Environmental Investigation and Remediation

1,4-Dioxane and Other Solvent Stabilizers

Environmental Investigation and Remediation

1,4-Dioxane and Other Solvent Stabilizers

Second Edition

By Thomas K.G. Mohr, William H. DiGuiseppi, Janet K. Anderson, and James W. Hatton

with chapters by Jeremy Bishop, Barrie Selcoe, and William B. Kappleman

CRC Press
Taylor & Francis Group
Boca Raton London New York

CRC Press is an imprint of the
Taylor & Francis Group, an **informa** business

CRC Press
Taylor & Francis Group
6000 Broken Sound Parkway NW, Suite 300
Boca Raton, FL 33487-2742

Printed on acid-free paper

International Standard Book Number-13: 978-1-138-39396-7 (Hardback)

Library of Congress Cataloging-in-Publication Data

A catalog record for this book has been requested.

Visit the Taylor & Francis Web site at
http://www.taylorandfrancis.com

and the CRC Press Web site at
http://www.crcpress.com

Dedications

In memory of my father, Eric Mohr, who always encouraged me to give back to my profession. In honor of my wife, Maria Kamau, whose support, vision, and encouragement made this book possible. This book is dedicated to the children, yours and mine (Kimathi, Makena, Njeri), who will inherit the resources we're protecting today.

– Thomas K.G. Mohr

In memory of my father, Jim DiGuiseppi, and in honor of my mother, Virginia DiGuiseppi, who always supported me, no matter what direction I chose to pursue. In appreciation of my wife, Cathy, who both tolerates and respects the commitments I've made to my career, the profession and this specific effort. For my children, Ginny and Mike, who I've encouraged to follow their passions and put in the effort it takes to achieve their dreams. And thanks to my management over the years, who's believed in supporting what I've found to be important and fulfilling work, especially John Mogge, Sharon Minchak, Paul Favara, and Jess Raphael-Ross.

– Bill DiGuiseppi

My work is dedicated in memory to my parents who were always my biggest cheerleaders; to my family for their support and understanding of the long hours spent tucked away in my office; and to my colleagues and friends who are in the trenches with me daily, who push me with their brilliance and dedication, and keep me (relatively) sane.

– Janet Anderson

Dedicated to my loving wife, Kelly, and my children, Emily and Nicholas.

– Jim W. Hatton

Contents

Acronyms

AAV	Acid acceptance value
ACGIH	American Conference of Governmental Industrial Hygienists
AES	Alcohol ethoxy sulfate
AGI	Amplified geochemical imaging
ALDH	Aldehyde dehydrogenase
ALP	Alkaline phosphatase
ALT	Alanine aminotransferase
AMO	Alkane monooxygenase
AOP	Advanced oxidation process
AOT	Advanced oxidation technology
AST	Aspartate aminotransferase
ASTM	American Society for Testing and Materials
ATSDR	Agency for Toxic Substances and Disease Registry
BCEE	*bis*-2-Chloroethylether
BCF	Bioconcentration factor
BHC	Benzene hexachloride (also Lindane)
BHT	Butylhydroxy toluene
BOD	Biological oxygen demand
BSM	Basal salt medium
BTEX	Benzene, Toluene, Ethylbenzene, and Xylene
CAR-PDMS	Carboxen-polydimethylsiloxane
CASRN	Chemical Abstracts Service Registry Number
CCL	Contaminant candidate list
CCl_4	Carbon tetrachloride
CCV	Continuing calibration verification
CDPHE	Colorado Department of Public Health and Environment
CEM	Conceptual exposure model
CFR	Code of Federal Regulations
CHO	Chinese hamster ovary (cells)
CHP	Catalyzed hydrogen peroxide
CLP	Contract Laboratory Program
CNS	Central nervous system
COC	Constituents of concern (or chemicals of concern)
COD	Chemical oxygen demand
CPK	Creatinine phosphokinase
CSF	Cancer slope factor
CSIA	Compound-specific isotope analysis
CSM	Conceptual site model
CT DPH	Connecticut Department of Public Health
CYP	Cytochrome P450
d, h, m, s	days, hours, minutes, seconds
DBCP	1,2-dibromo-3-chloropropane
DCE	1,1-Dichloroethylene
DCM	Dichloromethane (also methylene chloride)
DEG	Diethylene glycol
DEN	Diethylnitrosamine
DI	Deionized

DMBA	Dimethylbenzanthracene
DMT	Dimethyl terephthalate
DNAPL	Dense nonaqueous-phase liquid
DO	Dissolved oxygen
DOE	U.S. Department of Energy
DPE	Dual-phase extraction
DTSC	Department of Toxic Substances Control
DX	1,4-Dioxane
EC_{50}	Median concentration causing an effect in 50% of the test organisms
ECD	Electron capture detector
EDF	Ethylene glycol diformate (also ethylene-1,2-diformate)
EDTA	Ethylenediaminetetraacetic acid
EMD	Environmental molecular diagnostics
ERA	Ecological risk assessment
ESV	Ecological screening values
eV	Electron-volt
EVOS	Emulsified vegetable oil solution
FBR	Fluidized-bed reactor
FDA	U.S. Food and Drug Administration
FID	Flame ionization detector
f_{oc}	Fraction of organic matter
Freon 11	Dichlorodifluoromethane
Freon 113	1,1,2-Trichloro-1,2,2-trifluoroethane (also CFC-113)
Freon 12	Trichlorofluoromethane
FRI	Focused remedial investigation
GAC	Granular activated carbon
GAMA	Groundwater Ambient Monitoring and Assessment Program
GC	Gas chromatography
GC-IRMS	Gas chromatography-isotope ratio mass spectrometry
GCM	Group contribution model
GC–MS	Gas chromatography-mass spectrometry
GenX	Hexafluoropropylene Oxide Dimer Acid and Its Ammonium Salt
GGT	γ-Glutamyl transpeptidase
GO	Oxidized graphene
HCN	Health Council of the Netherlands
HEAA	Hydroxyethoxyacetic acid
HFC	Hydrofluorocarbon
HPV	High production volume
HSDB	Hazardous Substance Data Bank
i.p.	Intraperitoneal
i.v.	Intravenous
INEL	Idaho National Engineering Laboratory
IRIS	Integrated risk information system
ISCO	*In situ* chemical oxidation
ITRC	Interstate Technology & Regulatory Council
IUPAC	International Union of Pure and Applied Chemistry
IUR	Inhalation unit risk
KBV	Kauri-butanol value
K_d	Distribution coefficient
KIE	Kinetic isotope effect
K_{OC}	Organic carbon partition coefficient

K_i	Inhibition constant
K_{OW}	Octanol/water partition coefficient
K_p	Permeability coefficient
LC	Liquid chromatography
LC_{50}	Median concentration lethal to 50% of the test organisms
LD_{50}	Median lethal dose
LDPE	Low density polyethylene
LLNL	Lawrence Livermore National Laboratory
LNAPL	Light nonaqueous-phase liquid
LMS	Linear multi-stage (model)
LOAEL	Lowest observed adverse effects level
LOQ	Limit of quantitation
LPRO	Low pressure reverse osmosis
M	Molar
m/z	Mass-to-charge ratio
MBT	Molecular biological tools
MC	Methyl chloroform (also 1,1,1-trichloroethane)
MCI	Molecular connectivity index
MCL	Maximum contaminant level
MCLG	Maximum contaminant level goal
MDEQ	Michigan Department of Environmental Quality
MDL	Method detection limit
mM	Milli-moles
mm Hg	Millimeters of mercury
MMA	Michigan Manufacturers Association
MNA	Monitored natural attenuation
MoA	Mode of action
MOECC	Ontario Ministry of the Environment and Climate Change
MPP	Macro-porous polymer
MRE	Minimum relative entropy
MRL	Minimal risk level
MS	Mass spectrometry
MS/MSD	Matrix spike/matrix spike duplicate
MSDS	Material safety data sheet
MTBE	Methyl *tert*-butyl ether
MTD	Maximum tolerated dose
MW	Molecular weight
***n*-BGE**	*n*-Butyl glycidyl ether
NAPL	Non-aqueous phase liquid
ND	Not detected (also nondetect)
NDMA	Nitrosodimethylamine
NESHAP	National Emission Standards for Hazardous Air Pollutants
NHDES	New Hampshire Department of Environmental Services
NIOSH	National Institute for Occupational Safety and Health
NL	Notification level
nm	Nanometer
NMA	Natural mineral activation
NOAEL	No observed adverse effect level
NPDES	National Pollution Discharge Elimination System
OCWD	Orange County Water District
ODC	Ornithine decarboxylase (activity; toxicology context)

ODC	Ozone depleting compound (Chapter 1)
OEHHA	(California) Office of Environmental Health Hazard Assessment
ORP	Oxidation-reduction potential
OVIs	Organic volatile impurities
PA	Polyacrylate
Pa	Pascals (SI unit for pressure, 1 N/m^2)
PBPK	Physiologically based pharmacokinetic (model)
PCA	Potentially contaminating activity
PCB	Polychlorinated biphenyls
PCE	Perchloroethylene (also perchloroethene or tetrachloroethylene)
PCEMI	Perchloroethylene as a manufacturing impurity
PCR	Polymerase chain reaction
PDB	Passive diffusion bag
PDX	1,4-Dioxane-2-one
PEG	Polyethylene glycol
PET	Polyethylene terephthalate
PHG	Public health goal
PID	Photo-ionizing detector
PBPK	Pharmacokinetic
pMMO	Particulate methane monooxygenase
PNEC	Predicted no effect concentration
POEA	Polyoxyethyleneamine
POTW	Publicly owned treatment works (wastewater treatment plants)
ppb	Parts per billion
PPG	Pittsburg plate and glass
ppm	Parts per million
ppt	Parts per trillion
PRG	Preliminary remediation goal
PRP	Potentially responsible party
PSMS	Polysulfone membrane sampler
PT	Purge-and-trap
PTFE	Polytetrafluoroethene
PWS	U.S. public water system
PZ	Prohibition zone (Re: Ann Arbor, MI)
QSARs	Quantitative structure-activity relationships
RAGS	Risk Assessment Guidance for Superfund
RAP	Remedial action plan
RBC	Rotating biological contactor
RCRA	Resource Conservation and Recovery Act
RCS	Regenerated cellulose sampler
REACH	Registration, Evaluation & Authorization of Chemicals
R_f	Retardation factor
RfC	Reference concentration
RfD	Reference dose
RIVM	Dutch National Institute for Health and Environment
RoD	Record of decision
RO	Reverse osmosis
RP	Responsible party
RPD	Relative percent difference
RPPS	Rigid porous polyethylene samplers
RRF	Relative response factor

RSC	Relative source contribution
RSD	Relative standard deviation
RSL	Regional Screening Levels
RTECS	Registry of Toxic Effects of Chemical Substances
RVD	Relative vapor density
SARA	Superfund Amendments and Reauthorization Act of 1986
SCE	Sister chromatid exchange
SDS	Safety Data Sheet
SDWA	Safe Drinking Water Act
SEE	Steam enhanced extraction
SiG	Silicon-doped graphene
SIM	Selective ion monitoring (less commonly: selected-ion mode)
SLAC	Stanford Linear Accelerator Center
SMILES	Simplified Molecular Input Line Entry System
sMMO	Soluble methane monooxygenase
SMZ	Surfactant-modified zeolite
SPDE	Solid phase dynamic extraction
SPE	Solid phase extraction
SPME	Solid phase micro-extraction
SVE	Soil vapor extraction
SVOC	Semivolatile organic compound
SWAP	Source Water Assessment Program
TBA	*tert*-Butyl alcohol
TCA	1,1,1-Trichloroethane (also methyl chloroform)
TCD	Thermal conductivity detector
TCE	Trichloroethylene (also trichloroethene)
TCH	Thermal conductive heating
TCLP	Toxicity characteristic leaching procedure
TDI	Tolerable daily intake
TEG	Triethylene glycol
THF	Tetrahydrofuran
THP	Tetrahydropyran
THQ	Target hazard quotient
TICs	Tentatively identified compounds
TK	Toxicokinetic
TLV	Threshold limit value
TOC	Total organic carbon
TPA	Tetradecanoylphorbol-13-acetate
TR	Target cancer risk
TRI	Toxics release inventory
TRV	Toxicity reference value
TSS	Total suspended solids
TWA	Time-weighted average
UCMR	Unregulated contaminant monitoring rule
UF	Uncertainty factor
UN	United Nations Number
USAF	United States Air Force
USEPA	United States Environmental Protection Agency
USP	*U.S. Pharmacopeia*
UV	Ultraviolet
UVT	Ultraviolet light transmittance

VOA	Volatile organic analysis
VOCs	Volatile organic compounds
VI	Vapor intrusion
VISL	Vapor intrusion screening level
WHO	World Health Organization
wt	Weight
ZVI	Zero-valent iron

Preface

1,4-DIOXANE – THE CONTAMINANT THAT KEEPS EMERGING

Since early 2001, when I was tasked with a small research project into the potential groundwater quality impacts from solvent stabilizers, I have researched, investigated, networked, published, and engaged conference audiences in discussion on the question of the importance of considering 1,4-dioxane and other solvent stabilizers in chlorinated solvent release site investigations. As I prepared the manuscript for the first edition, I was in suspense: the USEPA Toxicological Review of 1,4-dioxane had not yet been finalized, and the findings could have rendered this book to be of value only to forensic geochemists. Instead, 1,4-dioxane toxicity was found to be more significant than the previous review, and the Science Advisory Board recommended that 1,4-dioxane be included in the Unregulated Contaminants Monitoring Rule 3 (UCMR3). The nationwide drinking water sampling results from UCMR3 showed that 1,4-dioxane is ubiquitous, vindicating the effort to write this book, and validating many of the forecasts included in the first edition.

The knowledge bank on 1,4-dioxane, measured in peer reviewed publications and government publications on 1,4-dioxane in groundwater, has grown almost exponentially. There have been quantum leaps in our understanding of 1,4-dioxane occurrence, fate and transport, and biodegradation, as well as significant improvements in laboratory methods for analysis of 1,4-dioxane. The Department of Defense environmental research programs (SERDP and ESTCP) have funded extensive work into 1,4-dioxane subsurface behavior and remedial technologies, carried out by scientists at universities and leading consulting firms, which has made an enormous contribution to advancing our understanding of 1,4-dioxane. The task of preparing a second edition was daunting, in view of the great number of new publications and new experiences with 1,4-dioxane. While our review of new findings was extensive, it's inevitable that we missed some important contributions to 1,4-dioxane research. The second edition brings current the 1,4-dioxane state of the art across the full spectrum of disciplines engaged on environmental problem solving. We believe you will find the second edition more useful than the first, and we welcome your feedback and inquiries.

Thomas K.G. Mohr
Santa Clara, CA, May, 2019

Disclaimer

The views presented here are those of the chapter authors only and are not the official policies, positions, or opinions of their employers. The reviewers who generously volunteered their time and knowledge to read and comment on draft chapters bear no responsibility for any errors, omissions, or other possible problems in the final manuscript.

Contact the Authors

Your comments, questions, corrections, and criticisms are welcome. Please contact the authors by e-mail or telephone at the following addresses and phone numbers:

Thomas K.G. Mohr, Mohr HydroGeoScience LLC,
tkgmohr@gmail.com, 408-832-1978
Bill DiGuiseppi
Bill.Diguiseppi@jacobs.com (720) 286-0784
Janet Anderson
jkanderson@gsienv.com (713) 522-6300
Jim W. Hatton
James.Hatton@Jacobs.com (720) 286-2655

Acknowledgments

The first edition of this book was the product of some basic library, gray literature, and internet research, but a large component arose from networking with groundwater professionals and environmental scientists from many disciplines: our colleagues (Santa Clara Valley Water District, Earth Tech/AECOM) and peers in regulatory agencies, professional associations, industry, and of course the amazing scientists and students at universities and research institutions. The second edition would not have been possible without the material and intellectual support and leadership of Bill DiGuiseppi (Jacobs), and by the coauthors who developed new material and updated many sections: Janet K. Anderson (GSI Environmental Inc.) and Jim W. Hatton (Jacobs). Chapters were substantially revised, updated, and improved by chapter authors Barrie Selcoe (Jacobs), William B. Kappleman (Jacobs), and Jeremy Bishop (Jacobs). Special thanks go to the dedicated support team at Jacobs, without whose help, this book never would have come together so well, especially Alexandra Salter-Blanc, Jared Khattak, Julia Roth, Alex Wilson-Fallon, Nicole Badon, Demitria Wright, Collin Hall, Judy Bristow, Shannon Bartow, and Amanda Struse. Special thanks to Steve Woodard (ECT_2), Chase Holton (Geosyntec), and Scott Grieco (Jacobs) as technical contributors.

Authors

Thomas K.G. Mohr is a consulting forensic and contaminant hydrogeologist with Mohr HydroGeoScience LLC, working on litigation regarding supply well contamination and forensic investigations to determine cost allocation for dry cleaner release sites. He retired from the Santa Clara Valley Water District as senior hydrogeologist in 2017, where he oversaw regional groundwater monitoring, salt and nutrient management planning, indirect potable reuse feasibility analysis, and provided oversight for remediation of solvent release sites in Silicon Valley. He has published on the application of stable isotope data for forensic investigations of perchlorate and nitrate contamination, bioremediation of gasoline, a method for assessing and ranking the potential for legacy contamination from past dry cleaners, and on a new refinement to methods for the laboratory analysis of 1,4-dioxane. Mohr previously managed and monitored environmental impacts from landfills for the County of Yolo and the City of Sunnyvale, and worked as a consultant conducting site characterization, remediation of leaking fuel and solvent tank sites, bioremediation of gasoline, and a variety of related projects.

Mohr was the 2006/2007 president of the Groundwater Resources Association of California and served on GRA's Board of Directors for 10 years. He has chaired six major symposia for GRA on perchlorate, dry cleaners, and 1,4-dioxane, and has been an invited speaker or session chair at symposia hosted by the International Society of Environmental Forensics, the National Groundwater Association, American Water Works Association, Battelle International, the California State Water Resources Control Board, and USEPA. He also enjoys mentoring students and has lectured at his alma mater, the University of California at Davis, as well as at Stanford University, San Jose State University, and California State University, East Bay. Mohr is continuing his pursuit of chlorinated solvent forensics and related topics through research, writing, and field investigations, while also home-schooling his three children and enjoying world travel.

William H. DiGuiseppi, PG, is a principal hydrogeologist with Jacobs, a Fortune 500 provider of environmental, water, transportation, and infrastructure engineering services. In his more than 30 years of consulting, his career has focused on soil and groundwater investigation and remediation programs at industrial sites and federal facilities. As the leader of the Emerging Contaminants Community of Practice at Jacobs, he has played a pivotal role in global technology transfer, ensuring that worldwide project teams leverage best practices to provide innovative and cost-effective solutions to complex remedial challenges related to 1,4-dioxane and per- and polyfluoroalkyl substances (PFAS), as well as other emerging and conventional contaminants. DiGuiseppi has authored or coauthored more than 30 peer-reviewed papers, presented at over 100 conferences, cochaired dozens of sessions on 1,4-dioxane or other emerging contaminants at national and international conferences and has been an invited guest lecturer at Colorado School of Mines and University of Colorado – Boulder. DiGuiseppi is the vice chair for Emerging Issues for the National Environmental Committee of the Society of American Military Engineers and holds leadership positions on the Interstate Technology & Regulatory Council PFAS and 1,4-dioxane Teams. He is a member of the Working Group on Emerging Contaminants for the Network for Industrially Contaminated Lands in Europe and was an invited speaker on Emerging Contaminants at the USEPA's Annual Groundwater Summit.

DiGuiseppi holds a BS in geology from George Mason University and an MS in geology from the University of Utah.

Dr. Janet K. Anderson is a human health toxicologist and environmental risk assessor with GSI Environmental Inc. She has over 15 years of experience providing toxicology expertise and consultation to federal agencies, private industry and municipal clients and is a nationally recognized leader in unregulated and emerging chemicals such as per- and polyfluoroalkyl substances and 1,4-dioxane. She specializes in risk communication, including the translation of key findings from toxicology studies to inform state and federal regulatory and public health decisions. Dr. Anderson performs critical reviews of federal and state risk assessments, guidance, regulatory policy, and legislations, with a goal of having the correct interpretation of science and science policy guide decision-making. She has developed strategies to communicate chemical risks, mitigate their human health impacts, and address associated environmental liability for both private and public sector clients. She has also provided litigation consulting and expert services and is a Diplomate of the American Board of Toxicology. Dr. Anderson received a BA degree in Biology from Wittenberg University, and a PhD in Molecular and Cancer Biology from the University of Cincinnati. She held a postdoctoral fellow position with the USEPA Office of Research and Development National Center for Environmental Assessment, and served as a government civilian employee for the U.S. Air Force (USAF) where she managed the USAF Emerging Issues and Emerging Contaminants Program and provided toxicology support to the Office of Secretary of Defense Chemical and Material Risk Management Program.

James W. Hatton is a principal technologist and senior engineer in Jacobs' environmental remediation group in Englewood, Colorado. Hatton has 30 years of applied engineering experience at a wide variety of sites, addressing the investigation and remediation of emerging contaminants, chlorinated volatile organic compounds, metals, and petroleum contaminants. Jim began working with 1,4-dioxane in the late 1980s at manufacturing facilities and has supported numerous investigations and remedial actions for 1,4-dioxane over his career. Hatton supports Jacob's emerging contaminants program and is the lead remediation engineer for programs addressing 1,4-dioxane and treatment of per- and polyfluoroalkyl substances in drinking water and groundwater. In addition to his work with 1,4-dioxane, Jim is globally recognized as subject matter expert for PFAS conceptual site models and remediation technologies. Hatton received his BS in petroleum engineering from West Virginia University in Morgantown, West Virginia.

1 Chlorinated Solvents and Solvent Stabilizers

Thomas K.G. Mohr

CONTENTS

TABLE 1.1
Major Chlorinated Solvents: Formulas and Structures

	Solvent				
	Carbon Tetrachloride	**Methyl Chloroform**	**Dichloromethane**	**Trichloroethylene**	**Perchloroethylene**
CASRN[a]	56-23-5	71-55-6	75-09-2	79-01-6	127-18-4
Formula	CCl_4	$C_2H_3Cl_3$	CH_2Cl_2	C_2HCl_3	C_2Cl_4
Structure	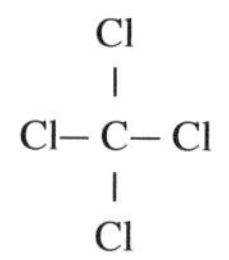	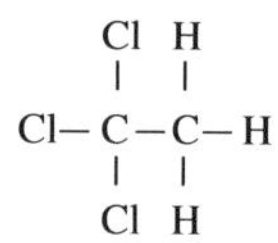	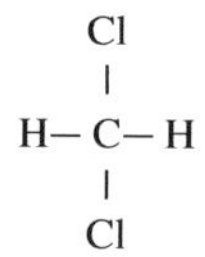	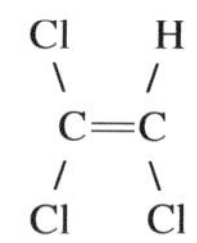	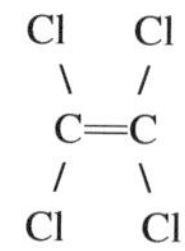

[a] CASRN, Chemical Abstracts Service Registry Number.

Solvent stabilizers have been added to a wide variety of solvents, including chlorinated and other halogenated solvents, petroleum solvents, ketone solvents, ether solvents, alcohol solvents, and others. This chapter deals with the stabilizers for the chlorinated solvents, which were by far the most frequently used solvents for degreasing, dry cleaning, cold cleaning, and dozens of other industrial applications. 1,4-Dioxane, a probable human carcinogen, has manifested as the most significant drinking water and environmental contaminant among the dozens of solvent stabilizers used for chlorinated solvents.

This chapter focuses primarily on solvent stabilizers added to the four major chlorinated solvents: methyl chloroform, trichloroethylene, perchloroethylene, and dichloromethane. The more common names for the four major solvents include methyl chloroform (1,1,1-trichloroethane or TCA), trichloroethylene (TCE), perchloroethylene (also called tetrachloroethylene, or PCE), and methylene chloride (also called dichloromethane). Tables 1.1 and 1.2 present identifying information and chemical structures for the four major chlorinated solvents, as well as lists of synonyms and trade names.

This chapter explores the uses of solvents in different applications, and the corresponding requirements for solvent stabilization, in substantial detail. While the environmental engineer or hydrogeologist's primary focus may be fate and transport of the contaminant in the subsurface, and the contaminants' physicochemical properties governing design of remediation systems, understanding the genesis of contaminants pays dividends for understanding both fate and transport and remedial design. A further benefit to focusing attention on how the chemicals were used prior to discharge is the forensic value that can be leveraged to sort out the timing of a release or to deconvolute commingled plumes originating from different potentially responsible parties. "Contaminant Archeology" is the term of art applied in this book to describe the exploration of a chemical's uses and composition before discharge. Chapter 1 therefore builds the foundation for exploring fate and transport aspects of solvents and their stabilizers (Chapter 3), treatment technologies suitable for removing both solvents and stabilizers (Chapter 8), and forensic investigation of contaminant release sites (Chapter 10).

1.1 HISTORY OF CHLORINATED SOLVENTS

Chlorinated solvents have played an integral role in the world's industrial societies for more than 70 years. The majority of manufactured consumer products made of metals or incorporating electronic components have undergone some form of cleaning process, most commonly using solvents to remove residuals from fabrication. Chlorinated solvents are nonflammable, easy to recover from

TABLE 1.2
Major Chlorinated Solvents: Identification

Carbon Tetrachloride	Methyl Chloroform	Dichloromethane	Trichloroethylene	Perchloroethylene
Tetrachloromethane	1,1,1-Trichloroethane, TCA, TCE	Methylene chloride, methane dichloride	TCE, trichloroethene, 1,1,2-trichloroethylene	PCE, tetrachloroethylene, tetrachloroethene, ethylene tetrachloride
Perchloromethane	Trichloro-1,1,1-ethane	Methylene dichloride	1,1-Dichloro-2-chloroethylene	Perc, Perclene™ (Diamond Shamrock/ DuPont)
Methane tetrachloride	Methyltrichloromethane	Freon 30	Ethylene trichloride	Persec (Vulcan Materials)
Carbon tet	Trielene	Aerothene MM™	Acetylene trichloride	DOWPER™ (Dow Chemical)
Freon 10	Aerothene TT™	DCM	TRI	Perchloroethylene SVG (Dow Chemical)
Halon 104	Chlorothane NU™		Tri-clene	Perchlor, Percosolv, Perklone™, Perstabil (Solvay et Cie)
Tetraform	Chlorothene SM™		Tri-plus	Soltene™ (Solvay et Cie)
Tetrasol	Chlorothene VG™		Triclene™	
Tetrachlorocarbon	Chlorothene NU™		Trilene™	
Carbona	Solvent 111		Trielin	
	"TRIC"		Narcogen	
	Inhibisol™		Dow-Tri	
	Genklene™		Neu-Tri™	
	1,1,1-TCE		Hi-Tri™	
			Solvent O-T-634C	

Sources: National Library of Medicine, 2006, Hazardous Substances Data Bank (HSDB). National Institutes of Health. http://toxnet.nlm.nih.gov/ (accessed March 15, 2006); Doherty, R.E., 2000a, *Journal of Environmental Forensics* 1: 69–81; Doherty, R.E., 2000b, *Journal of Environmental Forensics* 1: 83–93; Solvay, 2002a, Chlorinated solvents stabilisation. GBR-2900-0002-W-EN Issue 1 – 19.06.2002, Rue du Prince Albert, 33 1050 Brussels, Belgium. http://www.solvaychemicals.com (accessed November 9, 2003); and SDSs cited in Chapter 1 reference list.

Notes: Most commercial trade names in this list are registered trademarks; these are indicated by capitalization; the symbol ™ indicates that the solvent name was once registered with the U.S. Patent and Trademark office, but the name may no longer be registered or has since been assigned to another product.

the vapor phase, and easy to recycle by distillation when laden with oils and other debris from degreasing and dry cleaning processes. The relatively low boiling points of these chemicals make them economical for use in vapor degreasing and dry cleaning, and their high solvency makes them extremely effective in a wide variety of challenging cleaning applications, for example, cold cleaning and ultrasonic cleaning.

From the 1940s until recently, chlorinated solvents played a part in the production of most manufactured products. Today, many manufacturing processes continue to leverage the cleaning power and effectiveness of halogenated solvents. However, in compliance with regulations that prevent uncontrolled releases to the environment, these processes use equipment that has been engineered to eliminate or minimize emissions and spills. Chlorinated solvents provided an enormous benefit to the industry and consumers alike; however, they pose a threat to water quality where they were released to soil and groundwater. The unintended adverse consequences of past uncontrolled or under-controlled solvent uses have taught us valuable lessons, which have guided major improvements in equipment design, operator training, monitoring, regulation, and enforcement for continuing solvent use worldwide.

Substantial progress is being made toward environmentally safe uses of chlorinated solvents under controlled conditions. Concurrently, new solvents that pose a somewhat lower threat to drinking water, the atmosphere, and worker health and safety have been developed and introduced. For example, fluorinated solvents (hydrofluorocarbons – HFCs) are attracting increasing market attention due to their low boiling points, low volatility, and high stability, for which solvent stabilizers are not required. New regulatory restrictions are changing the marketplace for chlorinated solvents in the United States. In April 2016, the European Union banned all uses of TCE. Similarly, in January 2017, EPA proposed banning use of TCE in commercial vapor degreasing because of health risks identified in a 2014 Chemical Work Plan Chemical Risk Assessment for TCE. In 2007, the California Air Resources Board banned the installation of new PCE dry cleaning machines after 2007 and use of all California PCE dry cleaning machines will end in 2023, affecting 70% of California's ~3,400 PCE dry cleaning machines. The consequence of a changing marketplace favoring alternatives to chlorinated solvents, and regulatory restrictions for uses of chlorinated solvents, is that drinking water contamination with chlorinated solvents and their stabilizers is a primarily a legacy issue. However, as detailed in Chapter 2, drinking water contamination with 1,4-dioxane is ongoing where 1,4-dioxane is a by-product of production in processes that produce effluents discharged to surface water bodies that supply drinking water.

The enormous volume of chlorinated solvents used in the United States alone has left a monumental legacy that will keep contaminant hydrogeologists and environmental engineers occupied for decades. A compelling aspect of the historical use of chlorinated solvents is that most of the solvent used was ultimately released to the environment, primarily to the atmosphere, with a limited but significant portion to soil and groundwater. Solvents that were not emitted, discharged, or recycled were incinerated or placed in a landfill. When solvents were added to products such as paints, inks, or cleaning agents, the intended fate of the solvent was often evaporation. Degreasing and dry cleaning solvents were discarded when they were no longer usable. In earlier decades, waste solvents were sometimes dumped to the ground and often disposed in landfills, which, until the 1990s, were usually unlined or clay-lined landfills. Since the enactment of the Resource Conservation and Recovery Act in 1980, waste solvent has been subjected to "cradle-to-grave" management and hazardous waste manifesting. Waste minimization has emerged as an economical and regulatory compliance-driven solution to the challenge of safe solvent handling and disposal. Chlorinated solvent consumption in 2006 was about 10% of the consumption in 1976. The reduction was due to regulatory bans on ozone-depleting compounds (ODCs), elimination of open-top vapor degreasers, introduction of highly efficient degreasing and dry cleaning machinery, and replacement of chlorinated solvents by new "designer solvents" and other alternatives.

The history of chlorinated solvent production, use, and regulation has been thoroughly addressed in Richard E. Doherty's two landmark articles in the *Journal of Environmental Forensics* (Doherty, 2000a, 2000b). Doherty's articles profile which solvents were favored for different applications and how those preferences changed in response to the recognition of worker health and safety issues and new regulatory restrictions adopted to protect the environment. The sequence of regulations controlling solvent use, described further in Chapter 6, played a key role in the evolution of which solvents were put to use for various purposes. Economic and other forces also helped in determining specific solvent usage. For example, during World War II, chlorinated solvent markets were controlled by the government: the military received the first priority followed by essential uses such as grain fumigation and refrigeration; dry cleaning received the lowest priority (Doherty, 2000a). Figure 1.1 charts the U.S. production history of the major chlorinated solvents. An approximate U.S. production of chlorinated solvents in the early 1970s is represented by the following 1972 figures (Considine, 1974):

Methyl Chloroform	447 million pounds
Trichloroethylene	675 million pounds
Dichloromethane	500 million pounds
Perchloroethylene	500 million pounds

Methyl chloroform was substituted for TCE beginning in the mid-1960s. The change was largely due to the enactment of regulations that restricted the use of TCE, due to its role in the formation

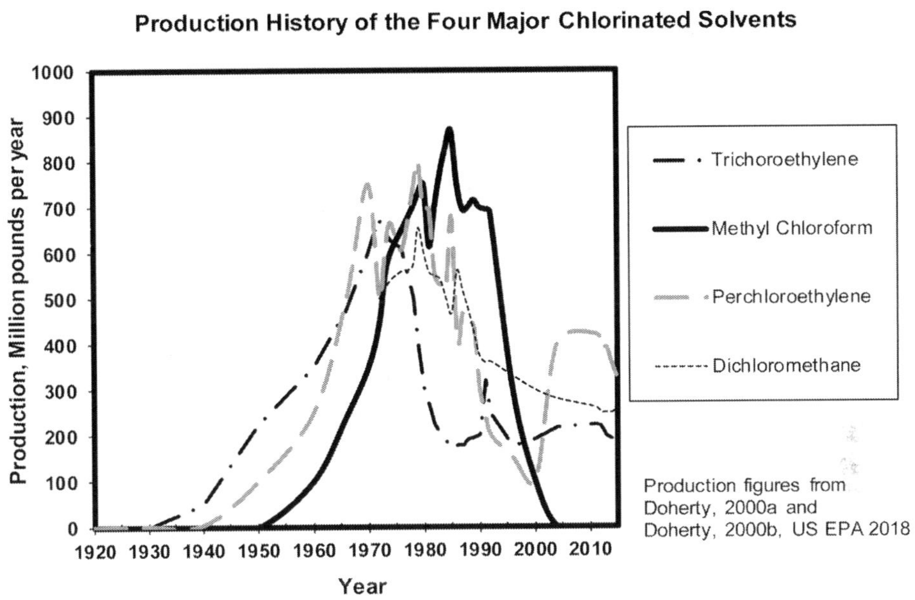

FIGURE 1.1 History of U.S. production of the four major chlorinated solvents. (After Doherty, R.E., 2000a. *Journal of Environmental Forensics* 1: 69–81. With permission.)

of ground level ozone (smog), and concern over worker exposure. Methyl chloroform production peaked in the mid-to-late 1980s. Reduced production was mandated by the Montreal Protocol on substances that deplete the ozone layer, an agreement among the world's nations to curtail the use of ozone-degrading compounds, including methyl chloroform. The Montreal Protocol went into effect on January 1, 1989; it called for 100% reduction in the 1989 methyl chloroform production levels by January 1, 1996 (UNEP, 1999).

1.1.1 Major Applications of Chlorinated Solvents

The major chlorinated solvents were previously put to the following primary uses:

- Degreasing (vapor degreasing and cold-cleaning)
- Extraction of oils or other food products
- Textile cleaning
- Wool processing
- Preparation of inks, paints, varnish, and lacquer
- Production of pharmaceuticals
- Cleaning of electronics, semiconductors, and precision instruments
- Cleaning of aircraft and other military hardware
- Minor uses in veterinary and medical applications
- A carrier for insecticides
- Applying paints to automobiles and other metal surfaces

Solvent boiling point, solvent power (solvency), stability, availability, and cost were factors that determined which solvents were used for particular applications. For example, PCE boils at 121°C, whereas dichloromethane boils at 39.7°C. Therefore, dichloromethane may have been preferred for cleaning electronics or other parts having plastics that are susceptible to melting, and PCE may have been preferred for removing waxes and resins whose melting points exceed the boiling point of other solvents.

1.1.1.1 Chlorinated Solvents in Historical or Current Use

1.1.1.1.1 Carbon Tetrachloride

Carbon tetrachloride is no longer produced for commercial use as an end product; it is still produced as an intermediate for the production of other chemicals. It was phased out because of its toxicity and its role as an ODC in the upper atmosphere. Carbon tetrachloride was the first of the major chlorinated solvents to be used in North America, beginning in 1898; its production in Europe began in the 1890s (Doherty, 2000a). It was used in fire extinguishers, in dry cleaning, as a grain fumigant, as a solvent for the extraction of oils from seeds and animal products, in rubber production, in metal degreasing, in rubber cement, in floor wax, as a refrigerant, as a household cleaning agent, in insecticides, and in numerous other applications (Irwin et al., 1997; Doherty, 2000a; European Chemicals Bureau, 2000b; ATSDR, 2005). Carbon tetrachloride was used to make Freon compounds that were widely used as aerosol propellants in the 1950s and 1960s (ATSDR, 2005; R.E. Doherty, personal communication, 2007). The majority of these uses of carbon tetrachloride were discontinued before 1970; some pesticide formulations included carbon tetrachloride until 1986. The regulation of ozone-depleting chemicals through the Montreal Protocol and the Clean Air Act Amendments of 1990, as well as bans by the U.S. Food and Drug Administration and other regulators on the basis of carcinogenicity, has led to the complete phase-out of carbon tetrachloride use (ATSDR, 2005; Morrison et al., 2006).

1.1.1.1.2 Trichloroethylene

TCE has been used in a multitude of applications (Shepherd, 1962; Irwin et al., 1997; Doherty, 2000a; European Chemicals Bureau, 2000b; HSIA, 2001; Mohr, 2001; ATSDR, 2005; Morrison et al., 2006):

- Dry cleaning, leather cleaning, and scouring wool
- Vapor degreasing of metal furniture, fabricated metal products, electric and electronic equipment, and military and transport equipment
- Cleaning liquid oxygen and hydrogen tanks in aerospace and other operations
- A carrier solvent for the active ingredients of insecticides
- A solvent in the rubber industry
- Adhesive formulations
- A solvent in printing inks, paints, lacquers, varnishes, adhesives, and paint strippers
- A solvent for dissolving and applying resin and rust-preventive coatings that dry as solvents vaporize
- A solvent in the pulp and paper industry
- A solvent in the textile industry for spotting fluids and as a solvent in waterless dyeing and finishing
- A solvent base for metal phosphatizing systems used to prepare metal surfaces for rust protection and painting
- A solvent in specialty glass production
- A component of household cleaners and consumer automotive products
- A component of drain cleaners and septic tank cleaners
- A component of metal polishes
- An extraction solvent for natural fats and oils such as palm, coconut, and soybean oils
- An extraction solvent for spices and hops and for removing caffeine from coffee
- A vapor-phase sterilizing agent for canned foods
- A solvent for the preparation of cosmetics
- A solvent for the preparation of pharmaceuticals
- An insecticidal grain fumigant
- An anesthetic in veterinary and medical procedures

The use of TCE as a skin, wound, and surgical disinfectant and as an inhalation anesthetic for childbirth and other short-duration medical procedures was banned in the regulations issued by the U.S. Food and Drug Administration in 1977 (ATSDR, 2005). In some developing nations, TCE continues to be used as an anesthetic even today. Because pharmaceutical grade TCE is no longer produced, industrial and reagent grades are used instead to substitute for more expensive anesthetic agents (Totonidis, 2005).

Metal degreasing consumed up to 80% of all TCE in European markets in the 1980s; similar consumption patterns were also present in North America in the 1970s (Doherty, 2000b; European Chemicals Bureau, 2004). In the 1950s and 1960s, TCE was used for aluminum degreasing in the United States, where chemists, for the major solvent producers, had worked out successful stabilization formulae. In Europe, however, reports of "spontaneous decomposition" of TCE, when used for vapor degreasing of aluminum parts, reflect a disparity in the art of stabilization of TCE between European and American solvent producers in that timeframe (Shepherd, 1962).

1.1.1.1.3 Perchloroethylene

The primary use of PCE is dry cleaning; about 60% of the 1991 PCE production was used for dry cleaning, whereas only 15% of that year's production was used for vapor degreasing (USEPA, 1994b). A minor use of PCE was as a dielectric fluid (e.g., to insulate electrical transformers). Improvements in dry cleaning machinery to minimize solvent losses, in response to increasing regulatory oversight of PCE use in dry cleaning operations, have greatly curtailed PCE consumption. Compared to other solvents, the solvent losses from PCE vapor degreasers were more easily controlled because of PCE's higher vapor density (Dow Chemical Company, 2006d). By 1998, the total annual U.S. production of PCE had declined to 341 million pounds; the total consumption by dry cleaners had fallen to 40% of the total PCE production, and only 10% was used for metal degreasing. The rest was used as a feedstock for the production of fluorocarbons (chlorofluorocarbons and HFCs) for refrigerant, blowing agent, and fluoropolymer applications.

PCE was preferred for cleaning alkali metals such as magnesium and aluminum, particularly before stabilized methyl chloroform became available, because PCE is more stable against reactions with metals than the other chlorinated solvents (Dow Chemical Company, 2001; Solvay, 2002a; Morrison et al., 2006). PCE has seen uses similar to those of TCE, including vapor degreasing of metals and electronics. PCE was preferred for removing wax and resin residues with high melting points, such as those used in the plating industry, because of its high boiling point. Its high boiling point results in a high-temperature vapor that completely dries wet work* because it vaporizes the moisture trapped in porous metals, deeply recessed parts, and blind holes (Dow Chemical Company, 2000). The high-temperature PCE vapor, compared with solvents that boil at lower temperatures, also provides a longer contact time because the room-temperature work – considered "cold" relative to the high-temperature solvent vapor – takes longer to warm to PCE's high boiling point (Morrison et al., 2006; Dow Chemical Company, 2006d). PCE was also used in ultrasonic cleaning.

PCE is used as an extractant in the production of pharmaceuticals and in veterinary formulations for deworming pets (NEMA worm capsules). It has also reportedly been used in the past as an anthelmintic† in the treatment of hookworm and some trematode‡ infestations in humans on the Indian subcontinent and other subtropical regions (USEPA, 2001).

In addition to dry cleaning, PCE was used for cleaning leather; as a scouring, sizing, and desizing agent for processing textiles; as a carrier solvent for fabric dyes and finishes; and as a water repellant in textile manufacture. It was used in mixtures with grain protectants and some liquid grain fumigants. PCE with a surfactant additive is preferred for vapor drying of glass products and

---

* In this book, the adjective "wet" is used in reference to water and not other liquids; the verb "wetting" takes on another meaning, as described in Chapter 2.

† Drugs that expel parasitic worms (helminthes) from the body.

‡ Parasitic flatworms, also called flukes. Some species chemically castrate their hosts: a capsule of perchloroethylene was most likely seen as a benign and preferred alternative!

some metal products requiring a high-quality surface finish. Aerosol formulations for the automotive aftermarket used PCE, particularly for brake cleaning, in which PCE content ranged from 50% to 100% (HSIA, 1999; Morrison and Murphy, 2013). Other applications were as an insulating fluid and cooling gas in electrical transformers; in paints, lacquers, and varnishes; in oil refineries for the regeneration of catalysts; in paint stripping; in cleaning printing equipment; in aerosol formulations; in polishes and lubricants; and as a solvent in various consumer products for household use (Doherty, 2000a; European Chemicals Bureau, 2004, 2005; Morrison et al., 2006).

1.1.1.1.4 Methyl Chloroform

Before being phased out in the mid-1990s owing to its listing as a stratospheric ODC, methyl chloroform was most commonly used for metal cleaning by cold cleaning, vapor degreasing, and ultrasonic cleaning baths. It reportedly first became widely used in the late 1950s as a cold-cleaning solvent for metal-cleaning applications that previously used carbon tetrachloride, because methyl chloroform was significantly less toxic (HSIA, 1994; Doherty, 2000b). Because of the inherent chemical instability of methyl chloroform vapors in the presence of aluminum and other alkali metals, it was not widely used for vapor degreasing until a suitable stabilized vapor-phase formulation became available. As discussed further in Section 1.2.6.6, the first stabilizer that enabled methyl chloroform use in vapor degreasing of aluminum and other metals was 1,4-dioxane. The addition of 1,4-dioxane to methyl chloroform was first patented in 1957 (Bachtel, 1957).

Methyl chloroform was commonly used as a solvent in aerosol formulations and in aerosol propellant formulations for various household products sold in spray cans. As an aerosol solvent, methyl chloroform was used to reduce the flammability of hydrocarbon propellants (HSIA, 1994). A number of consumer spray aerosol products contained methyl chloroform, including hair sprays, cosmetics, oven cleaners, spot removers, furniture polishes, automotive lubricants, automotive choke cleaners, water repellents, and adhesives (Aviado et al., 1976).

Methyl chloroform was promoted by Dow Chemical as a dry cleaning solvent in the early 1980s. It was used as a primary dry cleaning solvent (particularly in leather cleaning) by a very small number of dry cleaning operations – only 50 plants in the United States. The instability of methyl chloroform led to problems with machine and equipment corrosion; therefore, it is no longer used as a dry cleaning solvent, but is still found in some spot-removal solutions at dry cleaners that use PCE, hydrocarbon, or other solvents (Linn et al., 2004). Methyl chloroform has also been used in spray products to condition suede and for glossing and weatherproofing leather products.

In addition to its major application as a metal cleaner, methyl chloroform was also used in the following products and processes, among others (World Health Organization, 1990; USEPA, 1994a; Doherty, 2000a):

- Automotive products
- Coatings and finishes
- Consumer and industrial adhesives
- Various paints, coatings, and inks
- A textile-scouring agent
- A cleaner for machinery and tools used in the textile industry
- A dye carrier
- An electronics cleaner to remove flux and dry-film photoresist developer from printed circuit boards
- A grease cutter in drain cleaners and septic tank cleaners
- A cleaner for plastic molds
- A pharmaceutical extractant
- A plastic film cleaner in the film and television industry
- A solvent for various insecticides
- A degreening product for citrus fruits

- A postharvest fumigant for strawberries
- A solvent for natural and synthetic resins, oils, waxes, tars, and alkaloids
- A coolant in cutting oils
- An asphalt-extracting agent for paving aggregate
- A solvent for the application of water and oil repellents to paper and textiles

Methyl chloroform was banned from use as the active ingredient in pesticides in 1998. It was considered excellent for cleaning plastics, polymers, and resins because it is less likely to degrade these materials than PCE or TCE (Tarrer et al., 1989). Figures 1.2 through 1.4 show the end-use pattern for TCE, PCE, and methyl chloroform, respectively, for the United States.

1.1.1.1.5 Dichloromethane

Dichloromethane became an attractive replacement solvent as regulations were promulgated to protect worker's health and safety in the 1970s and the ozone layer in the 1980s and 1990s. It has no flash point under conditions of normal use, and therefore reduces the flammability of other solvents when mixed as an azeotrope. Among the benefits promoted for the use of dichloromethane in the solvents industry are its atmospheric characteristics. Compared to other chlorinated solvents, dichloromethane is not a major contributor to atmospheric pollution through the formation of smog, to the depletion of the stratospheric ozone layer, or to global warming (HSIA, 2003).

Dichloromethane provides a higher solvency than TCE, PCE, or methyl chloroform. Its aggressive solvency makes it an ideal paint remover, and it does not harm wood in the removal process (Dow Chemical Company, 2002; HSIA, 2003). Paint-stripping formulations make up the largest end use of dichloromethane; it has a unique ability to penetrate, blister, and lift a wide variety of paint coatings. Dichloromethane is found in industrial and commercial furniture strippers and in home paint removers, and it is used extensively in both flow-over and dip tanks in furniture-refinishing operations. It is also used in aircraft-maintenance formulations to inspect surfaces for damage (ATSDR, 2000).

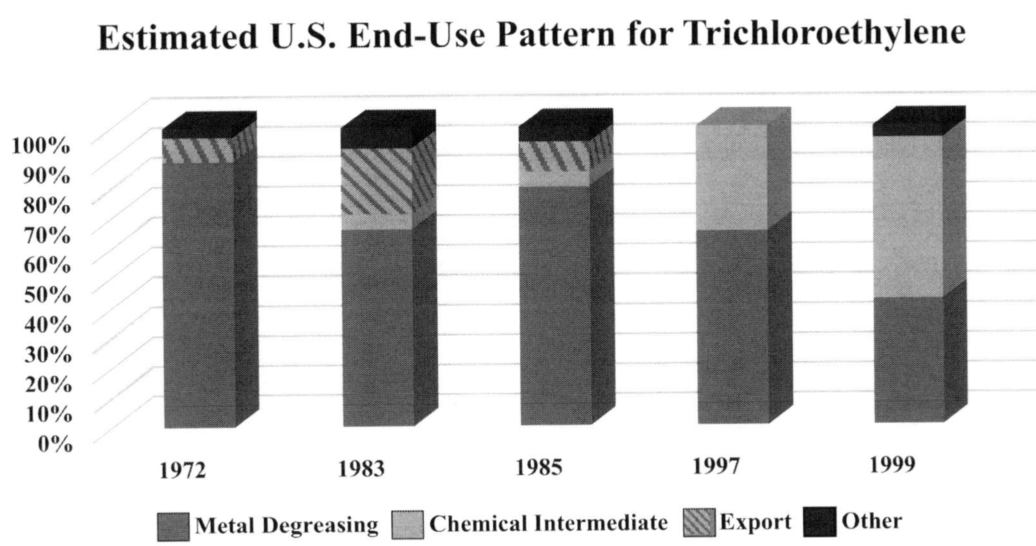

FIGURE 1.2 Estimated U.S. End-Use Pattern for Trichloroethylene. (*Sources:* National Library of Medicine, 2006, Hazardous Substances Data Bank (HSDB). National Institutes of Health. http://toxnet.nlm.nih.gov/ (accessed March 15, 2006); Halogenated Solvents Industry Alliance (HSIA), 2001, Trichlorethylene white paper. http://www.hsia.org/white_papers/paper.html (accessed April, 2002); U.S. Environmental Protection Agency (USEPA), 1989a, Locating and estimating air emissions from trichloroethylene and perchloroethylene. U.S. Environmental Protection Agency, Office of Air Quality Planning and Standards, Research Triangle Park, NC. EPA/450/2-89/013; and Doherty R.E., 2000a, *Journal of Environmental Forensics* 1: 69–81.)

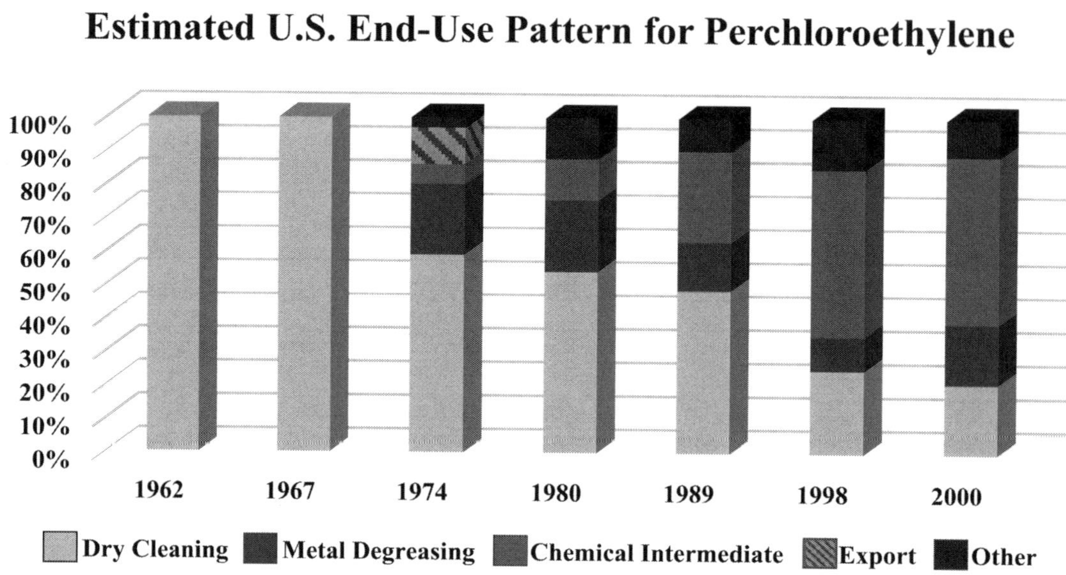

FIGURE 1.3 Estimated U.S. End-Use Pattern for Perchloroethylene. (*Sources:* National Library of Medicine, 2006, Hazardous Substances Data Bank (HSDB). National Institutes of Health. http://toxnet.nlm.nih.gov/ (accessed March 15, 2006); Halogenated Solvents Industry Alliance (HSIA), 1999, Perchlorethylene white paper. http://www.hsia.org/white_papers/paper.html (accessed April, 2002); U.S. Environmental Protection Agency (USEPA), 1989a, Locating and estimating air emissions from trichloroethylene and perchloroethylene. U.S. Environmental Protection Agency, Office of Air Quality Planning and Standards, Research Triangle Park, NC. EPA/450/2-89/013; and Doherty R.E., 2000a, *Journal of Environmental Forensics* 1: 69–81.)

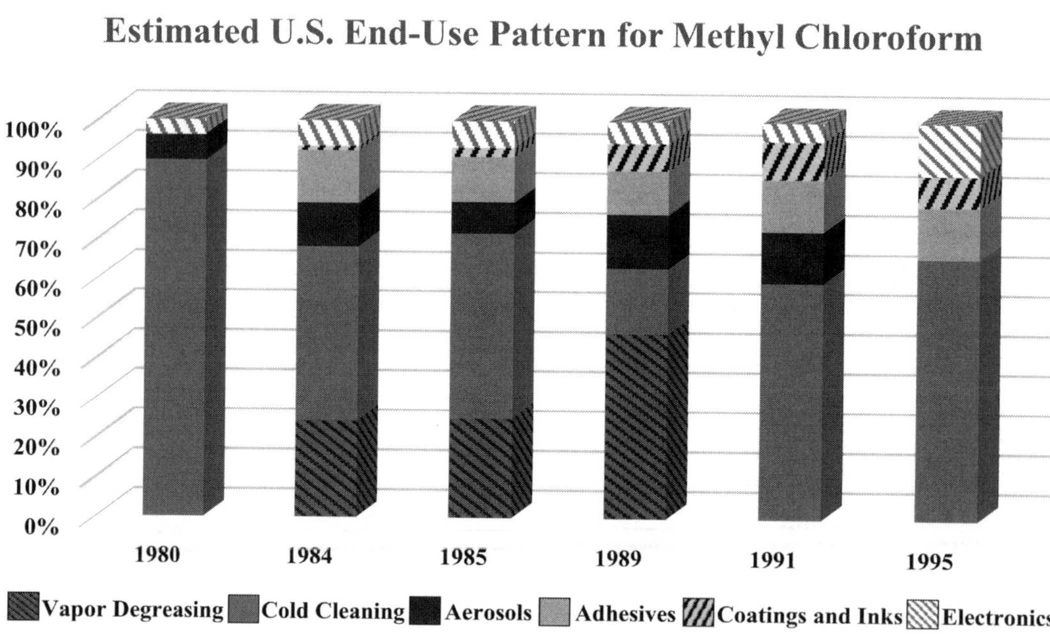

FIGURE 1.4 Estimated U.S. End-Use Pattern for Methyl Chloroform. (*Sources:* U.S. Environmental Protection Agency (USEPA), 1994a, Locating and estimating air emissions from methyl chloroform. U.S. Environmental Protection Agency, Office of Air Quality Planning and Standards. EPA/454/R-93/045 and Kavaler, A.R. (Ed), 1995, *Chemical Marketing Reporter* 247 (February 27): 9, 37.)

Dichloromethane is also used extensively as a foam-blowing agent for the production of softer grades of flexible polyurethane foams for furniture and bedding industries. Evaporation of the solvent during production causes expansion of the cells in urethane polymer foam, which reduces the urethane density while maintaining its flexibility. Dichloromethane controls the temperature of the reaction, which may otherwise become sufficiently high to burn or scorch the foam interior (HSIA, 2003).

Dichloromethane use as a propellant in aerosols such as insecticides, hair sprays, and spray paints increased significantly after the regulation of chlorofluorocarbons. Dichloromethane was a replacement for fluorocarbons in aerosols (World Health Organization, 1984b). Dichloromethane is used in aerosols as a strong solvent, a flammability suppressant, a vapor-pressure depressant, and a viscosity thinner. However, labeling requirements and health and environmental concerns led to a decline in the use of dichloromethane in consumer aerosol products in the late 1980s. Dichloromethane was used in hair sprays before it was banned by the U.S. Food and Drug Administration in 1989. It continues to be used in specialized spray paints and lubricants (ATSDR, 2000).

Figure 1.5 provides the approximate breakdown of the major uses of dichloromethane from the 1970s through the 1990s; usage figures vary widely by information source and year.

Dichloromethane is used as a reaction and recrystallization solvent in the extraction of several pharmaceutical compounds and in the production of many antibiotics and vitamins. It has been used as a carrier to coat tablets. Dichloromethane is eliminated from the coating of the finished tablet in order to comply with U.S. Food and Drug Administration regulatory standards.

A wide variety of nonpharmaceutical applications have been found for dichloromethane (World Health Organization, 1984b; USEPA, 1994c; ATSDR, 2000; European Chemicals Bureau, 2000d; HSIA, 2003; National Library of Medicine, 2006). It has been used in cold and vapor degreasing of metals. It is involved in the manufacture of synthetic fibers and polycarbonate resin used for the production of thermoplastics. Dichloromethane is a component in fire-extinguishing products and

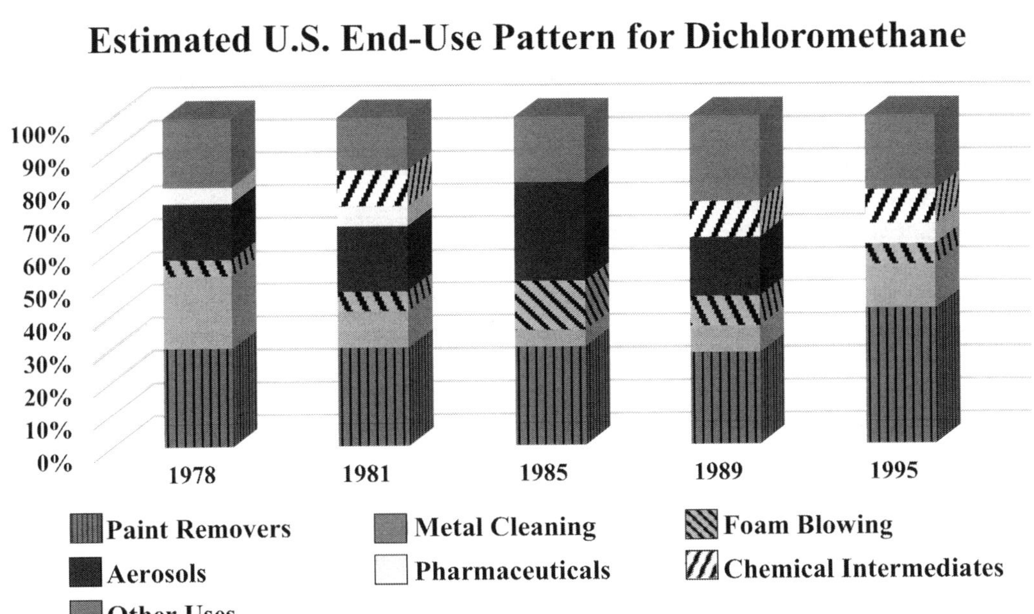

FIGURE 1.5 Estimated U.S. End-Use Pattern for Dichloromethane. (*Sources:* Compiled in National Library of Medicine, 2006, Hazardous Substances Data Bank (HSDB). National Institutes of Health. http://toxnet.nlm.nih.gov/ [accessed March 15, 2006].)

[a] **Includes vapor degreasing and cold cleaning combined.**

serves as a coolant and refrigerant. It is used as a solvent in the production of cellulose triacetate as a base for photographic film, in the solvent welding of plastic parts, as a metal-finishing solvent, and as an extraction solvent for naturally occurring, heat-sensitive products including edible fats, cocoa butter, caffeine, and beer flavoring in hops. Dichloromethane serves as a releasing agent to prevent the manufactured part from permanently bonding to the mold and for cleaning electronic components. In the agricultural sector, the chemical has applications as an insecticidal postharvest fumigant for grains, as a fumigant for strawberries, and as a ripening or "degreening" agent for citrus fruits. The dental profession makes use of dichloromethane in pour molding of dental material; a 50:50 mixture of dichloromethane and methyl methacrylate forms a cold-curing monomer to treat acrylic teeth used to improve bonding. Using dichloromethane to make decaffeinated coffee met with U.S. Food and Drug Administration approval; however, most coffee processors no longer use it for this purpose because of concerns over the consumer perception that a minor amount of dichloromethane may remain in the coffee as a residue (ATSDR, 2000). Nevertheless, improvements to making decaffeinated coffee with dichloromethane now produce coffee that meets the requirements of the U.S. Food and Drug Administration for the elimination of residues (HSIA, 2003).

1.1.1.2 Cold Cleaning

The manufacture of parts and products using metals and other materials often leaves them soiled with residues of cutting oils, machine oils, buffing compounds, soldering flux, rosins, lubricating grease, finely divided metal grindings, drill turnings, and residual powders. These soils must be removed before coatings such as rust preventives, primers, paints, waterproofing agents, varnishes, polishes, or other finishes can be applied. Many of these soils are insoluble in water. The two main methods for cleaning parts are alkali detergent washes and solvents in both liquid and vapor phases. The general categories of solvents used in washing manufactured parts include petroleum solvents, alcohols, ketones, and chlorinated aliphatic hydrocarbons.* Chlorinated solvents have been preferred for degreasing parts because they have high solvency for both cold cleaning and vapor degreasing. Their higher vapor density makes it easier to control the vapor level in a vapor degreaser, and their lower heats of vaporization make them more energy and cost effective than other solvents with higher solvency such as acetone and methanol. Chlorinated solvents also have low flammability compared to acetone and methanol (Archer and Stevens, 1977).

"Cold cleaning" refers to several common parts-washing technologies that use tanks for dipping soiled parts in a solvent at room temperature. Cold cleaning also refers to wiping or spraying parts or large equipment such as aircraft wings and fuselages, locomotive engines, and automotive engines. Cold-cleaning systems most often consist of simple tanks fitted with a lid and filled with a solvent. Cold cleaning in open tanks usually employed a layer of water atop the solvent to prevent solvent losses from evaporation (Gregersen and Hansen, 1986). Baskets filled with parts are immersed in the solvent tank and allowed to soak long enough to loosen and dissolve oils, grease, rosins, or flux. In processing printed circuit boards, a photoresist stripper employs cold cleaning by using an in-line design to dip parts on a conveyor rack (USEPA, 1989b).

Physical agitation of parts in a solvent bath aids cleaning. Operators use brushes or spray wands fitted with nozzles to clean parts, or they may use mechanical agitators or flow provided by circulation pumps. Circulation pumps produce solvent flow at the desired velocity. Agitation is achieved with oscillators that move parts back and forth through the cleaning solution or with ultrasonic units that use sound waves to induce high-frequency vibrations in and around the part (AFCEE, 1999; PRO-ACT, 1999; Dow Chemical Company, 1999b).

In ultrasonic cleaning, the cleaning action of the solvent is supplemented by sonically induced cavitation, which produces intense physical cleaning. Electricity at approximately 20 Hz powers a

* Chlorinated aliphatic hydrocarbons are straight-chained hydrocarbon compounds in which chlorine atoms are substituted for hydrogen atoms—for example, carbon tetrachloride, methyl chloroform, dichloromethane, TCE, and perchloroethylene.

crystal or magnetic transducer, which expands and contracts with each cycle to generate shock waves in the solvent. Shock waves cause the rapid formation and collapse of low-pressure bubbles throughout the solvent, which creates a vigorous scrubbing action. Ultrasonic cleaning is often utilized in the manufacture of electronic components and printed circuits as part of a cold-cleaning or vapor-degreasing operation. Methyl chloroform is a common ultrasonic cleaning agent (Considine, 1974). Ultrasonic cleaning is especially useful for cleaning metal chips from blind holes or removing small insoluble materials from fine machined surfaces, such as particulate residue from honing operations and some buffing compounds. Ultrasonic cleaning is most efficient in non-agitated liquids at temperatures 25–40 C below the solvent's boiling point (Dow Chemical Company, 1999b).

Methyl chloroform was preferred for cold cleaning because its worker-exposure threshold was significantly higher than that of other chlorinated solvents. Cold cleaning applications for methyl chloroform included maintenance cleaning of electronic components and precision instruments, small parts, aircraft parts, and other parts that require a high degree of cleanliness (USEPA, 1989b). TCE, PCE, and dichloromethane, while being very effective as cold-cleaning solvents, have lower worker-exposure thresholds and are regulated as hazardous air pollutants and as volatile organic compounds (USEPA, 1994a). In 1987, 70 million pounds of chlorinated solvents were used by cold cleaners in the United States. However, the amount of chlorinated solvents used for cold cleaning decreased dramatically as methyl chloroform was phased out following the Montreal Protocol in the 1990s. Methyl chloroform was replaced primarily with dichloromethane, but also with 1-bromopropane, limonene, terpenes, and other "designer" solvents (Petroferm, Inc., 1997; Dow Chemical Company, 1999b). Solvents used in immersion tanks were usually recycled on-site by filtering or distilling the solvent (AFCEE, 1999; PRO-ACT, 1999).

1.1.1.3 Vapor Degreasing

Degreasers have played a vital role in industrial manufacturing, particularly in the automobile, heavy equipment, aircraft, appliance, railroad, and electronics industries. Approximately 25,000 vapor degreasers operated in the United States in 1979 (Tarrer et al., 1989). Degreasing is favored for the removal of oils imparted to the item being cleaned ("the work") from cutting, molds, dies, welding, soldering, stamping, and processing, as well as for the removal of soil carried in cutting oils, such as metal fines and fluxes. Parts that have been die cast, stamped, machined, welded, soldered, or molded carry residues that can be cleaned by vapor degreasing. Vapor degreasers introduce solvents to small inaccessible crevices or "blind holes," allowing quick and efficient cleaning and drying (Mertens, 2000a). Vapor degreasing is used for removing mineral oil contamination prior to electro-cleaning and subsequent electroplating.

Vapor degreasing produces very clean, dry parts, suitable for subsequent finishing work such as the application of rust protection, primer, or paint. Clean dry parts are needed to facilitate quality control inspection and for subsequent assembly of machines. Parts requiring further metal work, such as welding, heat treatment, or machining, also benefit from vapor degreasing between steps to remove all traces of processing oils. Welding oily metal parts form carbonaceous deposits because of pyrolysis of organic matter at high temperatures; such deposits are extremely difficult to remove. Vapor degreasing oily metal parts before welding greatly reduces the subsequent cleaning requirements. Electron-beam welding requires that parts be completely free of oily wastes and is therefore preceded by high-quality vapor degreasing (Reuter, 2002). Vapor degreasing is also a final step before packing to deliver clean parts and before shipping copper, aluminum, and plated chromium, zinc, and silver products in a clean, bright, and shiny condition [American Society of Testing and Materials (ASTM), 1962].

1.1.1.3.1 Types of Vapor Degreasers

Three types of vapor degreasers are used for solvent cleaning: batch vapor, in-line machines, and non-air-interface machines. Most of the degreasing machines operating since about 1960 fall into the batch vapor category. Conveyor systems such as monorail degreasers are in-line machines.

Non-air-interface machines – vacuum degreasers that use subatmospheric pressure in the cleaning chamber – were introduced in the late 1990s to eliminate emissions and worker exposure. Energy costs in vacuum degreasers are reduced because parts can be dried at temperatures lower than the boiling point of the solvent. With a solvent such as PCE that has a high boiling point, it is possible to clean heat-sensitive parts at lower temperatures and still get them completely dry under vacuum, but the cycle time is longer (Dow Chemical Company, 1999b).

1.1.1.3.2 Vapor Degreaser Operations

In the 1960s through 1990s, vapor degreasing systems typically consisted of an open tank that is partially filled with solvent. A heater in the bottom of the tank heats the solvent to its boiling point, which generates vapors. Solvent vapors fill the tank to a level controlled by the elevation of the degreaser's cooling or condenser coils, positioned around the top of the tank's perimeter. These coils circulate cold water or a refrigerant, creating a layer of cold air that confines vapors to the tank and minimizes vapor escape. The condenser coils chill the air in the freeboard zone above the desired solvent vapor elevation. National Emission Standards for Hazardous Air Pollutants (NESHAP) require that refrigeration must create a cool air zone that is 30% or less of the solvent's boiling point. Table 1.3 lists the minimum temperatures needed for the major chlorinated solvents (Thomas, 1995).

The work requiring cleaning enters the degreaser at room temperature. When the work enters the heated vapor zone inside the tank, cool surfaces cause condensation of solvent vapors on the soiled metal surface. The condensed solvent then dissolves the oil, grease, flux, or rosin and loosens metal chips and other debris, which drip off into the bottom of the tank. Solvent vapor condensed by cooling coils is returned to a clean solvent reservoir, which overflows back into the boiling solvent compartment (Petering and Aitchison, 1945).

The cleaning action of a vapor degreasing system is dependent on the temperature differential between the vapor and the work. The efficiency of the process decreases as the work becomes warmer. Some degreaser operations use cool liquid solvent to presoak the work before bringing it into contact with solvent vapor; thereby extending the cleaning time, and promoting good solvent penetration from vapor condensation on the work (Petering and Aitchison, 1945). Some degreasing operations use multiple dips, where the work is moved from a dirty tank to subsequently cleaner tanks, followed by a final cleaning in clean solvent vapor (Dow Chemical Company, 1999b). The higher boiling point of PCE prolongs the condensation effect in a continuous degreaser because it takes more time to bring the cold work to PCE's higher vapor temperature. Compared to the solvents that have lower boiling points, PCE is therefore considered to have better cleaning power per unit of time that the work spends in a continuous vapor degreaser (Johnson, 1973).

Some degreaser systems are equipped with a wand that can be used to spray warm solvent onto the work to complete the cleaning process. The physical action of the spray pressure knocks off the

TABLE 1.3
NESHAP Temperature Requirements for Freeboard Refrigeration Devices in Vapor Degreasers

Solvent	Boiling Temperature [°C (°F)]	Required Air Blanket Temperature [°C (°F)]
Dichloromethane	40 (104)	0 (31)
Methyl chloroform	74 (165)	10 (50)
Trichloroethylene	87 (189)	13 (57)
Perchloroethylene	121 (250)	24 (75)

Sources: Thomas, T., 1995, Clean Air Act compliance for solvent degreasers – regulatory strategies for manufacturers affected by Clean Air Act amendments NESHAP for halogenated solvent cleaners. University of Tennessee Center for Industrial Services, Tennessee Department of Environment and Conservation.

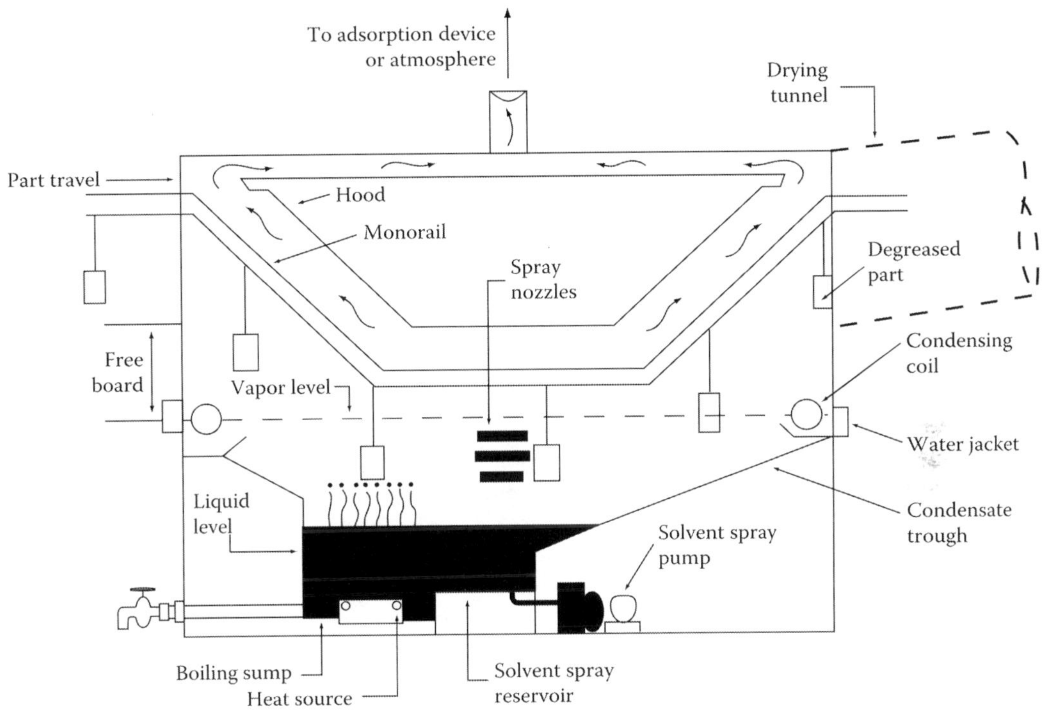

FIGURE 1.6 Configuration of typical offset vapor degreaser. (Adapted from University of Tennessee Center for Industrial Services, 1995.)

strongly bound soils (AFCEE, 1999; PRO-ACT, 1999; Dow Chemical Company, 1999b). However, high-pressure sprays can disrupt the vapor blanket, leading to solvent vapor losses, unless care is taken to spray the work beneath the vapor blanket or inside an enclosed compartment of a vapor degreaser. A diagram portraying the features of a typical vapor degreaser is shown in Figure 1.6.

Solvent vapors usually remain below the level of the condensing coils; nevertheless, open-top degreasers routinely incurred substantial solvent losses. Solvent vapors diffuse from the dense solvent vapor blanket into the overhead air mass at a rate dependent on solvent vapor density and boiling point and the temperature differential with the ambient air. Tests have shown that vapor loss from an idling open-top vapor degreaser can reach 0.20 (lb ft^2)/h for TCE, 0.26 (lb ft^2)/h for dichloromethane, and 0.29 (lb ft^2)/h for PCE (Dow Chemical Company, 1999b).*

1.1.1.3.3 Solvent Losses

Understanding the rate of solvent losses from degreasers gives an appreciation of the frequency at which solvents were replenished. The rate of solvent handling provides information about the volume of frequent minor losses from spills, drips, container waste, and other factors contributing to solvent releases associated with topping off or changing out solvents.

Solvent loss occurred in open-top vapor degreasers when the vapor blanket was disrupted by drafts, movement of the work through the degreaser on a conveyor system, drag-out by the work as it exits the vapor blanket, incomplete drying of the work, and frequent opening of closed degreasers during operation. Other causes of solvent losses included diffusion, discarded solvent in degreaser and still sludges, and leaks from solvent pipes and valves in the degreaser. Operator guidance since the 1960s has advised that the work should move through the solvent vapor blanket

* Based on tests in an idling open-top degreaser having a 24 × 58 in. (~10 ft^2) opening; units are pounds of solvent per square feet of exposed solvent per hour of idling.

no faster than 11 ft/min (ASTM, 1962; Thomas, 1995). Where problematic solvent levels are measured in the operator zone around the degreaser, operator guidance calls for slowing the conveyor rate to 3 ft/min (Howell and Tarrer, 1994). Solvent may also be lost when the condensed solvent fails to drain because of "cupping" attributes in the shape of the work, leading to liquid solvent carryout. Estimates for solvent losses due to these combined factors are as high as 70% of the solvent added to a degreaser over the course of a year (Mertens, 2000a). A description of degreasing practices from the early 1960s detailed an approach to solvent vapor management in which no hood is needed; in this approach, a large portable blower and roof ventilator are sufficient to vacate solvent vapors (Stoddard and Wells, 1962).

Solvent consumption rates due to emissions have been estimated at 1–2 gallons solvent per ton of small parts cleaned when a covered, in-line conveyor system is used, whereas small parts cleaned at a similar rate in manually operated open-top degreasers consumed 20% more (ASTM, 1962). The solvent and manufacturing industries substantially decreased solvent emissions from degreasing equipment over the decades during which chlorinated solvents were used for vapor degreasing. Improvements were motivated first by cost savings and later by regulation and enforcement of emissions standards. Modern vapor degreasing equipment is designed with vapor locks, vapor-detection devices, and other controls, so that solvent emissions are substantially reduced. Even so, vapor degreasers equipped with emission-control devices can still lose 5–25% of the solvent through vapor escape. In enclosed (non-air-interface) degreasers, almost all of the solvent is captured and condensed for reuse. Solvent losses in enclosed degreasers are less than 5% (Reuter, 2002).

Table 1.4 summarizes annual solvent loss to emissions and annual generation of waste solvent in degreasers studied at several Air Force bases in the southeastern states. The study shows that methyl chloroform degreasing operations generally produced more liquid waste than PCE degreasing operations. Differences in emissions and waste generation were also the result of differences in the quantity of parts cleaned and in the types of cleaning operation conducted in each vapor degreaser.

Table 1.5 provides a manufacturer's estimate of solvent losses at a major printed circuit board operation where Freon 113 was the primary solvent used. Although evaporative losses would be higher for Freon 113 than for chlorinated solvents, the values provide a general estimate of quantities released because of spills and leaks.

1.1.1.3.4 Water Separators

A smaller but significant source of solvent losses and possible releases to soil and groundwater involves the management of condensate water produced in vapor degreasers. Water enters degreasers when atmospheric water vapor condenses on the degreaser cooling coils, from moisture present

TABLE 1.4

Annual Solvent Waste Generation and Emission Losses in Vapor Degreasers at Air Force Bases in Southeastern States in the Late 1980s and Early 1990s

Solvent	Reservoir Volume (gallons)	Batch Changes	Annual Usage (gallons)	Annual Waste (gallons)	Emissions Loss (%)
Perchloroethylene	35	6	1,780	480	73
Perchloroethylene	50	17	1,824	420	77
Perchloroethylene	110	4	2,740	968	65
Methyl chloroform	50	10	654	275	58
Methyl chloroform	110	6	3,850	990	74
Methyl chloroform	110	8	4,180	1,320	65

Source: Howell, S.G. and Tarrer, A.R., 1994, Minimizing pollution in cleaning and degreasing operations. EPA/600/SR-93/191.

TABLE 1.5
Typical Solvent Losses in a Major Printed Circuit Board Manufacturing Facility

Solvent Handling Process	Cause of Loss	Solvent Lost (%)
Storage in holding tanks	Evaporative losses	1
	Leaks	1
Transport within facility	Spills	8
Vapor degreasing	Evaporative losses, leaky seals, etc.	12
	Drag-out	40
	Spills and leaks	8
On-site recovery by distillation	Evaporative losses, leaky seals	2
Hand cleaning	Evaporative losses	15
Off-site for recycling of waste solvent	—	18

Source: U.S. Environmental Protection Agency (USEPA), 1991, Conservation and recycling practices for CFC-113 and methyl chloroform. U.S. Environmental Protection Agency, Office of Air and Radiation. EPA/400/1-91/017.

Note: The primary solvent used at the facility was Freon-113; loss rates for other solvents may have lower evaporative losses.

on the work being cleaned, from leaks in the cooling water coils or steam heating lines in the solvent boiler, and from water-soluble cutting oils and buffing compounds (ASTM, 1962; Dow Chemical Company, 1999b). As discussed further in Section 1.2, the presence of water in the degreaser may lead to hydrolysis of the solvent, removal of water-soluble solvent stabilizers, and corrosion of the metal surfaces in the degreaser and on the work. Presence of water in the solvent is indicated by spots or even rust on the metal parts after cleaning. Another indication of excess water in the solvent is the formation of a dense white fog in the vapor zone, called "ghosting."

Vapor degreasers are equipped with water separators. Condensed water and solvent drop into a trough below the condenser coils and drain to the separator. Because water floats atop the solvent and does not dissolve significantly in the solvent, the water can be siphoned off. A relatively moisture-free solvent passes through the solvent return line at the bottom of the water separator and is then discharged to the degreaser or the storage tank (ASTM, 1962; Dow Chemical Company, 1999b). Operator guidance called for flushing with fresh water and keeping the water separator full by periodically adding solvent and water. Flushing prevented the buildup of corrosive acids from hydrolysis of solvents and corrosion of water separator pipes and tanks. The guidance specified that the pH of water in the separator should be monitored. If operators found the water layer to be acidic, it was replaced or buffered with an addition of a small amount of sodium bicarbonate to maintain the pH above 7. Some operations flushed the water layer weekly (Dow Chemical Company, 1999b). Operators were also directed to open the valve that drained the water off the top of the water separator frequently to allow accumulated water to drain out (ASTM, 1962). Wastewater from degreasers can be expected to contain solvents at concentrations equal to the solvent's solubility in hot water, well above its commonly cited solubility at 25°C. For example, at 75°C: PCE, ~250 mg/L; TCE, ~1,485 mg/L; methyl chloroform, ~1,575 mg/L; however, dichloromethane's solubility decreases with increasing temperature.

The volume of water condensed in a vapor degreaser is largely a function of the humidity at the degreaser location. The fate of water removed from degreasers varied with regulations in effect while the degreaser was operated and with the degree of operator compliance, as well as regulatory outreach and enforcement. Prior to regulation, degreaser wastewater was discharged to sewers, dry wells, storm drains, evaporation ponds, or the ground, or placed in drums and sent to unregulated landfills. Under actively enforced regulation, degreaser wastewater is managed through pretreatment using carbon filtration or on-site distillation, and waste discharges are subjected to inspection and sampling by local agency industrial wastewater pretreatment inspectors. After the mid-1980s,

TABLE 1.6
Solvent Boiling Temperature for Oil-Laden Solvent

Solvent	Normal Boiling Temperature [°C (°F)]	Boiling Temperature at 25% Oil [°C (°F)]
Dichloromethane	39.7 (104)	42.2 (108)
Methyl chloroform	73.8 (165)	78.8 (174)
Trichloroethylene	87 (189)	89.4 (193)
Perchloroethylene	121 (250)	125 (257)

Sources: ASTM, 1962, Handbook of Vapor Degreasing. American Society for Testing and Materials Special Technical Publication No. 310; Johnson, J.C. and Wedmore, L.K., 1983, Metal cleaning by vapor degreasing. Metal Finishing 81(9): 59–63; and Dow Chemical Company, 1999b, Degreasing: Economical and efficient degreasing with chlorinated solvents from DOW.

solvent-contaminated wastewater was managed as an RCRA (Resource Conservation and Recovery Act) waste, and discharge to sewer lines was prohibited. However, a common release mechanism for solvents from degreasing operations before and after RCRA implementation was through spills onto cracked concrete on shop floors or from poorly maintained sumps beneath the degreasers.

1.1.1.3.5 Heating Solvents

Vapor degreasers heat solvents to their boiling points with electric, steam, or gas heaters. Accumulation of sludge and metal fines on the heating elements may cause "hot spots" that result in solvent decomposition and burnout of the heating elements (Campbell, 1962). Dow Chemical advised operators to remove degreasing sludge when the solvent reached boiling temperatures listed in Table 1.6. To remove sludge, operators boiled solvent down to within 2 inches of the heating coils while diverting the solvent return line to a clean drum. The solvent sludge mixture was then drained and placed in drums for further distillation or disposal. Solvent cannot be recovered from solvent sludge mixtures that approach 60–70% oil because distillation becomes ineffective (Dow Chemical Company, 1999b).

1.1.1.3.6 On-Site Solvent Recovery in Degreasing Operations

The primary costs for operating vapor degreasers include solvent, energy for heating, and labor. New designs of degreaser covers and fume hoods to recover solvents from vapors were motivated by worker protection, rising solvent costs, and by solvent losses of up to 70% in open-top vapor degreasers. Solvent recovery from degreaser vapors became required with the introduction of standards for permissible worker exposure and restrictions on air emissions. The most common method for recovering solvent from vapors is carbon adsorption. Carbon cartridges designed to allow high air flows adsorb most of the solvent vapor. Periodically, steam is passed through the carbon bed to carry the vapors to a condenser. A water separator removes the water from the solvent, which is then recovered and returned to storage tanks. As discussed in Section 1.2.7.5, carbon adsorption also removes some solvent stabilizers from the solvent, while introducing water contamination to the solvent. Carbon bed desorption is also accomplished by using hot air flow on a daily 6–8-h cycle; solvent vapors are condensed and returned to the solvent reservoir in the degreaser, and the carbon is regenerated for further use (Reuter, 2002). Cryogenic solvent-recovery techniques are also used (Douthitt, 1990).

1.1.1.4 Dry Cleaning

Dry cleaning is the term of art for a number of textile-cleaning processes applied to clothing, fabric, rugs, draperies, and other textile articles, which uses petroleum solvents and halogenated solvents.

The term "dry" refers to the absence of water as a primary means for laundering clothes. Clothes obviously need to be cleaned regularly; laundering with water and detergent is insufficient for some fabrics and some soils and does not produce the same degree of cleanliness as can be obtained with chlorinated solvents. The details of how PCE was used, filtered, recovered, discharged, and vented from dry cleaning operations relate to the nature and volume of PCE releases and to the partitioning and release of solvent stabilizers, as discussed in Section 1.2.7.

PCE became the solvent of choice for the dry cleaning industry after surpassing petroleum solvents approximately in 1960. Carbon tetrachloride, TCE, methyl chloroform, and chlorofluorocarbons have also been used in the dry cleaning industry. In 2000, more than 80% of dry cleaners used PCE; some still used petroleum solvents, and a growing number of operators are switching over to machines that use carbon dioxide or other solvents or cleaning agents deemed to be "green," with fewer regulatory restrictions and environmental consequences.

The dry cleaning industry is composed of two main sectors: the commercial sector and the industrial sector. Previously, the coin-operated dry cleaning sector was economically important, but machine inefficiency and regulation led to its elimination. Coin-operated dry cleaning machines were banned in December 1994. In the early 1990s, the U.S. dry cleaning industry included about 30,500 commercial dry cleaners, 1,400 industrial cleaners, and 3,000 coin-operated dry cleaners (USEPA, 1995). Industrial dry cleaners are large operations that clean uniforms, aprons, overalls, shop towels, or other items. Approximately 50% of the industrial dry cleaners use petroleum solvents. A typical industrial cleaner of clothing may have a 500-pound capacity washer-extractor and 3–6 dryers, each with 100-pound capacity. The annual solvent usage ranges from 200 to 1,300 gallons per machine (USEPA, 1995).

1.1.1.4.1 Dry Cleaning Process

The descriptions in this section apply primarily to dry cleaning operations that used earlier-generation machines from the 1950s through the early 1990s. Operations from this timeframe contributed substantially more PCE to the subsurface than current or recent operations. Since 1986, there have been substantial improvements in solvent stewardship and dry cleaner operator training, adoption and enforcement of regulations, and improvements in equipment (Mohr et al., 2007).

The PCE dry cleaning process consists of washing clothes in liquid solvent, extracting solvent from the clothes, and drying the clothes. Two general categories of dry cleaning machines were used: transfer machines and dry-to-dry machines. Transfer machines performed the washing and extraction steps; clothes were then manually transferred to a separate dryer. Dry-to-dry machines perform all the three steps in the same machine (USEPA, 1980). Transfer machines have been prohibited, but their past operations had greater potential to contribute to PCE releases to soil and groundwater than dry-to-dry machines.

Dry cleaning involves two wash cycles, each followed by drying. In the first wash, soiled garments are loaded into a rotating, perforated, stainless-steel cylindrical basket with a capacity varying from 20 to 100 pounds of clothes or fabrics. Clothes are completely soaked and tumbled in detergent-charged solvent for 8–12 min at a temperature ranging from 70–85°F. A typical machine may recirculate PCE into the rotating basket filled with soiled clothing at a rate of 1,500 gallons per hour; for an 8-min cleaning cycle, soiled clothes would be soaked in 200 gallons of solvent (Cantin, 1992). The detergent-charged solvent dissolves greases and oily stains and is continually filtered during the wash cycle. Soil-laden solvent is first pumped through a button trap to remove pins and buttons from the solvent to prevent damage to the pump. Solvent charged with detergent is then pumped through either a disposable cartridge filter, a spin-disk filter, or a regenerative or flex-tube filter to remove soil, lint, and other insoluble materials (CARB, 1996).

A typical dry cleaning machine has one or two PCE solvent tanks: one for the pure solvent and the other for the "charged" solvent, made up of distilled solvent with a small amount of detergent and water to help clean water-soluble stains. Solvent is recycled through continuous distillation. Fresh solvent is added to the solvent tanks to replace the solvent lost to emissions, filters, and water separation.

To extract PCE from washed clothes, the solvent bath is drained; then clothes are spun at high speed to wring out excess solvent. Both transfer and dry-to-dry machines combined the washer-extractor functions in a single unit. In transfer machines, an operator moved clothes from the washer-extractor to the dryer-reclaimer. In the solvent-reclaiming dryer, heated air is blown over the clothes in a closed, recirculating loop to vaporize the solvent, which is recovered in a water-cooled condenser. After a timed drying step lasting 15–20 min, fresh air is blown over the clothes for approximately 5 min to complete the drying cycle and aerate the clothes. The air stream and PCE vapors are vented to a carbon adsorption system (USEPA, 1980).

In the second wash, clothing is rinsed in fresh, pure solvent for 4–5 min to remove detergent residue. Like the charged solvent, the pure solvent is filtered continuously by one of the three filtration systems described in Section 1.1.1.4.9. Rinsed clothing is rapidly spun to extract the solvent. The spent solvent is drained into the charged solvent tank and is used for the first wash for the next load of soiled clothing.

Freshly cleaned clothing is heated to about 160°F for approximately 40 min. The remaining PCE fumes and solvent are vaporized by warm air and then condensed over cooling coils to recover the solvent. The residual amount of PCE removed from clothing following the second wash ranges from 30 to 60 pounds for 100 pounds of clothing (Cantin, 1992). Most of the PCE is removed from the clothing via the last two extraction and drying steps; however, trace levels of PCE may remain in dry cleaned clothing.

1.1.1.4.2 First-Generation Dry Cleaning Machines

Transfer machines first appeared on the market in the 1930s and were used exclusively until the 1960s (Schmidt et al., 2001). In 1995, 34% of the dry cleaning machines in the United States were transfer machines. Transfer machines were banned in California on October 1, 1998 (Izzo, 1992). Transfer machines usually included a washer-extraction unit and a separate dryer unit. Clothing and solvent are agitated in the drum, which is then spun at high speed to extract the solvent. An attendant must manually transfer the clothing to a dryer. The clothing is tumbled in recirculating warm air to vaporize the residual solvent. After the drying cycle, cool air is circulated through the clothing to reduce wrinkling. Some transfer machines comprised a washer unit, a separate spin-dry extraction unit, and a tumbler dryer or reclaimer to vaporize the remaining solvent and fumes.

Transfer machines allowed increased productivity because a new load of clothing can be washed while the previous load is drying. Compared with dry-to-dry machines, maintenance of transfer machines was easier because of their basic construction with less automation. There was also less fabric damage because the drum remains cool after the previous load is removed (USEPA, 1995). The disadvantages of transfer machines included additional labor needed to manually transfer the solvent-soaked clothing from the washer-extractor to the dryer, worker exposure to solvent vapors during transfer of clothing, and loss of solvent during transfer.

To minimize worker exposure to PCE vapors during transfer of clothing from the washer to the dryer, a hamper enclosure made of impervious plastic was used to enclose the clothing when the clothing was transferred from the washer to the dryer. In a second approach, a room enclosure, composed of a metal frame covered with an impervious plastic, surrounded both the washer and dryer units and drew solvent vapors through a device to control vapor emissions. First-generation machines consumed and emitted more solvent than later-generation machines. PCE consumption in first-generation machines was typically in the range of 78–100 kg per 1,000 kg of clothes cleaned, depending on the primary control technology as well as on operator and maintenance practices (CARB, 1996).

1.1.1.4.3 Second-Generation Vented Dry-to-Dry Machines

Second-generation dry cleaning machines were the vented dry-to-dry machines introduced in the late 1960s. Washing, extracting, and drying were done in one unit, thereby avoiding the transfer of solvent-soaked clothing from the washer to the dryer. Second-generation machines were originally designed to release residual vapors to the atmosphere, but they could be retrofitted with a vapor-control device such as carbon absorbers or refrigerated condensers. In 1995, of U.S. dry cleaners, 66% had vapor-control devices consisting of refrigeration units, and 32% had vented units without vapor-control devices (USEPA, 1995).

Carbon adsorbers were installed on both transfer and vented dry-to-dry machines as retrofits to improve solvent recovery. Instead of venting solvent vapors directly to the atmosphere, retrofitted machines recover solvent by directing the solvent-laden vapors through an activated carbon bed that adsorbs the solvent. The adsorbed solvent is recovered by passing low-pressure steam or hot air through the carbon bed; then the mixed steam and solvent vapor are passed through a water-cooled condenser and collected in a phase separator. The desorbed activated carbon bed is dried and reused, and the recovered solvent is returned to the solvent tank. The removal efficiency is about 85% compared to noncontrolled vented dry-to-dry machines (USEPA, 1995).

1.1.1.4.4 Third-Generation Closed-Loop (Nonvented) Dry-to-Dry Machines

Third-generation or closed-loop dry-to-dry machines with built-in refrigerated condensers came on the market in the late 1970s. A third-generation machine is essentially a second-generation machine retrofitted with a refrigerated condenser. Closed-loop machines do not vent solvent vapors to the atmosphere; instead they recycle it continually throughout the dry cleaning cycle. In 1995, 34% of dry cleaning facilities used third-generation machines.

1.1.1.4.5 Fourth-Generation Closed-Loop (Nonvented) Dry-to-Dry Machines

In the early 1990s, fourth-generation machines came on the market. They feature a nonvented closed-loop dry-to-dry approach and have an additional internal vapor-recovery device, usually a built-in carbon adsorber installed in series with the refrigerated condenser. Another common feature of fourth-generation machines is secondary containment below the machine to capture accidental releases of PCE. Increased vapor controls in fourth-generation machines provided nominally higher solvent mileage than third-generation machines (CARB, 1996).

1.1.1.4.6 Fifth-Generation Closed-Loop (Nonvented) Dry-to-Dry Machines

Fifth-generation machines were introduced in the late 1990s. These machines are fitted with a single-beam infrared photometer to monitor the PCE concentration in the machine cylinder. An interlock on the machine door prevents the attendant from opening the door until PCE concentration in the cylinder falls below 290 ppm.

While fourth- or fifth-generation machines can achieve a factory solvent-mileage design rating of 10 kg of solvent per 1,000 kg of clothes, operator and maintenance practices are critically important for achieving the design performance. During sustained commercial operation, however, an average solvent mileage of 20 kg of PCE per 1,000 kg of clothing cleaned is a reasonable target for the dry cleaning industry as a whole. Therefore, fifth-generation machines provide a 5–10-fold improvement in solvent mileage compared to first-generation machines (USEPA, 1995). Table 1.7 summarizes the PCE mileage for five generations of dry cleaning machines.

1.1.1.4.7 Self-Service Coin-Operated Dry Cleaning Machines

Self-service coin-operated dry cleaners were introduced to the market in the early 1960s. Coin-operated machines were usually located in self-service laundries and were operated by consumers. The coin-operated machines are small dry-to-dry machines with 8–10 pounds clothing capacity that used PCE. These machines were inefficient, and maintenance issues invariably led to spills and emissions violations. Coin-operated dry cleaning machines were prohibited beginning in 1994.

1.1.1.4.8 Chlorinated Solvents Used for Dry Cleaning

By 1915, carbon tetrachloride provided a nonflammable alternative to the hydrocarbon solvents used for dry cleaning at the time. It was used by the majority of dry cleaners in the United States in

TABLE 1.7
Solvent Mileage by Machine Type

Machine Type	Perchloroethylene Consumption per 1,000 kg of Clothes Cleaned
First-generation transfer	78–100 kg (8–10%)
Second-generation dry-to-dry	77–94 kg (8–10%)
Third-generation closed loop	20–40 kg (2–4%)
Fourth-generation closed loop with refrigeration	10–20 kg (1–2%)
Fifth-generation closed loop with vapor locks	10–20 kg (1–2%)

Source: Mohr, T.K. et al., 2007, *Study of Potential for Groundwater Contamination from Past Dry Cleaner Operations in Santa Clara County*. Santa Clara Valley Water District: San José, CA.

the years prior to World War II. Carbon tetrachloride was replaced by PCE and TCE as dry cleaning and degreasing solvents because they are less toxic, easier to recover, and easier to ship. By 1950, three times as many dry cleaners were using PCE compared to those using carbon tetrachloride. Carbon tetrachloride is no longer used in the dry cleaning industry.

TCE was first used as a dry cleaning solvent in Germany in the 1920s and was introduced in the United States for this purpose in about 1930. TCE was more easily reused and cleaned faster and more safely than carbon tetrachloride. However, TCE damaged certain types of fibers and synthetic substances. Because of human toxicity and photochemical reactivity concerns, TCE is no longer used in the dry cleaning industry (Johnson, 1973).

PCE was introduced to the U.S. dry cleaning industry in the late 1930s. PCE became the solvent of choice for most small dry cleaners because it is not flammable, it safely cleans virtually all types of fabrics without damaging the material or bleeding dyes, and it is less toxic than carbon tetrachloride. As of 1952, the dry cleaning industry used 80% of the PCE produced. However, from that point forward, the rate of PCE used for dry cleaning decreased. Between 1972 and 1975, PCE used for dry cleaning decreased from 75% to 63% of the total production (Doherty, 2000a). The decline in PCE consumption corresponded to improvements in dry cleaning equipment and vapor-recovery systems, as well as to trends toward wash-and-wear clothing. In 1975, one 55-gallon drum of PCE cleaned 8,000 pounds of clothing, whereas in 1993, the same amount cleaned 16,000 pounds of clothing. By 1990, however, only 50% of the PCE produced was used for dry cleaning. Improved efficiencies in handling and recovery have caused dry cleaner PCE consumption to substantially decrease although PCE still dominates the U.S. dry cleaning industry. In spite of 10-fold improvement in solvent mileage and major progress with vapor recovery, dry cleaner operators are increasingly pressured to find alternatives to PCE. By January 1, 2023, all dry cleaners operating in California must convert to alternative solvents that are less toxic and do not contribute to the formation of ground-level ozone (smog).

In 1964, E. I. du Pont de Nemours and Company, introduced 1,1,2-trichloro-1,2,2-trifluoroethane, also known as Freon 113 or CFC-113, under the trade name Valclene™ (Johnson, 1973). Valclene™ is less aggressive than PCE and evaporates from fabrics at lower temperatures, so it was promoted for cleaning delicate fabrics. Valclene™ was preferred over PCE for rubber, painted fabrics, polystyrene, vinyl-coated fabrics, polyvinyl chloride fibers, some pigment colors, furs, and leathers (Johnson, 1973). Although Freon 113 had favorable attributes for delicate clothing, it was not widely used because it was more expensive and evaporated much faster than PCE (Morrison and Murphy, 2013). Use of Freon 113 for dry cleaning was banned in 1996 in accordance with the Montreal Protocol because of its potential to deplete ozone and contribute to global warming (Johnson, 1973; CARB, 1996). In 1975, 2 million pounds of Freon 113 were used for dry cleaning, whereas 3,345 million pounds of PCE were used for dry cleaning in the same year. In 1986, 2.2% of dry cleaning facilities used Freon 113, according to data from the U.S. Department of Commerce. A survey of 300 southern California dry cleaners in 1988 and 1989 also found that 2% used Valclene™ (Freon 113) (Morrison and Murphy, 2013). One consequence of Valclene™ use at a dry cleaner is it precludes the use of CFC-113 as an atmospheric tracer for determining of the age of a PCE release at the same facility. Where Freon 113 was used and spilled at dry cleaners (and at electronics manufacturers), methane reducing geochemical conditions can promote biodegradation and formation of chlorotrifluoroethylene, a toxic homolog to vinyl chloride that is often overlooked in site investigations.

In the early 1980s, Dow Chemical introduced methyl chloroform as a dry cleaning solvent (Linn et al., 2004). It was not widely used in the dry cleaning industry as a primary dry cleaning solvent because it damages delicate fabrics. Methyl chloroform is also somewhat unstable when mixed with water, where it breaks down and forms acids that corrode dry cleaning machines (CARB, 1996). Methyl chloroform was also restricted from most uses following the passage of the Montreal Protocol and Clean Air Act Amendments (Johnson, 1973; CARB, 1996).

1.1.1.4.9 Filtering Perchloroethylene

Three types of disposable cartridge filters are used by 90% of dry cleaners: carbon filters, paper filters, or combination paper and carbon filters. Disposable cartridges are changed after cleaning 450–1,000 pounds of clothing. Operators observe the solvent line pressure to determine when filters need to be changed, as indicated by increased pressure in the solvent line behind the filter (Cantin, 1992; CARB, 1996). Operators typically drain used or spent cartridges for 24 hours for paper filters or 48 hours for carbon cartridges prior to reuse or disposal at a facility that conducts hazardous waste treatment. Spent cartridges can retain as much as 1 gallon of PCE before draining, and from 0.13 to 0.25 gallon PCE remains in the cartridge after draining (CARB, 1996). The filter cartridge became prominent in the dry cleaning industry in about 1961.

A smaller number of dry cleaners use spin-disk filters. Spin-disk filters are made up of multiple high-surface-area polyester disks mounted in series to filter the dirty solvent. Spin-disk filtration systems may incorporate powdered activated carbon and diatomaceous earth. Filters are regenerated by spinning the disks and directing the solvent and the "caked" impurities through a drain valve into a distillation still. Spin-disk filters last longer and produce less waste than the cartridge systems, but skilled labor and frequent maintenance are required for proper operation (Hygnstrom, 1995).

A third type of filter is the regenerative or flex-tube filters, made of braided tubes coated with a combination of diatomaceous earth and activated carbon. The filter is cleaned by resuspending the powder and dirt in solution by a countercurrent of air and backwashing; the resuspension creates a mixture that is then redeposited back on the filter to form a highly porous "cake" and uniform filter media for the proper removal of contaminants. The regeneration action is repeated with each cleaning cycle. Regenerative filter systems are more expensive than spin-disk filters and require skilled labor to operate properly (CARB, 1996).

1.1.1.4.10 On-Site Distillation for Perchloroethylene Recovery

Approximately 90% of dry cleaners use distillation as a companion process to purify and recover used solvent (Cantin, 1992). Distillation stills are built into most modern dry cleaning machines; stand-alone stills, called "muck cookers," were used for earlier-generation machines. Steam is used to heat the PCE to its boiling point (121°C/250°F). Solvent and water vaporize and rise to the top of the still, where vapors are cooled back into a liquid that then goes to a water separator. Sludge or "still bottoms" – composed of nonvolatile residues such as detergents, waxes, dyestuffs, sizing, oils, and grease – accumulate in the sump and are periodically remove and disposed of as hazardous waste. The sludge or still bottoms can contain as much as 50–70% PCE. A "muck cooker" is used to recover most of the PCE. A cooker or still can recover as much as 95% of the solvent from the spent sludge (Morrison, 2002).

1.1.1.4.11 Water Separator

Water separators have been used in solvent distillation stills for dry cleaners and vapor degreasers since the early 1900s. The distilled solvent and water mixture sits in a gravity separator unit. Heavier PCE drains from the bottom of the unit to the solvent tank, and excess water is drained off the top of the separator. In earlier decades, separated water was drained to floor drains or onto the ground; later it was routed to the sanitary sewer system (Morrison and Murphy, 2013). In the past, it was legal to discharge wastewater to the sanitary sewer system, and high levels of PCE in the separator water were routinely so discharged. For example, levels of PCE up to 1,120 mg/L, with an average of 152 mg/L, were detected in samples of separator water; as much as 30% of the samples were pure solvent (Cohen and Izzo, 1992). Among older dry cleaning machines (ca. 1986), discharges from water separators commonly caused soil and groundwater contamination. Leaking sewer lines also facilitated PCE release to the subsurface (Cohen and Izzo, 1992).

1.1.1.5 Off-Site Solvent Recycling

Industrial plant managers and industrial engineers found various ways of dealing with the costs of solvent consumption and disposal of waste solvent. Improvements in degreasing and dry cleaning equipment significantly reduced consumption. As solvent costs rose, operators switched to more efficient equipment. When regulations prohibited land disposal of waste solvent, the cost of permitted disposal was a further incentive for improving the efficiency of solvent use in various operations and to use on-site solvent-recovery operations. Table 1.8 summarizes the relative ranks of solvents recycled in the United States in 1981.

Prohibitions of land disposal of waste solvent redirected most of this waste stream to the solvent recycling industry, which collected still bottoms from degreasing operations for distillation and solvent reclamation. Solvent recycling was already a viable business before the adoption of land disposal prohibitions because it offered attractive economic incentives for the generators of waste solvent. Recovered solvents were sold, and distillation residues were discarded or treated as hazardous waste. Prior to the adoption and enforcement of the Resource Conservation and Recovery Act in 1985, still bottoms and other waste solvent from recycling operations were generally sent off-site; the waste went into landfills and deep injection wells or was burned as fuel for blast furnaces in the steel industry or as fuel for cement kilns. In 1985, because steel mills abruptly stopped taking waste solvent to avoid the burden of obtaining permits as hazardous waste storage facilities, cement kilns became a primary destination for waste solvent (Douthitt, 1990). The oil residue in waste solvent provided enough heat energy (8,000–9,000 Btu) to make the waste attractive as fuel for kilns and boilers, provided the metal and chlorine contents were sufficiently low. Cement kilns burn waste at

TABLE 1.8
Ranks of Chlorinated Solvents by Relative Quantity Recycled in 1981

Solvent	Solvent Recycling Rank
Methyl chloroform	2
Dichloromethane	5
Trichloroethylene	7
Perchloroethylene	8

Source: Douthitt, C.A., 1990, Resource recovery through solvent reclamation. Presentation to the U.S. Department of Energy Industrial Solvent Recycling Conference October 15, 1990, Charlotte, NC. NASR (National Association of Solvent Recyclers). www.p2pays.org/ref/26/25692.pdf (accessed February, 2006).

Note: Other recycled solvent ranks: methyl ethyl ketone (#1), acetone (#3), xylene (#4), and toluene (#6).

temperatures above 2,600°F, assuring destruction efficiencies comparable to those of an incinerator (NCDHS, 1983; Dawson, 1989).

Recovered solvents were considered to be of good quality, meeting the same technical specifications required for virgin solvents, although solvent producers cautioned customers about the impurities in and instability of recycled solvents (NCDHS, 1983). Recycling companies restabilized reclaimed solvents and sold "restabilizers" to companies operating on-site solvent distillation units (NCDHS, 1983; see Section 1.2.7.6 for more information on replenishing stabilizers in reclaimed solvents).

Commercial solvent-recycling operations used two pricing models: custom toll recycling, which served generators of large quantities of waste solvent, and open-market recycling. In custom toll recycling, the generator's spent solvents are kept segregated, batch processed separately to the generator's specification, and then returned for reuse. The minimum batch size for toll recycling, determined by processing equipment capacity, was generally 1,000–2,000 gallons (Dawson, 1989). The toll recycling approach provided assurance to the generator that clean solvent returned for reuse would not contain any contaminants foreign to the generator's operation. The reclaimed solvent is returned to the generator at a price similar to or slightly lower than virgin solvent purchase, which was a cheaper alternative to the off-site disposal of the waste solvent (Dawson, 1989).

In open-market recycling, spent solvents are commingled with similar wastes from many generators and processed for resale as refined solvents. Because the recycler derives revenue from the resale of recycled solvents, the generator may pay a price lower than the alternative disposal cost, while obtaining assurance that the waste is managed in an environmentally acceptable manner.

The prevailing method for solvent reclamation is distillation or vaporization. Many solvents can be processed into a reusable state by simple vaporization in either a pot-still or a thin-film evaporator. In a pot-still evaporator, the spent solvent is heated to boiling by steam or hot oil heat exchangers. Solvent vapors exit the system to a water-cooled heat exchanger and are then condensed to provide clean solvent. The nonvolatile residues, or still bottoms, are pumped to storage where they await final disposal or transport to a cement kiln or boiler for burning.

In thin-film evaporators, waste solvent is spread in a thin layer on an internal surface by rotating wiper blades. The internal surface is heated by a steam or hot oil jacket, causing organic solvents to undergo flash evaporation. Vapors are condensed to produce clean solvent liquid. Nonvolatile residues continuously exit the bottom of the evaporator and are handled in the same manner as still bottoms. Thin-film evaporators are fed from an external storage tank and can therefore operate continuously. In thin-film evaporators, nonvolatile residues are removed without prolonged heating, thereby avoiding chemical breakdown that can cause odor or color problems in the reclaimed solvent (Dawson, 1989).

Both pot-still and thin-film evaporators can be operated under vacuum to reduce the boiling point of the solvent and the amount of energy required. Neither system is capable of separating different chlorinated solvents. If the waste solvent is properly segregated from the waste solvent of a different composition, the recovered solvent will be relatively pure. If volatile organic impurities are present in the waste, they will also be present in the reclaimed solvent. Some solvent recyclers install fractionating columns – including packed columns, bubble-cap columns, sieve-plate columns, and valve-tray columns – to separate solvents and ensure good quality control. Waste solvents are heated, and vapors enter the fractionating column. Components having lower boiling points migrate to the top of the column, whereas those with higher boiling points are condensed internally and subsequently migrate to the lower part of the column. As vapors with lower boiling points exit the top of the column, they are condensed to a liquid. Fractionating columns perform better when the boiling-point difference between the mixed solvents is large, generally at least 20–30°C (Dawson, 1989).

Solvent recyclers produce refined products used in almost all phases of industry, including pharmaceutical applications to manufacture drugs. Some solvent grades require further processing following reclamation, including extraction, filtering, restabilization, and drying (water removal).

The greatest challenge faced by solvent recyclers is the removal of excessive moisture. The reuse of solvent with moisture contamination creates corrosion problems in degreasing equipment, coating problems in paint solvents, and hydrolytic breakdown problems in chlorinated solvents. Solvent recyclers remove water by using fractional distillation, if available, or physical or mechanical removal methods, such as filtering the refined solvent through a desiccant. Anhydrous calcium chloride is very effective in the removal of trace water contamination. Refined solvents contaminated by less than 0.5% moisture can be dried by calcium chloride down to 200 ppm moisture, but chloride contamination of the refined solvent may occur, which can cause corrosion problems and solvent breakdown. Ion-exchange resins are also used for the removal of trace water contamination. In solvents with less than 0.5% moisture, water molecules attach to the resin, and dry solvent passes through, reducing moisture levels to less than 100 ppm in the best possible case. Ion-exchange resins used to remove water from the solvent can be regenerated. The best available technology for moisture removal is the molecular sieve bed, which traps water molecules in the interstices of the molecular sieve, allowing the dry solvent to pass through. Molecular sieves can dry solvents contaminated with 5% water to less than 1% and can dry solvents with trace moisture from 0.5% down to less than 200 ppm. Molecular sieves can also be regenerated for continued use (Dawson, 1989).

1.2 ROLE OF SOLVENT STABILIZERS

Chlorinated solvents can become unstable when subjected to environmental stresses in the broad variety of industrial applications for which they are used. Solvents are exposed to a wide range of physical conditions and a wide range of materials and substances that place unusual demands on solvent performance. Chlorinated solvents must perform equally well in both liquid and vapor phases and must remain inert to impurities. Solvents must remain stable under challenging physical and chemical conditions, including high temperatures, ultraviolet light, moisture and water vapor, exposure to alkali and acidic substances, and exposure to reactive metals.

As this section demonstrates, solvents usually need to be fortified against the severe physical and chemical conditions of the various environments in which they are used. "Solvent stabilizers" is the term of art chosen for the group of chemicals added to chlorinated solvents to ensure that they will not break down during their intended industrial applications. Unstabilized solvents will deteriorate – some very quickly with dramatic and dangerous reactions, others slowly – over repeated use cycles and in storage. In the most demanding applications, operators must regularly replenish stabilizers to replace those consumed or lost during use. Stabilizer replenishment is commonly accomplished by adding fresh, stabilized solvent to make up for solvent losses during operation (USEPA, 1989b). The need to replenish spent stabilizers was documented as early as 1937 (Dinley, 1937). As detailed in Section 1.1, solvents are lost from the operation when vapors escape, when solvent dissolves into condensed water, and when solvent remains on parts or in textiles. In dry cleaning, stabilizers with higher affinity for sorption to carbon may get filtered out in filters that use granular activated carbon. Improvements to machines that use solvents, such as vapor degreasers and dry cleaning machines, have minimized or completely eliminated solvent losses, as required by increasingly restrictive emissions regulations. Stabilizers are nevertheless consumed in newer equipment. Stabilizer replenishment is now accomplished with specialty stabilizer packages sold by the solvent manufacturer.

1.2.1 Types of Solvent Stabilizers and Solvent Breakdown

Solvent stabilizers, also called inhibitors (and in the early literature, "anticatalysts"), are selected for their ability to prevent the three main types of reactions that cause decomposition of chlorinated solvents, which leads to acid formation: (1) hydrolysis; (2) oxidation, initiated by exposure to air, by thermal breakdown, or by ultraviolet light; and (3) condensation reactions with alkali metal salts.* The main categories of solvent stabilizers are acid acceptors, antioxidants, and metal inhibitors. Ultimately, all stabilizers serve to mitigate the formation of acids, because each type of reaction produces acids. Antioxidants prevent reactions that form acids. Acid acceptors act to neutralize acids once they form in the solvent or are introduced into the operation from the work. Metal inhibitors prevent reactions between the solvent and alkali metals and their salts.

Solvents stored in drums may eventually deteriorate and form enough acid to corrode the drums. This happens when drums were not stored with airtight bungs or seals. Water vapor entering a drum will condense and form liquid water, which floats atop the solvent. Hydrolysis of impurities in the solvent carried over from production will form acids that then progressively deteriorate the solvent. A ring of pinholes at the solvent-water line near the top of the drum is a telltale sign of solvent deterioration. Various stabilizing formulations were developed to mitigate acid formation during solvent storage. Early solutions that stabilized TCE, PCE, and dichloromethane against acid formation during storage included addition of gasoline; however, large quantities were required, and the benefit of using a relatively nonflammable cleaning agent was negated (Pitman, 1933). Amylene, also known as 1-pentene, was added to TCE as an acid inhibitor more than 75 years ago (Harris, 1933). A 1959 patent notes that metal tanks corroded and failed within months because of the formation of acid that corroded tanks (Beckers, 1959). Such corrosion leads to leaks and contamination of underlying soil and groundwater.

PCE is susceptible to autoxidation when exposed to ultraviolet light in the presence of moisture. And in the absence of a metal stabilizer, methyl chloroform exposed to aluminum salts will undergo a violent reaction that quickly reduces the solvent to a tarry mass and evolves phosgene gas with potentially harmful or even fatal consequences to the equipment operator.

Carbon tetrachloride corrodes metals and brass in particular. Early fire extinguishers used carbon tetrachloride in brass containers. Valves on these carbon tetrachloride fire extinguishers corroded after lengthy storage and prevented them from discharging in the moment of need (Snelling, 1914). Solvent stabilizers were proposed to remedy the problem in a 1913 patent application filed in Pittsburgh, Pennsylvania.

To understand the challenges of solvent stabilization, some familiarity with the mechanics of industrial applications using chlorinated solvents is necessary. Section 1.1.1 provides an overview of the major industrial applications employing chlorinated solvents, with an emphasis on how these industrial processes may have contributed to the composition of waste solvents and the contamination patterns found in the investigation of contaminated sites.

1.2.2 Consequences of Solvent Breakdown

Degreasing operations require periodic shutdowns for the cleaning of degreasing equipment to remove accumulated metal fines, to recycle spent solvent through distillation, and to maintain pumps, seals, valves, nozzles, and other parts. When solvents become unstable, the degreasing equipment itself may be damaged or require intensive cleaning, which results in additional labor, production downtime, and loss of solvent. In early open-top vapor degreasers, operators faced potential hazards from exposure to solvents during routine operations, even when the solvent was performing as intended. The hazards increase considerably when solvents breakdown from a runaway acid formation reaction. Degreaser maintenance practices have a bearing on the nature and composition of the waste solvent that was stored in tanks, leaked, spilled, sent to landfills, or sent off to facilities for solvent recycling. Accordingly, the degreaser operating practices prescribed in guidance from

* A condensation reaction joins two molecules to create a complex molecule and a much simpler molecule as a by-product of the reaction. The simple molecule eliminated from the two reactants can be water, ammonia, or alcohol. Each reactant contributes to the eliminated molecule (Carey, 1987). The two combining molecules each contribute a single moiety (a portion of a molecule having a characteristic chemical property) to the eliminated molecule.

degreasing equipment manufacturers such as Baron-Blakeslee and from solvent providers such as Dow Chemical may provide clues to the expected composition of waste solvent.

1.2.2.1 Operator Safety

Reactions of acids and alkali metals with chlorinated solvents can produce several hazardous byproducts. Foremost among these is phosgene gas, also known as carbonyl chloride, or $COCl_2$. Phosgene forms during thermal decomposition of chlorinated solvents when they are overheated in a vapor degreaser. Subjecting chlorinated solvent vapors to intense ultraviolet light from arc welding will also form phosgene gas. Health and safety guidance for arc welding calls for no welding within 200 ft of a degreaser using chlorinated solvents (AFSCME, 2006).

Several studies have been performed for measuring the by-products that form when TCE is photocatalytically decomposed because of arc welding in proximity to a vapor degreaser. The major products are phosgene and carbon dioxide; small quantities of dichloroacetic acid and trichloroacetaldehyde remain at the end of the reaction (Blake et al., 1993; Rice and Raftery, 1999). PCE is known to break down to form phosgene, chloroacetic acid, and trichloroacetic acid (Skeeters, 1960a). Of the chlorinated solvents, PCE gives off the most phosgene when heated by flame, when in contact with a very hot surface, or when subjected to intense ultraviolet light (Smallwood, 1993).

Dichloromethane and carbon tetrachloride can produce phosgene; methyl chloroform also forms phosgene as a photocatalytic breakdown product, but to a lesser degree than TCE (Dreisbach, 1987; European Chemicals Bureau, 2000a). Phosgene is a particularly dangerous toxin. It was used as a chemical warfare agent in World War I.

Illnesses and fatalities have been attributed to degreasing operations in which destabilized solvents have evolved phosgene gas. In one incident, TCE was overheated when it was mistakenly used in a PCE degreaser. The thermal breakdown of TCE caused the death of an operator (Spolyar et al., 1951). The literature holds many other examples of similar fatalities, such as the chain-smoking proprietor of a dry cleaning establishment who smoked while operating dry cleaning machinery in an atmosphere laden with PCE vapors. Phosgene gas formed as he drew the vapors through his burning cigarette (Derrick and Johnson, 1943).

Several of the metal inhibitors used for the normally nonflammable methyl chloroform are highly flammable and can change the flash point associated with this solvent. Under severe conditions, stabilizers in methyl chloroform solvent can render it flammable, a fact that proved fatal to a welder working on the solid rocket boosters for the space shuttle. On August 10, 1983, an accident killed a welder who was working in a vapor degreaser pit filled with methyl chloroform vapors at a Utah facility that cleaned and inspected the used casings for the solid rocket boosters used on the space shuttle. The welder had been told that methyl chloroform would not burn. While in a basket suspended from an overhead crane inside the pit, he was attempting to weld a bracket into place when a ball of flame developed by his chest and then expanded; he jumped or fell into the pit. A fireball erupted from the pit, and the welder was burned to death or asphyxiated by the dense methyl chloroform vapors. Vapor combustibility tests showed that the liquid solvent would not burn, but the vapors would burn after the liquid was gone. Testing showed that degreasing grades of methyl chloroform were easier to ignite than vapors of pure or technical grade methyl chloroform. A high proportion of solvent stabilizers remaining in the solvent after the accident made it easier to ignite a sample (de Nevers, 1986).

Several stabilizers of methyl chloroform, such as nitromethane and 1,4-dioxane, may impart flammability to methyl chloroform if concentrated; these two compounds may have caused the Utah welding accident. The properties of stabilizers most commonly added to the major solvents and the industrial operations that lead to their partitioning are discussed in Chapter 3. Operator safety for industrial processes using chlorinated solvents is the subject of a large body of regulation. Occupational safety and health regulations for worker exposure to solvents and toxic stabilizer compounds are discussed in Chapter 6. Proper operation of degreasers and periodic testing of solvent stability prevents solvent-breakdown reactions and toxic reaction products such as phosgene gas.

1.2.2.2 Off-Spec Work

A solvent that has become acidic may stain or damage the work. Parts from which grease, soldering flux, cutting oils, buffing oils, and other oily wastes and debris must be removed can become discolored or receive spots and insoluble deposits from a solvent that has deteriorated. A contaminated solvent will discolor or stain metals, particularly brass and aluminum. Sensitive instruments, printed circuit boards, thermal switches, or other parts may become damaged if the solvent has accumulated too much oil that consequently vaporizes at a higher temperature. Temperature-sensitive parts may melt when the solvent boils too hot. In addition to spots and stains, the work may become corroded from the acid evolved during solvent-breakdown reactions.

A solvent that has deteriorated because of loss or imbalance of the stabilizer package may also become malodorous, colored, or too acidic for dry cleaning clothing and textiles. Fabrics cleaned with decomposing solvent appear "dull and lifeless," especially fabrics in pastel colors, and the odor of the residual solvent in the fabric and cleaning plant becomes increasingly unpleasant. PCE that has become too acidic from decomposition reactions can damage the clothing by leaching dyes and fabric colors. The leached dye discolors the PCE and makes it unsuitable for further use; its recovery is also difficult (Skeeters, 1960b).

In order to clean dirt and grime from uniforms, clothing, shop overalls, and other items, dry cleaning solvent is often infused with detergent and water to remove both water-soluble and water-insoluble fraction of the dirt (Skeeters, 1960a). However, some stabilized compositions are not fully compatible with soaps and detergents, which lead to solvent failure in dry cleaning applications.

Management of waste solvents has been the subject of a series of increasingly stringent regulations. Problems with spotting, staining, discolored textiles, and malodorous solvents often led operators to distill or discard the offensive solvent. Disposal or recycling of waste solvents was usually done in the most cost-effective and expedient manner available or was not actively regulated. For example, until 1986, dry cleaners could legally discharge condensate water to sanitary sewers. Today, the connection of dry cleaning machinery to sewers is prohibited; all waste must be inventoried, manifested, and sent to a licensed facility for recycling. Chapter 6 provides a full discussion on the evolution of waste-handling regulations and their effect on the nature and chemical composition of waste discharges.

1.2.2.3 Production Line Shutdown

Chlorinated solvents have played an integral role in supporting production lines for a wide variety of products such as automotive parts, armaments, aircraft, electronics, precision instruments, metal parts, textiles, semiconductors, and printed circuit boards. The value of goods produced from these assembly lines can be very high. The cost of a degreaser shutdown due to solvent that has lost its stability and has become acidic can be enormous: the loss of production and the cost of repairing the damaged degreaser far outweigh the cost of the solvent. The solvent must be drained from the degreaser, and all the surfaces must be cleaned with an alkaline detergent. The tarry material may have deposited in inaccessible components and surfaces, requiring the replacement of pipes and other equipment.

In addition to damage to products cleaned in degreasers from staining or spotting, the runaway reaction of an out-of-balance solvent can damage the degreasing equipment itself. To prevent this, operators test solvents frequently for indicators of stability versus impending breakdown. (Testing protocols for determining solvent stability are described in Section 1.2.7.) Preventing loss of production and damage to degreasing equipment was a strong incentive to discard and replace deteriorated solvents. Prior to regulation and wider understanding of the environmental and/or economic consequences of releasing solvent to the ground, operators faced with the issue of discarding waste solvent may have opted to dump it on-site or to contract with haulers to remove the waste to landfills, recyclers, or unknown destinations.

Solvents can rapidly deteriorate to the point where the work and the degreasing equipment are damaged. For example, methyl chloroform and methylene chloride readily undergo a decomposition reaction in the presence of aluminum chloride; to a lesser extent, TCE and PCE may also react with aluminum chloride. Solvent degradation is accompanied by a rapid rise in temperature and a discoloration of the solvent. The reaction advances until a black tarry mass forms and deposits on the work and the degreasing equipment (Petering and Aitchison, 1945; Starks and Kenmore, 1960a). Unstabilized methyl chloroform exposed to aluminum chloride or other alkali metal salt catalysts completely decomposes to tar in 20 min (Hardies, 1966). In addition to the tarry by-product of solvent decomposition, acids formed in the reaction may corrode the metal surfaces in a degreaser or dry cleaner. The first sign of acidity in a degreaser is often corrosion at the vapor-atmosphere interface, where rust appears on the walls of the unit around the cooling coils and condensate trough (Dow Chemical Company, 1999b).

The overall consequences of solvent decomposition – worker exposure to harmful or fatal gases, damage to the products being cleaned, and damage to the degreasing equipment with corresponding loss in production – provide a strong reason to keep solvents in good condition. Operators dealing with solvent that has become acidic owing to loss of stabilizers or other factors seek either to recover the solvent through on-site distillation or to replace the solvent. Solvent replacement could involve on-site recycling or collection and hauling to off-site solvent recyclers or to a hazardous waste incinerator or landfill; more expedient on-site disposal would have been in burn trenches, by drum burial, or directly disposing to the ground, before such practices were prohibited by actively enforced regulation.

The consequences of solvent breakdown were also strong motivators for the solvent producers to develop "stabilizer packages" (formulations of different solvent additives) capable of withstanding the numerous adverse conditions encountered in the industrial uses of chlorinated solvents. Equipment manufacturers for degreasing and dry cleaning were also challenged to design solutions to ensure that solvents were not subjected to extremes of temperature and stabilizer loss. Examples include increasing the freeboard in the degreaser to minimize vapor loss and the resulting imbalance of stabilizers in the solvent, as well as improvements to water separators and carbon adsorption systems for vapor recovery.

1.2.3 Stability of the Major Solvents

Among the four major chlorinated solvents, PCE is often listed as the most stable, whereas methyl chloroform is considered the least stable. Mixtures of chlorinated solvents, such as PCE and methyl chloroform, tend to decompose at an accelerated rate, compared to PCE alone (Goodner et al., 1977). Table 1.9 summarizes the relative stability of the major chlorinated solvents to aggressive agents encountered in operating environments.

TABLE 1.9
Solvent Stability in Harsh Operating Environments

Solvent	Oxidation	Hydrolysis	Alkali Metals	Pyrolysis	UV Light
Dichloromethane	Stable	Stable	Slightly unstable in vapor phase	Slightly unstable	Stable
Methyl chloroform	Slightly unstable	Slightly unstable	*Unstable*	*Unstable*	Stable
Perchloroethylene	Slightly unstable	Stable	Stable	Stable	*Unstable*
Trichloroethylene	*Unstable*	Stable	Slightly unstable	Slightly unstable	Slightly unstable

Sources: Solvay, S.A., 2002a, Chlorinated solvents stabilisation. GBR-2900-0002-W-EN Issue 1 – 19.06.2002, Rue du Prince Albert, 33 1050 Brussels, Belgium. http://www.solvaychemicals.com (accessed November 9, 2003) and Irani, 1977, United States Patent 4,032,584: Stabilized methylene chloride. Assignee: Stauffer Chemical Company, Westport, CT.

TABLE 1.10
Purity of Solvents Sold for Industrial Uses

Solvent	Grade	Purity (%)	References
Methyl chloroform	Degreasing	95	Ashland Chemical Company (1995)
	Technical	96.5	Dow Chemical Company (1990)
	Technical	94–98	Unocal (1989)
Dichloromethane	Technical	98	Vulcan Chemicals (1987)
	Technical	99.5	Dow Chemical Company (1995)
	Technical	99.9	Dow Chemical Company (1999a)
	Aerothene MM™	99.4	Dow Chemical Company (2006a)
Trichloroethylene	Neu-Tri™	99.4	Dow Chemical Company (1981, 2006a)
	Technical, O-T-634	> 95	Dow Chemical Company (1985a)
Perchloroethylene	Perchloroethylene-SVG™	99.5	Dow Chemical Company (1986)
	DOWPER CS solvent™, 1985	97	Dow Chemical Company (1985b)
	DOWPER solvent™, 2006	99[a]	Dow Chemical Company (2006a)
	Perclene™ dry cleaning grade	99–100	Occidental Chemical Corporation (1991)
	SVG double-stabilized	99[b]	Dow Chemical Company (2006a)

[a] 1% stabilizer.
[b] 0.1% stabilizer.

The relative stability of the major solvents can be inferred from the quantity of stabilizers required. Safety Data Sheets (SDSs) and other documents indicate the purity of solvents sold for industrial applications in recent years, as presented in Table 1.10.

Solvent purity listed in Table 1.10 reflects both the addition of stabilizers and impurities from manufacturing. Technical grade TCE (also called "industrial grade") has been known to include about 0.15% carbon tetrachloride as the major chlorinated hydrocarbon impurity (Willis and Christian, 1957). The compound 1,1,2-trichloroethane can be an impurity in the manufacture of methyl chloroform and is more reactive than methyl chloroform (World Health Organization, 1990). PCE and TCE produced from symmetric tetrachloroethane (1,1,2,2-tetrachloroethane) may retain detectable levels of the feed stock. The acetylene process used to produce TCE also carried over trace levels of symmetric tetrachloroethane. In the 1970s, production of TCE and PCE shifted from the acetylene process to an ethylene feedstock process, in which 1,2-dichloroethane was a common impurity (Morrison and Murphy, 2013).

Solvent impurities may also be introduced from cross-contamination from residues of other solvents in improperly cleaned tank cars, tank trucks, drums, piping, and hoses. Mixtures of chlorinated solvents, even if stabilized, may have substantially reduced stability when in contact with aluminum, compared to the stability of the individual solvents. For example, stabilized TCE can become unstable in the presence of as little as 2% of stabilized dichloromethane. When replacing one chlorinated solvent with another, it is therefore necessary to remove all traces of the first solvent before using the new solvent (EuroChlor, 2003). Most facilities would not frequently change solvents. However, solvent-recycling facilities that use the same distillation equipment for different solvents must clean out drum-filling pipes between batches. Several anecdotes describe drum-filling pipes that were flushed to the ground or storm drain sumps between batches of solvents at a solvent recycler in Silicon Valley, California.

1.2.3.1 Performance Criteria for Solvent Stabilizers in Various Applications

The industrial chemists who worked to solve the solvent-breakdown problem overcame the challenge of solvent stabilization through much experimentation. Additives were needed that would inhibit a variety of reactions with the aggressive substances found in industrial processes, especially where physically destructive conditions are present. The chemists' pursuit was to find stabilizers that would meet most of the following criteria (Levine and Cass, 1939; Klabunde, 1949; Kauder, 1960; Richtzenhain and Stephan, 1975; Irani, 1977; Blum, 1984; Smallwood, 1993). The ideal stabilizer must

- Adequately inhibit deleterious reactions to prevent solvent breakdown caused by the action of light, air, or temperature on the solvent
- Protect the solvent from the action of small amounts of strong acid or from metals and their salts
- Correct any incipient localized breakdown of the solvent
- Not impart increased toxicity to the solvent or endanger worker health and safety, particularly after 1984, when a federal specification was issued prohibiting use of stabilizers more toxic than the solvent, for solvents used in military and other federal projects (O-T-634C, April 25, 1984).
- Remain stable through use in the expected industrial process
- Stand up to repeated phase transitions from liquid to vapor and back to liquid
- Remain stable at high temperatures
- Partition to the vapor phase to provide sufficient solvent stability in vapor degreasing
- Be sufficiently volatile to exist in both the liquid and the vapor phases of the solvent and to be recoverable from distillation residues
- Boil at temperatures not more than 15°C higher or lower than the solvent boiling point
- Not be consumed by reactions
- Prevent the solvent from taking on the character of a mixture of solvents
- Not require addition in large quantities
- Not stain the metal part, textile, or clothing being cleaned
- Not leave deposits on the parts being cleaned
- Not promote the formation of corrosion products or insoluble sludge
- Be fully soluble in the solvent
- Be at least more soluble in the solvent than in water so that water would not extract the stabilizer
- Be inexpensive
- Not react with other stabilizers
- Be compatible with the metals used in the construction of the degreaser and with the many and varied contaminants encountered in the cleaning of metal parts from modern fabricating operations
- Not weaken textile fibers
- Not shrink textile fibers
- Not bleed dyes
- Not be malodorous or leave residual odor
- Be compatible with soaps and detergents if used in dry cleaning applications
- Not react with free fatty acids
- Not be so strongly alkaline that its volatility is reduced in the presence of acids
- Be easily measurable to confirm its concentration during use

1.2.3.2 Selection of Solvent Stabilizers

Selection of stabilizers was determined by the conditions of the particular application. The many permutations of the various requirements placed on solvents led to decades of competitive inventions of solvent stabilizer packages that would improve solvent performance in specific applications. For example, operations that cut and polish metals may use buffing compounds and drawing oils that must be removed from the metal parts by a vapor degreaser. Buffing compounds and drawing oils are sources of fatty acids. Buffing compounds contain 50% stearic acid, whereas drawing oils can contain 5–10% oleic acid (Klabunde, 1949). The stabilizer package must therefore not react with fatty acids.

As detailed in the first half of this chapter, PCE has been used in a great variety of applications, for example, in dry cleaning, in electrical transformers as a heat-transfer medium, in ultrasonic cold cleaning, in vapor degreasing, and as a paint thinner. PCE with a minor complement of stabilizers may remain stable for multiple uses when degreasing large pieces of stainless steel. The same formulation may become unstable when cleaning a more reactive metal or the same metal in which fine turnings, grindings, and other metal fines create a substantially increased surface area over which reactions may occur, or which may cause hot spots near heating coils, leading to pyrolysis of the solvent. Determination of the preferred stabilizer formulation therefore depends on the severity of the environment to which the solvent is subjected.

Among the most demanding applications requiring stabilization of PCE is its use as a dielectric fluid. A dielectric fluid must remain stable for up to 30 years without breaking down to form electrically conductive or corrosive materials. Breakdown of PCE by dehydrochlorination forms hydrogen chloride, which is conductive and therefore deleterious to the dielectric fluid and the electrical device used (EPO, 1989).

Another application in which PCE and TCE are subjected to severe stress is in phosphatizing baths. Nonaqueous phosphatizing baths are used to apply phosphate coatings to metallic surfaces, for example, automobile exteriors, to reduce corrosion and improve paint adhesion. Besides orthophosphoric acid and an agent to solubilize the acid, the phosphatizing bath is mostly TCE or PCE. The solubilizing agent is necessary because phosphoric acid is insoluble in TCE and PCE. While a wide range of compounds including alcohols and 1,4-dioxane have been identified as suitable solubilizing agents, 1,4-dioxane has not been claimed in patents for stabilizing phosphatizing baths (Campbell, 1966). Trichloroethylene and PCE can undergo rapid decomposition by a mechanism different from the type of decomposition for which most stabilizers were developed; thus, an application-specific group of stabilizers is required. For TCE and PCE, azobenzene compounds were used to stabilize solvents used for phosphatizing baths (Fullhart and Swalheim, 1962). Other stabilizer packages for phosphatizing baths included hydroquinones (simple, di-t-amyl-, and 2,5-di-t-butyl-), 4-t-butylpyrocatechol, dihydroxynaphthalene (1,6- and 1,5-), thiourea, and n-butanol (Vullo and Campbell, 1965; Campbell, 1966).

1.2.4 Causes of Solvent Breakdown

The primary causes of solvent breakdown – oxidation, hydrolysis, pyrolysis, ultraviolet light, and reaction with alkali metals – have been introduced. This section examines the chemical reactions that cause solvent deterioration. The structures of the chlorinated solvents (Table 1.1) play a decisive role in the nature of their decomposition.

1.2.4.1 Oxidation

TCE and PCE are susceptible to oxidative attack on the ethylene carbon-carbon double bond. Oxidative attack is a problem for the ethylene compounds PCE and TCE, but is not a problem for the alkane compounds methyl chloroform, dichloromethane, and carbon tetrachloride, which have single carbon-carbon bonds or only a single carbon atom (Levine and Cass, 1939). The initial deterioration of the ethylene bond in TCE is due to the oxidation initiated by light (photolysis, in which light energy is absorbed by a molecule). The reaction is strongly accelerated by ultraviolet rays from natural light, fluorescent lighting, and arc welding, by heat, and by the presence of catalysts, including metals such as iron and aluminum, and metallic salts such as ferric chloride and aluminum chloride. TCE is relatively immune to photolysis, except in the presence of oxygen (Shepherd, 1962). The products resulting from the oxidation of TCE also serve to promote further degradation of TCE. Humidity does not appear to affect the rate of oxidation of TCE (Solvay, 2002b). Air, moisture, and acid favor further deterioration once oxidation has been initiated. Because acid is evolved, the deterioration of TCE is autocatalytic (Pitman, 1933). As the breakdown of TCE progresses, the reaction takes the form of a self-accelerating Friedel-Crafts reaction. Once started, the reaction will proceed rapidly and, in some cases, explosively with the evolution of heat and large quantities of hydrogen chloride.

The acceleration of TCE degradation by ultraviolet light when oxygen is present suggests a free radical mechanism. A variety of products are formed, including phosgene gas, carbon monoxide, dichloroacetic acid, formic acid, glyoxylic acid, hydrochloric acid, chlorine gas, and a variety of polymer compounds (Shepherd, 1962).

Most halogenated solvents are resistant to attack by oxygen until a chlorine atom is removed. Removing a chlorine atom and an adjacent hydrogen atom leaves a double bond, which is more susceptible to oxidation, particularly in the presence of metals such as aluminum, copper, and vanadium (Howell and Tarrer, 1994).

The mechanics of solvent breakdown are profiled in a study by Arthur Tarrer of Auburn University, with researchers from Tyndall Air Force Base and from the U.S. Army Construction Engineering Research Laboratory (Tarrer et al., 1989; Howell and Tarrer, 1994). An initiator, such as heat or ultraviolet light, removes hydrogen from an unsaturated molecule, RH, forming a free radical, R. Oxygen then combines with the free radical to form a peroxide radical, ROO^-, and that then removes hydrogen from a new unsaturated molecule, R'H, thereby propagating the chain reaction (Tarrer et al., 1989).

Trichloroethylene is oxidized in a decomposition mechanism involving formation of a highly reactive epoxide which is readily isomerized to acetyl chloride. The susceptibility of TCE to oxidation alone, without an ultraviolet light initiator and free radical mechanism, was summarized by Shepherd, citing work in the German literature by Erdmann in 1911, as follows (Erdmann, 1911; Shepherd, 1962):

$$\overset{\text{Trichloroethylene}}{Cl_2C{=}CHCl} + \tfrac{1}{2}O_2 \rightarrow \overset{\text{Ethyl oxide}}{\underset{\diagdown\!\diagup \atop O}{Cl_2\tilde{C}\ \ CHCl}}$$

$$\underset{\diagdown\!\diagup \atop O}{Cl_2\tilde{C}\ \ CHCl} \rightarrow \overset{\text{Acetyl chloride}}{Cl_2CHCOCl}$$

In the presence of moisture, the acetyl chloride is hydrolyzed to dichloroacetic acid and hydrogen chloride:

$$\overset{\textit{Acetyl chloride}}{Cl_2CHCOCl} + H_2O \rightarrow \overset{\textit{Dichloroacetic acid}}{Cl_2CHCOOOH} + HCl$$

The epoxide has been verified experimentally by McKinney et al. (1955). PCE also forms an epoxide (Shepherd, 1962).

Commercial grades of PCE have been subject to free radical attack initiated by ultraviolet light; it is the least stable solvent in the presence of ultraviolet light. High-purity PCE is very inert to the action of air, light, heat, moisture, and the metal surfaces it contacts during commercial storage and use. However, most PCE contains a trace amount of impurities carried over from production. PCE obtained from the chlorination or chlorinolysis of lower aliphatic hydrocarbons may contain small amounts of saturated and unsaturated chlorinated hydrocarbons such as dichloroethylene, TCE, methyl chloroform, and unsymmetrical tetrachloroethane (i.e., 1,1,1,2-tetrachloroethane) (Skeeters, 1960a). These compounds are more susceptible to free radical attack than PCE and may initiate and catalyze the breakdown of PCE. The presence of the lower-molecular-weight chlorinated hydrocarbons in PCE leads to the formation of oxidation products that corrode metal surfaces. These lower-molecular-weight chlorinated hydrocarbons are thought to be the principal initial source of chloro-oxygen-containing impurities such as phosgene, chloroacetic acid, and trichloroacetic acid. It is these impurities that catalyze the decomposition of PCE (Skeeters, 1960b). Methyl chloroform was widely used as a spotting agent to remove spots from clothing prior to dry cleaning. Its introduction to the PCE vapor environment in a dry cleaner could also contribute to acid formation in PCE (Linn et al., 2004). Experimental data show that heat accelerates the deleterious effects of light and oxygen on the decomposition of solvents (Starks and Kenmore, 1960b).

1.2.4.2 Ultraviolet Light Breakdown

PCE shows poor light stability. Fluorescent light and ultraviolet light cause PCE to decompose, producing a measurably increased acidity and a corresponding change in pH (Stevens, 1955). Tests for stability of solvents often involved a multiday or multiweek reflux test. To show the instability of unstabilized PCE to light, a typical test might proceed for 72 hours, during which PCE with water and copper strips is heated to boiling and the vapor is exposed to fluorescent light. A change in pH of 1 unit or more is taken as an indicator of light instability. Without a light stabilizer, PCE can change from pH 6.8 to less than pH 2 in 72 hours in the presence of light (Stevens, 1955). The decomposition of PCE is more pronounced when it is in contact with iron or copper. Products of PCE decomposition cause metal corrosion. PCE will also degrade in the presence of light from a welder's arc (Dow Chemical Company, 1999b).

1.2.4.3 Thermal Breakdown

The major chlorinated solvents are stable and resistant to thermal breakdown at their normal boiling temperatures, but are vulnerable to thermal breakdown and acid formation if operating temperatures exceed their stability range (Petering and Aitchison, 1945; Archer and Stevens, 1977). The order of increasing thermal stability does not directly correspond to the order of the solvents' boiling points. The order, from least to most thermally stable, is methyl chloroform (b.p. = 74°C) < dichloromethane (b.p. = 41°C) < TCE (b.p. = 86.7°C) < PCE (b.p. = 121.4°C) (Archer and Stevens, 1977). TCE is stable up to 130°C; at higher temperatures, it is subject to pyrolysis. PCE is stable up to 150°C in the presence of air and moisture (Solvay, 2002b). In the absence of any catalysts, such as in the sealed environment of an electrical transformer, PCE may be stable up to 500°C (World Health Organization, 1984a). Decomposition in the presence of air and moisture begins above 125°C for TCE and above 150°C for PCE. For safety reasons, equipment operators are advised to avoid exceeding temperatures of 110°C and 140°C for TCE and PCE, respectively (Solvay, 2002b). Table 1.11 summarizes the susceptibility of the major solvents to pyrolysis in the presence of metal catalysts and air.

When chlorinated solvents are overheated, a chloride ion gets split off, leading to the formation of hydrochloric acid. Prolonged contact of solvents with oils, grease, and other soils at elevated temperatures also causes solvent instability problems (Archer, 1984). For example, thermal breakdown of TCE under conditions of normal use and in the absence of stabilizers probably proceeds according to the following reactions (Shepherd, 1962):

$$2\mathrm{ClCH}=\mathrm{CCl_2} \rightarrow (\mathrm{ClCH}=\mathrm{CCl_2})_2, \quad (1.1)$$

$$(\mathrm{ClCH}=\mathrm{CCl_2})_2 \rightarrow \mathrm{HCl}+\mathrm{Cl_2C}=\mathrm{CHCC}=\mathrm{CCl_2}. \quad (1.2)$$

1.2.4.4 Acid Breakdown

All halogenated solvents in contact with free-phase moisture and air will slowly react with water in a process called hydrolysis. Rates of hydrolysis for the chlorinated solvents vary widely. Methyl chloroform has a hydrolysis half-life of a little more than 2 years, whereas TCE and PCE are resistant to hydrolysis, with immeasurably slow hydrolysis half-lives. Hydrolysis produces hydrogen

TABLE 1.11
Pyrolysis Test Results for the Major Chlorinated Solvents

Solvent	Metal	Vapor Condition	Temperature at Which Slow Increase in Pyrolysis Begins (°F)	Temperature at Which Rapid Increase in Pyrolysis Begins (°F)
Dichloromethane	Black iron	Vapor only	525	700
	Black iron	With dry air	<400	600
	Aluminum	Vapor only	376	525
	Copper	Vapor only	475	525
Methyl chloroform	Black iron	Vapor only	525	625
	Black iron	With dry air	—	325
	Aluminum	Vapor only	—	675
	Copper	Vapor only	~350	675
Trichloroethylene	Black iron	Vapor only	—	825
	Black iron	With dry air	—	550
	Aluminum	Vapor only	~550	725
	Copper	Vapor only	—	575
Perchloroethylene	Black iron	Vapor only	—	>900
	Black iron	With dry air	—	>900
	Aluminum	Vapor only	—	>750
	Copper	Vapor only	—	>850

Source: Archer, W.L. and Stevens, V.L., 1977, *Industrial & Engineering Chemistry, Product Research and Development* 16(4): 319–326.

Note: Pyrolysis test consisted of passing chlorinated solvent vapor (0.2 mol/h), with or without air, through a heated metal pipe whose internal surface area was 200 in.^{2} at several elevated temperatures. Acidity of heated vapors was then determined by collecting 40 mL of condensed solvent from a cold trap; acid generated was determined by titrating with caustic soda.

chloride in chlorinated solvents. If exposed to water, hydrogen chloride produces hydrochloric acid. If not eliminated when formed, hydrochloric acid will catalyze and increase the rate of hydrolysis and produce acid at a faster rate. Hydrochloric acid also removes the protective oxide coating on metal surfaces, exposing fresh metal to the solvent, and enabling metal-catalyzed solvent deterioration. In the earliest study of problems with hydrolysis of chlorinated solvents, Levine and Cass (1939) described carbon tetrachloride as decomposing as a result of hydrolysis, "splitting of acid and reacting with water."

Acid can also be introduced into the degreaser solvent from oils, greases, soldering flux, and other material present on the work. The oxidation and decomposition of cutting oils dissolved from the work form acid, usually hydrochloric acid or acetic acid, which then reacts with the solvent (Starks and Kenmore, 1960b). Acids formed by oxidation of cutting oils or hydrolysis of solvents can corrode the work and the cleaning equipment itself. In dry cleaning applications, acids that form in PCE can leach dyes and fabric colors; this leaching damages the clothing and makes solvent recovery difficult (Cormany, 1977). A source of acid in dry cleaning is sebaceous oils and other bodily soil on the clothing. These soils contain short-chain, free fatty acids such as butyric acid and valeric acid that acidify moisture in the cleaning process, which leads to machine corrosion, odors on the clothing, and undesirable textile effects such as swales. Acidity cannot be tolerated in dry cleaning, where TCE was used to a limited extent in the 1930s through the 1950s. Free acidity was also undesirable for TCE when it was used to make decaffeinated coffee (Pitman, 1943).

TCE and PCE are relatively immune to hydrolysis, however, compared to carbon tetrachloride or methyl chloroform. Dichloromethane hydrolyzes very slowly (Solvay, 2002b). Table 1.12 provides

TABLE 1.12
Generalized Abiotic Hydrolysis Half-Lives of the Major Chlorinated Solvents

Solvent	Abiotic Hydrolysis Half-Life (years)	References
Perchloroethylene	1.3×10^6	Jeffers et al. (1989)
Trichloroethylene	1.3×10^6	Jeffers et al. (1989)
Dichloromethane	704	Mabey and Mill (1978)
Carbon tetrachloride	41	Jeffers et al. (1989)
Methyl chloroform	2.5	Vogel and McCarty (1987)

Source: Data and citations from Pankow, J.F. and Cherry, J.A., 1996, *Dense Chlorinated Solvents and Other DNAPLs in Groundwater.* Waterloo, Ontario, Canada: Waterloo Press.

commonly cited values for hydrolysis half-lives of the major solvents under ambient conditions in soil and groundwater.

In the absence of air and metals, solvents will undergo hydrolysis at slow rates and form only minor quantities of acids. TCE and PCE both hydrolyze at a rate of 0.01 mg/L per day in the absence of air and metals (6.8×10^{-7} mol% and 6.2×10^{-7} mol%), respectively (Carlisle and Levine, 1932). Citations of Carlisle and Levine's work record the rates of solvent decomposition listed in Table 1.13 when boiling TCE and PCE are subjected to the specific conditions given.

Through hydrolysis, methyl chloroform reacts abiotically to form 1,1-dichloroethylene:

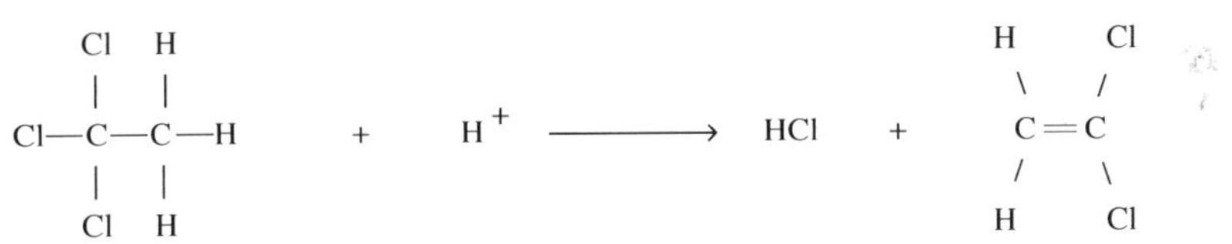

(1.3)

Methyl chloroform — Hydrogen — Hydrogen chloride — 1,1-dichloroethylene

Abiotic degradation of methyl chloroform has also been described as follows (Dubilier, 2003):

$$\underset{\textit{Methyl chloroform}}{CH_3CCl_3} \rightarrow \underset{\textit{1,1-dichloroethylene}}{CH_2CCl_2} + \underset{\textit{Acetic acid}}{CH_3COOH} \rightarrow \underset{\textit{Mineralization}}{CO_2 + H_2O + Cl^-} \tag{1.4}$$

TABLE 1.13
Acidity Generation by Boiling Trichloroethylene and Perchloroethylene

Test Condition	Trichloroethylene	Perchloroethylene
Dry solvent under nitrogen	10.2[a]	10.5[a]
Wet solvent under nitrogen	7.7	12.5
H_2O-saturated solvent exposed to soft steel under nitrogen	11.0	11.6
Wet solvent exposed to soft steel under nitrogen	5.6	—
Dry solvent in oxygen	820	782

Note: Reported as cubic centimeters of 0.01 *N* hydrochloric acid per 25 mL solvent (from Shepherd, 1962).
[a] Air leakage introduced oxygen.

1.2.4.5 Solvent Deterioration Catalyzed by Metals and Metal Salts

Methyl chloroform is most vulnerable to reaction with alkali metals, a problem also faced by dichloromethane and carbon tetrachloride when these two compounds are in the vapor phase. TCE and PCE are comparatively less vulnerable to reaction with alkali metals (Smallwood, 1993). The absence of hydrogen atoms in PCE makes it less sensitive to the action of bases and light metals (Solvay, 2002a). Chlorinated organic products in contact with a light or alkali metal such as aluminum or magnesium produce some hydrochloric acid and a Lewis acid catalyst, for example, anhydrous aluminum trichloride ($AlCl_3$). $AlCl_3$ will catalyze condensation of chlorinated solvents onto themselves or onto other organic substances, especially chloride ions or other nucleophiles.* The products of this reaction include a brownish-black, tarry condensate and more hydrochloric acid, which in turn reacts further with aluminum. Such a reaction increases the amount of $AlCl_3$ present; hence, the reaction is perpetuated. In some cases, the reaction can be violent and produce a considerable amount of hydrochloric acid and heat. Heat accelerates the reaction of condensation and evaporates some of the remaining solvent. $AlCl_3$ has a catalytic effect on TCE, dichloromethane, and especially methyl chloroform (Kauder, 1960). $AlCl_3$ is generally considered to have no catalytic effect on pure PCE (Solvay, 2002a). However, in earlier decades, impurities in PCE may have been carried over from production and reacted with alkali metals.

Methyl chloroform reacts with aluminum ($AlCl_3$), splitting off hydrochloric acid and leaving 1,1-dichloroethylene, shown here in a generalized reaction:

$$\underset{\textit{Methyl chloroform}}{CH_3CCl_3} \quad \underset{\textit{deydrodehalogenation}}{\overset{AlCl_3}{\longrightarrow}} \quad \underset{\textit{1,1-dichloroethane}}{CH_2CCl_2} \quad + \quad \underset{\textit{Hydrochloric acid}}{HCl} \tag{1.5}$$

Aluminum surfaces react with oxygen to form a protective aluminum oxide coating. If a machining, abrasive, or cleaning process removes this coating, the "virgin" metal is available to react with a solvent. The aluminum reaction with methyl chloroform is most easily observed in the "aluminum scratch test," as described in Section 1.2.5.2. Hydrochloric acid is formed during the scratch test and dissociates easily. The chlorine ions react with aluminum in the following sequence:

Hydration of oxide film:

$$2AlO_3 + 3H_2O \rightarrow 2Al(OH)_3 + \frac{3}{2}O_2. \tag{1.6}$$

Reaction of hydrated oxide film and chlorine (dissolution of film):

$$Al + 3HCl \rightarrow AlCl_3 + \frac{3}{2}H_2. \tag{1.7}$$

Reaction of aluminum and hydrochloric acid (dissolution of aluminum):

$$Al + 3HCl \rightarrow AlCl_3 + \frac{3}{2}H_2. \tag{1.8}$$

Precipitation of aluminum hydroxide:

$$AlCl_3 + 3H_2O \rightarrow Al(OH)_3 + 3HCl. \tag{1.9}$$

* A nucleophile, or nucleus-seeking compound, is typically negatively charged with unshared electron pairs that can be used to form a covalent bond with a carbocation (Carey, 1987).

The compounds produced by the reactions are aluminum hydroxide and hydrochloric acid; hydrochloric acid is not consumed and acts as a catalyst (Dubilier, 2003).

TCE, dichloromethane, and methyl chloroform react with aluminum to give aluminum chloride and oligomers of the solvent (a few solvent molecules bonded together). For TCE, the aluminum chloride goes on to react with the solvent to form hydrogen chloride, which acts as a catalyst and is not itself consumed by the reaction. Once initiated, this self-accelerating, "autocatalytic" dehydrochlorination (loss of hydrogen chloride) reaction produces a vigorous subsequent reaction between TCE and aluminum. Aluminum chloride does not have a similar reaction with dichloromethane (EuroChlor, 2003).

Aluminum chloride promotes a dimerization reaction with TCE to produce hydrochloric acid:

$$\underset{Trichloroethylene}{Cl_2C = CHCl} \overset{AlCl_3}{\rightarrow} Cl_2CHCCl_2CH = CCl_2 \quad (1.10)$$

$$Cl_2CHCCl_2CH = CCl_2 \underset{\text{Dimerization}}{\rightarrow} CCl_2 = CClCH = CCl_2 + HCl \quad (1.11)$$

The direct reaction between a chlorinated solvent, such as methyl chloroform, and a metal, such as aluminum, iron, or zinc, produces the metal chloride and the saturated dimer of the chlorinated reactant (Archer, 1984):

$$\underset{\text{Methyl chloroform}}{Cl_2CHCCl_2CH = CCl_2} \underset{\text{Dimerization}}{\rightarrow} 3CH_3CCl_2CCl_2CH_3 + 2AlCl_3 \quad (1.12)$$

The saturated dimer can in turn react with the metal to give the unsaturated *cis* and *trans* dimers, $CH_3CCl = CClCH_3$.

Reactions of stabilized and unstabilized solvents with various metal alloys have been the subject of extensive study. Numerous patents describe the rapid devolution of methyl chloroform refluxed with aluminum, which proceeds to a black tarry mass in as little as 20 min. The presence of a separate water phase floating atop the solvent accelerates the rate of metal corrosion for all solvents (Archer, 1984). Results of additional reflux (boiling and condensing solvent) and high-temperature exposure corrosion tests are listed in Tables 1.14 and 1.15.

1.2.5 Testing Solvent Stability and Performance

The balancing act of maximizing solvent utility while controlling and preventing solvent deterioration requires early detection of changes in solvent characteristics. Visual inspection of the solvent is

TABLE 1.14
Average Metal Corrosion Rates of Chlorinated Solvents

		Corrosion Rate[a] (mils/year)[b]		
Solvent	**Metal Inhibitor**	**2024 Aluminum**	**1010 Iron**	**Zinc**
Dichloromethane, technical grade	No	<1	<1	<0.1
Methyl chloroform	Yes	<3	<1	<1
Trichloroethylene with antioxidant	No	<0.1	<0.1	<0.1
Perchloroethylene with antioxidant	No	<0.1	<0.1	<1

Source: Archer, W.L. and Stevens, V.L., 1977, *Industrial & Engineering Chemistry, Product Research and Development* 16(4): 319–326.

[a] Penetration per year, determined in 7-day reflux tests.

[b] 1 mil = 0.001 in. = 0.0254 mm.

TABLE 1.15
Metals Corroded by Trichloroethylene and Dichloromethane at More Than 50 mils Penetration per Year

Solvent	Metal	Temperature (°C)
Trichloroethylene	Austenitic Cr–Ni stainless steel (18-8; 304/304L/347)	80–107
	Lead	–4–24
	Mo superalloy (Ni–Cr–Fe–9Mo; 625/725)	107–135
	Mo superalloy (Ni–16Cr–16Mo; C276)	107–135
	Mo superalloy (Ni–20Cr–16Mo–4W; 686)	107–135
	Mo superalloy (Ni–22Cr–16Mo; C22/59)	107–135
	Mo superalloy (Ni–23Cr–16Mo–1Cu; C20000)	107–135
	Nickel (200)	107–135
	Nickel-copper (400)	107–135
	Steels, carbon/low alloy	80–107
Dichloromethane	Austenitic Cr–Ni stainless steel (18-8; 304/304L/347)	88–139

Source: NACE International, The Corrosion Society, 2002, Corrosion survey database (COR·SUR). http://www.knovel.com/knovel2/Toc.jsp?BookID=532&VerticalID=0 (accessed 2006).
Note: 1 mil = 0.001, in. = 0.0254 mm.

not a reliable means for determining its condition. A solvent that is perfectly clear can be depleted of its stabilizers, whereas one that is badly discolored and cloudy can still be well stabilized. The most reliable method for confirming the presence and quantity of stabilizers is laboratory analysis by gas chromatography-mass spectrometry; however, the associated delay and expense are too impractical for most operations, so surrogate tests were developed (Howell and Tarrer, 1994). Operators of equipment using chlorinated solvents – such as degreasers, phosphatizing lines, and dry cleaners – routinely test solvents for the presence of acid. A variety of testing approaches have been developed since 1970.

The main approaches for testing solvents check for acids by titrating a base (the acid acceptance value [AAV] test) and test for reaction with alkali metals (the aluminum scratch test). Tests with more specific endpoints – such as reaction with copper (ASTM D 3316), zinc, or brass, and dye-bleed tests – have also been developed for specific applications. Light transmittance has been used in dry cleaning operations as an indicator of when a solvent is due for replacement. Solvents usually transmit light in the range from 450 to 600 nm; a 50% reduction in transmittance in solvent that has been filtered is often used as the rule of thumb for when the solvent must be replaced (Tarrer et al., 1989). Transmittance is measured by using a visible light spectrometer or a colorimeter. Color, odor, presence of dirt and grease, and boiling temperature have also been used by dry cleaner operators to gauge when the solvent is due for filtration, distillation, or replacement. All the solvent performance tests are temperature sensitive; a water bath is usually used to hold solvent temperatures constant to produce comparable results.

Solvent power is a measure of the solvent's effectiveness at dissolving oils, soldering flux, and buffing compounds. The most common test for judging changes to solvent power yields the Kauri-butanol value (KBV) (ASTM D 1133). The test uses Kauri gum,* which is very soluble in butanol but less soluble when butanol is diluted with a solvent that does not dissolve the resin. The test uses a burette, a flask, and a precision balance and provides a relative ranking of solvency. The solvent tested is added in small amounts until the solution becomes cloudy because of the precipitation

* Kauri gum is formed when resin exudes from a crack in the bark of the kauri tree (*Agathis australis*), found in New Zealand, and hardens on exposure to air. It also exists in fossil form. Its uses range from chewing gum to industrial applications.

of Kauri gum from the butanol solution. The more the solvent added before the solution becomes cloudy, the greater the solvent power and the higher the corresponding KBV. When tested as new solvent, uncontaminated by cleaning waste, water, or other impurities, the major chlorinated solvents have the following lower to higher order of solvent power and KBVs: PCE (KBV = 93) < methyl chloroform (KBV = 124) < TCE (KBV = 130) < dichloromethane (KBV = 178)[*] (Tarrer et al., 1989; Dow Chemical Company, 2002, 2006c; Solvay, 2002b). The solvent power of a solvent decreases with continued use and an increasing fraction of oily waste, grease, or soil.

Another test used to determine solvent stability involves the Acid Number. The Acid Number establishes the amount of fatty acid in a solvent, determined as the number of milligrams of potassium hydroxide to neutralize 1.28 mL of the solvent, by using a burette, a pipette, and a flask with potassium hydroxide, methanol, and phenolphthalein (Tarrer et al., 1989). Fatty acids such as stearic acid and oleic acid are introduced to the solvent when the work is coated with buffing compounds and drawing oils (Klabunde, 1949).

Additional measures of solvent stability and solvent oily waste content include viscosity (measured in a viscometer), electrical conductivity, and specific gravity. A hydrometer is often used for measuring differences in specific gravity for nonaqueous materials whose density is markedly higher or lower than water, such as fuels, or dense nonaqueous phase liquids including the major chlorinated solvents. However, a hydrometer is not sensitive enough to detect the small variations in specific gravity that occur when a solvent becomes contaminated with oily waste. The mixtures of lighter oils and heavy solvents approach the specific gravity of water and may become lighter than water as increasing amounts of oil are dissolved into the solvent. PCE, TCE, and dichloromethane have initial specific gravities of 1.62, 1.46, and 1.36, respectively; as the oil content approaches 10% by volume, the specific gravities of all three solvents converge on 0.92 (Solvay, 2002b; see Chapter 3 for further discussion). To determine the specific gravity of used solvents, a pycnometer or an electronic specific gravity meter is used. A pycnometer is a bottle that holds a specific volume of liquid and is weighed on a balance (Tarrer et al., 1989).

Solvent performance tests were generally run on samples obtained from the condensate tank on a vapor degreaser, because the condensate solvent was expected to hold the lowest level of stabilizers for dichloromethane, TCE, and PCE. Maintaining the stabilizers at recommended levels in the condensate ensures proper stabilizer levels throughout the vapor degreaser environment (Dow Chemical Company, 1999b).

1.2.5.1 Acid Acceptance Value

The total AAV test has been the most common method for determining whether acid in a solvent was sufficiently inhibited with acid-acceptor compounds. The standard for this test was ASTM D 2942, "Test Method for Total Acid Acceptance of Halogenated Organic Solvents – Nonreflux Method."

The AAV test gauges solvent condition by monitoring the concentration of neutral acid acceptors in the solvent. An aliquot of acid is added, and the solvent is titrated with sodium hydroxide to determine the amount of base required to neutralize the acid. Various degreasing equipment manuals advise operators to maintain AAV above 0.03% sodium hydroxide. For methyl chloroform, solvent with an AAV of less than 0.08% sodium hydroxide was considered to have a "borderline" condition, and solvent having an AAV of less than 0.04% was considered to be "unacceptable." Guidelines such as those mentioned in Table 1.16 informed operators when to distill, remove for recycling, or discard waste solvent.

[*] Dichloromethane's Kauri Butanol value (KBV) is listed as 87 for unstabilized methylene chloride and 178 for Solvaclene, metal-stabilized dichloromethane, both sold by Solvay SA. Dow lists a KBV for Dow Methylene Chloride (dichloromethane) as 136 expressed as cubic centimeters of solvent per 20 g of Kauri-Butanol solution. Dow's KBV for perchloroethylene is 90, for methyl chloroform, 124, and for NEU-TRI trichloroethylene 129 (Dow Chemical Company, 2002, 2006c).

TABLE 1.16
European Guidance for Acid Acceptance Values Requiring Solvent Management Action

Solvent	Fresh	Add Fresh Solvent	Replace or Distill Solvent
Tovoxene (trichloroethylene)	>0.16	0.05	0.02
Soltene™ (perchloroethylene)	>0.10	0.03	0.02
Solvaclene™ (dichloromethane)	>0.30	0.10	0.08

Source: Solvay, S.A., 2002a, Chlorinated solvents stabilisation. GBR-2900-0002-W-EN Issue 1 – 19.06.2002, Rue du Prince Albert, 33 1050 Brussels, Belgium. http://www.solvaychemicals.com (accessed November 9, 2003).
Note: AAVs are expressed in wt% NaOH.

Dow Chemical provides its own acid acceptance test kits to monitor stabilizer concentration for Dow Methylene Chloride Vapor Degreasing Grade, TCE Neu-Tri solvent, and PCE SVG (Dow Chemical Company, 1999b). Dow's test kits incorporate alkalinity testing as well as acid acceptance monitoring. In some cases, where the acid acceptance was maintained at a satisfactory level, corrosion could still be found in the degreaser equipment when there had been a drop in the pH of the solvent before the loss of the solvent stabilizer that performs as an acid acceptor. Solvent pH is normally greater than 8.5, primarily because of the reserve alkalinity from the antioxidants in the stabilizer system. Consequently, testing is needed for both reserve alkalinity and acid acceptance (Dow Chemical Company, 1999c).

Other acid tests and related solvent tests include the following:

- ASTM D 21-6-78 – an amine acid acceptance method that measures the concentration of an amine (basic) inhibitor by titration with standard acid.
- ASTM D 2942-74 – a test that determines the total AAV and measures the total concentration of an amine and neutral-type (α-epoxide) inhibitors in a solvent.
- ASTM D 1364-90 – a test that measures the water content of volatile solvents. The presence of water promotes corrosion of metals and hydrolysis of the chloride-carbon bond. The hydrolysis releases hydrogen chloride, which immediately combines with water to form hydrochloric acid. This test uses an analysis of Karl-Fischer type, which involves placing a sample into a drying oven at a predetermined temperature for a predetermined period of time. The water in the sample is vaporized and carried into a reaction vessel with methanol. The methanol traps the water, which is titrated to an endpoint with a Karl-Fischer reagent to determine the amount present.

1.2.5.2 Aluminum Scratch Test

The aluminum scratch test, ASTM D 2943-76, determines whether sufficient metal inhibitor (a compound that inhibits the catalyzing action of metal surfaces) is present to prevent reaction between aluminum and methyl chloroform. The test involves holding a clean coupon (a coupon is the metallurgical term for a thin strip of metal used for testing metal properties) of aluminum beneath the surface of the liquid solvent and scratching away the protective oxide coating. If the metal stabilizers in methyl chloroform (most commonly 1,4-dioxane) are depleted, a reaction will ensue that produces a blood-red color emitted from the metal surface. The color and bubbling are attributed to the complex formed between aluminum chloride and methyl chloroform, following dehydrochlorination and the production of hydrogen chloride gas. The degree of solvent discoloration, the amount of dark residue material formed, and the presence or absence of hydrogen chloride bubbling are used to gauge the inhibitor strength (Archer, 1984; Tarrer et al., 1989).

Other tests for inhibition of solvent breakdown by metals are as follows:

- The National Institute of Cleaning and Dying Standard 3–50 tests for an 18 mg or smaller loss in three uniformly sized copper strips (Skeeters, 1960b)
- Federal Specification OT-634A, "Trichloroethylene, Technical, 4/17/56" tests quantitatively for the extent to which metal-catalyzed oxidation decomposition of TCE has progressed

1.2.6 Identifying Solvent Stabilizers Used in Chlorinated Solvents

Determining the right mix of additives that will successfully stabilize a chlorinated solvent for a particular application was the subject of intensive research by industrial chemists working for the major solvent producers. Obtaining a successful formulation was a definite advantage in a competitive market. Proprietary formulations were therefore held as trade secrets, making difficult the task of establishing which stabilizers were used in commercial formulations of the major chlorinated solvents. Nevertheless, there are several ways of determining the identity and composition of stabilizers in commercial solvent formulations. This book documents the presence of specific stabilizer compounds in commercial solvent formulations using the following resources:

- SDSs (formerly called Material Safety Data Sheets)
- Chemical Safety Data Sheets from the Manufacturing Chemists Association (predating SDSs)
- Marketing literature by the major solvent producers
- Patents
- Industrial literature from technical journals and trade magazines
- Toxicology studies that analyzed and identified the complete range of toxic substances present in solvents

SDSs may not disclose all additives, indicating that formulations are proprietary, that quantities are less than 1%, or that components are either nontoxic or nonvolatile. Disclosure rules varied through time, and different states required disclosure of different information. For example, California's Proposition 65 rule requires listing reproductive toxins. Worker health and safety rules and transportation regulations require disclosure of some regulated compounds in the solvent formulation.

Product fact sheets and consumer guidance from the major solvent producers provide a rich resource for understanding the nature of solvent use, demands for stabilization, and the operator's challenge for maintaining solvent stability. The literature from Dow Chemical, available online, is particularly informative; however, with a few exceptions, the solvent marketing literature does not disclose the ingredients of solvent-stabilizer formulations. The approximate proportion of a given solvent formulation including its stabilizers can be inferred by contrasting the density of virgin, unstabilized solvent with that of a stabilized formulation, because stabilizers are in general substantially less dense than chlorinated solvents and most are less dense than water.

Hundreds of patents have been issued for stabilizing chlorinated solvents in the major industrial countries. The first patent for stabilizing a solvent was issued in 1914 to Walter Snelling of Pittsburgh, Pennsylvania, for stabilizing carbon tetrachloride fire extinguishers against corrosion of brass containers. Snelling used aliphatic hydrocarbons such as ethylene and acetylene (ethyne) compounds added at 0.5–3% to combine with "free chlorine" to inhibit corrosion.

Patents for stabilizing chlorinated solvents and new "designer solvents" continue to be issued today. The listing of a stabilizer formulation in a patent does not prove that it was deployed for commercial use; many patented formulations were never brought to market. In demonstrating how a particular new formulation presents an improvement over the prior art, some patents present the composition of degreasing grades of solvents in use at the time, determined by laboratory analysis. This information can be used to piece together the history of typical solvent formulations by the classes of compounds used to achieve stabilization of solvents against attack by acids, metals, and ultraviolet light. In some instances, patents list the specific compounds and the quantities of stabilizers of the leading formulations at the time the patent was issued. Patents also reveal the relative proportions of stabilizer compounds that work together to prevent reactions with the solvent. Some stabilizers work synergistically with other stabilizer compounds, and the patent literature reveals the stoichiometric relationships between them.

The industrial literature – including that from institutions promulgating specifications and performance standards for solvent use in military and industrial applications – is another source of information on identity and quantities of stabilizers in chlorinated solvents. Promulgation of air-quality regulations pursuant to the Clean Air Act led to an industry-wide effort to minimize solvent losses in vapor degreasing operations. Similarly, the land ban on disposal of liquid waste solvent led to increased attention to maximizing solvent recovery through recycling operations. The literature on solvent recycling includes information on replenishment of solvent stabilizers lost during distillation or other solvent-recovery processes and thereby reveals typical stabilizer formulations in the 1980s.

Studies on the toxicity of the major chlorinated solvents distinguished between the toxicity of pure solvents and that of technical grade solvents. A few solvent stabilizers are considerably more toxic than the solvents they stabilize. A number of articles on toxicology reveal the composition of the solvents assayed; some endeavor to separate the toxic effects of epoxide stabilizers from those of the chlorinated aliphatic solvents.

Tables 1.17 through 1.20 present a compilation of the documented presence of stabilizers in different solvent formulations obtained from cited sources, as identified at the beginning of this section. The patent literature is cited where a particular patent discloses stabilizer compositions in use at the time the patent was issued, but this practice is not intended to imply that the stabilizers claimed in the patent are present in the solvent. Patent *claims* on stabilizer packages are nevertheless useful and informative as an indirect means of understanding the empirical art of stabilizing solvents.

1.2.6.1 Solvent Stabilizer Chemicals

The major solvents have been stabilized with groups of chemicals identified as having an inhibiting effect on the reactions or catalysts that breakdown solvents. Each chemical added serves a distinct purpose, such as neutralizing acids, supplying an antioxidant, and inhibiting reactions with metal chloride salts. Finding the right chemicals to perform these tasks was the subject of much research and experimentation. For many years, the nature of the reactions that cause solvent breakdown was not identified or well understood by the industrial chemists whose task it was to prevent these reactions. In 1943, an inventor wrote in his patent, "The nature of deteriorative changes being unknown, the operative mechanism is also unknown. Stabilizing action must therefore be determined empirically" (Pitman, 1943).

The patent literature shows that industrial chemists understood which classes of compounds had the capacity to inhibit reactions. These chemists experimented with many candidate compounds before arriving at the right chemical to reliably stop solvent breakdown. Far from trial and error, industrial chemists applied sophisticated methods to select from thousands of candidate stabilizers. A number of patents claim groups of three or more stabilizer chemicals that work synergistically, such that the absence of one compound precludes the stabilizing action of the others. The author has compiled a database of more than 500 stabilizer compounds named in the patent literature. The summaries of important solvent stabilizer patents are presented in Appendix 1. Overall, the patent literature testifies to the enormous effort carried out by industrial chemists since 1914 to enable use of the major chlorinated solvents in demanding applications.

In the early decades of solvent stabilization, some inventors made claims for single additives to solve several stabilization problems. As the art of solvent stabilization evolved, industrial chemists developed solvent stabilizer packages commonly consisting of four or more chemicals to address the multiple reactions contributing to solvent breakdown. In recent decades, solvents were sold with

TABLE 1.17
Examples of General Citations of Stabilizers for the Major Solvents, by Year

Methyl Chloroform

Archer et al. (1977)

a. 1,3-Dioxolane; nitromethane; 1,2-butylene oxide; isobutyl alcohol; and toluene
b. 1,2-Butylene oxide; acetonitrile; 1,3,5-trioxane; and nitromethane
c. *tert*-Butyl alcohol; nitromethane; and methyl butynol
d. 1,2-Butylene oxide; *tert*-amyl alcohol; methyl ethyl ketone; and nitromethane
e. Nitromethane; acetonitrile; 1,2-butylene oxide; and isopropyl nitrate

Goodner et al. (1977)

DOW Chlorothene VG™: nitromethane; 1,4-dioxane; and acetonitrile

IARC (1979)

Nitromethane; *n*-methylpyrrole; 1,2-butylene oxide; 1,3-dioxolane; and secondary butyl alcohols

Dow Chemical Company (1980)

2-Methyl-3-butyn-2-ol; nitromethane; nitroethane; *t*-amyl alcohol; and 1,2-butylene oxide

Archer (1984)

United States: 0.5–0.8% 1,2-butylene oxide; 0.4–0.7% nitromethane; 2.0–3.5% 1,4-dioxane; 1.0–2.0% sec-butanol; and 1.0% 1,3-dioxolane

Europe: 0.6–1.0% 1,2-butylene oxide; 0.4–1.0% nitromethane; 3.5% 1,4-dioxane; 2.0–6.5% *tert*-butynol; 2–3% methyl butynol; 2% isopropyl nitrate; and 3% acetonitrile

Japan: 0.1–0.6% 1,2-butylene oxide; 0.1–0.7% nitromethane; and 3.5% 1,4-dioxane

Gregersen and Hansen (1986)

Epoxides; 1,4-dioxane; nitro compounds; carbonyl compounds; and *tert*-alcohols

Quast et al. (1988)

Methyl chloroform has a purity of 94%, with 5% stabilizers, including 1,2-butylene oxide, *tert*-amyl alcohol, methyl butynol, nitroethane, nitromethane, and <1% minor impurities

Dow Chemical Company (1992)

Dow Aerosol Grade: 1.5% *tert*-butyl alcohol; 0.5% 1,2-butylene oxide; and 1.9% dimethoxymethane

Meike (1993)

A typical cutting/tapping fluid based on a chlorinated hydrocarbon solvent was composed of 76% methyl chloroform, 16.3% 1,4-dioxane, 4.2% dichloromethane, and 3.5% 1,2-epoxyethane

Michigan ORR (2000)

tert-Butyl alcohol, *sec*-butyl alcohol, methylal (dimethoxymethane), and 1,2-butylene oxide

INEL (2006)

Dowclene EC: 1.9% 1,4-dioxane

Trichloroethylene

Stauffer (1956)

Triethylamine, *p*-*t*-amyl phenol, and diisobutylene

1976 and 1978 compositions cited in European Chemicals Bureau (2000c)

1976: 0.19% 1,2-butylene oxide; 0.04% ethyl acetate; 0.02% *n*-methylpyrrole; 0.03% diisobutylene; and 0.09% epichlorohydrin.

1978: 0.023% diisobutylene; 0.024% 1,2-butylene oxide; 0.052% ethyl acetate; 0.008% n-methylpyrrole; and 0.148% epichlorohydrin

Gregersen and Hansen (1986)

0.001–2% of bis(2-propyl)amine, 2-isopropyl-5-methyl-phenol, butylene oxide, and ethyl acetate

(Continued)

TABLE 1.17 (Continued)
Examples of General Citations of Stabilizers for the Major Solvents, by Year

Arena and Drew (1986)

Pentanol-2-triethanolamine; 2,2,4-trimethylpentene-1; and isobutanol

Dichloromethane

Gregersen and Hansen (1986)

Up to 1% phenol compounds, amines, epoxides, 1,4-dioxane, nitromethane, and methanol

20 ppm amylene, 50 ppm amylene, 0.2 wt% ethanol, or 150 ppm cyclohexane

Huntingtons Chemicals (2006)

Commercial methylene chloride is normally inhibited with small quantities of methanol, ethanol, amylene (2-methyl-but-2-ene), cyclohexane, or tertiary butylamine (typically 0.005–0.2% by weight) to prevent acidification and corrosion

Perchloroethylene

Gregersen and Hansen (1986)

0.1%, phenols, or mixtures of epoxides and esters

Morrison (2003)

Morpholine derivatives and mixtures of epoxides and esters

MDEQ (2004)

Dry cleaning grade: 4-methylmorpholine, diallylamine, tripropylene, cyclohexene oxide, benzotriazole, and β-ethoxypropionitrile

Linn (2002)

DOWPER™ Solvent: 4-methylmorpholine

DOWPER™ CS Solvent ("charged solvent"): nonylphenol – 4-mol ethylene oxide adduct (this was a solvent charged with detergent; it is no longer on the market)

Vulcan Materials' PerSec: diallylamine and tripropylene compounds

PPG's dry cleaning grade PCE: cyclohexene oxide and β-ethoxypropionitrile

PPG Perchlor Type 236 Stabilizer Concentrate SDS: cyclohexene oxide, β-ethoxypropionitrile, n-methylmorpholine, and 4-methoxyphenol

***n*-Propyl Bromide**[a]

Sclar (1999)

<0.5% 1,2-butylene oxide, <2.5% 1,3-dioxolane, and <0.25% nitromethane

Cheap Solvents (2002)

20–30% dimethoxymethane, 15–20% 2-methyl-2-propanol, 5–8% 1,2-butylene oxide (Stabilizer Solvent is a concentrate to refortify *n*PB for depleted stabilizers)

Petroferm, Inc. (2005)

10–15% 1,2-butylene oxide; 1–3% 2-methyl-2-propanol; and 1–3% dimethoxymethane

Swanson et al. (2002)

1,3-Dioxolane and 1,2-butylene oxide in *n*PB adhesive; contains 3% stabilizer

[a] *n*PB is n-propyl bromide, a recently introduced replacement solvent for methyl chloroform.

stabilizer packages customized for very specific applications particular to each solvent, such as vapor degreasing parts with iron and zinc but not aluminum or magnesium.

As with any endeavor involving organic chemistry, compiling lists of chemicals can be difficult because of the many acceptable synonyms for organic compounds as well as the larger list of trade names for some compounds. Appendix 1 lists some of the stabilizers claimed in the selected patents cited.

A number of articles, product fact sheets, and other sources provide numerous nonspecific lists of stabilizer compounds that name the class of compounds used or provide a longer list of chemicals that might have been used. Some listings of stabilizers for a given solvent include all the stabilizer

TABLE 1.18
Antioxidant Compounds Used in Chlorinated Solvents (Partial List)

Solvent and List of Antioxidant

Compounds	Quantity (vol%)[a]	References
Methylene chloride		
2-Methyl-2-butene	0.5	Irani (1977)
Phenol	0.05–0.5	Daras (1960); U.N. Environment Programme (UNEP) (1984)
Thymol	0.001–0.5	Pray and Chisholm (1960)
Cresol	Present in vapors	Ramos (1974)
Stabilized trichloroethylene		
Thymol	0.001–0.5, 0.28	Shepherd (1962); World Health Organization (1985); Pray and Chisholm (1960)
p-*t*-Amyl phenol	1–5	Watson and Rapp (1959); Shepherd (1962); Fullhart and Swalheim (1962)
Phenol	0.001–0.5 (0.05–0.5 g/L)	Pray and Chisholm (1960)
Pyrrole		Smallwood (1993)
n-Methyl pyrrole	0.022–0.028	Starks (1957); Copelin (1959); Shepherd (1962); Tarrer et al. (1989); OEHHA (1999)
1-Ethoxy-2-imino-ethane	0.01–0.02	Smallwood (1993)
Perchloroethylene		
2,6-di-*tert*-butyl-*p*-cresol[b]	30 ppm, 80 ppm	Howell and Tarrer (1994); Dempf et al. (1977)
Thymol	0.01% wt/wt	Reynolds and Prasad (1982)
Methyl chloroform		
Resorcinol	0.0001–1	World Health Organization (1990)

[a] Except as noted.
[b] Also known as Ionol™.

formulations used by the leading solvent producers; giving the incorrect impression that one manufacturer's grade of solvent might contain all the listed compounds. These listings are useful to obtain a sense of which classes of compounds were required to neutralize acid, inhibit acid formation, and impede solvent reactions with alkali metals. The examples in Table 1.17 are ordered by chemical species and year cited.

1.2.6.2 Antioxidants

Antioxidants reduce the solvent's potential to form oxidation products (Archer, 1984). The degree to which a solvent can undergo oxidation depends on its chemical structure, its exposure to ultraviolet light, and, if used in a vapor degreaser, its boiling point. Solvents with lower boiling points – dichloromethane (40°C) and methyl chloroform (74°C) – are less susceptible to oxidation than solvents with higher boiling points, such as PCE (121°C). The ethylene bond in TCE and PCE increase their vulnerability to oxidative attack.

Antioxidants usually fall into three chemical groups: phenols, amines, and amino-phenols, all of which contain an unsaturated benzene ring with either an amine group or a phenol group. Antioxidants suppress the free radical chain reaction that decomposes unsaturated solvents by forming stable resonance hybrids after losing a hydrogen atom to an oxidation-free radical and slowing the propagation step of autoxidation (Joshi et al., 1989).

Table 1.18 provides a list of commonly cited antioxidant compounds added to the major chlorinated solvents.

TABLE 1.19
Acid-Acceptor Compounds Used in Chlorinated Solvents

Solvent and Trade Name and List of Acid-Acceptor Compounds	Quantity (vol%)	References
Methylene chloride		
Amylene	<1.0	Solvay (2002a)
Methylene chloride, Technical Grade		
Propylene oxide	0.5	Dow Chemical Company (1995)
Dichloromethane, Technical MIL-D-6998		
Propylene oxide	2.0	Vulcan Chemicals (1987)
Methylene chloride, Technical		
Cyclohexane	<0.10	Dow Chemical Company (1999a)
Neu-tri solvent: Trichloroethylene #56530		
1,2-Butylene oxide	0.60	Dow Chemical Company (1981)
Stabilized trichloroethylene[a]		
1,2-Butylene oxide	0.053	Watson and Rapp (1959)
Trimethylpentene (diisobutylene)	0.30	Watson and Rapp (1959); Pray and Chisholm (1960); Fullhart and Swalheim (1962)
	0.03	
Stabilized trichloroethylene		
Isopropyl alcohol	0.25	Monroe and Rapp (1959)
Pyridine	0.001–0.1	Ferri and Patron (1959)
Triethylamine	0.05–0.20	McKinney et al. (1955); Shepherd (1962); Graham (1967); Tsuruta and Fukuda (1983)
Epichlorohydrin	0.31 0.09–0.8	Tsuruta and Fukuda (1983); Henschler et al. (1984); World Health Organization (1985); Tarrer et al. (1989)
Perchloroethylene		
Cyclohexene oxide	0.002 0.25 0.01–0.5	Pray and Chisholm (1960); Dempf et al. (1980); Blum (1984); Joshi et al. (1989)
Butoxymethyl oxirane (*n*-butyl glycidyl ether, BGE)	0.004 0.5	Dow Chemical Company (1987); Joshi et al. (1989)
4-Methylmorpholine	0.003 0.005 0.008	Blum (1984)
n-Methylmorpholine	0.0053	53 ppm in DOWPER: NIH (1986)
Epichlorohydrin	0.10 0.25	Beckers (1974); Van Waters and Rogers, Inc. (1990)
Methyl chloroform		
1,2-Butylene oxide	0.47 0.3–0.6	Dow Chemical Company (1990); Occidental Chemical Corporation (1989)
n-Methyl pyrrole	Present	Smallwood (1993)
Nitromethane	0.34 0.2–0.5	Dow Chemical Company (1990); Occidental Chemical Corporation (1989)

[a] Patent citation of a typical formulation of TCE in 1959 was 300 ppm 1,2-butylene oxide, 2,500 ppm isopropylacetate, and 1,500 ppm trimethylpentene (also called diisobutylene).

TABLE 1.20
Common Metal-Inhibitor Compounds Used in Chlorinated Solvents

Solvent and Trade Name and List of Metal-Inhibitor Compounds	Quantity (vol%)	References
Methylene chloride		
1,4-Dioxane	<1	Beckers (1973); World Health Organization (1984b)
	0.8	
Cyclohexane	Present	Rowe and Cawley (1977)
1,3-Dioxolane	~2	Beckers (1973)
Stabilized trichloroethylene		
Isoamyl alcohol		Petering and Aitchison (1945)
Perchloroethylene		
Cyclohexane	<1	Stevens (1955)
Ethyl acetate	0.1–0.5	Stevens (1955); Copelin (1959)
Methyl chloroform		
1,4-Dioxane	2–2.7	Occidental Chemical Corporation (1989)
1,4-Dioxane	3	Ashland Chemical Company (1995)
1,4-Dioxane	0–4	Unocal (1989)
1,3-Dioxolane	2	Great Western Chemical Company (1990)
1,3-Dioxolane	1–5	Ashland Chemical Company (1996)
Formaldehyde dimethyl hydrazone	Present	Tarrer et al. (1989)
2-Methyl-3-butyn-2-ol	Present	Ishibe and Metcalf (1982)

Although some sources list antioxidants in methyl chloroform, they are largely unnecessary. Methyl chloroform is not prone to oxidation reactions because it does not have a double bond (Archer, 1984). The use of phenolic compounds as antioxidants for TCE has caused problems for extracting essential oils from flavorings and for making decaffeinated coffee. For example, *p*-*t*-amyl phenol is unsuitable because of toxicity reasons.

A 1957 patent mentions that *n*-methyl pyrrole is a widely used antioxidant for TCE and is effective against normal air, light, and heat decomposition (Starks, 1957). However, contemporaneous patents note that pyrroles used as antioxidants have a tendency to form a dark purple sludge and that TCE stabilized with pyrrole becomes discolored over time when stored for several months in drums (Willis and Christian, 1957; Kauder, 1960). One proposed remedy to prevent the decomposition of pyrrole-stabilized TCE is the addition of diisopropylamine (Ferri and Patron, 1959). Furthermore, *n*-methyl pyrrole itself was shown to have stability problems, and an organometallic chelate compound was required to stabilize it (Starks, 1957).

Motorola's industrial chemists capitalized on the propensity for ethylene compounds to undergo oxidative attack by proposing that these compounds be used as stabilizers for methyl chloroform. Motorola's patent notes that Chlorothene VG, a widely used grade of methyl chloroform sold by Dow Chemical, will withstand semiconductor degreasing conditions for only 1 or 2 hours. For methyl chloroform, Motorola's chemists proposed a stabilizer package that uses TCE, PCE, and fluorine as free radical scavengers to meet the stringent demands of degreasing semiconductor devices while also satisfying EPA regulations. Free radical scavengers combine with trichloromethyl radicals as follows (Goodner et al., 1977):

$$R_3CH + Cl_3C \rightarrow R_3C + Cl_3CH. \quad (1.13)$$

Use of TCE, PCE, and fluorine as free radical scavengers in methyl chloroform will trap trichloromethyl radicals as intermediates before the acidic decomposition of these radicals takes place.

1.2.6.3 Light Inhibitors

A variety of chemicals have been patented to stabilize PCE against the oxidative attack of ultraviolet light. "Light inhibitors" claimed in patents have included *n*-methylmorpholine, isoeugenol (4-hydroxy-3-methoxy-1-propenylbenzene), alkyl cyanide compounds such as alkylaminoalkylcyanide, and nitrile compounds such as β-ethoxyacetonitrile and β-methoxyacetonitrile (Stevens, 1955; Strain and De Witt, 1956; Dial, 1957; Starks, 1957; Skeeters, 1959, 1960b). The mechanism for inhibiting the breakdown of PCE from ultraviolet light is not discussed in these patents. It is likely that these inhibitor compounds interrupt the free radical chain reaction, but do not physically impede the action of the ultraviolet light catalyst on the carbon-carbon double bond.

1.2.6.4 Thermal Stabilizers

A few compounds are described as providing thermal stability to PCE and TCE. Such stabilizers include diisobutylene, cyclohexene, amylene, pyrrole, and 1,2-epoxide (Stauffer, 1956; Shepherd, 1962; Smallwood, 1993).

1.2.6.5 Acid Acceptors

Acid acceptors, also referred to as antacids in the older patent literature, react with and chemically neutralize trace amounts of hydrochloric acid or acetic acid formed by hydrolysis or introduced during degreasing operations. The hydrolysis reaction cannot be prevented; hence, acid acceptors – which react with the acid to produce a benign alcohol – are added to solvents (DeGroot, 1998). Acid-acceptor compounds are either neutral (e.g., epoxide compounds) or slightly basic (e.g., amine compounds); they react with acid in the solvent and form an alcohol in the process (Archer, 1984; Tarrer et al., 1989). Compounds commonly used to neutralize acids in chlorinated solvents are summarized in Table 1.19.

Equation 1.14 (from Tarrer et al., 1989) demonstrates a typical acid acceptance reaction affected by an acid-accepting epoxide stabilizer:

$$\underset{\text{Ethylene oxide}}{\overset{\quad O}{\overset{/\ \ \backslash}{CH_2 — CH_2}}} \quad + HCl \quad \longrightarrow \quad \underset{\text{1-Chloroethanol}}{\overset{OH \quad\ \ Cl}{\overset{| \qquad\ |}{CH_2 — CH_2}}} \tag{1.14}$$

Amines were among the more commonly used acid-acceptor compounds until about 1955, when DuPont introduced a nonalkaline formulation based on a pyrrole, and Westvaco introduced a neutral formulation. These essentially replaced amines by 1961 (Doherty, 2000a). Used alone, amines can be disadvantageous because they are dissipated by reaction with acids. Amines have the further disadvantage that amine salts (1) accelerate the corrosion of metals such as zinc in galvanized degreasers and (2) can complex with copper and copper compounds (Stauffer, 1956; Copelin, 1959).

Certain combinations of acid-acceptor compounds can work together to regenerate the stabilizer by returning it to its original state when the reaction terminates. An example of this is the synergistic combination of an amine and an organic epoxide, which together are many times more effective than either additive used alone. The regeneration of an amine acid inhibitor (such as triethylamine) or the organic nitrogen-ring amine compound groups (pyridines and picolines) with an epoxide (such as 1,2-butylene oxide, propylene oxide, or cyclohexene oxide) may proceed as follows (Copelin, 1957):

$$\text{Amine} + HCl \rightarrow \text{Amine} \bullet HCl \tag{1.15}$$

$$\text{Amine} \bullet HCl + \text{Epoxide} \rightarrow \text{Chlorohydrin} + \text{Amine.} \tag{1.16}$$

Because the amine compound is regenerated, less is needed; the mole ratio of epoxide to amine can be 150:1. Eventually, epoxides will need to be replenished because they are consumed during the acid acceptance reaction.

Early efforts to find good acid-acceptor compounds included additions of the following:

- Caffeine, quinine, and limestone (calcium carbonate) to TCE (Dinley, 1937)
- Resins such as gum mastic, sandarac, and rosin to protect carbon tetrachloride and PCE for drum storage in contact with iron and copper (Stewart and DePree, 1933)
- Gasoline (Pitman, 1933)
- Camphor and turpentine (Dinley, 1937)

1.2.6.6 Metal Inhibitors

Metal inhibitors deactivate the catalytic properties of metal surfaces and complex any metal salts that might form. Metals and their salts promote solvent breakdown through catalysis. Metal inhibitors may also terminate free radicals through hydrogen donation (Archer, 1984; Howell and Tarrer, 1994). Metal inhibitors are Lewis bases that inhibit solvent-degradation reactions in the presence of a metal and its chloride (e.g., aluminum and aluminum chloride).* 1,4-Dioxane is a Lewis base because the oxygen molecules in 1,4-dioxane have electrons available for sharing (a base is a proton acceptor; a Lewis base is an electron-pair donor). Other Lewis bases used as metal inhibitors include 1,3-dioxolane, *tert*-amyl alcohol, methyl ethyl ketone, isopropyl nitrate, and nitromethane (van Gemert, 1982).

Metal inhibitors compete with the solvent for the aluminum chloride produced at microcorrosion sites on the aluminum surface. They either react with the active aluminum site, forming an insoluble deposit, or complex with aluminum chloride, preventing degradation of the solvent. Successful inhibition involves complexing of the chemisorbed aluminum chloride product with electronegative chemical groups in the molecular structure of the stabilizer (Archer, 1982). The solubility of the complex formed by the stabilizer and the metal chloride in the solvent determines the effectiveness of the stabilizer. Highly soluble stabilizer-metal chloride complexes are undesirable, because, instead of forming a protective coating over the active reaction sites on the metal surface, the complexes are removed by dissolution (Archer, 1982). Excess water promotes corrosion at the solvent-water interface because the metal chloride reaction product dissolves easily from the metal surface into the water phase (Archer, 1982).

The main function of a metal stabilizer is to compete with the solvent for electron-deficient sites on aluminum chloride adsorbed to the metal or metal oxide surface. The Lewis base stabilizer converts aluminum chloride into an insoluble coating on the metal surface, following the series of steps presented by Archer (1982):

$$\mathrm{Al(OH)_3} \leftrightarrow \mathrm{Al(OH)_2^+} + \mathrm{OH^-}, \tag{1.17}$$

$$\mathrm{Al(OH)_2^+} + \mathrm{Cl^-} \rightarrow \mathrm{Al(OH)_2Cl}\ \text{(soluble aluminum hydroxychloride salt)}. \tag{1.18}$$

$$2\left[\mathrm{Al^{3+}}\right] \cdot \mathrm{Cl_{adsorbed}} + 3\mathrm{CH_3Cl_3} \rightarrow 2\left[\mathrm{Al^{3+}}\right] + 4\mathrm{Cl^-} + 3(\mathrm{CH_3Cl_2C}\bullet), \tag{1.19}$$

$$\mathrm{Al_{surface}} \; + \; \underset{\text{Methyl chloroform}}{3(\mathrm{CH_3Cl_2C}\bullet)} \; \rightarrow \; \underset{\text{2,2,3,3-Tetrachlorobutane}}{\frac{3}{2}(\mathrm{CH_3CCl_2CCl_2CH_3})}. \tag{1.20}$$

---

* A Lewis base is any molecule or ion that can form a new coordinate covalent bond, by donating a pair of electrons. Lewis bases are preferred as metal inhibitors because they have available electrons capable of complexing as electron donors with an electron-deficient atom such aluminum. In this context, aluminum chloride is a Lewis acid.

In addition to the radical dimer product 2,2,3,3-tetrachlorobutane, an equimolar amount of 1,1-dichloroethane and a minor amount of chloroethane have been identified as products of the aluminum + methyl chloroform reaction, which suggests a reductive reaction pathway confirmed by an absence of vinyl chloride (van Gemert, 1982).

Aluminum is a highly reactive metal; its standard potential is –1.66 V, that is, it is strongly reducing. The insoluble oxide coating on aluminum will by itself prevent reaction with methyl chloroform; clean but oxide-coated aluminum dipped into methyl chloroform produces no reaction. However, any scratches in the aluminum oxide surface produce sites for microcorrosion to begin, which leads to the degradation of the solvent and the consumption of aluminum in the catastrophic decomposition reaction.

Methyl chloroform is relatively immune to the free radical mechanism for oxidation reactions; however, methyl chloroform can react violently with exposed aluminum or aluminum chloride salts. Methyl chloroform will also react with iron or tin in a similar fashion, but the corrosion rates are much slower than the reaction with aluminum (Archer and Stevens, 1977). Dichloromethane and, to some degree, TCE are also susceptible to degradation in the presence of aluminum chloride salts or aluminum swarf (aluminum fines), particularly in the vapor phase. PCE does not appear to have any significant reactivity toward aluminum (EuroChlor, 2003). A variety of metal stabilizers that are Lewis bases are added to chlorinated solvents to prevent or inhibit these reactions.

For inhibition to occur, there should be one or preferably two electronegative functional groups in the molecular structure of the selected metal inhibitor. Nitrogen functional groups have greater activity than sulfide groups, which in turn have greater activity than ether compounds. The activity of the functional group in this hierarchy is in reference to the ability to complex with Lewis acids such as aluminum chloride. Cyclic structures such as 1,4-dioxane and 1,3-dioxolane are generally more active than their straight-chain analogues. Compounds commonly used as inhibitors incorporate ether linkages, sulfide linkages, carbonyl groups, nitriles, amines, and alcohols (Archer, 1982). Because 1,4-dioxane has two ether linkages, it is considered "difunctional" and exhibits greater activity than a compound with a single ether linkage such as tetrahydropyran.

Compounds found to be effective metal inhibitors for methyl chloroform are not necessarily effective for other chlorinated solvents. For example, the hierarchy of functional groups in a mixture of 90% dichloromethane and 10% toluene (a common aerosol spray formulation) is opposite the hierarchy for methyl chloroform metal inhibitors (Archer, 1982). Other solvents require larger quantities of the same stabilizers that are effective for methyl chloroform. For example, 1,1,2-trichloroethane is easily stabilized by furfuryl alcohol and dimethyl oxalate, whereas 1,1-dichloroethane can be stabilized by furfuryl alcohol, dimethyl oxalate, pyrazine, glycidol, or 1,4-dioxane. Stabilizers preferred for the commonly used dichloromethane-toluene aerosol aluminum paint formulation contain an oxygen functional group, such as dimethoxymethane, dimethyl carbonate, or 1,4-dioxane (Archer, 1979).

The ability of solvent stabilizers selected to inhibit the aluminum + methyl chloroform reaction to protect the solvent from breakdown, ordered from best to worst, is 1,4-dioxane > nitromethane > *tert*-amyl alcohol > 1,2-butylene oxide* (van Gemert, 1982). However, when aluminum powder is added to a commercial formulation of methyl chloroform containing all these stabilizers, the first stabilizer to react is 1,2-butylene oxide, followed by *tert*-amyl alcohol. The reaction of aluminum with 1,4-dioxane and nitromethane is slow in comparison, but all of these stabilizers are more reactive toward aluminum than toward methyl chloroform. An explanation for this difference notes that epoxides† and alcohols form reaction products with aluminum chloride and are consumed, whereas 1,4-dioxane and nitromethane form complexes and are preserved. The 1,4-dioxane-aluminum chloride complex is highly insoluble; however, 1,2-butylene oxide will dissolve it and release

* 1,2-Butylene oxide, the most widely used acid acceptor for methyl chloroform, can also be considered a Lewis base (van Gemert, 1982).

† 1,2-Butylene oxide is an epoxide.

the 1,4-dioxane in its original form and a soluble aluminum-epoxide reaction product. This reaction can be conceptually represented as follows (van Gemert, 1982):

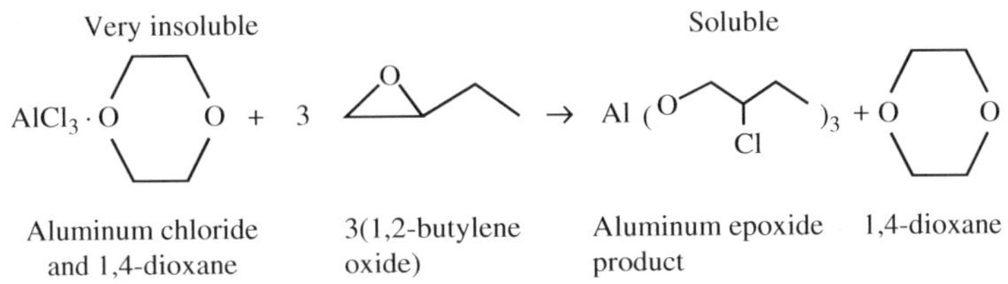

(1.21)

1,4-Dioxane is most often used in combination with 1,2-butylene oxide to stabilize vapor degreasing formulations. The electronegativity and molecular size of the complex formed by the stabilizer and the metal determine its solubility. In order to form an effective physical barrier at the reaction site, the complex should have only limited solubility in the solvent system (Archer, 1979).

Carbon tetrachloride also corrodes aluminum, after an initial induction period. The corrosion occurs by direct chemical attack and not by an electrochemical corrosion cell (Stern and Uhlig, 1952, 1953). The reaction of carbon tetrachloride with aluminum produces hexachloroethane and aluminum trichloride as follows (Minford et al., 1959):

$$2Al + 6CCl_4 \rightarrow 3C_2Cl_6 + 2AlCl_3. \tag{1.22}$$

PCE can react with fresh zinc surfaces, found in galvanized metals, to form toxic and flammable dichloroacetylene. The zinc surface eventually becomes passivated, and the reaction ceases, but this process can take up to 1 month. Since this reaction was discovered in the late 1970s, dry cleaning machine manufacturers have stopped making distillation units from galvanized metal and use stainless steel instead. Reactions of PCE with galvanized metals are also a potential problem for degreasing equipment and distillation equipment for solvent recycling (EuroChlor, 2003).

Inventors of stabilizer packages found that TCE can be stabilized against metals by adding 1 mol of an oxygen-containing compound per 99 mol of chlorinated solvent (Petering and Aitchison, 1945). Alcohol compounds are among several means of providing oxygen to stabilize solvents. The more alcohol or other oxygen compound present, the more effective the composition in restraining the metal-induced decomposition of solvents. For cleaning zinc, magnesium, and their alloys, as well as iron, steel, and copper, other metal stabilizers for TCE included oximes (Petering and Aitchison, 1945) and cyclohexene, which boils with TCE, allowing recovery on distillation. Cyclohexene is 17 times more effective as a metal stabilizer for TCE than cyclohexane (Larchar, 1950).

Among the earliest citations of solvent stabilizers to inhibit reactions with metals, corrosion of brass by carbon tetrachloride was inhibited by unsaturated hydrocarbons according to a 1914 patent, and the stabilization of carbon tetrachloride with a thiourea compound dissolved in methyl ethyl ketone was claimed in 1936 (Snelling, 1914; Missbach, 1936).

1,4-Dioxane provides two oxygen atoms to methyl chloroform. At the typical formulations listed below, the mole ratio of oxygen provided by 1,4-dioxane to methyl chloroform was 2:51.5. Oxygen-containing stabilizer compounds named in the patent literature include epoxides, oximes, ethers, and alcohols. Table 1.20 provides a listing of commonly utilized metal inhibitors for the major chlorinated solvents.

Stabilizing methyl chloroform against reaction with aluminum was a challenge that significantly delayed its commercial use in vapor degreasing and other metal-cleaning applications. These problems were memorialized in a 1957 patent to Dow Chemical, which tells us "No inhibitor heretofore used commercially has been capable of really stabilizing methyl chloroform in contact with aluminum" (Bachtel, 1957). Although the addition of 1,4-dioxane solved most aspects of the problem, it created others. Methyl chloroform stabilized with 1,4-dioxane, nitromethane, and *sec*-butyl alcohol does not react with aluminum, but its boiling vapor reacts with zinc. The compound 1,4-dioxane catalyzes the reaction with zinc, leading to damaged degreasing equipment and solvent distillation equipment made of galvanized steel sheeting. When 1,4-dioxane becomes concentrated in the degreaser boiling sump, the attack on zinc in galvanized metals can be severe (Spencer and Archer, 1981). Adding a vicinal* monoepoxide such as 1,2-butylene oxide prevents the reaction between methyl chloroform and zinc in the vapor phase. The most common stabilizer formulations for methyl chloroform contain 1,2-butylene oxide for this reason (Brown, 1962).

Pure 1,4-dioxane is susceptible to forming peroxides over time when stored. The compound 1,3-dioxolane will also combine with oxygen from the air to form explosive peroxides. Methyl chloroform stabilized with 1,4-dioxane or 1,3-dioxolane can develop peroxides or acidity during storage; however, 1,4-dioxane develops peroxides to a lesser extent than 1,3-dioxolane (Manner, 1977).

Efforts by Dow's competitors to stabilize methyl chloroform against the aluminum reaction involved other oxygen-containing compounds. Adding *tert*-butyl alcohol with 1,2-butylene oxide and nitromethane slows but does not completely inhibit the reaction, as shown by the formation of an amber precipitate. Pittsburgh Plate and Glass (later PPG), one of the producers of methyl chloroform since the 1960s, claimed that the addition of 1,2-dimethoxymethane prevents decomposition in the presence of aluminum and its salts (Hardies, 1966).

WHEN WAS 1,4-DIOXANE FIRST USED TO STABILIZE METHYL CHLOROFORM?

1,4-Dioxane was first identified in 1863 (Lourenço, 1863; Wurtz, 1863). It first became available for commercial use in the late 1920s (Reid and Hofman, 1929). Carbide and Carbon Chemicals Corporation was supplying 1,4-dioxane in the early 1930s for use as a solvent for cellulose acetate and plastics manufacturing (Jones et al., 1933). Reid and Hofman (1929) listed other uses motivating commercial production for 1,4-dioxane as a solvent:

- Degreasing
- Wool scouring
- Producing dyes
- Printing and staining
- Varnishes
- Paint and varnish removers
- Cosmetics
- Glues
- Shoe creams
- A preservative
- A fumigant
- A deodorant

The earliest mention of 1,4-dioxane as a solvent stabilizer was in 1945 in a patent for stabilizing solvents, in particular, TCE (Petering and Aitchison, 1945). However, the patent does not clearly indicate that 1,4-dioxane was preferred for use with TCE (see Box: Was 1,4-Dioxane a Stabilizer for Trichloroethylene?).

The first patent to claim 1,4-dioxane as a stabilizer for chlorinated solvents was filed in 1954 by Dow Chemical (Bachtel, 1957); however, Dow's chloroethene (methyl chloroform – technical grade) was formulated with 4.5% volume "diethylene ether" (1,4-dioxane) as early

(*Continued*)

* Vicinal means any two functional groups bonded to two adjacent carbon atoms; an epoxide is a cyclic ether with only three ring atoms (one oxygen and two carbons) in the shape of an equilateral triangle.

as 1951. Dow claimed 1,4-dioxane specifically to inhibit the reaction between methyl chloroform and aluminum and presented experimental data to show its effectiveness. Aluminum alloy coupons were scratched with a stylus to strip away the aluminum oxide surface coating. These were then immersed in unstabilized methyl chloroform, which led to immediate solvent breakdown and evolution of a carbonaceous, tarry mass. Methyl chloroform stabilized with 3.5% 1,4-dioxane remained water white with no visible reaction when in contact with scratched aluminum alloy coupons. Similar tests showed the effectiveness of 1,4-dioxane to inhibit the methyl chloroform reaction with aluminum in a boiling reflux chamber for 30 days and for storage in black iron drums for 2 months. Inclusion of *sec*-butyl alcohol or *tert*-amyl alcohol, in addition to 1,4-dioxane, prevented discoloration of the solvent.

Dow Chemical incorporated 1,4-dioxane into its next line of methyl chloroform products with the 1958 introduction of Chlorothene NU. However, as these improvements to make methyl chloroform more suitable for vapor degreasing were being developed, TCE remained the dominant vapor degreasing solvent, and was not surpassed by methyl chloroform until the late 1960s or early 1970s, depending on vapor degreasing market sector and local air pollution regulations (e.g., Los Angeles Rule 66).

WAS 1,4-DIOXANE A STABILIZER FOR TRICHLOROETHYLENE?

Petering and Aitchison's (1945) patent claims a variety of oxygen-containing compounds as stabilizers for chlorinated solvents, and TCE in particular. The patent (U.S. Patent 2,371,645) clearly lists 1,4-dioxane and even shows its molecular structure as an example of an "inner ether." However, the patent recommends propyl ether as a stabilizer for TCE, because it boils with the solvent, and emphasizes that the selection of the ether or oxide to stabilize TCE should be predicated upon its ability to distill with the solvent without loss of the stabilizer. In the claims section, the class of inner ethers is named, but the only compound specifically claimed by name is propyl ether. The patent also claims stabilizers that are "volatile with the chlorinated solvent" and stabilizers that are "simple aliphatic ethers" (1,4-dioxane is a cyclic ether).

The boiling-point difference between TCE and 1,4-dioxane is 14°C, a smaller difference than with methyl chloroform and 1,4-dioxane (which have a 26.9°C difference). The large boiling-point difference and tendency for 1,4-dioxane to partition and remain in the boiling sump did not prevent its widespread use for methyl chloroform; therefore, the smaller boiling-point difference would not by itself preclude the use of 1,4-dioxane to stabilize TCE.

In a 1960 patent filed in 1954 and issued to the Diamond Alkali Company of Cleveland, Ohio, an encyclopedic list of more than 100 compounds for use as stabilizers is provided, including "dioxan" (Skeeters, 1960b). This patent cites five other patents, none of which mention 1,4-dioxane. This patent is for stabilizing PCE, but the claims are for other compounds and do not specifically mention 1,4-dioxane.

A sampling survey of products made by processes that use organic solvents was completed in Japan in the 1980s. A large number of solvent samples (1,179) were collected and analyzed by gas chromatography. A small amount of 1,4-dioxane was detected as an additive in every sample that contained methyl chloroform; however, the study does not mention any detections of 1,4-dioxane in products produced with TCE (Inoue et al., 1983).

At a number of sites with monitoring wells, 1,4-dioxane is found together with TCE where no methyl chloroform is detected. Colocation of 1,4-dioxane with TCE does not in itself

(*Continued*)

prove that 1,4-dioxane was a stabilizer for TCE. At many facilities that operated over long periods of time, several different solvents were used in response to changing environmental and worker health and safety regulations. For example, in the late 1960s and early 1970s, TCE was replaced with methyl chloroform when it became apparent that TCE was contributing to smog in the Los Angeles basin and possibly to worker-exposure-related illness. In the mid-1980s, methyl chloroform fell out of favor when it was recognized as an ozone-depleting chemical. Many facilities switched from methyl chloroform to dichloromethane, PCE, or TCE in conjunction with adopting upgraded degreasers and ultrasonic cleaning equipment. A facility may have an old or recent release of TCE, followed or preceded by a 1,4-dioxane-containing methyl chloroform release. The abiotic half-life of methyl chloroform, approximately 2 years, is considerably shorter than either the abiotic or biodegradation half-life of TCE. Consequently, most of the methyl chloroform from an old spill will have converted to 1,1-dichloroethylene or 1,1-dichloroethane, while TCE will persist. This situation can result in the appearance that 1,4-dioxane was present in TCE. At some sites, the 1,4-dioxane was used separately for another purpose or was an ingredient of other products used in manufacturing operations equipped with TCE vapor degreasers (e.g., cutting and tapping fluids, some of which contained 16.3% 1,4-dioxane [Meike, 1993] and would have ended up in the TCE vapor degreaser sump). Statistical analysis of 1,4-dioxane and colocation with TCE has shown strong degrees of correlation, as expected from the successive use of TCE and methyl chloroform; however, the question of causation and the composition of the original solvent waste releases have not been demonstratively answered in a manner that can reliably conclude 1,4-dioxane was added to TCE in commercial formulations (Anderson et al., 2012; Karges, et al., 2018).

An independent and exhaustive review of the patent and industrial literature on solvent stabilizers provides an encyclopedic list of the stabilizers used with TCE, in which 1,4-dioxane is conspicuously absent (Morrison and Murphy, 2013).

Of greatest significance is the fact that TCE is inherently more stable toward reactions with aluminum and other alkali metals than methyl chloroform. Smaller quantities of stabilizers were required for TCE, and TCE was more easily stabilized against metal reactions by using other chemicals. Tests of TCE stabilization using 0.1% 1,4-dioxane did not measurably improve TCE stability against reactions with iron, zinc, copper, and aluminum (Willis and Christian, 1957). If TCE were stabilized with 1,4-dioxane, the mass strength of 1,4-dioxane releases at TCE sites would be markedly smaller than those measured in the aforementioned surveys of 1,4-dioxane colocation with TCE, because the smaller boiling-point difference between 1,4-dioxane and TCE would not have partitioned to the same degree as occurs for 1,4-dioxane in the boiling sump of a methyl chloroform degreaser.

Officials at Dow Chemical assertively emphasized that 1,4-dioxane was not a constituent of Dow's TCE grades (J.A. Mertens of Midland, Michigan, Dow Chemical Corporation, personal communication, December 20, 2000). Finally, none of the extensive material collected and reviewed for this book and its 2nd edition confirm that 1,4-dioxane was used to stabilize TCE.

1.2.7 Fate of Stabilizers in Industrial Applications and Solvent Recycling

The premise of "contaminant archeology" (as detailed in Chapter 10) is that understanding the nature of the industrial process in which the contaminant was used will inform the environmental scientist of the likely composition and properties of the chemicals were released to soil and groundwater. Understanding the demands on solvents for various uses will provide clues regarding the types of stabilizers that were needed for a particular process. Contrasting the physicochemical properties of stabilizers with those of their host solvents reveals which stabilizers would more likely end up in the vapor degreaser still bottoms and stored in waste tanks that ultimately leaked or discharged to unlined landfills. Contaminant archeology focuses on the clues that suggest the whole composition of the waste and how that composition may drive the subsurface behavior of the waste mixture, as well as the constituents of concern that should be considered for investigating a release of the waste solvent mixture. The properties that have the greatest bearing on the separation, concentration, or elimination of stabilizers in chlorinated solvents include boiling point, solubility in solvent and water, vapor pressure, organic carbon partition coefficient, thermal stability, and consumption in stabilizing reactions. These stabilizer and solvent properties and their consequences are discussed in more detail in this section.

1.2.7.1 Boiling-Point Differences

As described in Section 1.2.3.1, a stabilizer should not partition differentially: the stabilizer should have the same concentration in the solvent when in the vapor phase as it has in the liquid phase. Two or more organic solvent or stabilizer compounds, when brought to a boil, will distribute themselves differently between the two phases. The compound with a higher vapor pressure, or lower boiling point, will concentrate in the vapor phase. The compound with the lowest vapor pressure, or highest boiling point, will concentrate in the liquid phase. Stabilizers that concentrate in the vapor phase will eventually become depleted in open-top vapor degreasers (which were commonly used in earlier decades) because a substantial quantity of solvent vapor is lost to the atmosphere with evaporation. If a batch of stabilized solvent is evaporated under these conditions, the stabilizer will leave before the solvent is evaporated, resulting in lower stabilizer concentrations in the liquid solvent (DeGroot, 1998).

Stabilizers whose boiling points are higher or lower than the solvent to which they are added will partition differentially when the solvent is boiled for vapor degreasing or other applications.* For example, 1,4-dioxane boils at 101°C, whereas methyl chloroform boils at 78°C. Thus, 1,4-dioxane is a "high-boiler" relative to methyl chloroform. Although enough 1,4-dioxane will vaporize to stabilize methyl chloroform in the vapor phase, a larger proportion remains in the liquid phase in the vapor degreaser boiling sump. The proportion has been determined: 27% of the 1,4-dioxane will partition to the vapor phase, while 73% will remain in the liquid phase (Spencer and Archer, 1981). Through continued use, a vapor degreaser iteratively partitions 1,4-dioxane such that the proportion of 1,4-dioxane in the sump will increase over time. In several weeks of daily use, the liquid solvent can take on a composition that is as much as 10–20% 1,4-dioxane. Since 1,4-dioxane is flammable, this may pose a fire hazard (Archer et al., 1977).

The iterative degreasing cycle that partitions 1,4-dioxane in still bottoms presents a substantially greater mass strength for sources of 1,4-dioxane groundwater contamination – far greater than what might be inferred from the initial composition of stabilized methyl chloroform. As investigations of hundreds of 1,4-dioxane plumes mapped since 1995 have shown, the mass of 1,4-dioxane released in methyl chloroform degreasing waste or solvent reclamation waste was indeed enough to sustain large plumes.

* Partitioning stabilizers from boiling solvent may also depend on vapor pressure differences, vapor density contrasts, vapor removal, and other factors such as "drag-out." Vapor removal or drag-out is caused by retention of solvent on the parts exiting the vapor degreaser and by moving parts or drafts that stir the vapors and cause them to escape the cooling blanket. Although not accounting for uncontrolled factors such as drag-out, the tendency of a stabilizer compound to partition out of a mixture into the vapor phase can be determined by combining the Clausius-Clapeyron equation, Antoine's equation, and Raoult's law. This approach becomes difficult when consideration is given to the waste oil component of the mixture. Nevertheless, boiling-point difference is a good surrogate indicator of the potential for a stabilizer compound to become concentrated in the liquid phase or to become depleted.

TABLE 1.21
Partitioning of Trichloroethylene Stabilizers from New and Spent Solvent, Robbins Air Force Base

Solvent	1,2-Butylene Oxide[a]	Epichlorohydrin	Ethyl Acetate	Methyl Pyrrole
New trichloroethylene	1.64	1.66	34.6	15.9
Spent trichloroethylene	0.685	1.69	28.5	21.8
Percent change[b]	−58	+2	−18	+37

Source: Tarrer, A.R., Donahue, B.A., Dhamavaram, S., and Joshi, S.B., 1989, *Reclamation and Reprocessing of Spent Solvents*. Park Ridge, NJ: Noyes Data Corporation.

[a] Concentrations expressed as wt% × 10^3.

[b] Decreases are due to loss of stabilizer.

Laboratory analysis of new and spent methyl chloroform* sampled from a vapor degreaser at Hayes International Corporation showed a 68% increase in 1,4-dioxane concentration (Tarrer et al., 1989). Another vapor degreaser was operated under controlled conditions in a laboratory to measure changes in stabilizer composition over time. The starting composition of the stabilized methyl chloroform was 2.8% 1,4-dioxane. After operating the degreaser for 24 days, the 1,4-dioxane content increased to 7.5%. Other stabilizers of methyl chloroform were also partitioned in this experiment. *tert*-Amyl alcohol increased from 1.5% to 3.2%; nitromethane decreased from 0.5% to 0.25%; and 1,2-butylene oxide decreased from 0.75% to 0.5% (Spencer and Archer, 1981).

The decrease in nitromethane is due to its greater tendency to partition into the vapor phase. At 74°C, the boiling point of methyl chloroform, nitromethane partitions to 62% in the vapor phase and 38% in the liquid phase, even though its boiling point is 101°C (Spencer and Archer, 1981; Budavari, 1996). At 25°C, nitromethane has a vapor pressure of 35.8 mm Hg, whereas methyl chloroform has a vapor pressure of 127 mm Hg (Sunshine, 1969; Daubert and Danner, 1989). The relative loss of nitromethane to the vapor phase is compensated by including another nitroalkane as a stabilizer of methyl chloroform. Nitroethane has a vapor pressure of 20.8 mm Hg and so is more favorably retained in the liquid phase; adding nitroethane therefore inversely complements the behavior of nitromethane (Spencer and Archer, 1981).

In TCE, similar apportioning of stabilizer compounds has been measured. Samples of new and spent TCE from Robins Air Force Base outside Macon, Georgia, were analyzed by Tarrer et al. (1989); results are summarized in Table 1.21. Epichlorohydrin remained essentially unchanged because it is miscible in the solvent but has only limited solubility in water (66 mg/L), 1,2-butylene oxide decreased by 58%, ethyl acetate decreased by 18%, and methyl pyrrole increased by 37%.

Stabilizers used in PCE are also subject to partitioning because of boiling-point differences. Samples of new and spent PCE from a degreasing operation at Kelly Air Force Base near San Antonio, Texas, were analyzed and reported in Alfred Tarrer's book, *Reclamation and Reprocessing of Spent Solvents* (Tarrer et al., 1989). Cyclohexene oxide is negligibly diminished by about 6%, whereas *n*-BGE increases by about 75%. Table 1.22 presents data on the vapor partitioning of PCE stabilizers from new and spent solvents.

Aniline and morpholine, both claimed in patents as stabilizers for PCE, boil at higher temperatures than PCE. Amines such as aniline react to form anilidine, an undesired impurity that is retained in the distillation residue (Skeeters, 1960b). Aniline was one of the earliest stabilizers for

* "Spent" solvent is a relative term; degreasing operator guidance calls for removing and concentrating the still bottoms oily waste sludge when the oil content reaches 25–30% (Dow Chemical Company, 1999b). Temperature is a surrogate for measuring oil content: Dow recommends that still bottoms be removed when the boiling temperature reaches 42°C for dichloromethane (a 1°C increase from the normal boiling point for dichloromethane), 89°C for trichloroethylene (DOW NEU-TRI Solvent; a 2.7°C increase), and 125°C for perchloroethylene (a 3.6°C increase) (Dow Chemical Company, 1999b). Still bottoms for methyl chloroform should be removed when the boiling temperature reaches 77°C (a 3°C increase) (NZDL, 1981).

TABLE 1.22
Partitioning of Perchloroethylene Stabilizers from New and Spent Solvent, Robbins Air Force Base

Solvent	Cyclohexene Oxide	Butoxymethyl Oxirane[a]
New perchloroethylene	1.06	4.26
Spent perchloroethylene	0.998	7.45
Percent change (%)	–6	+75

Source: Tarrer, A.R., Donahue, B.A., Dhamavaram, S., and Joshi, S.B., 1989, *Reclamation and Reprocessing of Spent Solvents.* Park Ridge, NJ: Noyes Data Corporation.

Note: Concentrations expressed as wt% × 10^3.

[a] Also called *n*-BGE; 1-butoxy-2,3-epoxypropane; 1-butoxy-2,3-epoxypropane; and 2,3-epoxypropyl butyl ether.

chlorinated solvents (Ellis, 1925). It was also one of the earliest stabilizers for which toxicity and worker health and safety issues were raised (Dinley, 1937).

Boiling point and vapor pressure are integrally linked parameters. Both should be considered when weighing whether partitioning of a stabilizer from its solvent will likely occur.

Other impurities in solvents can also be partitioned because of boiling-point differences. For example, one of the methods of producing TCE also produces PCE (Shepherd, 1962). If PCE is carried over as a minor impurity into TCE, the PCE is concentrated in the vapor degreaser still bottoms owing to its much higher boiling point (121.4°C vs. 86.7°C for TCE).

Some investigators have concluded that the starting concentration of PCE as an impurity of TCE is probably too small to manifest as a detectable constituent at TCE waste release sites, even after iterative concentration in a vapor degreaser (Lane and Smith, 2006). Other investigators have drawn the opposite conclusion: PCE – if initially present at 1% in TCE – would reach a volume fraction of nearly 8% after multiple iterative cycles of boiling and condensation in a vapor degreaser, as calculated from the Clausius-Clapeyron equation. The operation of this process therefore allows enough PCE to be present in groundwater contaminated by TCE waste solvent to give the appearance that PCE was used for degreasing (Morrison et al., 2006). Moreover, direct chemical analysis of TCE of reagent grade, technical grade, and pharmaceutical grade has reportedly detected the presence of PCE in all three. The analysis was done to check for more toxic impurities in commercial grades of TCE, which are currently used to anesthetize patients for surgery as a cost-saving measure in hospitals that cannot afford more expensive agents of anesthesia (Totonidis, 2005). TCE produced in Europe had up to 0.03% PCE as an impurity (European Chemicals Bureau, 2004). Typical industrial grade TCE has been known to include about 0.15% carbon tetrachloride as the major chlorinated hydrocarbon impurity (Willis and Christian, 1957). See Section 10.1.2 for further discussion of PCE as an impurity of TCE.

Since the 1980s, the major producers of chlorinated solvents have sold increasingly reliable grades of TCE. In earlier decades, however, quality control was less stringent, and carryover of solvents from production led to the documented presence of impurities in the finished product sold for vapor degreasing. The patent literature from the mid-1950s notes that it was not then feasible to produce high-purity PCE for commercial use. The presence of lower aliphatic hydrocarbons introduces instability to PCE. PCE produced through chlorination and chlorinolysis* may contain trace impurities including dichloroethylene, TCE, methyl chloroform, and other compounds. Some patents from this era claim methods for restoring PCE to purity (Skeeters, 1960b). Analysis has shown that PCE with a purity of less than 50 ppm of chlorinated ethanes is still susceptible to decomposition, leading to a maximum acidity of 47 ppm (EPO, 1989).

* In the chlorinolysis method for producing perchloroethylene, hydrocarbons are chlorinated at or near pyrolytic conditions to produce a mixture of carbon tetrachloride and perchloroethylene.

The converse may also occur in which TCE is present in PCE as a degreasing blend designed to extend the life of the PCE. Solvent producers typically recommend replacing PCE in degreasing applications when the oil contamination reaches 25–30% or when the specific gravity drops to 1.4 from PCE's normal value of 1.6 (Mertens, 2000b). Producer guidance also advises that PCE should be replaced in degreasing operations when the AAV drops into the range 0.02–0.06%. TCE is sometimes added to PCE to allow continued use at higher acidity and higher oil content. The addition of up to 19% TCE to PCE extended its operational life. By lowering the boiling point from 121°C for PCE alone to 115°C for the PCE-TCE mixture, energy savings were also achieved (Roehl, 1981).

Partitioning of stabilizers from their host solvents due to boiling-point differences may increase as the boiling point of the solvent increases with continued use. A pure solvent has a defined boiling point. When oil, grease, or other compounds with higher boiling points are added to solvents during the vapor degreasing process, the boiling point of the mixture increases according to the molar concentration and the vapor pressure of each component. Solvent boiling temperature increases several degrees as the percentage of waste oil builds in the solvent sump. Changing out the solvent is recommended when the waste oil content in the degreaser approaches 25%; boiling temperature of the liquid solvent is used as a surrogate for determining the volumetric fraction of waste oil (Solvay, 2002b).

Solvent vapor losses from the open-top vapor degreasers used in earlier decades were substantial. The average emission of TCE from an open-top vapor degreaser was 2.6 g/min; about 10% of the total degreaser emissions escaped into the workplace. Solvents were frequently replenished by topping off the boiling sump with new solvent, sometimes daily (Wadden et al., 1989).

Tables 1.23 through 1.26 contrast the boiling points of stabilizers with the boiling points of their host solvents to estimate the tendency for stabilizers to be either concentrated in the still bottoms or

TABLE 1.23
Boiling-Point Contrasts for Stabilizers of Methyl Chloroform

Stabilizer	Maximum Stabilizer (wt%)	Boiling Point (°C)	Boiling-Point Difference[a] (°C)	Rank
Concentrated in Still Bottoms				
Resorcinol	—	280	205.9	1
n-Methyl pyrrole	—	115	40.9	2
1,3,5-Trioxane	1	114.5	40.4	3
Nitroethane	0.8	114	39.9	4
Toluene	—	110.6	36.5	5
Isobutyl alcohol	—	108	33.9	6
2-Methyl-3-butyn-2-ol	7	103.5	29.4	7
t-Amyl alcohol	3	102.4	28.3	8
Nitromethane	2	101.2	27.1	9
1,4-Dioxane	8	101.1	27	10
Isopropyl nitrate	—	101	26.9	11
sec-Butyl alcohol	1.22	99.5	25.4	12
Lost to Vapor				
tert-Butyl alcohol	—	82.4	8.3	13
Acetonitrile	—	81.6	7.5	14
Methyl ethyl ketone	—	80	5.9	15
1,3-Dioxolane	—	78	3.9	16
1,2-Butylene oxide	—	63.3	−10.8	17
Dimethoxymethane	—	44	−30.1	18

Source: See Appendix 2 for a tabulation of physical and chemical properties of solvent-stabilizer compounds with full citations.
[a] Methyl chloroform's boiling point is 74.1°C.

TABLE 1.24
Boiling-Point Contrasts for Stabilizers with Dichloromethane

Stabilizer	Maximum Stabilizer (wt%)	Boiling Point (°C)	Boiling-Point Difference[a] (°C)	Rank
Concentrated in Still Bottoms				
Thymol	0.5	233	193.25	1
Phenol	0.05	181.75	142	2
Ethanol	0.2	135	95.25	3
4-Cresol	0.5	129.5	89.75	5
Nitromethane	0.43	101.2	61.45	6
1,4-Dioxane	8	101.1	61.35	7
Cyclohexane	0.01	80.7	40.95	8
1,3-Dioxolane	1	78	38.25	9
Methanol	0.05	64.7	24.95	10
Lost to Vapor				
tert-Butyl amine	–	44.04	4.29	11
2-Methyl-2-butene	2.5	39	–0.75	12
Propylene oxide	–	34.23	–5.52	13
Methenamine	1	–6	–45.75	14

Source: See Appendix 2 for a tabulation of physical and chemical properties of solvent-stabilizer compounds with full citations.

[a] Dichloromethane's boiling point is 39.75°C.

TABLE 1.25
Boiling-Point Contrasts for Stabilizers with Trichloroethylene

Stabilizer	Maximum Stabilizer (wt%)	Boiling Point (°C)	Boiling-Point Difference[a] (°C)	Rank
Concentrated in Still Bottoms				
Triethanolamine	1	335.4	248.4	1
p-*t*-Amyl phenol	0.01	262.5	175.5	2
Thymol	0.5	233	146	3
2-Isopropyl-5-methyl-phenol		191	104	4
Phenol	0.5	181.75	94.75	5
Isoamyl alcohol	—	132.5	45.5	6
Epichlorohydrin	5	117.9	30.9	7
Pyridine	0.5	115.2	28.2	8
2,2,4-Trimethylpentene-1 (Diisobutylene)		101.4	14.4	10
Lost to Vapor				
Triethylamine	—	89.30	2.3	11
Ethyl acetate	—	77	–10	12
1,2-Butylene oxide	—	63.3	–23.7	13
Isobutyl alcohol	—	–88.6	–175.6	14

Source: See Appendix 2 for a tabulation of physical and chemical properties of solvent-stabilizer compounds with full citations.

[a] Trichloroethylene's boiling point is 87°C.

TABLE 1.26
Boiling-Point Contrasts for Stabilizers with Perchloroethylene

Stabilizer	Maximum Stabilizer (wt%)	Boiling Point (°C)	Boiling-Point Difference[a] (°C)	Rank
Concentrated in Still Bottoms				
2,6-di-*tert*-Butyl-*p*-cresol	0.00008	265	144	1
Thymol	0.003	233	112	2
Benzotriazole	—	204	83	3
Butoxymethyl oxirane	0.004	164	43	4
Cyclohexene oxide	—	129.5	8.5	5
Epichlorohydrin	—	117.9	−3.1	6
n-Methylmorpholine	—	113	−8	7
Lost to Vapor				
Diallylamine	—	111	−10	8
Cyclohexane	—	80.7	−40.3	9
Ethyl acetate	—	77	−44	10

Source: See Appendix 2 for a tabulation of physical and chemical properties of solvent-stabilizer compounds with full citations.

[a] Perchloroethylene's boiling point is 121.4°C.

driven off as vapor losses. Those stabilizers that tend to partition to the vapor phase will be retained in the waste solvent, but at a lower weight fraction than what was originally present. Stabilizers that are progressively depleted owing to vapor loss, consumption by reaction, or removal in the water may be replaced by periodically adding new stabilized solvent or by adding a stabilizer concentrate, as discussed in Section 1.2.7.6.

1.2.7.2 Solubility Differences

Most solvent stabilizer compounds are soluble in water; many are highly soluble or fully miscible in all proportions. Other stabilizers (e.g., epichlorohydrin) are relatively hydrophobic, with solubility of only 66 mg/L, but miscible in PCE and TCE. When water condenses in dry cleaning equipment and vapor degreasers, those stabilizers that are more soluble in water than the solvent may be extracted from the solvent by the layer of water floating atop the solvent in the sump. Water may also disperse within the liquid solvent as a mixture of fine free-phase water droplets. Dispersed water may be removed when solvents are filtered through membranes to remove waste in the solvent (Howell and Tarrer, 1994).

In dry cleaning operations, on-site distillation of PCE was commonly performed to remove impurities and water. Azeotropic reactions* typically allow for certain amounts of boil-over, causing moisture and other impurities to recirculate through the water separator in the solvent/water interphase and return to the clean tank. Distillation degrades and ultimately destroys most solvent stabilizers, which allows PCE to become acidic and corrosive (Laidlaw Company, 2004). The cause of stabilizer loss is most likely related to differences in water solubility between the stabilizers and PCE.

To minimize the effect of water on methyl chloroform, some solvent maintenance approaches capitalized on the miscibility of water in alcohol by using compounds known as polyhydroxy alcohols (i.e., propylene glycol). Methyl chloroform is treated by drawing off water accumulated atop the cooled liquid solvent and then adding a layer of propylene glycol, which also floats atop the solvent.

* An azeotrope is a liquid mixture of two or more substances that retains the same composition in the vapor state as in the liquid state when distilled or partially evaporated.

The remaining water and dissolved acids preferentially partition into the alcohol, which can then be easily removed because it remains as a free liquid phase atop the solvent (Cormany, 1977). Adding alcohols and ethers to solvents can change the solubility of substances in the solvent, sometimes with deleterious effects. For example, the addition of alcohols to TCE to stabilize metal reactions was known to increase the solubility of hydrochloric acid in TCE, which led to corrosion problems (Klabunde, 1949).

1.2.7.3 Vapor Pressure Differences

A stabilizer with a vapor pressure that is significantly lower than the host solvent will be partitioned into the still bottoms. Stabilizers whose vapor pressures are significantly higher than the host solvent will be progressively depleted as vapor is lost during operation of the degreaser or dry cleaner. Vapor pressure differences at ambient temperatures may contribute to slow loss or enrichment of stabilizers in cold cleaning operations where the solvent is used at room temperature. The literature usually has experimental vapor pressure values available for chemicals at standard temperature and pressure, that is, at 25°C and 1 atm pressure. The vapor pressures of substances change with temperature. The relevant indicator of the potential for stabilizer partitioning is the vapor pressure of the stabilizer at the boiling temperature of the solvent. Obtaining stabilizer vapor pressures at solvent boiling temperatures is more difficult as less experimental data are available.

Several methods are available for calculating vapor pressure at different temperatures, however. The most common method uses the Antoine equation:

$$\log_{10} P = A - \left[\frac{B}{T + C}\right], \tag{1.23}$$

where P is the vapor pressure (in mm Hg), T is the temperature (in °C), and A, B, and C are the Antoine constants for a particular chemical. The Antoine constants are available in *Yaws' Handbook of Antoine Coefficients for Vapor Pressure* (Knovel Corporation, 2006). The Antoine equation does not apply to all compounds and only applies to a specific temperature range for each compound. Figure 1.7 shows the vapor pressure of 1,4-dioxane and the vapor pressure of methyl chloroform; the plot graphically presents the tendency of 1,4-dioxane to concentrate.

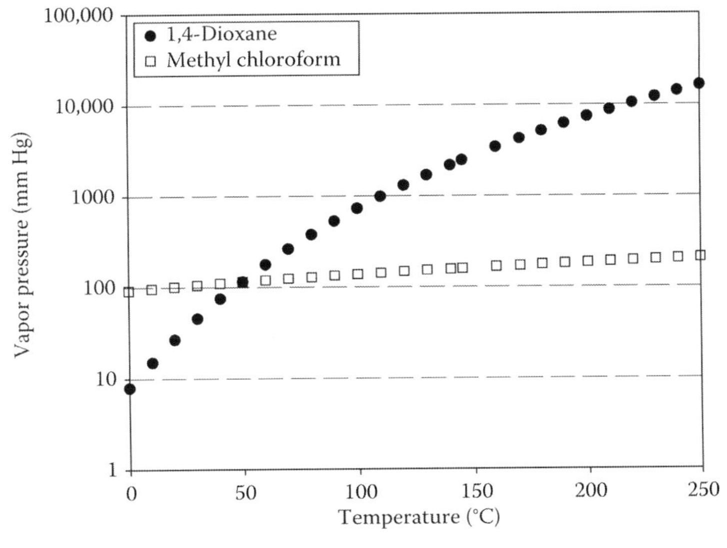

FIGURE 1.7 Vapor pressures of 1,4-dioxane and methyl chloroform. [1,4-Dioxane data from Vinson and Martin (1963). Methyl chloroform values calculated by using Antoine equation and constants from Knovel Corporation (2006).]

1.2.7.4 Carbon Adsorption

The use of activated charcoal filters for on-site purification of solvents to remove oily waste can contribute to the depletion of stabilizers, for example, in vapor degreasing operations and as a routine step in the recirculation of PCE in dry cleaning machinery after 1961 (Dow Chemical Company, 2006b; Morrison and Murphy, 2013). Acid acceptors in particular are prone to remain adsorbed to carbon filters used to filter spent solvents for reuse (Howell and Tarrer, 1994). The affinity of an organic compound to adsorb to carbon is commonly taken as the ratio of the chemical adsorbed per unit weight of organic carbon, referred to as a compound-specific parameter, K_{oc}. Tables 1.27 through 1.30 contrast the K_{oc} values for stabilizers with their host solvents, to predict which stabilizers would be lost to carbon filtration. These tables are also useful for predicting the fate of stabilizers in contact with soil organic matter as they migrate through the soil profile following a release. Fate and transport of stabilizers in the environment are described in detail in Chapter 3.

In addition to carbon filtration, physical filtration was also used to remove soil from solvents in degreasing, cold cleaning, dry cleaning, and solvent reclamation operations. Conventional filters, made of paper and other fibrous materials and often coated with granular activated carbon powder, were used to remove particles down to 0.5 μm in diameter. Solvent-resistant polymeric membranes were used to remove particles smaller than 1 μm. These filters remove high-molecular-weight matter, such as dissolved oil and grease, but allow molecules smaller than 150–200 Daltons to pass through (Howell and Tarrer, 1994). Membrane separation technology in solvent purification can remove stabilizers with higher molecular weights, such as triethanolamine, and impurities with higher molecular weights, such as PCE and carbon tetrachloride.

TABLE 1.27
Organic Carbon Partition Coefficient Contrasts for Stabilizers with Methyl Chloroform

Stabilizer	log K_{oc}	K_{oc} Ratio (Stabilizer:Solvent)	Rank
Retained in Filter			
n-Methyl pyrrole	4	54.95	1
Decreasing Retention in Carbon Filter			
Resorcinol	1.8	0.35	2
sec-Butyl alcohol	1.7	0.28	3
tert-Butyl alcohol	1.57	0.20	4
Toluene	1.56	0.20	5
2-Butanone	1.5	0.17	6
Nitroethane	1.48	0.17	7
1,4-Dioxane	1.23	0.09	8
Acetonitrile	1.2	0.09	9
1,3-Dioxolane	1.17	0.08	10
Isobutyl alcohol	0.95	0.05	11
1,2-Butylene oxide	0.90	0.04	12
2-Methyl-3-butyn-2-ol	0.54	0.02	13
t-Amyl alcohol	0.46	0.02	14
Nitromethane	0.28	0.01	15
Trioxane	−1.6	0.00	16

Source: See Appendix 2 for a tabulation of physical and chemical properties of solvent-stabilizer compounds with full citations.

Note: For methyl chloroform, log $K_{oc} = 2.26$.

TABLE 1.28
Organic Carbon Partition Coefficient Contrasts for Stabilizers with Dichloromethane

Stabilizer	Log	K_{oc} Ratio (Stabilizer:Solvent)	Rank
Retained in Filter			
Cyclohexane	2.2	6.61	1
Amine	2.16	6.03	2
2-Methyl-2-butene	1.83	2.82	3
Cresol	1.7	2.09	4
Decreasing Retention in Carbon Filter			
1,4-Dioxane	1.23	0.71	5
Phenol	1.2	0.66	6
Ethanol	1.2	0.66	6
1,3-Dioxolane	1.17	0.62	8
Thymol	0.52	0.14	9
tert-Butylamine	0.46	0.12	10
Nitromethane	0.28	0.08	11
Propylene oxide	0.03	0.04	12
Methanol	–0.77	0.01	13

Source: See Appendix 2 for a tabulation of physical and chemical properties of solvent-stabilizer compounds with full citations.

Note: For dichloromethane, log K_{oc} = 1.38.

TABLE 1.29
Organic Carbon Partition Coefficient Contrasts for Stabilizers with Trichloroethylene

Stabilizer	Log K_{oc}	K_{oc} Ratio (Stabilizer:Solvent)	Rank
Retained in Filter			
Diisobutylene	2.4	2.75	1
Pentanol-2-triethanolamine	2.2	1.45	2
Decreasing Retention in Carbon Filter			
Diisopropylamine	2.2	1.41	3
Isobutyl alcohol	1.8	0.63	4
Ethyl acetate	1.8	0.59	5
Styrene oxide	1.7	0.52	6
Pyridine	1.7	0.50	7
Epichlorohydrin	1.6	0.40	8
2-Methyl-3-butyn-2-ol	1.6	0.35	9
2-Butanone	1.5	0.32	10
Propylene oxide	1.4	0.25	11
Isopropyl alcohol	1.4	0.25	11
Phenol	1.2	0.16	13
Isopropyl acetate	1.2	0.15	14
1,2-Butylene oxide	0.90	0.08	15
Thymol	0.52	0.03	16
Triethanolamine	0.48	0.01	17
Triethylamine	0.48	0.03	17
t-Amyl alcohol	0.46	0.03	18

Source: See Appendix 2 for a tabulation of physical and chemical properties of solvent-stabilizer compounds with full citations.

Note: For trichloroethylene, log K_{oc} = 2.00.

TABLE 1.30
Organic Carbon Partition Coefficient Contrasts for Stabilizers with Trichloroethylene

Stabilizer	Log	K_{oc} Ratio (Stabilizer:Solvent)	Rank
Retained in Filter			
2,6-di-*tert*-Butyl-*p*-cresol	3.14	6.607	1
Decreasing Retention in Carbon Filter			
Cyclohexane	2.2	0.759	2
Benzotriazole	2.16	0.692	3
Triethylamine	2.16	0.692	3
Diisopropylamine	2.15	0.676	5
Diallylamine	1.98	0.457	6
Ethyl acetate	1.77	0.282	7
Epichlorohydrin	1.6	0.191	8
1,2-Butylene oxide	0.90	0.038	9
Thymol	0.52	0.016	10
n-Methylmorpholine	0.474	0.014	11
Butoxymethyl oxirane	0.24	0.008	12

Source: See Appendix 2 for a tabulation of physical and chemical properties of solvent-stabilizer compounds with full citations.

Note: For perchloroethylene, log K_{oc} = 2.32.

1.2.7.5 Thermal Stability

Stabilizers added to chlorinated solvents must be stable within the operating temperature range of the process for which the solvent is used. Stabilizers should be immune to pyrolysis at temperatures below the boiling point of the solvent and below the temperatures used for distilling solvents in reclamation processes. Solvent boiling temperatures may become slightly elevated when oily waste accumulates in the boiling sump; higher temperatures may be reached where metal fines build up around the heating coils in the boiling sump. In general, stabilizer chemicals are relatively immune to pyrolysis within the temperatures encountered in normal degreasing, dry cleaning, and other solvent applications. Higher temperatures may be encountered in transformers, for which the selection of suitable stabilizers requires special attention to their thermal stability.

1.2.7.6 Replenishing Solvent Stabilizers

Stabilizers were replenished in degreasers and in recycled solvent in two ways. For degreaser operators, the solvent distributors and degreasing equipment manufacturers recommended that the solvent be periodically refreshed by adding a quantity of fresh solvent. Operators were advised to purchase new solvent containing higher than usual levels of stabilizer to refortify the stabilizer-depleted solvent. The new solvent was added when the degreaser was not operating and the solvent was cold and provided enough stabilizers to restabilize the solvent (Howell and Tarrer, 1994).

In experiments conducted to simulate operating conditions in a degreaser for the purpose of testing a solvent maintenance system, 4 L of fresh solvent were added weekly to a degreaser whose solvent capacity was 45.4 L.* The weekly addition of almost 10% of the solvent volume was apparently a common practice (Cormany, 1977). Testing the solvent for acid acceptance and performing the aluminum scratch test before and after the addition of fresh solvent ensured the success of this approach.

In the process of recycling solvents, steam stripping was used to separate the oils and other materials solubilized in the solvent during the degreasing process. Many of the stabilizer compounds,

* A model MLW-120 stainless steel Baron-Blakeslee vapor immersion degreaser unit was used for the experiment.

particularly those added to methyl chloroform, are hydrophilic and are therefore siphoned off in steam cleaning or where water is used in the recovery process. Guidance for replenishing stabilizers calls for blending one part of reclaimed solvent with four parts of virgin solvent.

Stabilizers were also added directly back to solvents during ongoing solvent distillation in degreasing operations and in solvent-recycling operations. Operators were cautioned that (1) direct addition of stabilizers in pure form requires more skill and training to determine the correct amounts of acid acceptor, metal stabilizer, and antioxidant and (2) stabilizers are highly flammable, toxic, and possibly carcinogenic. Nevertheless, direct addition of solvent stabilizers was practiced. For example, technicians at Warner Robins Air Logistics Center in Georgia maintained a supply of 1,2-butylene oxide to replenish acid acceptors in reclaimed solvents (Howell and Tarrer, 1994).

Dow Chemical markets "MaxiStab," a series of stabilizer concentrates used to replenish stabilizers in different grades of solvents. In low-emission degreasing systems, stabilizers are depleted faster than they are replenished by the addition of fresh solvent. The MaxiStab line of stabilizer concentrates are tailored to each of Dow's chlorinated solvent lines, underscoring the specialization of stabilizer packages for specific applications (Dow Chemical Company, 2006b). A stabilizer concentrate is currently sold for newer lines of degreaser solvents, such as n-propyl bromide; the contents may include 1,2-butylene oxide, dimethoxymethane, and 2-methyl-2-propanol, as well as other compounds (Petroferm, Inc., 2005).

Another means of replenishing solvent stabilizers is accomplished through solvent conditioning. For dry cleaning, a deodorizing and antistatic powdered additive is used that also coagulates impurities to form a filterable residue, thereby decreasing the contact time with agents that could breakdown PCE. The powder also buffers the pH to increase the effectiveness of detergents added to PCE to remove water-soluble soil from clothing (Laidlaw Company, 2004).

The idea of adding stabilizers back to solvents predates vapor degreasing. A 1937 patent notes that solvent life may be indefinitely extended by repeated additions of stabilizing agents (Dinley, 1937). The concept of developing a concentrated stabilizer package was first described in a 1958 patent application (Kauder, 1960).

REFERENCES

Adamson, D.T., Mahendra, S., Walker, K.L., Rauch, S.R., Sengupta, S., and Newell, C.J., 2014, A multisite survey to identify the scale of the 1,4-dioxane problem at contaminated groundwater sites. *Environmental Science and Technology Letters* 1: 254–258.

AFCEE, 1999, Pro-ACT fact sheet: Parts washing technologies. TI#20835. Air Force Center for Environmental Excellence. http://www.afcee.brooks.af.mil/pro-act/factsheets.asp (accessed July 14, 2006).

AFSCME (American Federation of State, County and Municipal Employees), 2006, Welding hazards. http://www.afscme.org/health/faq-weld.htm (accessed July 28, 2006).

Anderson, R.H., Anderson, J.K., and Bower, P.A., 2012, Co-occurrence of 1,4-dioxane with trichloroethylene in chlorinated solvent groundwater plumes at US Air Force installations: Fact or fiction. *Integrated Environmental Assessment Management* 8(4): 731–737.

Archer, W.L., 1979, Comparison of chlorinated solvent-aluminum reaction inhibitors. *Industrial & Engineering Chemistry, Product Research and Development* 18: 131–135.

Archer, W.L., 1982, Aluminum – 1,1,1-trichloroethane: Reactions and inhibition. *Industrial & Engineering Chemistry, Product Research and Development* 21: 670–672.

Archer, W.L., 1984, A laboratory evaluation of 1,1,1-trichloroethane-metal-inhibitor systems. *Werkstoffe und Korrosion (Materials and Corrosion)* 35: 60–69.

Archer, W.L. and Stevens, V.L., 1977, Comparison of chlorinated aliphatic, aromatic, and oxygenated hydrocarbons as solvents. *Industrial & Engineering Chemistry, Product Research and Development* 16(4): 319–326.

Archer, W.L., Simpson, E.L., and Gerard, R.R., 1977, United States Patent 4,018,837: Stabilized methylchloroform. Assignee: Dow Chemical Company, Midland, MI.

Arena, J.M. and Drew, R.H. (Eds), 1986, *Poisoning – Toxicology, Symptoms, Treatments*, 5th ed. Springfield, IL: Charles C. Thomas Publishers.

Ashland Chemical Company, 1995, Material Safety Data Sheet for trichloroethane 111 degrs cold/V, issued October 28, 1995, Columbus, OH. MSDS No. 001466-014.001.

Ashland Chemical Company, 1996, Material Safety Data Sheet for trichloroethane 111 w/o dioxane, issued January 15, 1996, FSC: 6810, NIIN: 00072116.

ASTM, 1962, *Handbook of Vapor Degreasing*. American Society for Testing and Materials Special Technical Publication No. 310.

ATSDR, 2000, Toxicological profile for methylene chloride. U.S. Department of Health and Human Services Public Health Service, Agency for Toxic Substances and Disease Registry. http://www.atsdr.cdc.gov/2p-publications.html (accessed June, 2006).

ATSDR, 2005, Toxicological profile for carbon tetrachloride. U.S. Department of Health and Human Services Public Health Service, Agency for Toxic Substances and Disease Registry. http://www.atsdr.cdc.gov/2p-publications.html (accessed June, 2006).

Aviado, D.M., Zakhari, S., Simaan, J.A., and Ulsamer, A.G., 1976, *Methyl Chloroform and Trichloroethylene in the Environment*. Cleveland: CRC Press.

Bachtel, H.J., 1957, United States Patent 2,811,252: Methyl chloroform inhibited with dioxane. Assignee: Dow Chcmical Company, Midland, MI.

Beckers, N.L., 1959, United States Patent 2,903,488: Chemical method and composition. Assignee: Diamond Alkali Company, Cleveland, OH.

Beckers, N.L., 1973, United States Patent 3,923,912: Methylene chloride stabilized with ketones. Assignee: Diamond Shamrock Corporation, Cleveland, OH.

Beckers, N.L., 1974, United States Patent 3,796,755: Stabilization of chlorinated hydrocarbons. Assignee: Diamond Shamrock Corporation, Cleveland, OH.

Blake, D., Jacoby, W., Nimlos, M., and Noble, R., 1993, Identification and quantification of by-products and intermediates in the photocatalytic oxidation of gas phase trichloroethylene. In: M. Macias and M. Sanchez (Eds). *Proceedings, Volume 2, 6th International Symposium on Solar Thermal Concentrating Technologies*, pp. 1223–1231. Madrid, Spain.

Blum, K., 1984, United States Patent 4,469,906: Process for stabilizing epoxide-containing perchloroethylene. Assignee: Wacker-Chemie GmbH, Munich, Germany.

Brown, W.E., 1962, United States Patent 3,049,571: Stabilized degreasing solvent. Assignee: Dow Chemical Company.

Budavari, S. (Ed), 1996, *The Merck index – An Encyclopedia of Chemicals, Drugs, and Biologicals*. Whitehouse Station, NJ: Merck and Co., Inc.

California Air Resources Board (CARB), 1996, *Curriculum for the environmental training program for perchloroethylene dry cleaning operations*. California Environmental Protection Agency.

Campbell, D.H., 1962, United States Patent 3,040,108: Stabilization of Perchloroethylene. Assignee: Hooker Chemical Corporation, Niagara Falls, New York.

Campbell, D.H., 1966, United States Patent 3,281,285. Thiourea Stabilized Non-Aqueous Phosphatizing Composition and Method. Assignee: Hooker Chemical Corporation, Niagara Falls, New York.

Cantin, J., 1992, Overview of exposure pathways. In *Proceedings: International Roundtable on Pollution Prevention and Control in the Drycleaning Industry*. U.S. Environmental Protection Agency, Office of Pollution Prevention and Toxics. EPA/774/R-92/002.

Carey, F.A., 1987, *Organic Chemistry*. New York: McGraw Hill.

Carlisle, P.J. and Levine, A.A., 1932, Stability of chlorohydrocarbons. *Industrial & Engineering Chemistry* 24: 1164–1168.

Cheap Solvents, 2002, Material Safety Data Sheet for stabilizer solvent. http://www.cheapsolvents.com (accessed March 1, 2003).

Cohen, W.L. and Izzo, V.J., 1992, Degradation of ground water by tetrachloroethylene. In: *Proceedings of the Irrigation and Drainage Sessions at Water Forum '92, ed.*, pp. 63–69. New York: American Society of Civil Engineers.

Considine, D.M. (Ed), 1974, *Chemical and Process Technology Encyclopedia*. New York: McGraw-Hill Book Company.

Copelin, H.B., 1957, United States Patent 2,297,250: Stabilization of chlorinated hydrocarbons. Assignee: E. I. du Pont de Nemours and Company, Wilmington, DE.

Copelin, H.B., 1959, United States Patent 2,904,600: Stabilization of chlorinated hydrocarbons. Assignee: E. I. du Pont de Nemours and Company, Wilmington, DE.

Cormany, C., 1977, United States Patent 4,065,323: Degreasing process using stabilized methylchloroform solvent. Assignee: PPG Industries, Inc., Pittsburgh, PA.

Daras, N., 1960, United States Patent 2,935,537: Process for the stabilisation of chlorinated compounds. Assignee: Solvay et Cie, Brussels, Belgium.

Daubert, T.E. and Danner, R.P., 1989, *Physical and Thermodynamic Properties of Pure Chemicals Data Compilation*. Washington, DC: Taylor & Francis.

Dawson, B.R., 1989, Commercial off-site solvent reclamation, Chapter 11: U.S. Environmental Protection Agency (USEPA), September 1989. EPA/625/4-89/021.

de Nevers, N., 1986, A fatal fire with "nonflammable" methyl chloroform. *Archives of Environmental Health* 41(5): 279–281.

DeGroot, R., 1998, Stabilizing halogenated solvents thwarts undesirable reactions. *Precision Cleaning* 6(8): 25–27. http://www.p2pays.org/ref/02/01815.htm (accessed August 6, 2006).

Dempf, D., Fruhwirth, O., and Schmidhammer, L., 1977, United States Patent 4,034,051: Stabilization of perchloroethylene. Assignee: Wacker-Chemie GmbH, Munich, Germany.

Dempf, D., Knabl, R., Schmidhammer, L., and Mack, W., 1980, United States Patent 4,220,607: Stabilized perchloroethylene. Assignee: Wacker-Chemie GmbH, Munich, Germany.

Derrick, E.H. and Johnson, D.W., 1943, Three cases of poisoning by irrespirable gases – Phosgene from trichloroethylene, nitrogen dioxide, carbon dioxide with reduction of oxygen. *Medical Journal of Australia* 2: 355–358, as cited in NIOSH; Criteria Document: Phosgene, p. 33 (1976) DHEW Pub. NIOSH 76–137.

Dial, W.R., 1957, United States Patent 2,781,406: Stabilization of halogenated hydrocarbons. Assignee: Columbia Southern Chem Corp., Pittsburgh, PA.

Dinley, C.F., 1937, United States Patent 2,093,736: Stabilized chlorinated solvents and method of stabilizing such solvents. Assignee: James H. Bell, Philadelphia, PA.

Doherty, R.E., 2000a, A history of the production and use of carbon tetrachloride, tetrachloroethylene, trichloroethylene and 1,1,1-trichloroethane in the United States: Part 1 – historical background; carbon tetrachloride and tetrachloroethylene. *Journal of Environmental Forensics* 1: 69–81.

Doherty, R.E., 2000b, A history of the production and use of carbon tetrachloride, tetrachloroethylene, trichloroethylene and 1,1,1-trichloroethane in the United States: Part 2 – trichloroethylene and 1,1,1-trichloroethane. *Journal of Environmental Forensics* 1: 83–93.

Douthitt, C.A., 1990, Resource recovery through solvent reclamation. Presentation to the U.S. Department of Energy Industrial Solvent Recycling Conference October 15, 1990, Charlotte, NC. NASR (National Association of Solvent Recyclers), www.p2pays.org/ref/26/25692.pdf (accessed February, 2006).

Dow Chemical Company, 1980, Evaluation of Chloroethane VG and its components in the Ames Salmonella/Mammalian Microsome Mutagenicity Assay with attachments. EPA Doc. No. 86-890001189, Fiche No. OTS0520701.

Dow Chemical Company, 1981, Material Safety Data Sheet for Neu-Tri solvent: Trichloroethylene; #56530, December 14, 1981. FSC: 6810, NIIN: 00-924-7107.

Dow Chemical Company, 1985a, Material Safety Data Sheet for trichloroethylene-E; January 1, 1985. FSC: 6810, NIIN: 00-754-2813.

Dow Chemical Company, 1985b, Material Safety Data Sheet for 1/1/85. FSC: 6810, NIIN: LIIN: 00N003051 DOWPER CS SOLVENT.

Dow Chemical Company, 1986, Material Safety Data Sheet for perchloroethylene SVG, April 2, 1986. FSC: 6850, NIIN: LIIN: 00F049684.

Dow Chemical Company, 1987, Material Safety Data Sheet for perchloroethylene SVG, January 1, 1987. FSC: 6810, NIIN: 01-097-2020.

Dow Chemical Company, 1990, Material Safety Data Sheet for 1,1,1-trichloroethane, inhibited. FSC: 6810, NIIN: 00-476-5612.

Dow Chemical Company, 1992, Material Safety Data Sheet for 1,1,1-trichloroethane, aerosol grade, wide-spec.

Dow Chemical Company, 1995, Material Safety Data Sheet for dichloromethane, technical, December 7, 1995. FSC: 6810, NIIN: 00-616-9188.

Dow Chemical Company, 1999a, Material Safety Data Sheet for methylene chloride, technical grade, cyclohexane-inhibited. FSC: 6810, NIIN: 00-616-9188.

Dow Chemical Company, 1999b, Degreasing: Economical and efficient degreasing with chlorinated solvents from DOW.

Dow Chemical Company, 1999c, Maintaining stabilization of chlorinated solvents. *Choices & Solutions Newsletter* 3: 2. http://www.dow.com/gco/na/lit/n_letter.htm (accessed December 14, 2002).

Dow Chemical Company, 2000, Features and benefits for surface cleaning – Product Information Sheet. Form No. 100-06959. http://www.chlorinatedsolvents.com (accessed May 29, 2006).

Dow Chemical Company, 2001, Chlorinated solvents: Aluminum cleaning. Form No. 100-07024. http://www.chlorinatedsolvents.com (accessed May 29, 2006).

Dow Chemical Company, 2002, Chlorinated solvents solubility parameters. Form No. 100-07003 (revised). http://www.chlorinatedsolvents.com (accessed May 29, 2006).

Dow Chemical Company, 2006a, Chlorinated solvents product information: VOC content. Form No. 100-07056-0702. http://www.chlorinatedsolvents.com (accessed May 29, 2006).

Dow Chemical Company, 2006b, Maxistab Stabilizers Product Information Sheet. Form No. 100-06825-198QRP. http://www.chlorinatedsolvents.com (accessed May 29, 2006).

Dow Chemical Company, 2006c, Chlorinated solvents – physical properties. Form No. 100-06358. http://www.chlorinatedsolvents.com (accessed May 29, 2006).

Dow Chemical Company, 2006d, Perchloroethylene SVG double stabilized solvent Product Information Sheet. Form No. 100-07079-0902. http://www.chlorinatedsolvents.com (accessed May 29, 2006).

Dreisbach, R.H., 1987, *Handbook of Poisoning*, 12th ed. Norwalk, CT: Appleton and Lange.

Dubilier, Cornell, 2003, V-chip cleaning and coating guide. http://www.cornell-dubilier.com (accessed November 9, 2003).

Ellis, C., 1925, United States Patent 1,557,520: Cleaner. Assignee: Carlton Ellis, Montclair, NJ.

Erdmann, E., 1911, Autoxydation von trichloräthylen. *Journal für Praktische Chemie* 85(1): 78–89.

EuroChlor, 2003, Aluminium degreasing: An update on best practices. http://www.eurochlor.org/chlorsolvents/publications/digest20.htm (accessed November 17, 2003).

European Chemicals Bureau, 2000a, Evaluation and control of the risks of existing substances, IUCLID data set, substance ID 71-55-6, 1,1,1-trichloroethane. http://ecb.jrc.it/DOCUMENTS/Existing-Chemicals/IUCLID/DATA_SHEETS/71556.pdf (accessed July 29, 2006).

European Chemicals Bureau, 2000b, Evaluation and control of the risks of existing substances, IUCLID data set, substance ID 56-23-5, carbon tetrachloride. http://ecb.jrc.it/DOCUMENTS/Existing-Chemicals/IUCLID/DATA_SHEETS/56235.pdf (accessed July 29, 2006).

European Chemicals Bureau, 2000c, Evaluation and control of the risks of existing substances, IUCLID data set, substance ID 79-01-6, trichloroethylene. http://ecb.jrc.it/DOCUMENTS/Existing-Chemicals/IUCLID/DATA_SHEETS/79016.pdf (accessed July 29, 2006).

European Chemicals Bureau, 2000d, Evaluation and control of the risks of existing substances, IUCLID data set, substance ID 75-09-2, dichloromethane. http://ecb.jrc.it/DOCUMENTS/Existing-Chemicals/IUCLID/DATA_SHEETS/75092.pdf (accessed July 29, 2006).

European Chemicals Bureau, 2004, European Union risk assessment report: Trichloroethylene. 348. Environment Agency Chemicals Assessment Section Ecotoxicology and Hazardous Substances National Centre Isis House, Howbery Park, Wallingford, Oxfordshire.

European Chemicals Bureau, 2005, European Union risk assessment report: Tetrachloroethylene. Environment Agency Chemicals Assessment Section Ecotoxicology and Hazardous Substances National Centre Isis House, Howbery Park, Wallingford, Oxfordshire.

European Patent Office (EPO), 1989, European Patent Office Board of Appeals: Case number T 0131/87-3.3.2; Wacker-Chemie vs. Occidental Electrochemicals Company, European Patent No.0 041 220, 1985.

Ferri, A. and Patron, G., 1959, United States Patent 2,910,512: Method for stabilizing trichloroethylene. Assignee: Sicedison, S.P.A., Milan, Italy.

Fullhart, L. and Swalheim, D.A., 1962, United States Patent 3,051,595: Non-aqueous phosphatizing solution. Assignee: E. I. du Pont de Nemours and Company, Wilmington, DE.

Goodner, W.R., Smith, J.N., and Horvath, J., 1977, United States Patent 4,046,820: Stabilization of 1,1,1-trichloroethane. Assignee: Motorola, Inc., Schaumberg, IL.

Graham, G.W., 1967, United States Patent 3,314,892: Stabilization of halohydrocarbons. Assignee: Canadian Industries Ltd., Montreal, Quebec, Canada.

Great Western Chemical Company, 1990, Material Safety Data Sheet for 1,1,1-trichloroethane ("dioxane free"). FSC: 6810, NIIN: 00-476-5612.

Gregersen, P. and Hansen, T., 1986, Organic solvents – documentation of the neurotoxic effects in humans exposed to solvents. Miljøproject Nr. 72. National Agency of Environmental Protection, Denmark. http://www.mst.dk/homepage/ (accessed July 19, 2006).

Halogenated Solvents Industry Alliance (HSIA), 1994, White paper on methyl chloroform (1,1,1-trichloroethane). http://www.hsia.org/white_papers/paper.html (accessed April, 2002).

Halogenated Solvents Industry Alliance (HSIA), 1999, Perchlorethylene white paper. http://www.hsia.org/white_papers/paper.html (accessed April, 2002).

Halogenated Solvents Industry Alliance (HSIA), 2001, Trichlorethylene white paper. http://www.hsia.org/white_papers/paper.html (accessed April, 2002).

Halogenated Solvents Industry Alliance (HSIA), 2003, Methylene chloride white paper. http://www.hsia.org/white_papers/paper.html (accessed April, 2002).

Hardies, D.E., 1966, United States Patent 3,281,480: Stabilization of methyl chloroform. Assignee: Pittsburgh Plate and Glass Company, Pittsburgh, Pennsylvania.

Harris, C.R., 1933, United States Patent 1,904,450: Stabilization of halogenated hydrocarbons. Assignee: Roessler & Hasslacher Chemical Co., New York.

Henschler, D., Elasser, H., Romer, W., and Eder, E., 1984, Carcinogenicity study of trichloroethylene, with and without epoxide stabilizers in mice. *Journal of Cancer Research and Clinical Oncology* 107: 149–156.

Howell, S.G. and Tarrer, A.R., 1994, Minimizing pollution in cleaning and degreasing operations. EPA/600/SR-93/191.

Huntingtons Chemicals, 2006, MSDS for methylene chloride. http://www.huntingtons.fsworld.co.uk/tdsmecl2.htm (accessed August, 2006).

Hygnstrom, J., 1995, Pollution prevention: A tool kit for dry cleaners. University of Nebraska Cooperative Extension and Biological Systems Engineering, EC 95-741-S, No. 1.

INEL, 2006, Mass estimate of organic compounds in 743-series waste buried in the subsurface disposal area for operable units 7-08 and 7-13/14. U.S. Department of Energy, Idaho National Engineering Laboratory, Idaho Cleanup Project Report 23256.

Inoue, T., Takeuchi, Y., Hisanaga, N., et al., 1983, A nationwide survey on organic solvent components in various solvent products: 1. Homogeneous products such as thinners, degreasers and reagents. *Industrial Health* 21(3): 175–184.

International Agency for Research on Cancer (IARC), 1979, *Monographs on the Evaluation of the Carcinogenic Risk of Chemicals to Man*, 20, p. 516. Geneva: World Health Organization.

Irani, M.R., 1977, United States Patent 4,032,584: Stabilized methylene chloride. Assignee: Stauffer Chemical Company, Westport, CT.

Irwin, R.J., Van Mouwerik, M., Stevens, L., Seese, M.D., and Basham, W., 1997, *Environmental Contaminants Encyclopedia*. National Park Service, Water Resources Division, Fort Collins, CO.

Ishibe, N. and Metcalf, T.G., 1982, United States Patent 4,309,301: Methylchloroform stabilizer composition employing an alkynyl sulfide. Assignee: Dow Chemical Company, Midland, MI.

Izzo, V.J., 1992, Dry cleaners – a major source of PCE in ground water. Sacramento, California Regional Water Quality Board, Central Valley Region.

Jeffers, P.M., Ward, L.M., Woytowitch, L.M., and Wolfe, N.L., 1989, Homogeneous hydrolysis rate constants for selected chlorinated methanes, ethanes, ethenes, and propanes. *Environmental Science and Technology* 23: 965–969.

Johnson, J.C. and Wedmore, L.K., 1983, Metal cleaning by vapor degreasing. *Metal Finishing* 81(9): 59–63.

Johnson, K., 1973, *Dry Cleaning and Degreasing Chemicals and Processes*. Park Ridge, NJ: Noyes Data Corporation.

Jones, G.W., Seaman, H., and Kennedy, R.E., 1933, Explosive properties of dioxan-air mixtures. *Industrial & Engineering Chemistry* 25(11): 1283–1286.

Joshi, S.B., Donahue, B.A., Tarrer, A.R., Guin, J.A., Rahman, M.A., and Brady, B.L., 1989, Methods for monitoring solvent condition and maximizing its utilization. In: R.A. Conway, J.A. Frick, D.J. Warner, C.C. Wiles, and E.J. Duckett (Eds), *Hazardous and Industrial Solid Waste Minimization Practices: Proceedings of the 8th Symposium on Hazardous and Industrial Solid Waste Testing and Disposal held 12–13 November 1987 at Clearwater, FL*. Philadelphia, PA: American Society for Testing and Materials STP 1043, ASTM.

Karges, U., Becker, J., and Püttmann, W., 2018, 1,4-Dioxane pollution at contaminated groundwater sites in western Germany and its distribution within a TCE plume. *Science of the Total Environment*: 712–720.

Kauder, O.S., 1960, United States Patent 2,944,088: Stabilization of chlorinated solvents. Assignee: Argus Chemical Corporation, New York.

Kavaler, A.R. (Ed), 1995, Chemical profile: 1,1,1-Trichloroethane. *Chemical Marketing Reporter* 247 (February 27): 9, 37.

Klabunde, W., 1949, United States Patent 2,492,048: Stabilization of trichloroethylene and tetrachloroethylene. Assignee: E. I. du Pont de Nemours and Company, Wilmington, DE.

Knovel Corporation, 2006, *Yaws' Handbook of Antoine Coefficients for Vapor Pressure – Electronic Edition*. Available by subscription at http://www.knovel.com (accessed June, 2006).

Laidlaw Company, 2004, Technical product extra on LAIDLAW BUFF™. http://www.laidlawcorp.com/industrial/chemicals/buff.htm (accessed March, 2004).

Lane, V. and Smith, J., 2006, Fact or fiction: The source of perchloroethylene contamination in groundwater is a manufacturing impurity in chlorinated solvents. Presented to the National Groundwater Association Ground Water and Environmental Law Conference, July 7, 2006. http://www.ngwa.org (accessed August 2, 2006).

Larchar, A.W., 1950, United States Patent 2,517,894: Metal degreasing composition. Assignee: E. I. du Pont de Nemours and Company, Wilmington, DE.

Levine, A.A. and Cass, O.W., 1939, United States Patent 2,155,723: Stabilization of Trichloroethylene. Assignee: E. I. du Pont de Nemours and Company, Wilmington, DE.

Linn, W., 2002, Chemicals used in drycleaning operations. http://www.drycleancoalition.org/chemicals/ChemicalsUsedInDrycleaningOperations.htm (accessed 2002).

Linn, W., Appel, L., Davis, G., et al., 2004, Conducting contamination assessment work at drycleaning sites. State coalition for remediation of dry cleaners. http://www.drycleancoalition.org (accessed January, 2004).

Lourenço, M.A.-V., 1863, Recherches sur les composés polyatomiques [Research on the polyatomic compounds]. *Annales de Chimie et de Physique* (Paris) 67: 257–339.

Mabcy, W. and Mill, T., 1978, Critical review of hydrolysis of organic compounds in water under environmental conditions. *Physical Chemistry Reference Data* 7: 383–415.

Manner, J.A., 1977, United States Patent 4,026,956: Storage stabilized methylchloroform formulations. Assignee: PPG Industries, Inc., Pittsburgh, PA.

McKinney, L.L., Uhing, E.H., White, J.L., and Picken, J.C., Jr., 1955, Vegetable oil extraction: Auto-oxidation products of trichloroethylene. *Agricultural and Food Chemistry* 3(5): 413–419.

MDEQ, 2004, Michigan dry cleaning environmental compliance workbook. Michigan Department of Environmental Quality. http://www.deq.state.mi.us/documents/deq-ess-caap-drycleaners-workbook.pdf (accessed July 29, 2006).

Meike, A., 1993, Chemical and mineralogical concerns for the use of man-made materials in the post-emplacement environment. Lawrence Livermore National Laboratory, UCRL-ID-113383. http://www.osti.gov/energycitations/servlets/purl/139481-VIxX6s/139481.PDF (accessed August, 2006).

Mertens, J.A., 2000a, Vapor degreasing with chlorinated solvents. *Metal Finishing* 98(6): 43–51.

Mertens, J.A., 2000b, Keeping metal machines and the environment clean. *Machine Design* 72(5): 198.

Michigan Office of Regulatory Reform (ORR), 2000, Air Pollution Control, Natural Resources and Environmental Protection Act. Michigan Office of Regulatory Reform, Department of Environmental Quality, Air Quality Division, Air Pollution Control. *Michigan Register,* R 336.1122.

Minford, J.D., Brown, M.H., and Brown, R.H., 1959, Reaction of aluminum and carbon tetrachloride I. *Journal of the Electrochemical Society* 106(3): 185–191.

Missbach, E.C., 1936, United States Patent 2,043,258: Stabilized carbon tetrachloride. Assignee: Stauffer Chemical Company, Richmond, CA.

Mohr, T.K., 2001, *Solvent Stabilizers White Paper.* San José, CA: Santa Clara Valley Water District.

Mohr, T.K., Tulloch, C., Giri, S., Chan, S., Cook, G., and Crowley, J., 2007, *Study of Potential for Groundwater Contamination from Past Dry Cleaner Operations in Santa Clara County.* San José, CA: Santa Clara Valley Water District.

Monroe, R.F. and Rapp, D.E., 1959, United States Patent 2,906,783: Stabilization of chlorinated hydrocarbon solvents with azines. Assignee: Dow chemical company, Midland, MI.

Morrison, R., 2003, PCE contamination and the dry cleaning industry. *Environmental Claims Journal* 15(1): 93–106.

Morrison, R.D., 2002, Environmental toolbox: PCE contamination from dry cleaning operations. http://www.rmorrison.com/downloads (accessed 2002).

Morrison, R.D. and Murphy, B.L., 2013, *Chlorinated Solvents – A Forensic Evaluation.* Cambridge, United Kingdom: RSC Publishing.

Morrison, R.D., Murphy, B.L., and Doherty, R.E., 2006, Chlorinated solvents. In: R.D. Morrison and B.L. Murphy (Eds). *Environmental Forensics, Contaminant Specific Guide.* New York: Academic Press.

NACE International, The Corrosion Society, 2002, Corrosion survey database (COR-SUR). http://www.knovel.com/knovel2/Toc.jsp?BookID=532&VerticalID=0 (accessed 2006).

National Institutes of Health (NIH), 1986, *Toxicology and Carcinogenesis Studies of Tetrachloroethylene in F344/N Rats and B6C3F1 Mice (Inhalation Studies).* Research Triangle Park, NC: U.S. Department of Health and Human Services, National Institutes of Health, National Toxicology Program.

National Library of Medicine, 2006, Hazardous Substances Data Bank (HSDB). National Institutes of Health. http://toxnet.nlm.nih.gov/ (accessed March 15, 2006).

New Zealand Department of Labour (NZDL), 1981, Code of practice – vapour degreasing operations. http://www.osh.govt.nz/order/catalogue/archive/vapourdegrease.pdf (accessed August, 2006).

North Carolina Division of Health Services, Environmental Health Section (NCDHS), 1983, Alternatives for handling waste solvents in North Carolina. http://www.p2pays.org/ref/23/22202.pdf (accessed February, 2006).

Occidental Chemical Corporation, 1989, Material Safety Data Sheet for 1,1,1-trichloroethane vapor degreasing grade. FSC: 6810, NIIN: 00-476-5613.

Occidental Chemical Corporation, 1991, Material Safety Data Sheet for perclene drycleaning grade. FSC: 6850, NIIN: LIIN: 00D002823.

Office of Environmental Health Hazard Assessment California Environmental Protection Agency (OEHHA), 1999, Public health goal for trichloroethylene. Pesticide and Environmental Toxicology Section. http://www.oehha.ca.gov (see link for Public Health Goals) (accessed November 30, 2003).

Pankow, J.F. and Cherry, J.A., 1996, *Dense Chlorinated Solvents and Other DNAPLs in Groundwater.* Waterloo, Ontario, Canada: Waterloo Press.

Petering, W.H. and Aitchison, A.G., 1945, United States Patent 2,371,645: Degreasing process. Assignee: Westvaco Chlorine Products Corporation, South Charleston, WV.

Petroferm, Inc., 1997, Material Safety Data Sheet for Lenium CP, October 8, 1997. FSC: 6850, LIIN: 00N088108.

Petroferm, Inc., 2005, Material Safety Data Sheet for *n*PB Super Booster. http://www.petroferm.com/static/files/cleaning/msds (accessed July 30, 2006).

Pitman, A.L., 1933, United States Patent 1,910,962: Stabilized trichloroethylene. Westvaco Chlorine Products Corporation, South Charleston, WV.

Pitman, A.L., 1943, United States Patent 2,319,261: Stabilizing chlorinated hydrocarbons. Assignee: Westvaco Chlorine Products Corporation, South Charleston, WV.

Pray, B.O. and Chisholm, R.S., 1960, United States Patent 2,959,623: Stabilization. Assignee: Columbia-Southern Chemical Corporation.

PRO-ACT, 1999, Parts washing technologies (TI#20835). http://www.p2pays.org/ref/07/06053.htm (accessed July 14, 2006).

Quast, J.F., Calhoun, L.L., and Frauson, L.E., 1988, 1,1,1-Trichloroethane formulation: A chronic inhalation toxicity and oncogenicity study in Fischer 344 rats and B6C3F1 mice. *Fundamentals of Applied Toxicology* 11: 611–625.

Ramos, H., 1974, Health hazard evaluation toxicity determination report, No. HHE-73-5-110. Prepared by Cummins Northeastern Inc., Dedham, MA. National Institute for Occupational Safety and Health (NIOSH), Hazard Evaluations and Technical Assistance Branch, Cincinnati, OH.

Reid, E.W. and Hofman, H.E., 1929, 1,4-Dioxan. *Industrial and Engineering Chemistry* 21(7): 695–697.

Reuter, G., 2002, TCE and the right technology. *Clean Tech* September.

Reynolds, J.E.F. and Prasad, A.B. (Eds), 1982, *Martindale – The Extra Pharmacopoeia*, 28th ed., p. 106. London: The Pharmaceutical Press.

Rice, C.V. and Raftery, D., 1999, Photocatalytic oxidation of trichloroethylene using TiO_2 coated optical micro-fibers. *Chemical Communications* 1999: 895–896.

Richtzenhain, H. and Stephan, R., 1975, United States Patent 3,878,256: Stabilization of 1,1,1-trichloroethane with a four component system. Assignee: Dynamit Nobel AG, Troisdorf, Germany.

Roehl, E.O., 1981, United States Patent 4,289,542: Method of vapor degreasing. Assignee: Rho-Chem Corporation, Inglewood, CA.

Rowe, E.A., Jr. and Cawley, W.H., 1977, United States Patent 4,008,101: Methylene chloride phosphatizing. Assignee: Diamond Shamrock Corporation, Cleveland, OH.

Schmidt, R., DeZeeuw, R., Henning, L., and Trippler, D., 2001, State programs to clean up dry cleaners. State Coalition for Remediation of Dry Cleaners. http://www.drycleancoalition.org/survey/ (accessed 2002).

Sclar, G., 1999, Case report: Encephalomyeloradiculoneuropathy following exposure to an industrial solvent. *Clinical Neurology and Neurosurgery* 101: 199–202.

Shepherd, C.B., 1962, Perchloroethylene and trichloroethylene. In: S.J. Sconce (Ed), *Chlorine: Its Manufacture, Properties and Uses.* Washington, DC: American Chemical Society.

Skeeters, M.J., 1959, United States Patent 2,868,851: Stabilization of chlorinated hydrocarbons with acetylenic ethers. Assignee: Diamond Alkali Company, Cleveland, OH.

Skeeters, M.J., 1960a, United States Patent 2,958,711: Stabilization of chlorinated hydrocarbons with carboxylic acid esters of acetylenic alcohols. Assignee: Diamond Alkali Company, Cleveland, OH.

Skeeters, M.J., 1960b, United States Patent 2,947,792: Stabilization of tetrachloroethylene with a mixture of hydroxyl alkyne and isoeugenol. Assignee: Diamond Alkali Company, Cleveland, OH.

Smallwood, I., 1993, *Solvent Recovery Handbook*. New York: McGraw Hill.

Snelling, W.O., 1914, United States Patent 1,097,145: Fire extinguishing compound. Assignee: Walter O. Snelling, Pittsburgh, PA.

Solvay, S.A., 2002a, Chlorinated solvents stabilisation. GBR-2900-0002-W-EN Issue 1 – 19.06.2002, Rue du Prince Albert, 33 1050 Brussels, Belgium. http://www.solvaychemicals.com (accessed November 9, 2003).

Solvay, S.A., 2002b, Chlorinated solvents – physical and global properties. PROP-2900-0001-W-EN Issue 1 – 19.06.2002, Rue du Prince Albert, 33 1050 Brussels, Belgium. http://www.solvaychemicals.com (accessed November 9, 2003).

Spencer, D.R. and Archer, W.L., 1981, British Patent 1,582,803: Stabilization of 1,1,1-trichloroethane. Assignee: Dow Chemical Company, Midland, MI.

Spolyar, L.W., Harger, R.N., Keppler, J.F., and Bumsted, H.E., 1951, Generation of phosgene during operation of a trichloroethylene degreaser. *Archives of Industrial Hygiene and Occupational Medicine* 4: 156–160.

Starks, F.W., 1957, United States Patent 2,795,623: Stabilization of chlorinated hydrocarbons. Assignee: E. I. du Pont de Nemours and Company, Wilmington, DE.

Starks, F.W. and Kenmore, N.Y., 1960a, United States Patent 2,945,070: Stabilization of chlorinated hydrocarbons. Assignee: E. I. du Pont de Nemours and Company, Wilmington, DE.

Starks, F.W. and Kenmore, N.Y., 1960b, United States Patent 2,958,712: Stabilization of chlorinated hydrocarbons. Assignee: E. I. du Pont de Nemours and Company, Wilmington, DE.

Stauffer, W.O., 1956, United States Patent 2,751,421: Stabilization of trichloroethylene. Assignee: E. I. du Pont de Nemours and Company, Wilmington, DE.

Stern, M. and Uhlig, H.H., 1952, Effect of oxide films on the reaction of aluminum with carbon tetrachloride. *Journal of the Electrochemical Society* 99(10): 389–392.

Stern, M. and Uhlig, H.H., 1953, Mechanism of reaction of aluminum and aluminum alloys with carbon tetrachloride. *Journal of the Electrochemical Society* 100(12): 543–552.

Stevens, H.C., 1955, United States Patent 2,721,883: Stabilization of halogenated hydrocarbons. Assignee: Columbia-Southern Chemical Corp.

Stewart, L.C. and DePree, L., 1933, United States Patent 1,917,073: Method of stabilizing chlorinated aliphatic hydrocarbons. Assignee: Dow Chemical Company, Midland, MI.

Stoddard, D.L. and Wells, W.R., 1962, In-service solvent cleaning of electric motors. *Industrial Hygiene Association Journal* 23: 62–66.

Strain, F. and De Witt, B.J., 1956, United States Patent 2,737,532: Stabilization of perchloroethylene with alkoxyalkylnitriles. Assignee: Columbia-Southern Chemical Corporation.

Sunshine, I. (Ed), 1969, *CRC Handbook of Analytical Toxicology*. Cleveland: The Chemical Rubber Company.

Swanson, M.B., Geibig, J.R., and Kelly, K.E., 2002, Alternative adhesive technologies in the foam furniture and bedding industries: A cleaner technologies substitutes assessment. Volume 2: Risk screening and comparison. U.S. Environmental Protection Agency Office of Pollution Prevention Technology, Design for the Environment Program.

Tarrer, A.R., Donahue, B.A., Dhamavaram, S., and Joshi, S.B., 1989, *Reclamation and Reprocessing of Spent Solvents*. Park Ridge, NJ: Noyes Data Corporation.

Thomas, T., 1995, Clean Air Act compliance for solvent degreasers – regulatory strategies for manufacturers affected by Clean Air Act amendments NESHAP for halogenated solvent cleaners. University of Tennessee Center for Industrial Services, Tennessee Department of Environment and Conservation.

Totonidis, S., 2005, A role for trichloroethylene in developing nation anaesthesia. *Kathmandu University Medical Journal* 3(2,10): 181–190.

Tsuruta, H.I. and Fukuda, K., 1983, Analysis of trace impurities in reagent and technical grade trichloroethylene. *Industrial Health* 21(4): 293–295.

U.S. Environmental Protection Agency (USEPA), 1980, RACT compliance guidance for carbon adsorbers on perchloroethylene dry cleaners. U.S. Environmental Protection Agency, Office of General Enforcement. EPA/340/1-80/007.

U.S. Environmental Protection Agency (USEPA), 1989a, Locating and estimating air emissions from trichloroethylene and perchloroethylene. U.S. Environmental Protection Agency, Office of Air Quality Planning and Standards, Research Triangle Park, NC. EPA/450/2-89/013.

U.S. Environmental Protection Agency (USEPA), 1989b, Waste minimization in metal parts cleaning. U.S. Environmental Protection Agency, Office of Solid Waste. EPA/530/SW-89/049.

U.S. Environmental Protection Agency (USEPA), 1991, Conservation and recycling practices for CFC-113 and methyl chloroform. U.S. Environmental Protection Agency, Office of Air and Radiation. EPA/400/1-91/017.

U.S. Environmental Protection Agency (USEPA), 1994a, Locating and estimating air emissions from methyl chloroform. U.S. Environmental Protection Agency, Office of Air Quality Planning and Standards. EPA/454/R-93/045.

U.S. Environmental Protection Agency (USEPA), 1994b, Chemical summary for perchloroethylene. U.S. Environmental Protection Agency, Office of Pollution Prevention and Toxics. EPA/749/F-94/020a.

U.S. Environmental Protection Agency (USEPA), 1994c, Chemical summary for methylene chloride (dichloromethane). U.S. Environmental Protection Agency, Office of Pollution Prevention and Toxics. EPA/749/F-94/018a.

U.S. Environmental Protection Agency (USEPA), 1995, EPA Office of Compliance Sector Notebook Project: Profile of the dry cleaning industry. U.S. Environmental Protection Agency, Office of Enforcement and Compliance Assurance. EPA/310/R-95/001.

U.S. Environmental Protection Agency (USEPA), 2001, Sources, emission and exposure for trichloroethylene (TCE) and related chemicals. U.S. Environmental Protection Agency, National Center for Environmental Assessment, Office of Research and Development. EPA/600/R-00/099.

United Nations Environment Programme (UNEP), 1984, International Programme on Chemical Safety, Environmental health criteria 32: Methylene chloride. http://www.inchem.org/documents/ehc/ehc/ehc32.htm (accessed July, 2006).

United Nations Environment Programme (UNEP), Ozone Secretariat, 1999, 1998 Report of the Solvents, Coatings, and Adhesives Technical Options Committee. Nairobi, Kenya.

Unocal (Union Oil Company of California), 1989, Material Safety Data Sheet for 1,1,1-trichloroethane-15620. FSC: 6850, NIIN: 00011560.

van Gemert, B., 1982, Role of stabilizer-aluminum reactions in methylchloroform stabilization. *Industrial Engineering and Chemical Production Research and Development* 21: 296–299.

Van Waters and Rogers, Inc., 1990, Material Safety Data Sheet for tetrachloroethylene, technical, grade a, specification ASTM D 4081, 55 gallon drum. FSC: 6810, NIIN: 00-270-9982.

Vinson, C.G., Jr. and Martin, J.J., 1963, Heat of vaporization and vapor pressure of 1,4-dioxane. *Journal of Chemical Engineering Data* 8(1): 74–75.

Vogel, T.M. and McCarty, P.L., 1987, Abiotic and biotic transformations of 1,1,1-trichlorethane under methanogenic conditions. *Environmental Science and Technology* 21: 1208–1213.

Vulcan Chemicals, 1987, Material Safety Data Sheet for methylene chloride, degreasing grade. FSC: 6810, NIIN: 00-616-9188.

Vullo, W.J. and Campbell, D.H., 1965. United States Patent 3,220,890. Process and Composition for Phosphatizing Metals. Assignee: Hooker Chemical Corporation.

Wadden, R.A., Scheff, P.A., and Franke, J.E., 1989, Emission factors for trichloroethylene vapor degreasers. *American Industrial Hygiene Association Journal* 50: 496–500.

Watson, J.E. and Rapp, D.E., 1959, United States Patent 2,878,297: Stabilization of chlorinated solvents with aziridines. Assignee: Dow Chemical Company.

Willis, G.C. and Christian, C.A., 1957, United States Patent 2,803,676: Trichloroethylene stabilized with propargyl alcohol and pyrrole. Assignee: Dow Chemical Company, Midland, MI.

World Health Organization, 1984a, Perchloroethylene. International Program on Chemical Safety, Environmental Health Criteria 31. Geneva: Published under the joint sponsorship of the United Nations Environment Program, the International Labor Organization, and the World Health Organization. http://www.inchem.org/documents/ehc/ehc/ehc31.htm (accessed February 17, 2003).

World Health Organization, 1984b, Methylene chloride. International Program on Chemical Safety, Environmental Health Criteria 32. Geneva: Published under the joint sponsorship of the United Nations Environment Program, the International Labor Organization, and the World Health Organization. http://www.inchem.org/documents/ehc/ehc/ehc32.htm (accessed February 17, 2003).

World Health Organization, 1985, Trichloroethylene. International Program on Chemical Safety, Environmental Health Criteria 50. Geneva: Published under the joint sponsorship of the United Nations Environment Program, the International Labor Organization, and the World Health Organization. http://www.inchem.org/documents/ehc/ehc/ehc50.htm (accessed July 3, 2003).

World Health Organization, 1990, 1,1,1-Trichloroethane. International Program on Chemical Safety, Environmental Health Criteria 136. Geneva: Published under the joint sponsorship of the United Nations Environment Program, the International Labor Organization, and the World Health Organization. http://www.inchem.org/documents/ehc/ehc/ehc136.htm (accessed July 3, 2003).

Wurtz, A., 1863, Mémoire sur l'Oxyde d'Ethyléne et les Alcools Polyéthyleniques [Memorandum on the ethylene oxide and polyethylene alcohols]. *Annales de Chimie et de Physique* 69: 317–355.

2 1,4-Dioxane Chemistry, Uses, and Occurrence

Thomas K.G. Mohr

CONTENTS

1,4-Dioxane has many uses beyond its key role as a stabilizer for methyl chloroform. It is used directly in several industrial and commercial processes and is found in a wide range of consumer products. 1,4-Dioxane also occurs as a by-product in the production of certain surfactants, synthetic textiles, plastics, and resins. As a result of its widespread use, 1,4-dioxane is found to a limited extent in ambient air, surface water, and groundwater, but it occurs more commonly in wastewater, landfill leachate, and landfill gas. Documenting the uses, occurrence, and history of 1,4-dioxane production provides practical information for those investigating 1,4-dioxane releases to soil and groundwater. Knowledge of the many uses of 1,4-dioxane may also be helpful in reconstructing release histories by forensic investigative techniques, as discussed in Chapter 9.

1,4-Dioxane possesses some unique and surprising chemical properties, upon which its industrial utility and environmental fate depend. A comprehension of both the basic chemical structure of 1,4-dioxane and the many implications of that structure provides the foundation for (1) developing analytical methods capable of reliably separating low levels of this hydrophilic contaminant from water samples and (2) designing remedial technologies capable of removing the contaminant from affected water supplies. The chemical properties governing environmental fate and transport are described further in Chapter 3. Analytical methods for quantifying 1,4-dioxane in soil, water, air, and other media are described in detail in Chapter 4. Remedial engineering considerations for designing treatment technologies are reviewed in Chapter 8.

2.1 CHEMISTRY OF 1,4-DIOXANE

1,4-Dioxane was first described by A.V. Lourenço in 1863 as the product of reacting ethylene glycol and 1,2-dibromoethane (Stumpf, 1956; Flick, 1998). Lourenço published his discovery of 1,4-dioxane in *Annales de Chimie et de Physique* (Lourenço, 1863). In the same year, the derivation of 1,4-dioxane from ethylene oxide was described by A. Wurtz in a later issue of the same journal (Wurtz, 1863).

1,4-Dioxane is also known by the synonyms p-dioxane, diethylene oxide, 1,4-diethylene dioxide, and glycol ethylene ether. A complete list of synonyms is given in Table 2.1, and Table 2.2 lists identifiers for 1,4-dioxane as assigned by various organizations and agencies. 1,4-Dioxane is a cyclic ether that has four carbon atoms and two oxygen atoms, resulting in two ether functional groups in the same molecule. Cyclic ethers have a ring structure in which the oxygen has become part of the ring. In 1,4-dioxane, the oxygen atoms occur directly opposite each other to form symmetrical ether linkages. This structure makes 1,4-dioxane highly stable and relatively immune to reaction with acids, oxides, and oxidizing agents. The 1,4-dioxane ring does not rupture except in the presence of concentrated acids and strong oxidizing agents, under the influence of high temperature and pressure (Reid and Hoffman, 1942). The symmetry of the 1,4-dioxane ring should impart only a very small dipole moment to the molecule; nevertheless, it has good solvency (Stoye, 2005). The solvency and solubility of 1,4-dioxane are attributed to the polarity it acquires when a pair of 1,4-dioxane rings forms a dimer with two intermolecular hydrogen bonds (a dimer refers to a molecule composed of

TABLE 2.1
Synonyms, Translations, and Identification of 1,4-Dioxane

Dioxane	Glycol Ethylene Ether	Diethylene Dioxide	Diethylene Ether
p-Dioxane	Diethylene-1,4-dioxide	Dioxyethylene ether	1,4-Dioxacyclohexane
1,4-Diethylenedioxide	1,4-Dioxacyclohexane	Tetrahydro-1,4-dioxane	Tetrahydro-*p*-dioxane
Di(Ethylene oxide)	Tetramethylene 1,4-oxide	NE 220	Dioksan (Polish)
Diossano-1,4 (Italian)	Dioxaan-1,4 (Dutch)	Dioxan; dioxan-1,4 (German)	1,4-Dioksaani (Finnish)
1,4-Dioxano (Spanish)	Dioxane; dioxyethylene ether; *p*-dioxan (Czech)	1,4-Dioxanne; dioxane-1,4; dioxanne (French)	1,4-ジオキサン (Japanese)
1,4-Dioxan (Danish)	1,4-Dioxano (Portuguese)	1,4-διοξάνιο (Greek)	1,4-Dioxan (Swedish)

TABLE 2.2
Identification of 1,4-Dioxane

Identification Type	Identifier
Chemical Abstracts Service Registry Number (CASRN)	123-91-1
Registry of Toxic Effects of Chemical Substances (RTECS)	JG8225000
United Nations Number (UN)	1165
National Institute of Occupational Safety and Health (NIOSH)	JG8225000
EPA Hazardous Waste Identification Number	U108
Department of Transportation Number	1165; Guide 127
Standard Transportation Number	49 091 55; Dioxane
European Inventory of Existing Chemical Substances (EINICS Number)	204-661-8
Beilstein Registry Number	102551

two similar subunits or monomers linked together). The two remaining oxygen atoms are available for interaction with the water molecules (Mazurkiewicz and Tomasik, 2006).

1,4-Dioxane has a slight odor similar to butanol. It is fully miscible with water and most organic solvents, and water is fully soluble in dioxane. 1,4-Dioxane is favored as a solvent for cellulose derivatives, polymers, chlorinated rubber, and resins, and it has many other uses, as described further in this chapter. 1,4-Dioxane remains inert as a reagent in some chemical reactions, such as hydrogenation and sulfonation (Stoye, 2005).

There are two other dioxane isomers, 1,2-dioxane (CASRN: 5703-46-8) and 1,3-dioxane (CASRN: 505-22-6); in each designation, the two numbers separated by a comma indicate the oxygen positions in the ring structure, as shown in Figure 2.1. 1,3-Dioxane should not be confused with 1,3-dioxolane, which has a pentagonal ring structure, similar to tetrahydrofuran, and is also used as a stabilizer of methyl chloroform.

1,4-Dioxane is a Lewis base because its oxygen molecules have electrons available for sharing (a base is a proton acceptor; a Lewis base is an electron-pair donor). The molecular structure of 1,4-dioxane is shown in Figure 2.1. Its two oxygen atoms make it hydrophilic and infinitely soluble in water. Conventional rankings of solvency for different solvents predict that pure 1,4-dioxane should have a medium degree of solvency; however, its anomalously high dipole moment in aqueous solutions allows it to act as an efficient water-structure breaker, which gives 1,4-dioxane a higher solvency than expected. 1,4-Dioxane is most stable in the "chair conformation," shown in Figure 2.2 (Mazurkiewicz and Tomasik, 2006), but can shift between two boat conformations, occupying a different energy state in each conformation (Caulkins et al., 2006).

Table 2.3 summarizes structural properties of the 1,4-dioxane molecule, and Table 2.4 provides a list of chemical and physical properties compiled from a variety of sources. The key property controlling the fate and transport of 1,4-dioxane is its solubility. A tabulation of environmental fate and transport properties is presented in Chapter 3.

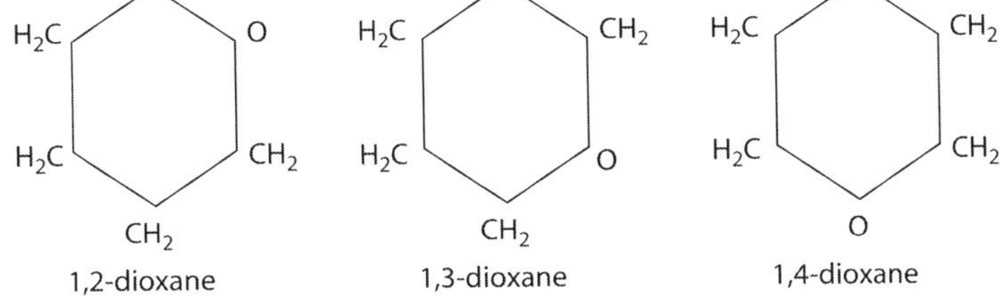

FIGURE 2.1 Structures of 1,2-dioxane, 1,3-dioxane, and 1,4-dioxane.

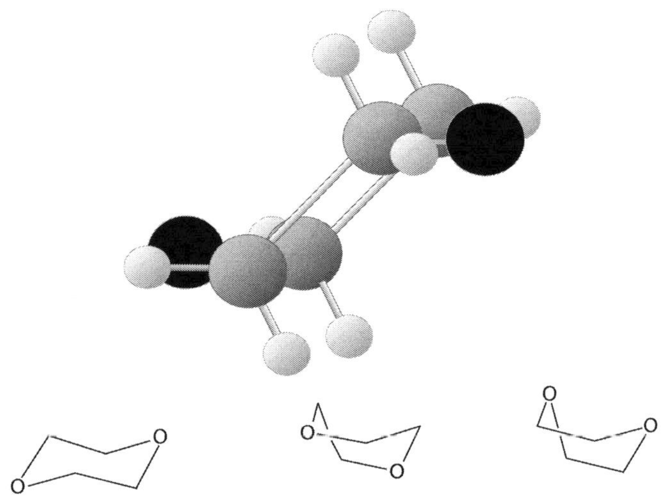

FIGURE 2.2 1,4-Dioxane in the chair conformation (black, oxygen; gray, carbon; white, hydrogen). The three conformations of 1,4-dioxane are shown below the ball and stick molecule. (Adapted from Mazurkiewicz, J. and Tomasik, P., 2006, *Journal of Molecular Liquids* 126: 111–116.)

2.2 HISTORY OF 1,4-DIOXANE PRODUCTION

1,4-Dioxane is produced from ethylene glycol. The most commonly used process, shown in Figure 2.3, involves heating ethylene glycol to 160°C and reacting it with concentrated sulfuric acid under a vacuum. 1,4-Dioxane forms when a catalyst – usually a strong acid – drives off water from ethylene glycol and reconfigures it to a ring structure. The reaction may have a 90% yield and is usually carried out on a continuous basis. The 1,4-dioxane/water-vapor azeotrope is collected and distilled with an acid trap to remove water and sulfuric acid (ATSDR, 2005; Surprenant, 2005). 1,4-Dioxane and its derivatives can also be produced with ethylene oxide as feedstock. A catalysis reaction on an acid ion exchange resin or on zeolites will dimerize ethylene oxide to diethylene dioxide, that is, 1,4-dioxane. Another method produces 1,4-dioxane and substitutes dioxane derivatives from bis(2-chloroethyl)ether (also called 2-chloro-2'-hydroxyethyl ether) by heating and treating with a 20% sodium hydroxide reactant (ATSDR, 2005; Surprenant, 2005).

TABLE 2.3
Structural Properties of the 1,4-Dioxane Molecule

Bond	Bond Length (Å)	Bond Angle Components	Bond Angle (°)	Structure
C–H	1.112			
		C–C–O	109.2	
C–O	1.423			Chair
		C–O–C	112.45	
C–C	1.523			

Source: Lide, D.R. (Ed.), 2007, Bond lengths and angles in gas-phase molecules. In: D.R. Lide (Ed.), *CRC Handbook of Chemistry and Physics, Internet Version* 2007, 8th Edition, Boca Raton, FL: Taylor & Francis.

Note: Determined on vapor phase 1,4-dioxane.

TABLE 2.4
Properties of 1,4-Dioxane

Property	Value	Notes
Molecular weight	88.106 Da	Daltons
Density	1.028 g/cm^3	~8.6 lb/gallons at 20°C
Composition	C 54.53%; H 9.15%; O 36.32%	
Boiling point	101.2°C	Range in citations: 101.1–101.3°C
Heat of vaporization	98.6 cal/g	
Freezing point	11.85°C	
Heat of fusion	33.8 cal/g	
Specific heat	36.01 cal/(mol K)	0.420 cal/g at 20°C
Vapor pressure	5.08 kPa at 25°C	38.09 mm Hg at 25°C
	15.8 kPa at 50°C	29 mm Hg at 20°C
	42.2 kPa at 75°C	
	97.2 kPa at 100°C	
	200 kPa at 125°C	
Antoine vapor coefficients	*A*: 7.94016	$P = 0.133322 10^{[A-(B/(C+T)]}$ *P* is vapor pressure (kPa) and *T* is temperature (°C); used for determining estimates of vapor pressure at different temperatures
	B: 1906.23	
	C: 275.577	
Vapor density	3.06	Air = 1; other citations: 3 and 3.3
Evaporation rate	2.42	Relative to butyl acetate = 1
Viscosity at 20°C	0.012 poise	
Surface tension (pure liquid)	32.8 mN/m	36.9 dyne/cm at 25°C; 40 dyne/cm at 20°C
Refractive index (at 20°C)	1.4224	
Auto-ignition temperature	180°C	
Flash point	12°C	Closed cup method
Lower explosive limit	2%	
Upper explosive limit	22%	
Dielectric constant	2.209	
Molar volume	85.8 cm^3/mol at 25°C	
Kinetic diameter	5.3 Å	
Dipole	0.4 D	D, debyes; 1 D, 1×10^{-18} esu cm; esu, electrostatic units of molecular charge; cm, intramolecular distance between positive and negative charge in the molecule in centimeters
Polarity	16.4	Water = 100
Hildebrand solubility parameter (δ)	10.13 cal$^{1/2}$ cm$^{-3/2}$ or 10 MPa	δ/MPa$^{1/2}$ = 2.0455 × ∂/cal$^{1/2}$ cm$^{-3/2}$
Solubility	Miscible	
Ultraviolet light absorption maximum	180 nm	
Acid dissociation constant	p**Ka** = –2.92	

Sources: Knovel, 2006, *Yaws' Handbook of Antoine Coefficients for Vapor Pressure – Electronic Edition.* http://www.knovel.com (accessed June 2006); Lide, D.R., 2007, Bond lengths and angles in gas-phase molecules. In: D.R. Lide (Ed.), CRC *Handbook of Chemistry and Physics, Internet Version* 2007, 87th Edition. Boca Raton, FL: Taylor & Francis; Mackay, D., Shiu, W.-Y., and Ma, K.-C., 1993, *Illustrated Handbook of Physical-Chemical Properties and Environmental Fate for Organic Chemicals. Volume III: Volatile Organic Chemicals.* Boca Raton, FL: Lewis Publishers (Taylor & Francis); Merck and Company, Inc., 1983, *The Merck Index.* Rahway, NJ: Merck and Company, Inc.; Monneyron, P.E., Manero, M.-H., and Foussard, J.-N., 2003, *Environmental Science and Technology* 37(11): 2410–2414; Smallwood, I.M., 1996, *Handbook of Organic Solvent Properties.* London: Arnold; Surprenant, K.S., 2005, Dioxane. In: W. Gerhartz, Y.S. Yamamoto, F.T. Campbell, et al. (Eds), *Ullmann's Encyclopedia of Industrial Chemistry.* Weinheim, Germany: Wiley Interscience; and Vinson, C.G. and Martin, J.J., 1963, Heat of vaporization and vapor pressure of 1,4-dioxane. *Journal of Chemical and Engineering Data* 8(1): 74–75.

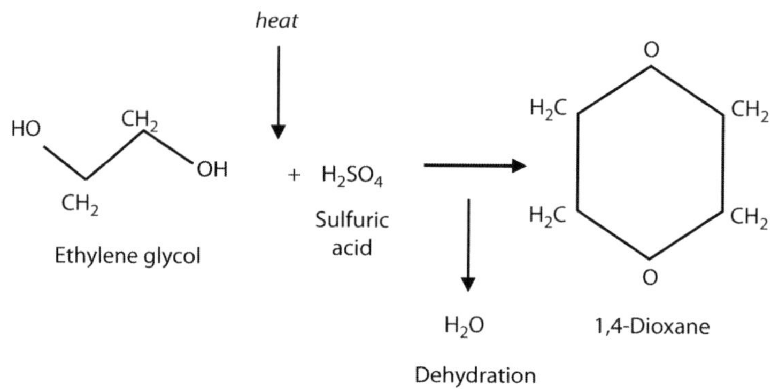

FIGURE 2.3 Conceptual representation of a common 1,4-dioxane production method involving the dehydration and ring closure of ethylene glycol with a strong acid catalyst (sulfuric acid).

2.2.1 1,4-Dioxane Production for Commercial Applications

1,4-Dioxane was first produced for commercial sale and use in 1929 (ATSDR, 2004). 1,4-Dioxane could be purchased in New York for $0.50/lb in 1929 (American Chemical Society, 1929); the price lowered to $0.25/lb by 1934 (American Chemical Society, 1935). The United States began larger-scale production of 1,4-dioxane in 1951 (National Cancer Institute, 1985). A company in Japan began production in 1958 (Osaka Organic Chemical Industry, Ltd., 2006). The largest demand for 1,4-dioxane arose in the late 1950s and early 1960s with its use in stabilizing methyl chloroform. Thereafter, 1,4-dioxane production closely tracked the production of methyl chloroform (Figure 2.4).

In 1985, the worldwide production of 1,4-dioxane was estimated at 30.8 million lbs (15,400 tons or 14,000 metric tons) (Surprenant, 2005). Production in the United States in 1985 was estimated

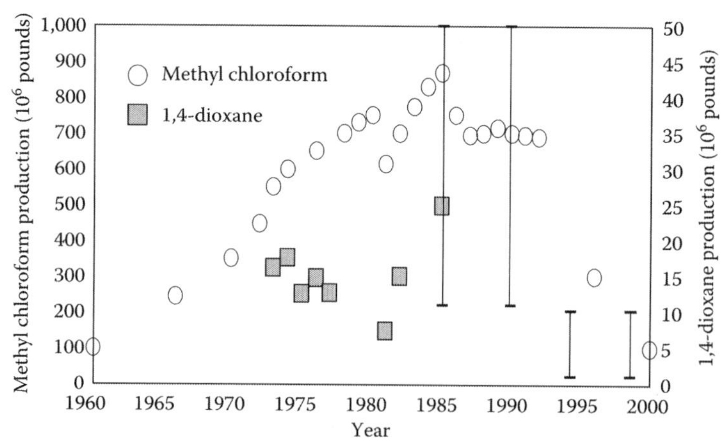

FIGURE 2.4 Annual U.S. production of methyl chloroform and 1,4-dioxane. Brackets indicate range of 1,4-dioxane production. (Adapted from Doherty, R.E., 2000, A history of the production and use of carbon tetrachloride, tetrachloroethylene, trichloroethylene and 1,1,1-trichloroethane in the United States: Part 1 – Historical background; carbon tetrachloride and tetrachloroethylene. *Journal of Environmental Forensics* 1: 11. With permission. Data from Stanford Research Institute, 1989 Directory of chemical producers – United States of America. Menlo Park, CA: SRI International; Hazardous Substances Data Bank (HSDB), 2006, Online database. National Library of Medicine, National Institutes of Health. http://toxnet.nlm.nih.gov/ (accessed July 2, 2005); and U.S. Environmental Protection Agency (USEPA), 2008, *Non-Confidential Inventory Update Reporting Production Volume Information.* http://www.epa.gov/oppt/iur/tools/data/2002-vol.htm (accessed January 2, 2008).)

TABLE 2.5
Producers of 1,4-Dioxane (ca. 1985)

Company	Location
Ferro Corporation/Grant Chemical Division	Baton Rouge, Louisiana
Union Carbide	South Charleston, West Virginia
CPS Chemical Company	Old Bridge, New Jersey
Dow Chemical Company	Freeport, Texas
Ugine-Kuhlmann	Frankfurt, Germany
BASF Aktiengesellschaft	Ludwigshafen, Germany
Dow Chemical Company	Terneuzen, Netherlands
Osaka Organic Chemical Industry, Ltd.	Osaka, Japan
Sanraku-Ocean Company	Tokyo, Japan
Toho Chemical Industry Company	Tokyo, Japan

Source: Surprenant, K.S., 2005. Dioxane. In: W. Gerhartz, Y.S. Yamamoto, F.T. Campbell, et al. (Eds), *Ullmann's Encyclopedia of Industrial Chemistry.* Weinheim, Germany: Wiley Interscience.

at 25 million pounds (12,500 tons or 11,300 metric tons). In the same year, the total methyl chloroform consumption in the United States – accounting for the total U.S. production, imports, and exports – was 450,000 tons. The approximate ratio of 1,4-dioxane production for domestic consumption to the comparable figure for methyl chloroform is about 3%. In the mid-1980s, about 90% of the 1,4-dioxane produced annually was used to stabilize methyl chloroform (National Cancer Institute, 1985). By 1995, the year in which ozone-depleting substance regulations severely curtailed the use of methyl chloroform, the production of 1,4-dioxane decreased to 22 million pounds (11,000 tons or 10,000 metric tons) (European Chemicals Bureau, 2002). The U.S. Environmental Protection Agency (USEPA) reports that the range of 1,4-dioxane production was from 10 to 50 million pounds between 1986 and the early 1990s; after 1994, production ranged from 1 to 10 million pounds (USEPA, 2008), as shown in Figure 2.4.

1,4-Dioxane was produced at ten locations in the 1980s, as summarized in Table 2.5. By 2000, fewer companies produced this chemical. The primary 1,4-dioxane production locations were Ferro Chemical Corporation (Baton Rouge, Louisiana), Dow Chemical (Freeport, Texas), BASF (Ludwigshafen, Germany), Osaka Yuki (Osaka, Japan), and Toho Chemical (Tokyo, Japan) (European Chemicals Bureau, 2002).

2.2.2 Impurities and Stabilizers of 1,4-Dioxane

1,4-Dioxane is available in several different grades that vary in purity, as well as the nature of impurities present due to different production methods. Table 2.6 lists the common impurities of 1,4-dioxane and their typical concentrations. Other impurities may also be present in 1,4-dioxane in unspecified amounts, such as acetaldehyde, acetic acid, glycol acetal paraldehyde, and crotonaldehyde (Wypych, 2001).

Low-molecular-weight ethers, including 1,4-dioxane, react with oxygen to form explosive peroxides. When 1,4-dioxane comes into contact with oxygen or ultraviolet light, it can form the peroxide compound 1,4-dioxanyl-2-hydroperoxide (Gierer and Pettersson, 1968). Prolonged exposure to air and sunlight may form unstable peroxides, particularly when water is absent. Peroxide compounds can form and accumulate in 1,4-dioxane, which may then explode when subjected to heat or shock. Handling 1,4-dioxane is most hazardous when peroxide levels are concentrated by distillation or evaporation. To prevent peroxide formation, 1,4-dioxane is stabilized by adding small amounts of

TABLE 2.6
Common Impurities of 1,4-Dioxane

Impurity	Concentration
bis(2-Chloroethyl) ether	<0.1%
Water	<0.1%
2-Methyl-1,3-dioxane	<0.1%
2-Ethyl-1,3-dioxane	<0.03%
Hydrogen peroxide	<0.001%
Lead	<0.25 ppm
Iron	<0.25 ppm

Sources: European Chemicals Bureau, 2002, European Union risk assessment report: 1,4-Dioxane. Luxembourg, European Union, Institute for Health and Consumer Protection, Report 21; DeRosa, C.T., Wilbur, S., Holler, J., et al., 1996, Health evaluation of 1,4-dioxane. *Toxicology and Industrial Health* 12: 43; U.S. Environmental Protection Agency (USEPA), 1981, Phase I risk assessment of 1,4-dioxane. U.S. Environmental Protection Agency, EPA Contract No. 68-01-6030; and VROM, 1999, Risk assessment: 1,4-Dioxane. Bilthoven, Netherlands: S.P. Netherlands Ministry of Housing and the Environment (VROM), Chemical Substances Bureau 153.

reducing agents such as stannous chloride or ferrous sulfate. In addition, it can be stabilized with 25 mg/kg 2,6-*tert*-butyl-*p*-cresol, a phenolic antioxidant also used to stabilize chlorinated solvents, or with 25 mg/kg butyl hydroxy toluene (BHT), an antioxidant commonly found in baked goods (BASF, 1986, 2005; European Chemicals Bureau, 2002). Another method to prevent peroxide formation is filling container headspace with nitrogen. When stabilized and stored under nitrogen in original containers, 1,4-dioxane has a shelf life of 24 months (BASF, 2005).

2.3 DIRECT USES OF 1,4-DIOXANE

The leading U.S. producer of 1,4-dioxane, Ferro Corporation, lists the following uses of 1,4-dioxane on its web site (Ferro Corporation, 2006):

- Wetting and dispersing agent in textile processing
- Dye baths and stain and printing compositions
- Some cleaning and detergent preparations, adhesives, cosmetics, deodorants, fumigants, emulsions, and polishing agents
- Some lacquers, paints, varnishes, and paint and varnish removers
- Solvent for fats, oils, waxes, and natural and synthetic resins
- Purification of drugs
- Reaction media in various organic synthesis reactions
- Stabilizer for chlorinated solvents

1,4-Dioxane is also used for making photosensitive resins and in magnetic tape production as a base for applying magnetic pigments to tape for audio and data recording (Royal Society of Chemistry, 1992). The solvents used in the manufacturing of certain varieties of magnetic tapes contain as much as 40% 1,4-dioxane. During formulation of some products, the pure substance is used (European Chemicals Bureau, 2002). The California Air Resources Board (CARB) lists manufacturing processes that use 1,4-dioxane directly or produce 1,4-dioxane as a by-product in regulatory guidance

TABLE 2.7
Emissions Sources for 1,4-Dioxane in Industrial Production Operations from Direct Use or Formation of 1,4-Dioxane as a By-Product

Emissions Sources for 1,4-Dioxane	
Vapor degreasing	Surface coating
Production of adhesives and sealants	Production of dyes
Production of miscellaneous printing inks	Production of polishes and waxes
Production of surface active agents	Reagent in university research facilities
Textile finishing	Paint and varnish removal
Wood finishing	Drug and pharmaceutical purification
Production of inorganic pigments	Paint production
Detergent manufacturing	Manufacture of membrane filters
Textile mills production	Weaving mills – synthetics, cotton, and wool

Source: California Air Resources Board (CARB), 2006, "Hot spots" inventory guidelines. California Air Resources Board, Health and Safety Code Section 44300 et seq.

documentation for the Air Toxics "Hot Spots" Information and Assessment Act of 1987 (California Assembly Bill 2588), as summarized in Table 2.7 (CARB, 2006).

A study of solvent uses by USEPA included applications of 1,4-dioxane (USEPA, 1996c). In addition to the uses just listed, possible uses of 1,4-dioxane mentioned in the study but not expressly confirmed included petroleum refining, petrochemicals, pulp and paper, explosives, electroplating, polishing, semiconductors, electronic components, photographic equipment, polymers, plastics, rubber manufacture, and organic and inorganic chemical manufacture. Use of 1,4-dioxane was confirmed in some industries in the USEPA study by questionnaire responses. Sixteen pharmaceutical facilities used 1,4-dioxane, primarily in small amounts for laboratory operations such as quality control of finished goods and laboratory experiments. Two pharmaceutical firms used 1,4-dioxane as a medium for crystallization and for the distillation and dissolution of an intermediate product. One laminated-paper firm identified 1,4-dioxane as a chemical to dissolve polymers to produce a coated paper. One business in the office-machines industry used 1,4-dioxane for the dissolution of pigment for an interface spray. Two photographic industry facilities surveyed in the USEPA study used 1,4-dioxane. One facility used it to dissolve resins and polymers for film coating. The other facility used 1,4-dioxane as a photochemical reaction medium and for miscellaneous research and development projects (USEPA, 1996c). Other surveys of 1,4-dioxane uses list pulping of wood, deodorants, production of fumigants, reagents to provide the mobile phase in some laboratory chromatography applications, reagents for measuring optical activity, reagent for cryoscopic analysis, and as a polymerization catalyst (ATSDR, 2005).

This chapter explores the various uses of 1,4-dioxane and its occurrence as a by-product in different industries by providing examples of how 1,4-dioxane was used or generated, with discussion of the relative quantities involved. Appreciation for the many ways 1,4-dioxane has been used or generated and released to the environment can assist with site investigations in locations where multiple industries have operated.

2.3.1 Cellulose Acetate Membrane Production for Reverse Osmosis and Kidney Dialysis Filters

Cellulose acetate membranes and cellulose triacetate membranes are used to filter small particles from water in reverse osmosis water treatment applications, from blood in kidney dialysis applications, and from water and other liquids in a variety of additional technological applications. Austrian chemist Karl Weissenberg recognized that cellulose acetate will flow when dissolved in 1,4-dioxane in rheological studies published in 1929 (Philippoff, 2006). 1,4-Dioxane was used to produce artificial silk in England in 1932 by treating cellulose acetate yarn with 1,4-dioxane (Barber, 1934, in USEPA, 2002).

The concept of reverse osmosis was first demonstrated in the late 1950s with cellulose acetate membranes. Reverse osmosis membranes are produced via solution deposition and hollow fiber extrusion. Solution deposition involves spreading a mixture of cellulose, a solvent, and a swelling agent over a smooth surface. As the solvent evaporates, the air-dried surface hardens and pores form within the cellulose layer. Reverse osmosis membranes are made with a very thin layer of polymer film applied on top of the porous cellulose acetate substrate. After the solvent evaporates from the cellulose membrane surface, the membrane is quenched by immersing in water, which the polymer precipitates (Klein and Smith, 1972).

The solvents used in this process are selected for their solubility in water. The solvent was commonly a mixture of 1,4-dioxane and acetone in ratios ranging from 1.5:1 to 2.5:1. Swelling agents include methanol and ethanol, used at between 2 wt% and 10 wt% of the final solution (Yashushi, 1977). Wastewater effluent from cellulose acetate and triacetate membrane production facilities has often included 1,4-dioxane. Several 1,4-dioxane groundwater contamination incidents have been caused by effluent from membrane production facilities, as described further in the case studies presented in Chapter 8.

Cellulose acetate and triacetate membranes produced by using 1,4-dioxane as a deposition solvent may contain residual 1,4-dioxane and thus require a California Proposition 65 warning label. Manufacturers acknowledge that trace amounts of 1,4-dioxane may be present and recommend flushing with water to purge 1,4-dioxane from the system. Because 1,4-dioxane is infinitely soluble, it is quickly removed from filters by flushing with water. Nearly all residual 1,4-dioxane is now flushed out in the manufacturing process, though this may not have always been the case. Most manufacturers no longer use 1,4-dioxane for producing reverse osmosis or kidney dialysis filters.

Cellulose acetate fibers are also used to produce some varieties of cigarette filters. 1,4-Dioxane is among the solvents listed in the patent literature for the process used to create the fiber bundles that remove tar from cigarette smoke (Matsumura et al., 1997).

2.3.2 Liquid Scintillation Cocktails

Numerous experiments and routine measurements in the medical and life sciences are made by injecting or orally administering a radio-labeled solution containing β-emitting radionuclides and then counting the radioactivity in blood, tissue, excreted urine, or other media. The process of counting levels of radiation often involves the use of liquid scintillation "cocktails." Samples selected for radiation measurement are dissolved in a solvent mixture, called a liquid scintillation cocktail, which includes fluorescent compounds or "fluors." Energy from emitted β-particles is transferred to solvent molecules, which then achieve an excited state and transfer excess energy to adjacent solvent molecules. The solvents selected for liquid scintillation cocktails remain in an excited state and return to the zero state without emitting light. Energy from excited solvent molecules is captured by the fluorescent solutes, and excited fluors dissipate the energy by emitting light photons, which permits radiation counting in the visible light range. Ideally, each β emission results in a pulse of light, though β-counting efficiencies are generally below 100% and vary by the strength of the β emitter. The liquid scintillation counter instrument consists of two photomultiplier tubes in a circuit. Only those light pulses reaching both tubes are counted. Each fluorescence event is proportional to a radioactive decay event, and the frequency of these events is directly proportional to the number of ^{14}C, ^{3}H, ^{32}P, ^{35}S, or other radionuclides present in the sample.

1,4-Dioxane was favored for liquid scintillation cocktails because it is fully miscible and holds a large amount of water in a homogeneous solution. Therefore, 1,4-dioxane is an ideal solvent for samples containing various amounts of water. Impurities in 1,4-dioxane, such as peroxides formed during storage, will quench the energy transmitted from β particles and must be avoided. The first widely used liquid scintillation formulation, today known as Bray's solution, was a mixture of naphthalene and 1,4-dioxane, with ethylene glycol as an antifreeze to allow long-term storage of counting vials in freezers (Bray, 1960). Bray's solution consisted of the following mixture dissolved in 1,4-dioxane; the 1,4-dioxane comprised more than 80% of the solution (Elmer, 2006):

Naphthalene	60 g/L
Methanol	100 mL/L
Ethylene glycol	20 mL/L
PPO = diphenyloxazole	4 g/L
POPOP = 1,4-bis-(5-phenyloxazolyl)benzene (note that 5-phenyloxazolyl = phenyloxazolylphenyl-oxazolyl-phenyl)	0.2 g/L

Material Safety Data Sheets (MSDSs, now referred to as Safety Data Sheets [SDSs]) indicate that scintillation-grade 1,4-dioxane used to make the cocktail has a composition of 95–99% 1,4-dioxane (National Diagnostics, Inc., 1987a, 1987b; Fisher Scientific, 2006). There are many other formulations for liquid scintillation cocktails; toluene and xylene are the common components. Larger laboratories have been known to recycle their scintillation wastes by distillation in a rotary evaporator and desiccation with excess potassium hydroxide. These steps were also followed to reduce waste volumes prior to disposal (Miyatake and Saito, 1984). Because the half-lives of the radioactive components of scintillation wastes are relatively short, wastes are often stored long enough for sufficient decay to occur until radioactivity is no longer measurable. Scintillation wastes have also been managed by incineration and by incorporating flammable liquid scintillation fluids into fuels; however, this practice releases the remaining radioactive components to the environment. The University of Illinois once burned liquid scintillation cocktails as a fuel supplement at its campus power plant with approval by the Nuclear Regulatory Commission, which conducted inspections to confirm that the levels of radioisotopes at the point of release were within regulatory limits. Because only small amounts were burned, the practice was exempt from Resource Conservation and Recovery Act (RCRA) Part B permit requirements. A firm in Gainesville, Florida, has processed liquid scintillation wastes by removing enough radioactive components to allow burning of the residual liquid in cement kilns (USEPA, 1996b).

A 1990 survey of more than 1,000 facilities producing liquid scintillation fluid wastes showed that the leading producers were industrial facilities, followed by academic facilities, government facilities, and medical facilities. The total quantity of all liquid scintillation fluids produced at the facilities surveyed was approximately 100,000 ft^3 (approximately 748,000 gallons) (USEPA, 1996b).

In the 1960s, landfills serving universities that performed biological research involving liquid scintillation counting were highly likely to have been contaminated with 1,4-dioxane. At Duke University in North Carolina, the campus landfill predates hazardous waste regulations. Laboratory wastes were buried in the landfill in the 1960s, and releases of 1,4-dioxane to groundwater at problematic levels (up to 2,800 μg/L) have been reported. At the University of North Carolina, a campus dump received scintillation counting wastes between 1963 and 1970, resulting in 1,4-dioxane detections in landfill monitoring wells as high as 22,000 μg/L. In 1964, the University of Kansas began operating a landfill for disposal of low-level radioactive wastes, including radionuclides contained in toluene and 1,4-dioxane; the landfill produced 1,4-dioxane contamination in off-site groundwater (Shamberg, 2002). Kansas State University also operated a landfill that received biological research waste, resulting in 1,4-dioxane detections as high as 14,000 μg/L. Additional examples of university landfills receiving scintillation counting wastes with 1,4-dioxane detections include the University of California at Davis, Stanford (DOE Linear Accelerator), University of Michigan, the University of Nebraska, University of Arkansas, University of Missouri, University of Florida, Dartmouth College, Cornell University, and the Gloucester Landfill located south of Ottawa in Ontario, Canada. Municipal landfills receiving hospital wastes and commercial medical and veterinary laboratory wastes may also be sources of 1,4-dioxane contamination. As discussed in Section 2.7, landfills of all types have been found to be sources of 1,4-dioxane, though most landfill groundwater-monitoring programs do not analyze groundwater, landfill leachate landfill gas, or landfill gas condensate for 1,4-dioxane. Further discussion of 1,4-dioxane in landfill leachate and landfill gas is included in Section 2.8.2.

2.3.3 Tissue Preservative in Histology and Other Laboratory Microscopy Methods

1,4-Dioxane has been used as a laboratory solvent for a variety of purposes, including study of tissue samples, as a carrier for biocides to eliminate biodegradation of analytes in water samples, for analysis of plant and wood samples, for purification/isolation of DNA elements, and for drying soil samples prior to analysis. Human and animal tissue samples are prepared as thin sections on microscope slides for examination by medical technologists and research scientists. A wide variety of methods are used, depending on the purpose for which the slide is prepared. Dehydration and staining are common steps in preparing tissue slides. 1,4-Dioxane has been used as a dehydrating agent for tissues in the preparation of histological slides since the 1930s (Mossman, 1937). It is also used by biologists to impregnate tissue sections with paraffin, and it has advantages in the preparation of histological slides for electron microscopy (Shearer and Hunsicker, 1980). 1,4-Dioxane is preferred over ethanol and xylene for dehydrating tissues because it removes water without causing substantial shrinkage or hardening in the tissue (Ralph, 1938).

1,4-Dioxane is also useful for dissolving and removing mercuric chloride, a biocide formerly used to preserve tissue samples. Tissue slides are also treated with stains or fixatives including dichromate (orange) and chromium trioxide (sometimes called "chromic acid" – red). When treated with a 70% 1,4-dioxane solution, both chrome-based fixatives reduce to chromic oxide, which is green-brown (Winsor, 2006). Because 1,4-dioxane is an expensive reagent (about $75/L for histological grade in 2006), some laboratories reclaim it by drying the 1,4-dioxane solution over a layer of calcium oxide or anhydrous cupric sulfate or by freezing the 1,4-dioxane with water in a sparkproof refrigerator (Winsor, 2006).

1,4-Dioxane is used for extraction and analysis of plant material, for example, lignin in wood samples. Analysis of wood samples involves hydrolyzing lignins and cellulose with a 1,4-dioxane solution. Some procedures use 100% 1,4-dioxane for lignin extraction (Agarwal and Ralph, 1997).

The study of soil samples by x-ray diffraction and other applications requires drying. Drying can be accomplished by introducing a hydrophilic solvent such as acetone or 1,4-dioxane. In some procedures, 1,4-dioxane is used to displace water; the 1,4-dioxane is subsequently displaced with a resin to allow study of friable soil structures. 1,4-Dioxane is preferred for soil drying because it preserves the soil structure better than acetone (Venediktova and Rudnyi, 1965; Moran et al., 1986, 1989; Chartres et al., 1989).

Preparation of water samples for the analysis of biodegradable analytes by ion-selective electrodes may involve treating with a biocide carried by 1,4-dioxane. One recommended protocol uses phenyl mercuric acetate as a biological inhibitor to treat water samples; 0.1 g is added to 20 mL of 1,4-dioxane, which is then diluted with deionized water to 100 mL of solution. A few drops of this solution are added to the sample prior to storage and analysis (NICO2000, Ltd., 2006). 1,4-Dioxane has been used as a reagent in various laboratory analytical methods, including detection of iodide in iodized salt (Saifer and Hughes, 1937). Its use for other iodide analysis methods dates back to 1926 (Anschutz and Broeker, 1926).

2.3.4 Printing Inks, Paints and Related Uses

2.3.4.1 Inks and Printing Operations

1,4-Dioxane has been used as a solvent in printing inks since the early 1950s (West, 1959). Although many general references cite 1,4-dioxane as a constituent of inks and paints, few citations were found to document the direct use of 1,4-dioxane in printing inks. The presence of 1,4-dioxane in printing inks may be due to the minor use of methyl chloroform as a solvent in solvent-based printing inks and in printing operations. Inks and coatings were listed as the destination for up to 5% of the 715 million pounds of methyl chloroform produced in 1989. The most common solvent component of ink formulations was toluene. Methyl chloroform was also used for contact cleaning of printing equipment. The USEPA's 1993 Toxic Releases Inventory (TRI) reported that 37 printing operations released a total of about 1.4 million pounds of methyl chloroform that year. Toxic Release Inventory data are inherently problematic; for example, the printing facilities reporting represented less than 0.15% of the 70,000 printing operations in the United States at the time (USEPA, 1995a). A 1997 update to USEPA's Sector Notebook Series (which gives surveys of chemical releases in different industries or sectors) provides 1995 figures for 1,4-dioxane releases from two printing facilities. The average release per facility was 8,500 lb per year (USEPA, 1997a).

Another possible source of 1,4-dioxane in inks and printing is the use of propylene glycol in both paint and ink formulations. Formulations for both paints and inks commonly contain up to 20% propylene glycol (Flick, 2005). 1,4-Dioxane is a known production by-product of propylene glycol (Johnson, 2001; Pundlik et al., 2001).

1,4-Dioxane is used in the office machines industry for the dissolution of pigment via an interface spray in photocopying equipment. 1,4-Dioxane is used to help inks adhere to or distribute in plastics. Some inks used to color thermoplastic fluoropolymers, such as electrical wiring insulation plastics, are dissolved in 1,4-dioxane (West, 1959). Inks containing resins to ensure durability – for example, to apply permanent ink markings to golf balls – use a variety of solvents that may include 1,4-dioxane (Kikuchi and Tanaka, 2003). 1,4-Dioxane has been used as an ingredient of deinking solvents for cleaning fingerprint rollers (Sirchie Finger Print Labs, 1989).

2.3.4.2 Historical Restoration

1,4-Dioxane is used to remove fungal stains from paper to preserve historical papers and art works. The solvent is also used in the conservation of oil-on-canvas and other paintings.

2.3.4.3 Painting, Coating, and Paint Stripping

Solvent-borne paints and coatings are made with solvents, binders, pigments, and additives. The combination of binder and solvent is referred to as the paint "vehicle"; pigment and additives are dispersed within the vehicle. Solvents comprise about 60% of the total formulation; binders, 30%; pigments, 7–8%; and additives, 2–3%. The primary paint solvents include toluene, xylene, methyl ethyl ketone, and methyl isobutyl ketone (Northeast Waste Management Officials Association, 1998). Chlorinated solvents such as trichloroethylene and perchloroethylene were used in phosphatizing applications to prepare car bodies and other metal surfaces for coating and painting, whereas methyl chloroform was often used as a solvent for paints formulated with alkyd and polyamide resins.

1,4-Dioxane can occur as a by-product of resin production and may be carried with resins into paint as an impurity. Resins commonly used in paints and coatings have included vegetable oils, alkyds, polyesters, polyamides, phenolics, polyurethanes, epoxies, silicones, acrylics, vinyls, cellulosics, and fluorocarbons. Coatings for metal surfaces are dominated by alkyd resins; however, water-based acrylics, epoxies, polyurethanes, and polyesters also are used for certain applications. Methyl chloroform was the favored solvent for alkyd resin coatings and for cleaning paint and coating spray guns and fluid hoses (USEPA, 1997b).

Methyl chloroform and dichloromethane could contribute 1,4-dioxane to paint compositions in minor amounts when 1,4-dioxane is present as a stabilizer. 1,4-Dioxane has been documented as a trace ingredient in paints that used methyl chloroform as a drying solvent. For example, polyurethane coatings contained less than 1% 1,4-dioxane (Glidden Paint, 1992). Resin formulations used large amounts of solvents to facilitate application and drying. A polyamide resin formula included 70% methyl chloroform with less than 5% 1,4-dioxane (DeSoto, Inc., 1991). Some lacquer formulations contained only 8% solids; the balance was composed of solvents and thickeners, including as much as 60% 1,4-dioxane (American Cyanamid, 1987). Pervo Paint Company (1993) stated that white traffic paints for striping lanes contained as much as 24% methyl chloroform and 0.5% 1,4-dioxane; yellow and red traffic paints had less than 0.1% 1,4-dioxane. However, according to the USEPA Toxic Release Inventory database, Pervo Paint Company removed 1,4-dioxane from its products in 1994.

Methyl chloroform is also used as a diluent for formulating and thinning paints used on military aircraft. The military specifications for aircraft coatings that used methyl chloroform as a diluent include MIL-P-23377F (Class 3 epoxy primer), TT-P-2760 (Class 3 polyurethane primer), and MIL-C-22750E (Type II epoxy topcoat) (Pulley, 1994).

Paint manufacturing plants produce wastewater that presumably contained 1,4-dioxane during the timeframe that methyl chloroform was used in paint formulations. Raw wastewater from four paint production plants in 1986–1987 ranged from non-detect to 1,444 μg/L methyl chloroform, while treated wastewater ranged from non-detect to 946 μg/L methyl chloroform (USEPA, 1989).

1,4-Dioxane has been found in the low parts per million (ppm) range in European felt tip pen inks and fabric dyes (Hansen, 2005). More recent measurements have documented maximum emission rates from children's marker pens procured in the United States at 0.64 μg/h among five pen types in which 1,4-dioxane was detected (Castorina et al., 2016). A European survey of art supply volatile organic compound (VOC) emissions noted the presence of polyethylene glycol (PEG) compounds, in which 1,4-dioxane is an impurity (Germinario et al., 2018). The U.S. Consumer Product Safety Commission recommends that parents purchase only art materials that conform with ASTM D-4236 and that do not bear a "CL" seal or other cautionary warnings.

Mixtures of cyclohexanone with 1,4-dioxane or tetrahydrofuran and 1,4-dioxane are preferred for painting the primer layer on magnetic tapes; the outer magnetic layer is applied by using xylene and toluene. 1,4-Dioxane comprises about 10% of the magnetic tape paint mix (Roller et al., 1977). Magnetic tape producers have historically discharged 1,4-dioxane in wastewater.

The dominant paint-removal solvent is dichloromethane, which in some instances includes 1,4-dioxane as a stabilizer. The most common paint-stripper formulation contains 60–65% dichloromethane. This solvent is particularly effective at penetrating the coating and causing it to swell and separate from the substrate (USEPA, 1996a). In 1989, more than 100 million pounds of dichloromethane was used for paint stripping, which was about a quarter of all dichloromethane produced in that year. About 40% of dichloromethane applied for paint stripping was used in the maintenance sector for stripping aircraft and other large vehicles and parts (SRRP, 1991). Because paint stripping often includes a pressure-washing step following treatment with the paint-stripping solvent, water may become contaminated with dichloromethane. USEPA estimates that the discharge of dichloromethane to water from paint-stripping operations in 1989 exceeded 8 million pounds (SRRP, 1991). Several citations in publications focusing on toxicology, risk, or summary fact sheets for 1,4-dioxane have listed 1,4-dioxane as a solvent used directly for paint stripping; however, none of the sources checked corroborated this assertion. Methyl chloroform has been included as an ingredient of some paint-stripping formulations, e.g., 8–12% in "Oops-Aerosol and Liquid" manufactured by New York Bronze Powder Company (Wollbrinck, 1993). However, the great majority of paint-stripping formulations were prepared using dichloromethane

2.3.5 Flame Retardant Production

1,4-Dioxane is used as a solvent in the production of brominated fire retardants. These are produced by reacting cyclododecatriene with bromide in the presence of a solvent, generally an alcohol such as isobutanol or 1,4-dioxane. 1,4-Dioxane is favored over alcohol because alcohol causes reaction intermediates to precipitate out of solution before they can be completely reacted and because alcohols directly consume the brominating agent. Using 1,4-dioxane eliminates the problem of intermediates remaining at the end of the reaction, and 1,4-dioxane does not react with the brominating agent (Business Communications Company, 2002). Waste 1,4-dioxane is recovered for off-site use as a fuel (USEPA, 2006).

2.3.6 Rubber and Plastics Industry

The USEPA's surveys of chemical releases in different industries, known as Sector Notebooks, group the rubber and plastics industries together. The Rubber and Plastics Industry Sector Notebook lists 1,4-dioxane as a waste emitted from four facilities out of more than 400 surveyed; however, the Toxics Release Inventory data cited are probably inconsistent in reporting 1,4-dioxane if it was not directly used. The Toxics Release Inventory data cited for 1993 include about 200 facilities releasing about 11 million pounds of methyl chloroform to the atmosphere and transferring about 1 million pounds to waste-handling facilities (USEPA, 1995b). The quantities of 1,4-dioxane reported in the 1993 Toxics Release Inventory, about 11,000 lb, probably understate the releases associated with its likely presence as a stabilizer of methyl chloroform. The 1995 Toxics Release Inventory also reports 1,4-dioxane for only four facilities, but lists more than 100,000 lb of 1,4-dioxane transferred and about 5,000 lb released.

2.3.7 Aircraft Deicing Fluid and Antifreeze

Before 2000, 1,4-dioxane was an ingredient or impurity in aircraft deicing fluids and antifreeze solutions. The Air Transport Association reportedly listed the following components of deicing fluid in 1994: ethylene glycol or propylene glycol, water, surfactants (wetting agents), corrosion inhibitors (including flame retardants), pH buffers, dyes, 1,4-dioxane, and complex polymers as thickening agents (USEPA, 2000). Glycol deicing formulations were previously used to treat airport runways. Because 1,4-dioxane was an impurity as residual solvent carried over from production of glycol, past uses of glycol deicing agents could leave trace levels of 1,4-dioxane in soil and groundwater near runways. For removing hoarfrost and ice from an aircraft, a mixture of glycols, urea, and hot water is used, which is sprayed at high pressure onto the fuselage. Three of the four types of deicing formulations include 90%, 65%, and 65% glycol, and may contain additives (2%). Quantities of spray range from as low as 40 liters/plane (10 gallons) to as high as 15,000 liters/plane (3,960 gallons) for large aircraft during bad weather. Runoff concentrations have been measured as high as 160,000 mg/L propylene glycol at an airport in Connecticut, and 20,000 mg/L of ethylene glycol and diethylene glycol (DEG) at Dallas Fort Worth Airport and Kansas City International Airport (Sulej et al., 2012).

MSDS for aircraft deicing fluids list a trace presence of 1,4-dioxane (0.0022% or 22 mg 1,4-dioxane per liter of deicing fluid) as an impurity of glycol compounds (Union Carbide, 1989). 1,4-Dioxane was also used as a wetting and dispersing agent but composed less than 0.5 mg/L of deicing fluids (USEPA, 2000). Others report that past levels of 1,4-dioxane in deicing fluids were on the order of 3 mg/L (Gelman Sciences, 1989a). 1,4-Dioxane can be detected in glycols in the low ppm range (Pundlik et al., 2001).

To reduce groundwater contamination, deicing fluid manufacturers have removed 1,4-dioxane from their formulations (USEPA, 2000). The main deicing agents used today are urea and calcium–magnesium acetate, more commonly known as "road salt."

Many brands of automotive antifreeze list 1,4-dioxane as an ingredient or a component at less than 0.0086% (86 mg 1,4-dioxane per liter of antifreeze fluid) (Old World Automotive Products, Inc., 2001). 1,4-Dioxane is also listed as a trace component of recycled antifreeze at less than 0.004% (40 mg/L) (FPPF Chemical Company, Inc., 2006). Analysis of consumer antifreeze products found 1,4-dioxane present at concentrations ranging from 0.1 to 3.4 mg/L (Gelman Sciences, 1989c). Analysis of radiator fluid boil-over verified the presence of 1,4-dioxane at up to 22 mg/L (ATSDR, 2005). The higher concentration of 1,4-dioxane in radiator boil-over compared to the concentration in antifreeze products suggests that heating ethylene glycol in a radiator, coupled with possible acidification from metal oxides, could lead to formation of 1,4-dioxane, as described in Figure 2.3.

In an effort to draw attention to the potential widespread contamination by 1,4-dioxane from antifreeze and deicing fluids, Charles Gelman, the CEO of Gelman Sciences, testified at a State of Michigan hearing that sample results from puddles of car radiator boil-over contained high levels of 1,4-dioxane (Judge, 1988). Gelman's letters cautioned that 1,4-dioxane could pollute groundwater at rest stops and wherever winterization practices for plumbing systems in summer cottages included filling pipes with ethylene glycol to prevent freezing during the winter. Flushing out the lines the following summer would cause 1,4-dioxane to flow through septic systems. The same issue was raised for winterizing recreational vehicles and large boats (Gelman Sciences, 1989b).

2.3.8 Adhesives

A number of glues and adhesives were prepared with a chlorinated solvent base to facilitate spreading and drying. 1,4-Dioxane is present in some glues as a solvent and in others as a stabilizer of methyl chloroform. The National Library of Medicine Household Products Database lists a brand of wood parquet floor paste as containing 0.5% 1,4-dioxane; an adhesive for affixing trim and detailing to automotive surfaces is listed as containing 1–3% 1,4-dioxane (DeLima Associates, 2004; HSDB, 2006). Adhesives containing methyl chloroform can also be expected to contain trace amounts of 1,4-dioxane. One formulation of wood glue contained from 1% to 5% 1,4-dioxane (DAP, Inc., 1994). Some formulations of contact cements had ratios of 1,4-dioxane to methyl chloroform at considerably higher proportions than those found in degreasing compositions of stabilized methyl chloroform (Wilke et al., 2004). DAP's Smooth Spread Contact Cement had 2.5% 1,4-dioxane and only 15–20% methyl chloroform (DAP, Inc., 1990).

Adhesives used in architectural coatings are thought to be the primary cause of 1,4-dioxane detections in indoor air samples. In offices equipped with cubicles, adhesives are used to join the many components of cubicle furniture; coatings and varnishes may also contain compounds bearing 1,4-dioxane. Testing of indoor air in offices using cubicles detected 1,4-dioxane as well as naphthalene, benzene, trichloroethylene, and perchloroethylene. Newer office furniture is designed to avoid emissions of these compounds (Betts, 2005). Draperies and drapery linings also emit measurable concentrations of 1,4-dioxane to indoor air (Smith and Bristow, 1994).

The use of epoxy adhesives for joining plastic parts required special cleaning and roughing preparation using a process referred to as "satinizing," which used abrasion followed by solvent cleaning. Abrasion was achieved by sanding with 280-grit emory cloth. The cleaning step, recommended by the epoxy adhesive manufacturer, involved immersing the sanded plastic part in a mixture of 96.85% perchloroethylene and 3% 1,4-dioxane at 80–120°C for 30 seconds, followed by air drying prior to a hot water rinse, air drying, then applying the epoxy adhesive (Petrie, 2006).

Film cements for joining or mounting sections of cellulose triacetate movie film use acetone, methylene chloride, and 1,4-dioxane as solvents. 1,4-Dioxane was a significant component in commercial film cement, exceeding 10% in some formulations (Thompson-Hayward Chemical Company, 1993). Another variety of film cement is composed of 50% 1,4-dioxane (Eastman Kodak Company, 1998). 1,4-Dioxane is used to dissolve resins and polymers into solution for film coating. It is also used as a photochemical reaction/synthesis medium as well as for miscellaneous photographic research and development projects (USEPA, 1996c).

2.3.9 Polyurethane Materials for Medical Devices

Polyurethanes are durable elastomers used for a wide variety of applications, including flexible and rigid foams, adhesives, seals, gaskets, carpet padding, and hard plastic parts. Polyurethanes are biocompatible, which makes them suitable for medical devices. Medical devices made with polyurethanes include pacemaker leads, catheters, feeding tubes, angioplasty balloons, condoms, surgical gloves, wound dressings, and many other products. Medical devices are made by injection molding, extrusion molding, and solution processing. Mold-release solvents include 1,4-dioxane, tetrahydrofuran, dichloromethane, methyl ethyl ketone, *N,N*-dimethylformamide, *N*-methylpyrrolidone, cyclohexanone, and chloroform. These solvents contribute to highly transparent polyurethanes, a desired feature for medical devices. 1,4-Dioxane and other polyurethane solvents dissolve polymers and form clear solutions. Criteria for solvent selection include ease of solvent removal from the product; all traces of solvents must be eliminated because even trace amounts may interfere with the treatment and the patient's health (Wypych, 2001).

Spray polyurethane foams, commonly used for insulating buildings, can be a source of 1,4-dioxane emissions during application. Spray polyurethane foams are produced at the construction site by mixing two sets of chemicals, including methylene diphenyl diisocyanate, flame retardants, catalysts, blowing agents, and polyols. The mixture emits 1,4-dioxane as a reaction by-product. In closed-cell measurement chambers, initial 1,4-dioxane emission rates from spray polyurethane foams were between 175 and 315 $\mu g/m^2/h$ immediately after application. Emission rates decreased to 50 $\mu g/m^2/h$ after 144 hours (6 days) (Poppendieck and Gong, 2016). A later study found higher and longer duration 1,4-dioxane emission rates: 900 $\mu g/m^2/h$ after 10 hours, and 150 $\mu g/m^2/h$ after 180 hours (Poppendieck et al., 2017).

2.3.10 Solvent-Based Cleaning Agents

Where methyl chloroform was a constituent of various household and automotive consumer and commercial contact cleaning agents, 1,4-dioxane was often present in the low percentage range as a stabilizer. Some cleaning agents are better known for their main ingredient, which may mask the presence of methyl chloroform and 1,4-dioxane. For example, brake-cleaning formulations used in automotive shops use tetrachloroethylene as the primary solvent (60–65%), but also contain methyl chloroform (30–35%) and 1,4-dioxane (2.5%) (Valvoline Oil Company, 1991). Some aerosol-spray brake-cleaner compositions were primarily methyl chloroform (89%), with higher percentages of 1,4-dioxane than normally required for vapor degreasing (up to 6.8%) (CRC Chemicals, 1987).

Other solvent-based consumer products contained only small amounts of methyl chloroform but larger amounts of 1,4-dioxane. Some consumer rust-removing agents contained up to 10% 1,4-dioxane, but only 4% methyl chloroform, with butyl alcohol and butyl cellosolve as the primary ingredients (Watsco Components, Inc., 1991). Rust inhibitors have been formulated with as much as 50% methyl chloroform and 0.8% 1,4-dioxane (Crown Industrial Products Company, 1989). Commonly used solvents for loosening frozen hardware such as nuts and bolts contained 1–3% 1,4-dioxane (Loctite Corporation, 1995).

2.3.11 Ether Supplements in Fuels

Cyclic ethers including 1,4-dioxane and 1,3-dioxolane have been considered for addition to fuels to enhance the octane value and to reduce the formation of carbon monoxide in automobile exhaust (Maurer et al., 1999). Ether compounds are sometimes added to racing fuel by karters (hot rodders) to increase fuel octane. Karters report that it takes 10–15% 1,4-dioxane added to racing fuel to produce increased octane and yield (Pro-Systems, 2002). Racing rules now prohibit using 1,4-dioxane, dinitrotoluene, propylene oxide, or nitropropane in racing fuels. 1,4-Dioxane is also banned as a racing motor oil additive in "XTREME Stock Car" and "Women on Wheels" racing (Susquehanna Speedway Park, 2006). However, ethers elude detection by the digital fuel meters commonly used at race tracks to verify fuel composition compliance. 1,4-Dioxane, propylene oxide, nitromethane, and nitropropane have all been used as "illegal" racing fuel additives. The quantities added vary according to the whims of the racer.

Amateur racers in the karting community have cautioned against the use of 1,4-dioxane because it is listed as a carcinogen and is easily absorbed through the skin (Copeland, 2004). Racers developed test methods to check racing fuels for the presence of banned substances including 1,4-dioxane, which is not detected by the widely used Godman DT-15 Digatron meter (a dielectric testing device used to check racing fuels for polar substances). The method involves adding an acid reagent and checking for the presence of a white precipitate (the "acid drop test").

1,4-Dioxane is a flammable liquid and tends to form explosive peroxides. Its development in glycols (low ppm levels), which are used as dehumidifying agents in refineries, may take place by condensation. 1,4-Dioxane thus formed gets distilled over with benzene in the refinery process. Therefore, it is necessary to identify and determine the levels of 1,4-dioxane in glycols as well as benzene. Gas chromatography (GC) is probably the best technique for this purpose. GC analysis may be carried out with a flame ionization detector. Results show that 1,4-dioxane can be reliably determined down to 2 ppm in glycols and benzene and 1 ppm in toluene (Cortellucci and Dietz, 1999).

2.4 1,4-DIOXANE AS A BY-PRODUCT OF MANUFACTURING

1,4-Dioxane occurs as a reaction by-product in several chemical processes used to produce polyester, soaps, and plastics. Consequently, waste streams from facilities that produced these products have included 1,4-dioxane, and some have caused contamination of soil and groundwater in past decades.

2.4.1 Ethoxylated Surfactants

1,4-Dioxane is produced as a by-product during the sulfonation reaction with alcohol ethoxylates, a process used to produce surfactants included in a wide variety of soaps and detergents. Ethoxylated alcohols are used as surfactants, detergents, foaming agents, emulsifiers, and wetting agents. During alcohol ethoxylation, ethylene oxide is combined and rearranged to form the polymer of ethylene oxide. This process allows ethylene oxide to dimerize to form diethylene dioxide, that is, 1,4-dioxane. If no effort is made to control formation of 1,4-dioxane, levels may approach 500 ppm or higher. By controlling mixing ratios, temperature, and other reaction parameters, formation of 1,4-dioxane has been limited to 30–200 ppm (Stepan Company, 2006). Since 1990, MSDS have reported 1,4-dioxane at an "upper bound concentration" or "typical maximum" of less than 15 ppm (Talmage, 1994). Analyses of surfactants and products containing surfactants have shown rather high levels of 1,4-dioxane associated with surfactants and associated products in the 1980s and 1990s. More recent analyses of consumer products containing surfactants show lower levels of 1,4-dioxane residual. Not all surfactant products contain detectable concentrations of 1,4-dioxane. This section provides a historical view of surfactants in consumer products, beginning with the earliest reported data and ending with a 2018 sampling survey of products in New York State.

Today, most manufacturers remove the 1,4-dioxane impurity in ethoxylated surfactants using a stripping process employed in many production lines (ATSDR, 2004). Between 1992 and 1997, the average concentration of 1,4-dioxane in finished cosmetic products was reported to fluctuate from 14 to 79 ppm (mg/kg) (ATSDR, 2004). As described in the Section 2.4.1.1, producers of nonionic surfactants also used 1,4-dioxane for routine quality-control testing, which could produce a laboratory waste stream with a substantial volume of 1,4-dioxane over time.

2.4.1.1 Personal Care Products: Detergents, Shampoos, and Sundries

1,4-Dioxane was historically present in the ppm range in alcohol ethoxy sulfate (AES) compounds. Sodium laureth sulfate [the compound sodium 2-(2-dodecyloxyethoxy) ethyl sulfate, CASRN 3088-31-1; also called sodium lauryl ether sulfate] in particular had elevated levels of 1,4-dioxane, as high as 500 ppm. Sodium laureth sulfate is a component of anionic surfactants in detergents and shampoos. Powdered detergents typically used as much as 12% AES, whereas liquid detergents used as much as 18% AES, and dishwashing detergents may contain as much as 27% AES (Greek and Layman, 1989; Stepan Company, 2006). In addition to sodium laureth sulfate, the

TABLE 2.8
Summary of 1,4-Dioxane Analysis of Ethoxylated Raw Materials as of September 30, 1981

Range of Dioxane Detections	Number of Samples in This Range (Out of 100)
Levels above 100 ppm	22
Levels between 10 and 100 ppm	36
Levels between 0.5 and 10 ppm	23
Not detected above 0.5 ppm	19

Source: Food and Drug Administration (FDA), 1981, Progress report on the analysis of cosmetics raw materials and finished cosmetics products for 1,4-dioxane. Division of Cosmetics Technology, Food and Drug Administration 15.

common shampoo, detergent, and dishwashing soap ingredients containing ammonium laureth sulfate and triethanolamine laureth sulfate may also contain 1,4-dioxane as a trace contaminant. In the 1980s, these ingredients contained elevated levels of 1,4-dioxane. Ammonium laureth sulfate contained 1,4-dioxane from 288 to 1,282 ppm; sodium laureth sulfate had from 69 to 340 ppm. In 1988, 400 million pounds of ethoxylated alcohol surfactants were used in household detergents in the United States (Talmage, 1994). Some earlier formulations of laundry presoak spray had particularly high levels of 1,4-dioxane. In one of six samples of laundry presoak spray analyzed, 1,4-dioxane was detected at a concentration of 15.0 wt%, i.e., 150,000 mg 1,4-dioxane per kilogram of laundry presoak (USEPA, 1992).

Since the 1950s, quality-control testing for surfactant production included determining the hydrophilic versus lipophilic balance or relative solubility number of nonionic surfactants. The relative solubility number is the volume in milliliters of distilled water necessary to produce persistent turbidity in a benzene and 1,4-dioxane solvent system consisting of 1 g of surfactant sample and 30 mL of solvent. To avoid using toxic solvents, the test now uses toluene and ethylene glycol dimethyl ether (Wu et al., 2004).

In 1979, the U.S. Food and Drug Administration (FDA) announced that hundreds of cosmetic products contained 1,4-dioxane. The announcement cautioned that the potential associated risk was unknown (Washington Post, 1979). In the early 1980s, the FDA tested 100 samples of ethoxylated surfactant ingredients used in personal care products and found that 81% of ingredients tested contained detectable levels of 1,4-dioxane. Results of the FDA study are summarized in Table 2.8 (FDA, 1981). In finished products in the sundries category, the same study found that 2 of 11 samples contained greater than 100 ppm 1,4-dioxane; 5 of 11 contained 1,4-dioxane between 10 and 100 ppm; one sample had less than 10 ppm; and three samples did not detect 1,4-dioxane at a reporting limit of 0.5 ppm (FDA, 1980, 1981). Table 2.9 summarizes levels of 1,4-dioxane in shampoos, bath, and sundry products.

Following the realization that 1,4-dioxane is present in cosmetics, shampoos, and detergents as an impurity of surfactants and emulsifiers, surfactant producers sought to reduce the amount of 1,4-dioxane formed. The degree of reduction in the levels of 1,4-dioxane present varied considerably in the 1980s and 1990s. A 1979 review of Colgate Palmolive products found levels of 1,4-dioxane up to 0.423 wt% (4,230 mg 1,4-dioxane per kilogram product); the permissible upper limit for by-products recommended by the National Institute of Occupational Safety and Health at that time was 1%, or 10,000 ppm (Belanger, 1980). In 1986, an analytical survey of German antidandruff shampoos for 1,4-dioxane found concentrations ranging from 10 to 390 mg/kg; a 1990 follow-on survey found the same German antidandruff shampoos free of 1,4-dioxane, a result that reflected improvements to surfactant production methods (Anonymous, 1987). However, a 1987 study of German personal care products including shampoos and liquid soaps reported 1,4-dioxane present in 22 products at concentrations ranging from less than 50 to 300 mg/kg. The same study found that 1,4-dioxane concentrations in German cosmetics were generally below 100 mg/kg and more commonly below 10 mg/kg.

TABLE 2.9
Partial Results for 1981 1,4-Dioxane Analyses in Single Samples of Personal Care Products

Cosmetic or Sundry Product	Manufacturer	Brand Name	1,4-Dioxane Concentration (ppm)
Shampoo	Estée Lauder	Azuree Natural Shampoo	66
Shampoo	Aramis	Aramis Malt Enriched Shampoo Concentrate	279
Bath gel	Ritz Group	Jean Naté Bath Gel	36
Bath gel	Peoples Drug Store	Peoples Peach Concentrated Foam Bath	64
Foundation	Cosmetic Sciences, Inc.	Liquid Jojoba Foundation	ND (<0.5)
Foundation or cream	Hasbro Industries, Inc.	Fresh'n Fancy Red/White Mixing Cream	ND (<0.5)
Skin cleanser	Winthrop Labs	Phisoderm	140
Lotion	Westwood Pharmaceuticals	Keri Lotion for Hands, Face, Body	ND (<0.5)
Lotion	Miller-Morton	Skin Quencher Hand and Body Lotion	30

Sources: FDA (1980, 1981; see also USEPA, 1989b, and Gelman Sciences, 1989c; the information presented here is from the summary tables in the FDA interim update memo included on EPA's OPPT TSCA (Office of Pollution Prevention and Toxics, and Toxic Substances Control Act) Web site (USEPA, 1989b).

Notes: ND, not detected. These data, produced 25 years before this publication, have little bearing on the safety of similar products produced today. All of the companies listed above and other companies producing similar products are regulated by the FDA and must comply with an extensive set of health-based protective standards. Technology to avoid formation of 1,4-dioxane and to strip any 1,4-dioxane and ethylene oxide formed during production before packaging has been employed for more than two decades. The information presented here is of interest primarily for the study of potential past modes of 1,4-dioxane introduction to the environment through discharge from septic tanks, leaking sewer lines, and treated wastewater. No laboratory test records or validated quality control records for the analyses presented were inspected or verified.

Shower gels, bubble baths, and hair care products containing alkyl ethers commonly contained up to 500 ppm 1,4-dioxane (Rümenapp and Hild, 1987). In 1989, West Germany produced 227,000 metric tons of surfactants in detergents and cleaners, of which 3 metric tons were estimated to be 1,4-dioxane, or a bulk concentration for all surfactant products of about 13 ppm.

A 1990 Danish study determined that 82% of the personal care products analyzed contained from 0.3 to 96 ppm 1,4-dioxane. The same study showed that 85% of dishwashing detergents contained from 1.8 to 65 ppm 1,4-dioxane. Of 76 products tested, four were in the range of 50–100 ppm; 27, from 20 to 50 ppm; 11, from 10 to 20 ppm; 22, from 0.3 to 10 ppm; and 12 did not detect 1,4-dioxane at a reporting limit of 0.3 ppm (Rastogi, 1990). A 1991 review found that 37 of 70 Japanese shampoos analyzed contained 1,4-dioxane greater than the study's 2 ppm detection threshold. The average concentration was 9 ppm, the maximum being 67 ppm (Fox, 1993). A review of shampoos in the U.S. market found 1,4-dioxane present in all 12 shampoos tested. The range was from 6 to 144 ppm (Italia and Nunes, 1991). A Polish study in 1998 found that low-quality ethoxylates were still being used to produce sundries, with 2–7 ppm 1,4-dioxane detected in shampoos, including baby shampoos, 2–8 ppm in body lotions, and 2–10 ppm in dishwashing soaps (Wala-Jerzykiewicz and Szymanowski, 1998).

An analytical survey of ethoxylated raw materials in the U.S. market found 1,4-dioxane at levels as high as 1,410 ppm and at levels as high as 279 ppm in personal care products. Children's shampoos contained levels of 1,4-dioxane in excess of 85 ppm (Black et al., 2001). In Taiwan, 1,4-dioxane residue was still present in 2005 in some nonionic surfactants, as well as in shampoos, liquid soaps, and dishwashing detergents (Fuh et al., 2005). Table 2.10 summarizes findings from analysis of surfactants and soaps in Taiwan.

Liquid laundry soaps often include nonylphenol ethoxylate (CASRN 127087-87-0), which may include trace residuals of 1,4-dioxane at levels sufficiently high to require reporting under California

TABLE 2.10
Detections of 1,4-Dioxane in Taiwanese Surfactants and Soaps

Sample Type	1,4-Dioxane (ppm)
Surfactant: Polyethylene oxide	72
Surfactant: Poly(ethylene/propylene) oxide	12
Surfactant: Polyhydric alcohol	65
Shampoo[a]	12–41 (in 3 of 9 samples; 6 ND)
Liquid soap[b]	8 (in 1 of 9 samples; 8 ND)
Dishwashing detergent[c]	6 (in 1 of 9 samples; 8 ND)

Source: Fuh, C.B., Lai, M., Tsai, H.Y., and Chang, C.M., 2005, Impurity analysis of 1,4-dioxane in nonionic surfactants and cosmetics using headspace solid-phase microextraction. *Journal of Chromatography A* 1071(1): 141–145.

Note: ND, not detected.

Proposition 65 and 40 CFR Section 302.4 rules. 1,4-Dioxane remaining in detergents and personal care products is released through use to publicly owned treatment works (POTWs) to produce a diffuse presence of 1,4-dioxane in wastewater and POTW effluent.

In 2007, the Campaign for Safe Cosmetics released results from a survey of 1,4-dioxane in children's shampoos and bubble baths. More than two dozen baby and children's consumer products were analyzed by a commercial laboratory by isotope-dilution, headspace/GCMS and a reporting limit of 0.2 ppm. Four of the products tested exceeded 10 ppm, the FDA recommendation for the maximum 1,4-dioxane content in cosmetic products. The median-detected 1,4-dioxane value (excluding eight samples with no detection) was 5 ppm; the average was 7 ppm, and the standard deviation was 7 ppm. Table 2.11 provides a list of 2007 product testing results. More recently, the Citizens Campaign for the Environment, conducted a survey of commercially available surfactant-based products in Long Island, New York (Schwartz, 2019). The results (Table 2.12) show

TABLE 2.11
2006 Analysis of Children's and Adult Shampoo and Bath Products for 1,4-Dioxane

Product	ppm	Product	ppm
Clairol Herbal Essence Rainforest Flowers Shampoo	23	Rite-Aid Tearless Baby Shampoo	4.3
Olay Complete Body Wash with Vitamins (normal skin)	23	Huggies Natural Care Baby Wash Extra Gentle and Tear Free	4.2
Hello Kitty Bubble Bath (Kid Care)	12	Huggies Baby Wash Shea Butter	4.0
Johnson's Kids Shampoo Watermelon Explosion (Johnson & Johnson)	10	Lil' Bratz Mild Bubble Bath (Kid Care)	3.7
Disney Clean as Can Bee Hair & Body Wash (Water Jel Technologies)	8.8	Scooby-Doo Mild Bubble Bath (Kid Care)	3.0
Gerber Grins & Giggles Gentle & Mild Aloe Vera Baby Shampoo	8.4	Disney Pixar Cars Piston Cup Bubble Bath (MZB Personal Care)	2.2
Johnson's Kids Tigger Bath Bubbles (Johnson & Johnson)	5.6 to 7.9	Suave Naturals Passion Flower	2.0
Sesame Street Wet Wild Watermelon Bubble Bath (The Village Company)	7.4	L'Oreal Kids Orange Mango Smoothie Shampoo	2.0
Johnson's Head-to-Toe Baby Wash (Johnson & Johnson)	5.3 to 6.1	Mr. Bubble Bubble Bath Gentle Formula with Aloe	1.5

Source: Published in David Steinman's *Safe Trip to Eden*, 2007; used with permission.

TABLE 2.12
2019 Testing of Consumer Sundries for 1,4-Dioxane (mg/kg)

Baby Products			
Dreft (Stage 1/Newborn)	10	Baby Magic Hair and Body Wash	7.6
Sesame Street Baby Wash	2.1	Angel of Mine Baby Bath	1.4
Mustela Gentle Cleansing Gel	0.9	Johnson's Baby Shampoo	0.87
Disney Frozen Bubble Bath	0.81	Original Bubble Mr. Bubble	0.59
Not detected (<0.025 ppm): Aveeno Baby Gentle Wash and Shampoo; Babyganics Bubble Bath; The Honest Company Shampoo and Body Wash			
Body Washes and Gels			
Victoria's Secret (Bombshell Fragrance Wash)	17	Victoria's Secret (Love Fragrance Wash)	16
Olay Moisture Ribbons Plus Body Wash (Shea + Lavender Oil)	3.5	Dr. Teal's Pink Himalayan Foaming Bath	3.4
Inspire Fresh Water Moisturizing Body Wash (Dollar Tree)	2.6	Neutrogena Rainbath Shower and Bath Gel	2.6
Soft Whisper by Power Stick (Pomegranate Lemon Verbena) (Dollar Tree)	2.2	Suave Essentials Body Wash (Wild Cherry Blossom)	2.0
St. Ives Oatmeal and Shea Butter	2.0	Caress Daily Silk Body Wash	1.7
Philosophy Love Body Wash	1.5	Modesa Coconut Ginger Body Wash (Family Dollar)	1.4
Smartly Meadow Scented Body Wash (Target Brand)	1.2	Nivea Care and Hibiscus Body Wash	1.0
Bath and Body Works Shower Gel (Sonoma Weekend Escape)			0.63
Not Detected (<0.025 ppm) Burt's Bees Lavender Honey Body Wash; Dove Go Fresh Body Wash (Cucumber and Green Tea); Dr. Bronner's Hemp Peppermint Soap; Method Body Wash (with Avocado Extract)			
Dish Soaps			
The Home Store Lemon Scented (Dollar Tree)	7.7	Up & Up Free and Clear (Target Brand)	6.4
Dawn	3.6	Palmolive Raspberry Coconut	2.8
Ajax	2.5		
Hand Soaps			
Soft Soap Fresh Breeze	1.9	Dial Coconut Water and Mango	0.99
RJ Walkins (Aloe and Green Tea)	0.34		
Laundry Detergents			
Tide Original	14	2X Ultra Ivory Snow (Gentle Care)	11
Gain Original	10	Tide Simply plus Oxi (Refreshing Breeze)	8.3
Persil Original (Family Dollar)	6.1	Homeline Original (Family Dollar)	5.4
All with Stain Lifters (Fresh Rain)	5.2	Up & Up Fresh Linen (Target Brand)	4.3
Arm and Hammer (Clean Burst, 2× Concentrated):	2.8	LA's Totally Awesome (Dollar Tree)	2.8
Purex plus Oxy Stain Removers (Fresh Morning Burst)	2.0	Xtra (Tropical Passion)	0.89
Method (Ginger Mango):	0.22		
Not Detected (<0.025 mg/kg): Mrs. Meyer's (Lavender Scent); Nature's Promise Free and Clear; Seventh Generation (Free and Clear); Suavitel (Field Flowers); Ultra Downy April Fresh (Fabric Softener)			
Men's Products			
Selsun Blue Dandruff Shampoo	3.1	Irish Springs Shower Gel	3.0
Old Spice Swagger Shower Gel/Shampoo	2.5	Axe Phoenix Shower Gel/Shampoo	2.4
Suave Sport Body Wash	2.0	Inspire Premium Surge Body Wash (Dollar Tree)	1.4
Power Stick Cool Blast Shower Gel/Shampoo (Dollar Tree)	0.8		
Shampoos			
Pantene Pro-V Nature Fusion	5.5	Herbal Essences Platinum	4.2
VO5 Passion Fruit Smoothie	3.4	Love Beauty and Planet Argan Oil and Lavender	2.9

(Continued)

TABLE 2.12 ***(Continued)***
2019 Testing of Consumer Sundries for 1,4-Dioxane (mg/kg)

John Frieda Brilliant Brunette	2.4	OGX Lavender Platinum	2.4
Dove Nutritive Solutions (Coconut and Hydration)	2.2	Head and Shoulders Classic Clean	1.7
Tresemme Moisture Rich with Vitamin E	1.5	L'Oreal Elvive Total Repair	1.4
Suave Professionals Moroccan Infusion (Color Care)	1.4	White Rain Ocean Mist (Dollar Tree)	1.1
Silkience Vitamin B5 and Avocado Oil (Dollar Tree)	0.98	Garnier Fructis with Active Fruit Protein	0.4
Not detected (<0.025 mg/kg): Aveeno Active Minerals Pure Renewal; Cantu Shea Butter for Natural Hair; Shea Moisture Coconut and Hibiscus Curl & Shine			

Source: Schwartz, D., 2019. Environmental group says 65 of 80 household products contain 1,4-dioxane. David M. Schwartz. *Newsday*, 3/26/2019. https://www.newsday.com/long-island/environment/1-4-dioxane-household-products-1.28968339.

Notes: Data listed directly from NewsDay article; underlying certificates of analysis (ALS Laboratory in Rochester, New York) were not obtained or inspected to check laboratory quality assurance, blanks analysis, and other quality control data. The data represent only a single analysis of each product, and do not account for possible analytical or product variability. This snapshot nevertheless shows improvement in reducing 1,4-dioxane content.

an across the board reduction of 1,4-dioxane content in commercially available, surfactant-based products; however, 1,4-dioxane concentrations remain sufficiently high to potentially cause releases of 1,4-dioxane to groundwater through septic tank effluent, leaking sewer lines, and wastewater treatment plant effluent.

2.4.1.2 Commercial Surfactant Products

While consumer groups have conducted market basket sampling of off-the-shelf products made from surfactants, similar sampling surveys have not been systematically completed to learn the 1,4-dioxane content of commercial, industrial, and agricultural products. The few data points available suggest that quantities of significant 1,4-dioxane may be present in nonconsumer products that use surfactants. For example, the New Hampshire Department of Environmental Services (NHDES) investigated the unexpected detection of 1,4-dioxane in fuel leak site monitoring wells. The gas station in question also operated a car wash. NHDES staff met with representatives of a car wash soap supplier, and learned that product analysis showed 12 to 20 mg/kg in soaps that were typically diluted 500-fold in the car wash. The manufacturer also analyzed the surfactant oil used to produce the car wash soap, and found it contained 760 mg/kg (Roy, 2013). Depending on how the car wash rinsate is contained and recycled, car washes may be a source of low to moderate levels of 1,4-dioxane in groundwater. Landfills receiving sediment from car wash filters may also be a source of 1,4-dioxane releases to groundwater.

2.4.1.3 Personal Care Products: Cosmetics

The *Cosmetic Handbook* (FDA, 1992) advises that cosmetics containing ethoxylated surfactant agents that may be contaminated with 1,4-dioxane – such as detergents, foaming agents, and emulsifiers – are identifiable by the words, acronyms, or suffixes "PEG," "polyethylene," "polyethylene glycol (PEG)," "polyoxyethylene," "-eth-," or "-oxynol-." A survey of cosmetic products in 1979 found that 31 of 65 products analyzed contained greater than 10 ppm 1,4-dioxane (Fishbein, 1981). An analytical survey of Belgian cosmetic products found 1,4-dioxane concentrations from less than 2 to 613 mg/kg (Beernaert et al., 1987). A survey of Italian cosmetic products found that 48% contained from 7.3 to 85.9 ppm 1,4-dioxane (Scalia and Menegatti, 1991; Scalia et al., 1992). The German IKW (German Cosmetic, Toiletry, Perfumery and Detergent Association) recommends that cosmetic products should not contain 1,4-dioxane at concentrations greater than 10 ppm (mg/kg) (Fruijtier-Pölloth, 2005). The FDA conducted periodic surveys of 1,4-dioxane levels in ethoxylated

TABLE 2.13
FDA Analytical Survey Results for 1,4-Dioxane in U.S. Ethoxylated Raw Materials and Finished Cosmetics

	1,4-Dioxane Concentration (ppm)			
	Raw Ethoxylated Surfactants/Alkyl Sulfate Surfactants		Finished Cosmetic Products	
Year	Average/Number of Analyses	Range	Average	Range
1979	49/229	71–580	—	—
1980	280/226	6–1,410	—	—
1981	—	—	50	2–279
1982	—	—	19	2–36
1983	—	—	2	1–8
1991	—	—	—	3–108
1992	—	—	41	5–141
1993	71/80	16–243	79	50–112
1994	—	—	45	20–107
1995	—	—	74	42–90
1996	180/188	20–653	14	6–34
1997	348/NA	45–1,102	19	6–34

Sources: Food and Drug Administration (FDA), 1981, Progress report on the analysis of cosmetics raw materials and finished cosmetics products for 1,4-dioxane. Division of Cosmetics Technology, Food and Drug Administration 15; Italia, M.P. and Nunes, M.A., 1991, Gas chromatographic determination of 1,4-dioxane at the parts-per-million level in consumer shampoo products. *Journal of the Society of Cosmetics Chemistry* 42: 97–103; Black, R.E., Hurley, F.J., and Havery, D.C., 2001, Occurrence of 1,4-dioxane in cosmetic raw materials and finished cosmetic products. *Journal of AOAC [Association of Official Analytical Chemists] International* 84(3): 666–670; and ATSDR, 2004, Draft toxicological profile for 1,4-dioxane. Division of Health Assessment and Consultation, U.S. Department of Health and Human Services.

Note: NA, not available.

raw materials; results are summarized in Table 2.13. The first edition of this book stated that 10 mg/kg was also an FDA-recommended concentration in cosmetic products, citing ASTDR (2004); however, the ASTDR's 2012 update to the 1,4-dioxane toxicity profile clarifies that FDA did not have a policy setting a limit for 1,4-dioxane in cosmetic products. The presence of 1,4-dioxane in cosmetic products is of potential concern for dermal toxicity, although the skin is not a significant pathway for 1,4-dioxane toxicity. The presence of 1,4-dioxane in cosmetics, soaps, detergents, and shampoos is of greater concern for wastewater and groundwater quality, and has in some locations caused domestic wells to become contaminated with low concentrations of 1,4-dioxane from septic leach field recharge.

Upon learning that 1,4-dioxane was a contaminant of nonionic surfactants, the U.S. Cosmetic, Toiletry and Fragrance Association commissioned a pharmacology and toxicology committee review of 1,4-dioxane contamination in cosmetics and concluded that the levels found are below a "virtually safe dose." The association nevertheless conducted a survey of suppliers of ethoxylated products to obtain data and information on specifications and processing. The goal was to work toward achieving acceptable levels and controls for 1,4-dioxane and to establish a standard for 1,4-dioxane-containing ingredients (Jass, 1989).

AES compounds and related surfactants have many varieties and many applications beyond detergents and personal care products. Secondary alcohol ethoxylate is used in crop protection agents, metal working and processing, textile processing, paints and coating compositions, and as a conditioning agent for pulp and paper processing (Kosswig, 2002). 1,4-Dioxane residuals may be present in ethoxylated C_{12}-C_{14} secondary alcohols, according to available MSDS. 1,4-Dioxane is also reportedly used in the dissolution of polymers to produce a coating for laminated paper products (USEPA, 1996c). Soil surfactants use ethoxylated surfactant compounds to improve irrigation of golf courses, athletic fields, and other turf and agricultural applications. MSDS listings for some varieties of soil surfactants note the trace presence of 1,4-dioxane.

Manufacturers of ethoxylated surfactants were challenged to both optimize surfactant production and minimize formation of 1,4-dioxane. The amount of 1,4-dioxane formed in AES compounds is a function of the mole ratio of sulfur trioxide fed to the reactor to the ethoxylated alcohol stock (e.g., alcohol) in the reactor. 1,4-Dioxane forms when two molecules of ethylene oxide are cleaved from the parent ethoxylated alcohol. 1,4-Dioxane formation is favored by an excess of sulfur trioxide, high temperatures, and moisture and longer ethylene oxide chains in the ethoxylated alcohol feedstock (Kosswig, 2002; Dado et al., 2006). The level of 1,4-dioxane in the product remains relatively low at 20–30 ppm until a critical point of oversulfation occurs. Once oversulfation occurs and the mole ratio exceeds 1.04, 1,4-dioxane production increases rapidly to values measured in hundreds of parts per million (Foster, 1997).

Vacuum stripping, steam stripping, drying, and other solutions were eventually successful in achieving substantial reductions in the 1,4-dioxane content of finished surfactant products (Sachdeva and Gabriel, 1997). The recovered 1,4-dioxane is often condensed and used as a fuel for on-site boilers or in other energy-recovery operations. Equipment specifically designed for removal of 1,4-dioxane in the production of lauryl ether sulfate achieves an eightfold decrease. Additional equipment is designed to subsequently destroy 1,4-dioxane in the condensate. Destruction methods include a catalyzed reaction with hydrogen peroxide to oxidize 1,4-dioxane, producing carbon dioxide and water (Chemithon Corporation, 2006).

2.4.1.4 1,4-Dioxane in Contraceptive Sponges and Spermicidal Lubricants

The active spermicide nonoxynol-9 has been commonly used in contraceptive products since the 1970s. Nonoxynol-9 is among the surfactants that may contain 1,4-dioxane as an impurity of production. The levels of 1,4-dioxane present were deemed to be less than the 10 ppm limit recommended for polysorbates in food and applied to contraceptives by the FDA (FDA, 1997a). Protracted debate over the safety of a contraceptive sponge in the 1980s and 1990s focused on the presence of 1,4-dioxane in nonoxynol-9 in the "Today" sponge. The FDA determined that there is no appreciable risk because of lack of vaginal absorption of nonoxynol-9 (Woodcock, 1997). Approximately one-sixth of the nonoxynol-9 contained in the sponge is released during use. The process used to create the contraceptive sponge is thought to drive off 1,4-dioxane, reducing its concentration sevenfold (FDA, 1997b). The level of 1,4-dioxane present in nonoxynol-9, 7 ppm, would probably not appear in the sponge (Medical Economics Publishing, 1983). Nonoxynol-9 is also used as a spermicidal lubricant in condoms and related products.

2.4.1.5 1,4-Dioxane in Polyethylene Glycol

PEG compounds are used in a wide range of products (Union Carbide, 2001):

- Pharmaceuticals
- Cosmetics such as deodorant sticks, lipsticks, shaving creams, toothpastes, and lotions
- Detergents in laundry soaps and dishwashing soaps
- Textiles and leather processing
- Plastics and resins
- Paper
- Printing inks
- Lubricants
- Mold-release agents in the rubber industry

- Metal corrosion inhibitors in the petroleum industry
- Anticracking and preservation in woodworking
- Brake fluid lubricants
- Fracking fluids for hydraulic fracture drilling

Pharmaceutical grades of PEG typically contain less than 10 ppm 1,4-dioxane.

Dow Chemical tested the 1,4-dioxane content of various grades of PEG. Some grades showed no detection; all grades tested less than 1 part 1,4-dioxane per million parts PEG (Dow Chemical Corporation, 1989). In 1988, Union Carbide notified its customers that its Carbowax line of PEG products contained no more than 5 ppm 1,4-dioxane and was therefore in compliance with California's Safe Drinking Water and Toxics Enforcement Act of 1986 (Proposition 65) and Section 313 of Title III of the Superfund Amendments and Reauthorization Act of 1986 (SARA Title III) (Union Carbide, 1988).

The PEG compounds of interest in the context of personal care products include ethers of propylene glycol, propylene glycol stearate, propylene glycol oleate, and propylene glycol cocoate. These compounds serve as cleansing, solubilizing, and emulsifying agents, skin-conditioning agents such as humectants and emollients, and solvents in cosmetic formulations. Only a minor percentage of cosmetic formulations use these compounds; some are not currently used. PEG, propylene glycol cocoate, and propylene glycol oleate are produced by the esterification of polyoxyalkyl alcohols with lauric acid and oleic acid, respectively. Impurities include ethylene oxide (maximum 1 ppm), 1,4-dioxane (maximum 5 ppm), polycyclic aromatic compounds (maximum 1 ppm), and the heavy metals lead, iron, cobalt, nickel, cadmium, and arsenic (maximum 10 ppm combined) (Johnson, 2001). PEG compounds have been added to fracking fluids as corrosion inhibitors, surfactants, and friction reducers.

2.4.1.6 1,4-Dioxane in Glyphosate Herbicides (Accord®, Roundup®, Rodeo®, Vision®)

1,4-Dioxane may also be present as an impurity of polyoxyethyleneamine (POEA), the salt of glyphosate [N-(phosphonomethyl)glycine], which is the major component in widely used herbicide formulations such as Roundup®. The United Nations Food and Agriculture Organization has established a limit of 1 ppm for 1,4-dioxane in POEA. The USEPA considers 1,4-dioxane as an "inert of toxicological concern" in pesticide and herbicide formulations. The most popular herbicide, Roundup®, contained less than 0.03% 1,4-dioxane in 1990 (<300 mg 1,4-dioxane per kilogram of Roundup®) (Diamond and Durkin, 1997). Monsanto, the producer of Roundup®, reported in 1997 that 1,4-dioxane contamination had been further reduced to 23 ppm (U.S. Forest Service, 1997). Although 1,4-dioxane content was probably too low to cause groundwater contamination from application to fields, locations where glyphosate-based herbicide was loaded into or washed from spray equipment could be sources of low-level 1,4-dioxane contamination. The practice of rinsing herbicide containers prior to recycling may also cause detectable 1,4-dioxane contamination in groundwater.

Some urea herbicides could potentially contain 1,4-dioxane as a residue of production. 1,4-Dioxane is reacted with 4-chloroaniline, anhydrous hydrogen chloride, and phosgene at 70–75°C to produce *p*-chlorophenyl isocyanate, an intermediate used for the production of urea herbicides (Roig et al., 2003). For example, Monuron, a uracil herbicide, is among the products formulated with 1,4-dioxane and chloroaniline, and may contain 1,4-dioxane as a residual (USEPA, 1978). Another herbicide used in plantings of tomatoes and other crops, rimsulfuron, could contain 1,4-dioxane as an impurity of production. Technical rimsulfuron, also called Shadeout, is manufactured in a two-step batch process in which a carbamic acid and an aniline compound are reacted in a 1,4-dioxane solution (California Environmental Protection Agency, 1997). Rimsulfuron is 1-(4,6-dimethoxypyrimidin-2-yl)-3-(3-ethylsulfonyl-2-pyridylsulfonyl)urea (CASRN 122931-48-0; formula: $C_{14}H_{17}N_5O_7S_2$). Agricultural products that contain ethoxylated surfactants may also contain traces of 1,4-dioxane. 1,4-Dioxane is listed as present but below reportable quantities in the insecticide Dursban, which includes surfactants among its components (Dow Chemical Corporation, 1985).

2.4.1.7 1,4-Dioxane in Pesticides

Some insecticidal formulations claimed in patents have used a solvent mixture comprising 30–50% of the formulation. An example solvent mixture claimed in the patent literature is 20% 1,4-dioxane along with methyl chloroform, methanol, and ethanol (Chang, 1999). It is unlikely that 1,4-dioxane was widely used in commercial production as a carrier in insecticides in such high percentages because of its high cost. 1,4-Dioxane was present in some pesticide formulations as a stabilizer of methyl chloroform, which has been documented as an "inert" ingredient in pesticides. In 1989, USEPA's Office of Pesticide Programs issued a notice concerning inert ingredients of pesticide products, in which 1,4-dioxane was included in the list of "inerts of toxicological concern." Methyl chloroform is also a carrier solvent for pesticides; it was included on a list of "potentially toxic inerts" (USEPA, 1989a). Trace amounts of 1,4-dioxane are often found in pesticides because of the inclusion of ethoxylated surfactants, which include 1,4-dioxane as an impurity (see Section 2.4.1). For example, a solid mosquito larvicide designed to float atop ponds and release its active biocidal ingredient, *Bacillus thuringiensis*, lists 1,4-dioxane as present at less than 20 ppm. Other formulations list 1,4-dioxane present at less than 10 ppm as a by-product with ethoxylated alcohols (Cognis Corporation, 2001). Some agency risk assessments and other reviews that list 1,4-dioxane use as a pesticide may confuse 1,4-dioxane derivatives for 1,4-dioxane. For example, dioxathion [CASRN 78-34-2, is *O,O,O',O'*-tetraethyl *S,S'*-(1,4-dioxane-2,3-diyl) diphosphorodithioate], a pesticide for deciduous fruit, is a derivative of 1,4-dioxane that could be mistaken for 1,4-dioxane used as a pesticide.

1,4-Dioxane was used together with benzene to extract and distill the pesticide hexachlorocyclohexane (also called benzene hexachloride, BHC, Lindane), which was widely used as a contact pesticide for head lice, as a wood preservative, and for other purposes. The solvent mixture was recovered and reused; however, losses to the waste by-products and product were routinely replaced with new solvent. Waste solvents were thermally destroyed in well-controlled production operations; however, there have been examples of extraordinary volumes of waste BHC isomers littering production sites in Europe (Vijgen, 2006). Disposal of BHC wastes could also be a source of 1,4-dioxane releases to soil and groundwater.

2.4.2 Terephthalate Esters (Polyester) and Resins

Polyethylene terephthalate (PET) polyester was first introduced in 1952 (USEPA, 1997b). Polyester fibers are produced from the polycondensation reaction of a dicarboxylic acid, such as terephthalic acid, and a dihydroxy alcohol, such as ethylene glycol. 1,4-Dioxane forms as a by-product during esterification (Ellis and Thomas, 1998). To manufacture polyester fiber, terephthalic acid and ethylene glycol are first passed through primary and secondary esterifiers to form the monomer. The melt is then passed to a polymerizer equipped with a high vacuum to allow excess ethylene glycol to escape. Wastes generated during polymerization may include emissions of VOCs from leaks, spills, and vents, solid wastes from off-specification polymers, and spent solvent from incomplete polymerization (USEPA, 1997b).

Polymers of PET always contain a certain amount of incorporated DEG. DEG is formed in a side reaction during the ester interchange of dimethyl terephthalate (DMT) with ethylene glycol. Alternatively, during direct esterification of terephthalic acid with ethylene glycol, DEG is formed via an unusual type of reaction: ester + alcohol ether + acid = DEG + 1,4-dioxane + methyl cellosolve + methyl carbitol (Hovenkamp and Munting, 1970).

Factories producing PET are generally equipped with wastewater treatment plants to remove waste chemicals; however, 1,4-dioxane has proved difficult to remove. Treated effluent from an industrial wastewater treatment plant at a PET manufacturing facility contained 100 mg/L of 1,4-dioxane in 1995 (European Chemicals Bureau, 2002). Nearly 50% of effluent samples from a PET resin plant in Spain had 1,4-dioxane concentrations exceeding 1,000 μg/L; the remainder showed 1,4-dioxane present from 100 to 1,000 μg/L. The mean concentration of 35 samples was 6,400 μg/L; the range

TABLE 2.14
USEPA 1995 Toxic Release Inventory Data for 1,4-Dioxane from EPA Sector Notebook

Fugitive Air	Point Air	Water Discharges	Total Releases	Average Releases per Facility
3810	1763	17,246	22,841	4568
Disposal transfers	Treatment transfers	Energy recovery transfers	Total transfers	Average transfer per facility
271	12,655	11,990	24,916	4983

Source: U.S. Environmental Protection Agency (USEPA), 1995b, Profile of the rubber and plastics industry. Office of Enforcement and Compliance Assurance, USEPA.
Note: Values in pounds per year for five plastic-resin manufacturing facilities.

was 100 to 31,400 μg/L (Romero et al., 1998). Waste streams from polyester production plants in the eastern United States routinely contained as much as 2,000 mg of 1,4-dioxane per liter (Grady et al., 1997). Air emissions created during polyester production are estimated to include 1 mg of 1,4-dioxane per kilogram of polyester fiber produced (Laursen et al., 1997).

The equipment used to produce polyester must be cleaned periodically. The cleaning process consists of dipping parts in triethylene glycol (TEG) to remove accumulated polyester and by-products. Spent TEG is recovered and transported off-site for recycling. Parts are then rinsed with water to remove residual TEG and polyester ingredients. This rinsed water represents the primary wastewater stream at a typical polyester production facility and may total as much as 2 million gallons annually. Rinse-water constituents typically include TEG, 1,4-dioxane as a by-product of heating TEG, and polyester by-products such as methanol. Rinsed waters may be processed in wastewater treatment plants or shipped by rail for off-site treatment (North Carolina Department of Environment and Natural Resources, 1998).

In 1992, chemical manufacturing operations produced 2.56 billion kg of terephthalic acid (2.82 million tons). A summary of 1,4-dioxane emissions, releases, and off-site transfers for five plastic-resin manufacturing facilities in 1995 is given in Table 2.14.

Many photographic films are produced from DMT to create a polyester film base. Polyester waste material, such as old x-rays and used movie film, is commonly recycled to make DMT available for reuse. The processing of polyester waste produces 1,4-dioxane as a by-product, which has been discharged from operations as wastewater. For example, in 1994, 140,000 lb of 1,4-dioxane were discharged to the industrial sewer in Rochester, New York, from Kodak, the leading U.S. film products company. Because the industrial wastewater treatment plant was inefficient for 1,4-dioxane removal, plant engineers have since devised and deployed several innovative means to separate and collect 1,4-dioxane for on-site energy recovery so that the chemical can be used as a fuel for plant boilers. The plant successfully eliminated the discharge of 1,4-dioxane (Eastman Kodak Corporation, 2001).

Another resin-related application for 1,4-dioxane is in the production and use of phenolic resins, e.g., in printed circuit boards. Cresol novolac resins are used for etching ultrafine circuit line patterns into the photoresist surface coatings of semiconductor silicon wafers, as well as in printing plate production. For epoxy resins used in electronic applications, such as cresol epoxy novolacs, more powerful polar aprotic solvents, such as 1,4-dioxane or dimethyl formamide, have been used to hydrolyze chlorohydrin and organically bound chlorides (Jahn and Goetzky, 1988).

2.4.3 1,4-Dioxane as a By-Product of Petroleum Refining and Gasoline Antiknock Agent

The literature contains reference to the use and formation of 1,4-dioxane during petroleum refining; however, the information that exists is insufficient to suggest that 1,4-dioxane was present in gasoline at levels that can be detected or cause environmental impacts. The main impurities in refined benzene are nonaromatics with nine or fewer carbon atoms, toluene, 1,4-dioxane, and aromatics containing eight carbon atoms (Wypych, 2001). 1,4-Dioxane may also occur as an impurity of toluene. Some petroleum refining methods for isolating and purifying toluene have reportedly used oxygenated reagents. Analytical methods have been specifically developed to address the possible residual presence of 1,4-dioxane in toluene (Cortellucci and Dietz, 1999).

Universal Oil Products Company of Chicago obtained a patent in 1938 for a petroleum refining process to produce 1,4-dioxane from the ethylene by-product of petroleum refining and then to use the 1,4-dioxane to serve as an antiknock agent in gasoline. Ethylene is first converted to ethylene glycol by adding hydrogen peroxide; the ethylene glycol is then converted to 1,4-dioxane by using concentrated sulfuric acid. 1,4-Dioxane acts as a polymerizing agent in the refining process, and excess 1,4-dioxane remaining in the gasoline acts as an antioxidant to preserve the antiknock rating in gasoline stored for 6 months or more (McCaffrey, 1938). 1,4-Dioxane was thus apparently considered a viable oxygenate for gasoline at levels up to 0.05 wt%, although the extent to which this invention was put into practice has not been determined.

2.5 1,4-DIOXANE IN FOOD

The first systematic study designed to determine the levels of 1,4-dioxane in food was conducted in 2003 by a team of Japanese scientists led by Tetsuji Nishimura of Japan Food Research Laboratories in Tokyo (Nishimura et al., 2004). A number of risk assessments and drinking water fact sheets have cited earlier literature to assert that 1,4-dioxane is commonly found in fried chicken, tomatoes, shrimp, and other foods. However, none of the studies cited actually quantified the amount of 1,4-dioxane present, and the studies were all performed in the 1970s and 1980s, before analytical methodologies for 1,4-dioxane detection were improved (ATSDR, 2004). Table 2.15 summarizes early citations of 1,4-dioxane in various food products at unquantified levels.

In Nishimura's landmark study, a market-basket approach was taken to analyze groups of food by blending them together to assay 1,4-dioxane content. Food was prepared in the same manner in which consumers cook food for routine consumption, using purified water and clean utensils. Raw and cooked foods in 12 dietary groups were blended together with other foods from the same group and then analyzed by using active carbon solid-phase extraction cartridges (described further in Chapter 4), coupled with GC and mass spectrometry. 1,4-Dioxane was found in all foods except rice and rice noodles; detections were in the low part per billion (ppb) range. Corresponding dietary intake was determined to be in the nanogram-per-day range, far below the applicable risk thresholds. The study does not specifically explain the cause of the nearly ubiquitous presence of 1,4-dioxane in food; however, the widespread presence of 1,4-dioxane from

TABLE 2.15
Studies Reporting the Identification of 1,4-Dioxane in Food Products at Unquantified Levels Prior to Japanese Study by Nishimura et al. (2004)

Food Product	Reference
Chicken flavor and meat volatiles	Shahidi et al. (1986)
Fried chicken volatile flavor compounds	Tang et al. (1983)
Tomato fruit juice volatiles	Chung et al. (1983)
Fat oil compound (tillnoloin) formed in deep frying	Chang et al. (1978)
Cooked shrimp volatile flavor compounds	Choi et al. (1983)
Patis (Philippine fermented fish sauce)	Sanceda et al. (1984)
Brazilian coffee	Spadone et al. (1990)

TABLE 2.16
1,4-Dioxane in Japanese Dietary Food Groups

Dietary Group	Foods	1,4-Dioxane Content (μg/kg)	Total Dietary Intake (μg)
I	Rice, rice noodles	ND	0
II	Oatmeal, wheat flour, bread, buckwheat, cornflakes, sweet potato, almonds, instant noodles, potato, taro root	6	0.057
III	Sugar, jam, caramel, rice cracker, pound cake, chocolate, others	6	0.010
IV	Butter, margarine, soybean oil, lard, dressing	8	0.008
V	Soybean paste, tofu, frozen tofu, deep fried tofu	3	0.006
VI	Orange, apple, banana, strawberry, tomato juice, watermelon, apricot, loquat	4	0.0026
VII	Carrot, spinach, green pepper, tomato, broccoli, celery, okra	3	0.015
VIII	Daikon radish, cucumber, onion, cabbage, bean sprouts, pickles, pickled relish, mushroom, green seaweed	8	0.076
IX	Soy sauce, ketchup, salt, sauce, sake, beer, wine, tea, soft drinks	7	0.070
X	Salmon, tuna, mackerel, flatfish, scallop, salted cod, bonito, sardine	5	0.022
XI	Beef, pork, chicken, lamb, ham, chicken, eggs	6	0.055
XII	Milk, cheese, yoghurt, sake sediment	13	0.095

Source: Nishimura, T., Iizuka, S., Kibune, N. et al., 2004, Study of 1,4-dioxane intake in the total diet using the market-basket method. *Journal of Health Science* 50(1): 101–107.
Note: ND, not detected.

industrial discharges to river water and groundwater used for irrigation and food processing are suggested as possible reasons. Table 2.16 summarizes the findings from the analysis of 1,4-dioxane in different food groups.

Food additives have been regulated in the United States and Europe for their 1,4-dioxane content. The U.S. Food Chemicals Codex restricts residual 1,4-dioxane content in polysorbates to 10 mg/kg. Polysorbate 60 [polyoxyethylene (20) sorbitan monostearate] is an emulsifier used in ice cream, frozen custard, ice milk, fruit sherbet, and other frozen desserts. Polysorbate 60 is used alone or in combination with polysorbate 65 and/or polysorbate 80. Polysorbates have been found to contain trace levels of 1,4-dioxane and ethylene oxide as impurities, which result from the production processes used for polysorbates (FDA, 1999). Polysorbate 60 and polysorbate 80, which are produced from the polymerization of polyoxyethylene, have historically been found to contain 1,4-dioxane (Birkel et al., 1979). Levels of 1,4-dioxane in these compounds have been reported to range from 5 to 6 ppm (mg/L) (ATSDR, 2004).

1,4-Dioxane was detected in Tween 20 and Tween 80 (polysorbate 20 and polysorbate 60) at approximately 100 and 200 ppm, respectively (Guo and Brodowsky, 2000). Table 2.17 summarizes the varieties of polyoxyethylene-based emulsifiers in which 1,4-dioxane may be present in trace amounts.

Eating ice cream and other frozen desserts that use polysorbates as an emulsifier could cause a maximum exposure to 1,4-dioxane of 19 ng per person per day (FDA, 1999). For a 70 kg person, this daily value equates to 2.7×10^{-7} mg/kg, which is about 40,000 times less than the suggested Integrated Risk Information System daily oral slope factor for estimated

TABLE 2.17
Emulsifiers in Which 1,4-Dioxane May Form at Trace Levels

Emulsifier Name	Chemical Description	Uses
Polysorbate 20	Polyoxyethylene sorbitan monolaurate	Emulsifier
Polysorbate 80	Polyoxyethylene sorbitan mono-oleate	Emulsifier, flavoring, surfactant, defoaming agent
Polysorbate 40 (BAN)	Polyoxyethylene sorbitan monopalmitate	Emulsifier (from animal fatty acids)
Polysorbate 60	Polyoxyethylene sorbitan monostearate	Emulsifier
Polysorbate 65	Polyoxyethylene sorbitan tristearate	Emulsifier

Source: Food and Drug Administration (FDA), 1999, *Federal Register* 64: 57974–57976.

carcinogenic risk, that is, 1.1×10^{-2} mg/kg. The FDA concluded that there is reasonable certainty that no harm from exposure to 1,4-dioxane would result from the daily consumption of ice cream (FDA, 1999).

Glycerides and polyglycides of hydrogenated vegetable oils, also known as PEG esters, may also contain 1,4-dioxane as an impurity. Glycerides include mixtures of mono-, di-, and triglycerides, whereas polyglycides include PEG mono- and diesters of fatty acids. The Code of Federal Regulations Title 21, Food and Drugs, sets limits for allowable 1,4-dioxane in food and excipients of dietary supplements at no greater than 10 mg/kg (FDA, 2006). Consumption of glycerides and polyglycides in dietary supplement tablets, capsules, and liquid formulations produces an exposure to 1,4-dioxane that equates to not more than 800 ng per person per day (FDA, 2006). This exposure level is about 1,000 times lower than the suggested oral slope factor for cancer risk for 1,4-dioxane.

2.6 1,4-DIOXANE USE IN THE PHARMACEUTICAL INDUSTRY

The pharmaceutical industry fabricates, processes, and formulates medicinal chemicals and pharmaceutical products and grinds, grades, and mills botanical products. Pharmaceutical products may include agents of natural origin, hormonal products, basic vitamins, and isolated alkaloids from botanical drugs and herbs (USEPA, 1997c). The majority of pharmaceutical production facilities in the United States are located in California, New Jersey, New York, and Puerto Rico. Formulated pharmaceutical products are manufactured by chemical synthesis, fermentation, or isolation and recovery from natural sources, or a combination of these processes. Fermentation is economically the most important process; however, by tons produced, chemical synthesis dominates. Fermentation processes are used to produce antibiotics such as penicillin and tetracycline. Chemical synthesis processes are used to manufacture psychotropic and antihistamine drugs. Extracts from animal products provide hormone-based drugs, biological processes are used to produce vaccines and serums, and vegetable extraction processes yield steroids and alkaloids (Chenier, 2002). Production of some pharmaceutical products incorporates the use of various solvents, including 1,4-dioxane.

Solvents are used primarily as a vehicle or reagent and to clean residues from equipment between production runs of different drugs. Solvent wastes, primarily as vapors, may be generated at most steps in the production of pharmaceuticals, including

- Cleaning reaction vessels and condensers
- Separation by extracting, decanting, centrifugation, filtration, or crystallization
- Purification by recrystallization, centrifugation, or filtration
- Product drying

Solvent-laden wastewater may be generated in the reaction, separation, extraction, and purification steps. Devices to control air emissions and equipment for wastewater treatment are designed to remove residual solvent and comply with air quality and discharge elimination regulations (USEPA, 1997c).

Many bulk pharmaceutical reactions require organic solvents to dissolve chemical intermediates and reagents. To prevent reaction with reagents and intermediates, nonreactive solvents are preferred. Dichloromethane is often the optimum choice for pharmaceutical reactions. The most commonly used solvents in the pharmaceutical industry include methanol, ethanol, acetone, isopropanol, and other oxygenated organic solvents (USEPA, 1997c). 1,4-Dioxane is used in some processes but has not been a major solvent used in production pharmaceuticals. In addition to 1,4-dioxane, other compounds used in the pharmaceutical industry that also serve as solvent stabilizers include acetonitrile, *n*-butyl alcohols, cyclohexane, pyridine, and tetrahydrofuran. Specific uses for 1,4-dioxane include chemical synthesis and biological and natural product extraction (USEPA, 1997c). 1,4-Dioxane, toluene, xylene, and pyridine are routinely monitored by the pharmaceutical industry (Kauffman, 2003).

In 1995, TRI participation involved 200 out of the 916 pharmaceutical production facilities listed in the 1992 census of manufacturers. Of those 200 facilities, two reported release of 1,4-dioxane (USEPA, 1997c). This low percentage may be because very few facilities used 1,4-dioxane, but it is more likely the result of a tendency for TRI responses to exclude chemicals used in only minor quantities. The two facilities that did report 1,4-dioxane releases in 1995 summarized the quantities released, in pounds per year, as follows:

Fugitive Air Emissions	Point Air Emissions	Water Discharges	Total Releases	Average Releases per Facility
270	260	0	530	265

The same two facilities reported larger quantities transferred for treatment at wastewater facilities (4,170 lb) and energy-recovery facilities such as cement kilns (47,916 lb) (USEPA, 1997c). Following the Pollution Prevention Act of 1990, the pharmaceutical industry engaged in a large effort to minimize solvent use and emissions; the industry developed alternative, solvent-free methods for producing pharmaceuticals. In earlier decades (i.e., the 1980s and earlier), pharmaceutical facilities *potentially* released 1,4-dioxane from on-site handling of wastewater, burial of solvent-laden sludge from the solid residues of botanical extraction processes (or landfills that received such sludge), and energy-recovery operations that handled solvents from pharmaceutical facilities.

An indirect means of establishing the use of 1,4-dioxane in the pharmaceutical industry is found in the literature for analytical methods developed to detect organic volatile impurities (OVIs) in finished drug products. U.S. Pharmacopeia (USP) analytical method 467 for analysis of OVIs also specifies the maximum permissible levels of five solvents of interest, including methylene chloride, benzene, trichloroethylene, 1,4-dioxane, and chloroform. Literature from the commercial laboratory sector, the analytical instrument sector, peer-reviewed journals on laboratory analysis, and the pharmaceutical industry literature all cite methods for improving the detection of 1,4-dioxane as an OVI in finished pharmaceutical products to ensure compliance with U.S. Food and Drug restrictions on residual solvents. For example, one gas chromatograph manufacturer's technical literature notes that analysis of ascorbic acid (vitamin C) and acetaminophen (a common pain killer) showed all solvents well below the regulated levels. Vitamin C had benzene at 35 ppb, and acetaminophen had dichloromethane at 407 ppb; neither product had 1,4-dioxane above a method detection limit of just less than 1 ppm (Rankin, 1996).

In the United States and Europe, the permitted daily exposure to 1,4-dioxane through consumption of pharmaceuticals is 3.8 mg per person. The permissible concentration limit in the finished drug product is 380 ppm (FDA, 1997a; European Medicines Agency, 1998). The same levels are considered permissible in U.S. veterinary medicinal products (FDA, 2001)

2.7 DETECTIONS OF 1,4-DIOXANE IN AMBIENT SURFACE WATER, GROUNDWATER, AND AIR

The Third Unregulated Contaminant Monitoring Rule 3 (UCMR 3) list included 1,4-dioxane analysis using EPA Method 522, a single-analyte method using solid phase extraction and tandem mass spectroscopy with a reporting limit of 0.07 μg/L. The UCMR is designed to collect occurrence data from a predetermined number of U.S. public water systems (PWSs) as part of a 5-year review cycle. As part of an "assessment monitoring" program, 21 of these contaminants were required to be monitored by all large PWSs (serving >10,000 connections) and 800 pre-selected smaller systems. UCMR3 monitoring began in January 2013 and was completed in December 2015. In all, 4,864 different public water supply systems (PWSs) were tested, representing 73,453 individual sources (41,111 groundwater sources, 27,800 surface water sources, 3,708 mixed sources and 834 samples of groundwater under the influence of surface water). Results were compared to the 10^{-6} cancer risk level, using the EPA 2012 Health Advisory Level of 0.35 μg/L. Among the 63 states, territories, and tribal areas tested, 50 had PWSs with 1,4-dioxane detections.

An analysis of UCMR3 results for 1,4-dioxane analyses found a higher frequency of detection in larger water systems (11.7%) than small (5.8%), as well as a higher reference concentration exceedance rate in large water systems (3.1%) than small (1.0%) (Adamson, et al., 2017). 1,4-Dioxane was detected above the minimum reporting level of 0.07 μg/L in 22% of PWSs tested (7% > 0.35 μg/L) (USEPA, 2017). Samples from a large system were 2.18 times more likely to be associated with a 1,4-dioxane detection than a sample from a smaller system ($p < 0.0001$). Higher 1,4-dioxane detection frequency was observed in samples from groundwater sources (12.0%) than in samples from surface water sources (10.0%). Groundwater samples also exhibited a higher exceedance frequency (3.5%) than surface water samples (2.0%). Groundwater samples were 1.25 times more likely to result in a 1,4-dioxane detection than surface water sources ($p < 0.0001$) and 1.8 times more likely to result in a 1,4-dioxane exceedance of the 0.35 μg/L reference concentration than surface water sources ($p < 0.0001$) (Adamson et al., 2017).

The expected sources of 1,4-dioxane in groundwater are profiled in this chapter. While many sources are possible, the predominant source, based on 90% of 1,4-dioxane production used for stabilizing methyl chloroform, is various uses of methyl chloroform, particularly for vapor degreasing. As such, 1,4-dioxane is primarily a legacy groundwater contaminant, with ongoing releases from methyl chloroform uses tapering off in the 1990s, following the 1996 ban on methyl chloroform because of its ozone-depleting properties.

Detections of 1,4-dioxane in surface water represent ongoing discharges from wastewater treatment plants, and industries that either use 1,4-dioxane or produce it as a by-product. Effluent limits for 1,4-dioxane have generally been set rather high, in the mg/L range, because 1,4-dioxane is not toxic to aquatic organisms at low concentrations. Hence industrial discharge of 1,4-dioxane concentrations that may be problematic for downstream drinking water outtakes may be permitted under the narrow restrictions for aquatic toxicity. UCMR3 surface water results show that this approach has been shortsighted, with surface water sources for several large water systems impaired by 1,4-dioxane, particularly in the Cape Fear Watershed in North Carolina. In addition, where groundwater treatment systems are approved to discharge to the sanitary or storm sewer after removing chlorinated solvents, 1,4-dioxane discharges may continue where 1,4-dioxane testing has not been required. Where water reuse is contemplated, particularly indirect potable reuse, 1,4-dioxane contamination of wastewater presents treatment challenges for compliance with recycled water regulations.

The concentration of 1,4-dioxane in municipal wastewater has been studied in detail, profiling 1,4-dioxane content in the effluent from 40 municipal wastewater treatment plants located across the United States. The median measured 1,4-dioxane concentration was 1.13 μg/L, while the maximum measured concentration was 3.3 μg/L, and 90% of results were less than 1.75 μg/L (Simonich et al., 2013). 1,4-Dioxane concentrations were independent of treatment plant flow rates.

The dilution factor for wastewater effluent prior to the nearest downstream drinking water outtake was also assessed, with a median dilution factor of 245, and 10% of drinking water outtakes diluted less than 44.5-fold, and a maximum dilution of 48,113-fold. A Monte Carlo analysis was completed to account for variability in combined 1,4-dioxane effluent content and dilution factors. The predicted median, 90th percentile and 99th percentile dioxane concentrations at drinking water intakes, at mean stream flow, were 0.0041, 0.030, and 0.16 μg/L, respectively. The authors concluded that municipal wastewater treatment plant effluent, in which 1,4-dioxane generally originates from detergents and other surfactant-based products, does not pose a significant risk to drinking water quality from surface water sources (Simonich et al., 2013).

The high incidence of 1,4-dioxane detection in United States's drinking water is a clear indication that a similarly high percentage of drinking water sources worldwide, both surface and groundwater, may also be impaired by 1,4-dioxane. However, most other countries do not require public water supply testing for 1,4-dioxane. The European Union has not yet required universal drinking water testing for 1,4-dioxane, but state and country-level initiatives have shown that 1,4-dioxane is a drinking water problem in Europe as well.

A national survey of "micro-organic" contaminant data in 2,605 groundwater production wells in England and Wales was conducted, with most of the data collected in annual sampling since 2003, which included analysis for 800 compounds with reporting limits for most compounds at 0.1 μg/L. The sampling effort and data review identified methyl chloroform in 9.6% of samples, with maximum 1,4-dioxane concentration at about 95 μg/L (Manamsa et al., 2016). Frequency of detection varied by aquifer type and land use, with 1.5% in Permo-Triassic Sandstone aquifers, 2.5% in the Chalk aquifers, 2.6% in the Lower Greensand aquifers, and 3% in Jurassic Limestone aquifers, and only in areas classified as urban-industrial. 1,4-Dioxane was detected at a maximum concentration of about 70 μg/L (Manamsa et al., 2016).

Several rivers in Germany have been profiled in a detailed 1,4-dioxane sampling campaign using solid phase extraction coupled with GC/MS, with a limit of detection at 0.016 μg/L, and a limit of quantitation of 0.052 μg/L on river water. The average 1,4-dioxane load in the Main, Oder, and Rhine Rivers were determined to be 6.5 kg/day, 34.1 kg/day, and 134.5 kg/day, respectively (Stepień, 2013). The corresponding concentrations of 1,4-dioxane were 2.2 μg/L in the Oder River, and 0.86 μg/L in both Main and Rhine rivers. Raw water (following bank filtration) contained 0.650 μg/L to 670 μg/L 1,4-dioxane, whereas the concentration in the finished drinking water fell only to 0.6 and 0.490 μg/L, respectively (Stepień, 2013). In Germany, 70% of drinking water is obtained from groundwater, and about 17% from riverbank filtrate (Karges et al., 2018). The German Umweltbundesamt (Environment Agency) has suggested a precautionary limit for 1,4-dioxane in drinking water at 0.1 μg/L.

Levels of 1,4-dioxane between 0.2 and 1.5 μg/L were detected in tap water samples collected during 1995 and 1996 in Kanagawa Prefecture, Japan, an 870-square-mile province adjacent to and southwest of Tokyo (Abe, 1997). An additional comprehensive survey of the Kanagawa Prefecture included multiple sampling events in three rivers and 20 wells and found 1,4-dioxane to be ubiquitous at low levels; a few locations had relatively elevated levels. 1,4-Dioxane was detected in 90% of the wells, with two-thirds of the detections falling below 1 μg/L, 20% between 1 and 10 μg/L, and 10% above 10 μg/L; the highest detections ranged from 50 to 95 ppb (Abe, 1999). All of the seven sampling stations in the three rivers had 1,4-dioxane detections in each of five sampling events. Three river sampling stations had all five analyses falling in the range 0.1–0.99 ppb; three fell in the range 0.2–3.3 ppb, and one river sampling station fell in the range 2–16 ppb (Abe, 1999). Sources of 1,4-dioxane were attributed to legacy contamination by methyl chloroform and industrial and domestic wastewater discharges.

1,4-Dioxane has been detected in ambient air monitoring samples in a few isolated studies. USEPA compiled 1,4-dioxane measurements of ambient air performed at 45 locations in 12 cities taken between 1979 and 1984. The results indicated concentrations ranging from below detection limits to 30 μg/m^3. The mean ambient concentration of 1,4-dioxane was 0.44 μg/m^3 (0.12 parts per billion volume - ppbv) (USEPA, 1993). Air samples in industrial areas in Newark, Elisabeth, and

Camden, New Jersey, were collected from July 6 to August 16, 1981. The geometric mean 1,4-dioxane concentrations ranged from 0.01 to 0.02 ppbv (0.04 to 0.07 μg/m^3). 1,4-Dioxane was detected in 51% of the samples (Harkov et al., 1981).

In the 1980s, USEPA compiled a National Ambient Air Database of indoor and outdoor measurements of 320 VOCs. The database includes 585 analyses for 1,4-dioxane, for which the average concentration was 1.03 ppbv; 75% of measured values were less than 0.09 ppbv, and more than half showed no detection (Shah and Singh, 1988).

Obviously, the ambient atmospheric concentration of 1,4-dioxane will vary with distance from potential sources. As noted in Chapter 3, 1,4-dioxane has a relatively short atmospheric half-life because it is quickly photo-oxidized. Therefore, detections in air samples reflect recent, nearby emission sources. The industrial facilities that use 1,4-dioxane or produce it as a by-product (e.g., pharmaceuticals, polyethylene teraphthalate plastics, textiles, surfactant oils, and surfactant-based products, etc.) can constitute sources of atmospheric emissions, as well as wastewater discharges. Air samples collected downwind of a major petrochemical complex in Tarragona, Spain, detected elevated levels of 1,4-dioxane. At two locations, the mean 1,4-dioxane concentrations were 42 and 43.8 μg/m^3, while the maxima and minima at the two sites were 105 and 14, and 63 and 21, respectively (Gonzales, 2011). A primary source for 1,4-dioxane emissions was identified as an industrial wastewater treatment plant.

2.8 ESTIMATED RELEASES OF 1,4-DIOXANE TO THE ENVIRONMENT

The amount of 1,4-dioxane released to the environment since 1988 can be estimated by tracking the quantity of 1,4-dioxane reported in USEPA's TRI database. 1,4-Dioxane is reportable in waste streams at concentrations above 0.1% because it is listed by the Occupational Safety and Health Administration as a probable carcinogen, based on reports from the National Toxicology Program and the International Agency for Research on Cancer. For each year, the TRI database presents data compiled from hundreds of thousands of reports from hazardous waste generators; the data are categorized into a variety of sectors of treatment and disposal. It is not surprising that the data available from USEPA's TRI database, particularly in the first few years of TRI data collection and reporting, are sometimes imprecise, given the monumental scope of the task. The TRI data sets are nonetheless useful for providing the approximate quantities, locations, and final disposition of 1,4-dioxane releases.

Because 90% of all 1,4-dioxane produced in the United States was reportedly used as a stabilizer for methyl chloroform, tracking chemical market reports for methyl chloroform provides a means of corroborating TRI data for 1,4-dioxane. Prior to 1988, 1,4-dioxane production, consumption, and releases to the environment can only be inferred from chemical market reports. The release of 1,4-dioxane as a by-product of manufacture can be estimated from total production in the plastics, synthetic textiles, and ethoxylated surfactants industries. The trends apparent from TRI data for 1,4-dioxane releases to air, surface water, landfills, and injection wells are presented in Figure 2.5. Trends in the TRI database for transfers of 1,4-dioxane wastes for treatment and energy recovery are shown in Figure 2.6.

During the 16 years of waste handling records depicted in Figures 2.5 and 2.6, 28 million pounds were transferred to treatment facilities, and 15 million pounds were released. Releases of 1,4-dioxane to air and surface water have decreased substantially, while recovery and transfer to treatment has increased. For example, U.S. facility emissions of 1,4-dioxane to air reported in the TRI database have decreased from a high of about 840,000 lb in 1989 to approximately 115,000 lb in 2004. Discharges to surface water reported in the TRI database increased from about 200,000 lb in 1988 to 650,000 lb in 1993, and then decreased to about 90,000 lb in 2004. Overall discharges have decreased approximately sixfold from 1993 to 2004. Some of the apparent trends in the TRI database may be reporting artifacts, but the overall trend shows decreases in discharges and increased recovery and treatment of 1,4-dioxane from facilities that use and generate it. A cursory review of

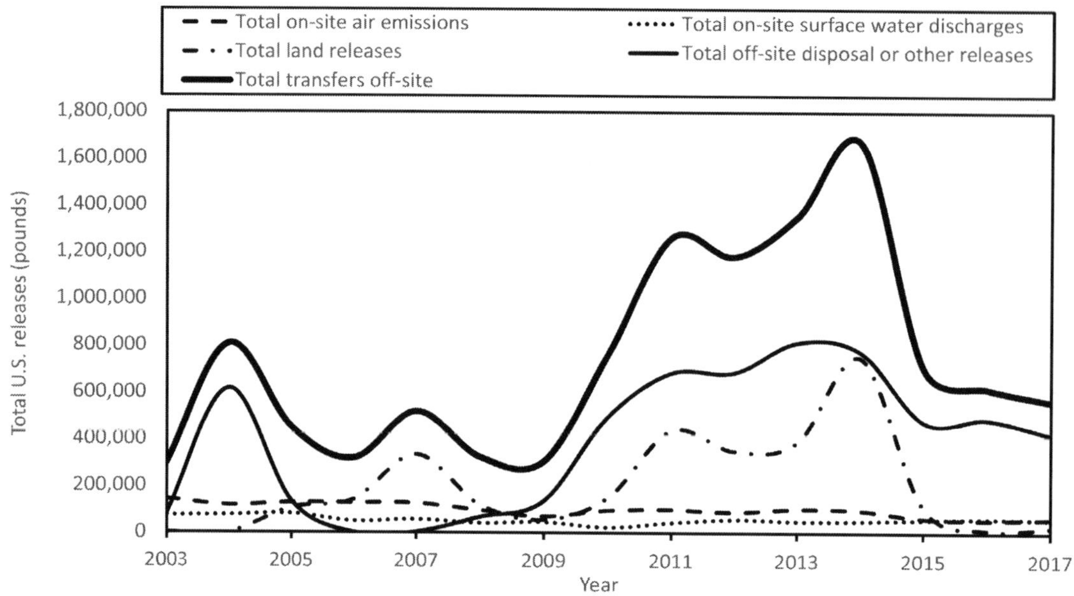

FIGURE 2.5 Estimated releases of 1,4-dioxane from 2003 through 2017. [U.S. Environmental Protection Agency (USEPA), 2019, 2017 TRI Factsheet: Chemical – 1,4-Dioxane, 000123911, Toxic Release Inventory Program, USEPA.]

records for facilities generating 1,4-dioxane shows that since methyl chloroform was banned by the Montreal Protocol Clean Air Act Amendments, 1,4-dioxane emissions and releases to surface water have been eliminated or significantly decreased. 1,4-Dioxane occurrence at hazardous waste sites is the focus of Chapter 8.

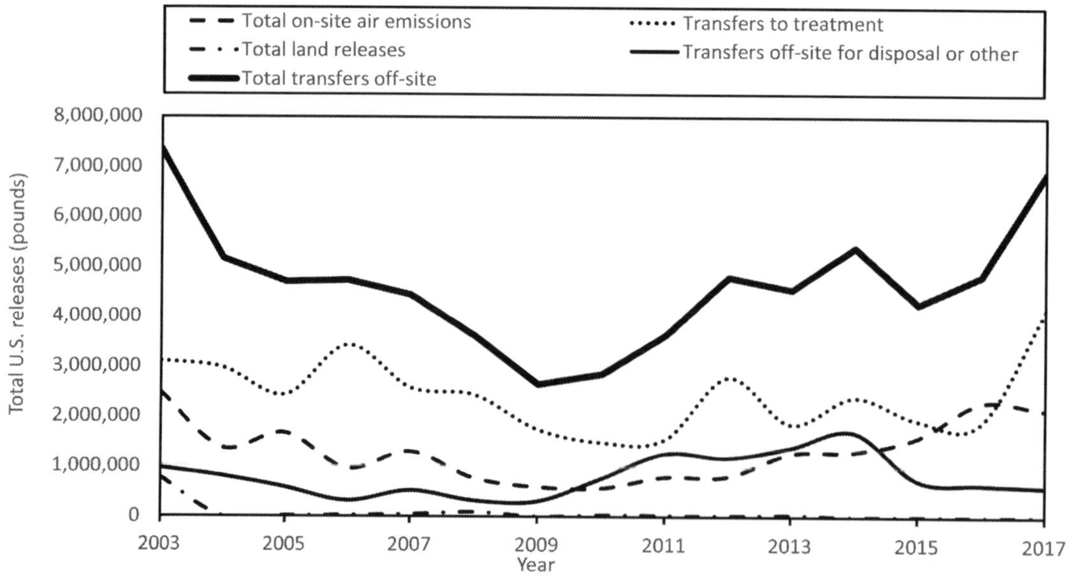

FIGURE 2.6 Estimated transfers and treatment of 1,4-dioxane from 2003 through 2017. [U.S. Environmental Protection Agency (USEPA), 2019, 2017 TRI Factsheet: Chemical – 1,4-Dioxane, 000123911, Toxic Release Inventory Program, USEPA.]

2.8.1 1,4-Dioxane in Wastewater

1,4-Dioxane has been detected in raw wastewater in Michigan and Japan. Treated effluent from a wastewater treatment plant serving several apartment complexes in Japan's Kanagawa Prefecture was found to contain an average of 0.25 mg 1,4-dioxane per person per day. Sources of 1,4-dioxane in households discharging to the treatment plant were presumed to be shampoos and liquid dish-washing and laundry soaps, which in Japan contain from 0.2 to 0.56 mg/L 1,4-dioxane (Abe, 1999).

The City of Ann Arbor, Michigan, tested raw wastewater and treated wastewater effluent for 1,4-dioxane. In each of three sampling events, 1,4-dioxane was present in influent at an average concentration of 3 ppb; 1,4-dioxane remained present in all the three corresponding treated wastewater effluent samples at an average concentration of 2 ppb. The source of 1,4-dioxane in raw wastewater was not identified in the study (Skadsen et al., 2004). A similar comparison of wastewater influent and treated wastewater effluent from Japan showed 16 ppb in influent and 11 ppb in effluent for one sampling event; a subsequent sampling event showed 3 ppb in influent and 3 ppb in effluent. The treatment plant used an activated sludge process. Sources of 1,4-dioxane in the Japanese study were indicated as industrial pollution and surfactants (Abe, 1999). A 1988 survey of organic compounds in U.S. sewage sludge samples detected 1,4-dioxane in three samples out of more than 100 samples analyzed (USEPA, 1990). The ability of conventional aerobic treatment technology to remove 1,4-dioxane is discussed in Chapter 7. A case study of 1,4-dioxane in recycled water is presented in Chapter 8.

2.8.2 1,4-Dioxane in Landfills

1,4-Dioxane is frequently detected in landfill leachate, groundwater beneath municipal and industrial landfills, and in landfill gas and landfill gas condensate. Landfills receiving vapor degreasing solvent wastes, paint filters, scintillation and other laboratory wastes, ink sludge, pesticide containers, household products with methyl chloroform as an ingredient, and industrial sludges from textile production, resin production, and cellulose acetate membrane production have a higher likelihood of 1,4-dioxane presence in leachate, gas, and affected groundwater. 1,4-Dioxane is routinely detected in landfill gas, as noted in Table 2.18.

An USEPA survey of U.S. landfills reviewed the chemistry of leachates. The mean concentration of 1,4-dioxane in leachate from all municipal landfills, including landfills not regulated under

TABLE 2.18
Landfill Leachate Concentrations of 1,4-dioxane

Facility	Maximum Leachate 1,4-Dioxane, µg/L	Analytical Method	Notes	Reference
Hanford, Washington DOE Solid Waste Landfill	180 µg/L	NA	Average ~85 µg/L	Del Mar (2001)
West Contra Costa Sanitary Landfill, CA	2,100 µg/L	NA	Groundwater sample beneath hazardous waste landfill	DTSC (2003)
Operating Industries, Inc., Monterey Park, CA	19,000 µg/L	NA	Municipal and industrial landfill; received hazardous waste	USEPA (1987)
UNC Hazardous Waste Dump	22,000 µg/L	NA	Scintillation Counting Waste – University Landfill	The Hill (2002)
Three municipal landfills in Göteborg, Sweden	11–36 µg/L	SPE-GC/MS		Paxéus (2000)
New Hampshire Survey of 31 Landfills	1–11,000 µg/L	NHDES list		NHDES

Subtitle D, was 118 ppb. The mean concentration of 1,4-dioxane at a construction and demolition landfill was 49 ppb; four other construction and demolition landfills had no detections. The median concentration of 1,4-dioxane in municipal landfill leachate at Subtitle D landfills was 11 ppb, the mean was 118 ppb, and the maximum detected concentration was 323 ppb (USEPA, 2000). At hazardous waste landfills, the median concentration was 466 ppb and the maximum was 7,611 ppb (USEPA, 2000). Landfill leachate concentrations obtained from an ad hoc survey of the literature and regulatory compliance reports are presented in Table 2.18.

In Japanese landfills, 1,4-dioxane was detected in 87.5% of leachate samples analyzed. Concentrations ranged from 1.1 to 109 ppb with a reporting limit of 0.9 ppb (Yasuhara et al., 1997). 1,4-Dioxane was detected in leachates from hazardous wastes landfills in Japan in the range 20.7–1370 μg/L (Yasuhara et al., 1999).

A study of the origins of 1,4-dioxane in leachate at a Japanese landfill accepting only fly ash and bottom ash from two incineration facilities sought to determine what chemical processes during incineration might cause 1,4-dioxane concentrations to be as high as 134 μg/L. Investigators concluded that the thermal treatment of foam plastics, such as incineration of polyurethane foam upholstery material in automobile-shredder wastes, may produce 1,4-dioxane as a by-product (Watanabe et al., 2006). Bench-scale laboratory tests partially demonstrated that this synthesis can occur. A leaching test on the incineration residue of isolated foam plastic waste produced 0.2 ppm 1,4-dioxane, whereas leachate from unincinerated foam waste released no 1,4-dioxane (Fujiwara et al., 2006). These findings were corroborated by a second study that analyzed 38 leachate samples from Japanese municipal solid-waste landfills. The maximum detected concentration of 1,4-dioxane in leachate was 340 ppb. Of the leachate samples from landfills that received incinerator wastes and shredded solid waste mixed with incinerator waste (as opposed to noncombustible waste), 70% had 1,4-dioxane detections. However, just 38% of the leachate samples from landfills receiving only noncombustible waste had 1,4-dioxane detections (Kurata et al., 2006).

REFERENCES

Abe, A., 1997, Determination method for 1,4-dioxane in water samples by solid phase extraction GC/MS. *Journal of Environmental Chemistry* 7: 95–100.

Abe, A., 1999, Distribution of 1,4-dioxane in relation to possible sources in the water environment. *The Science of the Total Environment* 227: 41–47.

Adamson, D., Piña, E.A., Cartwright, A.E., Rauch, S.R., Anderson, R.H., Mohr, T., and Connor, J., 2017, 1,4-dioxane drinking water occurrence data from the third unregulated contaminant monitoring rule. *Science of the Total Environment* 596–597: 236–245.

Agarwal, U.P. and Ralph, S.A., 1997, FT-Raman spectroscopy of wood: Identifying contributions of lignin and carbohydrate polymers in the spectrum of black spruce *(Picea mariana). Applied Spectroscopy* 51(11): 1641–1648.

American Chemical Society, 1929, Market report – October, 1929. *Industrial and Engineering Chemistry* 21(11): 1153–1154, DOI: 10.1021/ie50239a051.

American Chemical Society, 1935, Market report. *Industrial and Engineering Chemistry* 27(6): 733–734, DOI: 10.1021/ie50306a028.

American Cyanamid, 1987, Material Safety Data Sheet for FM 47 identification lacquer, 8% solids, issued January 1, 1987, Wayne, NY. FSC: 8010, LIIN: 00F003660.

Anonymous, 1987, Testergebnisse: Von appretur bis zahnbürste [test results: From glaze to toothbrushes]. *Der Okotest* 2: 129–135.

Anschutz, L. and Broeker, W., 1926, Dioxan als Lösungsmittel, insbesondere bei Molekulargewichts-Bestimmungen [dioxane as a solvent, particularly for determining molecular weights]. *Berichte der Deutschen Chemische Gesellschaft* 59: 2844–2847.

ATSDR, 2004, Draft toxicological profile for 1,4-dioxane. Division of Health Assessment and Consultation, U.S. Department of Health and Human Services.

ATSDR, 2005, Public health implications of 1,4-dioxane-contaminated groundwater at the Durham Meadows Site, Durham, CT. Division of Health Assessment and Consultation, Agency for Toxic Substances and Disease Registry, U.S. Department of Health and Human Services.

Barber, H., 1934, Haemorrhagic nephritis and necrosis of the liver from dioxane poisoning. *Guy's Hospital Reports* 84: 267–280.

BASF, 1986, Datenblatt Dioxan [Dioxane datasheet]. BASF (Badische Anilin- & Soda-Fabrik).

BASF, 2005, Dioxane – Stab Technical Data Sheet. Ludwigshafen, Germany: BASF.

Beernaert, H., Herpol-Borremans, M., and De Cock, F., 1987, Determination of 1,4-dioxane in cosmetic products by headspace gas chromatography. *Belgian Journal of Food Chemistry and Biotechnology* 42(5): 131–135.

Belanger, P.L., 1980, Health hazard evaluation determination, Report No. HE 80-21-721. National Institute for Occupational Safety and Health, U.S. Department of Health and Human Services, Center for Disease Control.

Betts, K.S., 2005, The changing chemistry of office cubicles. *Environmental Science and Technology Online – Technology News* 39(15):319A–320A, June 22, 2005. http://pubs.acs.org/subscribe/journals/esthag-w/2005/jun/tech/kb_cubicles.html (accessed July 5, 2005)

Birkel, T.J., Warner, C.R., and Fazio, T., 1979, Gas chromatographic determination of 1,4-dioxane in polysorbate 60 and polysorbate 80. *Journal of the Association of Official Analytical Chemists* 62(4): 931–935.

Black, R.E., Hurley, F.J., and Havery, D.C., 2001, Occurrence of 1,4-dioxane in cosmetic raw materials and finished cosmetic products. *Journal of AOAC [Association of Official Analytical Chemists] International* 84(3): 666–670.

Bray, G.A., 1960, A simple efficient liquid scintillator for counting aqueous solutions in a liquid scintillation counter. *Analytical Biochemistry* 1: 279–285.

Business Communications Company, 2002, Process makes purer plastics retardant. *Flame Retardancy News* 12(8): 1.

California Air Resources Board (CARB), 2006, "Hot spots" inventory guidelines. California Air Resources Board, California Air Resources Board, Health and Safety Code Section 44300 et seq.

California Environmental Protection Agency, 1997, Assessment of the tolerances for Section 3 registration of Shadeout (rimsulfuron) on tomatoes. Department of Pesticide Regulation.

Castorina, R., Tysman, M., Bradman, A., Hoover, S., Iyer, S., Russell, N., Sultana, D., and R. Maddalena, 2016, Volatile organic compound emissions from markers used in preschools, schools, and homes. *International Journal of Environmental Analytical Chemistry*, 96:13, 1247–1263. November, 2016.

Caulkins, J., Labbe, N., Luth, C., Valleries, P., Wilcox, J., and Thompson, R., 2006, Using computational chemistry to understand effective adsorption strategies for separating contaminants from water. American Society of Environmental Engineers Student Poster Competition, Worcester Polytechnic Institute, March 17, 2006.

Chang, B., 1999, United States Patent 5,912,003: Spray-type insecticidal paint and manufacturing process thereof. Assignee: Kukbo Pharma Co., Ltd., Seoul, Korea.

Chang, S.S., Peterson, R.J., and Ho, C.-T., 1978, Chemical reactions involved in the deep-fat frying of foods. *Journal of the American Oil Chemist's Society* 1: 718–727.

Chartres, C.J., Ringrose-Voase, A.J., and Raupach, M., 1989, A comparison between acetone and dioxane and an explanation of their role in water replacement in undisturbed soil samples. *Journal of Soil Science* 40(4): 849–850.

Chemithon Corporation, 2006, DRS™ dioxane reduction system. http://www.chemithon.com/drs_diox_reduction.html (accessed March 22, 2006).

Chenier, P.J., 2002, *Survey of Industrial Chemistry*. New York: Kluwer Academic Publishing.

Choi, S.H., Kobayashi, A., and Yamanishi, T., 1983, Odor of cooked small shrimp, *Acetes japonicus* kishinouye: Difference between raw material and fermented product. *Agricultural and Biological Chemistry* 47(2): 337–342.

Chung, T.-Y., Hayase, F., and Kato, H., 1983, Volatile components of ripe tomatoes and their juices, purees and pastes. *Agricultural and Biological Chemistry* 47: 343–351.

Cognis Corporation, 2001, Material Safety Data Sheet for Agnique MMF mosquito larvicide & pupicide, issued September 4, 2002, Cincinnati, OH.

Copeland, J., 2004, The question of fuel – Part 4. Oxygenators are the "hottest" topic in fuel chemistry. http://www.foxvalleykart.com/fuel4.html (accessed March, 2004).

Cortellucci, R. and Dietz, E.A., Jr., 1999, Determination of 1,4-dioxane in toluene by gas chromatography. *Journal of High Resolution Chromatography* 22(4): 201–204.

CRC Chemicals, 1987, Material Safety Data Sheet for 5089 Brakleen Aerosol, issued April 10, 1987, Warminster, PA. FSC: 6850, LIIN: 00F024407.

Crown Industrial Products Company, 1989, Material Safety Data Sheet for rust inhibitor 6007, issued February 28, 1989, Hebron, IL. FSC: 6850, LIIN: 00N003634.

Dado, G.P., Knaggs, E.A., and Nepras, M.J., 2006, Sulfonation and sulfation. *Kirk-Othmer Encyclopedia of Chemical Technology Online Edition*. http://mrw.interscience.wiley.com/emrw/9780471238966/home/WileyInterscience (accessed October 3, 2007).

DAP, Inc., 1990, Material Safety Data Sheet for smooth spread contact cement, issued February 22, 1990, Dayton, OH. FSC: 8040, LIIN: 00F017978.

DAP, Inc., 1994, Material Safety Data Sheet for weldwood contact cement, issued January 4, 1994, Dayton, OH. FSC: 8040, LIIN: 00N056641.

Del Mar, R.A., 2001, Leachate and soil gas monitoring at the solid waste landfill. https://www.pnnl.gov/main/publications/external/technical_reports/PNNL-13788.pd

DeLima Associates, 2004, Consumer product information database. https://hpd.nlm.nih.gov/cgi-bin/household/list?tbl=TblBrands&alpha=A (accessed July 2, 2005).

Department of Energy, 2000, Hanford Annual Environmental Report 2000. Hanford, Washington, DC: Pacific Northwest National Laboratory.

DeRosa, C.T., Wilbur, S., Holler, J., Richter, P., and Stevens, Y., 1996, Health evaluation of 1,4-dioxane. *Toxicology and Industrial Health* 12: 43.

DeSoto, Inc., 1991, Material Safety Data Sheet for polyamide resin component 910X746, issued February 21, 1991, Berkeley, CA. FSC: 8010, LIIN: 00N019667.

Diamond, G.L. and Durkin, P.R., 1997, Effects of surfactants on the toxicity of glyphosate, with specific reference to RODEO. U.S. Department of Agriculture Animal and Plant Health Inspection Service Report No. 53-3187-5-12, prepared by Syracuse Research Corporation, Syracuse, NY.

Doherty, R.E., 2000, A history of the production and use of carbon tetrachloride, tetrachloroethylene, trichloroethylene and 1,1,1-trichloroethane in the United States: Part 1 – Historical background; carbon tetrachloride and tetrachloroethylene. *Journal of Environmental Forensics* 1: 11.

Dow Chemical Corporation, 1985, Material Safety Data Sheet for Dursban, issued October 16, 1985, Midland, MI. FSC: 6840, LIIN: 00D001801.

Dow Chemical Corporation, 1989, TSCA 8(e) letter: Analyses of polyethylene glycol for 1,4-dioxane, January 20, 1989. Office of Toxic Substances Report No. 89-890000079, USEPA.

DTSC, 2003. Docket HWCA 20061079 West Contra Costa Sanitary Landfill Inc. Enforcement Order. Department of Toxic Substances Control, California Environmental Protection Agency. https://dtsc.ca.gov/wp-content/uploads/sites/31/2018/01/WCCSL_ENF_CO_EXHIBITA_B.pdf

Eastman Kodak Company, 1998, Material Safety Data Sheet for Kodak film cement, issued June 24, 1998, Rochester, NY. MSDS No. CKRRX.

Eastman Kodak Corporation, 2001, 2000 Kodak Park Environmental Update. Rochester, NY: Eastman Kodak Corporation 8.

Ellis, R.A. and Thomas, J.S., 1998, United States Patent 5,817,910: Destroying 1,4-dioxane in by-product streams formed during polyester synthesis. Assignee: Wellman, Inc., Shewsbury, NJ.

Elmer, P., 2006, The evolution of liquid scintillation counter cocktails and safety. Formerly located at http://www.npl.co.uk/lsuf/2003/the_evolution_of_lsc_cocktails_and_safety.pdf (accessed August 3, 2006).

European Chemicals Bureau, 2002, European Union risk assessment report: 1,4-Dioxane. Luxembourg, European Union, Institute for Health and Consumer Protection, Report 21.

European Medicines Agency, 1998, ICH harmonised tripartite guideline: ICH topic Q 3 C (R3) impurities: Residual solvents. In: *Note for Guidance on Impurities: Residual Solvents*. London, UK: European Medicines Agency. Table 2, Page 6: https://www.ich.org/fileadmin/Public_Web_Site/ICH_Products/Guidelines/Quality/Q3C/Q3C__R6___Step_4.pdf (accessed May 10, 2006).

Ferro Corporation, 2006, Ferro fine chemicals: 1,4-Dioxane. Formerly located at http://www.ferro.com/Our+Products/Fine+Chemicals/Products+and+Markets/1+4-Dioxane/ (accessed November 5, 2006).

Fishbein, L., 1981, Carcinogenicity and mutagenicity of solvents: I. Glycidyl ethers, dioxane, nitroalkanes, dimethylformamide and allyl derivatives. *Science of the Total Environment* 17(2): 97–110.

Fisher Scientific, 2006, Material Safety Data Sheet for OptiDry ACC# 26970, issued June 6, 1999, Fair Lawn, NJ. https://www-s.mechse.uiuc.edu/cleanroom/files/Current%20MSDS/1,4-Dioxane.pdf (accessed October 16, 2006).

Flick, E.W., 1998, *Industrial Solvents Handbook*. Westwood, NJ: Noyes Data Corporation.

Flick, E.W., 2005, *Paint and Ink Formulations Database*. Norwich, NY: William Andrew Publishing. http://www.knovel.com/knovel2/Toc.jsp?BookID=1197&VerticalID=0 (accessed November 5, 2006).

Food and Drug Administration (FDA), 1980, Progress report on the analysis of ethoxylated ingredients for 1,4-dioxane. Division of Cosmetics Technology, Food and Drug Administration.

Food and Drug Administration (FDA), 1981, Progress report on the analysis of cosmetics raw materials and finished cosmetics products for 1,4-dioxane. Division of Cosmetics Technology, Food and Drug Administration 15.

Food and Drug Administration (FDA), 1992, *Cosmetic Handbook*. Center for Food Safety and Applied Nutrition. https://books.google.com/books/about/Cosmetic_handbook.html?id=SrI_ZBK46bwC

Food and Drug Administration (FDA), 1997a, Guidance for industry: Q3C impurities: Residual solvents. Rockville, MD: U.S. Department of Health and Human Services, Food and Drug Administration, Center for Drug Evaluation and Research ICH 16.

Food and Drug Administration (FDA), 1997b, Letter from FDA regarding dioxane in contraceptive sponges. Center for Drug Evaluation and Research 10.

Food and Drug Administration (FDA), 1999, Food additives permitted for direct addition to food for human consumption: Polysorbate 60. *Federal Register* 64: 57974–57976.

Food and Drug Administration (FDA), 2001, Impurities: Residual solvents in new veterinary medicinal products, active substances and excipients. Rockville, MD: Center for Veterinary Medicine, U.S. Food and Drug Administration, https://www.ema.europa.eu/en/documents/scientific-guideline/vich-gl18r-impurities-residual-solvents-new-veterinary-medicinal-products-active-substances_en.pdf

Food and Drug Administration (FDA), 2006, Food additives permitted for direct addition to food for human consumption: Glycerides and polyglycides of hydrogenated vegetable oils. *Federal Regulations* 21:172.736.

Foster, N.C., 1997, *Sulfonation and Sulfation Processes*. Seattle, Washington, DC: Chemithon. www.chemithon.com (accessed November 5, 2006).

Fox, C., 1993, Hair care literature and patent review. *Cosmetics and Toiletries* 108(3): 29–50.

FPPF Chemical Company, Inc., 2006, Material Safety Data Sheet for glyclean recycled antifreeze 6–6. FPPFChemical Company, Inc., issued February 22, 1999, Buffalo, NY. FSC: 6850, NIIN: LIIN: 00N092811.

Fruijtier-Pölloth, C., 2005, Safety assessment on polyethylene glycols (PEGs) and their derivatives as used in cosmetic products. *Toxicology* 214: 1–37.

Fuh, C.B., Lai, M., Tsai, H.Y., and Chang, C.M., 2005, Impurity analysis of 1,4-dioxane in nonionic surfactants and cosmetics using headspace solid-phase microextraction. *Journal of Chromatography A* 1071(1): 141–145.

Fujiwara, T., Fukuda, N., Tamada, T., Ono, Y., Kurata, Y., Ono, Y., Nishimura, F., Ohtoshi, K., 2006, Investigation of the origins of 1,4-dioxane in leachate from municipal solid waste landfills. In: *Proceedings of the 4th Intercontinental Landfill Research Symposium*, June 14–16, 2006, Gällivare, Sweden.

Gelman Sciences, 1989a, Information on the dioxane content of polyethylene glycol, raw materials and consumer products with cover letter dated 011789 [submitted to USEPA under TSCA Section 8E, OTS0516624-1].

Gelman Sciences, 1989b, Reports on 1,4-dioxane and polyethylene glycol in products with cover letter dated 011489 [submitted to USEPA under TSCA Section 8E, OTS0516624-1].

Gelman Sciences, 1989c, Letter from Gelman Sciences to USEPA regarding contamination of 1,4-dioxane at rest stops with attachments [submitted to USEPA under TSCA Section 8E, OTS0516624].

Germinario, G., Garrappa, S., D'Ambrosio, V., Dorothé van der Werf, I., and Sabbatini, L., 2018. Chemical composition of felt-tip pen inks. *Analytical and Bioanalytical Chemistry* 410:1079–1094.

Gierer, J. and Pettersson, I., 1968, On the preparation and characterization of p-dioxanyl hydroperoxide. *Acta Chemica Scandinavica* 22: 3183–3190.

Glidden Paint, 1992, Material Safety Data Sheet for Glid-Thane One polyurethane coating 6124, issued June 1, 1992, Cleveland, OH.

Gonzales, N.R., 2011. Organic Contaminants in Environmental Atmospheres and Waters. Dissertation. Universitat Rovina Virgili, Tarragona, Spain.

Grady, C.P.L., Jr., Sock, S.M., and Cowan, R.M., 1997, Biotreatability kinetics. In: G.S. Sayler, J. Sanseverino, and K.L. Davis (Eds), *Biotechnology in the Sustainable Environment* Vol. 54, p. 14. New York: Plenum Press.

Greek, B.F. and Layman, P.L., 1989, Higher costs spur new detergent formulations. *Chemical and Engineering News* 67(4): 29–38.

Guo, W. and Brodowsky, H., 2000, Determination of the trace 1,4-dioxane. *Microchemical Journal* 64: 174–180.

Hansen, O.C., 2005, Screening for health effects from chemical substances in textile colorants. In: F. Jensen and O.C. Hansen (Eds), *Survey of Chemical Substances in Consumer Products*. Danish Ministry of the Environment, Danish Technological Institute, Report No. 57. http://miljoestyrelsen.dk/homepage/ (accessed October 9, 2006).

Harkov, R., Katz, R., Bozzelli, J., and Kebbekus, B., 1981, Toxic and carcinogenic air pollutants in New Jersey – volatile organic substances. In: J.J. McGovern (Ed.), *Proceedings of the International Technical Conference of Toxic Air Contaminants*. Pittsburgh, PA: Air Pollution Control Association.

Hazardous Substances Data Bank (HSDB), 2006, Online database. National Library of Medicine, National Institutes of Health. http://toxnet.nlm.nih.gov/ (accessed July 2, 2005).

Hovenkamp, S.G. and Munting, J.P., 1970, Formation of diethylene glycol as a side reaction during production of polyethylene terephthalate. *Journal of Polymer Science: Part A-1. Polymer Chemistry* 8(3): 4.

Italia, M.P. and Nunes, M.A., 1991, Gas chromatographic determination of 1,4-dioxane at the parts-per-million level in consumer shampoo products. *Journal of the Society of Cosmetics Chemistry* 42: 97–103.

Jahn, H. and Goetzky, P., 1988, Chemical analysis of epoxides and epoxy resins. In: C.A. May and Y. Tanaka (Eds), *Epoxy Resins Chemistry and Technology*, 2nd Edition, pp. 1049, 1087. New York: Marcel Dekker, Inc.

Jass, H.E., 1989, CTFA today – despite numerous personnel changes, the workload continues heavy. *Cosmetics and Toiletries* 1989(9): 10–12.

Johnson, W., Jr., 2001, Final report on the safety assessment of PEG-25 propylene glycol stearate, PEG-75 propylene glycol stearate, PEG-120 propylene glycol stearate, PEG-10 propylene glycol, PEG-8 propylene glycol cocoate, and PEG-55 propylene glycol oleate. *International Journal of Toxicology* 20(4): 13–26.

Judge, P., 1988, Gelman cites other dioxane sources. *Ann Arbor News* October 13, 1988.

Karges, U., Becker, J., and Püttmann, W., 2018. 1,4-Dioxane pollution at contaminated groundwater sites in western Germany and its distribution within a TCE plume. *Science of the Total Environment* 619–620: 712–720.

Kauffman, J.S., 2003, Trends in impurity analysis: Determination of extractables, leachables, residual solvents, and unknowns by mass spectrometry. *Pharmaceutical Technology North America* 2003: 24–27.

Keel, L. and Franzmann, A., 2000, Downtown Bellingham air toxics screening project: 1995–1999. Northwest Air Pollution Authority (NWAPA).

Kikuchi, M. and Tanaka, H., 2003, United States Patent 6,646,021: Golf ball printing ink, golf ball bearing a mark printed with the ink, and method of manufacturing the golf ball. Assignee: Sumitomo Rubber Industries, Ltd., Kobe, Japan.

Klein, E. and Smith, J.K., 1972, Asymmetric membrane formation: Solubility parameters for solvent selection. *Industrial Engineering and Chemical Production Research and Development* 11(2): 207–212.

Knovel, 2006, *Yaws' Handbook of Antoine Coefficients for Vapor Pressure – Electronic Edition*. http://www.knovel.com (accessed June 2006).

Kosswig, K., 2002, Surfactants. In: W. Gerhartz, Y.S. Yamamoto, and F.T. Campbell (Eds), *Ullmann's Encyclopedia of Industrial Chemistry*. Weinheim, Germany: Wiley-VCH Verlag GmbH & Co. KGaA, DOI: 10.1002/14356007.a25_747.

Kurata, Y., Watanabe, Y., Kawasaki, M., Ono, Y., and Ono, Y., 2006, Occurrence of trace organic compounds in leachates from municipal solid waste landfill sites in Japan. *4th Intercontinental Landfill Research Symposium*. Gällivare, Sweden, Swedish Waste Management Association.

Laursen, S.E., Hansen, J., Bagh, J., Jensen, O., and Werther, I., 1997, Environmental assessment of textiles – life cycle screening of textiles containing cotton, wool, viscose, polyester or acrylic fibres. Danish Environmental Protection Agency, Copenhagen, Denmark. http://www.defra.gov.uk/science/project_data/DocumentLibrary/SD14010/SD14010_3064_FRP.doc (accessed November 5, 2006).

Lide, D.R., 2007, Bond lengths and angles in gas-phase molecules. In: D.R. Lide (ed.), *CRC Handbook of Chemistry and Physics, Internet Version 2007*, 8th Edition. Boca Raton, FL: Taylor & Francis.

Loctite Corporation, 1995, Material Safety Data Sheet for RP 701 LV Anti-Seize 70175, issued January 3, 1995, Rocky Hill, CT. FSC: 8030, NIIN: LIIN: 00N070591.

Lourenço, M.A.-V., 1863, Recherches sur les composés polyatomiques [Research on the polyatomic compounds]. *Annales de Chimie et de Physique* (Paris) 67: 257–339.

Mackay, D., Shiu, W.-Y., and Ma, K.-C., 1993, *Illustrated Handbook of Physical-Chemical Properties and Environmental Fate for Organic Chemicals: Volume III. Volatile Organic Chemicals*. Boca Raton, FL: Lewis Publishers (Taylor & Francis).

Manamsa, K., Crane, E., Stuart, M., Talbot, J., Lapwortha, D., and A. Hart, 2016. A national-scale assessment of micro-organic contaminants in groundwater of England and Wales. *Science of the Total Environment* 568 (2016) 712–726.

Matsumura, S., Shimamoto, S., and Shibata, T., 1997, United States Patent 5,692,527: Tobacco smoke filter materials, fibrous cellulose esters, and production processes. Assignee: Daicel Chemical Industries, Ltd., Himeji, Japan.

Maurer, T., Hass, H., Barnes, I., and Becker, K.H., 1999, Kinetic and product study of the atmospheric photo-oxidation of 1,4-dioxane and its main reaction product ethylene glycol diformate. *Journal of Physical Chemistry A* 103: 8.

Mazurkiewicz, J. and Tomasik, P., 2006, Why 1,4-dioxane is a water-structure breaker. *Journal of Molecular Liquids* 126: 111–116.

McCaffrey, J.W., 1938, United States Patent 2,119,110: Treatment of hydrocarbon oils. Assignee: Universal Oil Products Company, Chicago, IL.

Medical Economics Publishing, 1983, Contraceptive sponge holds up under criticism. *Mark Drug Topics* 127 (August 15), 23.

Merck and Company, Inc., 1983, *The Merck Index*. Rahway, NJ: Merck and Company, Inc.

Miyatake, H. and Saito, K., 1984, Distillation of used liquid scintillation cocktail and the possibility of re-utilization of the recovered solvent. *Radioisotopes* 33(3): 146–149.

Monneyron, P.E., Manero, M.-H., and Foussard, J.-N., 2003, Measurement and modeling of single- and multicomponent adsorption equilibria of VOC on high-silica zeolites. *Environmental Science and Technology* 37(11): 2410–2414.

Montgomery, J.H., 1996, *Groundwater Chemicals Desk Reference*, 2nd Edition. Chelsea, MI: Lewis Publishers.

Moran, C.J., McBratney, A.B., and Ringrose-Voase, A.J., 1986, A method for the dehydration and impregnation of clay soil. *European Journal of Soil Science* 40(3): 569–575.

Moran, C.J., McBratney, A.B., Ringrose-Voase, A.J., and Chartres, C.J., 1989, A method for the dehydration and impregnation of clay soil. *Journal or Soil Science* 40: 569–585.

Moran, J.E., Hudson, G.B., Eaton, G.F., and Leif, R., 2003, Ambient groundwater monitoring and assessment: A contamination vulnerability assessment for the Sacramento area groundwater basin. Lawrence Livermore National Laboratory internal report UCRL-TR-203258.

Mossman, H.W., 1937, The dioxan technic. *Stain Technology* 12(4): 147–156.

National Cancer Institute, 1985, *Monograph on Human Exposure to Chemicals in the Workplace: 1,4-Dioxane*. Bethesda, MD: National Cancer Institute.

National Diagnostics, Inc., 1987a, Material Safety Data Sheet for Bray's solution, issued October 7, 1987, Manville, NJ. FSC: 6810, NIIN: LIIN: 00N067365.

National Diagnostics, Inc., 1987b, Material Safety Data Sheet for Dioxscint LS-171, issued October 9, 1987, Manville, NJ. FSC: 6810, NIIN: LIIN: 00N067374.

NICO2000, Ltd., 2006, Ion selective electrode analysis. http://www.nico2000.net/index.htm (accessed October 28, 2006).

Nishimura, T., Iizuka, S., Kibune, N., and Ando, M., 2004, Study of 1,4-dioxane intake in the total diet using the market-basket method. *Journal of Health Science* 50(1): 101–107.

North Carolina Department of Environment and Natural Resources, 1998, Focus: Waste minimization. *The North Carolina Division of Pollution Prevention and Environmental Assistance Newsletter* 7(3). http://www.p2pays.org (accessed November 5, 2006).

Northeast Waste Management Officials Association, 1998, Metal painting and coating operations. http://www.p2pays.org/ref/01/00777/toc.htm (accessed July 18, 2007).

Old World Automotive Products, Inc., 2001, Material Safety Data Sheet for full force antifreeze and coolant. Northbrook, IL.

Osaka Organic Chemical Industry, Ltd., 2006, History of Osaka Organic Chemical Industry, Ltd. http://www.ooc.co.jp/00english/03company/company_01.html (accessed July 18, 2007).

Paxéus, N., 2000. Organic compounds in municipal landfill leachates. *Water Science & Technology* 42(7):323–332.

Pervo Paint Company, 1993, Material Safety Data Sheet for White R D traffic paint, 5950, issued January 22, 1993, Los Angeles, CA. FSC: 8010, LIIN: 00N065622.

Pesticide Action Network, 2006, Glyphosate fact sheet. Pesticide News.

Petrie, E.M., 2006, *Epoxy Adhesive Formulations*, p. 554. New York: McGraw-Hill.

Philippoff, W., 2006, Weissenberg's contributions to rheology. Karl Weissenberg – The 80th birthday celebration essays. http://innfm.swan.ac.uk/bsr/frontend/home.asp (accessed October 29, 2006).

Poppendieck, D., and Gong, M., 2016, The Influence of Chamber Size on Chemical Emissions from Closed Cell Spray Polyurethane Foam. Indoor Air 2016: The 14th International Conference of Indoor Air Quality and Climate, July 3–8, 2016, Ghent, Belgium. National Institute of Standards and Technology.

Poppendieck, D., and Gong, M., Emmerich, S., 2017, Characterization of Emissions from Spray Polyurethane Foam Final Report to U.S. Consumer Product Safety Commission. NIST Technical Note 1921.

Pro-Systems, 2002, *Under the Scoop: Carburetors and Concepts*. Spring Lake, MI. http://www.pro-system.com/pjames072702.htm (accessed November 25, 2006).

Pulley, D., 1994, Revision of Navy Paint specifications. Proceedings of the Tri-Service Conference on Corrosion, June 21–23, 1994, Orlando, Florida.

Pundlik, M.D., Sitharaman, B., and Kaur, I., 2001, Gas chromatographic determination of 1,4-dioxane at low parts-per-million levels in glycols. *Journal of Chromatographic Science* 39(2): 4.

Ralph, P.A., 1938, Comparative study of some dehydration and clearing agents. *Stain Technology* 13: 9–15.

Rankin, T., 1996, Residual solvents in drugs by static headspace analysis. In: *Pharmaceutical Analysis: GC Technical Report No. TR9115/99.04*. Austin, Texas: ThermoQuest. http://www.thermo.com (accessed November 25, 2006).

Rastogi, S.C., 1990, Headspace analysis of 1,4-dioxane in products containing polyethoxylated surfactants by GC–MS. *Chromatographia* 29(9–10): 441–444.

Reid, E.W. and Hoffman, H.E., 1942, 1,4-Dioxan. *Industrial and Engineering Chemistry* 21(7): 695–697.

Roig, B., Allan, I.J., and Greenwood, R., 2003, A toolbox of existing and emerging methods for water monitoring under the water framework directive. Main uses of the priority compounds (Appendix 1) In: B. Roig, I.J. Allan, and R. Greenwood (Eds), *Water Framework Directive*. St.-Eticnnc, Francc: Ecologic Institute for International and European Environmental Policy. www.ecologic-events.de/swift-germany/documents/Emerging_tools_manual.pdf (accessed November 5, 2006).

Roller, H., Senkpiel, W., Wunsch, J.W., et al., 1977, United States Patent 4,018,967: Magnetic recording tape. Assignee: BASF Aktiengesellschaft, Ludwigshafen, Germany.

Romero, J., Ventura, F., Caixach, J., Rivera, J., Godé, L.X., and Niñerola, J.M., 1998, Identification of 1,3-dioxanes and 1,3-dioxolanes as malodorous compounds at trace levels in river water, groundwater, and tap water. *Environmental Science and Technology* 32(2): 206–216.

Roy, Stephen, 2013, Personal Communication re: Approximate levels of 1,4-dioxane in car wash soaps. October 7, 2013. New Hampshire Department of Environmental Services.

Royal Society of Chemistry, 1992, Union Carbide sells dioxane business to division of Ferro. *American Paint and Coatings Journal* 21: 1.

Rümenapp, J. and Hild, J., 1987, Dioxan in kosmetischen Mitteln [Dioxane in cosmetic preparations]. *Lebensmittelchemie und Gerichtliche Chemie* 41(3): 59–62.

Sachdeva, Y.P. and Gabriel, R.P., 1997, Apparatus for decontaminating a liquid surfactant of dioxane. United States Patent and Trademark Office, Pharm-Eco Laboratories, Lexington, Massachusetts.

Saifer, A. and Hughes, J., 1937, Dioxane as a reagent for qualitative and quantitative determination of small amounts of iodide: Its application to the detection of iodide in iodized salt. *The Journal of Biological Chemistry* 118: 241–245.

Sanceda, N.G., Kurata, T., and Arakawa, N., 1984, Fractionation and identification of volatile compounds in patis, a philippine fish sauce. *Agricultural and Biological Chemistry* 48: 3047–3052.

Scalia, S. and Menegatti, E., 1991, Assay of 1,4 dioxane in commercial cosmetic products by HPLC. *Farmaco* 46(11): 5.

Scalia, S., Testoni, F., Frisina, G., and Guarnerij, M., 1992, Assay of 1,4-dioxane in cosmetic products by solidphase extraction and GC–MS. *Journal of the Society of Cosmetic Chemists* 43: 6.

Schwartz, D., 2019, Environmental group says 65 of 80 household products contain 1,4-dioxane. David M. Schwartz. Newsday, 3/26/2019. https://www.newsday.com/long-island/environment/1-4-dioxane-household-products-1.28968339.

Shah, J.J. and Singh, H.B., 1988, Distribution of volatile organic chemicals in outdoor and indoor air – a national VOCs data base. *Environmental Science and Technology* 22(12): 1381–1388.

Shahidi, F., Rubin, L.J., and D'Souza, L.A., 1986, Meat flavor volatiles: A review of the composition, techniques of analysis, and sensory evaluation. *CRC Critical Reviews in Food Science and Nutrition* 24: 141–252.

Shamberg, J.B., 2002, State of Kansas settles litigation over pollution from KU landfill. http://www.sjblaw.com/headlines/landfill.html (accessed March 2002).

Shearer, T.P. and Hunsicker, L.G., 1980, Rapid method for embedding tissues for electron microscopy using 1,4-dioxane and polybed 812. *Journal of Histochemistry and Cytochemistry* 28(5): 465–467.

Simonich, S.M., Sun, P., Casteel, K., Dyer, S., Wernery, D., Garber, K., Carr, G., and Federle, T., 2013, Probabilistic analysis of risks to us drinking water intakes from 1,4-Dioxane in domestic wastewater treatment plant effluents. *Integrated Environmental Assessment and Management* 9(4):554–559.

Sirchie Finger Print Labs, 1989, Material Safety Data Sheet for lab/roller cleaner, 234T, issued June 8, 1989, Raleigh, NC. FSC: 7930, NIIN: 00N047242.

Skadsen, J.M., Rice, B.L., and Meyering, D.J., 2004, The occurrence and fate of pharmaceuticals, personal care products, and endocrine disrupting compounds in a municipal water use cycle: A case study in the city of Ann Arbor, Michigan. Ann Arbor, MI: Water Utilities and Fleis & VandenBrink Engineering, Inc. http://www.ci.annarbor.mi.us/PublicServices/water/WTP/EndocrineDisruptors.pdf (accessed October 9, 2006).

Smallwood, I.M., 1996, *Handbook of Organic Solvent Properties*. London: Arnold.

Smith, B., and Bristow, V., 1994, Indoor air quality and textiles: An emerging issue. *American Dyestuff Reporter*, 83(1): 37–48.

Spadone, J.C., Takeoka, G., and Liardon, R., 1990, Analytical investigation of Rio off-flavor in green coffee. *Journal of Agricultural and Food Chemistry* 38(1): 226–233.

SRRP, 1991, Paint removal. Source Reduction Research Partnership: Metropolitan Water District of Southern California and the Environmental Defense Fund.

Stanford Research Institute, 1989, *1989 Directory of chemical producers – United States of America*. Menlo Park, CA: SRI International.

Stepan Company, 2006, Test plan for sodium 2-(2-dodecyloxyethoxy) ethyl sulfate. High Production Volume Program, USEPA. http://www.epa.gov/chemrtk/pubs/summaries/sodium22/c16316tc.htm (accessed November 8, 2006).

Stepień, D.K., 2013. Occurrence, distribution and behavior of hydrophilic ethers in the aquatic environment. Dissertation, Johann Wolfgang Goethe Universität, Frankfurt am Main.

Stoye, D., 2005, Solvents. *Ullman's Encyclopedia of Industrial Chemistry*. Weinheim, Germany: Wiley-VCH Verlag GmbH & Co. KGaA.

Stumpf, W., 1956, Chemie und Anwendungen des 1,4-dioxans [Chemistry and applications of 1,4-dioxane]. *Monographien Zu Angewandte Chemie und Chemie-Ingenieur-Technik*. Weinheim/Bergstraβe GMBH: Verlag Chemie, Wiley-VCH.

Sulej, A.M., Polkowska, Z., and Namieśnik, J., 2012, Pollutants in Airport runoff waters. *Critical Reviews in Environmental Science and Technology* 42(16):1691–1734.

Surprenant, K.S., 2005, Dioxane, In: W. Gerhartz, Y.S. Yamamoto, F.T. Campbell, R. Pfefferkorn, and J.F. Rounsaville (Eds), *Ullmann's Encyclopedia of Industrial Chemistry*, pp. 1–6. Weinheim, Germany: Wiley Interscience.

Susquehanna Speedway Park, 2006, 2006 SSP XTREME Stock Car and Women on Wheels (W.O.W.) specification rules. http://www.sspracing.net/docs/2006xtreme.pdf (accessed November 28, 2007).

Talmage, S.S., 1994, *Environmental and Human Safety of Major Surfactants: Alcohol Ethoxylates and Alkylphenol Ethoxylates*. Boca Raton, FL: CRC Press.

Tang, J., Jin, Q.Z., and Shen, G.-H., 1983, Isolation and identification of volatile compounds from fried chicken. *Journal of Agricultural and Food Chemistry* 31: 1287–1292.

The Hill, 2002. UNC [University of North Carolina] Hazardous Waste Dump Ordered Removed Immediately. http://alumni.unc.edu/car/weekly/story.asp?sid=206

Thompson-Hayward Chemical Company, 1993, Material Safety Data Sheet for cement-o-film D-139, issued September 7, 1993, Chicago, IL. FSC: 6850, LIIN: 00N052426.

U.S. Environmental Protection Agency, 1981, Phase I risk assessment of 1,4-dioxane. U.S. Environmental Protection Agency, EPA Contract No. 68-01-6030.

U.S. Environmental Protection Agency (USEPA), 1978, Development Document for Effluent Limitations Guidelines for the Pesticide Chemicals Manufacturing Point Source Category. April, 1978. PB-285 480.

U.S. Environmental Protection Agency (USEPA), 1981, Phase I risk assessment of 1,4-dioxane.

U.S. Environmental Protection Agency (USEPA), 1987, Superfund record of decision: Operating Industries, Inc. Superfund Program, USEPA.

U.S. Environmental Protection Agency (USEPA), 1989a, Inert ingredients in pesticide products: Policy statement; revision and modification of lists. Office of Pesticide Programs, USEPA.

U.S. Environmental Protection Agency (USEPA), 1989b, Memo from Gelman Sciences to USEPA. Office of Pollution Prevention and Toxics, and Toxic Substances Control Act, USEPA. http://yosemite.epa.gov/oppts/epatscat8.nsf/ALLIDS/8FC76CFF09DF768485256F0800556E8A/$FILE/89890000081.pdf?OpenElement (accessed April 2, 2008).

U.S. Environmental Protection Agency (USEPA), 1989c, Preliminary Data Summary for the Paint Formulating Point Source Category. Office of Water Regulation and Standards. EPA 440/1-89/050.

U.S. Environmental Protection Agency (USEPA), 1990, National sewage sludge survey: Availability of information and data and anticipated impacts on proposed regulations. USEPA, *Federal Register* 55: 47210–47283.

U.S. Environmental Protection Agency (USEPA), 1992, Project summary: Indoor air pollutants from household product sources. Las Vegas, NV: USEPA, EPA Contract No. 600/S-49-1025: 4.
U.S. Environmental Protection Agency (USEPA), 1993, Final report on ambient concentration summaries for Clean Air Act Title III hazardous air pollutants. Research Triangle Park, North Carolina: USEPA, EPA Contract No. 600/R94/090.
U.S. Environmental Protection Agency (USEPA), 1995a, Profile of the printing and publishing industry. Office of Enforcement and Compliance Assurance, USEPA.
U.S. Environmental Protection Agency (USEPA), 1995b, Profile of the rubber and plastics industry. Office of Enforcement and Compliance Assurance, USEPA.
U.S. Environmental Protection Agency (USEPA), 1996a, Pollution prevention in the paints and coatings industry. National Risk Management Research Laboratory Office of Research and Development, Center for Environmental Research Information, USEPA.
U.S. Environmental Protection Agency (USEPA), 1996b, Profile and management options for EPA laboratory generated mixed waste. Center for Remediation Technology and Tools; Radiation Protection Division; Office of Radiation and Indoor Air, USEPA. EPA Contract No. 402-R-96-015:120.
U.S. Environmental Protection Agency (USEPA), 1996c, Solvents study. Office of Solid Waste, USEPA.
U.S. Environmental Protection Agency (USEPA), 1997a, Sector notebook data refresh – 1997. Office of Enforcement and Compliance Assurance. USEPA.
U.S. Environmental Protection Agency (USEPA), 1997b, Profile of the plastic resin and manmade fiber industries. Office of Enforcement and Compliance Assurance, USEPA.
U.S. Environmental Protection Agency (USEPA), 1997c, Profile of the pharmaceutical manufacturing industry. Office of Enforcement and Compliance Assurance, USEPA.
USEPA, 2000, Development document for final effluent limitations guidelines and standards for the landfills point source category. Office of Water, EPA-821-R-99-019.
USEPA, 2000, Preliminary data summary: Airport deicing operations. Office of Water, USEPA
U.S. Environmental Protection Agency (USEPA), 2002, 1,4-Dioxane: Proposed acute exposure guideline levels (AEGLs). Office of Pollution Prevention and Toxics, USEPA 76.
USEPA, 2006, Envirofacts data warehouse. http://www.epa.gov/enviro/ (accessed July 22, 2007).
U.S. Environmental Protection Agency (USEPA), 2008, Inventory update reporting. *Non-Confidential Inventory Update Reporting Production Volume Information*. http://www.epa.gov/oppt/iur/tools/data/2002-vol.htm (accessed January 2, 2008).
U.S. Environmental Protection Agency (USEPA), 2017, Unregulated contaminant monitoring rule 3 (UCMR3) occurrence data. Available from: www.epa.gov/dwucmr/occurrence-data-unregulated-contaminant-monitoring-rule#3.
U.S. Environmental Protection Agency (USEPA), 2019, TRI Chemical Factsheet: Chemical-1,4- Dioxane, 000123911, Toxic Release Inventory Program. Available from: https://iaspub.epa.gov/triexplorer/chemical.html?pYear=2017&pLoc=000123911&pParent=TRI&pDataSet=TRIQ1 (accessed May 17, 2019).
U.S. Environmental Protection Agency (USEPA), 2019, TRI Chemical Factsheet: Chemical-1,4- Dioxane, 000123911, Toxic Release Inventory Program. Available from: https://iaspub.epa.gov/triexplorer/chemical.html?pYear=2017&pLoc=000123911&pParent=TRI&pDataSet=TRIQ1 (accessed May 17, 2019).
U.S. Forest Service, 1997, Glyphosate herbicide information profile. U.S. Forest Service.
Union Carbide, 1988, Right-to-know information for polyethylene glycol products. USEPA.
Union Carbide, 1989, Material Safety Data Sheet for UCAR aircraft deicing fluid 2-D, PM-6435, issued November 14, 1989. Bound Brook, NJ. FSC: 6850, NIIN: LIIN: 00F015044.
Union Carbide, 2001, *Polyethylene Glycols Fact Sheet*. Belgium: INEOS Oxides.
University of North Carolina, 2002, UNC hazardous waste dump ordered removed immediately. The Hill Online. http://alumni.unc.edu/car/weekly/story.asp?sid=206 (accessed July 2007).
Valvoline Oil Company, 1991, Material Safety Data Sheet for Mac's Brake and Electric Motor Cleaner 4700, Valvoline Oil Company. LIIN: 00F029811.
Venediktova, R.I. and Rudnyi, N.M., 1965, Study of mass transfer in the extraction of moisture from capillary-porous materials. *Journal of Engineering Physics and Thermophysics* 9(3): 273–274.
Vijgen, J., 2006, The legacy of lindane HCH isomer production – main report: A global overview of residue management, formulation and disposal. Copenhagen, Denmark: International HCH & Pesticides Association. www.ihpa.info/library_access.php (accessed November 28, 2007).
Vinson, C.G. and Martin, J.J., 1963, Heat of vaporization and vapor pressure of 1,4-dioxane. *Journal of Chemical and Engineering Data* 8(1): 74–75.
VROM, 1999, Risk assessment: 1,4-Dioxane. Bilthoven, Netherlands: S.P. Netherlands Ministry of Housing and the Environment (VROM), Chemical Substances Bureau 153.

Wala-Jerzykiewicz, A. and Szymanowski, J., 1998, Headspace gas chromatography analysis of toxic contaminants in ethoxylated alcohols and alkylamines. *Chromatographia* 48(3/4): 299–304.

Washington Post, 1979, A Suspect Chemical Is Found in Cosmetics. July 20, 1979, p. A16.

Watanabe, Y., Kurata, Y., Kawasaki, M., and Ono, Y., 2006, Leaching of the chemical substance from MSW incineration ashes and the shredded residues. *4th Intercontinental Landfill Research Symposium*. Gällivare, Sweden: Swedish Waste Management Association.

Watsco Components, Inc., 1991, Material Safety Data Sheet for 6404 rust eliminator DR-1, issued August 1, 1991, Hialeah, FL. LIIN: 00F025818.

West, F.W., 1959, United States Patent 2,915,416: Ink composition. Assignee: Minnesota Mining and Manufacturing, St. Paul, MN.

Wilke, O., Jann, O., and Brodner, D., 2004, VOC- and SVOC-emissions from adhesives, floor coverings and complete floor structures. *Indoor Air* 14(s8): 98–107.

Winsor, L., 2006, Tissue processing. Formerly located at http://home.primus.com.au/royellis/tp/tp.htm (accessed October 28, 2006).

Wollbrinck, T., 1993, The composition of proprietary paint strippers. *Journal of the American Institute for Conservation* 32(1): 43–57.

Woodcock, J., 1997, Letter to associated pharmacologists and toxicologists, Washington, D.C., and to Empire State Consumer Association, Rochester, NY. Rockville, MD: Center for Drug Evaluation and Research, U.S. Food and Drug Administration.

Wu, J., Xu, Y., Dabros, T., and Hamza, H., 2004, Development of a method for measurement of relative solubility of nonionic surfactants. *Colloids and Surfaces A: Physicochemical and Engineering Aspects* 232: 229–237.

Wurtz, A., 1863, Mémoire sur l'oxyde d'ethyléne et les alcools polyéthyleniques [Memorandum on the ethylene oxide and polyethylene alcohols]. *Annales de Chimie et de Physique* 69: 317–355.

Wypych, G. (Ed.), 2001, *Handbook of Solvents*. Toronto, Ontario: Chemtec Publishing.

Yashushi, J., 1977, United States Patent 4,323,627: Hollow fiber and method of manufacturing the same. Assignee: Nippon Zeon Co., Ltd., Tokyo, Japan.

Yasuhara, A., Shiraishi, H., Nishikawa, M., Yamamoto, T., Nakasugi, O., Okumura, T., Kenmotsu, K., Fukui, H., Nagase, M., and Kawagoshi, Y., 1999, Organic components in leachates from hazardous waste disposal sites. *Waste Management and Research* 17(3): 186–197.

Yasuhara, A., Shiraishi, H., Nishikawa, M., Yamamoto, T., Uehiro, T., Nakasugi, O., Okumura, T., Kenmotsu, K., Fukui, H., Nagase, M., Ono, Y., Kawagoshi, Y., Baba, K., and Noma, Y., 1997, Determination of organic components in leachates from hazardous waste disposal sites in Japan by gas chromatography–mass spectrometry. *Journal of Chromatography A* 774(1–2): 11.

3 Environmental Fate and Transport of Solvent-Stabilizer Compounds

Thomas K.G. Mohr and James Hatton

CONTENTS

Industrial waste discharges to the environment are universally associated with degradation of water quality and risk to human and ecological health. Many chemicals in industrial effluents do not persist because of physical removal, chemical reaction, or biodegradation to less harmful by-products. The factors that differentiate compounds that are recalcitrant and compounds that are assimilated into environmental systems without appreciable consequences are called *contaminant fate and transport properties*. Fate and transport properties describe the propensity for a chemical to volatilize, photo-oxidize, adsorb, hydrolyze, or biodegrade and, if the chemical survives these fates, how it may advect, diffuse, or otherwise migrate through soil, groundwater, surface water, and air. These processes not only eliminate some compounds and retard the movement of others, but also affect the phase in which the chemical is present (gas, liquid, or solid). Some compounds are relatively immune to degradation and will persist in the environment for long periods of time. If released in sufficient mass, those contaminants that are persistent, mobile, and also toxic constitute the greatest potential threat to drinking water resources.

Chapter 1 profiled the industrial fate of solvent-stabilizer compounds. The environmental fate of the stabilizers discussed in Chapter 1 is summarized here in brief descriptions and tables.

In this chapter, the environmental fate and transport of stabilizer compounds is examined by profiling their potential to persist in air, surface water, soil, and groundwater. The focus is on those stabilizer compounds that become concentrated by the industrial processes that precede their discharge. A more detailed review is provided of the fate and transport of 1,4-dioxane, a solvent stabilizer that poses an elevated threat to drinking water resources because of its persistence and mobility.

3.1 FATE AND TRANSPORT PROCESSES

The processes governing the fate and transport of contaminants following their release to the environment are the subject of a number of books and thousands of peer-reviewed papers. Authoritative works on this subject include Freeze and Cherry (1979), Lyman (1990), Dragun (1988), Howard et al. (1991), Chappelle (1993), Mackay et al. (1993), Schwarzenbach et al. (1993), Thibodeaux (1996), Hemond and Fechner (1994), Verschueren (1996), Domenico and Schwarz (1997), Suthersan (2002), Fetter et al. (2017), and many others. This section briefly reviews the key processes that determine the environmental fate and transport properties for 1,4-dioxane and some of the other solvent-stabilizer compounds.

Fate and transport properties of stabilizer compounds are estimated through a variety of approaches where literature citations of measured or estimated values are not available. The leading approach to property estimation is quantitative structure-activity relationships (QSARs); related approaches include quantitative property-property relationships (QPPRs), quantitative structure-property relationships (QSPRs), group contribution models (GCMs), and similarity-based models (Reinhard and Drefahl, 1999). These methods relate chemical structure to chemical behavior. For example, the measured relationship between molecular weight and boiling point for many compounds permits estimation of boiling points for compounds whose boiling points have not been measured. The EPIWIN Suite* from USEPA and Syracuse Research Corporation was used in this

* Estimations Programs Interface for Windows (EPI Suite) (USEPA, 2007a).

chapter to calculate some physicochemical properties for solvent-stabilizer compounds for which experimental values could not be located.

3.1.1 Atmospheric Fate and Transport Processes

Solvent-stabilizer compounds can enter the atmosphere through the following industrial processes:

- Direct vapor emission from vapor degreasers, cold cleaning, dry cleaners, solvent-charged steam cleaners, and other applications in which solvents are vaporized
- Volatilization from open-top vapor degreasers, vented solvent storage vessels such as tanks and drums, waste lagoons or from solvent wastes poured on the ground for disposal, and aeration units in wastewater treatment plants receiving solvent-laden wastewater
- Volatilization from aqueous solutions discharged as condensation water in degreaser or dry cleaner water separators, solvent-laden steam-cleaning wastewater, surface-water bodies to which solvent wastes were discharged, or air-stripping towers used to treat solvent-contaminated groundwater
- Emissions of uncombusted stabilizer compounds from solvent incineration operations at cement kilns and waste incinerators

As discussed in the next section, contaminated soil and groundwater can serve as secondary release sources to the atmosphere.

3.1.2 Volatilization from Dry Soil

Estimating the rate of chemical transfer from soil to air is complicated by the wide range of possible fates for the stabilizer compound when discharged to soil. In addition to volatilizing directly into air, the compound may be subjected to some or all of the following abbreviated list of possible fates:

- Sorption to soil mineral and organic matter surfaces
- Migration under matric suction into the soil
- Penetration into the soil as a liquid and then volatilization into a vapor that sinks or rises depending on vapor density and soil-air pressures
- Dissolution into soil moisture and downward migration to groundwater
- Absorption into plant roots and translocation into plant tissue or transpiration to the atmosphere
- Microbial metabolic processes

An effective strategy for estimating the transfer of a particular chemical from dry soil to air alone is to compartmentalize the system and assess the degree to which partitioning between the two media is expected to occur. This simplification assumes that chemicals in the environment are present in a pure state; however, contaminant releases usually occur as complex mixtures (Kinerson, 1987). This approach forms the basis for broad estimates of volatilization and other contaminant fate and transport processes.

The rate of evaporation of a chemical from dry soil is a function of soil, chemical, and air temperatures, wind speed, air turbulence and surface roughness, air humidity, solar radiation, and the relevant properties of the chemical, including total mass released, molecular weight, vapor pressure, vapor density, and diffusivity in air (Mackay et al., 1993; Thibodeaux, 1996). In particular, vapor pressure describes the concentration of a chemical as its partial pressure in the gas phase in air above a liquid sample, usually measured at 20°C or 25°C and reported in pressure units (e.g., mm Hg). Vapor density is the mass per unit volume of a chemical in the vapor phase at a fixed temperature, usually expressed as a ratio to air density. Vapors heavier than air are reported as a multiple of air density and measured at 25°C (e.g., perchloroethylene, vapor density = 5.7). Diffusion is the average rate of migration of a chemical in air in response to temperature, pressure, and concentration gradients exclusive of any chemical movement in response to advection. The air diffusion constant is sometimes called air diffusivity, often denoted as D_a (expressed in units of cm^2/s). Temperature affects the air diffusion constant, which affects the volatilization rate.

Values for the air diffusion constant, D_a, are not widely available from experimental data. Methods to estimate D_a include empirical equations based on a compound's molecular weight and specific gravity. The following equation has been used to estimate D_a (USEPA, 2001a):

$$D_a = \frac{0.0029(T+273.16)^{3/2}\sqrt{0.034+(1/\text{MW})}\left(1-0.00015\text{MW}^2\right)}{\left[(\text{MW}/2.5\rho)^{1/3}+1.8\right]^2}, \tag{3.1}$$

where D_a is the diffusion coefficient of the chemical in air (in cm^2/s), T is the temperature (in °C), MW is the molecular weight (in g/mol), and ρ is the density (in g/cm^3).

A useful index for contrasting vapor densities of single compounds is relative vapor density (RVD). A compound's RVD is the ratio of the density of dry air saturated with that compound at 25°C and 1 atm total pressure to the density of dry air. The RVD may be calculated by using Equation 3.2 (Pankow and Cherry, 1996):

$$\text{RVD} = \frac{(p^\circ/760)\text{MW}+\left[(760-p^\circ)29.0\right]/760}{29}, \tag{3.2}$$

where p° is the saturated vapor pressure, MW (in g/mol) is the molecular weight, and 29 g/mol is the mean molecular weight of dry air. RVD is dependent on temperature, so the temperature at which the RVD is calculated must also be stated (usually at 25°C). The saturated vapor pressure p° can be obtained from the Antoine equation.* At a compound's boiling point, RVD = MW/29.

Table 3.1 provides vapor pressure values cited in the literature and estimated values of vapor pressure, vapor density, and air diffusion constants for 1,4-dioxane and other commonly used solvent stabilizers likely to be present in vapor degreasing and other solvent wastes. For comparison, data for the major chlorinated solvents are included.

Chemicals on dry soil partition to the vapor phase according to the parameters in Table 3.1. Once in the vapor phase, advection causes vapor to move from soil to turbulent air, and diffusion causes upward movement of vapor into still air (Dragun, 1988). A number of empirical equations are available to estimate the rate of volatilization of a pure chemical from dry soil. The rates at which chemicals diffuse at a given temperature are inversely proportional to the square roots of their molecular weights:

$$\frac{D_{a_1}}{D_{a_2}} = \left(\frac{\text{MW}_1}{\text{MW}_2}\right)^{1/2}. \tag{3.3}$$

An equation for estimating the rate of vapor generation of a pure chemical from dry soil under steady-state conditions was proposed by Shen (1981):

$$E = 2P_V W_A \left(\frac{L_A D_a v}{(\pi f)}\right)^{1/2} \left(\frac{W_c}{W}\right), \tag{3.4}$$

* The Antoine equation describes the relationship between saturated vapor pressure of pure substances and temperature by the relationship $P = 10[A-(B/(C+T))]$, where by historic convention P is the pressure (in mm Hg), T is the temperature (in °C), and A, B, and C are the Antoine equation coefficients available from the National Institute of Standards and Technology (http://webbook.nist.gov/chemistry/).

TABLE 3.1
Properties Affecting Volatilization Rates from Dry Soil for Chlorinated Solvents and Selected Stabilizer Compounds

Compound	Molecular Weight (g/mol)	Vapor Pressure (mm Hg, 25°C)	Vapor Density Relative to Air (g/cm³)	Diffusivity (cm²/s)	Evaporation Rate Relative to Evaporation Rate of Butyl Acetate
Stabilizers					
1,4-Dioxane	**88.1**	**38.09 [1]**	**3.03 [2]**	**0.229 [3]**	**2.7; 2.42 [4]**
1,3-Dioxolane	74.08	70 at 20°C [5] 79 at 20°C [6]	2.6 [7]	*0.146*[a]	0.29 [18]
tert-Butyl alcohol	74.12	40.7 [1]	2.55 [8]	*0.115*[a]	1.05 [19]
sec-Butyl alcohol	74.12	18.3 [1]	2.6 [7]	*0.101*[a]	1.3 [20]
tert-Amyl alcohol	88.15	13.8 [1]	3	*0.102*[a]	0.93 [21]
Epichlorohydrin	92.52	16.4 [1]	3.29 [8]	0.086 [3]	1.35 [6]
Nitroethane	75.08	20.8 [1]	2.58 [8]	*0.165*[a]	1.2 [9]
Nitromethane	61.0	35.8 [1]	2.11 [8]	*0.116*[a]	1.39 [10]
Cyclohexane	84.18	96.9 [11]	2.9 [12]	*0.089*[a]	6.1 [12]
Triethanolamine	149.19	3.59×10^{-6} [1] < 0.01 [13]	5.1 [13]	*0.121*[a]	<1 [13]
1,2-Butylene oxide	72.11	207 [3] 180 [14]	2.2 [7]	0.135 [3]	6.05 [16]
Solvents					
Methyl chloroform	133.42	100 [15]	4.63 [2]	0.078 [3]	4.6 [15]
Dichloromethane	84.93	355 [15]	2.93 [16]	0.101 [3]	7.0 [15]
Trichloroethylene	131.39	60 [15]	4.53 [17]	0.079 [3]	3.0 [15]
Perchloroethylene	165.83	14 [15]	5.7 [17]	0.072 [3]	1.5 [15]

Sources: [1] Daubert and Danner (1985); [2] Verschueren (1996); [3] USEPA (1998a); [4] USEPA (1981); [5] Riddick et al. (1985); [6] Sax and Lewis (1987); [7] NFPA (1991); [8] Sax (1984); [9] Mackison et al. (1981); [10] NLM (2006); [11] Chao et al. (1983); [12] Fisher Scientific (1999); [13] Baker (2006); [14] Osborn and Scott (1980); [15] Dow Chemical (2006); [16] Kirk and Othmer (1982); [17] Budavari (1996); [18] Beaujean (2007); [19] NIOSH (1978); [20] Smallwood (1996); and [21] Cheremisinoff and Archer (2003).

[a] Air diffusion constants displayed in italic font were estimated by using Equation 3.1 at 25°C.

where E is the emission rate (in cm³/s), P_v is the equivalent vapor pressure (in percent), where P_v = vapor pressure (in mm Hg) divided by 760 mm Hg, W_A is the width of the area occupied by the chemical (in cm), L_a is the length of the area occupied by the chemical (in cm), D_a is the diffusion coefficient of the chemical in air (in cm²/s), generally at 25°C, v is the wind speed (in cm/s), W_c/W is the weight fraction of the chemical in soil (in g/g), and f is the correction factor, where $f = (0.985–0.00775P_v)$ and the range of P_v is 0–80%. The above volumetric emission rate can be converted to a mass emission rate:

$$Q = \frac{E(\mathrm{MW})}{G}, \tag{3.5}$$

where Q is the mass emission rate (in g/s), E is the volumetric emission rate (in cm³/s), MW is the molecular weight (in g/mol), and the empirical conversion factor G is 24,860 cm³/mol (Dragun, 1988).

Many other empirical equations are available for estimating rates of volatilization; however, given the many unknown aspects of ground surface conditions, the best use of such equations is to determine the relative volatility and transfer rates among compounds instead of seeking absolute

TABLE 3.2
Calculated Relative Mass Flux Estimates (Chemical Volatilization Rates) of Chlorinated Solvents and Stabilizer Compounds from Dry Soil for a 10-L Spill in a 1.5 m² Area

Compound[a]	Molecular Weight (g/mol)	Vapor Pressure (mm Hg, 25°C)	Density (g/cm³)	10 L Spill Mass (g)	Mass Fraction in Soil[b] (%)	Diffusion Constant[c] D_a	Mass Flux[d] Q(g/s)
Dichloromethane	84.93	355	1.3255	13,255	0.79	0.101	22.0
1,2-Butylene oxide	72.11	207	0.8297	8297	0.50	0.135	9.1
Methyl chloroform	133.42	100	1.3376	13,376	0.80	0.078	7.6
Trichloroethylene	131.39	60	1.4642	14,642	0.88	0.079	5.0
1,3-Dioxolane	74.08	70	1.06	10,600	0.63	*0.146*	4.4
1,4-Dioxane	**88.1**	**38.09**	**1.0329**	**10,329**	**0.62**	**0.229**	**4.3**
Cyclohexane	84.18	96.9	0.779	7790	0.47	*0.089*	3.1
Nitromethane	61	35.8	1.1322	11,322	0.68	*0.116*	1.6
tert-Butyl alcohol	74.12	40.7	0.78581	7858	0.47	*0.115*	1.5
Perchloroethylene	165.83	14	1.6227	16,227	0.97	0.072	1.5
Nitroethane	75.08	20.8	1.0448	10,448	0.63	*0.165*	1.5
Epichlorohydrin	92.52	16.4	1.175	11,750	0.70	0.086	0.8
sec-Butyl alcohol	74.12	18.3	0.8063	8063	0.48	*0.101*	0.6
tert-Amyl alcohol	88.15	13.8	0.8096	8096	0.48	*0.102*	0.5
Triethanolamine	149.19	3.59×10^{-6}	1.1242	11,242	0.67	*0.121*	4×10^{-7}

[a] Nonitalic compounds are stabilizers. Italic compounds are chlorinated solvents; these are listed for comparison to show whether the stabilizer is likely to evaporate with the solvent or remain behind in the soil where it can be leached down to groundwater.
[b] Mass fraction = grams of compound per gram of soil, if soil bulk density is assumed to be 1.85 g/cm³.
[c] Air diffusion constants shown in italic font were estimated by using Equation 3.1, with temperature set to 25°C.
[d] Mass flux was calculated by using the Shen (1981) equation (Equation 3.4 in the text).

transfer rates for an individual compound. Several equations for estimating rates of volatilization from soil are found in Dragun (1988). Calculated estimates of the relative mass flux of chlorinated solvents and stabilizer compounds from dry soil to air are provided in Table 3.2.

The Shen estimate of relative volatilization rates from dry soil isolates one mode of transfer, but discharge of a single chemical to dry soil is an uncommon occurrence. Generally, wastes are discharged as complex mixtures of chlorinated solvents with concentrated stabilizers, oil, grease, soldering flux, acids, and water. Soil-moisture plays an important role in the fate and transport of chemicals discharged to soil. Mixtures of wastes will partition from moist soil to air according to their solubilities, vapor pressures, chemical stability, and capacity to adsorb to soil minerals and organic matter. As discussed in the next section, the interaction of the chemical at the water-air interface plays an important role in fate and transport processes.

3.1.3 Volatilization from Water

Several possible fate and transport mechanisms act on a chemical discharged to wet soil, groundwater, or surface water. The propensity for a chemical to partition between aqueous and vapor phases is a well-studied phenomenon, as Henry's law mathematically describes.

3.1.3.1 Henry's Law

The constant describing the equilibrium concentration of a compound in both the aqueous and vapor phases at the liquid-air interface at a fixed temperature is called the Henry's law constant.

The general principle is that the higher that constant, the greater the potential for the compound to volatilize from the water surface. The Henry's law constant is a function of molecular structure and air and water temperatures (Goss, 2006).

The Henry's law constant, H, is usually defined at a particular temperature and at equilibrium as

$$H = \frac{C_a}{C_w}, \tag{3.6}$$

where C_a is the concentration of a compound in air and C_w is the concentration of that compound in water. H is conventionally defined in terms of gas concentrations (in atm) and liquid concentrations (in mol/m^3); therefore, typical units for H are (atm · m^3)/mol. H is usually reported from measurements made at 25°C or 20°C and 1 atm pressure. It is important to take note of the temperature at which H was measured to ensure that values for different compounds can be compared.

The dimensionless form of the Henry's law constant, noted as H_C, is obtained by converting gas concentrations from partial pressures in atmospheres to moles per cubic meter. The ideal gas law relates pressure, volume, temperature, and number of moles:

$$\frac{n}{V} = \frac{P}{RT}, \tag{3.7}$$

$$H_C = \frac{H}{RT}, \tag{3.8}$$

where R is the universal gas constant and is equal to 8.2×10^{-5} (m^3 atm)/(mol K); T is the temperature (in K). Note that the dimensionless H_C actually has the units of concentration of the gas (in mol/m^3) per concentration of the liquid (in mol/m^3); the units can also be volume (in m^3 of liquid) per volume (in m^3 of gas).

The Henry's law constant is often stated as the simple ratio of a compound's vapor pressure to its aqueous solubility. A limitation to this interpretation of Henry's law is the inherent assumption that water does not dissolve into the compound. The vapor pressure of the pure compound is used in this statement of Henry's law, but the solubility used is for the compound when saturated with water. The Henry's law constant estimates that are calculated from the compound's vapor pressure and solubility fail where the solubility of water in the chemical exceeds a few percent (Corsi, 1998). An exhaustive compilation of measured Henry's law constants is found in Sander (1999) and Sander (2015).

The Henry's law constant describes the tendency of a compound to escape the liquid phase and move into the gaseous phase. The term for the potential to transfer from one medium to another is *fugacity,* or escaping tendency, and is a function of a compound's activity coefficient (Arbuckle, 1983). Compounds such as dichloromethane with high vapor pressures (i.e., with high fugacity) and high aqueous-phase activity coefficients will tend to partition to the gas phase to equalize the chemical potentials for the compound in water and air; thus, such compounds have high values of H. Conversely, compounds such as alcohols with low vapor pressures and high solubilities (i.e., low activity coefficients in water) will have low values of H and will tend to partition into the aqueous phase (Schwarzenbach et al., 1993).

Where the ionic strength of a solution increases, as may occur when polluted water is discharged to marine waters, a compound's solubility in the saline solution will be lower, while its vapor pressure remains constant, leading to higher values of H and increased rates of volatilization. This phenomenon, called electrostriction or "salting out," is leveraged in laboratory analysis, as described in Chapter 4. Table 3.3 provides Henry's law constants in two forms for solvent-stabilizer and chlorinated solvent compounds.

TABLE 3.3
Henry's Law Constants for Stabilizer and Chlorinated Solvent Compounds

Compound[a]	*H* [(atm m³)/mol]	*HC* (dimensionless)[b]	References
Cyclohexane	1.50×10^{-1}	6.14	Bocek (1976)
Perchloroethylene	1.77×10^{-2}	7.24×10^{-1}	Gossett (1987)
Trichloroethylene	9.85×10^{-3}	4.03×10^{-1}	Leighton and Calo (1981)
Methyl chloroform	8.00×10^{-3}	3.27×10^{-1}	Dilling et al. (1975); Lyman (1990)
Dichloromethane	3.25×10^{-3}	1.33×10^{-1}	Leighton and Calo (1981)
tert-Amyl alcohol	7.30×10^{-4}	2.99×10^{-2}	Butler et al. (1935)
1,2-Butylene oxide	1.80×10^{-4}	7.37×10^{-3}	Bogyo et al. (1980)
Nitroethane	4.76×10^{-5}	1.95×10^{-3}	Gaffney et al. (1987)
Epichlorohydrin	3.00×10^{-5}	1.23×10^{-3}	Lyman (1990)
Nitromethane	2.59×10^{-5}	1.06×10^{-3}	Lyman (1990)
1,3-Dioxolane	2.40×10^{-5}	9.82×10^{-4}	Hine and Mookerjee (1975)
sec-Butyl alcohol	9.06×10^{-6}	3.71×10^{-4}	Snider and Dawson (1985)
tert-Butyl alcohol	9.05×10^{-6}	3.70×10^{-4}	Altschuh et al. (1999)
1,4-Dioxane	$\mathbf{4.80 \times 10^{-6}}$	$\mathbf{1.96 \times 10^{-4}}$	Park et al. (1987)
1,4-Dioxane	$\mathbf{4.29 \times 10^{-6}}$	$\mathbf{1.75 \times 10^{-4}}$	Hiatt (2013) ***preferred value***
Triethanolamine	1.00×10^{-7}	4.09×10^{-6}	Hine and Mookerjee (1975)

[a] Nonitalic compounds are stabilizers. Italic compounds are chlorinated solvents; these are listed for comparison to show whether the stabilizer is likely to evaporate with the solvent or remain behind in the soil where it can be leached down to groundwater.
[b] Dimensionless values of Henry's law constants calculated from literature values by using Equation 3.8.

3.1.3.2 Published Measurements of Henry's Law Constants for 1,4-Dioxane

The 2015 compendium of Henry's law constants compiled by Rolf Sander of the Max Planck Institute for Chemistry in Mainz, Germany, lists published Henry's law constants for hundreds of compounds, including miscible compounds (Sander, 2015). The published experimental Henry's law constants for miscible compounds show a range of results, which suggests that careful consideration should be given to the context, methods, and precision by which Henry's law constants were measured. In general, experimental values are found to over-predict the actual volatilization rates measured under environmental conditions (Mackay and Shiu, 1981). Precision in volatility measurements is easier to achieve with the hydrophobic and more volatile chlorinated solvents. However, even among replicate measurements of Henry's law constant for hydrophobic, highly volatile compounds by different researchers using the same methods (EPICS – equilibrium partitioning in closed space), there was little concordance. The measured Henry's law constants of hydrophobic, highly volatile compounds in distilled water in four different studies are listed in Table 3.4 (Yurteri et al., 1987).

The low agreement among published values for Henry's law constants for hydrophobic, highly volatile compounds shown in Table 3.4 should decrease one's confidence in the published values for Henry's law constants for low volatility compounds, because the experiments to measure Henry's law constant in fully miscible compounds are challenged by a number of factors. These include:

- Molar strengths of solution must be increased to high levels to produce measurable levels in the gas phase (e.g., 0.5 mol/m³ for alcohols and ketones, compared to 0.01–0.1 mol/m³ for chlorinated solvents);
- Experiments must be run up to 5 times longer (e.g., 50 hours) for hydrophilic compounds as compared to hydrophobic compounds; and
- Analytical difficulties (in the 1980s) due to low stripping efficiencies for compounds at room temperature (Mackay and Yeun, 1983).

TABLE 3.4
Reproducibility of Henry's Law Constant Measurements in Four Studies Using the EPICS Method for Seven Highly Volatile VOCs in Distilled Water at 20°C

Compound	HLC 1	HLC 2	HLC 3	HLC 4	Average	Range	Maximum % Difference
Carbon Tetrachloride	0.936	1.264	0.971	1.202	1.093	0.328	26
1,1,1-trichlorethane	0.645	0.715	0.552	0.712	0.656	0.163	45
1,1-dichloroethylene	3.089	1.085	0.862	1.054	1.523	2.23	209
cis-1,2-dichloroethylene	0.181	0.182	0.123	0.186	0.168	0.063	88
trans-1,2-dichloroethylene	0.375	0.390	0.302	0.386	0.363	0.088	70
Trichloroethylene	0.430	0.398	0.299	0.417	0.386	0.131	70
Tetrachloroethylene	0.535	0.540	0.549	0.695	0.58	0.16	47

Henry's Law Constants (HLCs) listed are dimensionless

Source: Yurteri, C., Ryan, D.F., Callow, J.J., and Gurol, M.D., 1987. The effect of chemical composition of water on Henry's law constant. *Journal of the Water Pollution Control Federation* 59(11): 950–956.

The Sander compilation (2015) lists five experimental values for the 1,4-dioxane Henry's law constant. The most recent measurement comes from USEPA's Las Vegas laboratory in 2013, while the other four come from studies published in 1973, 1979, 1987, and 1992. These studies were all published in the field of analytical chemistry for the primary purpose of improving estimates of partitioning coefficients of standards used for resolving retention and selectivity of chromatographic columns used in gas chromatography. The 2013 study by Hiatt stated its objective was to leverage vacuum distillation technology to improve the precision of Henry's law constant measurements. The five publications on experimental determinations of the Henry's law constant for 1,4-dioxane are listed in Table 3.5.

The wide range in experimental 1,4-dioxane Henry's law constants among published studies raises the question of which value is best, and why most regulatory vapor intrusion (VI) guidance documents and other lists of fate and transport properties are inclined to cite Park et al. (1987). Inspection of the studies yields the following observations:

1. Until the Hiatt (2013) study by USEPA's National Exposure Research Laboratory was published, no experimental Henry's law constant papers for 1,4-dioxane had been published for 20 years.
2. The four papers published in 1973, 1979, 1987, and 1992 all use gas chromatography with flame-ionizing detectors (GC-FID), which is reliable for compound identification, but is less robust than GC-MS for compound quantitation. The 2013 study by Hiatt uses GC-MS and internal standards (1,4-dioxane-d8), and is therefore considerably more reliable.
3. The range of concentrations for the infinite dilution criterion of Henry's law to be valid should be <0.1% (Tipler, 2016). However, the concentration range in Park et al. (1987) was 0.56–1.9%.
4. Kolb et al. (1992) and Friant and Suffett (1979) published Henry's law constants at higher than standard temperatures (40°C and 30°C). Kolb provides a temperature dependency relationship and reports good accuracy; however, the measurements were made at concentrations up to 1.7%, which exceeds the infinite dilution assumption of Henry's law.

Among the five available studies, the most recent study by Hiatt is the most reliable, in view of its use of internal standards for analysis, even though it suffers from low recovery due to 1,4-dioxane's very low partitioning rate from water to air (even when vacuum distillation is used). Hiatt also

TABLE 3.5
Experimental Values for 1,4-Dioxane Henry's Law Constants

Study	Method	Concentration (mg/L)	Temperature (°C)	Henry's Law Constant	Error
Rohrschneider, 1973	Headspace analysis by GC-FID	2,256	25	5,396 L:G **4.54×10^{-6} atm-m³/mol**	< ±15%
Friant & Suffett, 1979	Headspace analysis by GC-FID Dioxane detection limit = 0.74 mg/L	93.3	*30*	0.000278 G:L **6.92×10^{-6} atm-m³/mol**	±19.4%
Park et al., 1987	Headspace analysis by GC-FID	5,590–18,600	25	5096 L:G **4.8×10^{-6} atm-m³/mol**	<2%
Kolb et al., 1992	Headspace analysis by GC-FID	680–17,070	25	3486 L:G (extrapolated)* **7.05×10^{-6} atm-m³/mol**	±0.7%*
Hiatt, 2013	Headspace + vacuum extraction + GC-MS	1	25	232.1 mol/kg-bar **4.29×10^{-6} atm-m³/mol**	19%

* *The estimated error for the Kolb et al. study was not reported for 1,4-dioxane; the listed value is for methyl ethyl ketone. The Henry's law constant listed for the Kolb study at 25°C was extrapolated from measurements at higher temperatures using the 1,4-dioxane specific temperature correction equation derived in the study.*

evaluated the Henry's law constant at several temperatures, and reports constants for 1,4-dioxane and 90 other compounds for the van't Hoff temperature correction equation.

3.1.3.3 Mass Transfer Rates from Water to Air: Flux Density

Henry's law values can be used to estimate the rate of mass transfer, or flux density, from a surface-water body by evaluating the volatility of an aqueous solution in isolation from other factors. The air-water interface can be idealized as a thin layer of static air above the water surface, above which turbulent air flows, and a thin layer of stagnant water beneath the water surface, below which water flows. In "thin-film theory," the thickness of the static air layer above the water surface is approximated as 1 mm, whereas the stagnant water layer is taken to be 0.1 mm thick (Schwarzenbach et al., 1993). Molecular diffusion is the dominant transport mechanism in the thin films above and below the water surface, whereas turbulent diffusion is active in the fluids above and below the thin films. Equation 3.9 provides an expression for flux density (Hemond and Fechner, 1994):

$$J = -\left[\frac{1}{\delta_w / D_w + \delta_a / (D_a \cdot H)}\right]\left[C_a - \frac{C_a}{H}\right], \tag{3.9}$$

where J is the flux density(M/L²T), δ_w is the thickness of the hypothetical thin water layer (L), D_w is the molecular diffusion coefficient for the chemical in water (L^2/T), δ_a is the thickness of the hypothetical thin air layer (L), D_a is the molecular diffusion coefficient for the chemical in air (L^2/T), H is the Henry's law constant (dimensionless), C_w is the chemical concentration in water (M/L^3), C_a is the chemical concentration in air (M/L^3), and L, M, and T represent any consistently applied units of length, mass, and time.

Because wind speed affects the thickness of the air film above water and water flow or circulation affects the thickness of the water film below the interface, this flux density equation should only be used for rough approximations or to compare relative mass flux rates. The air film is 10 or more times thicker than the water film, and the molecular diffusion coefficient for a compound in air is about 10,000 times higher than its molecular diffusion coefficient in water. If the air concentration of the compound of interest is zero, then the flux density is directly proportional to the magnitude of the compound's Henry's law constant and its concentration in water (Thomas, 1990). Table 3.3 can therefore be interpreted to provide an approximation of the relative magnitude of mass transfer from aqueous solutions to air for fixed concentrations. Comparison of relative mass transfer rates inferred from ratios of the Henry's law constants for chlorinated solvents and their stabilizers suggests that water-to-air partitioning after a waste is discharged to a surface-water body may favor the ether, epoxide, nitroalkane, and alcohol stabilizers remaining in water, whereas the solvents and cyclohexane volatilize to the atmosphere. The Henry's law constant is also a key parameter for designing groundwater and wastewater treatment systems where air stripping is viable. Chapter 8 discusses the unsuitability of air stripping for remediation of 1,4-dioxane.

A more tangible means of relating the Henry's law constant to the environmental fate of stabilizers discharged to surface water is to estimate their rates of volatilization from rivers and lakes. Models are available to predict the volatilization half-life of a compound in water. Modeled half-lives for stabilizer and chlorinated solvent compounds are profiled in Table 3.6. The model used was Estimation Programs Interface Suite, EPI Suite™ (a trademark of ImageWare Systems®, Inc.), which predicts physical-chemical properties of compounds entered by their Chemical Abstracts Service Registry Number or by their SMILES notation.* The model within EPI Suite™ for estimation of volatilization from water, WVOLNT, or Water Volatilization Program is based on the method outlined in *Handbook of Chemical Property Estimation Methods* (Thomas, 1990).

The accuracy of the properties estimated by EPI Suite™ and the associated estimates of chemical phase partitioning depends on the chemical's class, the quality of the available chemical data used by the model as a training set, and whether the chemical's properties fall within the range of the 353 chemicals in the training data set. In general, EPI Suite™ is able to predict the measured property value within an order of magnitude of measured values, provided the chemical is appropriate for the types of regression and other estimation techniques used (USEPA, 2007a).

3.1.4 Vapor Intrusion

VI is the migration of vapor-forming chemicals, including volatile organic compounds (VOCs) and some semi-volatile organic compounds (SVOCs), from contaminated soil or groundwater sources to indoor air in overlying buildings. Assessment of the VI pathway and potential inhalation risk is challenging due to the number of factors that influence subsurface, atmospheric, and building processes. To counter this complexity, VI assessment often relies on multiple lines of evidence to determine whether the pathway is complete and significant (i.e., whether indoor air concentrations of chemicals of concern are due to VI and are at or above risk-based thresholds).

VI guidance and related regulations generally define volatile chemicals as those with Henry's law constants greater than 1×10^{-5} atmosphere-meter cubed per mole (atm m^3 mol^{-1}) or vapor pressures greater than 1 mm mercury (mm Hg) at room temperature (USEPA, 1991, 2015, 2017). Common VI chemicals of potential concern, including trichloroethylene (TCE), tetrachloroethylene, benzene, toluene, ethylbenzene, and xylene (BTEX), meet both volatility criteria. Some solvent stabilizer compounds also meet these criteria, including acetonitrile, nitromethane, and cyclohexane, cyclohexene, and pyridine. Conversely, 1,4-dioxane, methanol, *sec*-butyl alcohol, and isobutyl alcohol

* SMILES is "Simplified Molecular Input Line Entry System," a notation system for describing molecular structure. See Reinhard and Drefahl (1999) for a brief introduction.

TABLE 3.6
Modeled Volatilization Half-Lives and Volatilization Residence Times for Solvent-Stabilizer Compounds in a Model River and a Model Lake

Chemical Name[a]	Estimated Volatilization from Model River (Half-Life)	Time for 99% Removal from Model River	Estimated Volatilization from Model Lake (Half-Life)	Time for 99% Removal from Model Lake
Cyclohexane[b]	0.9 h	6.3 h	3.6 days	25 days
Isopentane	0.9 h	6.3 h	3.4 days	24 days
Isoprene	0.9 h	6.3 h	3.3 days	23 days
Diisobutylene	1.1 h	7.7 h	4.2 days	29 days
Dichloromethane[b]	1.1 h	7.7 h	3.7 days	26 days
Methyl chloroform[b]	1.2 h	8.4 h	4.6 days	32 days
Trichloroethylene	1.2 h	8.4 h	4.6 days	32 days
Carbon tetrachloride	1.3 h	9.1 h	4.9 days	34 days
Perchloroethylene	1.4 h	9.8 h	5.1 days	36 days
1,2-Butylene oxide[b]	3.6 h	1.1 days	4.6 days	32 days
Cyclohexene oxide	4.5 h	1.3 days	5.5 days	39 days
Triethylamine	5 h	1.5 days	5.8 days	41 days
Ethyl acetate[b]	5.1 h	1.5 days	5.6 days	39 days
Isopropyl acetate[b]	6.2 h	1.8 days	5 days	35 days
Diisopropylamine	7.2 h	2.1 days	6.8 days	48 days
Propylene oxide[b]	7.2 h	2.1 days	5.9 days	41 days
Acetonitrile	12 h	3.5 days	7.5 days	53 days
2-Butanone[b]	16 h	4.7 days	7.3 days	51 days
Nitromethane	17 h	5.0 days	10 days	70 days
Nitroethane	19 h	5.5 days	8.3 days	58 days
Epichlorohydrin	20 h	5.8 days	12 days	84 days
1,3-Dioxolane	1.4 days	9.8 days	13 days	91 days
Styrene oxide	1.7 days	12 days	23 days	161 days
Nonylphenol	1.8 days	13 days	18 days	126 days
tert-Amyl alcohol	2.6 days	18 days	22 days	154 days
Methanol	3.1 days	22 days	35 days	245 days
Pyridine	3.1 days	22 days	25 days	175 days
sec-Butyl alcohol[b]	3.6 days	25 days	29 days	203 days
tert-Butyl alcohol[b]	3.6 days	25 days	29 days	203 days
1,4-Dioxane[b]	**4.8 days**	**34 days**	**56 days**	**1.1 years**
Butoxymethyl oxirane	6.4 days	45 days	74 days	1.4 years
2-Methyl-3-butyn-2-ol	8.7 days	61 days	66 days	1.3 years
2,6-di-*tert*-Butyl-p-cresol	16 days	112 days	123 days	2.4 years
p-tert-Amyl phenol	23 days	161 days	173 days	3.3 years
Diepoxybutane	28 days	196 days	303 days	5.8 years
Thymol	62 days	1.2 years	1.9 years	13.3 years
Phenol	71 days	1.4 years	2.1 years	15 years

Notes: The times predict only volatility; biodegradation, photolysis, sorption, hydrolysis, etc. are not addressed. Half-lives shown are exclusive of other fate processes such as photodegradation, hydrolysis, and biodegradation. Model assumptions: river = 1 m deep, flow is 1 m/s, wind speed is 5.0 m/s; lake = 1 m deep, wind velocity is 0.5 m/s, and circulation or current is 0.05 m/s. Modeled using EPI Suite™ version 3.20 (USEPA, 2007a).

[a] Nonitalic compounds are stabilizers. Italic compounds are chlorinated solvents; these are listed for comparison to show whether the stabilizer is likely to evaporate with the solvent or remain behind in the soil where it can be leached down to groundwater.

[b] Chemical is included in the EPI Suite™ training set.

only meet the vapor pressure volatility criteria, meaning their Henry's law constants are lower and, thus, their tendency to stay in water is higher.

The 1,4-dioxane VI potential is of interest due to its persistence and mobility in the subsurface, vapor pressure greater than 1 mm Hg, and the availability of inhalation toxicity data (see Chapter 5). The USEPA vapor intrusion screening levels (VISLs) include screening concentration values for 1,4-dioxane in groundwater, soil gas, and indoor air. As states update their VI guidance to be more aligned with USEPA's final VI guidance (2015), new screening levels for 1,4-dioxane and other solvent stabilizer compounds are expected. For example, Michigan Department of Environmental Quality (MDEQ) issued an emergency rule on VI of 1,4-dioxane in October 2016. The rule was initiated due to concerns of impacted groundwater intrusion into subgrade structures (e.g., cellars, basements) and set a groundwater VI screening level at 29 μg/L, which is over three orders of magnitude lower than the USEPA 1,4-dioxane groundwater VISL. MDEQ updated their groundwater VI screening level for 1,4-dioxane to 1,900 μg/L in April 2017 perhaps after consideration of the importance of Henry's law and the tendency of 1,4-dioxane to stay in soil moisture or groundwater.

Despite the prevalence of potential subsurface and indoor sources of 1,4-dioxane, there is limited information on the occurrence of 1,4-dioxane in indoor air due to VI. This may be related to the analytical challenges associated with 1,4-dioxane or because it has not been viewed as a potential VI concern until recently. In recent presentations on the topic (Holton et al., 2017; Bell et al., 2019), practitioners have noted the limited data on this subject and the likelihood that other comingled compounds are more likely to drive the VI risk unless certain conditions are met (e.g., relatively dry soils, shallow vadose zone source, etc.).

IS 1,4-DIOXANE A POTENTIAL INDOOR AIR CONCERN?

The Michigan Department of Environmental Quality ("MDEQ") proposed R 299.27 ("Rule 27") that focuses on the potential for miscible or hydrophilic compounds (i.e., compounds that are soluble in water in all proportions) to volatilize and complete the VI pathway. It is the authors' view that regulation of hydrophilic compounds for the VI pathway, particularly in groundwater, is not scientifically supported. Several salient points contribute to this view, including the following:

- Definition of Volatility: MDEQ relied on USEPA's 2015 definition of VOCs: compounds with Henry's law constants greater than 1×10^{-5} atm-m^3 mole^{-1} OR a vapor pressure greater than 1 mm Hg. This definition is inappropriate for hydrophilic compounds, for which low Henry's law constants and miscibility with water exert far more control over volatility than do pure phase vapor pressures. The vapor pressure of hydrophilic compounds is only relevant in the environmental setting where water, including soil moisture, is absent, which is rare. A better definition of volatility is a compound with a Henry's law constant greater than 1×10^{-5} atm-m^3/mole^{-1} AND a vapor pressure greater than 1 mm Hg.
- Temperature Dependency of Properties Governing Vapor Partitioning: MDEQ has improved its method for adjusting Henry's law constants measured at 25°C to the system temperature of 10°C. However, several other factors (e.g., air and water diffusivity, water evaporation rate) are also temperature dependent, but are not adjusted for the system temperature. The method could be further improved for estimating the Henry's law constants for 1,4-dioxane and other contaminants if these factors were corrected to the proper system temperature.
- "Potential for Evaporation from Flooded Basements" Vapor Intrusion Scenario: This scenario poses limited risk of unacceptable indoor air exposures to 1,4-dioxane

(*Continued*)

because: (a) as a hydrophilic compound, 1,4-dioxane will remain dissolved in water and recede back to the groundwater table, or it will be pumped/vacuumed out and not evaporate; (b) to the extent such evaporation occurs, the acceptable indoor air levels for 1,4-dioxane are based on chronic exposure (30 years, 350 days/year), and the potential for evaporation from infrequent and brief flooding events would not create the chronic exposure that could result in unacceptable indoor air exposures.

- Soil Gas Detections of 1,4-Dioxane: Rarely is 1,4-dioxane detected in soil gas, and when it is, there is little evidence for concluding that soil gas detections arose from volatilization from groundwater. Among the few datasets reviewed, the more likely source of the soil gas detections is uninvestigated sources of 1,4-dioxane that likely contributed to soil contamination. The soil gas detections alone do not demonstrate that 1,4-dioxane is sufficiently volatile in groundwater to complete the VI pathway.
- Challenges with Measuring Vapor Intrusion of 1,4-Dioxane: There is potential ambiguity and confusion that could result from measurement of 1,4-dioxane in indoor air from interior sources, including household uses of products containing 1,4-dioxane and emissions from architectural coatings. For example, residential exposure to airborne 1,4-dioxane from household uses of laundry detergent, dishwashing detergent, or shampoos is likely orders of magnitude higher than the negligible, if measurable, concentrations of 1,4-dioxane that could volatilize from groundwater into a residence.

3.1.5 Atmospheric Fate of Stabilizer Compounds

Chemical reactions in the atmosphere may occur in the vapor phase as gas-phase collisions between molecules, on the surfaces of airborne solid particulate matter, and in aqueous solution in water droplets. Reactions in water droplets are predominately of acid-base type. Reactions on particle surfaces are of minor importance because of their short residence time in the atmosphere. Gas-phase reactions are the dominant chemical reaction that transforms chemicals released to the atmosphere (Seinfeld, 1986).

Physical reactions—including condensation and precipitation, dissolution into water droplets followed by rain, and reactions with sunlight—can also remove, transform, or eliminate stabilizer compounds emitted to the atmosphere. The longevity of a chemical in the atmosphere is referred to as its atmospheric half-life, that is, the length of time before half the mass of the emitted compound is removed, transformed, or eliminated by chemical or physical reactions. Most studies of atmospheric half-lives for chemicals involve laboratory bench-top testing of compounds reacting with simulated sunlight or atmospheric oxidants such as hydroxyl radicals, nitrate, ozone, and chlorine. Modeling studies are also used to predict the atmospheric fate of compounds subjected to chemical and physical reactions.

The most important atmospheric reactions affecting stabilizer compounds are photolysis, in which ultraviolet (UV) light reacts directly with the compound to break it down, and photo-oxidation, a reaction with hydroxyl radicals, chlorine, and other oxidants in the atmosphere. These reactions are described in more detail in the following sections.

3.1.5.1 Photolysis

Photolysis occurs when a chemical absorbs light energy and undergoes chemical transformation. Light is said to excite electrons and destabilize bonds in some compounds. Absorption of light does not result in degradation in the majority of compounds; instead, absorbed energy is released through fluorescence, and the compounds return to their beginning energy state (Hemond and Fechner, 1994). Photolysis, also called direct photodegradation, is more likely to occur in compounds that have double carbon bonds, such as alkenes (e.g., perchloroethylene and TCE) and aromatic rings (e.g., benzene and toluene). Shorter-wavelength light has higher frequency and energy and is therefore the primary agent in photolysis. Visible light in the wavelength range from 280 to 730 nm (nanometers) is primarily responsible for photolysis of chemicals in the upper atmosphere (Seinfeld, 1986). Photolysis of some compounds may occur in both the atmosphere and in surface-water bodies; however, because light in the lower atmosphere has longer wavelengths, many compounds are not directly photolyzed.

The intensity of sunlight passing through the atmosphere is decreased through absorption by ozone and other atmospheric gases and by molecular and aerosol scattering. Essentially no light is transmitted to the Earth's surface atmosphere (the troposphere) at wavelengths less than 295 nm, and there is a sharp decrease in intensity in the 280–320 nm wavelength range due mainly to ozone absorption. Sunlight in this wavelength interval, often called UV-B radiation, causes sunburn as well as direct photolysis of many pollutants residing in the upper layers of surface-water bodies (Zepp and Cline, 1977).

3.1.5.1.1 Photolysis of 1,4-Dioxane

1,4-Dioxane is photolyzed at wavelengths shorter than the wavelengths of light penetrating the troposphere, that is, less than 290 nm; therefore, direct photolysis is inconsequential to the atmospheric fate of 1,4-dioxane. Nevertheless, photolysis experiments hold interest for developing treatment technologies involving UV light (see Chapter 8).

Laboratory experiments on photolysis of 1,4-dioxane produced volatile solid by-products, including ρ-formaldehyde and trioxane. Proposed pathways (Hentz and Parrish, 1971) for direct photolysis of 1,4-dioxane vapor at 1470 Å (147 nm) include

$$C_4H_8O_2 \xrightarrow{hv} C_2H_4 + 2CH_2O, \tag{3.10}$$

$$C_4H_8O_2 \xrightarrow{hv} C_2H_4 + CH_2O + H_2 + CO, \tag{3.11}$$

$$C_4H_8O_2 \xrightarrow{hv} H_2 + [\text{Unidentified product}]. \tag{3.12}$$

1,4-Dioxane absorbs UV light in the wavenumber range* 52,200–60,510 cm^{-1} (165–191 nm), producing an absorption curve with a single peak, which may be the superposition of absorption associated with the two symmetrical oxygen atoms on the ether ring (Pickett et al., 1951). A more detailed pathway for direct photolysis of 1,4-dioxane exposed to light at 190 nm for 200 hours was elicited in later studies, as shown in Figure 3.1.

The hydrogen abstraction from 1,4-dioxane ([B] in Figure 3.1) is disputed; other researchers find the primary mechanism for 1,4-dioxane photolysis to be scission of the CO bond rather than cleavage of the CH bond (Houser and Sibbio, 1977). Photolysis experiments on liquid 1,4-dioxane conducted at 185 nm produced "an exceedingly complex mixture" including six major components and 50 minor components. The products included ethylene glycol, glycolaldehyde, dioxanone, hydroxymethyldioxane, and a variety of ethers, alcohols, aldehydes, and some carbonyl compounds (Houser and Sibbio, 1977). Vapor-phase photolysis of 1,4-dioxane at 147 nm produced formaldehyde and ethylene (Hentz and Parrish, 1971).

Some experiments have sought to replicate the light range to which 1,4-dioxane would be exposed in the near-surface atmosphere. One experiment simulated UV light with a short-wavelength cutoff of 290 nm and an intensity 2.6 times greater than the noonday sun found in Freeport, Texas (Dilling et al., 1976). Half the mass of 1,4-dioxane disappeared in 3.4 hours, giving a corrected half-life of

* Wavenumber (in cm^{-1}) is the inverse of wavelength.

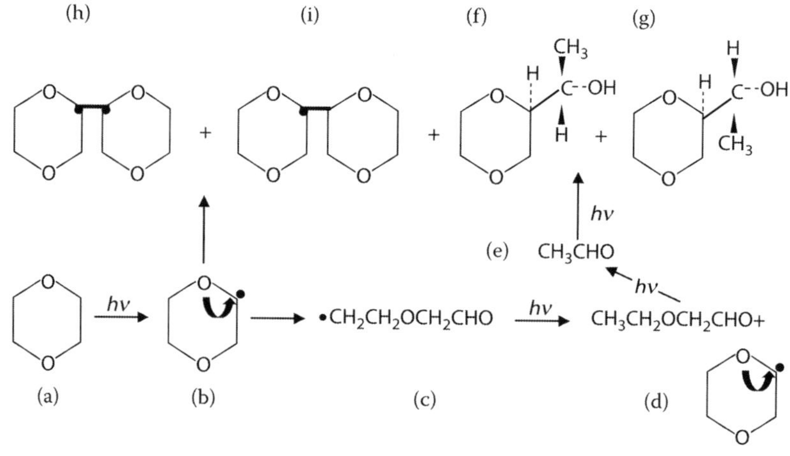

FIGURE 3.1 Laboratory photolysis of 1,4-dioxane [a] produces a dioxyl radical [b] by abstracting a hydrogen from dioxane (indicated by the arrow and dot in [b]). A pair of dioxane dimers ([h] and [i]) is formed by the dimerization of dioxyl radicals. The dioxyl radical also undergoes bond cleavage to open the ring and form radical [c], which then forms ethoxyacetaldehyde [d] and, with further light exposure, acetaldehyde [e]. Acetaldehyde can be photo-reduced to the alcohol isomers [f] and [g]. *Note:* These reactions are not expected to occur in the natural environment. (From Mazzocchi, P.H. and Bowen, M.W., 1975, Photolysis of dioxane. *Journal of Organic Chemistry* 40(18): 2689–2690. With permission.)

8.8 hours to account for the artificially enhanced light intensity. However, this experiment was conducted in the presence of 5 ppm nitrous oxide, an oxidant involved in photo-oxidation.

3.1.5.1.2 Photolysis in Surface Water

The degree to which light penetrates surface water governs the degree to which chemicals in solution are photolyzed. The rates of photolytic reactions in a water body are affected by the solar spectral irradiance at the water surface, radiative transfer of light from air into water, and the transmission of sunlight in the water body.

Indirect or "sensitized" photolysis occurs in surface water when light-absorbing natural organic compounds absorb photons. These compounds, called chromophores, can transfer energy in the form of electrons or hydrogen ions to another compound. Titanium dioxide and other inorganic compounds can also serve as an intermediate in the indirect photolysis of organic contaminants (Hemond and Fechner, 1994). The degree to which contaminants in surface water may be susceptible to photolysis or indirect photodegradation from chromophores is limited by the depth of light penetration in the surface-water body. Light penetration in water is governed by light intensity,* the angle of light entry, turbidity, and dissolved organic matter in water. Light intensity at the water surface decreases with the angle of sunlight entering the water; therefore, intensity decreases from midday to sunset, from summer to winter, and from the tropics to higher latitudes, particularly in the UV-B range (280–320 nm), but also for visible light (UV-A, 320–400 nm) (Zepp and Cline, 1977). Intensity also varies by seasonal changes in atmospheric ozone content and cloud cover. Less than 10% of light from the sun or scattered light from the sky arriving at the water surface is reflected.

Light intensity diminishes with water depth because of scattering and absorption by dissolved humic substances such as humic and fulvic acids and by suspended sediment. The depth range in which photolysis is an active agent for chemical degradation may be on the order of a few centimeters in highly turbid water to several meters in water with greater clarity and lower levels of dissolved organic matter. In the ocean, light absorption is due primarily to water itself. Water is most transparent in the blue region, and scattering is relatively independent of wavelength; hence, solar

* Light intensity = the number of photons per unit surface area per unit time.

radiation in clear ocean water acquires a blue hue at great depths. Sunlight penetrates deeper into ocean waters than into inland surface waters, where absorption is due mainly to dissolved organic matter (Zepp and Cline, 1977; Hemond and Fechner, 1994).

A compound in solution in a surface-water body will undergo photolysis if its molecules can absorb UV light in the wavelength range penetrating water and if the bond energy is lower than the energy of the absorbed light. 1,4-Dioxane and other ethers are weak absorbers of UV light in the wavelength range that persists through the troposphere and penetrates surface waters (Houser and Sibbio, 1977; Hill et al., 1997). Therefore, direct photolysis of 1,4-dioxane in surface water is not expected to occur. In addition to ethers, the alcohol, amine, and nitrile stabilizer compounds also do not absorb light in the solar spectral region, that is, between 290 and 700 nm (Harris, 1990a).

3.1.5.2 Photo-Oxidation in the Atmosphere

Photo-oxidation is the reaction of an atmospheric or aqueous compound with hydroxyl radicals and other oxidants. Hydroxyl radicals, $OH^{\bullet}$, are highly reactive and can degrade a wide range of organic contaminants (Edney and Corse, 1987; Atkinson, 1988).* They are produced from the reaction of sunlight with water vapor in the atmosphere or in surface water. In air, hydroxyl radicals may form from four related reactions:

$$H_2O \xrightarrow{hv} H^+ + OH^{\bullet}, \tag{3.13}$$

$$H^+H_2O \xrightarrow{hv} H_2 + OH^{\bullet}, \tag{3.14}$$

$$H^+O_2 \xrightarrow{hv} OH^{\bullet} + O, \tag{3.15}$$

$$O + H_2O \xrightarrow{hv} 2OH^{\bullet}. \tag{3.16}$$

In reaction 3.13, the hydroxyl radical is most commonly produced by breaking the O-H bond of the water molecule to form a hydrogen atom (H^+) and a hydroxyl radical ($OH^{\bullet}$). In reaction 3.14, the resulting hydrogen atom then reacts with another water molecule to form a hydrogen molecule (H_2) and a second hydroxyl radical. Alternatively, via reaction 3.15, the hydrogen atom reacts with an oxygen molecule (O_2) to form a second hydroxyl radical and an oxygen atom. The new oxygen atom can then react with another water molecule via reaction 3.16 to form two new hydroxyl radicals. Hydroxyl radical formation by these reactions supplies a concentration of about 5–10 million hydroxyl radicals per cubic centimeter of air at ground level (Seinfeld, 1986).

Photons impinging on hydrogen peroxide can cause it to dimerize into two hydroxyl radicals, $OH^{\bullet}$. Hydroxyl radicals constitute the most important catalyst for atmospheric reactions. The potential for photodegradation of stabilizer compounds likely to be released to the atmosphere is summarized in Table 3.7.

Hydroxyl radicals are produced in the presence of sunlight. At night, there is a 10-fold decrease in the concentration of hydroxyl radicals. Atmospheric residence times exceeding 12 hours may result in wet deposition of soluble airborne contaminants at night, leading to a low-concentration distribution to surface water and groundwater. Rain also washes out soluble and particulate airborne contaminants. Numerous studies have been conducted to elicit the photo-oxidation pathway for 1,4-dioxane (Maurino et al., 1997; Platz et al., 1997; Geiger et al., 1999; Li and

* In the notation $OH^{\bullet}$, the dot represents a free electron; hv represents photon energy, where h is Planck's constant, $h = 6.626 \times 10^{-34}$ J s $= 4.136 \times 10^{-15}$ eV s, and v is the light frequency (in Hz).

TABLE 3.7
Calculated and Experimental Atmospheric Half-Lives of Stabilizer Compounds

	Photo-Oxidation in Presence of 5 ppm NO		Photo-Oxidation	
Compound	$t_{1/2}$	**Reference**	$t_{1/2}$	**Reference**
n-Methyl pyrrole	0.4 h	1	—	
2,4,4-Trimethyl-1-pentene	1.6 h	1	—	
Cyclohexene	2.3 h	1	—	
Triethanolamine	—		4 h	2
sec-Butyl alcohol	17 h	1	—	
1,4-Dioxane	**8.8 h**	**1**	**22.4 h**	**3**
			1–2 days	4
Nitromethane	24 h	1	100 days	5
1,3-Dioxolane	—		26.5 h	6
Cyclohexane	18 h	1	45 h	7, 8
tert-Amyl alcohol	—		3.3 days	7, 9
1,2-Butylene oxide	—		8.4 days	9
tert-Butyl alcohol	90 h	1	14 days	9
Epichlorohydrin	42 h	1	45 days	9
Nitroethane	—		107 days	9

Sources: [1] Dilling et al. (1976); [2] Dow Chemical (1980); [3] Maurer et al. (1999); [4] Platz et al. (1997); [5] Nielsen et al. (1989); [6] Buxton et al. (1988); [7] Bidleman (1988); [8] Chao et al. (1983); and [9] Meylan and Howard (1993).

Notes: Values cited in the literature from the Dilling et al. (1976) study often mistakenly cite the 1,4-dioxane half-life as 3.4 hours. The Dilling study used light intensity much higher than sunlight; measured half-lives are multiplied by 2.6 to estimate the half-life under bright sunlight conditions. Values from the Dilling study were measured in the presence of 5 ppm of nitrous oxide (NO); therefore, photo-oxidation is likely the dominant process for values from Dilling et al.

Pirasteh, 2006). The pathway by which 1,4-dioxane reacts with hydroxyl radicals is summarized in Figure 3.2.

In the photo-oxidation pathway for 1,4-dioxane (mapped in Figure 3.2), one of the eight hydrogen atoms on the 1,4-dioxane molecule [a] reacts with $OH^\bullet$ followed by addition of oxygen to form a 1,4-dioxyl radical [b]. The dioxyl radical in turn converts NO to NO_2, producing an alkoxyl radical [c]. The alkoxyl radical is rapidly converted to the $HC(O)O(CH_2)_2OCH_2$ radical [d] via a very fast ring-opening reaction. Addition of oxygen forms another peroxyl radical $HC(O)O(CH_2)_2OCH_2O_2$ [e]. This peroxyl radical will convert NO to NO_2, forming $HC(O)O(CH_2)_2OCH_2O$ [f], which reacts with molecular oxygen to produce ethylene-1,2-diformate (EDF—also called ethylene glycol diformate) [g]. EDF will degrade from successive reactions with hydroxyl radicals, as delineated by Maurer et al. (1999).

Atmospheric oxidation of 1,4-dioxane in the presence of nitrogen oxide (NO) produces EDF with an approximately 100% molar yield. If the annual average tropospheric $OH^\bullet$ concentration of 1×10^6 molecules per cubic centimeter is used, the total residence times of 1,4-dioxane and EDF in the atmosphere will be 22.4 hours and 24 days, respectively. EDF can therefore be expected to travel far from its source; however, because EDF is highly soluble in water, removal by wet precipitation or "rain-out" will shorten its atmospheric lifetime in wet climatic regions (Maurer et al., 1999).

Although the $OH^\bullet$ concentration in the atmosphere varies with location, time of day, season, and meteorological conditions, a reasonable 24-hour global average atmospheric lifetime for 1,4-dioxane

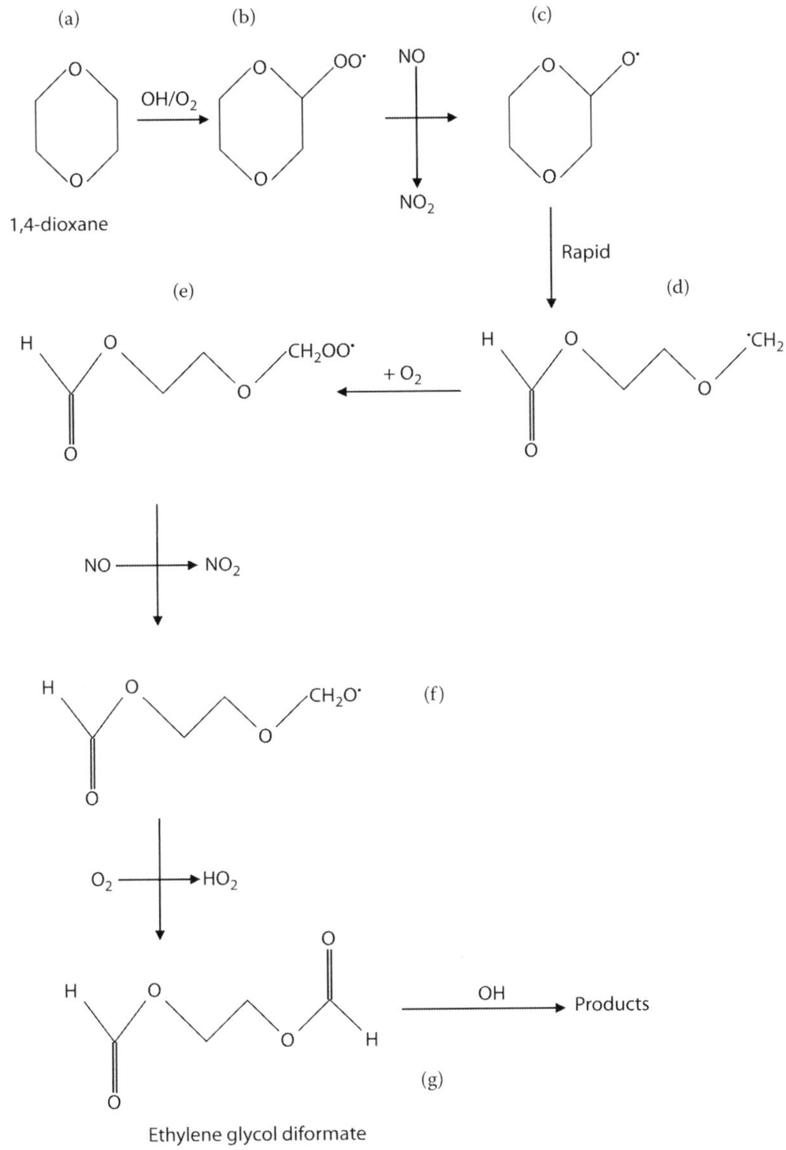

FIGURE 3.2 Photo-oxidation pathway for 1,4-dioxane in the presence of NO_x. Note: The product of step (f), ethylene glycol diformate, is also called ethylene-1,2-diformate (EDF). (From Geiger, H., Maurer, T., and Becker, K.H., 1999, OH-initiated degradation mechanism of 1,4-dioxane in the presence of NO_x. *Chemical Physics Letters* 314: 465–471. With permission.)

is 1–2 days (Platz et al., 1997). 1,3-Dioxolane follows a similar hydroxyl radical oxidation pathway in which ethylene carbonate and methylene glycol diformate are produced (Freitas-Dinis et al., 2001).

Atomic chlorine is another important atmospheric oxidant capable of oxidizing 1,4-dioxane. Chlorine atoms are introduced into the atmosphere through reactions of HCl with OH and by photolysis of chlorine molecules from industrial emissions and sea salt. The chlorine atoms play a significant role in removal of VOCs from the atmosphere in coastal regions where marine boundary layers occur (Li and Pirasteh, 2006). The reaction of atomic chlorine and 1,4-dioxane follows the general form

$$Cl + C_4H_8O_2 \rightarrow HCl + C_4H_7O_2\cdot \quad (3.17)$$

TABLE 3.8
Ranges of Measured and Calculated Aqueous Photo-Oxidation Half-Lives for Solvent-Stabilizer Compounds

	Aqueous Photo-Oxidation, $t_{1/2}$		
Compound	**High**	**Low**	**References**
1,4-Dioxane	**9.1 years**	**67 days**	Anbar and Neta (1967), Dorfman and Adams (1973)
sec-Butyl alcohol	23 years	129 days	Anbar and Neta (1967), Dorfman and Adams (1973)
1,2-Butylene oxide	55 years	1.4 years	Howard et al. (1991)
tert-Butyl alcohol	64,500 years	2.1 years	Anbar and Neta (1967), Dorfman and Adams (1973)
Ethyl acetate	110 years	2.75 years	Dorfman and Adams (1973), Howard et al. (1991)
Pyridine	24.4 years	14.7 years	Dorfman and Adams (1973), Howard et al. (1991)
Acetonitrile	12,560 years	314 years	Dorfman and Adams (1973), Howard et al. (1991)
Methyl ethyl ketone	81.4 years	48.8 years	Anbar and Neta (1967), Howard et al. (1991)
Cyclohexane	7,800,000 years	160,000 years	Howard et al. (1991)

The product of the above reaction, $C_4H_7O_2$, is a dioxane alkyl peroxy radical, which degrades to EDF, a toxic by-product with an atmospheric residency of 24 days (Maurer et al., 1999).

3.1.5.3 Photo-Oxidation in Water

Hydroxyl radicals are present at concentrations on the order of 10^{-17} mol/L in illuminated surface waters (Hemond and Fechner, 1994), but can be much higher in direct, intense sunlight, e.g., 6.7×10^{-15} to 4.0×10^{-12} mol/L (Allen et al., 1996). Hydroxyl radicals form when an organic chromophore absorbs light and reacts with water to form hydrogen peroxide (H_2O_2). The formation and accumulation of H_2O_2 in natural waters are correlated with the concentration of naturally occurring humic substances (Cooper et al., 1988). Hydrogen peroxide results from a reaction with superoxide, O^-, which forms when oxygen is reduced by free electrons generated by light-absorbing substances through photoionization or energy transfer. Superoxide can also form by electron transfer from reduced metals such as titanium dioxide (Cooper et al., 1988). As a source of hydroxyl radicals, hydrogen peroxide plays a significant role in photo-oxidation of contaminants in natural surface waters, and it may react directly with pollutants. The measured hourly photochemical accumulation rate of H_2O_2 in surface water exposed to midday sunlight ranges from 2.7×10^{-7} to 48×10^{-7} mol/L in waters with dissolved organic carbon ranging from 0.53 to 18 mg/L, respectively. The measurements were made at 24.3° northern latitude, where the midday sunlight intensity is 0.4 W/m^2 in the wavelength range 295–385 nm (Cooper et al., 1988). H_2O_2 will break into two hydroxyl radicals ($OH^\bullet$) if it absorbs a sufficiently energetic photon.

The aqueous photo-oxidation half-life of 1,4-dioxane in water has been estimated to be as low as 67 days and as high as 9.1 years (Dorfman and Adams, 1973; Howard et al., 1991). Estimated photo-oxidation half-lives for additional stabilizer compounds and chlorinated solvents are tabulated in Table 3.8.

3.2 SURFACE-WATER FATE AND TRANSPORT PROCESSES

Solvent wastes discharged to surface water enter a dynamic environment dominated by moving water, biological processes, sedimentation, sunlight, and a variety of chemical reactions. Although discharge of solvent wastes to surface water has been effectively eliminated with the adoption and successful implementation and enforcement of regulations to protect streams, rivers, and lakes, the discharge of 1,4-dioxane to surface water continues today. Untreated 1,4-dioxane may enter surface-water bodies as the effluent of groundwater treatment systems designed to remove chlorinated solvents such as air strippers and granular activated carbon (GAC) treatment vessels, which are not effective at removing 1,4-dioxane. Other stabilizer compounds that are not effectively removed by conventional pump-and-treat technologies employed at solvent release sites may also be discharged to surface-water bodies. Therefore, the surface-water fate and transport processes for solvent-stabilizer compounds may warrant evaluation for managing the remediation of chlorinated solvent sites. In addition, 1,4-dioxane is present in municipal wastewater from surfactants used in household detergents, shampoos, and soaps, and is not removed by conventional sewage treatment plants before discharge to rivers.

The transport processes governing contaminants in surface water are advection, diffusion, and sorption to suspended sediment. Diffusion, wind-induced currents, and thermal stratification are important in lakes, whereas advective flow governed by gravity and water velocity is important in streams and rivers. Advection of low to moderate concentrations of dissolved organic compounds generally proceeds independently of chemical properties; however, diffusion is limited by compound-specific properties. Surface-water transport processes are well studied and will not be described further; a concise summary is given in Hemond and Fechner (1994), and a detailed, analytical treatment is given in Thibodeaux (1996). The potential for photolysis and photo-oxidation of stabilizer compounds and their relative potential rates of volatilization from surface-water bodies is discussed in the previous sections of this chapter. This section focuses on the physical and chemical fate of stabilizer compounds in surface-water bodies. Most of these processes are common to the groundwater fate and transport of stabilizer compounds; adsorption and biological processes are discussed in Section 3.3.

3.2.1 Hydrolysis

Hydrolysis is a form of nucleophilic substitution reaction in which water is the nucleophile (Harris, 1990b; Hemond and Fechner, 1994). More specifically, hydrolysis is a chemical reaction in which an organic molecule, RX,* reacts with hydroxide from the ionization of a water molecule, forming a new carbon—oxygen bond and cleaving a carbon—X bond in the organic molecule, producing H^+X^-. Both the organic compound and the water molecule are split in a hydrolysis reaction, which is essentially a displacement of X by a hydroxyl group (–OH):

$$RX + H_2O \rightarrow ROH + X^- + H^+ \cdot \tag{3.18}$$

Hydrolysis acts on organic molecules that possess electrophiles, that is, electron-seeking atoms such as carbon and phosphorous. The nucleophile (water or hydroxide) attacks the electrophile (e.g., H^+) and displaces the functional group that leaves the molecule, often chloride or phenoxide. Some groups of chemicals are resistant to hydrolysis, including many of the solvent-stabilizer compounds, such as ethers, alcohols, alkanes, alkenes, aromatics, and heterocyclic compounds. Compounds susceptible to hydrolysis include amines, amides, epoxides, nitriles, esters, and several other compounds not found among the stabilizer or chlorinated solvent compounds (Harris, 1990b). There are no hydrolyzable functional groups on the 1,4-dioxane molecule; therefore, 1,4-dioxane is not expected to hydrolyze significantly (NICNAS, 1998).

Hydrolysis is a first-order reaction whose rate varies directly with the concentration of the organic species, RX, as shown:

$$\frac{d[RX]}{dt} = -k[RX], \tag{3.19}$$

* In the RX notation, R usually represents a hydrocarbon, such as an alkyl group, and X represents an organic functional group, such as an ester, amine, or aldehyde.

where k is the first-order hydrolysis rate constant in units of inverse time (in s^{-1}), giving a disappearance rate of moles per liter per second. The first-order rate constant describes the loss of RX at any concentration of RX and at constant pH, pressure, and temperature. At any time t, the concentration of RX, $[RX]_t$, is given by

$$[RX]_t = [RX]_0 \, e^{-kt}, \tag{3.20}$$

where $[RX]_0$ is the starting concentration of RX, t is time in seconds, k is the first-order rate constant, and e is the natural logarithm (USEPA, 1998a; Bennett, 2004a). The time needed to decrease the concentration by one-half, that is, the half-life ($t_{1/2}$), is

$$t_{1/2} = \frac{\ln 2}{k} = \frac{0.693}{k}, \tag{3.21}$$

The hydrolysis reaction described above is the unimolecular form. A bimolecular form of the hydrolysis reaction involves kinetics that depend on both the concentration of the organic molecule RX and the nucleophile, H_2O, H^+, and OH^-. Hydrolysis reactions are therefore sensitive to the pH of the water, so that a difference of 2 pH units would cause the reaction rate to change 100-fold. Depending on whether hydroxide (OH^-) or hydrogen (H^+) is the dominant species at a given pH, bimolecular hydrolysis reactions are characterized as base-catalyzed hydrolysis or acid-catalyzed hydrolysis (Hemond and Fechner, 1994). Catalysis by microbial or other agents may further complicate the kinetics of hydrolysis; therefore, citations of hydrolysis rate constants from the literature or from predictive modeling must be used with caution, as natural systems are likely to be a great deal more complex than the system measured or modeled.

Equations and software to estimate hydrolysis rate constants for those compounds susceptible to acid-catalyzed or base-catalyzed hydrolysis are freely available (USEPA, 2000a). The HYDROWIN module in the EPIWIN Suite only estimates aqueous hydrolysis rate constants for esters, carbamates, epoxides, halomethanes, and selected alkyl halides (USEPA, 2000a). HYDROWIN does not estimate neutral hydrolysis rate constants, which may be the dominant hydrolysis rate for some epoxide compounds, leading to underestimation of the rate constant. Molecular structure features are used to estimate the acid- and base-catalyzed rate constants, which are then used to calculate hydrolysis half-lives (Mill et al., 1987; USEPA, 2000a). Table 3.9 summarizes estimates for those solvent-stabilizer compounds susceptible to hydrolysis. Literature values for two chlorinated ethanes are included to indicate their resistance to hydrolysis. Ethers are generally resistant to hydrolysis; HYDROWIN does not estimate hydrolysis rate constants for ether compounds, including 1,4-dioxane.

3.2.2 Acid Dissociation Potential

Organic compounds that are prone to form ions by dissociating when dissolved in water will behave differently when in pure form. Ionization of an organic compound will affect its solubility and adsorption. This section reviews the potential for stabilizer compounds' participation in acid-base interactions in aqueous solution.

The primary chemical attribute describing a compound's potential to participate in acid-base chemical reactions is its acid dissociation constant, K_a, the equilibrium constant for the reaction

$$HA + H_2O \rightarrow H_3O^+A^-, \text{(j)}$$

TABLE 3.9
Estimated Surface-Water Hydrolysis Half-Lives and Residency Times (Time for >99% Removal by Hydrolysis Acting Alone) for Chlorinated Solvents and Solvent Stabilizers Prone to Hydrolysis in Water

Chemical Name	Hydrolysis Half-Lives (Literature Values)	References	Hydrolysis Half-Lives (Estimated)	Residency Time
Solvents				
Dichloromethane	260,000 days	5	Very long	Persistent
Methyl chloroform	2.9 years at 15°C	4	Very long	Persistent
Stabilizers				
Epichlorohydrin	8.2 days in distilled water	3	350 years	Persistent
Diepoxybutane[a]	—		107 years	Persistent
Glycidol[a]	—		63 years	Persistent
Butoxymethyl oxirane[a]	—		62 years	Persistent
1,2-Butylene oxide[a]	11.6 days; converts to ethylene glycol	1	3 years	21 years
Propylene oxide[a]	11.6 days	1	2.9 years	20 years
Isopropyl acetate	2.4 years at pH 7; 88 days at pH 8	2	2.4 years	17 years
Ethyl acetate	2 years at pH 7 and 25°C	3	1.82 years	13 years
Cyclohexene oxide[a]	—		37.5 days	260 days
Styrene oxide[a]	—		9 days	63 days

Sources: [1] Bogyo et al. (1980) and Hoechst Celanese (2006); [2] Harris (1990b); [3] NLM (2006) (note: the Hazardous Substances Data Bank includes errors in citation and transcription of values from journal articles to online summaries; where available, original sources were checked); [4] Wing (1977); and [5] Capel and Larson (1995).

Notes: Estimates made with HYDROWIN (USEPA, 2000a), a module in the EPIWIN Suite (USEPA, 2007a). Stabilizer compounds not appearing in this table (e.g., 1,4-dioxane) are not susceptible to hydrolysis. "Very long" describes modeled estimates in excess of 1,000 years.

[a] Epoxides estimated for only acid- or base-catalyzed hydrolysis, not for neutral hydrolysis; underestimation may result.

where HA is an organic chemical associated with a proton H, H_3O^+ is hydronium, and A^- is the organic compound remaining after splitting off one or more hydrogen atoms. The equilibrium constant K_a is the ratio of the activities of the reactants as follows (Lyman et al., 1990):

$$K_a = \frac{a_{H_2O^+} a_{A^-}}{a_{HA} a_{H_2O}},\ a_{H_2O} = 1 \therefore K_a = K_a = \frac{a_{H_3O^+} a_{A^-}}{a_{HA}} = \frac{\left(\gamma_{H_3O^+} M_{H_3O^+}\right)\left(\gamma_{A^-} M_{A^-}\right)}{\gamma_{HA} M_{HA}}. \tag{3.22}$$

The activity of water is taken as one by convention, and activity is defined as the molar activity coefficient, γ, multiplied by the molar concentration, M. The convention for expressing acid dissociation constants is similar to pH; pK_a is the negative log of the acid dissociation constant. The chlorinated solvents and most of the solvent-stabilizer compounds are stable with respect to acid-base reactions, and few have literature values for the acid dissociation constant.

Weak organic acids in aqueous solutions are strongly affected by the pH of the aquatic system:

$$\log \frac{M_{A^-}}{M_{HA}} = \text{pH} - \text{p}K_a \cdot \tag{3.23}$$

In the normal pH range of aquatic systems, the relevant range of pK_a is 3–10; pK_a values less than 3 are expected to be 99% dissociated, whereas pK_a values greater than 10 are expected to remain

TABLE 3.10
Acid Dissociation Constants for Stabilizer Compounds

Chemical Name	pK_a	Temperature (°C)	Percent Dissociated at pH 7	References
Pyridine	5.25	25	Complete dissociation	Perrin (1972)
Nitroethane	8.46	25	3.47	Perrin (1972)
Phenol	9.994	20	0.13	Perrin (1972)
Nitromethane	10.21	25	0.062	Perrin (1972)
p-*tert-Amyl* phenol	10.43	25	0.0167	Schultz (1986)
Triethylamine	10.778	25	0.017	Riddick et al. (1985)
Nonylphenol	10.778	25	0.017	Lipnick et al. (1986)
Diisopropylamine	11.05	25	0.009	Perrin (1972)
2,6-di-*tert*-Butyl-p-cresol	12.23	25	5.89×10^{-6}	Serjeant and Dempsey (1979)
tert-Butyl alcohol	19.2	25	6.31×10^{-13}	Serjeant and Dempsey (1979)

Note: The majority of stabilizer compounds, including 1,4-dioxane, do not contain functional groups capable of dissociation.

completely undissociated (Lyman et al., 1990). The higher the pH, the more a compound with a pK_a between 3 and 10 will dissociate (Schultz, 1986). Most of the pK_a measurements available are made at room temperature (25°C), which is much warmer than temperatures expected in most surface waters and groundwaters. Therefore, laboratory values of pK_a will typically overestimate the tendency for an organic compound to ionize under ambient subsurface conditions. Table 3.10 provides acid dissociation constants for the few stabilizer compounds whose acid dissociation constants have been measured and reported in the literature. The majority of stabilizer compounds do not contain functional groups capable of dissociation.

3.3 SUBSURFACE FATE AND TRANSPORT PROCESSES

Contaminants in the subsurface migrate slowly. The most rapid rates of migration in the unsaturated zone occur by vapor transport through relatively dry soils, particularly when vapors are denser than air. The most rapid migration rates in the saturated zone are equal to the groundwater flow velocity. Contaminant migration is retarded by chemical, physical, and biological processes contributing to the contaminant's elimination, transformation, or otherwise delayed arrival at the groundwater interface or at an observation well down-gradient from the point of release. The following sections focus on the subsurface fate and transport processes in groundwater that favor migration of some stabilizer compounds while eliminating others.

3.3.1 Aqueous Solubility of Stabilizer Compounds and Stabilizer-Solvent-Waste Mixtures

Solubility, the maximum mass of a compound that will dissolve in pure water at a specific temperature before phase separation occurs, is for many compounds the most important physical property governing subsurface fate. Highly soluble compounds are less prone to partition into the vapor phase and are generally less likely to be adsorbed to organic and mineral surfaces. Consequently, hydrophilic and highly soluble compounds are not significantly retarded in groundwater flow and will migrate rapidly in groundwater. Relative retardation of contaminant migration is discussed further in Section 3.4.

In general, liquids obey the "like dissolves like" rule; polar molecules (e.g., alcohols, ketones, and some ethers) are soluble in polar solvents (e.g., water), and nonpolar molecules (e.g., benzene) are soluble in nonpolar solvents (e.g., oil).

Natural water contains mineral salts that may cause a minor change in the capacity of water to dissolve organic compounds. In most groundwater, ionic strength contributed by dissolved minerals will have a negligible effect on organic compound solubility. For groundwater having higher ionic strength, a correction factor derived from experimental data for aromatic compounds can be applied (Hashimoto et al., 1984):

$$\log_{10} S_{SW} = \left[(0.0298I + 1)\log_{10} S_{W}\right] - 0.004I, \tag{3.24}$$

where S_{sw} is the solubility in seawater (in mol/L), S_w is the freshwater solubility (in mol/L), and I is the ionic strength, defined by

$$I = \frac{1}{2}\sum\left(C_i Z_i^2\right) \tag{3.25}$$

where C is the concentration and Z_i is the charge of the ith ionic species, summed over all ionic species in solution (Lewis and Randall, 1921).

Chemical properties of organic compounds and the geochemical characteristics of groundwater are the primary determinants of pollutant solubility, and molecular structure is the main determinant of most chemical properties. Molecular structures that include branching have increased water solubility for many compounds. Ring formation increases water solubility. Inclusion of a double bond in the molecule, ring, or chain also increases water solubility. The presence of a second or third double bond in a hydrocarbon proportionately increases water solubility, and a triple bond in a chain molecule imparts greater solubility than two double bonds (Verschueren, 1996).

The technical literature lists a wide range of solubilities for the same compounds. Older literature typically lists higher solubility values because older products were not as pure as today's chemicals and because analytical methods have improved (Verschueren, 1996). In addition to laboratory experiments, a number of estimation methods have been developed to calculate the expected compound solubility based on molecular structures (QSARs) and a variety of other approaches.

The WATERNT program uses the atom/fragment contribution (AFC) estimation methodology developed by Syracuse Research Corporation. The AFC estimation methodology is based on a "fragment constant" method wherein coefficients for individual fragments and groups were derived by multiple regressions of 1,000 reliably measured water solubility values. WATERNT retrieves experimental water solubility values from a database containing more than 6,200 organic compounds with measured values. When a compound structure matches a database structure via an exact atom-to-atom connection match, the experimental water solubility value is retrieved (USEPA, 2007a). Table 3.11 lists literature values for measured and calculated solubilities for solvent-stabilizer compounds and chlorinated solvents. The large differences between commonly cited solubility values and values estimated by using QSAR techniques (shown in Table 3.11) underscore the importance of reviewing the basis for solubility values listed in data compilations.

Table 3.12 lists estimated and measured solubilities for individual compounds. Stabilizer compounds in a solvent waste mixture dissolve differently than individual compounds when released to groundwater. Each component in the mixture will partition between the aqueous phase and the mixture (Verschueren, 1996). Mixtures of organic compounds will dissolve into water according to their effective solubilities, defined by Raoult's law:

$$S_{\text{eff}} = X_i S_{\text{water}}, \tag{3.26}$$

where S_{eff} is the effective solubility of a compound in a mixture (lower than the compound's solubility when measured alone in water), X_i is the mole fraction of compound i in the mixture, and S_{water} is the compound's single-component aqueous solubility (Banerjee, 1984; Jackson and Mariner, 1995; Jackson and Dwarakanath, 1999). Equation 3.26 applies to mixtures without significant cosolvent

TABLE 3.11
Estimated and Frequently Cited Solubilities of Solvent-Stabilizer Compounds and Chlorinated Solvents

Stabilizer and Solvent Compounds[a]	Commonly Cited Solubility[b] (mg/L)	Estimated Solubility, WATERNT[c] (mg/L)	Temperature for Commonly Cited Value (°C)	References
Ethanol	Miscible	1,000,000	25	Budavari (1989)
Glycidol	Miscible	1,000,000	25	Sax and Lewis (1987)
Methanol	Miscible	1,000,000	25	Riddick et al. (1985)
W-Methylmorpholine	Miscible	1,000,000	—	Fisher Scientific (2007)
Triethanolamine	Miscible	1,000,000	25	Budavari (1996)
Pyridine	Miscible	729,800	20	Kirk and Othmer (1982)
1,3-Dioxolanc	Misciblc	276,900	25	Lide (2000)
2-Methyl-3-butyn-2-ol	Miscible	239,500	20	Ahlers (1998)
Diepoxybutane	Miscible	220,500	25	Dean (1985)
tert-Butyl alcohol	Miscible	217,500	25	Riddick et al. (1985)
1,4-Dioxane	**Miscible**	**213,900**	**25**	Reid and Hoffman (1929)
Acetonitrile	Miscible	136,700	25	Riddick et al. (1985)
sec-Butyl alcohol	Miscible	130,400	25	Hefter (1984)
Trichloroacetic acid	1,200,000	11,990	25	Willis and McDowell (1982)
Propylene oxide	590,000	129,300	25	Bogyo et al. (1980)
2-Butanone	353,000	76,100	10	Verschueren (1996)
1,3,5-Trioxane	212,000	289,800	25	Budavari (1996)
tert-Amyl alcohol	110,000	70,700	25	Hefter (1984)
1,2-Butylene oxide	95,000	24,650	25	Bogyo et al. (1980)
Phenol	82,800	26,160	25	Southworth and Keller (1986)
Hydroquinone	72,000	129,500	25	Granger and Nelson (1921)
Resorcinol	71,700	85,710	25	Yalkowsky and Dannenfelser (1992)
Epichlorohydrin	65,900	50,630	25	Yalkowsky and Dannenfelser (1992)
Ethyl acetate	64,000	29,930	25	Wasik et al. (1981)
Nitroethane	45,000	38,130	20	Yalkowsky and Dannenfelser (1992)
Isopropyl acetate	29,000	9,268	25	Yalkowsky and Dannenfelser (1992)
Dichloromethane	*20,000*	*10,950*	*25*	Pankow and Cherry (1996)
Butoxymethyl oxirane	20,000	26,590	20	Patty (1963)
Triethylamine	15,000	68,260	20	Verschueren (1996)
Styrene oxide	13,000	4,228	25	Lapkin (1965)
Nitromethane	11,100	111,200	25	Riddick et al. (1985)
Diisopropylamine	11,000	75,310	25	Kirk and Othmer (1982)
Methyl chloroform	*1334*	*666*	*25*	Banerjee et al. (1980)
Trichloroethylene	*1280*	*779*	*25*	Horvath et al. (1999)
Thymol	1000	437	25	Sax and Lewis (1987)
1-Naphthol	866	1,126	25	Hassett et al. (1980)

(Continued)

TABLE 3.11 *(Continued)*
Estimated and Frequently Cited Solubilities of Solvent-Stabilizer Compounds and Chlorinated Solvents

Stabilizer and Solvent Compounds[a]	Commonly Cited Solubility[b] (mg/L)	Estimated Solubility, WATERNT[c] (mg/L)	Temperature for Commonly Cited Value (°C)	References
Carbon tetrachloride	*757*	*280*	25	Banerjee et al. (1980)
Isoprene	642	339	25	McAuliffe (1966)
Perchloroethylene	*200*	*80*	25	Coca and Diaz (1980)
p-tert-Amyl phenol	168	113	25	Barton (1984)
Cyclohexane	55	43	25	McAuliffe (1966)
Isopentane	48	185	25	Riddick et al. (1985)
Nonylphenol	6	2	25	Shiu et al. (1994)
2,6-di-*tert*-Butyl-p-cresol	0.4	6	20	Verschueren (1996)
Diisobutylene	Insoluble in water	12	—	Lide (2000)

[a] Nonitalic compounds are stabilizers. Italic compounds are chlorinated solvents; these are listed for comparison to show whether the stabilizer is likely to evaporate with the solvent or remain behind in the soil where it can be leached down to groundwater.

[b] Solubility values commonly cited in ChemFate database (SRC, 2007b) and the Hazardous Substances Databank (NLM, 2006).

[c] WATERNT™ (USEPA, 2007a) in "Estimations Programs Interface for Windows" (EPI Suite™) v. 3.20; http://www.epa.gov/oppt/exposure/pubs/episuitedl.htm

effects (Payne et al., 2008). Calculations using an assumed solvent waste composition (based on information in the solvent recovery and patent literature) yield the effective solubility of 1,4-dioxane relative to methyl chloroform, nitromethane, *sec*-butyl alcohol, 1,3-dioxolane, 1,2-butylene oxide, and waste cutting oil (Table 3.12).

Table 3.12 shows that the effective solubility of components in a mixture is considerably reduced from the individual compound's solubility when measured as a single-component system in pure water. Also, the effective solubility of 1,4-dioxane is 10,000 times greater than the effective solubility of methyl chloroform. As a result, 1,4-dioxane will be preferentially removed from the waste mixture; in effect, groundwater will "suck" the dioxane out of the waste mixture because of this large difference in effective solubilities. Because 1,4-dioxane will dissolve in groundwater earlier in time and at a much faster rate, the 1,4-dioxane is likely to lead the other components in the waste mixture in the migration front emanating from the point of release. Consequently, 1,4-dioxane, if released in sufficient quantities to a sand or gravel aquifer, can be expected to migrate farthest and can be found at distances several times the length of the methyl chloroform plume. This pattern has been observed in many plumes, as discussed further in Chapters 3 and 9.

3.3.2 Adsorption

Adsorption is the attachment of a compound in the aqueous or vapor phase onto a solid surface. A related process, *ab*sorption, involves the transfer of a compound from the aqueous phase *into* a volume of solid material. Because adsorption and absorption are usually not distinguishable in natural systems, the term *sorption* is used to indicate that the specific mechanism of attachment to or movement into a solid surface or volume is not known (Schwarzenbach et al., 1993).

TABLE 3.12
Effective Solubilities of Components in a Waste Mixture of Methyl Chloroform Degreasing Solvent from a 1970s-Era Aerospace Industry Degreaser

Component	Molecular Weight (g/mol)	Specific Gravity (g/cm³)	Initial Mass Fraction[a] (%)	Final Mass Fraction[b] (%)	Mass in 1 mol of Mixture (g/mol)	*Xi* (Mole Fraction)	Single-Component Aqueous Solubility[c] (mg/L)	S_{eff} (Effective Solubility)[d] (mg/L)
Methyl chloroform	113.40	1.35	93.25	53.62	62.71	0.55	1334	740
1,4-Dioxane	88.12	1.03	3.00	15.00	17.54	0.20	10,000,000	1,991,600
sec-Butyl alcohol	74.10	0.80	1.50	7.06	8.25	0.11	10,000,000	1,114,000
1,3-Dioxolane	74.09	1.06	1.00	1.27	1.48	0.02	10,000,000	200,000
1,2-Butylene oxide	72.12	0.83	0.70	0.31	0.36	0.01	95,000	480
Nitromethane	61.04	1.14	0.55	2.75	3.22	0.05	11,100	590
Cutting oil[e]	400.00	0.95	0.00	20.00	23.39	0.06	50	3
Overall mixture[f]	116.96	1.13	100	100	116.96	1.00		

[a] Initial mass fraction represents the composition of methyl chloroform-containing spent-solvent waste listed in Archer (1984).
[b] Final mass fraction represents the roughly estimated composition of spent-solvent waste after 6 weeks of intensive vapor degreasing during which volatile vapors are lost to the atmosphere, "high-boilers" are concentrated in the still bottoms, and cutting oil accumulates. The assumed oil composition and hypothetical aerospace industry setting are based on Jackson and Dwarakanath (1999).
[c] Miscible compounds assigned a solubility of 10^6 mg/L for calculation; values are from frequently cited solubilities in Table 3.11.
[d] The effective solubility is calculated by using the adaptation of Raoult's law explained in Section 3.3.1 and in Jackson and Mariner (1995) (values are rounded).
[e] Molecular weight, specific gravity, and solubility of cutting oil are assumed values.
[f] Properties of overall mixture are for final composition of the degreasing waste at the point it was removed from the degreaser and replaced with new solvent.

Adsorption of molecules, while technically a partitioning phenomenon, can be represented as a chemical reaction:

$$A + B \leftrightarrow A{\cdot}B, \tag{3.27}$$

where A represents the adsorbate, B the adsorbent, and AB the adsorbed compounds. The "adsorption reaction" involves a variety of intermolecular forces and is often reversible: molecules continue to accumulate onto the soil particle surface until equilibrium is reached and the rate of forward reaction (adsorption) equals the rate of reverse reaction (desorption) (Snoeyink, 1999).*

A key determinant of whether a contaminant will be mobile in the subsurface is the kinetics of the sorptive process compared to the rate of chemical transport by advection and dispersion. Chemicals that are adsorbed more quickly than they are advected will be retained in the aquifer matrix (Schwarzenbach and Westall, 1981).

Adsorption of chemicals to organic and mineral surfaces controls retardation of contaminant migration in surface water and groundwater (Brusseau and Reid, 1991).† Compounds may also sorb onto micrometer-sized colloids that are advected with groundwater flow. Low-solubility hydrophobic compounds such as chlorinated solvents can be substantially more mobile when sorbed to

* The technical term "soil" is typically reserved for the root-zone material; the geologic material comprising aquifers is referred to as "aquifer solids." For convenience, this chapter uses the less precise practice of calling all subsurface solids by the term "soils."
† Retardation is the rate of chemical migration in a saturated aquifer relative to groundwater velocity; see Section 3.3.2.3 for further discussion of retardation.

colloids than they are in the aqueous phase (Huling, 1989). Adsorption of contaminants to carbon surfaces also has bearing on treatability using liquid or granular activated-carbon filters.

The forces governing sorption include electrostatic interactions, van der Waal's interactions, hydrogen bonding, charge transfer, ligand exchange, direct and induced dipole-dipole interactions, chemisorption, and hydrophobic bonding (Suthersan, 2002). Organic compounds can adsorb to both organic matter and mineral surfaces. The degree to which sorption occurs depends on both soil properties such as soil organic matter content, clay type, and properties of the organic compound, including water solubility, molecular volume, the octanol/water partition coefficient (K_{ow}), and the organic carbon partition coefficient (K_{oc}). These terms and their role in adsorption of stabilizer compounds to aquifer solids and suspended solids are described further in this section.

Contaminant residence time in the subsurface also plays an important role in sorption processes, as contaminants tend to become more strongly bound to soil over time. The age of a release is important as it has bearing on enhanced sorption of organic compounds to inorganic soil particles. Generic fate and transport models rely on published K_{oc} values and assumed total organic carbon content to estimate contaminant partitioning between sorbed and aqueous phases, as well as to predict retardation (see Section 3.4). This approach can significantly underestimate the sorption potential of a contaminant and overpredict its mobility in the soil or groundwater.

3.3.2.1 Soil Properties Affecting Adsorption

Organic matter in soil or sediment is the primary factor in the sorption of nonionic organic contaminants from water by soils. Organic compounds also adsorb to soil minerals but only weakly, at least initially, owing to the strong competitive adsorption of water on polar mineral surfaces. Sorption to soil organic matter has an approximately inverse proportionality to a compound's aqueous solubility (Chiou and Kile, 1994).

Adsorption is a surface phenomenon: it increases with higher surface area, which is in cubic proportion to particle size and is also related to pore size. Micropores increase the adsorption capacity. The distribution of pore sizes in organic matter is relevant to sorption processes, but it is difficult to quantify and is not usually measured. Large molecules whose hydrated radii exceed the diameter of pore sizes will be excluded from absorption into soil organic matter ("steric exclusion") and will be limited to surface adsorption (Bennett, 2004b).

Most organic matter in soil is bound to clay as a clay-organic complex. Two major types of adsorbing surfaces are available to an organic compound: clay-organic and clay alone. The relative contribution of organic and inorganic surface areas to adsorption will depend upon the extent to which the clay is coated with organic matter. The influence of clay on organic chemical adsorption is significant in soils with organic matter contents below 1% (Dragun, 1988). Active agents governing adsorption of organic compounds onto mineral surfaces include silicates and aluminum and iron oxides and hydroxides. Oxide surfaces may be amphoteric hydroxyl sites* capable of both donating and accepting a proton, providing anionic, neutral, and cationic sites, depending on pH and the hydroxyl reaction rate constant, pK. Amphoteric hydroxyl sites strongly attract water, creating a layer of tightly bound water at the surface (vicinal water). Organic molecules interact first with the charged surface of the vicinal water layer; direct adsorption of the compound onto the mineral surface must be preceded by displacement of the water (Bennett, 2004b).

3.3.2.2 Molecular Properties Affecting Adsorption

The size of a molecule influences its potential to be adsorbed to solid surfaces. Large molecules have more sites where van der Waal's forces† contribute to attraction of the molecule to charged soil particles, due to temporary dipoles resulting from time-varying electron distribution or polarity

* Amphoteric substances can react either as acids or bases.

† Van der Waal's forces, also called London dispersion forces, are the weak electrostatic charges within an otherwise neutral molecule resulting from movement of electrons forming temporary dipoles or charge concentrations.

induced by charges on soil particles. The positive regions in the temporary dipole attract electrons in the adjacent soil surface, resulting in a net attraction of the compound to the soil surface (Carey, 1987; Dragun, 1988). For organic chemicals with molecular weights greater than 400–500, van der Waal's forces become the dominant adsorption mechanism in nonsandy soils (Dragun, 1988).

Solubility in water, exemplified by hydrophilic compounds, also controls the degree to which organic compounds are adsorbed to soil organic matter. Molecular fragments including carbon, hydrogen, and the halogens chlorine, bromine, and iodine are hydrophobic, whereas fragments including nitrogen, sulfur, oxygen, and phosphorus are generally hydrophilic. The solubility of compounds with both hydrophilic and hydrophobic groups depends on which fragments are dominant and the geometric arrangement of the molecule (Dragun, 1988). In general, highly soluble and miscible compounds are less prone to adsorption.

The molecular charge of an organic chemical can also play an important role in adsorption onto soil surfaces if the molecule is polar. Many of the stabilizer compounds do not carry a sufficient charge to be affected by electrostatic sorption. Some stabilizer compounds including amines and quaternary ammonium compounds may react in acidic waters to form functional groups that carry a positive charge to make them prone to adsorption. Compounds with hydroxyl or carboxyl groups may acquire a negative charge in alkaline waters and become adsorbed by oxide surfaces on alumina-silicate clay minerals (Dragun, 1988).

Hydrogen bonding can also affect organic chemical adsorption onto soil surfaces. Hydrogen bonding occurs where a hydrogen atom serves as a bridge between two electronegative atoms, with a covalent bond on one end and an electrostatic bond on the other. For example, hydrogen bonding may join a polar organic molecule to an adsorbed cation through a water molecule in the cation's primary hydration shell. This type of hydrogen bonding occurs with pyridine as well as with ketones and amides in soils containing smectite clays (Dragun, 1988). As discussed in Chapter 2, 1,4-dioxane's three conformations and propensity to form dimers may result in varying polarity, which affects its capacity to adsorb to clay surfaces or move through the monomolecular layer of water molecules and penetrate the clay mineral structure.

3.3.2.3 Distribution Coefficients and Sorption Isotherms

Distribution coefficients provide a quantitative description of partitioning between chemicals in solution and soils, bottom sediments, suspended sediments, and other solid media. The distribution coefficient, K_d, describes a linear relationship between the concentration of a chemical in solution and its concentration adsorbed to soil surfaces. At low concentrations, adsorption is a linear process, where the amount adsorbed is directly proportional to the amount in solution (Bennett, 2004b). Sorption potential increases with increasing values of K_d. Sorption isotherms become nonlinear and sorption decreases as the concentration increases, particularly for polar and ionizable compounds and for soils low in organic carbon or high in clay (Suthersan, 2002). The value of K_d is restricted to the system for which that value was obtained; it will vary in different soils. Subsurface heterogeneity creates a range of K_d values even within a single boring; accordingly, estimation of adsorption from K_d values obtained from soil cores should be coupled with a geostatistical realization of soil texture distribution, or other probabilistic approaches, rather than applying a single value of K_d.

The relationship between the quantity of a chemical absorbed at constant temperature per quantity of adsorbent, q_e, and the equilibrium concentration of adsorbate in solution, C_e, is called the adsorption isotherm (Snoeyink, 1999). The distribution coefficient K_d is the slope of the sorption isotherm at the concentration of interest (a constant for linear isotherms). The use of isotherms for estimating mass of contaminant adsorbed to soil assumes that equilibrium between the adsorbent and the adsorbate is attained instantaneously and that the isotherm is reversible.

The Freundlich equation describes adsorption data for organic compounds adsorbing to organic carbon, particularly GAC in a water treatment setting:

$$q_e = KC_e^{1/n}, \quad (3.28)$$

or in linear form

$$\log q_e = \log K + \frac{1}{n}\log C_e \tag{3.29}$$

where K and n are constants.

The units of K and n are determined by the units of q_e, the quantity of chemical adsorbed per quantity of adsorbate, and C_e, the concentration of adsorbate in solution (Snoeyink, 1999). The constant K primarily describes the capacity of the adsorbent for the adsorbate, and $1/n$ is a function of the strength of adsorption. When the adsorbent has become saturated with respect to the adsorbent, q_e is constant. The Freundlich isotherm is more applicable at lower concentrations.

The distribution coefficient is described as the ratio of the mass of a given compound sorbed to soil mineral and organic surfaces to the mass remaining in solution:

$$K_d \frac{q_e}{C_e}, \tag{3.30}$$

$$q_e \frac{(C_e - C_i)V}{m}, \tag{3.31}$$

where K_d is the distribution coefficient (in mL/g), q_e is the mass of chemical on the solid phase (in mg/g), and C_e is the mass of chemical in solution (in mg/L). The mass of chemical on the solid phase, q_e, is taken as the difference between the mass of chemical in solution and the initial aqueous-phase concentration C_i (in mg/L) multiplied by the volume of solution and divided by the mass of the solid, m (in g).

In general, compounds with lower values of K_d indicate that those compounds will have higher mobilities. A screening-level approach for estimating K_d involves multiplying K_{oc}, the organic carbon partition coefficient, by f_{oc}, the fraction of organic matter: $K_{oc} \times f_{oc}$. As a rule of thumb, compounds whose values of K_d are higher than 20 are considered to be relatively immobile; these compounds are unlikely to leach from soils or migrate through aquifer materials because of their strong potential for sorption to soil surfaces.

K_d Value of Compounds	Mobility (Fetter, 1993)
>20	Relatively immobile
10–20	"Slight"
1–10	"Low"
0–1	"High"

A related parameter, the retardation factor, R_f or R, expresses the rate of chemical migration in a saturated aquifer relative to groundwater velocity:

$$R_f = 1 + \frac{\rho_b K_d}{\Theta_e} = \frac{V_w}{V_c}, \tag{3.32}$$

where ρ_b is the bulk density of the soil (in g/cm^3), K_d is the distribution coefficient, Θ_e is the effective porosity (unitless because it is vol/vol), V_w is the velocity of water, and V_c is the velocity of the contaminant. In the literature, K_d is sometimes represented as K_p, the mass-specific partition (or distribution) coefficient. The retardation factor assumes that sorption is reversible, instantaneous, and linear (i.e., not proportional to solute concentration or organic matter) (Freeze and Cherry, 1979).

3.3.2.4 Organic Carbon Partition Coefficient, K_{oc}

The sorption of hydrophobic organic compounds is strongly controlled by the presence of soil organic material. K_{oc} is the partitioning coefficient for a solute between the aqueous and organic phases and is defined as the ratio of the amount of chemical adsorbed (in micrograms) per unit mass of organic carbon (oc, in grams) in the soil or sediment to the concentration (in micrograms per milliliter of solution) of the chemical in solution at equilibrium (Lyman et al., 1990). K_{oc} is unitless* and is represented as

$$K_{OC} = \frac{K_d}{f_{OC}}, \tag{3.33}$$

where f_{oc} is the fraction of organic carbon in soil (mass ratio).

K_{oc} is not a primary physicochemical compound property because the degree to which a compound is adsorbed is in relation to its K_{oc} and dependent upon the soil organic matter content, organic matter properties, and groundwater chemistry. Measured values of K_{oc} are obtained for particular conditions of the measured soil and solution; however, because K_{oc} is the K_d value normalized for an aquifer's organic carbon content, f_{oc}, use of K_{oc} values to contrast contaminant mobility reduces the variability of partition-coefficient measurements in different soils (Dragun, 1988). K_{oc} values derived from linear isotherms are therefore considered to be largely independent of the properties of soil or sediment. Nevertheless, variability in soil properties including clay content, the particular clay mineral(s), surface area, soil pH, soil temperature, and type of soil organic matter may contribute to different measured organic carbon partition coefficients. Values of K_{oc} for the same compound measured on different soils have uncertainties ranging from 10% to 140% (Lyman, 1990). When estimating the distribution coefficient K_d from the organic carbon partition coefficient K_{oc} and soil organic carbon content f_{oc} (i.e., $K_d = K_{oc} \times f_{oc}$), the following assumptions are made (Suthersan, 2002):

- Sorption is exclusively to the organic component of the soil.
- All soil organic matters have the same sorption capacity per unit mass.
- The sorption—desorption process is in equilibrium.
- The sorption—desorption isotherms are identical, that is, hysteresis does not occur.

Of course, soil organic matter varies in type and sorption capacity among different soils. A significant portion of soil organic matter is composed of humic and fulvic acids. Fulvic acid is composed of organic polymers with molecular weight greater than 2,000, whereas humic acid is composed of organic polymers with higher molecular weights, up to 300,000 Daltons.† Some sources report that compounds have a greater propensity to sorb to one type of organic matter than another (Dragun, 1988), whereas others find that there is only a small variability of the sorption properties of soil organic matter (Nguyen et al., 2005). For humic acid, absorption of organic compounds into the organic matrix plays a greater role than adsorption onto its surfaces (Niederer et al., 2006b). Sorption by amorphous soil organic matter involves absorption *into* instead of adsorption *onto* the organic matter and is referenced as the solubility of the organic compound into the soil organic matter, which is generally proportional to the inverse of aqueous solubility (Chiou and Kile, 1994).

Estimated values of K_{oc} are often used in environmental fate assessment because measurement of K_{oc} is expensive and values are generally not transferable to other settings. Most published values for K_{oc} are estimated by using a variety of QSARs. More than 200 different correlations for estimating K_{oc} have been published. The correlations used to estimate K_{oc} are based on compound solubilities, octanol/water partition coefficients, molecular factors, topological indices, solvation energy relationships, and others (Reinhard and Drefahl, 1999). Table 3.13 lists values for K_{oc} estimated by using the PCKOCWIN algorithm as well as values published in various chemical fate and transport property compilations.

The variation among predicted values of K_{oc} listed in Table 3.13 underscores the uncertainty inherent to estimation of retardation due to sorption onto soil organic matter. Calculated values of

* Note that K_{oc} is unitless, which assumes that 1 mL of solution = 1 g of solution, that is, organic compounds are present at low concentrations. Adsorption isotherms become nonlinear at high concentrations.

† Dalton (Da) is an atomic mass unit equal to one-twelfth of the mass of an unbound atom of carbon-12, i.e., the approximate mass of a hydrogen atom, a proton, or a neutron.

TABLE 3.13
Estimated K_{oc} Values for Stabilizer Compounds

Stabilizer Compound	log K_{oc} Estimated by PCKOCWIN	Commonly Cited log K_{oc}	Estimated[a] K_d ($K_{oc} \times f_{oc}$)	Mobility[b]	References
Nonylphenol	4.79	4.49	6,089	Immobile	Swann et al. (1983)
p-*tert*-Amyl phenol	3.58	2.41	380		Lyman et al. (1990)
1-Naphthol	3.48	4.4	304		Deneer et al. (1988)
Resorcinol	2.64	1.81	43.4		Swann et al. (1983)
Hydroquinone	2.64	1.325	43.4		Swann et al. (1983)
Diisobutylene	2.44	2.44	27.55		Swann et al. (1983)
Phenol	2.43	1.48	26.8		Briggs (1981)
Cyclohexane	2.22	2.2	16.55	Low mobility	Swann et al. (1983)
Diisopropylamine	2.04	2.15	10.9		Swann et al. (1983)
Triethylamine	2.03	2.16	10.72		Swann et al. (1983)
Isopentane	1.83	3.0	6.77		MSDS for isopentane
Isoprene	1.83	2.69	6.77		Lyman et al. (1990)
Styrene oxide	1.73	1.72	5.42		Lyman et al. (1990)
Pyridine	1.52	1.7	3.3		Lyman et al. (1990)
Nitroethane	1.20	1.48	1.58		Swann et al. (1983)
Triethanolamine	1.0	0.477	1.0	High mobility	Hansch and Leo (1985)
Isopropyl acetate	0.98	1.18	0.95		Swann et al. (1983)
Nitromethane	0.91	0.28	0.82		Lyman et al. (1990)
Ethyl acetate	0.79	1.77	0.613		Swann et al. (1983)
Epichlorohydrin	0.65	1.6	0.45		Swann et al. (1983)
Acetonitrile	0.65	1.2	0.45		Swann et al. (1983)
1,2-Butylene oxide	0.65	0.65	0.45		Swann et al. (1983)
2-Butanone	0.58	1.5	0.383		Walton et al. (1992)
2-Methyl-3-Butyn-2-Ol	0.47	1.55	0.292		UNEP (1998)
tert-Amyl alcohol	0.47	0.46	0.292		Swann et al. (1983)
Butoxymethyl oxirane	0.45	0.24	0.283		SPI (2006)
Trichloroacetic acid	0.44	0	0.274		Weber et al. (1981)
Propylene oxide	0.37	1.4	0.232		Swann et al. (1983)
sec-Butyl alcohol	0.31	1.7	0.205		Swann et al. (1983)
tert-Butyl alcohol	0.17	1.57	0.16		Swann et al. (1983)
tert-Butyl phenol	0	2.77	0.10		NICNAS (2003)
1,4-Dioxane	**0**	**1.23**	**0.10**		Lyman et al. (1990)
1,3-Dioxolane	0	1.18	0.10		Lyman et al. (1990)
Methanol	0	−0.74	0.10		USEPA (1996)
Methyl chloroform	1.687	2.18	15.2	Low	Pankow and Cherry (1996)
Dichloromethane	1.375	0.944	0.88	High	
Trichloroethylene	1.831	2.1	12.6	Low	
Perchloroethylene	2.029	2.56	36.4	IMM	

Notes: PCKOCWIN is a soil adsorption coefficient estimation program and part of the *Estimation Programs Interface Suite* (EPI Suite™) (USEPA, 2007a).

[a] To calculate K_d, the contaminant distribution coefficient, K_{oc} is multiplied by the fraction of soil organic matter, f_{oc}, where f_{oc} is assumed to be 0.1.

[b] Mobility classes from classification scheme proposed by Fetter (1993); IMM = immobile.

K_{oc} can provide an order of magnitude estimate of retardation when combined with measured values of the weight fraction of soil organic matter in aquifer solids. However, measured values of K_{oc} must address organic matter variability due to aquifer heterogeneity. In general, less soil organic matter is present in higher-conductivity strata, and sorption to organic matter does not substantially contribute to retardation in sands and gravels. K_{oc} values in clays and silts are similar, but K_{oc} values for sands are generally 25% lower than for clays (McCarty and Reinhard, 1981).

When reviewing literature values of K_{oc}, it is important to distinguish whether experiments used whole soils or grain-size segregates, given the strong influence of particle size. Fine-grained, low-conductivity sediments such as clays and silts are richer in organic matter and play a greater role in retarding contaminant migration. The type of soil organic matter also makes a difference. Organic matter in consolidated sediments that has been altered by compression and elevated temperature (i.e., diagenesis) may behave differently than unaltered soil organic matter (Patterson et al., 1985).

PCKOCWIN (a soil adsorption coefficient program) estimates K_{oc} for organic compounds. Earlier methods for estimating K_{oc} relied upon relationships with the octanol/water partition coefficient. PCKOCWIN uses a molecular connectivity index (MCI) to predict K_{oc} values for hydrophobic organic compounds, but the MCI approach does not reliably estimate K_{oc} values for hydrophilic polar aprotic compounds such as 1,4-dioxane.* PCKOCWIN uses a series of group contribution factors to predict K_{oc} for polar compounds (USEPA, 2007a). The group contribution method outperforms the traditional estimation methods based on octanol/water partition coefficients and water solubility (Meylan et al., 1992). A comparison of PCKOCWIN estimates with experimentally determined distribution coefficients shows that (1) the MCI-based prediction model provides unreliable and inconsistent K_{oc} values, for both polar and nonpolar compounds, and (2) the divergence between experimental and calculated distribution coefficients becomes larger with increasing molecular size (Niederer et al., 2006b). K_{oc} values generated by QSAR models should be used with caution. The similarity of the compounds in the calibration data set for the model and the compound of interest should be confirmed (Niederer and Goss, 2007).

3.3.2.5 Predicting K_{oc} from K_{ow}, the Octanol/Water Partition Coefficient

The octanol/water partition coefficient, K_{ow}, is often used to predict the potential for a compound to adsorb to the aquifer matrix. K_{ow} is the ratio of the concentration of a chemical in octanol and in water at equilibrium and at a specified temperature. Octanol is an organic solvent that is used as a surrogate for natural organic matter. The relationship between the mass-specific distribution coefficient, K_d, and the octanol/water partition coefficient, K_{ow}, has been firmly established for nonpolar organic compounds present in concentrations at less than half their solubility limit and in aquifers whose organic carbon fraction is 0.1% or greater (McCarty and Reinhard, 1981; Schwarzenbach and Westall, 1981). The total organic carbon content of the aquifer matrix has less bearing on the potential for sorption in compounds with low octanol/water partition coefficients (Suthersan, 2002).

For hydrophobic compounds, K_d and K_{ow} are highly correlated for all sorbents except pure mineral substrates, in which micropores may play a role. For nonpolar organic compounds such as chlorinated solvents, the correlation between the logarithms of K_{oc} and K_{ow} values was determined to follow the relationship (Schwarzenbach and Westall, 1981)

$$\log \mathrm{K_{OC}} = 0.72(K_{OW} + 0.49),\ R^2 = 0.95. \quad (3.34)$$

Another commonly used relationship (Karickhoff, 1981) is

$$K_{OC} = 0.41K_{OW}. \quad (3.35)$$

* A polar aprotic solvent is one that does not contain an O-H or N-H bond.

These two relationships are not reliable for estimating partition coefficients of polar organic compounds such as alcohols and aliphatic ethers. A regression analysis of about two dozen experimentally determined K_{oc} values produced the following correlation (Nguyen et al., 2005):

$$\log K_{oc} = 0.73\ (\log K_{ow} + 0.52), R^2 = 0.83. \quad (3.36)$$

Equation 3.36 has a rather low correlation coefficient (R^2). The compounds used to derive Equation 3.36 do not include any ether compounds. A more dynamic and reliable approach to estimating K_{oc} uses linear free-energy relationships (LFERs)* to correlate compound structure and sorption to soil organic matter (Niederer et al., 2006b).

Numerous other relationships have been developed for different classes of compounds. In general, compounds with octanol/water partition coefficients less than 1,000 will migrate easily through the subsurface without appreciable retardation due to sorption (McCarty and Reinhard, 1981).

K_{ow} values in the literature are prone to considerable uncertainty, sometimes approaching two orders of magnitude or more. Experimental determination of K_{ow} values may incorporate method errors; for example, small quantities of emulsified octanol in the aqueous phase can carry high concentrations of the organic compound being tested, leading to erroneously high determinations of the concentration in water (Mackay et al., 1993). K_{ow} sorption models may be unreliable for polar compounds even if experimental K_{ow} values are used for input data (Niederer et al., 2006b).

Using physicochemical data without carefully assessing its reliability can lead to significant error in estimating contaminant fate. Fate and transport experts address poor data quality of K_{ow} values in the literature by using structure—activity estimation methods to compute K_{ow}, by using carefully selected data, or by applying probabilistic assessments to fate and transport modeling (Renner, 2002).

3.3.2.6 Effect of Surface Oxides on Sorption

Increased amounts of oxygen on granular activated-carbon surfaces can decrease the affinity of organic compounds for sorption to carbon. Oxidation of carbon by acids increases the presence of oxygen surface functional groups and decreases the adsorption capacity of organic compounds (Snoeyink, 1999). Decreased adsorptive capacity was observed in activated carbon for several stabilizer compounds (phenol, nitromethane, methyl ethyl ketone, *n*-butanol, and 1,4-dioxane) as a result of an increase in acidic surface oxides (Vidic and Suidan, 1991).

3.3.2.7 Competitive Sorption

Solvent-stabilizer compounds are usually released as mixtures composed primarily of chlorinated solvents and oily wastes. The rate of migration of individual components of the mixture will depend primarily on their solubility, but also on their capacity to adsorb to soil organic matter and mineral surfaces. In mixtures, competitive sorption occurs, in which compounds with greater affinities for soil organic matter are preferentially adsorbed and those compounds with lower affinities stay in solution. Competitive sorption occurs between organic compounds in anthropogenic wastes and natural dissolved organic substances such as humic and fulvic acids. Competitive sorption in soil can cause compounds initially adsorbed to organic matter surfaces to become dislodged and replaced by compounds with stronger affinities for sorption (Snoeyink, 1999). Consequently, compounds that migrate more rapidly because of higher solubility and weak capacity to adsorb to soil surfaces may be displaced as compounds with higher K_{oc} values arrive at the soil surface.

The potential adsorption of a compound determined from a pure-component test does not necessarily predict its degree of removal from a dynamic, multicomponent mixture (Verschueren, 1996). The presence of other organic compounds in solution, whether natural or anthropogenic, usually reduces the adsorption capacity of activated carbon or soil organic matter for a particular organic compound.

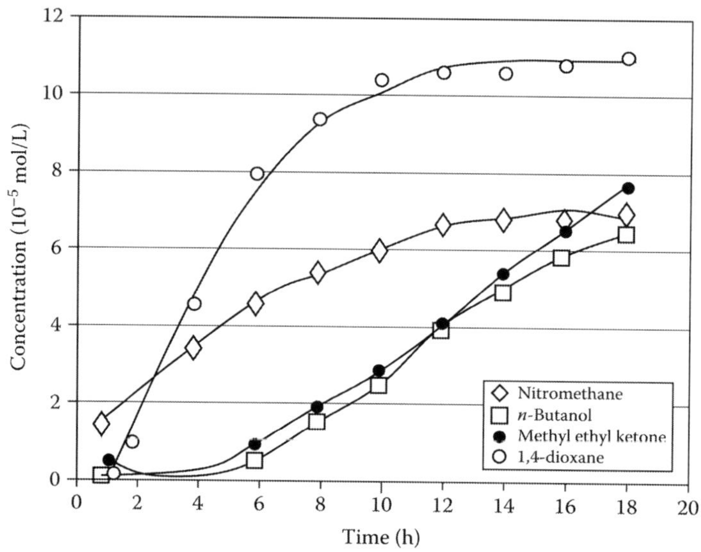

FIGURE 3.3 Breakthrough curves for a mixture of nitromethane, methyl ethyl ketone, *n*-butanol, and 1,4-dioxane on granular activated carbon (GAC). Influent concentrations were 10^{-4} mol/L each; flow rate = 23 mL/min, pH = 8.0, detention time = 2.1 min. (From McGuire, M.J. and Suffett, I.H., 1978, Adsorption of organics from domestic water supplies. *Journal of the American Water Works Association* 70(11): 621–635. With permission.)

In a study of competitive sorption of aqueous-phase solvent-stabilizer compounds onto GAC, nitromethane was found to have an adsorption capacity similar to methyl ethyl ketone (2-butanone), both of which have substantially larger capacities than *n*-butanol, which in turn has a larger sorption capacity than 1,4-dioxane (McGuire and Suffett, 1978). The adsorption isotherms for the four-compound mixture were significantly lower than those for each individual compound in solution. As the equilibrium concentration was reduced, the mixture isotherms for *n*-butanol and 1,4-dioxane became identical to the single-compound isotherms. 1,4-Dioxane was progressively desorbed by the three other compounds in the mixture. Nitromethane, *n*-butanol, and methyl ethyl ketone continued to adsorb to the GAC after 1,4-dioxane had completely broken through, indicating that they are more successful competitors for adsorption sites than 1,4-dioxane. Initial adsorption followed by displacement by compounds with stronger affinity to adsorb to carbon is referred to as the "chromatographic effect" (McGuire and Suffett, 1978). Figure 3.3 shows the effect of competitive sorption on the retention on GAC of 1,4-dioxane and its displacement by other solvents.

3.3.2.8 Use of Laboratory Batch Tests to Measure K_d in Soil

Contaminants residing in the vadose zone may leach to underlying groundwater from percolation of rainfall, flood water, or applied irrigation. The degree to which contaminants are prone to dissolution, desorption, leaching, and migration to the saturated zone is the subject of the various estimation methods described in this section. Fate and transport models often rely on published values of K_{oc} and assumed organic matter contents (f_{oc}) in soil to estimate contaminant retardation and partitioning between sorbed and aqueous phases. This approach can significantly underpredict the sorption potential of a contaminant and overpredict its mobility in soil or groundwater, when compared with soil-column leach testing.*

Laboratory batch tests can be used to quantify the mobility of contaminants in soil and to estimate the initial concentration of the contaminant in soil leachate and the final concentration as the

* Personal communication with Roger Brewer, PhD, State of Hawaii Department of Health.

contaminant enters the saturated zone. Batch tests involve placing a small quantity of soil in buffered, deionized water, agitating the soil slurry for a set period of time, and measuring the soluble fraction of the contaminant. A minimum of three soil samples is generally needed to validate batch test data for each area investigated (HDOH, 2007). The ratio of the mass of a contaminant that remains sorbed to the contaminant mass that goes into solution is referred to as the contaminant's desorption coefficient or K_d value (see additional discussion of K_d—also called the distribution coefficient, but most often called the partition coefficient—in later sections).

The contaminant concentration that partitions into leachable soil moisture can be calculated by using the soil-specific K_d value in an equilibrium partitioning equation:

$$C_{\text{total}} = C_{\text{leachate}} \times \left[K_d \left(\frac{\Theta_w + (\Theta_a \times H')}{\rho_b} \right) \right] \times \left(\frac{1 \text{ mg}}{1{,}000 \text{ μg}} \right), \tag{3.37}$$

where C_{total} is the total concentration of chemical in the soil sample (in mg/kg), C_{leachate} is the dissolved-phase concentration of chemical (in μg/L), K_d is the estimated or measured distribution coefficient (in L/kg), Θ_w is the water-filled porosity (in L/L or volume of water per volume of soil), Θ_a is the air-filled porosity (in L/L or volume of air per volume of soil), H' is the Henry's law constant at 25°C [in (μg/L)/μg/L), that is, micrograms of chemical per liter of vapor divided by micrograms of chemical per liter of water], and ρ_b is the soil bulk density (in kg/L).

The final concentration of the contaminant in groundwater as a consequence of its leaching is estimated by assuming a groundwater–leachate dilution factor, DF, usually assumed to be 20 for sites less than 0.5 acre in size (USEPA, 2001b). The DF is the volume of affected groundwater divided by the volume of leachate.

Where site-specific hydrogeological data are available, a site-specific DF can be calculated according to the following equation (USEPA, 2001b):

$$\text{DF} = 1 + \left(\frac{K \times i \times d}{I \times L} \right), \tag{3.38}$$

where K is the aquifer hydraulic conductivity (in m/year), i is the regional hydraulic gradient, d is the assuming mixing zone depth (default is 2 m), I is the surface-water infiltration rate (in m/year), and L is the length (in m) of the contaminated soil area that is parallel to groundwater flow (HDOH, 2007).

3.3.2.9 Sorption of Vapor-Phase 1,4-Dioxane

Because 1,4-dioxane is hydrophilic and has a very low Henry's law constant, it is unlikely to migrate in the vapor phase once it has dissolved into water. Vapor transport is often a concern for chlorinated solvent contamination source areas with thick unsaturated zones where VI into occupied buildings may occur (Little et al., 1992). The extensive literature on sorption includes several studies in which sorption of 1,4-dioxane vapor was measured. A 1994 study by Chiou and Kile measured the vapor adsorption of several solvent-stabilizer compounds dissolving into low-mineral-content samples of peat soil from the Florida Everglades, including nitroethane, 1,4-dioxane, and acetonitrile. The isotherm of acetonitrile was found to be about 4.2 times greater than the isotherm for 1,4-dioxane, and the isotherm for nitroethane was approximately 3.5 times greater than the isotherm for 1,4-dioxane (Chiou and Kile, 1994). The comparatively low sorption to peat, which is composed of 86.4% organic carbon,* suggests a lower potential for 1,4-dioxane vapors to be sorbed to the much smaller fraction of soil organic matter found in most vadose zone soils.

* Measured on a dry-weight basis. The Florida Everglades peat is a reference sample of the International Humic Substances Society.

TABLE 3.14
Humic Acid/Air Partition Coefficients at 15°C

	Relative Humidity			
	<0.01%	45%	70%	98%
Substance	$K_{HA\text{-}air}$ (L/kg)	$K_{HA\text{-}air}$ (L/kg)	$K_{HA\text{-}air}$ (L/kg)	$K_{HA\text{-}air}$ (L/kg)
Phenol	797,000	1,770,000	967,000	1,920,000
1,4-Dioxane	**4,680**	**12,200**	**7,000**	**16,900**
Nitroethane	2,390	3,360	3,430	3,040
Nitromethane	1,560	4,270	2,650	3,110
2-Butanone	2,400	1,730	1,340	24,40
n-Propyl Acetate	5,030	1,220	1,480	1,930
2-Butanone	2,400	1,730	1,340	2,440

Source: From Niederer, C., Goss, K.U., and Schwarzenbach, R.P., 2006a, Sorption equilibrium of a wide spectrum of organic vapors in leonardite humic acid: Experimental setup and experimental data. *Environmental Science and Technology* 40(17): 5368–5373. With permission.

Notes: The $K_{HA\text{-}air}$ partition coefficients are stated as the liters of air containing the given substance per kilogram of humic acid. Data for 1,4-dioxane are shown in boldface. Calculations are according to Equation 3.39.

Scientists at the Swiss Institute of Biogeochemistry and Pollutant Dynamics rigorously measured the humic acid/air partition coefficients for 188 polar and nonpolar organic compounds at different temperatures and relative humidities by using a dynamic flow-through technique. Humic acid/air partition experiment data reflect the concentration sorbed relative to the concentration in air. The measurements were made by using inverse gas chromatography with the humic acid sorbent serving as the chromatographic column stationary phase in a gas chromatograph and the organic compound of interest injected onto the column under isothermal conditions.* The measured retention reflects the tendency of the compound to sorb to this particular variety of humic acid at different temperatures and humidities.† The Swiss experiments provide the most comprehensive and reliable data set for sorption into humic material available (Niederer et al., 2006a). The humic acid/air partition coefficients for a few compounds that may be present in solvent wastes as solvent stabilizers are provided in Table 3.14, and Table 3.15 lists humic acid/soil-vapor distribution coefficients at 98% relative humidity. For example, the following equation describes the 1,4-dioxane partitioning between air and humic acid:

$$K_{\text{air/humic acid}-1,4-\text{dioxane}} = 4{,}680\left[\frac{\left(n_{1,4-\text{dioxane}} \,/\, m_{\text{humic acid}}\right)}{\left(n_{1,4-\text{dioxane}} \,/\, V_{\text{air}}\right)}\right] = 4{,}680\frac{V_{\text{air}}}{m_{\text{humic acid}}}, \tag{3.39}$$

where n is the number of moles, m is the mass (in kg), and V is the volume (in L).

The 1,4-dioxane aqueous-phase/soil organic carbon partition coefficient is comparatively low, but in the vapor-phase 1,4-dioxane partitions easily into humic acid, the major component of soil organic matter. The variation in humic acid/soil-vapor distribution coefficients with humidity may reflect changes in the polarity of the humic substance with water saturation (Chiou and Kile, 1994).

3.3.2.10 High-Strength 1,4-Dioxane Solutions May Cause Clay Swelling

At high concentrations, some organic compounds, including 1,4-dioxane, can induce swelling and flocculation of smectite (i.e., montmorillonite) clays (Bradley 1945; MacEwan, 1948; Brindley and

* See Chapter 4 for a brief explanation of chromatographic column stationary phases and gas chromatography.

† The Swiss study used the "leonardite humic acid" with an organic carbon content of 63.8%.

TABLE 3.15
Humic Acid/Soil-Vapor Partition Coefficients at 98% Relative Humidity

	Temperature		
	5°C	15°C	25°C
Substance	$K_{HA\text{-}air}$ (L/kg)	$K_{HA\text{-}air}$ (L/kg)	$K_{HA\text{-}air}$ (L/kg)
Phenol	2,080,000	990,000	416,000
2-Butanone	2,430	708	759
Ethyl acetate	1,340	247	—
Tetrahydrofuran	1,620	504	—
1,4-Dioxane	**15,100**	**6,550**	**4,380**
Acetonitrile	2,630	1,460	978
Nitromethane	2,900	1,880	1,070
Nitroethane	2,850	1,940	—
Pyridine	—	2,270	—

Source: From Niederer, C., Goss, K.U., and Schwarzenbach, R.P., 2006a, Sorption equilibrium of a wide spectrum of organic vapors in leonardite humic acid: Experimental setup and experimental data. *Environmental Science and Technology* 40(17): 5368–5373. With permission.

Notes: The $K_{HA\text{-}air}$ partition coefficients are stated as the liters of soil vapor containing the given substance per kilogram of humic acid. Data for 1,4-dioxane are shown in boldface.

Tsunashima, 1972). Chemical alteration of clay-mineral arrangement in natural soils and engineered clay liners can affect permeability and diffusive and advective transport (Wu et al., 1994).

Smectite clays have silica and aluminum layers assembled in a 2:1 ratio of silica to aluminum, that is, –Si-Al-Si-Si-Al-Si–. The lattice structure of smectite clays can expand in response to water and other agents, including sorption of organic compounds on both external and internal surfaces. Organic molecules and water can penetrate the layers of the clay-mineral crystal, changing its structure and swelling properties.

Wu et al. (1994) mixed 1,4-dioxane at concentrations of 0.5, 1, and 3 M [4.4%, 8.8%, and 26.4% (i.e., 264,000 mg/L)] in a suspension of deionized water and sodium-saturated smectite. The suspension was deposited on a ceramic filter within a sealed environmental chamber equipped with a window to allow x-ray diffraction measurements of spacing between clay layers and a drain to retrieve samples of drained solution. The 1,4-dioxane solution was expressed through the filter by applying pressure with helium; the expressed solution was drained at atmospheric pressure. Successive x-ray diffraction measurements at different helium pressures and for different concentrations of 1,4-dioxane showed that 1,4-dioxane converts fully expanded clay layers with spacing of 30 Å to partially expanded layers with a spacing of 15 Å (Wu et al., 1994).*

The researchers in this study postulate that 1,4-dioxane separates from solution and fills the space between the partially expanded layers in the sodium montmorillonite clay, that is, a pure-phase 1,4-dioxane accumulates between the clay layers. 1,4-Dioxane has also been observed to accumulate between layers of calcium montmorillonite (Zhang et al., 1990). 1,4-Dioxane apparently rearranges water molecules between the expanded clay layers to produce an open, coordinated structure that may force layers apart because of the larger volume of the inferred structure (Wu et al., 1994).

Zhang et al. (1990) studied 1,4-dioxane adsorption to sodium and calcium smectite suspensions in the <2 μm size fraction by using ^{14}C-labeled 1,4-dioxane. The 1,4-dioxane sorption isotherm for a calcium smectite followed the Langmuir pattern, wherein the rate of sorption is initially first-order and subsequently zero-order, producing a steep initial slope in the curve of quantity adsorbed

* Å = angstrom; 1 Å = 10^{-8} m = 0.1 nm.

versus 1,4-dioxane concentration, followed by an asymptotic tapering off to a zero-slope plateau.* This result was markedly different from the sorption isotherm for sodium smectite, for which the millimoles of 1,4-dioxane adsorbed per gram of clay increased monotonically with increasing concentration. The Zhang study found that the 1,4-dioxane adsorption reaction with calcium smectite was exothermic, but bonding was not indicated. Adsorption of most other organic compounds is usually endothermic. Zhang proposed that 1,4-dioxane adsorption onto calcium smectite clay being exothermic is attributable to its unique structure-breaking effect on the tightly held monolayer of water on the clay surface (Zhang et al., 1990). The behavior of 1,4-dioxane as a water-structure breaker has been confirmed; two 1,4-dioxane rings can form a dimer stabilized by two intermolecular hydrogen bonds, leaving two oxygen atoms available for interaction with the water molecules and creating a dipole moment (Mazurkiewicz and Tomasik, 2006).

In solutions with 1,4-dioxane, the interlayer distance in calcium smectite expands from 0.55 to 1.48 nm. In a 1.0 M solution of 1,4-dioxane, the interlayer concentrations in the calcium smectite were found to be 6.0 M (Zhang et al., 1990). Zhang concluded that 1,4-dioxane is not adsorbed because of any specific interaction with the surfaces of the clay layer; instead, 1,4-dioxane is distributed between the bulk phase and the interfacial phase, but prefers the interfacial phase, which facilitates its migration through clays. A different conclusion was reached in an earlier study by Brindley et al. (1969): calcium smectite clays in 1,4-dioxane solutions ranging from 1 to 100 mol% had constant basal spacings, leading to the conclusion that 1,4-dioxane is strongly and preferentially adsorbed by the clay. That study has been superseded by newer research under more controlled conditions, for example, Wu et al. (1994).

3.3.3 Subsurface Vapor-Phase Transport

The migration of contaminants in the vapor phase constitutes an important pathway by which releases of volatile chemicals to soil may manifest as groundwater contamination. Soil vapors are a concern when they migrate to and accumulate in enclosed spaces, where they may lead to either toxic exposure in occupied buildings or create an explosion hazard. Man-made migration pathways for soil vapors—such as permeable sand or gravel fill in trenches for buried electrical lines or rapid movement through sewer lines and storm drains—may provide the connection between sources of flammable vapors and sources of ignition.

Vapors accumulating beneath building foundations may intrude into occupied spaces, creating an inhalation exposure hazard. The degree to which vapors may intrude into enclosed spaces depends on the same factors as for volatilization from dry soil described in Section 3.1.2, but also includes factors specific to the building-soil interface. These include the surface area of the building slab or foundation, diffusion coefficients describing the efficiency of vapor transport through cracks in the slab, the thickness of the slab, air exchange rates for the structure, soil-gas flow rates, and other factors. The reader is directed to the authoritative references on the topic of VI†; this section addresses the intrinsic vapor transport characteristics of 1,4-dioxane and other solvent stabilizers.

The physical and chemical parameters controlling contaminant vapor transport are source-zone concentration, vapor pressure, vapor density, solubility, and sorption characteristics. These parameters determine contaminant equilibrium distribution between the vapor, sorbed, and dissolved phases. Contaminant biodegradability and chemical stability will determine contaminant persistence in each of these subsurface phases, as discussed in Section 3.3.4.

The soil characteristics governing contaminant vapor transport in a homogeneous soil include porosity, soil moisture content, soil-air water vapor content or humidity, and soil-air flow rates governed by pressure and temperature gradients. In the absence of advective flow of soil air, molecular

* In first-order processes, the rate is proportional to the concentration of a single reactant or substrate; in a zero-order process, the rate is independent of the substrate concentration.

† See Johnson and Ettinger (1991), Little et al. (1992), Johnson et al. (1999), American Petroleum Institute (1998), DTSC (2004), and USEPA (2015).

diffusion, characterized by contaminant diffusivity in air as discussed in Section 3.1.2, is the dominant property controlling soil-vapor transport. Soil is rarely homogeneous; soil characteristics such as stratigraphy, structures, organic matter, roots, biota, mineralogy, and density all present challenges to the reliable estimation of soil-vapor transport.

The vapor pressure of a liquid or solid is the pressure of the gas in equilibrium with the liquid or solid at a given temperature. Volatilization, the evaporative loss of a chemical from the liquid to the vapor phase, depends on the vapor pressure of the chemical and on environmental conditions that influence diffusion from the evaporative surface. Chemicals with relatively low vapor pressures, high sorption onto solids, or high solubility in water are less likely to vaporize and become airborne than are chemicals with high vapor pressures or less affinity for solution in water or adsorption to solids and sediments. Because most solvent-stabilizer compounds are highly soluble and have relatively low vapor pressures, they are generally unlikely to partition into the soil-vapor phase and migrate through the soil as vapors.

The probability that a compound occurs in the gas phase depends not only on its vapor pressure but also on its water solubility and adsorption/desorption behavior. Therefore, even substances that have a relatively low vapor pressure (down to 10^{-3} Pa) can be found in soil vapor in measurable quantities (Verschueren, 1996).

Vapor pressures, vapor densities, diffusivities, and relative evaporation rates for selected stabilizer compounds are summarized in Table 3.1. Appendix 3 provides a comprehensive listing of physicochemical parameters for a longer list of solvent-stabilizer compounds. Figure 3.4 contrasts vapor pressures with aqueous solubilities for the solvent and solvent-stabilizer compounds in Appendix 3. In Figure 3.4, compounds plotting in the box in the upper left corner have the highest likelihood of partitioning into the vapor phase and being transported.

An understanding of vapor transport is important for evaluating the potential for VI into living spaces and the associated risk from that exposure. Inhalation risk for humans is addressed in Chapter 7; however, the fate and transport factors affecting VI of 1,4-dioxane are more appropriately described in this section. Reliance on USEPA's (2015) definition of volatility based on numeric limits to Henry's law constants or vapor pressure is inappropriate for describing volatility of hydrophilic compounds, which are soluble in water in all proportions. A chemical's Henry's law constant is a measure of the chemical's ability to volatilize from water and, thus, is directly applicable to evaluating the potential for unacceptable VI risks from groundwater contamination. A vapor pressure greater than 1 mm Hg describes compounds that will volatilize from an open container of the pure compound. However, this measure is not a reliable indicator of volatility from groundwater or soil pore water when a compound is infinitely soluble in water. A binary system of a dissolved organic compound and water has its own vapor pressure, which for hydrophilic compounds is predominantly attributed to the vapor pressure of water (Hovorka et al., 1936). The vapor pressure of miscible compounds—if relevant at all to those chemicals' ability to present a VI risk—is only relevant when these chemicals are present in dry soil. Vapor pressure alone should not be used as an independent measure of a miscible compounds' ability to volatilize out of water.

The 2015 USEPA Vapor Intrusion Guidance relies on the 1991 Risk Assessment Guidance for Superfund (RAGS). At that time, the conventional thinking promulgated in the RAGS was that "risk from indoor inhalation of volatiles is assumed to be relevant only for chemicals that easily volatilize." (USEPA, 1991). On this basis, the 1991 RAGS recommends using a rule of thumb to determine whether a chemical is volatile as follows: "Chemicals with a Henry's law constant of greater than 1×10^{-5} atm-m^3/mole and a molecular weight of less than 200 g/mole." The 1991 RAGS notes that this guidance should be used to include chemicals that meet the criteria, but not necessarily to exclude chemicals that do not. Chemicals not meeting these criteria are recommended for site-specific evaluation during the feasibility study for a site. The 1991 RAGS definition of VOC would not explicitly include 1,4-dioxane as a VOC, because the most commonly cited Henry's law constant for 1,4-dioxane is 4.8×10^{-6} atm-m^3/mol (ATSDR, 2012; USEPA, 2014). The molecular weight of 1,4-dioxane is 88.106 g/mol (Lide, 2007).

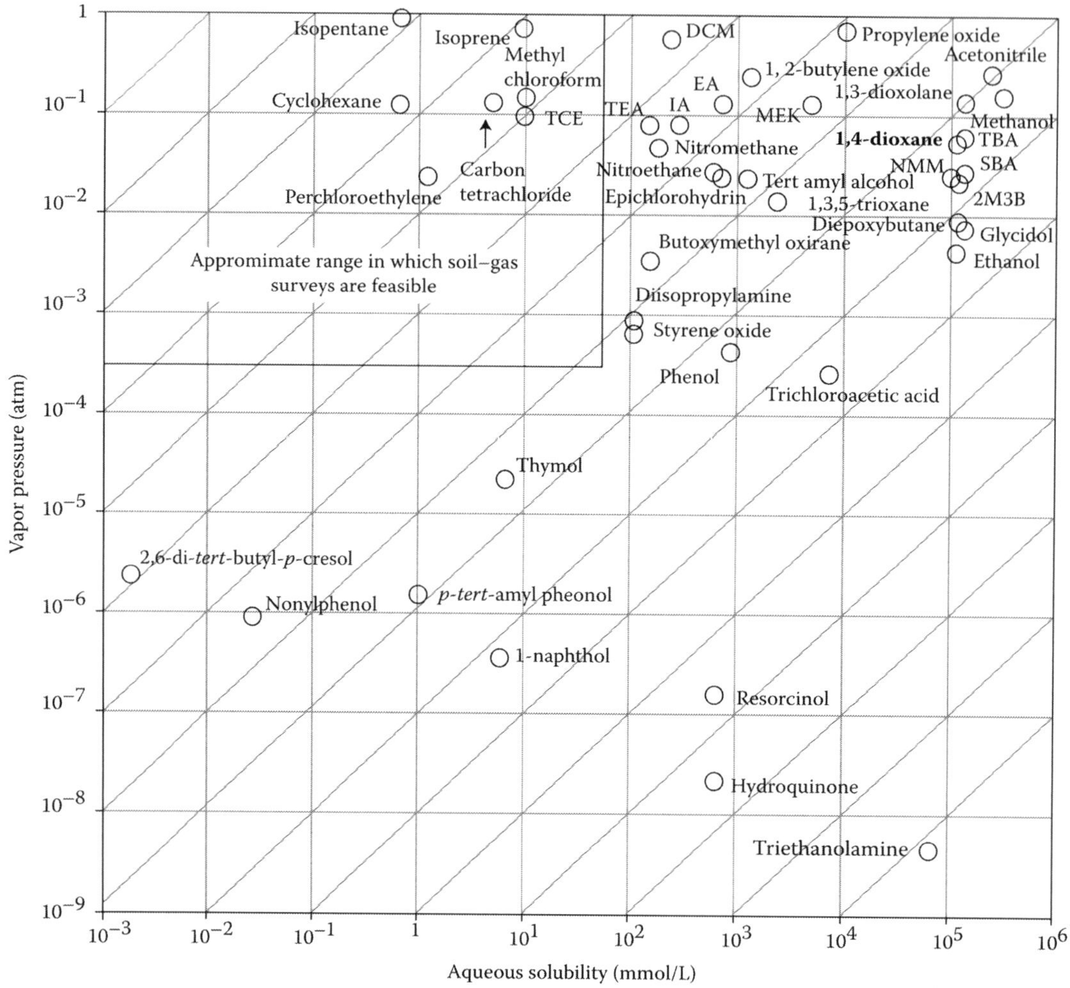

FIGURE 3.4 Plot of vapor pressure versus aqueous solubility for chlorinated solvents and solvent-stabilizer compounds. Diagonal lines represent Henry's law constant equivalents, increasing to the upper left (Boublik, 1984; Clayton, 1994; Hawley, 1977; Huntsman, 2005; Mallinckrodt, 2008; USEPA, 2000b, 2007c).

3.3.4 Biodegradability of Solvent-Stabilizer Compounds

Information presented in this chapter focuses on laboratory studies geared toward identifying organisms capable of 1,4-dioxane destruction or transformation, identification of degradation pathways, and theoretical considerations related to biodegradation. Chapter 8 further assesses various studies related to the possibility of field application of bioremediation as a technology to carry sites toward closure. Although there is some overlap between these two chapters, Chapter 3 primarily addresses the theoretical approaches, and Chapter 8, the applied approaches.

The ability of an organic compound to persist in soil and groundwater is inversely related to the potential for indigenous microflora to dissimilate it (White et al., 1996). Biodegradation refers to the general ability of indigenous soil microbes or fungi to acclimate to and metabolize a contaminant under ambient temperature conditions through structural transformation of the parent compound that changes its molecular integrity. Complete or ultimate biodegradation converts a compound to carbon dioxide and water ("mineralization"), whereas acceptable biodegradation is microbial transformation that results in reduced toxicity (Scow, 1982). Microbes can adjust their metabolism to degrade most anthropogenic organic compounds by induction of enzymes, by selecting new metabolic capabilities produced by genetic changes, or by increasing their population (Chappelle, 1993). Although fungi do not increase their population in response to introduction of a contaminant, some can use the contaminant as their sole carbon source.

The term "biodegradable" is qualified by the conditions under which biodegradation occurs. Biodegradable or nonbiodegradable are commonly used terms to indicate whether a compound will be susceptible to microbial destruction under ambient conditions. Some compounds widely considered to be nonbiodegradable under ambient conditions can be degraded under enhanced conditions, which may be created in *ex situ* bioreactors or *in situ* by injection of a substrate, inoculation of acclimated microbes, redox manipulation, temperature control, or adding nutrients essential to sustaining microbial populations.

The by-products of microbial metabolism can sometimes produce compounds that are more toxic and regulated at a lower threshold than the original contaminant. If this is the case, additional metabolic steps are necessary to detoxify the parent compound. The complete mineralization of the contaminant may involve multiple metabolic steps, possibly by different groups of microbes and at different metabolic rates. Biodegradation may be limited by the toxicity of the subject compound to microbes. The metabolic by-product may also be toxic to microorganisms, limiting further biodegradation after an initial step. "Complete biodegradation" occurs in six half-lives, after which only approximately 1.5% of initial contaminant concentration remains; after seven half-lives, less than 1% remains.

Testing for biodegradability often involves using pure cultures of a single strain of microbe to establish the identity of microbes responsible for biodegradation mechanisms and testing for intermediate and end products of biodegradation. Laboratory studies of biodegradation by mixed cultures of microbes are more relevant for determining biodegradability under ambient conditions; however, the conditions under which tests are performed are usually not representative of ambient conditions (Boethling et al., 1994). For example, temperatures in laboratory microcosm studies are elevated, flow rates are increased, growth media and nutrient solutions are enhanced, and high concentrations of the target contaminant are used to accelerate the process and complete the study within a reasonable time frame. However, caution is required when extrapolating microcosm results to field conditions. For example, while tests are often run on high concentrations of the contaminant, when chemicals are present at trace concentrations, microorganisms may not assimilate carbon from the chemicals, and they will not grow or produce the large, acclimated populations needed for enhanced biodegradation (Alexander, 1985). Tests representing ambient conditions may take a long time to complete. Data from different studies may not be comparable where different environmental conditions are used.

Microorganisms in groundwater are more often found attached to grains of aquifer material than in flowing groundwater, and most microbes are larger than the smallest pores in most clays; consequently, hydrophilic contaminants that have diffused into very small pores or intraparticle voids may not be directly subjected to biodegradation.

Microbes generally mediate biotransformation that is energetically favorable. Biotransformation reactions produce a net decrease in the free energy of the chemical system, and the microbes harvest some of the released energy for their own use, stimulating their growth (Hemond and Fechner, 1994). If a degradation pathway does not allow a microbe to harvest energy, the reaction is unlikely to proceed.

Microorganisms responsible for biodegradation include heterotrophic bacteria (including actinomycetes), some autotrophic bacteria, fungi including yeasts and basidiomycetes, and some protozoa. Fungi are less able to utilize recalcitrant compounds as growth substrates than are the bacteria, which possess greater metabolic capacity. Fungi function only aerobically, whereas bacteria may evolve to metabolize only in aerobic systems, only in anaerobic systems, or in both environments (i.e., facultative organisms) (Hemond and Fechner, 1994).

Most compounds become usable as an energy or nutrient source only after a period of acclimation during which the population capable of metabolizing the substance grows to a threshold where appreciable degradation becomes noticeable. The acclimation of a microbial community to a chemical may not occur below a threshold concentration; low concentrations of contaminants may be resistant to biodegradation (Alexander, 1994).

Acclimation also involves existing microbes adapting through induction of enzymes that catalyze biodegradation. The number of steps in the biodegradation sequence for a compound will affect the rate of biodegradation (Scow, 1982). Induction is the series of cellular processes by which microbes produce enzymes specific to a particular substrate when that substrate is actually present (Chappelle, 1993). Products generated during the catabolism* of one substrate may repress the synthesis of enzymes utilized for the degradation of a second substrate. This effect, called catabolite repression, may suppress the enzymatic activity of a population (Alexander, 1994).

The acclimation of a microbial community to one chemical substrate can result in acclimation to chemical substrates that are structurally similar. Cometabolism is the fortuitous degradation of a compound that microorganisms cannot utilize as growth substrates, that is, the compound does not provide a nutrient or energy source for the degrading organisms (Alexander, 1994; Hyman, 1999). For example, enzymes produced during the microbial degradation of the cyclic ether tetrahydrofuran (THF) also degrade 1,4-dioxane. If a synthetic organic compound is transformed by cometabolism, the mass of enzymes involved will be independent of the mass of the compound transformed and will be determined by the mass of the compound that induces the enzymes (Grady et al., 1997).

The rate of biodegradation is controlled by the availability and concentration of the compound to be degraded (e.g., adsorption favors availability) and by the microbe population and temperature of the system, as well as other environmental variables, such as pH, moisture content, and availability of soil organic matter or other organic substrates. Table 3.16 summarizes the conditions that control rates of biodegradation.

TABLE 3.16
Conditions Controlling Rates of Biodegradation

Substrate-Related	**Influence**
State	Liquid> solid> gas
Physicochemical properties	Sorption favors biodegradation; high solubility does not
Concentration	Upper and lower thresholds may restrict biodegradation
Size, shape, ionic charge	Permeation of cell membrane and interaction with enzymes
Toxicity	Specific or general growth-limiting toxicity
Molecular structure	Degree of branching and nature, number, and position of substituents; biodegradability of molecular structural units
Organism-Related	
Species composition of population	High populations favor biodegradation
History of acclimation to substrate	Affects rate; a lag period may be required
Inter- and intraspecies interactions	Competition for resources may favor biodegradation or prevent it if dominant strains do not degrade compound
Enzymatic composition and activity	Enzymes must catalyze contaminant breakdown
Environment-Related	
Temperature	Higher temperatures favor breakdown; upper limits apply
pH	Some pH ranges unfavorable for biodegradation
Moisture	Critical to nearly all biodegradation
Oxygen availability	Necessary to support aerobic biodegradation and fungi
Salinity	May suppress biodegradation above a threshold
Presence of other substances toxic to microbes	Can prevent sufficient population growth for biodegradation

Source: After Scow, K.M., 1982, Rate of biodegradation. In: W.J. Lyman, W.F. Reehl, and D.H. Rosenblatt (Eds), *Handbook of Chemical Property Estimation Methods*, pp. 9-1-9-85. Washington, DC: American Chemical Society and Knapp, J.S. and Bromley-Challenor, K.C.A., 2006, Recalcitrant organic compounds. http://www.bmb.leeds.ac.uk/mbiology/ug/ugteach/level3.html (accessed September 9, 2007).

* Catabolism is the metabolic process that breaks down molecules into smaller units.

TABLE 3.17
Index of Biodegradability According to BOD_5/COD Ratio

BOD_5/COD	Rating	Example Compounds
<0.01	Not degradable	1,4-Dioxane, propylene oxide, morpholine, triethanolamine
0.01–0.1	Moderately degradable	*n*-Butyl acetate, acetonitrile
>0.1	Relatively degradable	*n*-Propyl alcohol, *n*-butyl alcohol, pyridine, *n*-amyl alcohol, phenol

Source: Scow, K.M., 1982, Rate of biodegradation. In: W.J. Lyman, W.F. Reehl, and D.H. Rosenblatt (Eds), *Handbook of Chemical Property Estimation Methods*, pp. 9-1-9-85. Washington, DC: American Chemical Society.

The kinetics of microbial processes determine whether biodegradation can play an important role in reducing or eliminating a compound in the subsurface within the time frame of interest, for example, the proposed period of monitored attenuation following source removal in site remediation. The first-order kinetics follows a linear relationship between concentration reduction by microbial degradation and time, whereas the second-order kinetics follows a hyperbolic relationship (Scow, 1982).

A screening-level index of the biodegradability of organic compounds uses the ratio of biochemical oxygen demand from a 5-day test (BOD_5) to chemical oxygen demand (COD) according to the classification scheme shown in Table 3.17.

3.3.4.1 Predicting Biodegradability

Most of the thousands of organic compounds introduced to the marketplace and subsequently released to the environment are subjected to only screening-level tests for biodegradability. Reliable and consistently generated biodegradation data are not widely available. A goal in the assessment of the environmental persistence of organic compounds is the ability to use biodegradation models to predict whether the compounds will be mineralized by ambient soil microbes. Databases of consistently developed biodegradation data—such as the Japanese Ministry of International Trade and Industry (MITI) Test database, which includes more than 800 compounds—are gradually overcoming this limitation to biodegradation modeling, and predictions of biodegradability are improving (Pavan and Worth, 2006). One such biodegradation model is BIOWIN, a commonly used model developed by Syracuse Research Corporation and available as part of the EPIWIN Suite (USEPA, 2007a).

BIOWIN is an algorithm that predicts the potential for aerobic biodegradation from a compound's molecular structure and molecular weight. High-molecular-weight compounds are less prone to aerobic biodegradation. The BIOWIN classification of chemical compounds is based on the presence of structural fragments in a molecule that are also present in compounds whose biodegradability is known. A training set of 200 compounds whose biodegradation fates are well studied is used to predict the biodegradation potential for compounds that include the same structural fragments (Boethling et al., 1994). For example, ester, alcohol, and carboxylic acid groups usually enhance biodegradability, whereas the presence of halogens and nitro groups generally increases resistance to biodegradation (Scow, 1982). Alcohols, aldehydes, acids, esters, and amides are more susceptible to biodegradation than are the corresponding alkanes, olefins, ketones, dicarboxylic acids, nitriles, amines, and chloroalkanes. Ether functions are particularly resistant to biodegradation because of the high dissociation energy of the ether linkage (about 360 kJ/mol) (Kim and Engesser, 2004). Highly branched compounds are also more resistant to biodegradation than are straight chains, and short chains are not as quickly degraded as long chains. Halogenated compounds that resist aerobic degradation may be more rapidly degraded under anaerobic conditions (Aronson et al., 1999). Unsaturated compounds (alkenes or alkynes) are more susceptible to biodegradation than saturated compounds (alkanes). Table 3.18 lists output from the BIOWIN model for predicting biodegradability of solvent stabilizers and chlorinated solvents, and Table 3.19 gives ratings for biodegradability

TABLE 3.18
BIOWIN-Predicted Biodegradability of Chlorinated Solvents and Solvent-Stabilizer Compounds

Chemical Name[a]	Readily Biodegradable?	Time Frame	BIOWIN1 Ranking	MITI Probability[b]
Triethanolamine	Yes	Weeks	0.95	0.96
Ethyl acetate	Yes	Weeks	0.88	0.84
Isopropyl acetate	Yes	Weeks	0.87	0.70
2-Butanone	Yes	Weeks	0.72	0.67
Acetonitrile	Yes	Weeks	1.04	0.66
sec-Butyl alcohol	Yes	Weeks	0.87	0.65
1,2-Butylene oxide	Yes	Weeks	0.37	0.58
Cyclohexane	Yes	Weeks	0.71	0.58
Propylene oxide	Yes	Weeks	0.37	0.57
tert-Amyl alcohol	Yes	Weeks	0.52	0.57
tert-Butyl alcohol	Yes	Weeks	0.53	0.56
Butoxymethyl oxirane	Yes	Weeks	0.10	0.56
1,3-Dioxolane	Yes	Weeks	0.02	0.55
Nitroethane	Yes	Weeks	0.71	0.54
Phenol	Yes	Weeks	0.95	0.54
1,4-Dioxane	**Yes**	**Weeks**	**0.01**	**0.53**
Nitromethane	Yes	Weeks	0.72	0.53
2-Methyl-3-butyn-2-ol	Yes	Weeks	0.52	0.53
Diepoxybutane	Yes	Weeks	0.01	0.52
Epichlorohydrin	Yes	Weeks	0.24	0.52
Dichloromethane	*No*	*Months*	*0.48*	*0.51*
1,3,5-Trioxane	Yes	Weeks	−0.34	0.51
Pyridine	No	Weeks	0.56	0.48
Triethylamine	No	Months	0.49	0.48
Diisobutylene	No	Months	0.51	0.45
Nonylphenol	No	Weeks	0.92	0.44
p-*tert*-Amyl phenol	No	Months	0.60	0.44
Styrene oxide	No	Weeks	0.47	0.41
N-Methylmorpholine	No	Months	0.15	0.41
Methyl chloroform	*No*	*Months*	*0.17*	*0.39*
Diisopropylamine	No	Weeks	0.85	0.34
Trichloroethylene	*No*	*Months*	*0.35*	*0.33*
Carbon tetrachloride	*No*	*Months*	*0.04*	*0.33*
2,6-di-*tert*-Butyl-*p*-cresol	No	Months	0.45	0.32
Perchloroethylene	*No*	*Months*	*0.22*	*0.22*

[a] Nonitalic compounds are stabilizers. Italic compounds are chlorinated solvents; these are listed for comparison to show whether the stabilizer is likely to evaporate with the solvent or remain behind in the soil where it can be leached down to groundwater. Data for 1,4-dioxane are shown in boldface.

[b] "MITI Probability," also called "Inherent Biodegradability," is determined by the Modified MITI Test, an activated sludge assay of biochemical oxygen demand, from the Chemicals Inspection, and Testing Institute, Japan (OECD, 1981). MITI probability values are included in BIOWIN output.

TABLE 3.19
Summary of Biodegradability Test Results from the Literature in Syracuse Research Corporation's BIODEG Database

Compound[a]	Aerobic Summary[b]	Qual[c]	Anaerobic Summary	Qual[c]	Year Filed	Number of References
tert-Amyl alcohol	Slow + Acc	1	—	—	1985	5
1,4-Dioxane	**Slow**	**1**	**Slow**	**3**	**1984**	**3**
tert-Butyl alcohol	BST – 1	1	—	—	1984	15
Nonylphenol	BST – 1	1	—	—	1986	3
Pyridine	BST – 1	1	Fast + Acc	3	1986	6
Acetonitrile	Fast + Acc	1	—	—	1987	6
Triethanolamine	Fast + Acc	1	—	—	1985	9
Dichloromethane	Fast + Acc	1	—	—	1987	5
Diisopropylamine	Fast + Acc	2	—	—	1988	3
Carbon tetrachloride	Fast + Acc	2	Fast	1	1987	9
Phenol	Fast	1	BST – 1	1	1986	66
Isopropyl acetate	Fast	1	—	—	1986	2
2-Butanone	Fast	1	—	—	1986	10
sec-Butyl alcohol	Fast	1	—	—	1987	10
2,6-di-*tert*-Butyl-*p*-cresol	Fast	1	Fast	3	1986	6
Ethyl acetate	Fast	1	Fast	3	1987	9

[a] Nonitalic compounds are stabilizers. Italic compounds are chlorinated solvents; these are listed for comparison to show whether the stabilizer is likely to evaporate with the solvent or remain behind in the soil where it can be leached down to groundwater. Data for 1,4-dioxane are shown in boldface.

[b] BIODEG Evaluation Reliability Codes ("Qual"): Fast = biodegrades fast; Fast + Acc = biodegrades fast with acclimation; Slow = biodegrades slowly; Slow + Acc = biodegrades slowly with acclimation; and BST = biodegrades sometimes.

[c] Reliability ratings: Qual = 1: chemical tested in three or more tests, consistent results. Qual = 2: chemical tested in two tests, or results in more than two tests are interpretable, but conflicting data. Qual = 3: only one test or uninterpretable, conflicting data.

interpreted from literature studies and collated in the Syracuse Research Corporation's BIODEG database (SRC, 2007a).

Because of imperfections in the prediction algorithm, estimates of biodegradability predicted by using BIOWIN must be used conservatively. An evaluation of the reliability of BIOWIN predictions determined that the prediction "not *readily degradable*" is highly accurate (correctly predicted for more than 90% of compounds); however, the prediction "*readily degradable*" often *does not agree* with MITI test data. A linear regression of biodegradability predicted by BIOWIN and results from MITI tests had a rather low correlation coefficient, $R^2 = 0.58$ (Arnot et al., 2005). If BIOWIN predicts fast biodegradation, the estimate should be independently verified. If the program predicts slow biodegradation, this result can be reliably used as a confirmation that the compound is not readily biodegradable. The presence of a biodegradation-enhancing fragment generally indicates a *possible* metabolic step, which does not necessarily lead to complete mineralization (Pavan and Worth, 2006).

3.3.4.2 Biodegradability of 1,4-Dioxane: Laboratory Studies

Research into the biodegradability of 1,4-dioxane was conducted first in the context of treating industrial effluents from processes that produce 1,4-dioxane as a by-product, for example, polyethylene terephthalate plastic manufacturing, or processes that use 1,4-dioxane in the manufacture of goods, for example, magnetic tape and cellulose acetate membranes. Recent research into 1,4-dioxane biodegradability is motivated by the desire to develop effective *in situ* remedial technologies or viable *ex situ* treatments.

Engineered aerobic wastewater treatment systems in Japan decrease but do not completely remove elevated influent concentrations of 1,4-dioxane (1,000 μg/L) in activated sludge processes (Abe, 1999). Low influent concentrations of 1,4-dioxane (2–3 μg/L) at the City of Ann Arbor, Michigan, wastewater treatment plant are unchanged in effluent (Skadsen et al., 2004). The general conclusion of the growing body of research into biodegradation of 1,4-dioxane is that it does not biodegrade appreciably under ambient conditions; however, in enhanced laboratory-controlled environments and in engineered field systems, 1,4-dioxane can be degraded completely. The current state of the practice, which has developed substantially since the early 2000's, is biodegradation of 1,4-dioxane by cometabolism, as will be discussed in Chapter 8. Table 3.20 summarizes key research on the capacity of microbes to degrade 1,4-dioxane.

TABLE 3.20
Summary of Studies of Biodegradation of 1,4-Dioxane

Substrate: 1,4-Dioxane Alone

Pure Culture

Bernhardt and Diekmann (1991): Six *Rhodococcus* strains tested; none grew on 1,4-dioxane alone

Burback and Perry (1993): *Mycobacterium vaccae* Strain JOB5 catabolized 100 ppm 1,4-dioxane <50%, that is, growth not supported

Parales et al. (1994): Isolated Strain CB1190 from THF enrichments; 1,4-dioxane degraded at 30°C in 18 hours from 4 mM to <10 μM without THF

Kim and Engesser (2004): *Rhodococcus* Strain DEE5151 from diethyl ether enrichment cultures did not degrade 1,4-dioxane

Mahendra and Alvarez-Cohen (2005): Isolated *Pseudonocardia dioxanivorans,* the only known pure culture that grows on DX alone. Cells grew at 30°C for up to 28 days

Mahendra and Alvarez-Cohen (2006): 1,4-dioxane served as a growth substrate for *Pseudonocardia dioxanivorans* Strain CB1190 and *Pseudonocardia benzenivorans* Strain B5

Huang et al. (2014): *Acinetobacter baumannii DD1* (DD1), degraded 1,4-dioxane aerobically

Sei et al. (2013): Four soil bacteria (*actinomycetes sp. Strains D6, D11, and D17,* and *Afipia sp. Strain D1*) can use 1,4-dioxane as a sole carbon source

Mixed Culture

Fincher and Payne (1962): No oxygen demand on 1,4-dioxane at 30°C for 72–96 hours

Klecka and Gonsior (1986): No oxygen demand in a 20-day BOD test on 1,4-dioxane. No biodegradation of 1,4-dioxane by sewage cultures at 1 year; aerobic shaker bath at 30°C achieved 44.5% 1,4-dioxane removal in 32 days including a 10-day lag

Roy et al. (1994): 1,4-dioxane completely degraded at 150 mg/L; sludge microbes utilize 1,4-dioxane as sole source of carbon and energy. Incomplete biodegradation at concentrations of >150 mg/L; by-product toxicity indicated

Taylor et al. (1997): Soil microbes did not degrade 1,4-dioxane aerobically or anaerobically on representative groundwater samples

Aoyagi et al. (2018): Isolates from activated sludge treatment at a petrochemical plant degraded 1,4-dioxane; ^{14}C tagged carbon identified in 10 organisms

Fungi and Other

Patt and Abebe (1995), Deshpande (1992): *Aureobasidium pullulans* grew on 1,4-dioxane after lag of 36 hours, 90% of 50 ppm 1,4-dioxane was degraded in the next 36 hours

Taylor et al. (1997): Fungi from 1,4-dioxane-contaminated site soils did not degrade 1,4-dioxane

Nakamiya et al. (2005): *Cordyceps sinensis* (a fungal insect pathogen) cultivated with 1,4-dioxane as sole source of carbon; 1,4-dioxane degradation nearly complete (90%) after 3 days. Degradation product was ethylene glycol

Substrate: 1,4-Dioxane with THF as Cometabolite

Pure Culture

Bernhardt and Diekmann (1991): Six *Rhodococcus* strains tested; 1,4-dioxane completely degraded with 7.5 mM THF

Mahendra and Alvarez-Cohen (2006): See listing below

Vainberg et al. (2006): Biodegradation of 1,4-dioxane with THF by *Pseudonocardia* sp. Strain ENV478

(Continued)

TABLE 3.20 *(Continued)*
Summary of Studies of Biodegradation of 1,4-Dioxane

Mixed Culture

Zenker et al. (1999): Complete biodegradation of 1,4-dioxane and THF by sludge microbes at 35°C, biodegradation of 1,4-dioxane dependent upon THF presence

Zenker et al. (2000): Biodegradation of 1,4-dioxane with THF as cometabolite; 1,4-dioxane without THF failed to stimulate biodegradation by soil microbes

Fungi and Other

Kelley et al. (2001): Phytoremediation with hybrid poplars, CB1190 + THF; 100 mg/L 1,4-dioxane removed within 45 days with cosubstrates THF and 1-butanol

Substrate: 1,4-Dioxane with Other Cometabolites

Pure Culture

Hyman (1999): *Mycobacterium vaccae* Strain JOB-5 cometabolically degrades cyclic ethers after growth on straight-chain/branched alkanes; catalyzed by alkane monooxygenases

Mahendra and Alvarez-Cohen (2006): Cometabolic transformation of 1,4-dioxane observed for monooxygenase-expressing strains induced with methane, propane, tetrahydrofuran, and toluene: *Methylosinus trichosporium* Strain OB3b, *Mycobacterium vaccae* Strain JOB5, *Pseudonocardia* Strain K1, *Pseudomonas mendocina* Strain KR1, *Ralstonia pickettii* Strain PKO1, *Burkholderia cepacia* Strain G4, and *Rhodococcus* Strain RR1

Vainberg et al. (2006): Biodegradation of 1,4-dioxane with sucrose, lactate, yeast extract, 2-propanol, and propane, by *Pseudonocardia* sp. strains ENV478 and ENV425

Deng et al. (2018a): *Azoarcus sp.* strain DD4 degraded 1,4-dioxane using propane as the substrate

Deng et al. (2018b): *Mycobacterium dioxanotrophicus PH-06* degraded 1,4-dioxane using propane as the substrate

Hatzinger et al. (2017): *Mycobacterium sphagni* degraded 1,4-dioxane using ethane as a substrate

Mixed Culture

Hatzinger et al. (2017): An ethane oxidizing culture from Myrtle Beach AFB degraded 1,4-dioxane using propane as a substrate

Fungi and Other

Skinner (2007): *Graphium* sp., a filamentous fungus, degrades DX by cometabolism after growth on THF or gaseous alkanes, catalyzed by cytochrome P450 monooxygenase Kelley et al. (2001): phytoremediation with 1-butanol

Notes: DX = 1,4-dioxane; THF = tetrahydrofuran.

Most of the research into 1,4-dioxane biodegradability has focused on aerobic biodegradation. Anaerobic biodegradation of cyclic ethers is not widely reported. Microcosm studies on soils from dioxane-contaminated sites in Maryland and New York showed no degradation under anaerobic conditions (Steffan, 2006). USEPA's guidance document (1999) for implementation of monitored natural attenuation remedies describes protocols for conducting microcosm tests for anaerobic biodegradation. To dissolve hydrophobic or nonaqueous-phase liquids, the guidance recommends using 1,4-dioxane instead of alcohol, because alcohols will biodegrade anaerobically whereas 1,4-dioxane will not (USEPA, 1999).

Aliphatic (straight, hydrocarbon chain) ethers such as methyl *tert*-butyl ether (MTBE) can be degraded anaerobically (Pruden et al., 2005). MTBE will mineralize in anoxic conditions at slow but measurable rates, beginning with cleavage of the ether linkage (Zenker, 2006). Cyclic ethers like 1,4-dioxane may also be cleavable under strongly anaerobic conditions. Tests of anaerobic biodegradation of the cyclic ether THF with methanotrophic bacteria showed less than 30% of the theoretical methane production (Battersby and Wilson, 1989). Humic acids have many ether linkages in their structure and may be a candidate for stimulating cometabolic breakdown of 1,4-dioxane in both oxic and anoxic conditions (Zenker, 2006). Anaerobic biodegradation of 1,4-dioxane under humic and iron-reducing conditions has been reported (Pan and Chen, 2006; Shen et al., 2008), although newer research demonstrates hydroxyl radicals produced as iron cycles between oxidation and reduction can degrade 1,4-dioxane abiotically (Sekar et al., 2016), suggesting the observed degradation may be abiotic.

The ether linkage, -C-O-C-, is commonly encountered in nature (lignins) and in anthropogenic compounds (agrochemicals, detergents) and is generally resistant to biological mineralization (White et al., 1996). The intrinsic refractivity of the ether linkage to biological destruction was recognized early in the history of biodegradation research (Alexander, 1965). Bacterial scission of ether bonds is not thermodynamically favored, as noted in a seminal paper surveying the topic (White et al., 1996): "Microbial cleavage of the ether bond is a remarkable phenomenon, since the C-O bond energy* is 360 kJ/mol and necessarily demands an appreciable investment of energy to effect its fission in relation to the sometimes relatively small yield of assimilable carbon." The enzymes responsible for bacterial scission of ether bonds were called etherases in early work and were believed to catalyze a hydrolysis reaction that cleaves the bond; however, catalysis of the ether bond division is now attributed to a heterogeneous group of enzymes that exhibit a variety of mechanisms to break the ether bond. Hydrolysis does not play an important role in cleaving the ether bond (White et al., 1996).

Table 3.20 summarizes the research into metabolism and cometabolism of 1,4-dioxane by pure and mixed cultures of bacteria and fungi. Where biodegradation has been documented, the rate and sustainability must also be determined. Biodegradation that occurs too slowly will not have a significant impact on 1,4-dioxane concentrations in the subsurface environment if migration rates substantially exceed degradation rates. Intermediate products that are toxic to and inhibit the growth of microflora, such as 2-hydroxyethoxyacetic acid (HEAA, described below) will also limit the degree to which biodegradation lowers concentrations of 1,4-dioxane in groundwater.

Rates of biodegradation of 1,4-dioxane can be described using the microbial growth equation (Monod, 1949) as presented in the literature (Chappelle, 1993; Evans et al., 2006):

$$r = \frac{\mu_{max} XS}{Y(K_s + S)}, \tag{3.40}$$

where r is the dioxane biodegradation rate, μ_{max} is the maximum growth rate, X is the bacteria concentration, S is the dioxane concentration, Y is the growth yield on dioxane, K_s is the half-saturation constant, and r/X is the specific dioxane biodegradation rate.

The minimum concentration that will maintain growth in a biological treatment system under steady-state conditions ($S_{S,min}$; Alexander, 1994) is

$$S_{S,min} = \frac{K_s b}{\mu_{max} - b}. \tag{3.41}$$

Table 3.21 summarizes published parameters of biodegradation kinetics for bacteria capable of metabolizing 1,4-dioxane.

Information on kinetics is extremely important because it characterizes the concentration of the chemical remaining at any time, permits prediction of the levels likely to be present at some future time, and allows assessment of whether the chemical will be eliminated before it is transported to a site at which exposure will occur (Alexander, 1994).

The relationship between bacterial growth rate and the concentration of the carbon substrate increases, but only at low substrate concentrations. A concentration is reached above which the growth rate falls as the substrate level rises further. This decline is a result of the antimicrobial action (i.e., toxicity to microbes) of the chemical substrate at high concentrations (Alexander, 1994).

* kJ = kilojoules; 1 kJ = 0.239 kilocalories (kcal); 360 kJ/mol = 86 kcal/mol. Alexander (1994) lists the ether linkage bond energy as 85.5 kcal/mol.

TABLE 3.21
Kinetics of 1,4-Dioxane Biodegradation

Culture	Temperature (°C)	Maximum Growth Rate (μ_{max}/h)	Yield (mg/mg)	Half-Saturation Constant, K_s (mg/L)	Specific Rate, r/X [mg/(g h)]	References
Soil Bacteria	25		0.179-1.223	20.6–69.8	52–263 at 100 ppm	Sei et al. (2013)
CB1190	30	0.0230	—	—	9.9 at 500 ppm	Adamus et al. (1995)
Mixed	25	0.0100	0.33	13.51	27 at 100 ppm	Grady et al. (1997)
Mixed	35	0.0430	0.64	1.04	66 at 100 ppm	Grady et al. (1997)
Mixed	—	0.141	0.12	9.9	0.12	Grady et al. (1997)
Industrial activated sludge	—	0.062	0.44	1.65	—	Sock (1993) (cited in Zenker et al., 2002)
Nocardioform actinomycete	30	0.033	0.02	—	1.98	Parales et al. (1994)
Industrial activated sludge	—	0.0053	0.22	182	—	Roy et al. (1995)
Pseudonocardia strain ENV478	30				21 at 100 μg/L DX	Vainberg et al. (2006)
Mixed	30	0.083	0.75	13 ± 7.6	0.05 ± 0.003 at 200 ppm	Zenker et al. (2000)
Pseudonocardia dioxanivorans CB1190	30	0.10	0.09	160 ± 44	1.1 ± 0.008 at 100 ppm	Mahendra and Alvarez-Cohen (2006)
Pseudonocardia benzenivorans B5	30	0.07	0.03	330 ± 82	0.1 ± 0.006 at 100 ppm	Mahendra and Alvarez-Cohen (2006)
Amycolata sp. CB1190	~24	—	0.01	—	0.92 ± 0.29 at 100 ppm	Kelley et al. (2001)
Pseudonocardia strain ENV478	30	—	—	—	19	Vainberg et al. (2006)

Source: From Grady, C.P.L., Jr., Sock, S.M., and Cowan, R.M., 1997, Biotreatability kinetics. In: G.S. Sayler, J. Sanseverino, and K.L. Davis (Eds), *Biotechnology in the Sustainable Environment*. New York: Plenum Press. With permission.

Note: The last entry, for *Pseudonocardia* Strain ENV478, is for biodegradation of 1,3-dioxolane. The other entries are for biodegradation of 1,4-dioxane.

The Monod equation (Monod, 1949) describes the growth of a population on a chemical substrate constrained by the toxicity to the population of that substrate in high concentrations:

$$\mu = \frac{\mu_{\max} S}{K_s + S + \left(S^2 / K_1\right)} \approx \frac{\mu_{\max}}{K_s + S}, \tag{3.42}$$

where μ is the specific growth rate of the bacterium, μ_{max} is its maximum specific growth rate, S is the concentration of the chemical substrate, K_s is a constant that represents the concentration of the chemical substrate at which the rate of growth is half the maximum rate, and K_1 is the inhibition constant that represents the suppression of the growth rate by the toxic substrate (Alexander, 1994). This equation may apply to 1,4-dioxane, whose metabolic by-product, 2-hydroxyethoxyacetic acid (HEAA), is toxic to microorganisms, as described in research by Vainberg et al. (2006) (see Chapter 8).

Bacterial growth requires that the substrate be present at sufficiently high concentrations relative to the cell population that acts on the substrate to allow the population to continue doubling. When cell density is higher than the substrate concentration can sustain, growth is limited.

Fungal biomass increases through hyphal lengthening* and branching instead of by the splitting and doubling (binary fission) mechanism used by bacteria. Fungal growth occurs primarily at the ends of the filaments, which grow at a constant rate. Because fungi and other filamentous organisms such as actinomycetes may grow radially in three dimensions, growth may follow cubic kinetics; however, growth in the subsurface is constrained by the porous matrix.

Recent studies explore the inhibition of 1,4-dioxane degradation in water by the presence of chlorinated VOC co-contaminants (Mahendra et al., 2013; Zhang et al., 2016). Inhibition is expressed as a constant (K_i) within the microbial growth equation, that divides the starting concentration of the inhibiting compound (Mahendra et al., 2013; Zhang et al., 2016). Therefore, inhibition is proportionate to the co-contaminant concentration and inversely proportionate to K_i. Because studies of inhibition involve determining kinetic constants with varying concentrations of 1,4-dioxane and co-contaminants, many kinetic constants are available for the many cases tested. The reader is referred to the referenced studies for the constants (Mahendra et al., 2013; Zhang et al., 2016). To summarize the results of these studies, methyl chloroform was found to not inhibit degradation of 1,4-dioxane at concentrations up to 50 mg/L, while TCE and cis-1-2-dichloroethylene (cDCE) did not inhibit degradation of 1,4-dioxane at 5 mg/L but did inhibit degradation at 50 mg/L. 1,1-Dichloroethylene (1,1-DCE) inhibited 1,4-dioxane degradation at 5 mg/L, with no degradation of 1,4-dioxane observed at 50 mg/L. However, a more recent study (Zhao et al., 2018) found less inhibition in column studies under flow-through conditions and suggested microcosm studies exacerbate inhibition.

A microcosm study noted that the presence of non-cometabolism stimulating substrates also hindered methane degradation (Li et al., 2017). The presence of more readily degradable compounds prevented cometabolism from degrading 1,4-dioxane, at least until the more stimulating substrates were depleted.

In addition to co-contaminant inhibition, 1,4-dioxane degradation is impacted by the types of enzymes generated by the cometabolic process. Methanotrophic degradation of 1,4-dioxane is mediated by soluble methane monooxygenase (soluble MMO or sMMO), but not by particulate MMO (pMMO) (Mahendra and Alvarez-Cohen, 2006). sMMO is only produced under copper-deficient conditions (Stanley et al., 1983; Prior and Dalton, 1985). If adequate copper is available to the microorganisms, pMMO will be produced in lieu of sMMO. Since pMMO does not degrade 1,4-dioxane, methane oxidation will not lead to 1,4-dioxane degradation under excessive copper conditions (Mahendra and Alvarez-Cohen, 2006). This effect is not reported for other common co-contaminants, and potentially makes treatment of 1,4-dioxane by methane cometabolism a more difficult process.

The pathway for aerobic degradation of 1,4-dioxane parallels that of THF, which involves oxygenase-mediated hydroxylation of the carbon present in the no. 2 position on the dioxane ring and its subsequent dehydrogenation to form the lactone ring. Hydroxylation is any process that introduces one or more hydroxyl groups (-OH) into a compound (or radical), thereby oxidizing it. In biochemistry, hydroxylation reactions are often facilitated by enzymes called hydroxylases. A lactone is a cyclic ester. After opening the lactone ring, intermediary metabolism pathways may begin (Zenker, 2006).

3.3.4.3 Synopsis of 1,4-Dioxane Biodegradability Studies

The key studies of 1,4-dioxane biodegradation listed in Table 3.20 are briefly summarized in this section.

One of the earliest studies of the degradation of ethers was by Fincher and Payne (1962). The study was conducted as part of a waste-stream treatability test for effluent from manufacturing

* A hypha (plural hyphae) is a long, branching filamentous cell of a fungus and also of unrelated *Actinobacteria*. In fungi, hyphal lengthening is the main mode of growth.

processes that produced ether glycols as a by-product, such as in the manufacturing of explosives, glues, cosmetics, pharmaceuticals, shampoos and sundries, and synthetic detergents. In this study, the author's evaluated utilization of ethers for growth by using a presumed *Pseudomonas-Achromobacter* member (designated TEG-5). The test assayed cells for oxidative activity in cultures maintained at 30°C for 72–96 hours in a glycol-basal salt medium. No oxidative activity or other symptoms of growth were observed in the assay for 1,4-dioxane. Fincher and Payne determined that 1,4-dioxane was not utilized by the bacteria for growth and therefore was not directly degraded by the bacteria studied.

Klecka and Gonsior (1986) observed negligible oxygen consumption in a 20-day test for biochemical oxygen demand for 1,4-dioxane. They also did not observe any biodegradation of 1,4-dioxane by mixed cultures of sewage microorganisms exposed for 1 year in a wastewater treatment plant where influent concentrations of 1,4-dioxane ranged from 100,000 to 900,000 μg/L. No toxicity to microorganisms from 1,4-dioxane concentrations up to 300,000 μg/L was observed. Klecka and Gonsior also conducted an aerobic shaker bath test at 30°C and measured 44.5% removal of 1,4-dioxane after 32 days of treatment, including an initial 10-day lag time during which there was little respirometric response. Their study confirmed that aerobic bacteria in sludge could degrade 1,4-dioxane as a primary source of carbon and energy. However, instead of complete destruction, their data suggested conversion of 1,4-dioxane to intermediates and not complete mineralization; they inferred that the metabolic by-products may be toxic to microorganisms and inhibit their growth.

Bernhardt and Diekmann (1991) studied degradation of 1,4-dioxane, THF, and other cyclic ethers by the actinobacterium strain *Rhodococcus* from forest soil or sludge that was obtained from the settling basin of an aerobic wastewater purification plant at a chemical firm. Despite considerable effort, no strains were enriched or isolated when 1,4-dioxane or cyclohexane was the sole carbon substrate. Six strains belonging to the genus *Rhodococcus* that degrade THF were isolated and classified. A strain that degraded dioxane instead of or in combination with THF was further characterized. *Rhodococcus* Strain 219 grows fast and degrades 1,4-dioxane at an optimal temperature of 30°C. The maximum growth rate, μ_{max}, of Strain 219 on 1,4-dioxane in the presence of 7.5 mM THF was 0.019 hours^{-1}. Growth occurred in as little as 0.22 mM (16,000 μg/L) THF. In the shaker test, at the end of the logarithmic growth phase, no THF or 1,4-dioxane could be detected by gas chromatography; COD measurement in the supernatant after centrifugation was zero, which was interpreted to mean that no intermediates had accumulated and the substances were totally destroyed. A total of 34 compounds containing nitrogen or oxygen, including the stabilizers 1,3-dioxolane, morpholine, and *n*-methyl morpholine, were tested. Oxygen demands were 0.37 and 2.32 for 1,3-dioxolane and 1,4-dioxane, respectively, and zero for the morpholine compounds. The presumed primary products of THF and 1,4-dioxane degradation were THF-2-ol and 1,4-dioxane-2-ol, but these compounds were not analytically identified.

Burback and Perry (1993) used a pure culture of the propane-oxidizing actinobacterium *Mycobacterium vaccae* [ATCC 29678 (JOB-5)] to demonstrate the biodegradability of 1,4-dioxane. A total of 100 ppm 1,4-dioxane was added to 100 mL of a 1 mg/mL cell suspension and incubated at 30°C in a rotating shaker. Vials were removed at 12, 24, and 48 hours, with zero hours as the control, and extracted with ethyl ether for analysis. The bacterium was able to use propane, acetone, and toluene as a sole carbon source, but 1,4-dioxane and the other contaminants evaluated could not sustain growth and were therefore determined to be only cometabolically degraded.

Roy et al. (1994) published a study on the biodegradation of 1,4-dioxane and diglyme in industrial waste. They recorded complete degradation of 1,4-dioxane at 150,000 μg/L based on an oxygen-uptake curve. Roy et al. interpreted the data to mean that microbes utilized 1,4-dioxane as the sole source of carbon and energy below 150,000 μg/L and that there was incomplete biodegradation at concentrations greater than 150,000 μg/L, indicating production of an inhibitory toxic by-product of 1,4-dioxane.

In a follow-on paper (Roy et al., 1995), the growth of microorganisms on 1,4-dioxane was modeled to estimate biodegradation kinetic parameters by using respirometric data. The growth data were obtained from experiments that used microbes from an acclimatized industrial waste to investigate the biodegradability of pure 1,4-dioxane. Growth of the microbes was measured by monitoring oxygen uptake in a respirometer. Experiments were performed on a pure solution of 1,4-dioxane (300,000 and 670,000 μg/L) in electrolytic respirometers as part of a 1989 Master's degree thesis (Anagnostu, 1989). The modeling assumed no stable by-products, no toxicity to microorganisms from dioxane, and Monod kinetics. From a best fit to experimental data of 1,4-dioxane depletion and oxygen uptake, the maximum specific growth rate, μ_{max}, was found to be 0.127 day^{-1}, the saturation constant, K_s, was 182,000 μg/L, and the cell yield was 0.218.

Grady et al. (1997) reported that 1,4-dioxane was very difficult to biodegrade because of the stability of the two ether bonds in the 1,4-dioxane saturated heterocyclic structure. Their study confirmed that 1,4-dioxane could serve as a sole source of carbon and energy through a series of bioreactor experiments using 1,4-dioxane concentrations up to 150,000 μg/L. Growth on 1,4-dioxane alone was sustained for 5 weeks within a complex bacterial community with several bacterial genera present; however, they were unable to isolate a pure culture capable of converting carbon to energy from 1,4-dioxane alone. Biodegradation of 1,4-dioxane followed Monod kinetics with no indication that either substrate or product inhibition occurred at 1,4-dioxane concentrations up to 2,100,000 μg/L. Grady et al. determined that growth is particularly sensitive to temperature, as profiled in Figure 3.5. At low temperatures, the growth rate of microbes on 1,4-dioxane would be too slow to maintain itself without getting flushed out of the bioreactor. This kinetics-based failure to biodegrade 1,4-dioxane occurred especially in winter when bioreactor temperatures decreased to 20°C; it was attributable neither to an inability of bacteria to degrade 1,4-dioxane nor to any inhibitory toxic characteristics.

Hyman (1999) studied the structural features that influence the reactivity of a single, nonspecific monooxygenase enzyme toward a variety of ether compounds. Hyman noted that many ether-bonded compounds are oxidized by microorganisms that cannot further metabolize the products of those reactions and that relatively few microorganisms appear to be able to utilize ether-containing compounds as growth substrates. Hyman's study examined the rates of cometabolism of six ethers

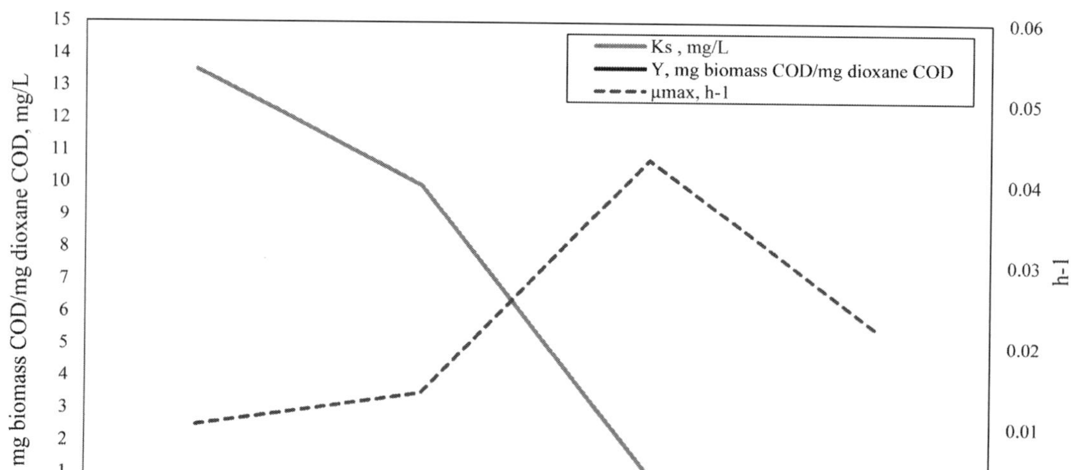

FIGURE 3.5 Effect of temperature on 1,4-dioxane biodegradation kinetics in bioreactor experiments. (From Grady, C.P.L., Jr., Sock, S.M., and Cowan, R.M., 1997, Biotreatability kinetics. In: G.S. Sayler, J. Sanseverino, and K.L. Davis (Eds), *Biotechnology in the Sustainable Environment*. New York: Plenum Press. With permission.)

and the degradation products obtained by bacteria and fungi. The study also investigated the physiological consequences of ether cometabolism on bacteria. Hyman found that the alkane monooxygenase (AMO) enzyme is largely unreactive to cyclic ethers such as tetrahydropyran (THP) and dioxane isomers; even though this enzyme oxidizes several sulfur analogs of the cyclic ethers. The main focus of Hyman's study was ether oxidation by propane-oxidizing bacteria. AMO is a bacterial enzyme; its capacity for oxidizing specific carbon-hydrogen bonds in some organic compounds leads to cometabolic degradation of recalcitrant contaminants.

Hyman's study used *Mycobacterium vaccae* JOB-5 grown on straight-chain (C3–C8) and branched alkanes to cometabolically degrade the cyclic ethers THF, THP, hexamethylene oxide (HMO), and the dioxane isomers 1,3-dioxane and 1,4-dioxane. The oxidation of all of these compounds was apparently catalyzed by AMO, because all of the transformation reactions were inhibited by acetylene and only occurred in cells grown on alkane substrates. Hyman noted that *Mycobacterium vaccae* cannot utilize cyclic ethers (THF, THP) as growth substrates. However, their lactone (butyrolactone, valerolactone) and hydroxy acid derivatives all supported rapid growth of the bacterium. The cometabolism of ethers produced resultant products that can sustain growth of the organism.

Hyman's research demonstrated some of the potential limitations of cometabolic processes for the degradation of cyclic ethers, including compounds that exert toxic effects on the bacteria that can degrade these compounds. Hyman postulated that the reason ether-bonded compounds are often recalcitrant in the environment may be that latent toxic effects of ether-bonded compounds limit their rate of biodegradation even when appropriate organisms, cosubstrates, and environmental conditions are present.

A study by Kim et al. (2007) isolated and identified a 1,4-dioxane-degrading bacteria (*Mycobacterium sp.* PH-06) from 1,4-dioxane contaminated river sediments in Korea. This bacterial strain was able to degrade 90% of the 1,4-dioxane (at 900 mg/L) in 15 days, using 1,4-dioxane as the sole carbon and energy source, that has been confirmed by only a few researchers (e.g., Roy et al., 1994; Grady et al., 1997; Mahendra and Alvarez-Cohen, 2006). Degradation metabolites (i.e., 1,4-dioxane-2-ol and ethylene glycol) and isotope tagged 1,4-dioxane were used to confirm degradation. The authors identified an initial monooxygenation step that appears to be common to degradation identified by other researchers using different strains of aerobic bacteria.

The ability of the *Rhodococcus* sp. Strain DEE5151 to degrade MTBE and cyclic ethers was studied by Kim and Engesser (2004). *Rhodococcus* Strain DEE5151 was selected because it is able to utilize a broad range of alkyl ethers, including diethyl ether, di-*n*-propyl ether, di-*n*-butyl ether, di-*n*-pentyl ether, di-*n*-hexyl ether, di-*n*-heptyl ether, *tert*-butyl ethyl ether, and other ethers as sole carbon and energy sources. Kim and Engesser isolated 18 ether-degrading *Rhodococcus* strains and two ether-degrading *Sphingomonas* strains from sewage sludge from Stuttgart, Germany. None of the 20 isolates tested was able to utilize dimethyl ether, MTBE, isopropyl ether, THF, THP, or 1,4-dioxane as the sole carbon source.

Researchers with Japan's National Institute for Environmental Studies investigated degradation of 1,4-dioxane and other cyclic ethers by the fungus *Cordyceps sinensis* (Nakamiya et al., 2005). The genus *Cordyceps* is sometimes called the caterpillar fungus and resides in *Hepialus armoricanus* (Lepidoptera: Hepialidae). *C. sinensis* has been identified as mushrooms that grow on the *H. armoricanus* caterpillar in mountainous regions of China. It makes a fruiting body and is consumed as an herbal medicine in Asia.* *C. sinensis* exhibits antioxidant activity that could be useful in bioremediation.

* The herbal medicine is Dong Chong Tsia Tsiao (winter caterpillar summer grass). The caterpillar fungus consists of larvae of *H. armoricanus* (Lepidoptera: Hepialidae) infected with the fungus *C. sinensis* (Clavicipitales, Ascomycotina). The pharmacological properties of the caterpillar fungus are similar to those of ginseng. Dong Chong Tsia Tsiao is mainly used for weak lungs, coughing and shortness of breath, weak kidney, back pain, and other ailments. The wholesale price in China in 2007 was about \$700 kg^{-1} (Steinkraus and Whitefield, 1994; Huang, 2007).

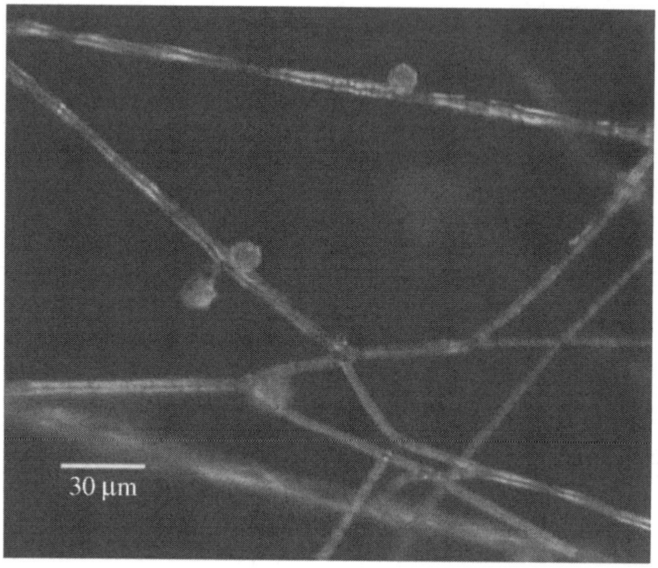

FIGURE 3.6 Light micrograph of *C. sinensis* Strain A. (From Nakamiya, K., Hashimoto, S., Ito, H., Edmonds, J.S., and Morita, M., et al., 2005, Degradation of 1,4-dioxane and cyclic ethers by an isolated fungus. *Applied and Environmental Microbiology* 71(3): 1254–1258. With permission.)

The Japanese researchers confirmed that *C. sinensis* (Strain A) can utilize 1,4-dioxane as the sole carbon and energy sources. Strain A has a septum and spores as shown in Figure 3.6 and was identified as the fungus *C. sinensis* from 18S rRNA gene homology (99.75% of a partial sequence of 1617 base pairs).

Nakamiya et al. (2005) found that Strain A can grow in solutions containing up to 0.09 M 1,4-dioxane, but optimal growth was obtained at 0.034 M 1,4-dioxane. The decrease of 1,4-dioxane in the medium was essentially complete after 3 days, and there was a concomitant increase in the concentration of the degradation product, ethylene glycol (0.97×10^{-3} M), after 3 days. After 4 days, the ethylene glycol was almost completely utilized. Rates of 1,4-dioxane degradation and product formation by *C. sinensis* are profiled in Figure 3.7.

Nakamiya et al. applied an innovative technique, reacting phenyl boronate with the degradation products, to identify the products of *C. sinensis* biodegradation of 1,4-dioxane. Phenyl boronate forms ether bonds with dihydroxy compounds in water-methanol; the derivatives are amenable to gas chromatography/mass spectrometry analysis. By analyzing the breakdown products, Nakamiya et al. were able to propose a degradation pathway for filamentous fungus degradation of 1,4-dioxane, as shown in Figure 3.7. The pathway follows the sequential production of ethylene glycol, glycolaldehyde, glycolic acid, and oxalic acid followed by incorporation of the glycolic acid and/or oxalic acid into the tricarboxylic acid cycle. This study proposed etherases and oxidases as the likely enzymes involved in *C. sinensis* degradation of 1,4-dioxane by Strain A, as shown in Figure 3.8. The authors note that ethylene glycol formation may occur via two hemiacetals at one side of the 1,4-dioxane molecule by etherase-type reactions.

Mahendra and Alvarez-Cohen (2005) identified *Pseudonocardia dioxanivorans* as a novel species of actinomycete that grows on 1,4-dioxane in pure culture. Cells of *Pseudonocardia dioxanivorans* Strain CB1190^T were grown at an optimal temperature of 30°C for up to 28 days. No growth was observed at temperatures higher than 37°C. The cells, 1–3 mm in size, appeared as powdery white colonies floating on the liquid surface and formed biofilms on the inside wall and the bottom of the

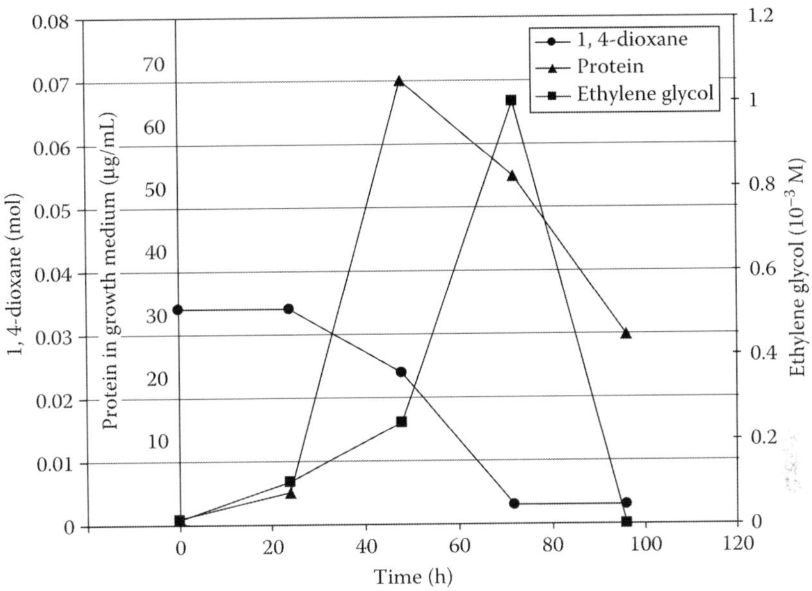

FIGURE 3.7 Rates of 1,4-dioxane degradation and product formation by *C. sinensis*. (From Nakamiya, K., Hashimoto, S., Ito, H., Edmonds, J.S., and Morita, M., 2005, Degradation of 1,4-dioxane and cyclic ethers by an isolated fungus. *Applied and Environmental Microbiology* 71(3): 1254–1258. With permission.)

flask. Figure 3.9 shows an electron micrograph of CB1190^T cells. The physiological and biochemical characteristics of the bacterium are unlike the other species identified in the *Pseudonocardia* genus. To identify the strain, the researchers extracted genomic DNA from CB1190^T to perform polymerase chain reaction (PCR) amplification of the 16S rRNA gene to obtain gene sequences for comparison with the GenBank database for phylogenetic analyses. Genotypic and phenotypic data support the formal recognition of *Pseudonocardia dioxanivorans* as a unique species.

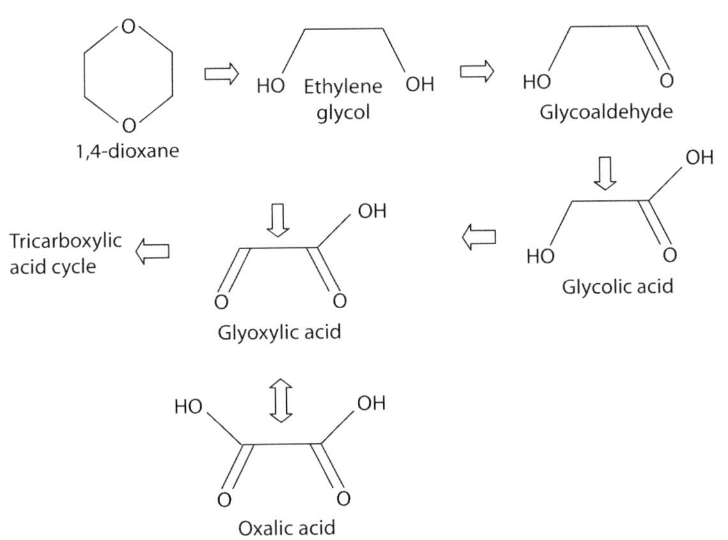

FIGURE 3.8 Proposed pathway for *C. sinensis* degradation of 1,4-dioxane. (From Nakamiya, K., Hashimoto, S., Ito, H., Edmonds, J.S., and Morita, M., 2005, Degradation of 1,4-dioxane and cyclic ethers by an isolated fungus. *Applied and Environmental Microbiology* 71(3): 1254–1258. With permission.)

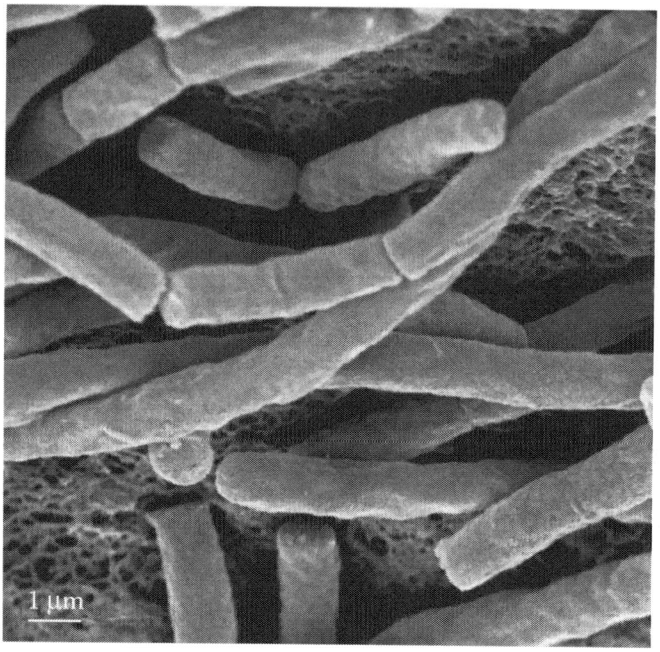

FIGURE 3.9 Scanning electron micrograph of cells of Strain CB1190^T. (From Mahendra, S., and Alvarez-Cohen, L., 2005, *Pseudonocardia dioxanivorans* sp. nov., a novel actinomycete that grows on 1,4-dioxane. *International Journal of Systematic and Evolutionary Microbiology* 55: 593–598. With permission.)

CB1190^T was grown aerobically in ammonium BSM at 30°C with 5 mM 1,4-dioxane. Cell growth was determined by measuring total protein and by assaying the enzyme activities of soluble monooxygenases. The observed maximum growth rate over 1 hours was 90 µg/L, and the maximum 1,4-dioxane degradation rate, 0.1 mg/mg protein/hour, is described by Monod kinetics. Cell yield was 0.09 ± 0.002 g of protein per gram of 1,4-dioxane. Mahendra and Alvarez-Cohen also investigated the ability of CB1190^T to grow on 1,4-dioxane under anaerobic conditions; however, their data did not show anaerobic growth.

Twenty bacterial isolates, 13 of which were capable of transforming 1,4-dioxane, were evaluated by Mahendra and Alvarez-Cohen (2006). 1,4-Dioxane served as a growth substrate for *Pseudonocardia dioxanivorans* CB1190 and *Pseudonocardia benzenivorans* B5; yields were 0.09 g of protein per gram of 1,4-dioxane and 0.03 g of protein per gram of benzene, respectively. The researchers observed cometabolic transformation of 1,4-dioxane for monooxygenase-expressing bacterial strains induced with methane, propane, THF, and toluene, including *Methylosinus trichosporium* OB3b, *Mycobacterium vaccae* JOB5, *Pseudonocardia* K1, *Pseudomonas mendocina* KR1, *Ralstonia pickettii* PKO1, *Burkholderia cepacia* G4, and *Rhodococcus* RR1. Many of the cometabolic reactions resulted in incomplete degradation of 1,4-dioxane because of product toxicity. Subjecting the bacteria to brief exposure to acetylene, a known monooxygenase inhibitor, prevented oxidation of 1,4-dioxane and toluene in all cases, which supported the hypothesis that monooxygenase enzymes expressed by these strains are the active agent for degrading 1,4-dioxane. The researchers also used a colorimetric assay of naphthol production to rapidly detect monooxygenase activity. The ability of the strains to oxidize naphthalene to naphthol independently confirms the strains' ability to express (i.e., utilize) monooxygenases and that dioxane degradation was positively correlated with monooxygenase activity. Bacterial monooxygenases are similar in structure, function, and reaction mechanisms to mammalian P450 enzymes, which have been documented in the metabolism of 1,4-dioxane during toxicological studies using rats (Woo et al., 1977).

Most of the cultures that acted on 1,4-dioxane cometabolically degraded dioxane incompletely, which suggests that their transformation capacity for dioxane is limited. The primary substrate (e.g., THF) was removed to prevent competition with dioxane during the cometabolic degradation reactions. Incomplete degradation of 1,4-dioxane could be caused by a limited reductant (NADH)* supply to fuel the monooxygenase reaction, by toxicity due to dioxane, and/or by toxicity due to dioxane metabolites. 1,4-Dioxane is not likely to exert direct toxicity; CB1190 is able to degrade 1,4-dioxane at concentrations above 1,000,000 µg/L and Strains JOB5, OB3b, and RR1 all tolerate 1,4-dioxane concentrations as high as 500,000 µg/L. Metabolite toxicity is more likely to explain incomplete 1,4-dioxane degradation: the major dioxane metabolite in mammalian cells, 1,4-dioxane-2-one, is more toxic than dioxane. In rats, the lethal dose of 1,4-dioxane-2-one is about one-tenth that of 1,4-dioxane (Woo et al., 1977).

Mahendra and Alvarez-Cohen (2006) found that two of the bacterial strains studied were capable of sustained growth on 1,4-dioxane as the sole carbon and energy source: *Pseudonocardia dioxanivorans* CB1190 and *Pseudonocardia benzenivorans* B5. Rates of oxidation of 1,4-dioxane at 50,000 µg/L were

- 0.01–0.19 mg/h for the metabolic processes
- 0.1–0.38 mg/h for cometabolism by the monooxygenase-induced strains
- 0.17–0.60 mg/h for the recombinant strains per milligram of bacterial protein present

1,4-Dioxane degradation did not occur in the absence of molecular oxygen. Stoichiometric uptake of dissolved oxygen accompanied 1,4-dioxane degradation by CB1190 within the first minute.

Mahendra and Alvarez-Cohen reported that Monod growth kinetics was observed for CB1190; the maximum dioxane degradation rate (k) and half-saturation concentration (K_s) were computed as 1.1 ± 0.008 mg of dioxane per hour per milligram of protein and 160,000 ± 44,000 µg/L, respectively, as calculated by a nonlinear regression fit of the model to the data. Similarly, k and K_s values for Strain B5 were calculated as 0.1 ± 0.006 mg of dioxane per hour per milligram of protein and 330,000 ± 82,000 µg/L, respectively. The cell yields for Strains CB1190 and B5 were 0.09 ± 0.002 mg of protein per milligram of dioxane and 0.03 ± 0.002 mg of protein per milligram of dioxane, respectively. The electron equivalents (e^- eq) of these yields were 0.07 e^- eq protein per e^- eq dioxane and 0.008 e^- eq protein per e^- eq dioxane, respectively.

Measured growth rates were relatively low because cell yields of all strains growing on 1,4-dioxane were low. Mahendra and Alvarez-Cohen (2006) noted that the yields were consistently low for all cultures reported in the literature, even on an e^- eq/e^- eq basis. The theoretical dioxane transformation capacity of a microbial consortium was reported to be 1.0 ± 0.36 mg of dioxane per milligram of protein (Zenker et al., 2000). Mahendra and Alvarez-Cohen's study was the first to definitively show the role of monooxygenases in dioxane degradation by using several independent lines of evidence. The complete biodegradation pathway of 1,4-dioxane by monooxygenase-expressing bacteria observed in the study by Mahendra et al. (2007) is presented in Figure 3.10.

Cai et al. (2008) isolated Strain D4 from a polyester production sewage aeration pond in China. They determined that Strain D4 could use 1,4-dioxane as the sole carbon source. Observations of morphology and physiology as well as biochemical tests and 16S rDNA genetic analysis were used to identify Strain D4 as *Bacillus pumilus*. Ideal conditions for 1,4-dioxane degradation were a pH of 7.0, a temperature of 30°C, and an inoculating quantity of 10%. Degradation rates of 83.7% and 85.6% were determined after 24 and 48 hours from an initial concentration of 1,000 mg/L.

Skinner (2007) completed an extensive study of biodegradation of ether compounds by the eukaryotic alkanotroph *Graphium* sp. (ATCC 58400). *Graphium* sp. does not grow directly on 1,4-dioxane; however, it is able to cometabolize 1,4-dioxane after growth on either THF or alkanes such as propane when mycelia are grown at 27 ± 3°C. Both n-alkane- and THF-grown mycelia oxidize

* NADH is nicotinamide adenine dinucleotide, an important coenzyme found in cells that serve as a carrier of electrons and a participant in metabolic redox reactions.

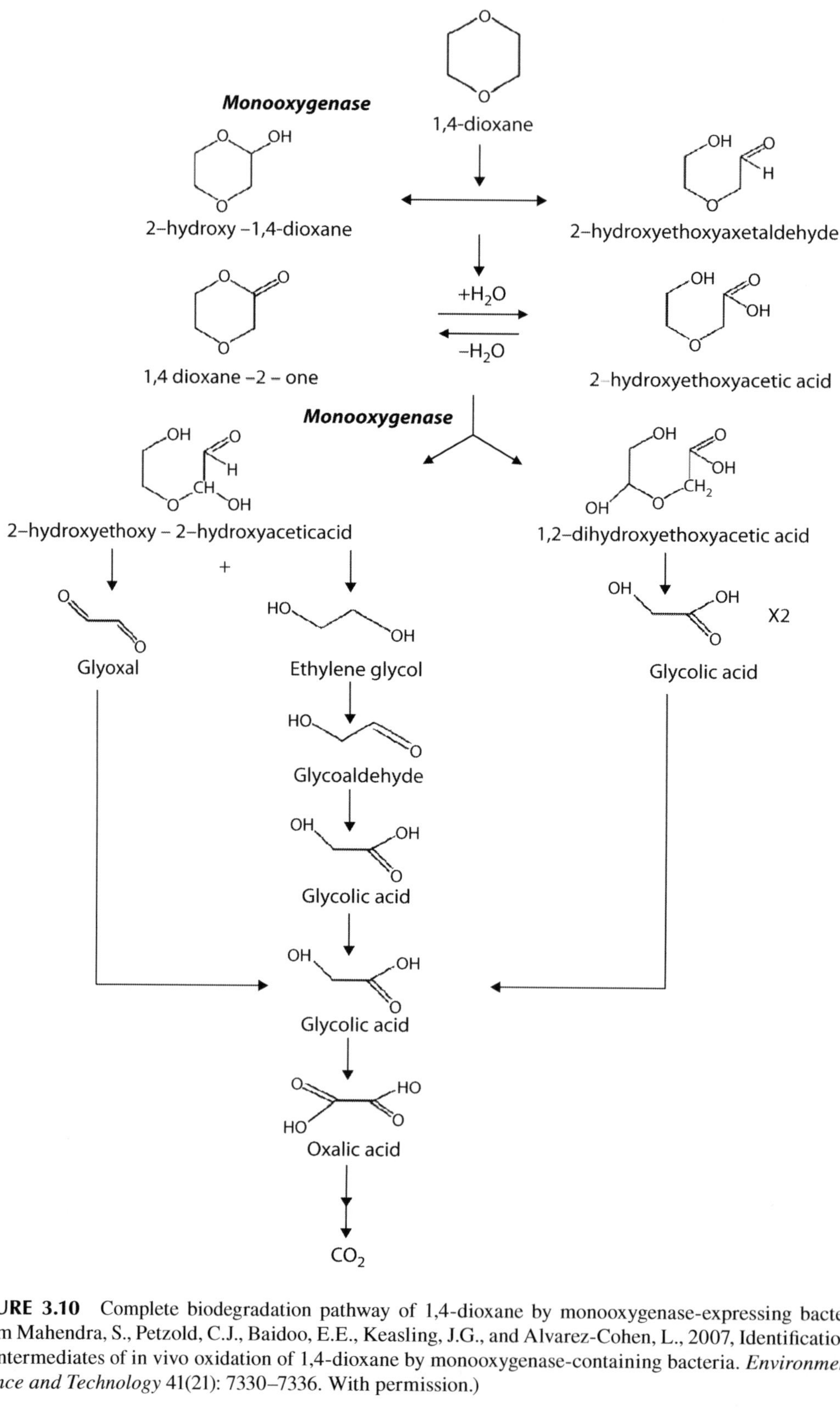

FIGURE 3.10 Complete biodegradation pathway of 1,4-dioxane by monooxygenase-expressing bacteria. (From Mahendra, S., Petzold, C.J., Baidoo, E.E., Keasling, J.G., and Alvarez-Cohen, L., 2007, Identification of the intermediates of in vivo oxidation of 1,4-dioxane by monooxygenase-containing bacteria. *Environmental Science and Technology* 41(21): 7330–7336. With permission.)

TABLE 3.22
Initial Rates of 1,4-Dioxane and THF Oxidation by Propane-, THF-, and Dextrose-Induced Mycelia of *Graphium* sp.

Induction Compound	1,4-Dioxane	THF
Propane	3.8 ± 1	115 ± 3
THF	9.2 ± 5	123 ± 22
Dextrose	0 ± 6	20.4 ± 25

Source: Skinner, K.M., 2007, Characterization of the molecular foundations and biochemistry of alkane and ether oxidation in a filamentous fungus, a *Graphium* species. PhD Dissertation, Oregon State University, Corvallis, OR.
Note: Units: nmol/h per milligram dry weight.

1,4-dioxane. *Graphium* sp. utilizes THF as a sole source of carbon and energy under aerobic conditions via the same THF metabolic pathway used by *Rhodococcus ruber* and two *Pseudonocardia* strains. The cometabolic growth rate of *Graphium* sp. on THF was approximately 10 times greater than on 1,4-dioxane, which itself supports growth about twice the rate of 1,3-dioxane. Growth rates of *Graphium* sp. from Skinner's study are presented in Table 3.22.

Skinner's dissertation explained that physiological and inhibitor studies suggest that THF degradation is initiated by the same cytochrome P450 responsible for oxidizing *n*-alkanes, diethyl ether, and MTBE. There may be an even greater overlap between the enzyme activities involved in *n*-alkane and THF oxidation. Growth of *Graphium* sp. on THF was fully inhibited by acetylene, ethylene, propylene, or propyne (0.5% vol/vol gas phase), whereas growth on either γ-butyrolactone or succinate was unaffected by the presence of these gases at these concentrations.

Although *Graphium* sp. was grown on cyclic ethers containing a single oxygen atom (e.g., THF), no growth was observed on cyclic ethers containing two oxygen atoms (e.g., 1,3-dioxane and 1,4-dioxane), and no growth was observed for any of the nitrogen heterocycles tested as potential growth substrates (morpholine, pyrrolidine, piperazine, and piperidine).

Evidence that the alkane and ether oxidation pathways are superimposed was determined by measuring rates of THF, MTBE, and 1,4-dioxane degradation by *Graphium* mycelia incubated under a spectrum of inductive conditions. Propane-induced mycelia were able to degrade ether compounds (and vice versa) without either a lag phase or the buildup of metabolic intermediates.

The oxidation of alkanes and ethers by *Graphium* sp. is linked through a common catalyst, a cytochrome P450 AMO enzyme that mediates the first step of these pathways. The study characterized this enzyme, designated GSPALK1, through molecular, genetic, and biochemical analyses. Skinner's characterization of GSPALK1 through gene encoding is the first description of a cytochrome P450 involved in the terminal oxidation of gaseous *n*-alkanes and cyclic ethers, as well as the first description of a cytochrome P450 involved in MTBE and 1,4-dioxane cometabolism.

Huang et al. (2014) isolated a gram-negative bacterial strain *Acinetobacter baumannii* DD1 (DD1), capable of degrading 1,4-dioxane aerobically as the sole carbon source in an activated sludge sample from an urban treatment sewage plant. The culture was able to tolerate and degrade concentrations of 1,4-dioxane up to 1,000 mg/L and "efficiently degraded" 100 mg/L of 1,4-dioxane in 42 hours. A partial degradation pathway was elucidated, with initial degradation by monooxygenase and the generation of ethylene glycol and oxalic acid.

In Sei et al. (2013), a soil sample impacted by drainage from a chemical factory yielded four bacteria found to be capable of aerobically degrading 1,4-dioxane as the sole carbon source. Three of the strains, *Mycobacterium* sp. D6, *Mycobacterium* sp. D11, and *Pseudonocardia* sp. D17, are gram-positive *actinomycetes* sp. while the fourth strain, *Afipia* sp. D1, is gram-negative. While most known 1,4-dioxane degrading bacteria are gram-positive *actinomycetes*, D1 is the first known gram-negative

1,4-dioxane degrading bacteria. Using batch experiments, all four strains exhibited higher degradation rates and cell yields than CB1190^T, suggesting they were highly adept at degrading 1,4-dioxane. All four strains also were capable of degrading a variety of cyclic ethers in addition to 1,4-dioxane.

Aoyagi et al. (2018) used stable isotope probing of rDNA and rRNA to link isolated bacterial species to 1,4-dioxane degradation. Working with a sample collected from the aeration tank of an activated sludge wastewater treatment process at a petrochemical industrial plant, microcosms were used to test degradation of 200 mg/L 1,4-dioxane in ^{13}C labeled and unlabeled experiments under aerobic conditions. 1,4-Dioxane concentrations decreased from the initial concentration of 200 mg/L by 49–51 mg/L in both the labeled and unlabeled tests. Concentrations did not decline in the killed controls. ^{13}C content of the carbon dioxide that accumulated in the microcosms increased over time, indicating the tagged carbon was being released by microbial activity. Results indicated aerobic degradation of 1,4-dioxane in the microcosms. Using the ^{13}C to tag genes activated by 1,4-dioxane revealed ten "phylogenetically diverse" cultures associated with 1,4-dioxane degradation. These cultures correlated to known microbes as follows:

- Culture 1: identical to *P. dioxanivorans*
- Culture 2: highly similar to *P. dioxanivorans*
- Culture 3: identical to *Micrococcus luteus*
- Culture 4: slight similarity to *Acidithrix ferrooxidans*
- Culture 5: betaproteobacterial related to *Nitrosomonas ureae*
- Culture 6 and 7: related to some extent to *Methyloversatilis universalis* and *Chitinimonas koreensis*
- Culture 8: alphaproteobacterial related to *Blastochloris viridis* and *Methylocystaceae bacterium*
- Culture 9: alphaproteobacterial identical to *Rhodopseudomonas palustris*
- Culture 10: cyanobacterial OTU 8532 slightly similar to *Vampirovibrio chlorellavorus*

Deng et al. (2018a) isolated a gram-negative motile, rod-shaped, non-filamentous, aerobic bacterium (*Azoarcus sp.* strain DD4) capable of co-oxidizing 1,4-dioxane and 1,1-DCE using propane as the primary substrate and an activated sludge sample. This bacterium is potentially more suitable for *in situ* bioremediation than the filamentous bacteria currently identified as 1,4-dioxane degraders because it is thought that the rod-shaped bacteria will be easier to disperse into an aquifer than filamentous organisms and because DD4, unlike other cultures, grows readily in a nutrient media. In microcosm testing, DD4 degraded both 1,4-dioxane (starting concentration 10.4 mg/L) and 1,1-DCE (3.28 mg/L), with degradation of 1,4-dioxane partially inhibited until the 1,1-DCE was depleted. Final 1,4-dioxane concentrations were less than 1 µg/L after 14 days incubation. The 1,4-dioxane daughter compound HEAA was detected while 1,4-dioxan-2-one was not detected (Deng et al., 2018a). This suggests this 1,4-dioxane degradation pathway shares an initial degradation step with other cometabolic degraders but deviated from that pathway once the ether ring was cloven. The study also found the process was inhibited by acetylene, which is known to inhibit monooxygenase activity. A gene associated with a family of known dioxane degraders (soluble di-iron monooxygenases) was discovered in the bacterium (Deng et al., 2018a). This gene was reportedly similar to genes associated with other known 1,4-dioxane degraders *Burkholderia sp.* H160 and *Rhodococcus sp.* RHA1.

Deng et al. (2018b) conducted laboratory tests on a propane oxidizing bacteria *Mycobacterium dioxanotrophicus* PH-06 (PH-06). This bacterium had previously been found to degrade 1,4-dioxane using a propane monooxygenase. Microcosm testing was used to culture the PH-06 in the presence of varying compounds including 1,4-dioxane and THF, as well as degradation products of 1,4-dioxane and THF (1-propanol, HEAA, pyruvate) and common substrates not typically associated with 1,4-dioxane and THF degradation (glucose and succinate) as the sole carbon sources, with and without propane amended to the headspace. PH-06 cells were harvested for total RNA extraction and reverse transcription PCR and reverse transcription quantitative PCR (qPCR) analysis.

Additional tests were run to assess the oxidation of 1,4-dioxane under various conditions. Testing included controls that were not inoculated. Samples were analyzed for the presence of 1,4-dioxane and production of metabolites.

Genes related to the propane monooxygenase were stimulated when exposed to 1,4-dioxane, THF, and propane, but were not affected by the other compounds (Deng et al., 2018b). Degradation products of 1,4-dioxane were identified in microcosms (HEAA and 1,4-dioxan-2-one). Results indicate that the PH-6 can initiate degradation of 1,4-dioxane using the propane monooxygenase. The authors note similarities in this process to that observed in wild-types of bacteria thought to contribute to natural attenuation. The authors also noted that the propane-monooxygenase-expressing genes are not located on the same genetic group as the THF-degrading genera, suggesting tests to detect THF degraders may not detect propane degraders.

Hatzinger et al., 2017, evaluated the potential to use methane and ethane to stimulate biodegradation of 1,4-dioxane in groundwater using microcosm experiments with varying conditions of growth media and availability of copper ions. During the experiments, ethane was found to stimulate degradation of 1,4-dioxane (using the *Mycobacterium sphagni* ENV482 culture and a soil inoculum from Myrtle Beach Air Force Base), but methane did not stimulate biodegradation under the same conditions. Methanotrophic cultures tested included soil inoculums (Myrtle Beach Air Force Base and Vandenberg Air Force Base), and pure cultures (*M. trichosporium* OB3b, *Methylocella palustris* ATCC 700799 Strain K and *Methylomonas methanica* strain 68-1). The presence or absence of copper made no difference in terms of methane stimulating 1,4-dioxane degradation. On the other hand, ethane stimulated 1,4-dioxane degradation using both pure culture and soil inoculum, although a significant lag period was observed using the soil inoculum. 1,1-DCE was degraded in both the methane and ethane-supplemented microcosms, suggesting chlorinated VOCs are more readily degraded by cometabolism. The results of the study confirmed that ethane can be a valuable substrate for the remediation of 1,4-dioxane by cometabolism and further suggests that ethane may be more useful for 1,4-dioxane biodegradation than methane under field conditions. In general, it calls into question the usefulness of methane in contributing to natural attenuation of 1,4-dioxane. It appears that unknown factors also can control the degradation of 1,4-dioxane under methanotrophic conditions and because of this, methane may not be a consistently effective substrate for 1,4-dioxane biological degradation.

3.3.4.4 Environmental Molecular Diagnostics for 1,4-Dioxane Biodegradability Studies

Environmental molecular diagnostics (EMDs) is a group of diagnostic techniques that can be used to analyze the biological and chemical characteristics of environmental samples (ITRC, 2013). Since publication of the first edition, great advances have been made in applying these tools to environmental clean-up and they have become readily available from commercial laboratories. A number of these tools are adapted to 1,4-dioxane and should be considered when remediating or managing 1,4-dioxane sites. While there are many techniques available to site investigators, this discussion will focus on two types of tools that are now readily available, molecular biological tools (MBTs) and compound-specific isotope analysis (CSIA).

MBTs often focus on the detection of biomarkers, compounds in samples that indicate a key microbe or enzyme is present. These indicators can be used to determine if the "right" microbes are present at a site, or if the desired biological process is active. They are indicators of either the potential for degradation to occur, or in an engineered system, an indication that biostimulation is having the desired effect. In the case of 1,4-dioxane, CSIA measures the ratios of stable isotopes of carbon and hydrogen in 1,4-dioxane isolated from a groundwater sample using synthetic media. Because many chemical and biological processes favor removing lighter isotopes over heavy, degradation often results in loss of lighter isotopes and accumulation of heavy isotopes. For example, carbon CSIA assessment capitalizes on the fact that ^{12}C is typically liberated by biological activity more readily than ^{13}C, leading to an accumulation of ^{13}C (enrichment) in samples that have been degraded by biological activity over time. The amount of ^{13}C enrichment and the rate at which it is enriched relative to the initial concentration not only provide direct evidence of degradation, but can be also be characteristic of the degradation process (USEPA, 2008).

3.3.4.4.1 Molecular Biological Tools

Mahendra and Alvarez-Cohen (2006) found that cometabolism of 1,4-dioxane could be induced by sMMO. Detection of sMMO can be an indicator of biodegradation of 1,4-dioxane, although as noted above (Hatzinger et al., 2017), this may not always be true and there may be other processes that degrade 1,4-dioxane that do not produce sMMO (Deng, 2018b). Gedalanga et al. (2014) identified dioxane monooxygenase and aldehyde dehydrogenase as markers of metabolic and cometabolic biodegradation of 1,4-dioxane. The DNA sequences in the genes that encode THF and 1,4-dioxane monooxygenases are virtually identical. Li et al. (2014) identified a tetrahydrofuran/dioxane monooxygenase also thought to initiate 1,4-dioxane catabolism and proposed it as a marker for 1,4-dioxane degradation. It is thought that the two markers are nearly identical (Gedalanga et al., 2016). Similarly, Deng et al. (2018b) identified a propane monooxygenase as a marker for biodegradation of 1,4-dioxane by propane cometabolism.

3.3.4.4.2 Compound-Specific Isotope Analysis

Pornwongthong (2011) monitored $^{13}C/^{12}C$ ratios in 1,4-dioxane degraded in *Pseudonocardia dioxanivorans* CB1190 microcosms under metabolic and cometabolic conditions, developing a kinetic carbon isotopic enrichment factor (ε). Enrichment values were −1.59 ± 0.09 and −2.34 ± 0.05 for the metabolic and cometabolic culture, respectively. Initially, CSIA was limited for 1,4-dioxane because it was difficult to analyze low concentrations and presumably, more degraded samples, but advancements in analytical techniques (Bennett, 2018) currently allow application of CSIA to low concentrations of 1,4-dioxane (as low as 1 μg/L for carbon isotope ratios, and as low as 20 μg/L for hydrogen isotope ratios). CSIA has been further developed to include two-dimensional analysis using carbon and hydrogen isotope analysis (Wang, 2016). Adding the hydrogen isotope analysis reportedly allows better differentiation of sources in addition to detecting degradation of 1,4-dioxane. CSIA is discussed further in the context of forensic assessment of 1,4-dioxane in Chapter 10.

Both MBTs and CSIA are now commercially available, offering detection of several potential 1,4-dioxane degrader markers by qPCR techniques (Microbial Insights, 2019) and both one-dimensional and two-dimensional CSIA (Pace Analytical Laboratories, 2019).

3.4 LABORATORY, FIELD, AND MODELING STUDIES OF 1,4-DIOXANE MOBILITY

The best evidence for 1,4-dioxane's high mobility comes from field studies. Laboratory studies of 1,4-dioxane breakthrough in soil columns are also informative. Modeling can also yield an understanding of 1,4-dioxane's high mobility relative to methyl chloroform, its host solvent, and 1,1-DCE and 1,1-dichloroethane (1,1-DCA), the abiotic and biodegradation products of methyl chloroform. The rapid migration of 1,4-dioxane occurs in the mobile phase of groundwater, i.e., in sands and gravels, which often comprise only a small fraction of the total aquifer volume. Where 1,4-dioxane migrates into silts and clays, 1,4-dioxane often is retained in the static or immobile phase of groundwater, due to its capacity to invade micropores. In soil cores retrieved from the unsaturated zone, silts and clays have been found to harbor two or more orders of magnitude higher 1,4-dioxane concentrations than found in sands or gravels. This phenomenon can be observed by centrifuging soil cores, or using the synthetic precipitation leaching procedure to quantify the mass of 1,4-dioxane retained in fine-grained aquitards. This section summarizes laboratory and field studies and presents some screening-level modeling calculations of the relative mobility of 1,4-dioxane at solvent release sites.

3.4.1 Laboratory Studies

A common approach to establishing the retardation rates of contaminants migrating in the subsurface is to perform column tests on saturated soil columns. This approach generally involves sterilizing a column containing a known mass of soil and saturating the soil with water, establishing a pressure gradient and flow rate for the water, then adding a solute to the water and measuring for its presence in the column effluent. Samples of column effluent water are taken at regular intervals and analyzed for the contaminant under study. Breakthrough occurs when the effluent concentration first yields a detection of the subject contaminant, and equilibrium is established when the effluent concentration equals the initial concentration.

Retardation factors (R_f – see Equation 3.32 in Section 3.3.2.3) vary with the flow velocity in column tests. In field tests, the duration of the observation period and the distance over which field measurements are made also affect the retardation factor value. Retardation factors for 1,4-dioxane predicted from field data were twice the values measured in laboratory tests of soil columns (Table 3.23). The difference was attributed to heterogeneities in the aquifer media encountered at the field scale. Highly soluble and miscible compounds are more likely to migrate into low-permeability zones, thereby increasing retardation values derived from field observations (Lesage, et al, 1990; Priddle and Jackson, 1991).

TABLE 3.23
Comparison of Retardation Factors (R_f) Based on Field Data, Equations, and Column Data

			Retardation Factor Value Based on Type of Data					
			Field Data			Laboratory Data		
Compound	**Solubility[a] (g/L)**	**log K_{ow}**	**Plume Length Estimate[b]**	**Purge Well Test Result[c]**	**Correlation Equation[d]**	**S and W Equation[b,e]**	**Center of Mass[f]**	**Maximum Concentration[g]**
1,4-Dioxane	M	−0.27	1.6	1.4	1.6	1.0	1.1	1.2
Tetrahydrofuran	M	0.46	2.2	2.2	2.5	1.0	NT	NT
1,2-Dichloroethane	8.7	1.48	7.6	n.p.	5.7	1.2	7.2	4–5
Benzene	1.78	2.04	8.8	n.p.	10.0	1.4	14.3	6–8
1,1-Dichloroethylene	0.40	2.13	n.m.	n.p.	11.0	1.5	10.7	6–7

Source: From Jackson, R.E. and Dwarakanath, V., 1999, Chlorinated degreasing solvents: Physical-chemical properties affecting aquifer contamination and remediation. *Ground Water Monitoring Review* 19(4): 102–110. With permission.

[a] M = Miscible.

[b] $R_f = L_{Cl}/L_{org}$, the ratio of the length of the chloride plume (L_{Cl}) to that of the organic plume (L_{org}) (Patterson et al., 1985); n.m. = not measured. The chloride originated from the same waste-disposal trench as the organic compounds.

[c] Tracer-labeled contaminants were injected at 140 L/min and extracted 5 m down gradient at the same rate for 6 days; samples were collected at multilevel sampling wells between injection and extraction wells. The retardation factor was determined from the ratio of the times at which $C/C_0 = 0.5$ for the organic compound and the tracer compounds (iodide and fluorinated benzoic acid) (Whiffin and Bahr, 1984); n.p. = contaminant not present.

[d] Correlation equation from field data at the Gloucester landfill site: $\log(R_f - 1) = 0.5 \log(K_{ow}) - 0.065$ (Patterson et al., 1985).

[e] $R_f = 1 + \rho_b K_d/n$, where $K_d = 3.2 f_{oc} \times 0.72(K_{ow})$ (S and W = Schwarzenbach and Westall, 1981); NT = not tested.

[f] In column test data, the center of mass of organic compound versus the center of mass of iodide (Priddle and Jackson, 1991); NT = not tested.

[g] In column test data, the ratio of the time at which $C/C_{max} = 0.5$ for organic plume versus the time at which $C/C_{max} = 0.5$ for iodide plume (Priddle and Jackson, 1991).

3.4.2 Field Studies

In a study of 1,4-dioxane migration relative to other contaminants at the Gloucester Landfill near Ottawa, Canada, researchers found that the rate of contaminant migration was inversely proportional to the octanol/water partition coefficient (K_{ow}). The retardation factors and relative mobilities of organic compounds released at the landfill were ascertained by using the ratio of the plume length for each organic compound to the length of the chloride plume, which presumably originated at the same location and time and which is assumed to have a retardation factor of 1.0 (Patterson et al., 1985). A comparison of retardation factors determined with the plume-length approach is given in Table 3.23.

GLOUCESTER LANDFILL SOIL-COLUMN STUDY

(PRIDDLE AND JACKSON, 1991)

Retardation factors measured from soil core data, presented in Table 3.23, were generated by injecting 1,4-dioxane, THF, diethyl ether, 1,2-dichloroethane, trichloromethane, benzene, and 1,1-dichloroethylene into soil cores at fixed and varying flow rates and using iodide as a conservative tracer to index relative breakthrough. The soil cores were from an aquifer at a well-studied landfill site in Gloucester, near Ottawa, Canada. The contaminants studied were selected because they are found in the aquifer at the site.

Iodide is considered an ideal tracer because it experiences little or no sorption. It has symmetrical breakthrough curves in soil-column tests, indicating equilibrium sorption behavior during transport, because any sorption is temporary and reversible and little or no sorption occurs. 1,4-Dioxane breakthrough curves display a similarly symmetrical pattern, whereas the other contaminants display long tailing, that is, the plot of relative concentration to pore volume rises sharply, then slowly decreases as an increasing number of pore volumes passes through the soil core.

For 1,4-dioxane, a retardation factor of 1.1 was obtained from soil core tests at pore-water velocities of 45 and 90 cm/day.

DUKE FOREST LANDFILL SOIL SORPTION STUDY

(BALL AND BARTLETT, 1992)

A screening-level soil sorption study of 1,4-dioxane transport in native soils was conducted at Duke University. A landfill near the university had received wastes containing 1,4-dioxane, and monitoring wells showed concentrations of several thousand micrograms per liter. Researchers performed soil sorption tests on sieved and homogenized cores of aquifer solids retrieved from below the landfill site to verify that 1,4-dioxane would not bind strongly to soil material. Soil samples sterilized with sodium azide were mixed with water and 1,4-dioxane. For 4 days, the slurry was rotated to maintain active mixing; then it was centrifuged to separate the aqueous-phase solution from the soil for analysis. Because of 1,4-dioxane losses in sample handling (13%) and low analytical recovery on some analyses (e.g., 44%), the distribution coefficient K_d could not be reliably estimated. Nevertheless, the study clearly demonstrated that the potential for 1,4-dioxane to sorb to soil is low. The maximum estimated retardation factor was 4.2, that is, 1,4-dioxane would not move slower than one-quarter the rate of groundwater flow. A related study summarized previous investigations of the Duke Forest landfill site and noted that a consultant report determined an average retardation factor of 1.2 (Liu et al., 2000).

TABLE 3.24
1,4-Dioxane and TCE Plume Characteristics[a]

	Dioxane (n = 108)	TCE (n = 1,212)
% of plumes >1 km in length	11	8
% of plumes with a majority of detections of <100 μg/L	86	91
% of plumes with a majority of detections of <50 μg/L	72	84
% of plumes >1 km in length with a majority of detections of <50 μg/L	7	6

[a] From Adamson, D.T., Mahendra, S., Walker, K.L., Rauch, S.R., Sengupta, S., and Newell, C.J., 2014, A multi-site survey to identify the scale of the 1,4-dioxane problem at contaminated groundwater sites. *Environmental Science & Technology Letters* 1(5): 254–258. With permission.

Table 3.23 shows that estimates of retardation factors calculated with the Schwarzenbach and Westall equation can be as little as one-sixth the field-measured values. The Schwarzenbach and Westall equation does not adequately account for adsorption and matrix diffusion; it was derived from data for hydrophobic polar compounds, which do not sufficiently represent the behavior of hydrophilic compounds such as 1,4-dioxane. Table 3.23 also demonstrates that migration of 1,4-dioxane is only minimally retarded. It therefore migrates substantially faster than most other organic contaminants.

Adamson et al. (2014) compared 1,4-dioxane and chlorinated VOC plume lengths using California's GeoTracker database, compiled by the State Water Resources Control Board. Data were evaluated from sites with groundwater concentration data for the period between February 2000 and July 2013, specifically omitting data earlier than 2010 to avoid spurious results from earlier analytical methods. Chlorinated VOCs assessed included methyl chloroform, TCE, and 1,1-DCE. 1,1-DCE was included as the primary degradation product of methyl chloroform; however, the biodegradation product of methyl chloroform, 1,1-DCA, and the biodegradation product(s) of TCE, cis- or trans-1,2-dichloroethylene, were not included. Plume lengths were estimated as the maximal distance of all possible well pairs over the reporting period for a given site. Not all sites could have plume lengths estimated because of limitations in monitoring well networks and monitoring results. Table 3.24 summarizes plume characteristics compiled from the sites meeting the evaluation criteria.

The length of the observed plumes was assessed to define the relationship between 1,4-dioxane plume length and chlorinated solvent plume length. Figure 3.11 breaks those plume lengths into three categories: instances where the 1,4-dioxane plume is the longest, instances where both plumes are approximately the same length, and instances where the chlorinated solvent plume is the longest. Only 21% of the sites had a 1,4-dioxane plume that was the longest, versus 62% where the chlorinated solvent plume was the longest; 17% of the sites had plumes of approximately the same length. At first glance, these results seem to contradict what would be expected when comparing mobility of a miscible compound with a very low sorption coefficient (1,4-dioxane) to compounds with limited solubility and higher sorption coefficients (chlorinated solvents). However, when taking into consideration the history and circumstances of release for 1,4-dioxane versus chlorinated solvents, some differences are apparent that could impact the resulting plume configuration. First, when 1,4-dioxane was released with chlorinated solvents, the stabilizer was estimated to be present at levels of less than 20% as described earlier in this book. The feedstock methyl chloroform typically had 3% 1,4-dioxane and with losses of the solvent base, the 1,4-dioxane would have concentrated in the waste materials. But even with that concentration effect, it was unlikely that 1,4-dioxane would have been present at levels greater than 20% because the performance of the degreasing solvent would have been compromised. Thus, the amount of 1,4-dioxane released would have been at most one-fifth of the amount of chlorinated solvent released in the waste product. Second, the duration

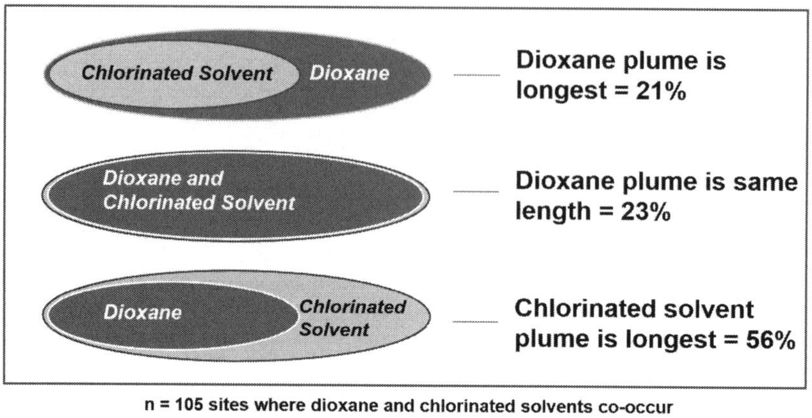

FIGURE 3.11 Plume length comparison between 1,4-dioxane and chlorinated solvents. (From Adamson, D.T., Mahendra, S., Walker, K.L., Rauch, S.R., Sengupta, S., and Newell, C.J., 2014, A multi-site survey to identify the scale of the 1,4-dioxane problem at contaminated groundwater sites. *Environmental Science & Technology Letters* 1(5): 254–258). With permission.)

of release, and thus the migration period, for the chlorinated solvent would likely have been much longer than for 1,4-dioxane. Chlorinated solvents, and TCE in particular, were generally put into widespread use in the 1950s and thus would have been released and initiated migration 20-years earlier than the 1,4-dioxane, which in general was not in widespread use as a methyl chloroform stabilizer until the 1970s when many industries switched over from TCE to methyl chloroform. These two factors (one-fifth the mass and 20 years less migration) could easily account for the shorter plume length for 1,4-dioxane at chlorinated solvent sites. Finally, with few exceptions, the monitoring well networks reviewed in the GeoTracker database for the Adamson 2014 survey were designed to delineate the lateral extents of chlorinated solvent plumes. 1,4-Dioxane was most often discovered well after monitoring well networks were installed, because monitoring requirements did not include 1,4-dioxane and awareness was limited until the early 2000s. If the objective of the monitoring well network designs had been to characterize the lateral extent of 1,4-dioxane migration, it's likely that in many cases, the network would need to be expanded, and that more instances of 1,4-dioxane migrating ahead of the chlorinated VOC plume would be identified. Where 1,4-dioxane has been the driver for monitoring well placement, this pattern has been observed many times. The 2014 Adamson survey nevertheless produces valuable data for the distribution of 1,4-dioxane within the known footprint of chlorinated solvent plumes.

3.4.3 Modeling 1,4-Dioxane Transport and Matrix Diffusion

Modeling tools can provide useful insights into the relative rates of migration of contaminants under assigned subsurface conditions such as groundwater flow rate, organic matter content of soils and aquifer solids, retardation factors, biodegradation rate constants, abiotic rate constants, source mass strength, release duration, and other factors. There are many analytical approaches and many models available to simulate contaminant transport. The subject of modeling contaminant transport is beyond the scope of this book. The first edition of this book included a simplistic application of BIOCHLOR (Aziz et al, 2000, 2002), which showed 1,4-dioxane migrating considerably farther than methyl chloroform or 1,1-dichloroethane. This is not surprising, given that BIOCHLOR does not account for matrix diffusion (Srinivasan, et al, 2007; USEPA, 2007), and that the exercise was constrained by using a rate constant of "zero" for 1,4-dioxane biodegradation. This assumption ignores data collected by Chiang et al. (2012) at Air Force Plant 44 which used stable isotope probing with ^{13}C-baited Bio-Traps® to confirm high levels of ^{13}C-dioxane incorporation into biomass, indicating that 1,4-dioxane serves as a growth supporting substrate for native bacteria. It also does not consider the observation by Feenstra et al. (1984), that diffusion slows the spread of contaminants in an aquifer, because the contaminants diffuse into the inaccessible porosity and are lost to the mobile porosity. Ignoring the impact of degradation and diffusion exaggerates the potential for 1,4-dioxane migration. As such, the first edition presentation of the BIOCHLOR screening level migration estimate was an example of confirmation bias, and the predicted migration of 1,4-dioxane far exceeding methyl chloroform and its breakdown products is not reflected in field evidence in the majority of sites studied in California's GeoTracker database (Adamson et al., 2014; Adamson et al., 2017). A further limitation to the BIOCHLOR analysis in the first edition was not adequately accounting for the long-term dissolution of the NAPL mixture of methyl chloroform and 1,4-dioxane, i.e., the source term duration was artificially curtailed. While the BIOCHLOR exercise in the first edition was not completely useless, and its limitations were acknowledged with ample caveats, it is not repeated here, as more careful and advanced analysis now supplants the simple and possibly misleading presentation of the BIOCHLOR results in the first edition.

The 2010 edition BIOCHLOR-predicted relative migration rates showed 1,4-dioxane migrating further than methyl chloroform and its breakdown products; however, the real world, with heterogeneous distributions of sands, gravels, silts, and clays, as well as variable biogeochemical conditions, is not well matched to the BIOCHLOR results. The 2014 survey by Adamson illustrates that at many sites, matrix diffusion into fine-grained deposits, which is not accounted for by BIOCHLOR, will cause 1,4-dioxane to be retained in the interstitial pore spaces of fine-grained deposits, particularly at the source zone. Matrix diffusion is an important process that governs 1,4-dioxane migration in particular, owing to its shifting polarity and ability to penetrate water structures and clay minerals. Adamson et al. (2017) conducted modeling of 1,4-dioxane penetration into and out of fine-grained deposits in the saturated zone, using the Square-Root Model in the Matrix Diffusion Toolkit (Farhat et al., 2012; Parker et al., 1994). The modeling results indicate that high 1,4-dioxane concentrations should be expected in the source zone immediately following a release, but can decline rapidly. Back-diffusion of 1,4-dioxane will persist for many years, resulting in 1,4-dioxane concentrations well above the range of regulatory cleanup thresholds.

The model scenario included 1,4-dioxane initially released as part of a chlorinated solvent NAPL spill, and used the effective solubility of 1,4-dioxane, approximated using Raoult's Law. Simulating a spill of 13 drums beginning in 1970, initial mole fractions of 0.20 for 1,4-dioxane and 0.55 for TCA were selected to represent the relative ratio of these compounds in spent solvent waste, with oily waste comprising the balance (see Table 3.12). The source loading concentrations for this scenario were 1,4-dioxane, 29,100 mg/L for 2.7 years, and methyl chloroform, 751 mg/L for 39.9 years. The source loading durations represent the time required to deplete 99.9% of the NAPL mixture (methyl chloroform, 1,4-dioxane, and cutting oil waste), determined using a dissolution model based on Charbeneau, 2000 (see Adamson et al., 2017 for details).

Figure 3.12 presents modeled source zone concentrations versus time (Adamson, et al., 2016). Note that while methyl chloroform NAPL is most often considered to be denser than water, the presence of 25% oily waste and 20% 1,4-dioxane can render the specific gravity of the mixture close to or even less than 1.0. In Figure 3.12, the mass fraction of oily waste (cutting oil) is shown increasing and persisting over the forty-year loading period. However, this assumes the cutting oil mass is conserved under site conditions and does not account for biodegradation, which can be considerable for petroleum under either aerobic or anaerobic source zone conditions, and which may result in the oil being degraded more quickly than methyl chloroform. If biodegradation were active, the density of the NAPL would not increase as predicted in the model.

1,4-Dioxane matrix diffusion was modeled with a coarse-grained layer (transmissive zone) overlying a fine-grained layer (low-K zone), using the equations from the Matrix Diffusion Toolkit (Adamson, et al., 2017). After the 1,4-dioxane dissolves out of the source zone, the modeled initial 1,4-dioxane concentration exceeded 100 mg/L, which reflects the substantial mass of 1,4-dioxane that has penetrated the low-permeability layer. During an 80-year modeled timeframe, during

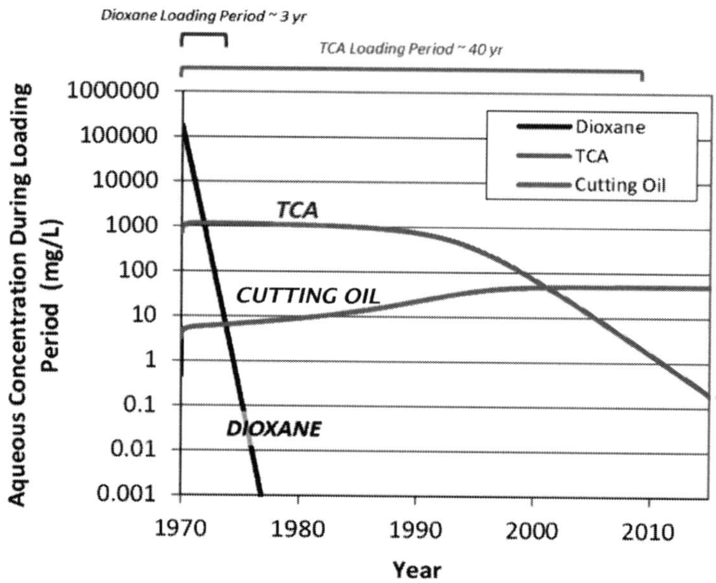

FIGURE 3.12 Modeled source concentration versus time for 1,4-dioxane and methyl chloroform (TCA) from NAPL dissolution. (From Adamson, D.T., de Blanc, P.C., Farhat, S.K., and Newell, C.J., 2016, Implications of matrix diffusion on 1,4-dioxane persistence at contaminated groundwater sites. *Science of the Total Environment* 562 (2016): 98–107. With permission.)

which back-diffusion from the low-K zone occurs, 1,4-dioxane concentrations in the transmissive zone declined by over two-orders of magnitude. Even with a >100-fold decrease, 1,4-dioxane concentrations remained significantly higher than drinking water action levels, even after 80 years (Adamson et al., 2017).

Matrix diffusion plays a critical role in the distribution of 1,4-dioxane in the subsurface. Accordingly, acquiring a detailed appreciation for the dynamics of matrix diffusion is essential for effective site investigation and remedial design. The reader is directed to the Adamson et al. (2017) report, *In Situ Treatment and Management Strategies for 1,4-Dioxane-Contaminated Groundwater* (SERDP Project ER-2307) and the Matrix Diffusion Toolkit (Farhat et al., 2012). Failure to account for 1,4-dioxane matrix diffusion can lead to failure to fully account for 1,4-dioxane mass location and longevity; hence, the time investment to acquire this knowledge is important for successful remedial investigation and design.

3.5 DIFFUSIVE TRANSPORT OF 1,4-DIOXANE AND STORAGE IN FINE-GRAINED SOILS

Transport in clays and silts plays an important role in the overall subsurface fate of groundwater contaminants. In clays found in aquitards and in the engineered clay liners formerly used as the primary liner for landfills and waste lagoons, contaminant migration by advective transport is very slow. Where the aquitards or liners have low hydraulic conductivity or where hydraulic gradients are insignificant, molecular diffusion through the clay pore fluid is the primary mode of contaminant transport. Predicting contaminant transport through fine-grained strata and engineered clay liners involves modeling diffusive transport (Barone et al., 1992; Hoffman et al., 1998; Young and Ball, 1998; Ball, 2000).

Diffusion-driven contamination is inherently slow. It is controlled by concentration gradients, solubility, and temperature. Limitations on diffusion-driven contamination include the occurrence and heterogeneity of fine-grained media within and adjacent to an aquifer and rates of adsorption, biodegradation, and abiotic degradation. Over long periods of time, contaminants with high concentrations in coarse-grained media will move into adjacent low-permeability zones, including aquitards, fine-grained lenses within an aquifer, and the micropores within individual grains of the aquifer matrix. The process of contaminant diffusion into fine-grained sediments is best articulated by Mackay and Cherry's classic 1989 article "Groundwater Contamination: Pump and Treat Remediation," paraphrased as follows:*

> *As dissolved contaminants spread through aquifers, they rapidly migrate through more permeable zones while slowly invading less permeable materials by diffusion. Over many years, diffusion can cause dissolved contaminants to occupy large volumes of low-permeability material. To obtain clean water from wells, the lower-permeability components of the aquifer system must be remediated as well as the higher-permeability zones.*
>
> *Consider a clay lens in the middle of a predominantly sandy hydrostratigraphic unit through which contaminants have migrated for decades. The porosity of the clay lens is often larger than that of the adjacent aquifer, promoting diffusion into the clay. Dissolved contaminants permeate the clay lens by molecular diffusion, which leads to delayed diffusive release. Also, the capacity of a clay lens to sorb contaminants is generally much greater per unit volume than the sorption capacity of the aquifer. When the aquifer is flushed by clean water, the only significant process for release of the contaminant from the clay will be a reversal of the diffusion direction. The relatively slow rate of release of contaminants from the clay by diffusion and the potentially appreciable contaminant mass contained in dissolved and sorbed form in the clay causes a long-term "bleed" or "back-diffusion" of contaminants into the aquifer during remediation. In many aquifers, there are numerous thin beds of silt and clay.*
>
> *Contaminant concentrations in water extracted for remediation or for beneficial uses often remain higher than low regulatory cleanup thresholds for very long periods of time, greatly increasing the duration and cost of cleanup. Most investigations fail to delineate site geology in sufficient detail to predict the effects of delayed diffusive release.*

Diffusion into fine-grained media is driven by a concentration gradient, away from sustained high concentrations in the higher-mobility aquifer zone supplied by a continuous source of contamination. When that source of contamination is terminated or remediated, the concentration gradient reverses, leading to back-diffusion out of the fine-grained media. The back-diffusion concept is often explained in the context of a sorption-desorption system; however, in the case of high-solubility or miscible hydrophilic compounds with low potential for sorption such as 1,4-dioxane, sorption may not play an active role in the process. Instead, diffusion can lead to long-term storage of 1,4-dioxane in fine-grained media near the source of the release, followed by long-term slow release (Suthersan, 2002; Payne et al., 2003, 2008; Nguyen et al., 2005). Remediation projects commonly cease extraction when the plot of mass removal over time becomes asymptotic; however, a subsequent rebound in concentrations is often observed. Figure 3.13 shows the relative distribution of hydrophilic contaminants in aquifer media.

The diffusion coefficient for a contaminant in water (D_w) can be predicted by using a variety of QSAR equations. A commonly used approach is the Wilke-Chang equation, which incorporates the molar volume of the solute, the viscosity and temperature of water, and several empirical constants (Wilke and Chang, 1955). For many organic compounds, D_w is in the range of 10^{-6}–10^{-4} m²/day. Through the use of the Wilke-Chang method, the calculated diffusion coefficient for 1,4-dioxane in pure water at 22°C is 10.6×10^{-6} cm²/s (Barone et al., 1992). The measured diffusion coefficient of 1,4-dioxane in pure water at environmentally relevant concentrations is 1.11×10^{-5} cm²/s at 25°C (Leaist et al., 2000). At a more commonly encountered field groundwater temperature, 15°C, 1,4-dioxane water diffusivity in pure water was measured at 8.99×10^{-5} cm²/s (Holz et al., 2000). As the concentration of 1,4-dioxane increases to 30%, water diffusivity initially decreases about twofold until concentration reaches about 60%, whereafter water diffusivity increases sharply (Leaist et al., 2000).

* See Mackay and Cherry (1989).

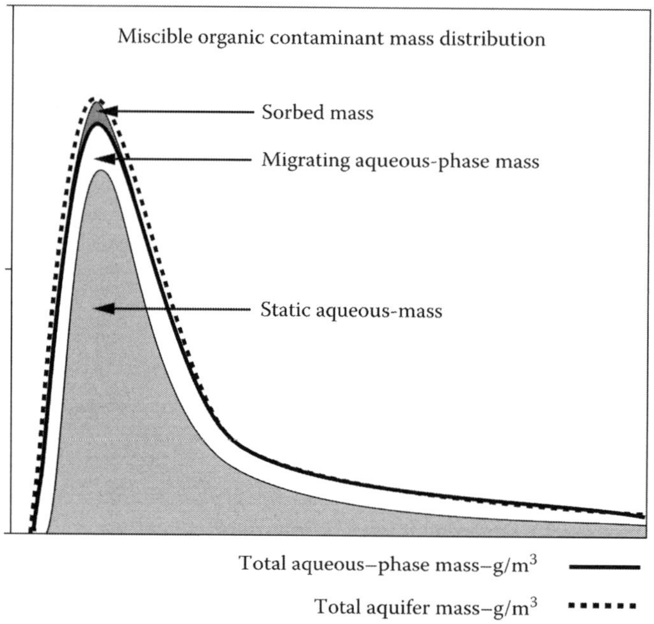

FIGURE 3.13 Distribution of mass of hydrophilic compounds as mass versus distance from site or time at a location down-gradient from the point of release. (From Payne, F.C., Suthersan, S.S., Molnaa, B., and Davis, S., 2003, Developing *in situ* reactive zone strategies for 1,4-dioxane. Presented at 1,4-dioxane and other solvent-stabilizer compounds in the environment. December 8, 2003, San Jose, CA, Groundwater Resources Association of California. http://www.grac.org. With permission.)

More relevant than diffusion into distilled, deionized water is the diffusion of 1,4-dioxane into saturated clay soils. A laboratory study to measure diffusion coefficients and adsorption coefficients for 1,4-dioxane and four other compounds in samples of a saturated clayey soil was conducted by Barone et al. (1992) at the University of Western Ontario.

Distribution coefficients (K_d) were estimated from laboratory diffusion tests in plug samples of a saturated and undisturbed clayey soil using five organic compounds, each tested alone at 300 mg/L (Barone et al., 1992). A K_d value of 0.17 mL/g was estimated for 1,4-dioxane, on the basis of a measured diffusion coefficient of 4×10^{-6} cm^2/s in the clayey soil. The characteristics of the soil tested are listed in Table 3.25. The results of the soil plug diffusion tests are listed in Table 3.26.

The high retardation factor, 2.19, for 1,4-dioxane calculated in Table 3.26 is twice the 1.1 value obtained by Priddle and Jackson (1991). The higher retardation factor calculated by using data from Barone et al. (1992) reflects the low hydraulic conductivity of the Sarnia clay used for diffusion and distribution coefficient testing.

The ratio of the measured effective diffusion coefficient for a water-borne contaminant in soil to the calculated or measured diffusion coefficient of that contaminant through pure water represents the tortuosity of the soil. As a compound diffuses in a porous medium, the rate of diffusion is reduced by the increased length of the molecule's path as it travels around solid particles in the soil. The length of the path traveled by a diffusing molecule relative to the straight-path distance is called the tortuosity factor, τ (Bear, 1972). Tortuosity can be derived from diffusivity as follows:

$$\tau = \frac{L}{L_e} = \frac{D_e}{D_w} \approx \phi^P, \tag{3.43}$$

where L is the length of a straight path through the porous media, L_e is the length of a tortuous path around solid grains, D_e is the effective diffusion coefficient of the water-borne contaminant,

TABLE 3.25
Characteristics of the Sarnia Clay in 1,4-Dioxane Diffusion Study

Property	Value
Specific gravity	2.73
Saturation	100%
Dry density	1.68 g/cm^3
Calcite/dolomite ratio	0.42
Moisture content	23%
Porosity	39%
Organic carbon content, f_{oc}	0.58%
Cation-exchange capacity	10 meq per 100 g dry weight
Minerals in <0.074 mm Fraction	
Calcite and dolomite	34%
Quartz and feldspars	15%
Illite	25%
Chlorite	24%
Smectite	2%
Grain-size Distribution	
Clay (<0.002 mm)	45%
Silt (0.074–0.002 mm)	43%
Sand (2.0–0.074 mm)	10%
Gravel (>2.0 mm)	2%

Source: Barone, F.S., Rowe, R.K., and Quigley, R.M., 1992, A laboratory estimation of diffusion and adsorption coefficients for several volatile organics in a natural clayey soil. *Journal of Contaminant Hydrology* 10(3): 225–250.

TABLE 3.26
Measured Fate and Transport Parameters of Selected Organic Compounds in Clay from Soil Plug Diffusion Tests

Compound	Distribution Coefficient (mL/g)	Calculated Aqueous Diffusion Coefficient in Pure Water[a] D_w (cm²/s)	Measured Effective Aqueous Diffusion Coefficient[b] D_e (cm²/s)	Relative Concentration C/C_0 after 14 days	Calculated Retardation Factor[c] R_f
Acetone	0.19	11.5×10^{-6}	5.6×10^{-6}	0.9	2.33
1,4-Dioxane	0.17	10.6×10^{-6}	4.0×10^{-6}	0.9	2.19
Aniline	1.3	9.8×10^{-6}	7.0×10^{-6}	0.7	10.1
Chloroform	4.5	10.6×10^{-6}	7.0×10^{-6}	0.5	43
Toluene	11	8.9×10^{-6}	5.8×10^{-6}	0.4	183

Source: From Barone, F.S., Rowe, R.K., and Quigley, R.M., 1992, A laboratory estimation of diffusion and adsorption coefficients for several volatile organics in a natural clayey soil. *Journal of Contaminant Hydrology* 10(3): 225–250. With permission.

[a] Diffusion constants calculated by Barone et al. (1992) by using the Wilke-Chang equation for water at laboratory temperature, 22°C.

[b] Barone et al. (1992) noted that the aqueous diffusion coefficients first obtained for chloroform and toluene were unreasonably high owing to the role of nonsettling particles in the soil column.

[c] Retardation factors calculated by the author using soil properties from Barone et al. (1992) and the familiar retardation equation (Equation 3.32).

and D_w is the calculated or measured diffusion coefficient of the contaminant in pure water (Bear, 1972; Hoffman et al., 1998). The last expression, ϕ^P, provides an empirical way to obtain tortuosity, where ϕ is porosity and P is an empirical factor selected for the geological medium under consideration (Millington and Quirk, 1961; Parker et al., 1996). Tortuosity, sometimes represented with the Greek letter χ, accounts for all factors that limit solute diffusion through the porous medium such as the tortuous nature of the diffusion path, dead-end pores, and steric hindrance (Young and Ball, 1998). Higher tortuosity factors indicate shorter flow paths. Well-sorted sands (i.e., those that have essentially uniform grain size) have higher tortuosity factors because smaller grains are not filling pore spaces. Tortuosity factors in well-sorted, fine-grained sands range from 0.6 to 0.8, whereas τ in well-sorted, coarse-grained sands averages 0.4; for poorly sorted sands, τ ranges from 0.2 to 0.8 (Fetter, 1993; Hoffman et al., 1998). Barone et al. (1992) obtained tortuosity factors for the subject Sarnia clay of 0.65 from the calculated and measured diffusivities of chloroform and toluene. A theoretical estimate of tortuosity in unconsolidated materials is 0.67 (Bear, 1972).

Another approach to estimating the tortuosity of a soil involves contrasting measured diffusion coefficients of a compound of interest to the diffusion coefficient of a nonretarded compound to obtain a relative approximation of tortuosity. This approach has been used to compare carbon tetrachloride, perchloroethylene, and TCE to chloroform, whose retardation is considered negligible (Hoffman et al., 1998). Knowledge of the tortuosity of a medium is useful for estimating the role of diffusion in retaining contaminants in fine-grained deposits.

Barone et al. (1992) estimated breakthrough times for vertical transport in clay soil. 1,4-Dioxane was calculated to advance more than 5 m in 100 years, while toluene advanced less than 1 m in the same time. The hypothetical calculation estimates that a leachate containing 1,4-dioxane at 300 mg/L discharged to the subject Sarnia clay would diffuse through a 1-m-thick clay landfill liner in 5 years and contaminate underlying groundwater to concentrations in excess of 30 μg/L (the Canadian drinking water guidance in effect during the 1989 study; it is now 50 μg/L).

To address the role of sorption in diffusive transport, scientists at Lawrence Livermore National Laboratory (LLNL) developed a hybrid contaminant transport term that accounts for both tortuosity and retardation by sorption (Hoffman et al., 1998). The LLNL scientists noted that effective diffusion is affected by sorption as follows:

$$D_a = \frac{D_e}{R} = \frac{D_w \times \tau}{R}. \tag{3.44}$$

Hoffman et al. (1998) combined the effects of diffusion and sorption into a single parameter, the diffusion reduction coefficient, Ω:

$$\Omega = \frac{\tau}{R}, \tag{3.45}$$

where Ω is the diffusion reduction coefficient, τ the tortuosity factor, and R the retardation factor. With these parameters, the rate of migration due to diffusion for a soil with known bulk properties can be estimated.

Diffusion rates in clays are both difficult to measure and difficult to predict. Measured values of diffusion coefficients in clayey materials may only be accurate to within a factor of four, whereas tortuosity factors may be accurate to within a factor of two if controls such as measurement of tritiated water are used. Estimation of diffusion coefficients in low-permeability porous media requires specific measurements for the system of interest (Young and Ball, 1998). Nevertheless, because diffusion plays a critical role in the transport of hydrophilic compounds such as 1,4-dioxane, diffusion coefficients are estimated here by using the approaches of Barone et al. (1992) and Hoffman et al. (1998), described previously in this section, to provide a *relative* sense of the role of molecular diffusion in the transport of solvent stabilizers in aquitards. Table 3.27 provides estimated rates of

TABLE 3.27
Relative Magnitudes of *Estimated* Effective Aqueous Diffusion Coefficients and Migration of Chlorinated Solvents and Solvent-Stabilizer Compounds in Sarnia Clay

Compound[a]	Distribution Coefficient[b] K_d (Dimensionless)	Aqueous Diffusion Coefficient[c] D_w (cm²/s)	Effective Diffusion Coefficient[d] D_e (cm²/s)	Calculated Retardation Factor[e] R_t (Dimensionless)	Diffusion Reduction Coefficient[f] Ω (Dimensionless)	Apparent Diffusion Coefficients[g] D_a (cm²/s)
1,3-Dioxolane	0.58	2.66×10^{-5}	5.31×10^{-6}	3.50	5.72×10^{-2}	1.52×10^{-6}
1,4-Dioxane	**0.58**	$\mathbf{2.62 \times 10^{-5}}$	$\mathbf{5.24 \times 10^{-6}}$	**3.50**	$\mathbf{5.72 \times 10^{-2}}$	$\mathbf{1.50 \times 10^{-6}}$
tert-Butyl alcohol	0.928	2.14×10^{-5}	4.28×10^{-6}	5.00	4.00×10^{-2}	8.56×10^{-7}
sec-Butyl alcohol	1.189	2.14×10^{-5}	4.28×10^{-6}	6.12	3.27×10^{-2}	6.98×10^{-7}
2-Methyl-3-butyn-2-ol	1.6936	2.19×10^{-5}	4.37×10^{-6}	8.30	2.41×10^{-2}	5.27×10^{-7}
Butoxymethyl oxirane	1.6414	2.10×10^{-5}	4.20×10^{-6}	8.07	2.48×10^{-2}	5.20×10^{-7}
Epichlorohydrin	2.61	2.52×10^{-5}	5.03×10^{-6}	12.24	1.63×10^{-2}	4.11×10^{-7}
Acetonitrile	2.61	2.32×10^{-5}	4.63×10^{-6}	12.24	1.63×10^{-2}	3.78×10^{-7}
1,2-Butylene oxide	2.61	2.20×10^{-5}	4.41×10^{-6}	12.24	1.63×10^{-2}	3.60×10^{-7}
Nitromethane	4.756	2.73×10^{-5}	5.45×10^{-6}	21.49	9.31×10^{-3}	2.54×10^{-7}
Diisopropylamine	63.22	1.94×10^{-5}	3.89×10^{-6}	273.33	7.32×10^{-4}	1.42×10^{-8}
Cyclohexane	95.99	2.13×10^{-5}	4.25×10^{-6}	414.50	4.83×10^{-4}	1.03×10^{-8}
Methyl chloroform	88.16	8.30×10^{-6}	1.66×10^{-6}	381	5.25×10^{-4}	4.36×10^{-9}
Dichloromethane	5.104	1.24×10^{-5}	2.48×10^{-6}	23	8.70×10^{-3}	1.08×10^{-7}
Trichloroethylene	73.08	8.60×10^{-6}	1.72×10^{-6}	316	6.33×10^{-4}	5.45×10^{-9}
Perchloroethylene	211.1	7.70×10^{-6}	1.54×10^{-6}	910	2.20×10^{-4}	1.69×10^{-9}

Source: From Barone, F.S., Rowe, R.K., and Quigley, R.M., 1992, A laboratory estimation of diffusion and adsorption coefficients for several volatile organics in a natural clayey soil. *Journal of Contaminant Hydrology* 10(3): 225–250. With permission.

[a] Nonitalic compounds are stabilizers. Italic compounds are chlorinated solvents; these are listed for comparison to show whether the stabilizer is likely to diffuse out of the solvent or remain with the solvent.

[b] K_d calculated by using $K_{oc} \times f_{oc}$ (see Section 3.3.2.3) and soil parameters from the Sarnia clay.

[c] Aqueous diffusion coefficients in pure water calculated by using the Wilke-Chang method ($T = 22°C$).

[d] Effective diffusion coefficient obtained by dividing aqueous diffusion coefficient by an assumed tortuosity factor, 0.2, for a poorly sorted fine-grained material.

[e] Retardation factor calculated by using soil parameters from Sarnia clay and Equation 3.32.

[f] Diffusion reduction coefficient obtained from Equation 3.45.

[g] D_a obtained as the product of the aqueous diffusion coefficient and the diffusion reduction coefficient.

migration by diffusion for several solvent-stabilizer compounds and chlorinated solvents through a 1-m-thick clay lens by using the bulk soil parameters from the Barone study Sarnia clay described in Table 3.25.

Table 3.27 profiles the potential for compounds to migrate into clays and silts by molecular diffusion and to back-diffuse into the mobile phase when the concentration gradient reverses. The stabilizer compounds in Table 3.27 that have higher apparent diffusion coefficients are candidates for long-term storage in fine-grained material. These stabilizer compounds include the cyclic ethers 1,4-dioxane and 1,3-dioxolane; the alcohols *sec*-butyl alcohol, *tert*-butyl alcohol, and 2-methyl-3-butyn-2-ol; and the epoxy compound butoxymethyl oxirane.

The aqueous and effective diffusivity coefficients presented in Table 3.27 span a relatively small range; therefore, the expected mass transfer by diffusion should be similar among these compounds. Accordingly, concentration ratios of stabilizer compounds in the mobile phase of groundwater should be primarily a reflection of their respective effective solubilities (Feenstra and Guiguer, 1996). However, the apparent diffusion coefficient derived by Hoffman et al., which incorporates sorption and tortuosity, shows a wider range. Hydrophilic and highly soluble ethers, alcohols, and some epoxides can be expected to participate to a greater degree in the aquitard storage and subsequent back-diffusion process than other stabilizers, and 100-fold more than methyl chloroform, TCE, and perchloroethylene.

The retardation factor for 1,4-dioxane in Table 3.27 is higher than the values measured by Priddle and Jackson (1991; Table 3.23) or Barone et al. (1992; Table 3.25) because it is calculated from the soil parameters for the Sarnia clay used in the Barone et al. study. Where aquifer sands are interbedded with silts and clays or include a clayey matrix, molecular diffusion will play a significant role in the transport, storage, and long-term secondary release back into the aquifer for several solvent-stabilizer compounds. Matrix diffusion from sands into clays and long-term release by back-diffusion can cause persistent presence of 1,4-dioxane at low concentrations, contrary to the expectation that it would quickly flush through the porous reaches of an aquifer owing to its high solubility and low K_{oc} (Payne et al., 2008).

The geometry of the interface between coarse- and fine-grained aquifer strata plays an important role in determining the degree to which matrix diffusion will occur. Compared to thick sand beds, thin sand beds interbedded with clays and silts provide a larger sand/clay contact surface area across which diffusion may occur. In strongly heterogeneous conditions, contaminant flow in sand lenses less than 10 cm thick will be retarded by aqueous-phase diffusion, causing a significant mass of solute to reside in immobile dissolved-phase storage. Most aquifers are characterized as highly heterogeneous and anisotropic and exhibit sharp contrasts of hydraulic conductivity, often on the scale of a few centimeters (Payne et al., 2008).

The concentration gradient that drives diffusion across interfaces from high to low hydraulic conductivities will rapidly dissipate as 1,4-dioxane or other solutes invade fine-grained pores. The flux density will decrease in response to a decreased concentration gradient, according to Fick's law:

$$J = \Theta_{\text{total}} \cdot D_{\text{eff}} \frac{dC}{dz}, \tag{3.46}$$

where Θ_{total} is total porosity (dimensionless), D_{eff} is effective diffusivity (in m²/day), dC/dz is the solute concentration gradient [i.e., the change in mass concentration over distance, in $(\text{kg/m}^3) \cdot \text{m}^{-1} = \text{kg/m}^2$], and J is the flux density [in $\mu\text{g/(cm}^2 \cdot \text{s)}$]. The diffusive flux density will rapidly decay as contaminant concentration increases in the "immobile porosity," that is, in the micropores in the clays and silts, thereby decreasing the solute concentration gradient, dC/dz. In calculations performed by Payne et al. (2008), the diffusive flux for a TCE gradient with an initial source-side concentration of 200 μg/L will decline 10-fold in 100 days and 1,000-fold in 1,000 days.

There are two main consequences of appreciable diffusive mass flux density (Payne et al., 2008):

- The diffusive loss of solute from the mobile porosity slows the solute velocity relative to groundwater velocity.
- Immobile porosity in clays and silts serves as a repository for dissolved-phase solute that enters via aqueous-phase diffusion.

A further consequence of diffusion is contaminant transport to sites unavailable to microorganisms, precluding biodegradation. For example, in a silt loam, half the measured pore volume consists of pores with radii less than 1 μm, whereas the mean diameter of most soil bacteria is 0.5–0.8 μm, and the mean diameter of soil pores occupied by bacteria is 2 μm (Alexander, 1994). Therefore, diffusion of chemical contaminants into pores smaller than those occupied by soil microbes is an important factor controlling the rate of biodegradation of these compounds. The Matrix Diffusion Toolkit (Farhat et al., 2012) now provides methods for quantitative analysis of the role of matrix diffusion, whose effect on 1,4-dioxane migration causes a surprisingly high mass retention in low-K zones and a very long period of back-diffusion at concentrations exceeding regulatory thresholds (Adamson et al., 2017).

REFERENCES

Abe, A., 1999, Distribution of 1,4-dioxane in relation to possible sources in the water environment. *The Science of the Total Environment* 227: 41–48.

Adamson, D.T., de Blanc, P.C., Farhat, S.K., and Newell, C.J., 2016, Implications of matrix diffusion on 1,4-dioxane persistence at contaminated groundwater sites. *Science of the Total Environment* 562(2016): 98–107.

Adamson, D.T., Mahendra, S., Walker, K.L., Rauch, S.R., Sengupta, S., and Newell, C.J., 2014, A multi-site survey to identify the scale of the 1,4-dioxane problem at contaminated groundwater sites. *Environmental Science & Technology Letters* 1(5): 254–258.

Adamson, D.T., Piña, E.A., Cartwright, A.E., Rauch, S.R., Anderson, R.H., Mohr, T., and Connor, J.A., 2017, 1,4-Dioxane drinking water occurrence data from the third unregulated contaminant monitoring rule. *Science of the Total Environment*. 236–245

Adamus, J.E., May, H.D., Paone, D.A., Evans, P.J., and Parales, R.E., 1995, United States Patent 5,474,934: Biodegradation of ethers. Assignee: Celgene Corporation, Warren, NJ.

Agency for Toxic Substances and Disease Registry (ATSDR), 2012, Toxicological Profile for 1,4-Dioxane. Atlanta, GA, April.

Ahlers, J., 1998, Screening information data set initial assessment report: 2-Methylbut-3-yn-ol; 8th SIDS2 Initial Assessment Meetings, Paris, France. Nairobi, Kenya: UNEP (United Nations Environment Programme) Organization for Economic Cooperation and Development (OECD).

Alarie, Y., Nielsen, G.D., Andonianhaftvan, J., and Abraham, M.H., 1995, Physicochemical properties of non-reactive volatile organic chemicals to estimate RD_{50}: Alternatives to animal studies. *Toxicology and Applied Pharmacology* 134(1): 92–99.

Alexander, M., 1965, Biodegradation: Problems of molecular recalcitrance and microbial fallibility. *Advances in Applied Microbiology* 7: 35–80.

Alexander, M., 1985, Biodegradation of organic chemicals. *Environmental Science and Technology* 19(2): 106–111.

Alexander, M., 1994, *Biodegradation and Bioremediation*. San Diego: Academic Press.

Allen, J.M., Lucas, S., and Allen, S.K., 1996, Formation of hydroxyl radical (OH) in illuminated surface waters contaminated with acidic mine drainage. *Environmental Toxicology and Chemistry* 15(2): 107–113.

Altschuh, J., Berggemann, R., Santl, H., Eichinger, G., and Piringer, O.G., 1999, Henry's law constants for a diverse set of organic chemicals: Experimental determination and comparison of estimation methods. *Chemosphere* 39(11): 1871–1886.

American Petroleum Institute (API), 1998, Assessing the Significance of Subsurface Contaminant Vapor Migration to Enclosed Spaces, Site-Specific Alternatives to Generic Estimates. Publication No. 4674. Health and Environmental Sciences Department. Washington D.C.

Anagnostu, G., 1989, Biological detoxification of industrial wastes containing dioxane and diglyme. Master of Science Thesis, Louisiana State University, Baton Rouge, LA.

Anbar, M. and Neta, P., 1967, A compilation of specific bimolecular rate constants for the reactions of hydrated electrons, hydrogen atoms, and hydroxyl radicals with inorganic and organic compounds in aqueous solution. *International Journal of Applied Radiation and Isotopes* 18(7): 493–523.

Aoyagi, T., Morishita, F., Sugiyama, Y., Ichikawa, D., Mayumi, D., Kikuchi, Y., Ogata, A., Muraoka, K., Habe, H., and Hori, T., 2018, Identification of active and taxonomically diverse 1,4-dioxane degraders in a full-scale activated sludge system by high-sensitivity stable isotope probing. *The ISME Journal* 12: 2376–2388.

Arbuckle, W.B., 1983, Estimating activity coefficients for use in calculating environmental parameters. *Environmental Science and Technology* 17(9): 537–542.

Archer, W.L., 1984, A laboratory evaluation of 1,1,1-trichloroethane-metal-inhibitor systems. *Werkstoffe und Korrosion (Materials and Corrosion)* 35: 60–69.

Arnot, J., Gouin, T., and Mackay, D., 2005, Development and application of models of chemical fate in Canada—practical methods for estimating environmental biodegradation rates. Report to Environment Canada, Peterborough, Ontario, Canada: Environment Canada Report No. 200503.

Aronson, D., Citra, M., Shuler, K., Printup, H., and Howard, P.H., 1999, Aerobic biodegradation of organic chemicals in environmental media: A summary of field and laboratory studies. Syracuse Research Corporation, North Syracuse, NY. Syracuse Research Corporation Report No. TR 99–002.

Atkinson, R., 1988, Estimation of gas-phase hydroxyl radical rate constants for organic chemicals. *Environmental Toxicology and Chemistry* 7: 435–442.

Aziz, C.E., Newell, C.J., Gonzales, J.R., and Jewett, D.G., 2000, *BIOCHLOR Natural Attenuation Decision Support System, User's Manual Version 1.0.* U.S. Environmental Protection Agency, Office of Research and Development, National Risk Management Research Laboratory.

Aziz, C.E., Newell, C.J., Gonzales, J.R., and Jewett, D.G., 2002, *BIOCHLOR Natural Attenuation Decision Support System Version 2.2, User's Manual Addendum.* San Antonio, TX: Groundwater Services, Inc., and Air Force Center for Environmental Excellence Technology Transfer Division.

Baker, J.T., 2006, Material Safety Data Sheet for triethanolamine, Phillipsburg, NJ.

Ball, W.P., 2000, Diffusion-limited contamination and decontamination in a layered aquitard: Forensic and predictive analysis of field data. Paper presented at Confinement and Remediation of Environmental Hazards meeting, January 15–19, 2000, University of Minnesota.

Ball, W.P. and Bartlett, L., 1992, *Evaluation of Para-Dioxane Sorption by Duke Forest Subsurface Solids.* Durham, NC: Duke University Department of Civil and Environmental Engineering.

Banerjee, S., 1984, Solubility of organic mixtures in water. *Environmental Science and Technology* 18(8): 587–591.

Banerjee, S., Yalkowsky, S.H., and Valvani, S.C., 1980, Water solubility and octanol/water partition coefficients of organics: Limitations of the solubility partition coefficient correlation. *Environmental Science and Technology* 14(10): 1227–1229.

Barone, F.S., Rowe, R.K., and Quigley, R.M., 1992, A laboratory estimation of diffusion and adsorption coefficients for several volatile organics in a natural clayey soil. *Journal of Contaminant Hydrology* 10(3): 225–250.

Barton, A.F.M., 1984, *Alcohols with Water,* Solubility Data Series, Vol. 15. Oxford: Pergamon Press.

BASF, 2006, Safety Data Sheet for 2-methylbut-3-yn-2-ol, Florham Park, NJ.

Battersby, N.S. and Wilson, V., 1989, Survey of the anaerobic biodegradation potential of organic chemicals in digesting sludge. *Applied and Environmental Microbiology* 55(2): 433–439.

Bear, J., 1972, *Dynamics of Flow in Porous Media.* New York: American Elsevier Publishing Company.

Beaujean, M., 2007, Acetals as new solvents in paint and coating formulations, paint strippers and graffiti removers. *Pitture E Vernici—European Coatings* 2007(9): 41–48.

Bell, C., Sage, S., Offenberger, S., and Forsberg, N., 2019, Vapor Intrusion of 1,4-Dioxane? Regulatory Myth or Real Issue? Fifth International Symposium on Bioremediation and Sustainable Environmental Technologies, Baltimore, MD, April.

Bennett, P., 2004a, Hydrolysis. http://www.geo.utexas.edu/courses/387E/387E_notes_list.htm (accessed April 2, 2007).

Bennett, P., 2004b, Sorption. http://www.geo.utexas.edu/courses/387E/387E_notes_list.htm (accessed April 2, 2007).

Bennett, P., 2017, Extending the Applicability of Compound-Specific Isotope Analysis to Low Concentrations of 1,4-Dioxane. SERDP Project ER-2535. Washington, DC. February.

Bennett, P., Hyman, M., Smith, C., El Mugammar, H., Chu, M., Nickelsen, M., and Aravena, R., 2018, Enrichment with carbon-13 and deuterium during monooxygenase-mediated biodegradation of 1,4-Dioxane. *Environmental Science and Technology Letters.* 5(3): 148–153.

Bernhardt, D. and Diekmann, H., 1991, Degradation of dioxane, tetrahydrofuran, and other cyclic ethers by an environmental *Rhodococcus* strain. *Applied Microbiology and Biotechnology* 36(1): 120–123.

Bhattacharya, S.K., Qu, M., and Madura, R.L., 1996, Effects of nitrobenzene and zinc on acetate utilizing methanogens. *Water Research* 30(12): 3099–3105.

Bidleman, T.F., 1988, Atmospheric processes—wet and dry deposition of organic compounds are controlled by their vapor-particle partitioning. *Environmental Science and Technology* 22(4): 361–367.

Bingham, E., Cohrssen, B., and Powell, C.H., 2001, *Patty's Toxicology*. New York: John Wiley and Sons, Inc.

Bocek, K., 1976, Relationships among activity coefficients, partition coefficients and solubilities. *Experientia Supplementum* 23: 231–239.

Boethling, R.S., Howard, P.H., Meyian, W., Stiteler, W., Beauman, J., and Tiradot, M., 1994, Group contribution method for predicting probability and rate of aerobic biodegradation. *Environmental Science and Technology* 28(3): 459–465.

Bogyo, D.A., Lande, S.S., Meylan, W.M., Howard, P.H., and Santodonato, J., 1980, Investigation of selected potential environmental contaminants: Epoxides—final technical report. Prepared by Syracuse Research Corporation for the U.S. Environmental Protection Agency Office of Toxic Substances.

Boublik, T., Fried, V., and Hala, E., 1984, *The Vapor Pressures of Pure Substances: Selected Values of the Temperature Dependence of the Vapor Pressures of Some Pure Substances in the Normal and Low Pressure Region*. New York: Elsevier Science Publishing Company.

Bradley, W.F., 1945, Molecular associations between montmorillonite and some polyfunctional organic liquids. *Journal of the American Chemical Society* 67(6): 975–981.

Briggs, G.G., 1981, Theoretical and experimental relationships between soil adsorption, octanol-water partition coefficients, water solubilities, bioconcentration factors, and the parachor. *Journal of Agricultural and Food Chemistry* 29 (5): 1050–1059.

Brindley, G.W. and Tsunashima, A., 1972, Montmorillonite complexes with dioxane, morpholine, and piperidine: Mechanisms of formation. *Clays and Clay Minerals* 20(4): 233–240.

Brindley, G.W., Wiewiora, K., and Wiewiora, A., 1969, Intracrystalline swelling of montmorillonite in some water-organic mixtures. *The American Mineralogist* 54(November-December): 1635–1644.

Brusseau, M.L. and Reid, M.E., 1991, Nonequilibrium sorption of organic chemicals by low organic-carbon aquifer materials. *Chemosphere* 22(3–4): 341–350.

Budavari, S. (Ed.), 1989, *The Merck Index—An Encyclopedia of Chemicals, Drugs and Biologicals*. Rahway, NJ: Merck and Co., Inc.

Budavari, S. (Ed.), 1996, *The Merck Index—An Encyclopedia of Chemicals, Drugs, and Biologicals*. Rahway, NJ: Merck and Co., Inc.

Burback, B.L. and Perry, J.F., 1993, Biodegradation and biotransformation of groundwater pollutant mixtures by *Mycobacterium vaccae*. *Applied Microbiology and Biotechnology* 59(4): 1025–1029.

Butler, J.A.V., Ramchandani, C.N., and Thompson, D.W., 1935, The solubility of non-electrolytes: Part I. The free energy of hydration of some aliphatic alcohols. *Journal of the Chemical Society, Faraday Transactions* 58: 280–285.

Buxton, G.V., Greenstock, C.L., Helman, W.P., and Ross, A.B., 1988, Critical review of rate constants for reactions of hydrated electrons, hydrogen atoms and hydroxyl radicals (•OH/•O$^-$) in aqueous solution. *Journal of Physical and Chemical Reference Data* 17: 513–886.

Cai, Z.-Q., Yang, G.-H., Li, E.-Y., and Zhao, X.-Y., 2008, Isolation, identification and degradation character of dioxane degradative strain D4. *China Environmental Science* 28: 49–52.

Capel, P.D. and Larson, S.J., 1995, A chemodynamic approach for estimating losses of target organic chemicals from water during sample holding time. *Chemosphere* 30(6): 1097–1107.

Carey, F.A., 1987, *Organic Chemistry*. New York: McGraw-Hill.

Chao, J., Lin, C.T., and Chung, T.H., 1983, Vapor pressure of coal chemicals. *Journal of Physical and Chemical Reference Data* 12(4): 1033–1063.

Chappelle, F.H., 1993, *Ground-water Microbiology & Geochemistry*. New York: Wiley.

Charbeneau, R.J., 2000, *Groundwater Hydraulics and Pollutant Transport*. Upper Saddle River, New Jersey, USA: Prentice Hall.

Cheremisinoff, N.P. and Archer, W.L., 2003, *Industrial Solvents Handbook*, 2nd ed. Boca Raton, FL: CRC Press.

Chiang, S.Y., Mora, R., DiGuiseppi, W.H., Davis, G., Sublette, K., Gedalanga, P., and Mahendra, S., 2012, Characterizing the intrinsic bioremediation potential of 1,4-dioxane and trichloroethene using innovative environmental diagnostic tools. *Journal of Environmental Monitoring*, 14: 2317–2326.

Chiou, C.T. and Kile, D.E., 1994, Effects of polar and nonpolar groups on the solubility of organic compounds in soil organic matter. *Environmental Science and Technology* 28(6): 1139–1144.

Clayton, G.D. and Clayton, F.E., 1994, *Patty's Industrial Hygiene and Toxicology*. New York: Wiley.

Coca, J. and Diaz, R., 1980, Extraction of furfural from aqueous solutions with chlorinated hydrocarbons. *Journal of Chemical & Engineering Data* 25(1): 80–83.

Cooper, W.J., Zika, R.G., Petasne, R.G., and Plane, J.M.C., 1988, Photochemical formation of H_2O_2 in natural waters exposed to sunlight. *Environmental Science and Technology* 22(10): 1156–1160.

Corsi, R., 1998, Environmental risk assessment. http://www.crwr.utexas.edu/gis/gisenv98/class/risk/lecture/Lect4/Fate.html#chemical#chemical (accessed April 2, 2007).

Daubert, T.E. and Danner, R.P., 1985, *Data Compilation Tables of Properties of Pure Compounds*. New York: American Institute of Chemical Engineers.

Dean, J.A., 1985, *Lange's Handbook of Chemistry*. New York: McGraw-Hill.

Deneer, J.W., Sinnige, T.L., Seinen, W., and Hermens, J.L.M., 1988, A quantitative structure-activity relationship for the acute toxicity of some epoxy compounds to the guppy. *Aquatic Toxicology* 13(3): 195–204.

Deng, D., Li, F., and Li, M., 2018b, A novel propane monooxygenase initiating degradation of 1,4-dioxane by mycobacterium dioxanotrophicus PH-06. *Environmental Science and Technology Letters*, 5(2): 86–91.

Deng, D., Li, F., Wu, C., and Li, M., 2018a, Synchronic biotransformation of 1,4-dioxane and 1,1-dichloroethylene by a gram-negative propanotroph azoarcus sp. DD4. *Environmental Science and Technology Letters* 5(8): 526–532.

Department of Toxic Substances Control (DTSC), 2004, Guidance for the evaluation and mitigation of subsurface vapor intrusion to indoor air. California Environmental Protection Agency, Department of Toxic Substances Control. http://www.dtsc.ca.gov (accessed November 11, 2006).

Deshpande, M.S., Rale, V.B., and Lynch, J.M., 1992, *Aureobasidium pullulans* in applied microbiology: A status report. *Enzyme and Microbial Technology* 14(7): 514–527.

Dilling, W.L., Bredeweg, C.J., and Tefertiller, N.B., 1976, Simulated atmospheric photodecomposition rates of methylene chloride, 1,1,1-trichloroethane, trichloroethylene, tetrachloroethylene, and other compounds. *Environmental Science and Technology* 10(4): 351–356.

Dilling, W.L., Tefertiller, N.B., and Kallos, G.J., 1975, Evaporation rates and reactivities of methylene chloride, chloroform, 1,1,1-trichloroethane, trichloroethylene, tetrachloroethylene, and other chlorinated compounds in dilute aqueous solutions. *Environmental Science and Technology* 9(9): 833–838.

Domenico, P. and Schwarz, F.W., 1997, *Physical and Chemical Hydrogeology*. New York: Wiley.

Dorfman, L.M. and Adams, G.E., 1973, Reactivity of the hydroxyl radical in aqueous solutions. National Bureau of Standards, National Standard Reference Data System.

Dow Chemical Company, 1980, *The Alkanolamine Handbook*. Midland, MI: Dow Chemical Company.

Dow Chemical Company, 2006, Chlorinated solvents—physical properties. Form No. 100–06358. www.chlorinatedsolvents.com (accessed May 29, 2006).

Dragun, J., 1988, *The Soil Chemistry of Hazardous Materials*. Silver Spring, MD: The Hazardous Materials Control Research Institute.

Edney, E.O. and Corse, E.W., 1987, *Hydroxyl Radical Rate Constant Intercomparison Study*. Research Triangle Park, NC: U.S. Environmental Protection Agency, Atmospheric Science Research Laboratory, EPA-600/S3–86/056.

Evans, P.J., Parales, R.E., and Parales, J.V., 2006, Bioreactors for treatment of 1,4-dioxane. Presented at Emerging Contaminants in Groundwater: A Continually Moving Target, June 8, 2006, Concord, CA. Groundwater Resources Association of California. http://www.grac.org.

Farhat, S.K., Newell, C.J., Seyedabbasi, M.A., McDade, J.M., Mahler, N.T., Sale, T.C., Dandy, D.S., and Wahlberg, J.J., 2012, Matrix Diffusion Toolkit. ER-201126. Environmental Security Technology Certification Program (ESTCP) by GSI Environmental Inc., Houston, Texas. Project ER-201126.

Feenstra, S. and Guiguer, N., 1996, Dissolution of dense non-aqueous phase liquids (DNAPLs) in the subsurface. In: J.F. Pankow and J.A. Cherry (Eds), *Dense Chlorinated Solvents and Other DNAPLs in Groundwater: History, Behavior, and Remediation*, pp. 203–232. Portland, OR: Waterloo Press.

Feenstra, S., Cherry, J.A., Sudicky, E.A., and Haq, Z., 1984, Matrix diffusion effects on contaminant migration from an injection well in a fractured sandstone. *Groundwater* 22(3): 307–316.

Fetter, C.W., 1993, *Contaminant Hydrogeology: First Edition*. Englewood Cliffs, NJ: Prentice-Hall.

Fetter, C.W., Boving, T., and Kreamer, D., 2017, *Contaminant Hydrogeology: Third Edition*. Long Grove, IL; Waveland Press.

Fincher, E.L. and Payne, W.J., 1962, Bacterial utilization of ether glycols. *Applied and Environmental Microbiology* 10(6): 542–547.

Fisher Scientific, 1999, Material Safety Data Sheet for cyclohexane 99%, University of Wisconsin Nanotechnology Department, ACC# 05870.

Fisher Scientific, 2007, Material Safety Data Sheet for n-methyl morpholine. https://fscimage.fishersci.com/msds/95430.htm (accessed February 4, 2007).

Freeze, R.A. and Cherry, J.A., 1979, *Groundwater*. Englewood Cliffs, NJ: Prentice-Hall.

Freitas-Dinis, C.M., Geiger, H., and Wiesen, P., 2001, Kinetics of the reactions of OH($X^2\Pi$) radicals with 1,3-dioxolane and selected dialkoxy methanes. *Physical Chemistry and Chemical Physics* 3: 2831–2835.

Friant, S.L., and Suffet, I.H., 1979, Interactive effects of temperature, salt concentration, and pH on head space analysis for isolating volatile trace organics in aqueous environmental samples. *Analytical Chemistry* 51(13): 2167–2172.

Gaffney, J.S., Streit, G.E., Spall, W.D., and Hall, J.H., 1987, Beyond acid rain. Do soluble oxidants and organic toxins interact with SO_2 and NO_x to increase ecosystem effects? *Environmental Science and Technology* 21(6): 519–524.

Gedalanga, P.B., Pornwongthong, P., Mora, R., Chiang, S.-Y.D., Baldwin, B., Ogles, D., and Mahendra, S., 2014, Identification of biomarker genes to predict biodegradation of 1,4-dioxane. *Applied and Environmental Microbiology* 80(10): 3209–3218.

Gedalanga, P., Madison, A., Miao,Y., Richards. T., Hatton, J., DiGuiseppi, W.H., Wilson, J., Mahendra, S., 2016, A Multiple Lines of Evidence Framework to Evaluate Intrinsic Biodegradation of 1,4-Dioxane. *Remediation Journal*, 27(1):93–114.

Geiger, H., Maurer, T., and Becker, K.H., 1999, OH-initiated degradation mechanism of 1,4-dioxane in the presence of NO_x. *Chemical Physics Letters* 314: 465–471.

Goss, K.U., 2006, Prediction of the temperature dependency of Henry's law constant using poly-parameter linear free energy relationships. *Chemosphere* 64(8): 1369–1374.

Gossett, J.M., 1987, Measurement of Henry's law constants for C1 and C2 chlorinated hydrocarbons. *Environmental Science and Technology* 21(2): 202–208.

Grady, C.P.L., Jr., Sock, S.M., and Cowan, R.M., 1997, Biotreatability kinetics. In: G.S. Sayler, J. Sanseverino, and K.L. Davis (Eds), *Biotechnology in the Sustainable Environment*. New York: Plenum Press.

Granger, F.S. and Nelson, J.M., 1921, Oxidation and reduction of hydroquinone and quinone from the standpoint of electromotive-force measurements. *Journal of the American Chemical Society* 43(7): 1401–1415.

Hansch, C. and Leo, A.J., 1985, Medchem Project Issue No. 26. Pomona College, Medchem Project, Claremont, CA.

Harris, J.C., 1990a, Rate of aqueous photolysis. In: W.J. Lyman, W.F. Reehl, and D.H. Rosenblatt (Eds), *Handbook of Chemical Property Estimation Methods: Environmental Behavior of Organic Compounds*, pp. 8-1–8-41. Washington, DC: American Chemical Society.

Harris, J.C., 1990b, Rate of hydrolysis. In: W.J. Lyman, W.F. Reehl, and D.H. Rosenblatt (Eds), *Handbook of Chemical Property Estimation Methods: Environmental Behavior of Organic Compounds*, pp. 7-1–7-48. Washington, DC: American Chemical Society.

Hashimoto, Y., Tokura, K., Kishi, H., and Strachan, W.M.J., 1984, Prediction of seawater solubility of aromatic compounds. *Chemosphere* 13(8): 881–888.

Hassett, J.J., Means, J.C., Banwart, W.L., and Wood, S.G., 1980, *Sorption Properties of Sediments and Energy-Related Pollutants*. Athens, GA: U.S. Environmental Protection Agency, Office of Research and Development, Environmental Research Laboratory, EPA-600/3-80-041.

Hatzinger, P.B., Banerjee, R., Rezes, R., Streger, S.H., McClay, K., and Schaefer, C.E., 2017, Potential for cometabolic biodegradation of 1,4-dioxane in aquifers with methane or ethane as primary substrates. *Biodegradation*; published online 11 October 2017.

Hawaii Department of Health (HDOH), 2007, Use of laboratory batch tests to evaluate potential leaching of contaminants from soil (April 2007). Hawai'i Department of Health, Office of Hazard Evaluation and Emergency Response. http://www.hawaii.gov/health/environmental/hazard/eal2005.html (accessed February 21, 2007).

Hawley, G.G., 1977, *The Condensed Chemical Dictionary*. New York: Van Nostrand Reinhold Company.

Hefter, G.T., 1984, Solubility data. 2-Butanol-water system. In: A.F.M. Barton (Ed.), *Alcohols with Water*, Vol. 15. Research Triangle Park, NC: The International Union of Pure and Applied Chemistry (IUPAC).

Hemond, H.F. and Fechner, E.J., 1994, *Chemical Fate and Transport in the Environment*. New York: Academic Press.

Hentz, R.R. and Parrish, C.F., 1971, Photolysis of gaseous 1,4-dioxane at 1470 angstroms. *Journal of Physical Chemistry* 75(25): 3899–3901.

Hiatt, M.H., 2013, Determination of Henry's law constants using internal standards with benchmark values. *Journal of Chemical & Engineering Data* 58: 902–908.

Hill, R.R., Jeffs, G.E., Morris, M.S., and Rebhun, M., 1997, Photocatalytic degradation of 1,4-dioxane in aqueous solution. *Journal of Photochemistry and Photobiology A: Chemistry* 108: 55–58.

Hine, J.H. and Mookerjee, P.K., 1975, Structural effects on rates and equilibriums. XIX. Intrinsic hydrophilic character of organic compounds. Correlations in terms of structural contributions. *Journal of Organic Chemistry* 40(3): 292–298.

Hoechst Celanese, 2006, Ethylene oxide user's guide. Celanese, Dow, Shell, Sunoco, Equistar. http://www.ethyleneoxide.com (accessed March 11, 2007).

Hoffman, F., Chiarappa, M.L., O'Boyle, J., Fox, K., and Bair, K., 1998, Diffusive transport of dissolved volatile organic compounds in sediments. U.S. Department of Energy, Lawrence Livermore National Laboratory Report UCRL-ID-129845.

Holton, C., DiGuiseppi, B., Mohr, T., 2017, Assessment of Vapor Intrusion of Volatile but Miscible Compounds. RemTEC Summit, Denver, CO, March.

Holz, M., Heil, S., and Sacco, A., 2000, Temperature-dependent self-diffusion coefficients of water and six selected molecular liquids for calibration in accurate ^{1}H NMR PFG measurements. *Physical Chemistry Chemical Physics* 2: 4740–4742.

Horvath, A.L., Getzen, F.W., and Maczynska, Z., 1999, IUPAC-NIST solubility data series 67: Halogenated ethanes and ethenes with water. *Journal of Physical and Chemical Reference Data* 28(2): 395–629.

Houser, J.J. and Sibbio, B.A., 1977, Liquid-phase photolysis of dioxane. *Journal of Organic Chemistry* 42(12): 2145–2151.

Hovorka, F., Schaefer, R.E., and Dreisbach, D., 1936, The system dioxane and water. *Journal of the American Chemical Society* 58(11): 2264–2267.

Howard, P.H., Boethling, R.S., Jarvis, W.F., Meylan, W.M., and Michalenko, E.M., 1991, *Handbook of Environmental Degradation Rates*. Chelsea, MI: Lewis Publishers.

Huang, H., Shen, D., Li, N., Shan, D., Shentu, J., and Zhou, Y., 2014, Biodegradation of 1,4-dioxane by a novel strain and its biodegradation pathway. *Water, Air, and Soil Pollution* 225: 2135.

Huang, Z., 2007, *Insects in Chinese culture*. http://www.cyberbee.net/~huang/pub/insect.html (accessed September 9, 2007).

Huling, S.G., 1989, *Facilitated Transport: Superfund Groundwater Issue*. Ada, OK: U.S. Environmental Protection Agency, Office of Research and Development, EPA/540/4-89/003.

Huntsman, 2005, *Technical Bulletin: N-Methylmorpholine (NMM) CAS 109-02-4*. The Woodlands, TX: Huntsman Corporation.

Hyman, M., 1999, Final report on aerobic cometabolism of ether-bonded compounds. North Carolina State University, U.S. Environmental Protection Agency, Office of Research and Development, National Center for Environmental Research.

Interstate Technology & Regulatory Council (ITRC), 2013, Environmental Molecular Diagnostics. New Site Characterization and Remediation Enhancement Tools. Technical and Regulatory Guidance. April.

Jackson, R.E. and Dwarakanath, V., 1999, Chlorinated degreasing solvents: Physical-chemical properties affecting aquifer contamination and remediation. *Ground Water Monitoring Review* 19(4): 102–110.

Jackson, R.E. and Mariner, P.E., 1995, Estimating DNAPL composition and VOC dilution from extraction well data. *Ground Water* 33(3): 407–414.

Johnson, P.C. and Ettinger, R.A., 1991, Heuristic model for predicting the intrusion of contaminant vapors into buildings. *Environmental Science and Technology* 25(8): 1445–1452.

Johnson, P.C., Kemblowski, M.W., and Johnson, R.L., 1999, Assessing the significance of subsurface contaminant vapor migration to enclosed spaces: Site-specific alternatives to generic estimates. *Journal of Soil Contamination* 8(33): 389–421.

Karickhoff, S.W., 1981, Semi-empirical estimation of sorption of hydrophobic pollutants on natural sediments and soils. *Chemosphere* 10(8): 833–846.

Kelley, S., Aitchison, E.W., Deshpande, M., Schnoor, J.L., and Alvarez, P.J.J., 2001, Biodegradation of 1,4-dioxane in planted and unplanted soil: Effect of bioaugmentation with *Amycolata* sp. CB1190. *Water Research* 35(16): 3791–3800.

Kim, Y.H. and Engesser, K.H., 2004, Degradation of alkyl ethers, aralkyl ethers, and dibenzyl ether by *Rhodococcus* sp. strain DEE5151, isolated from diethyl ether-containing enrichment cultures. *Applied and Environmental Microbiology* 70(7): 4398–4401.

Kim, Y.H., Engesser, K.H., Kim, S.J., 2007. Physiological, numerical, and molecular characterization of alkyl ether-utilizing *rhodococci*. *Environmental Microbiology* 9(6):1497–1510.

Kinerson, R.S., 1987, Modelling the fate and exposure of complex mixtures of chemicals in the aquatic environment. In: G.C. Butler, V.B. Vouk, A.C. Upton, D.V. Parke, and S.C. Asher (Eds), *Methods for Assessing the Effects of Mixtures of Chemicals*, pp. 409–421. New York: Wiley.

Kirk, R. and Othmer, D., 1982, *Kirk-Othmer Encyclopedia of Chemical Technology*. New York: Wiley.

Klecka, G.M. and Gonsior, S.J., 1986, Removal of 1,4-dioxane from wastewater. *Journal of Hazardous Materials* 13(2): 161–168.

Knapp, J.S. and Bromley-Challenor, K.C.A., 2006, Recalcitrant organic compounds. http://www.bmb.leeds.ac.uk/mbiology/ug/ugteach/level3.html (accessed September 9, 2007).

Kolb, B., Walter, C., and Bichler, C., 1992. Determination of partition coefficients by automatic equilibrium headspace gas chromatography by vapor phase calibration. *Chromatographia* 34(5–8): 235–240.

Lapkin, M., 1965, Epoxides. In: R.E. Kirk and D.F. Othmer (Eds), *Encyclopedia of Chemical Technology*, 2nd ed., Vol. 8, pp. 263–293. New York: John Wiley and Sons Inc.

Leaist, D.G., MacEwan, K., Stefan, A., and Zamari, M., 2000, Binary mutual diffusion coefficients of aqueous cyclic ethers at 25°C. Tetrahydrofuran, 1,3-dioxolane, 1,4-dioxane, 1,3-dioxane, tetrahydropyran, and trioxane. *Journal of Chemical & Engineering Data* 45(5): 815–818.

Leighton, D.T. and Calo, J.M., 1981, Distribution coefficients of chlorinated hydrocarbons in dilute air-water systems for groundwater contamination applications. *Journal of Chemical & Engineering Data* 26(4): 382–385.

Lewis, G.N. and Randall, M., 1921, The activity coefficient of strong electrolytes. *Journal of the American Chemical Society* 43(5): 1112–1154.

Li, M., Mathieu, J., Liu, Y., Van Orden, E.T., Yang, Y., Fiorenza, S., and Alvarez, P.J.J., 2014, The abundance of tetrahydrofuran/dioxane monooxygenase genes (thmA/dxmA) and 1,4-dioxane degradation activity are significantly correlated at various impacted aquifers. *Environmental Science and Technology Letters* 1: 122–127.

Li, M., Liu, Y., Jacques, Y.H., Mathieu, J., Hatton, J., DiGuiseppi, W., Alvarez. P.J.J., 2017. Hindrance of 1,4-dioxane biodegradation in microcosms biostimulated with inducing or non-inducing auxiliary substrates. Water Research, Volume 112, 1 April 2017, Pages 217–225.

Li, Z. and Pirasteh, A., 2006, Kinetic study of the reactions of atomic chlorine with several volatile organic compounds at 240–340 K. *International Journal of Chemical Kinetics* 38(6): 386–398.

Lide, D.R. (Ed.), 2000, *CRC Handbook of Chemistry and Physics*. Boca Raton, FL: CRC Press LLC.

Lide, D.R. (Ed.), 2007, *CRC Handbook of Chemistry and Physics*, Internet Version, 87th ed. Boca Raton, FL: Taylor & Francis Group.

Lipnick R.L., Watson K.R., Strausz, A.K., 1987, A QSAR study of the acute toxicity of some industrial organic chemicals to goldfish. Narcosis, electrophile and proelectrophile mechanisms. In: *Aquatic Toxicology and Environmental Fate* G.W. Suter and M.A. Lewis (Eds.), Eleventh Volume, American Society for Testing and Materials, Philadelphia. ASTM STP 1007. https://www.astm.org/DIGITAL_LIBRARY/STP/PAGES/STP33571S.htm.

Little, J.C., Daisey, J.M., and Nazaroff, W.M., 1992, Transport of subsurface contaminants into buildings. *Environmental Science and Technology* 26(11): 2058–2066.

Liu, W.H., Medina, M.A., Jr., Thoman, W., Piver, W.T., and Jacobs, T.L., 2000, Optimization of intermittent pumping schedules for aquifer remediation using a genetic algorithm. *Journal of the American Water Resources Association* 36(6): 1335–1348.

Lyman, W.J., 1990, Adsorption coefficients for soils and sediments. In: W.J. Lyman, W.F. Reehl, and D.H. Rosenblatt (Eds), *Handbook of Chemical Property Estimation Methods: Environmental Behavior of Organic Compounds*, Chapter 4. Washington, DC: American Chemical Society.

Lyman, W.J., Reehl, W.F., and Rosenblatt, D.H., 1990, *Handbook of Chemical Property Estimation Methods: Environmental Behavior of Organic Compounds*. New York: McGraw-Hill.

MacEwan, D.M.C., 1948, Complexes of clays with organic compounds. I. Complex formation between montmorillonite and hallyosite and certain organic liquids. *Transactions of the Faraday Society* 44: 349–367.

Mackay, D. and Shiu, W.Y., 1981, A critical review of Henry's law constants for chemicals of environmental interest. *Journal of Physical and Chemical Reference Data* 10(4): 1175–1199.

Mackay, D. and Yeun, T.K., 1983, Mass transfer coefficient correlations for volatilization of organic solutes from water. *Environmental Science and Technology* 17(4): 211–217.

Mackay, D., Shiu, W.Y., and Ma, K.C., 1993, *Illustrated Handbook of Physical-Chemical Properties and Environmental Fate for Organic Chemicals: Volume III. Volatile Organic Chemicals*. Boca Raton, FL: Lewis Publishers.

Mackay, D.M. and Cherry, J.A., 1989, Groundwater contamination: Pump and treat remediation. *Environmental Science and Technology* 23(6): 630–636.

Mackison, F.W., Stricoff, R.S., and Partridge, L.J., Jr. (Eds), 1981, *NIOSH/OSHA—Occupational Health Guidelines for Chemical Hazards*. Washington, DC: National Institute of Occupational Safety and Health (NIOSH).

Mahendra, S. and Alvarez-Cohen, L., 2005, *Pseudonocardia dioxanivorans* sp. nov., a novel actinomycete that grows on 1,4-dioxane. *International Journal of Systematic and Evolutionary Microbiology* 55: 593–598.

Mahendra, S. and Alvarez-Cohen, L., 2006, Kinetics of 1,4-dioxane biodegradation by monooxygenase-expressing bacteria. *Environmental Science and Technology* 40(17): 5435–5442.

Mahendra, S., Grostern, A., and Alvarez-Cohen, L., 2013, The impact of chlorinated solvent co-contaminants on the biodegradation kinetics of 1,4-dioxane. *Chemosphere* 91(1): 88–92.

Mahendra, S., Petzold, C.J., Baidoo, E.E., Keasling, J.G., and Alvarez-Cohen, L., 2007, Identification of the intermediates of *in vivo* oxidation of 1,4-dioxane by monooxygenase-containing bacteria. *Environmental Science and Technology* 41(21): 7330–7336.

Mallinckrodt Baker, 2008, Material Safety Data Sheet for *tert*-amyl alcohol—75-85-4, Phillipsburg, NJ.

Maurer, T., Hass, H., Barnes, I., and Becker, K.H., 1999, Kinetic and product study of the atmospheric photo-oxidation of 1,4-dioxane and its main reaction product ethylene glycol diformate. *Journal of Physical Chemistry A* 103: 5032–5039.

Maurino, V., Calza, P., Minero, C., Pelizzeti, E., and Vincente, M., 1997, Light-assisted 1,4-dioxane degradation. *Chemosphere* 35(11): 2675–2688.

Mazurkiewicz, J. and Tomasik, P., 2006, Why 1,4-dioxane is a water-structure breaker. *Journal of Molecular Liquids* 126: 111–116.

Mazzocchi, P.H. and Bowen, M.W., 1975, Photolysis of dioxane. *Journal of Organic Chemistry* 40(18): 2689–2690.

McAuliffe, C., 1966, Solubility in water of paraffin, cycloparaffin, olefin, acetylene, cycloolefin and aromatic hydrocarbon. *Journal of Physical Chemistry* 70(4): 1267–1275.

McCarty, P.L. and Reinhard, M., 1981, Trace organics in groundwater. *Environmental Science and Technology* 15(1): 40–51.

McGuire, M.J. and Suffett, I.H., 1978, Adsorption of organics from domestic water supplies. *Journal of the American Water Works Association* 70(11): 621–635.

Meylan, W.M. and Howard, P.H., 1993, Computer estimation of the atmospheric gas-phase reaction rate of organic compounds with hydroxyl radicals and ozone. *Chemosphere* 26(12): 2293–2300.

Meylan, W.M., Howard, P.H., and Boethling, R.S., 1992, Molecular topology/fragment contribution method for predicting soil sorption coefficients. *Environmental Science and Technology* 26(8): 1560–1567.

Michigan DEQ (Department of Environmental Quality). 2016. Establishment of Cleanup Criteria for 1,4-Dioxane: Emergency Rules. October.

Michigan DEQ. 2017. Rule 27(10)(b) for 1,4-Dioxane. April.

Microbial Insights, 2019, 1,4-Dioxane. Web Site accesses on 3/22/2019 at: https://www.microbe.com/14-dioxane/

Mill, T., Haag, W., Penwell, P., Pettit, T., and Johnson, H., 1987, *Environmental Fate and Exposure Studies Development of a PC-SAR for Hydrolysis: Esters, Alkyl Halides and Epoxides*. Menlo Park, CA: Stanford Research Institute, EPA Contract No. 68–02–4254.

Millington, R.J. and Quirk, J.P., 1961, Permeability of porous solids. *Transactions of the Faraday Society* 57: 1200–1207.

Monod, J., 1949, The growth of Bacterial Cultures, *Annual Review of Microbiology* 3: 371–394, https://doi.org/10.1146/annurev.mi.03.100149.002103.

Nakamiya, K., Hashimoto, S., Ito, H., Edmonds, J.S., and Morita, M., 2005, Degradation of 1,4-dioxane and cyclic ethers by an isolated fungus. *Applied and Environmental Microbiology* 71(3): 1254–1258.

NFPA (National Fire Protection Association), 1991, *National Fire Protection Guide: Fire Protection Guide to Hazardous Materials*. Quincy, MA: National Fire Protection Association.

Nguyen, T.H., Goss, K.U., and Ball, W.P., 2005, Polyparameter linear free energy relationships for estimating the equilibrium partition of organic compounds between water and the natural organic matter in soils and sediments. *Environmental Science and Technology* 39(4): 913–924.

NICNAS (National Industrial Chemicals Notification and Assessment Scheme), 1998, 1,4-Dioxane: Priority Existing Chemical No. 7. Canberra, Commonwealth of Australia: National Occupational Health and Safety Commission.

NICNAS (National Industrial Chemicals Notification and Assessment Scheme), 2003, *Full Report:* 2,6-di-*t*-Butyl-4-Methyl Phenol. Canberra, Commonwealth of Australia: N. I. C. N. A. S. (NICNAS), National Occupational Health and Safety Commission.

Niederer, C. and Goss, K.U., 2007, Quantum chemical modeling of humic acid/air equilibrium partitioning of organic vapors. *Environmental Science and Technology* 41(10): 3646–3651.

Niederer, C., Goss, K.U., and Schwarzenbach, R.P., 2006a, Sorption equilibrium of a wide spectrum of organic vapors in leonardite humic acid: Experimental setup and experimental data. *Environmental Science and Technology* 40(17): 5368–5373.

Niederer, C., Goss, K.U., and Schwarzenbach, R.P., 2006b, Sorption equilibrium of a wide spectrum of organic vapors in leonardite humic acid: Modeling of experimental data. *Environmental Science and Technology* 40(17): 5374–5379.

Nielsen, O.J., Sidebottom, H.W., O'Farrell, D.J., Donlon, M., and Treacy, J., 1989, Rate constants for the gas-phase reactions of OH radicals and Cl atoms with $CH_3CH_2NO_2$, $CH_3CH_2CH_2NO_2$, $CH_3CH_2CH_2CH_2NO_2$, and $CH_3CH_2CH_2CH_2CH_2NO_2$. *Chemical Physics Letters* 156(4): 312–318.

NIOSH, 1978, *Occupational Health Guideline for tert-Butyl Alcohol*. Washington, DC: National Institute for Occupational Safety and Health.

NLM (National Library of Medicine), 2006, Hazardous substance data bank: MEDLARS Online Information Retrieval System. National Library of Medicine. http://toxnet.nlm.nih.gov/ (accessed April 12, 2006).

OECD (Organisation for Economic Cooperation and Development), 1981, Inherent biodegradability: Modified MITI test (II). OECD Protocol 302C, adopted May 12, 1981. Paris, France: Organisation for Economic Cooperation and Development.

Osborn, A.G. and Scott, D.W., 1980, Vapor pressures of 17 miscellaneous organic compounds. *The Journal of Chemical Thermodynamics* 12(5): 429–438.

Pace Analytical Laboratories, 2019, 1,4-Dioxane. Web Site accesses on 3/22/2019 at: https://www.pacelabs.com/environmental-services/specialty-services/14-dioxane.html

Pan, S. and Chen, H., 2006, Anaerobic degradation of 1,4-dioxane under humic-reducing or Fe(III)-reducing conditions. Presented at the 231st ACS National Meeting, Atlanta, GA. American Chemical Society, Abstract ENVR 144.

Pankow, J.F. and Cherry, J.A., 1996, *Dense Chlorinated Solvents and Other DNAPLs in Groundwater: History, Behavior, and Remediation*. Portland, OR: Waterloo Press.

Parales, R.E., Adamus, J.E., White, N., and May, H.D., 1994, Degradation of 1,4-dioxane by an actinomycete in pure culture. *Applied Environmental Microbiology* 60(12): 4527–4530.

Park, J.H., Hussam, A., Couasnon, P., Fritz, D., and Carr, P.W., 1987, Experimental reexamination of selected partition coefficients from Rohrschneider's data set. *Analytical Chemistry* 59(15): 1970–1976.

Parker, B.L., Cherry, J.A., and Gilham, R.W., 1996, The effects of molecular diffusion on DNAPL behavior in fractured porous media. In: J.F. Pankow and J.A. Cherry (Eds), *Dense Chlorinated Solvents and Other DNAPLs in Groundwater: History, Behavior, and Remediation*, pp. 355–393. Portland, OR: Waterloo Press.

Parker, B.L., Gillham, R.W., and Cherry, J.A., 1994, Diffusive disappearance of immiscible-Phase organic liquids in fractured geologic media. *Groundwater* 32: 805–820.

Patt, T.E. and Abebe, H., 1995, United States Patent 5,399,495: Microbial degradation of chemical pollutants. Assignee: The Upjohn Company, Kalamazoo, MI.

Patterson, R.J., Jackson, R.E., Graham, B.W., Chaput, D., and Priddle, M.W., 1985, Retardation of toxic chemicals in a contaminated outwash aquifer. *Water Science and Technology* 17: 57–69.

Patty, F., 1963, *Industrial Hygiene and Toxicology: Volume II. Toxicology*. New York: Wiley-Interscience Publishers.

Pavan, M. and Worth, A.P., 2006, Review of QSAR models for ready biodegradation. European Commission Directorate General Joint Research Centre, European Chemicals Bureau.

Payne, F.C., Quinnan, J.A., and Potter, S.T., 2008, *Remediation Hydraulics*. Boca Raton, FL: Taylor & Francis Group.

Payne, F.C., Suthersan, S.S., Molnaa, B., and Davis, S., 2003, Developing *in situ* reactive zone strategies for 1,4-dioxane. Presented at 1,4-Dioxane and other solvent stabilizer compounds in the environment, December 8, 2003, San Jose, CA. Groundwater Resources Association of California. http://www.grac.org.

Perrin, D.D., 1972, *Dissociation Constants of Organic Bases in Aqueous Solution: Supplement*. London: Butterworths.

Pickett, L.W., Hoeflich, N.J., and Liu, T.C., 1951, The vacuum ultraviolet absorption spectra of cyclic compounds. II. Tetrahydrofuran, tetrahydropyran, 1,4-dioxane and furan. *Journal of the American Chemical Society* 73(10): 4865–4869.

Platz, J., Sehested, J., Mgelberg, T., Nielsen, O.J., and Wallington, T.J., 1997, Atmospheric chemistry of 1,4-dioxane: Laboratory studies. *Journal of the Chemical Society, Faraday Transactions* 93(16): 2855–2963.

Pornwongthong, P., Paradis, G.L., and Mahendra, G.L., 2011, Stable carbon isotope fractionation during 1,4-dioxane biodegradation. *Proceedings of the Water Environment Federation 2011* 18: 111–116.

Priddle, M.W. and Jackson, R.E., 1991, Laboratory column measurement of VOC retardation factors and comparison with field values. *Ground Water* 29(2): 260–266.

Prior, S. D. and Dalton, H., 1985, The effect of copper ions on membrane content and methane monooxygenase activity in methanol-grown cells of *Methylococcus capsulatus*. *Journal of General Microbiology* 131: 155–163.

Pruden, A., Sedran, M.A., Suidan, M.T., and Venosa, M.D., 2005, Anaerobic biodegradation of methyl *tert*-butyl ether under iron-reducing conditions in batch and continuous-flow cultures. *Water Research* 77(3): 297–303.

Reid, E.W. and Hoffman, H.E., 1929, 1,4-dioxan. *Industrial and Engineering Chemistry* 21(7): 695–697.

Reinhard, M. and Drefahl, A., 1999, *Handbook for Estimating Physicochemical Properties of Organic Compounds*. New York: Wiley.

Renner, R., 2002, The K_{ow} controversy: Doubts about the quality of basic physicochemical data for hydrophobic organic compounds could be undermining many environmental models and assessments. *Environmental Science and Technology* 36(21): 411A–413A.

Riddick, J.A., Bunger, W.B., and Sakano, T.K., 1985, *Techniques of Chemistry, 4th Edition: Volume II. Organic Solvents*. New York: Wiley.

Rohrschneider, L. 1973. Solvent characterization by gas-liquid partition coefficients of selected solutes. *Analytical Chemistry* 45(7):1241–1247.

Roy, D., Anagnostu, G., and Chaphalkar, P., 1994, Biodegradation of dioxane and diglyme in industrial waste. *Journal of Environment Science and Health, Part A: Environmental Science* 29(1): 129–147.

Roy, D., Anagnostu, G., and Chaphalkar, P., 1995, Analysis of respirometric data to obtain kinetic coefficients for biodegradation of 1,4-dioxane. *Journal of Environment Science and Health, Part A: Environmental Science* 30(8): 1775–1790.

Sander, R., 1999, Compilation of Henry's law constants for inorganic and organic species of potential importance in environmental chemistry. Max-Planck Institute of Chemistry, Air Chemistry Department, Mainz, Germany. http://www.mpch-mainz.mpg.de/~sander/res/henry.html (accessed May 6, 2007).

Sander, R., 2015, Compilation of Henry's law constants (Version 4.0) for water as solvent. *Atmospheric Chemistry and Physics* 15: 4399–4981.

Sax, N.I., 1984, *Dangerous Properties of Industrial Materials*. New York: Van Nostrand Reinhold Company.

Sax, N.I. and Lewis, R.J.S., 1987, *Hawley's Condensed Chemical Dictionary*. New York: Van Nostrand Reinhold Company.

Schultz, T.W., 1986, The use of the ionization constant (pK_a) in selecting models of toxicity in phenols. *Ecotoxicology and Environmental Safety* 14(2): 178–183.

Schwarzenbach, R.P. and Westall, J., 1981, Transport of nonpolar organic compounds from surface water to groundwater: Laboratory sorption studies. *Environmental Science and Technology* 15(11): 1360–1367.

Schwarzenbach, R.P., Gschwend, P.M., and Imboden, D.M., 1993, *Environmental Organic Geochemistry*. New York: Wiley.

Scow, K.M., 1982, Rate of biodegradation. In: W.J. Lyman, W.F. Reehl, and D.H. Rosenblatt (Eds), *Handbook of Chemical Property Estimation Methods*, pp. 9-1–9-85. Washington, DC: American Chemical Society.

Sei, K., Miyagaki, K., Kakinoki, T., Fukugasako, K., Inoue, D., and Ike, M., 2013, Isolation and characterization of bacterial strains that have high ability to degrade 1,4-dioxane as a sole carbon and energy source. *Biodegradation* 24, 665–674.

Seinfeld, J.H., 1986, *Atmospheric Chemistry and Physics of Air Pollution*. New York: Wiley.

Sekar, R., Taillefert, M. and DiChristina, T.J., 2016, Simultaneous transformation of commingled trichloroethylene, tetrachloroethylene, and 1,4-dioxane by a microbially driven Fenton reaction in batch liquid cultures. *Applied and Environmental Microbiology* 82(21). November.

Serjeant, E.P. and Dempsey, B., 1979, *Ionisation Constants of Organic Acids in Aqueous Solution*. New York: Pergamon Press.

Shen, T.T., 1981, Estimating hazardous air emissions from disposal sites. *Pollution Engineering* 13(1981): 31–35.

Shen, W., Chen, H., and Pan, S., 2008, Anaerobic biodegradation of 1,4-dioxane by sludge enriched with iron-reducing microorganisms. *Bioresource Technology* 99(7): 2483–2487.

Shiu, W.Y., Ma, K.C., Varhanickova, D., and Mackay, D., 1994, Chlorophenols and alkylphenols: A review and correlation of environmentally relevant properties and fate in an evaluative environment. *Chemosphere* 29(6): 1155–1224.

Skadsen, J.M., Rice, B.L., and Meyering, D.J., 2004, The occurrence and fate of pharmaceuticals, personal care products, and endocrine disrupting compounds in a municipal water use cycle: A case study in the City of Ann Arbor. Water Utilities, City of Ann Arbor, and Fleis & VandenBrink Engineering, Inc.

Skinner, K.M., 2007, Characterization of the molecular foundations and biochemistry of alkane and ether oxidation in a filamentous fungus, a *Graphium* species. PhD Dissertation, Oregon State University, Corvallis, OR.

Smallwood, I.M., 1996, *Handbook of Organic Solvent Properties*. New York: Halsted Press.

Snider, J.R. and Dawson, G.A., 1985, Tropospheric light alcohols, carbonyls, and acetonitrile: Concentrations in the southwestern United States and Henry's law data. *Journal of Geophysical Research* 90(D2): 3797–3805.

Snoeyink, V.L., 1999, Adsorption of organic compounds. In: F.W. Pontius (Ed), *Water Quality & Treatment Handbook*, pp. 781–875. New York: McGraw-Hill.

Sock, S.M., 1993, A comprehensive evaluation of biodegradation as a treatment alternative for the removal of 1,4-dioxane. Master of Science Thesis, Clemson University, Department of Environmental Engineering and Science, Clemson, SC.

Southworth, G.R. and Keller, J.L., 1986, Hydrophobic sorption of polar organics by low organic carbon soils. *Journal of Water, Air, & Soil Pollution* 28(3–4): 239–247.

SPI (Society of the Plastics Industry), 2006, Test plans for alkyl (C_{12}-C_{14}) glycidyl ether (CAS No. 68609–97–2) and n-butyl glycidyl ether (CAS No. 2426-08-6). Society of the Plastics Industry/Epoxy Resin Systems Task Group, Chemical Right-to-Know—HPV Challenge Program.

SRC (Syracuse Research Corporation), 2007a, BIODEG. Syracuse Research Corporation.

SRC (Syracuse Research Corporation), 2007b, ChemFate. Syracuse Research Corporation.

Srinivasan, V., Clement, T.P., and Lee, K.K., 2007, Domenico solution—is it valid? *Ground Water* 45(2): 136–146.

Stanley, S.H., Prior, S.D., Leak, D.J., and Dalton, H., 1983, Copper stress underlies the fundamental change in intracellular location of methane mono-oxygenase in methane-oxidizing organisms: studies in batch and continuous cultures. *Biotechnology Letters* 5: 487–492.

Steffan, R.J., 2006, *Biodegradation of Ether-Containing Pollutants*. Arlington, VA: Federal Remediation Technologies Roundtable. http://www.frtr.gov/ (accessed August 2, 2007).

Steinkraus, D.C. and Whitefield, J.B., 1994, Chinese caterpillar fungus and world record runners. *American Entomologist* 40(4): 235–239.

Suthersan, S.S., 2002, *Natural and Enhanced Remediation Systems*. Boca Raton, FL: Lewis Publishers.

Swann, R.L., Laskowski, D.A., McCall, P.J., and Vander Kuy, K., 1983, A rapid method for the estimation of the environmental parameters octanol/water partition coefficient, soil sorption constant, water to air ratio and water solubility. *Residue Reviews* 85: 17–28.

Taylor, S.W., Lange, C.R., and Lesold, E.A., 1997, Biofouling of contaminated ground-water recovery wells: Characterization of microorganisms. *Ground Water* 35(6): 973–980.

Thibodeaux, L.J., 1996, *Environmental Chemodynamics*. New York: Wiley.

Thomas, R.G., 1990, Volatilization from water. In: W.J. Lyman, W.F. Reehl, and D.H. Rosenblatt (Eds), *Handbook of Chemical Property Estimation Methods*, pp. 1–34. Washington, DC: American Chemical Society.

Tipler, A., 2016. An Introduction to Headspace Sampling in Gas Chromatography – Fundamentals and Theory. Perkin Elmer Inc. https://www.perkinelmer.com/lab-solutions/resources/docs/GDE_Intro_to_Headspace.pdf (accessed August, 2016).

U.S. Environmental Protection Agency (USEPA), 1981, Phase I risk assessment of 1,4-dioxane. U.S. Environmental Protection Agency Contract No. 68-01-6030.

U.S. Environmental Protection Agency (USEPA), 1991, Risk Assessment Guidance for Superfund (RAGS), Volume I: Human Health Evaluation Manual (Part B, Development of Preliminary Remediation Goals), Interim. Office of Emergency and Remedial Response, Washington, D.C. EPA-540-R-92-003. December. http://www.epa.gov/oswer/riskassessment/ragsb/index.htm.

U.S. Environmental Protection Agency (USEPA), 1996, Requirements for management of hazardous contaminated media: Appendix A. Summary of physical/chemical properties. 40 Code of Federal Regulations Parts 260, 261, 262, 264, 268, 269 and 271, Vol. 61, No. 83. Government Printing Office.

U.S. Environmental Protection Agency (USEPA), 1998a, *Industrial Waste Air Model Technical Background Document*. Washington, DC: U.S. Environmental Protection Agency, Office of Solid Waste, EPA 530-R-99-004.

U.S. Environmental Protection Agency (USEPA), 1999, Use of monitored natural attenuation at Superfund, RCRA corrective action, and underground storage tank sites. U.S. Environmental Protection Agency, Office of Solid Waste and Emergency Response, Directive 9200.4-17P.

U.S. Environmental Protection Agency (USEPA), 2000a, Aqueous Hydrolysis Rate Program (HYDROWIN). A module in the EPIWIN Suite, U.S. Environmental Protection Agency.

U.S. Environmental Protection Agency (USEPA), 2000b, High production volume challenge program submission for 1,3,5-trioxane by Trioxane Manufacturer's Consortium. Prepared by Toxicology and Regulatory Affairs, Flemington, NJ. http://www.epa.gov/hpv/pubs/summaries/triox/c12863.pdf (accessed January 4, 2009).

U.S. Environmental Protection Agency (USEPA), 2001a, *User's Guide for Water9 Software Version 2.0.0.* Research Triangle Park, NC: U.S. Environmental Protection Agency, Office of Air Quality Planning and Standards.

U.S. Environmental Protection Agency (USEPA), 2001b, Supplemental guidance for developing soil screening levels for Superfund sites (Peer Review Draft, May 2001). U.S. Environmental Protection Agency, Office of Solid Waste and Emergency Response, Publication 9355.4–24.

U.S. Environmental Protection Agency (USEPA), 2007a, Estimations programs interface for windows (EPISuite), v. 3.20. http://www.epa.gov/oppt/exposure/pubs/episuite.htm (accessed November 3, 2007).

U.S. Environmental Protection Agency (USEPA), 2007b, CSMoS comments on the potential limitations of the Domenico-based fate and transport models. Ground Water and Ecosystems Restoration Research. http://www.epa.gov/ada/csmos/domenico.html (accessed September 3, 2007).

U.S. Environmental Protection Agency (USEPA), 2007c, High Production Volume Information System (HPVIS) online database. http://www.epa.gov/chemrtk/hpvis/index.html (accessed November 3, 2007).

U.S. Environmental Protection Agency (USEPA), 2008, A Guide for Assessing Biodegradation and Source Identification of Organic Ground Water Contaminants using Compound Specific Isotope Analysis (CSIA). EPA 600/R-08/148, Office of Research and Development National Risk Management Research Laboratory, Ada, Oklahoma, December 2008.

U.S. Environmental Protection Agency (USEPA), 2014, Technical Fact Sheet – 1,4-Dioxane. https://www.epa.gov/sites/production/files/2014-03/documents/ffrro_factsheet_contaminant_14-dioxane_january2014_final.pdf

U.S. Environmental Protection Agency (USEPA), 2015, Technical Guide for Assessing and Mitigating the Vapor Intrusion Pathway from Subsurface Vapor Sources to Indoor Air. Office of Solid Waste and Emergency Response (OSWER) Publication 9200.2-154. Washington, DC, June 2015.

U.S. Environmental Protection Agency (USEPA), January 2017, Addition of a subsurface intrusion component to the hazard ranking system (EPA-HQ-SFUND-2010-1086; FRL-9956-58-OLEM).

UNEP (United Nations Environment Programme), 1998, *Screening Information Data Set (SIDS) for 2-Methyl-3-yn-2-ol.* Nairobi, Kenya: United Nations Environment Programme Report 115-19-5. http://www.chem.unep.ch/irptc/sids/OECDSIDS/115195.pdf (accessed June 3, 2007).

UNEP (United Nations Environment Programme), 2002, Screening information data set (SIDS) for 2,6-di-tert-butyl-p-cresol [betahydroxy toluene or BHT]: Initial assessment report for SIDS initial assessment meeting (SIAM), 14th meeting, Paris, France, March 26–28, 2002.

Vainberg, S., McClay, K., Masuda, H., Root, D., Condee, C., Zylstra, G.J., and Steffan, R.J., 2006, Biodegradation of ether pollutants by *Pseudonocardia* sp. strain ENV478. *Applied and Environmental Microbiology* 72(8): 5218–5224.

Verschueren, K., 1996, *Handbook of Environmental Data on Organic Chemicals*, 3rd ed. New York: Van Nostrand Reinhold Company.

Vidic, R.D. and Suidan, M.T., 1991, Role of dissolved oxygen on the adsorptive capacity of activated carbon for synthetic and natural organic matter. *Environmental Science and Technology* 25: 1612–1618.

Walton, B.T., Hendricks, M.S., and Anderson, T.A., 1992, Soil sorption of volatile and semivolatile organic compounds in a mixture. *Journal of Environmental Quality* 21: 552–558.

Wang, Y., 2016, Breakthrough in 2D-CSIA technology for 1,4-dioxane. *Remediation.* Winter 2016.

Wasik, S.P., Tewari, Y.B., Miller, M.M., and Martire, D.E., 1981, *Octanol/Water Partition Coefficients and Aqueous Solubilities of Organic Compounds.* Washington, DC: U.S. Department of Commerce, National Bureau of Standards, NBSIR 81–2406.

Weber, R., Parker, P., and Bowser, M., 1981, Vapor pressure distribution of selected organic chemicals. U.S. Environmental Protection Agency, Office of Research and Development, EPA/600/2-81/021.

West, M.R., Kueper, B.H., and Ungs, M.J., 2007, On the use and error of approximation in the Domenico, 1987 solution. *Ground Water* 45(2): 126–135.

Whiffin, R.B. and Bahr, J.H., 1984, Assessment of purge well effectiveness for aquifer decontamination. *Proceedings of the Fourth National Symposium on Aquifer Restoration and Ground Water Monitoring.* Las Vegas, NV: National Water Well Association, May 23–25.

White, G.F., Russell, N.J., and Tidswell, E.C., 1996, Bacterial scission of ether bonds. *Microbiology and Molecular Biology Reviews* 60(1): 216–232.

Wilke, C.R. and Chang, P., 1955, Correlation of diffusion coefficients in dilute solutions. *AIChE Journal* 1(2): 264–272.

Willis, G.H. and McDowell, L.L., 1982, Pesticides in agricultural runoff and their effects on downstream water quality. *Environmental Toxicology and Chemistry* 1: 267–279.

Windholz, M. (Ed.), 1983, *The Merck Index*. Rahway, NJ: Merck & Co., Inc.

Wing, M.R., 1997, Apparent first-order kinetics in the transformation of 1,1,1-trichloroethane in groundwater following a transient release. *Chemosphere* 34(3): 771–781.

Woo, Y.T., Argus, M.F., and Arcos, J.C., 1977, Metabolism *in vivo* of dioxane: Effect of inducers and inhibitors of hepatic mixed-function oxidases. *Biochemical Pharmacology* 26(16): 1539–1542.

Wu, J., Low, P.F., and Roth, C.B., 1994, Effect of 1,4-dioxane on the expansion of montmorillonite layers in montmorillonite/water systems. *Clays and Clay Minerals* 42(2): 109–113.

Yalkowsky, S.H. and Dannenfelser, R.M., 1992, *The AQUASOL Database of Aqueous Solubility*. Tucson: University of Arizona, College of Pharmacy.

Yaws, C.L., 1994, *Handbook of Vapor Pressure: Volume 2. C_5 to C_7 Compounds*. Houston, TX: Gulf Publishing Company.

Young, D.F. and Ball, W.P., 1998, Estimating diffusion coefficients in low-permeability porous media using a macropore column. *Environmental Science and Technology* 32(17): 2578–2584.

Yurteri, C., Ryan, D.F., Callow, J.J., and Gurol, M.D., 1987. The effect of chemical composition of water on Henry's law constant. *Journal of the Water Pollution Control Federation* 59(11): 950–956.

Zenker, M.J., 2006, 1,4-Dioxane biodegradation—a review. In: A.R. Gavaskar and A.S.C. Chen (Eds), *Remediation of Chlorinated and Recalcitrant Compounds*. Columbus, OH: Battelle Press.

Zenker, M.J., Borden, R.C., and Barlaz, M.A., 1999, Investigation of the intrinsic biodegradation of alkyl and cyclic ethers. *Proceedings of the Fifth International In Situ and On-Site Bioremediation Symposium*, San Diego, CA, April 19–22. Columbus, OH: Battelle Press.

Zenker, M.J., Borden, R.C., and Barlaz, M.A., 2000, Mineralization of 1,4-dioxane in the presence of a structural analog. *Biodegradation* 11(4): 239–246.

Zenker, M.J., Borden, R.C., and Barlaz, M.A., 2002, Modeling cometabolism of cyclic ethers. *Environmental Engineering Science* 19(4): 215–228.

Zepp, R.G. and Cline, D.M., 1977, Rates of direct photolysis in aquatic environment. *Environmental Science and Technology* 11(4): 359–366.

Zhang, S., Gedalanga, P.B., and Mahendra, S., 2016, Biodegradation kinetics of 1,4-dioxane in chlorinated solvent mixtures. *Environmental Science and Technology* 50(17): 9599–9607.

Zhang, Z.Z., Low, P.F., Cushman, J.H., and Roth, C.B., 1990, Adsorption and heat of adsorption of organic compounds on montmorillonite from aqueous solutions. *Soil Science Society of America Journal* 54(1): 59–66.

Zhao, L., Lu, X., Polasko, A., Johnson, N.W., Miao, Y., Yang, Z., Mahendra, S., and Gu, B., 2018, Co-contaminant effects on 1,4-dioxane biodegradation in packed soil column flow-through systems. *Environmental Pollution* 243 (Pt A): 573–581.

4 Sampling and Laboratory Analysis for Solvent Stabilizers

Thomas K.G. Mohr and Jeremy Bishop

CONTENTS

In the 1990s and early 2000s, solvent-release site investigators may have overlooked 1,4-dioxane because (1) they were not required to look for it and (2) they did not find a compelling reason to sample for chemical compounds that were not among the known or regulated threats to drinking water and human health. Awareness of the widespread occurrence of 1,4-dioxane as a co-contaminant of methyl chloroform is still growing and is beginning to be better recognized. Historically, an understanding that 1,4-dioxane was a co-contaminant was not widespread; consequently, most site investigators did not direct the laboratory to analyze for 1,4-dioxane. A sensitive and reliable laboratory analytical method for 1,4-dioxane was not available until 1997, when the California Department of Health Services developed a method for low-level detection of 1,4-dioxane (Draper et al., 2000). Other solvent-stabilizer compounds may also be overlooked because (1) most investigations of solvent-release sites focus on analyzing for familiar chlorinated aliphatic hydrocarbon compounds; (2) the methods used are unsuitable for trace detection and quantification of ethers, alcohols, and other highly soluble compounds that have been used as solvent stabilizers; and (3) the presence of much higher concentrations of solvents in environmental samples may obscure the detection of trace level stabilizers.

The challenge to analyze for 1,4-dioxane in water is a direct consequence of its miscibility with water, its low water-air partition coefficient, and its low affinity to organic carbon. Understanding the challenges faced by laboratory chemists helps remedial project managers make better use of available laboratory analytical methods. This chapter is written for the "layperson," that is, engineers, geologists, and other people interested in contaminated site remediation who are laboratory clients but who have relatively limited familiarity with the subtleties of laboratory analysis. This chapter also includes detailed information on 1,4-dioxane analysis for the benefit of laboratory chemists.

Sampling methods can make an important difference in the results of analyzing for hydrophilic compounds. The requirements of sampling for 1,4-dioxane by routine or innovative methods are summarized here. By way of review, the basic principles of gas chromatography (GC), mass spectrometry (MS), and sample preparation are briefly discussed here in the context of 1,4-dioxane analysis.

This chapter demonstrates that in laboratory analysis of environmental samples, the chemist is faced with myriad considerations for how the many available combinations of different options in each analytical step best serve to isolate and quantify the contaminants of interest. This universe of possibilities makes analytical chemistry a fascinating science; indeed, its skillful execution deserves the same appreciation as fine art!

4.1 THE FLAWED PARADIGM OF ANALYTE LISTS

To satisfy regulatory requirements and generate comparable results, project managers often require the use of standard methods for analysis of samples from sites contaminated by hazardous wastes. These methods are usually selected from the U.S. Environmental Protection Agency (USEPA) *Test Methods for Evaluating Solid Waste* [SW-846 (see USEPA, 2014a, and earlier editions)]. These methods are designed to reliably report accurate results for a limited list of compounds that behave similarly and that can be analyzed by using the same general approach and instrument parameters.

Analyzing for solvents from standard lists of priority pollutants will overlook important compounds that may be equally or more toxic and mobile than those that the site investigator seeks. Therefore, more information could be obtained by focusing the laboratory phase of an investigation on a wider range of organic compounds than is included on the standard method list. For example, the typical laboratory report for U.S. Environmental Protection Agency (USEPA) Method 8260D (USEPA, 2018a), "Volatile Organic Compounds by Gas Chromatography/Mass Spectrometry (GC/MS)," includes far fewer than the 128 compounds listed in Method 8260; often only 33 compounds, or an extended list of 62 compounds, will be reported by commercial laboratories. Historically, 1,4-dioxane was often excluded from target compound lists at commercial laboratories; however, as the importance of solvent stabilizers has been better understood, 1,4-dioxane has become a more common target analyte for commercial laboratories. The compounds reported often vary from laboratory to laboratory and project to project. Project managers may specify which target analytes to include for each method requested (Simmons, 1997). The choices of sample preparation, chromatographic column, detector, calibration standards, and mass spectrometer settings all affect the ability of a particular analysis to detect and quantify compounds. In order to favor detection of chlorinated solvents at low concentrations, other compounds are typically excluded from reporting. Additionally, at high concentrations chlorinated solvents may obscure the presence of solvent stabilizers or laboratory practices may focus on the highest concentration analytes in a sample and inadvertently exclude lower concentration compounds.

For laboratory methods using GCMS, the chemist has the option of looking at nontarget peaks and making tentative identifications by comparison with a library of mass spectra. Tentatively identified compounds (TICs) are usually reported with an estimated concentration, because the instrument is not calibrated using authentic standards for TICs. The USEPA's Contract Laboratory Program (CLP) requires that laboratories report up to 20 TICs per sample. Some laboratories have fixed prices for reporting the 10 or 20 most commonly encountered, or highest response, nontarget compounds (Simmons, 1997).

A variety of ancillary benefits can be obtained from knowing the full complement of anthropogenic compounds in groundwater samples at solvent-release sites. As discussed in Chapter 3, the chlorinated solvents targeted in an investigation are often recalcitrant and retarded relative to groundwater flow rates. Hydrophilic compounds introduced as solvent stabilizers can be helpful for mapping the full extent of a plume, as they are retarded to a much smaller extent and therefore behave like a conservative tracer. Knowing the migration pathways traversed by the more mobile contaminants can be useful for predicting the likely direction of migration of less mobile contaminants released at the same time. Further, parameters of plume dispersion such as dispersivity can be inferred from the distribution of the most mobile contaminant. Knowledge of these migration parameters aids modeling site history and remedial alternatives. Tracking molar ratios of chlorinated solvents (which biodegrade slowly under certain ambient conditions) to 1,4-dioxane (which does not biodegrade significantly under ambient conditions) can also be instructive for analysis of monitored natural attenuation or enhanced biodegradation as a remediation strategy for chlorinated solvents.

The consequences of *not* analyzing for a wider variety of analytes and *not* requesting that TICs be reported at solvent-release sites may include the following: (1) The site characterization may be incomplete (see Chapter 9); (2) the capture zone and treatment system design may be ineffective in protecting drinking water (see Chapter 9); (3) the risks to drinking water consumers may not be fully addressed (see Chapter 6); (4) adjudication of source allocation in legal disputes may be erroneous (see Chapter 10).

Unfortunately, 1,4-dioxane is not a benign tracer. At sufficiently high concentrations, it can cause health effects when consumed, as explained in Chapter 5. Therefore, an expanded analysis of samples for 1,4-dioxane at solvent-release sites should be considered.

In several conference presentations, Christian Daughton of the USEPA summarized the inherent filtering process that results from following the paradigm of restricting analyses by method analyte lists. Daughton has noted that nearly 28 million organic and inorganic substances had been documented as of April 2006 (as indexed by the American Chemical Society's Chemical Abstracts Service in their CAS Registry); 10 million of these are commercially available. Only 240,000 of these compounds are subjected to any form of regulation. Of these 240,000, a much smaller number of chemicals are subjected to routine laboratory analysis of drinking water or environmental samples. A still smaller number of chemicals have been subjected to sufficient study to provide reliable information about health effects (Daughton, 2005).

Daughton (2003, p. 758) has offered several deceptively simple but profoundly important insights into the limitations of environmental analyses:

- Not everything that can be measured is worth measuring, and not everything worth measuring is measurable
- What one finds usually depends on what one aims to search for

- Only those compounds targeted for monitoring have the potential for being identified and quantified
- Those compounds not targeted will elude detection
- The spectrum of pollutants identified in a sample represents but a portion of those present and they are of unknown overall risk significance

These axioms apply very well to solvent stabilizers and help to explain why it took more than 20 years of cleanup at solvent-release sites for 1,4-dioxane to be discovered. Figure 4.1 summarizes the manner in which the universe of possible chemicals present in the environment is narrowed to the target analyte list by USEPA methods.

The tendency for 1,4-dioxane to be overlooked by routine approaches for analyzing samples for organic compounds was first noted in 1977. In a review article by William T. Donaldson of the USEPA, the basis for establishing analyte lists was questioned. Donaldson noted that a list of 50 compounds *expected* to occur because of their known use in industry did not match well with a

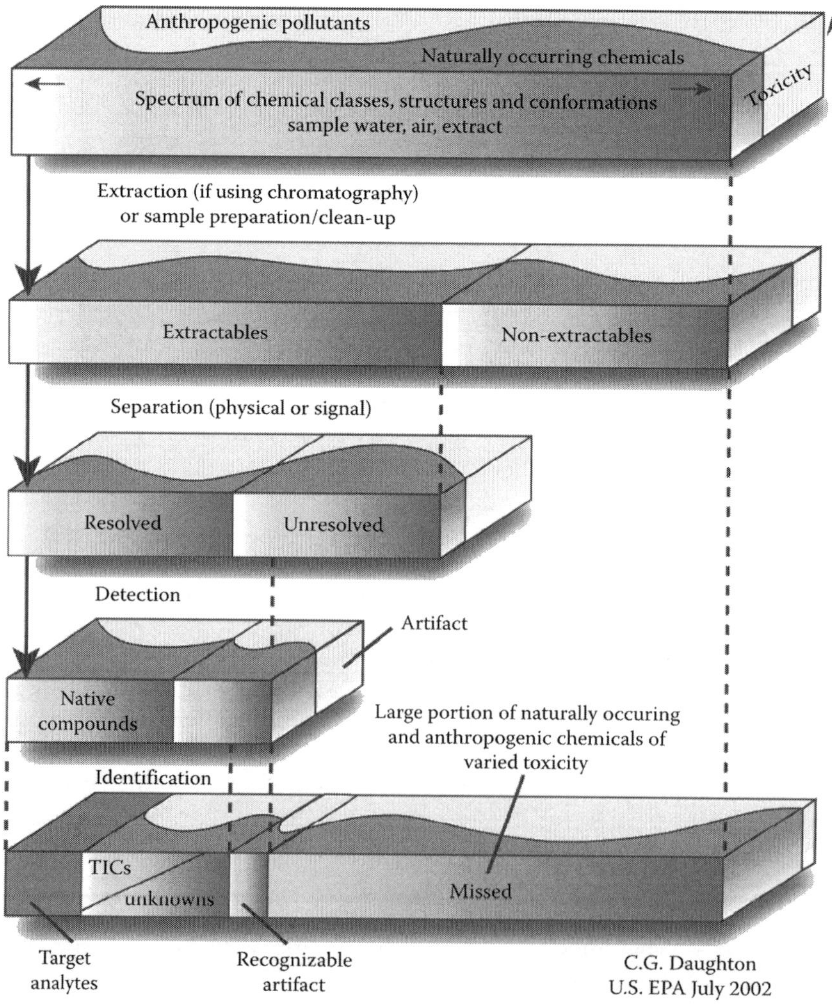

FIGURE 4.1 Limitations and complexity of environmental chemical analysis. TICs, tentatively identified compounds. (From Daughton, C.G., 2006, Pharmaceuticals as environmental contaminants: An overview of the science. Paper presented at the Water Education Foundation Pacific Conference. With permission.)

survey of 5,500 analyses of water samples. Whereas many of the compounds expected in water were not found, 1,4-dioxane was observed in 15 analyses, and nitromethane was observed in 10 samples (Donaldson, 1977).

4.2 SAMPLE COLLECTION, PRESERVATION, AND HANDLING FOR ANALYSIS OF 1,4-DIOXANE

USEPA methods that include 1,4-dioxane as an analyte include USEPA SW846 Method 8260, USEPA SW846 Method 8270, USEPA SW846 Method 1624, and USEPA drinking water Method 522. Sample-preservation requirements for these methods follow the general protocol for all analytes obtained from those methods, as summarized in Table 4.1.

To ensure data validity, it is best to follow the sample container, preservation, and holding time protocols for the USEPA method that will be used for 1,4-dioxane analysis or the USEPA method upon which the commercial laboratory's adaptation is based. Although acidic preservatives are required for USEPA methods for volatile organic compound (VOC) analysis, they are not necessary for 1,4-dioxane. 1,4-Dioxane extraction and recovery are independent of pH (Yoo et al., 2002). Biodegradation of 1,4-dioxane in samples is not likely to be a concern, given the absence of evidence for rapid biodegradation of 1,4-dioxane under ambient conditions (Vainberg et al., 2006).

4.2.1 Sampling for 1,4-Dioxane in Water

Conventional groundwater sampling approaches involve a one-stop visit to a well site for sample retrieval with a bailer or a submersible sampling pump, after purging three to five well-casing volumes or enough groundwater so that consecutive readings of pH and conductivity are constant.

TABLE 4.1
USEPA Recommended Sample Containers, Preservatives, and Holding Times for 1,4-Dioxane Analysis of Aqueous Samples

Method	Container	Preservation	Holding Time	Notes
8260	Standard 40-mL glass screw-cap VOA vials with PTFE-lined silicone septa	Cool to 4°C and adjust pH to less than 2 with H_2SO_4, HCl, or $NaHSO_4$	14 days	Aqueous samples should be stored with minimal or no headspace
8270	1-L amber glass with PTFE-lined silicone septa	Cool to 4°C	Extract within 7 days; analyze within 40 days of extraction	3 mL 10% sodium thiosulfate per gallon if sample is chlorinated
1624	20- to 40-mL glass with PTFE-lined lid	Cool to 4°C; acidify to pH 2 with HCl (1:1) for aromatics	Analyze within 14 days of collection	Sodium thiosulfate (10 mg per 40 mL) recommended if sample of drinking water is chlorinated
522	500- or 100-mL glass bottles with PTFE-lined screw caps	Cool to less than 6°C; acidify to pH 4 with sodium bisulfite Extract stored at less than −5°C	Extract within 28 days; analyze within 28 days of extraction	Sodium sulfite (50 mg/L) added to reduce disinfectant residual; sodium bisulfite added as microbial inhibitor and to adjust pH

Source: U.S. Environmental Protection Agency (USEPA), 2015, Test Methods SW-846 on line. https://www.epa.gov/hw-sw846 (accessed March 15, 2019).

Note: VOA, volatile organic analysis; PTFE, polytetrafluoroethene.

Other systems include dedicated tubing that pulls samples directly from depth-discrete zones in a monitoring well through the use of a peristaltic pump or a nitrogen lift bladder pump. Another approach for sampling soluble organic compounds is the passive-diffusion concept, which provides the basis for a variety of no-purge sampling devices (Parsons, 2005; O'Neill, 2006). These include the polyethylene diffusion bag (PDB; also called passive-diffusion bag), rigid porous polyethylene (RPP), polysulfone membrane, regenerated cellulose, Snap Sampler™, Hydrasleeve™, and Amplified Geochemical Imaging (AGI) Universal Sampler. All of these approaches cost less than conventional sampling in the long run. Performance data for sampling 1,4-dioxane are available for most of these samplers. PDB samplers have been studied extensively and are widely used. Performance data for PDBs can be found online.*

This section briefly examines how some of these passive-diffusion samplers are used and how their results compare for 1,4-dioxane sampling. Diffusion methods determine the concentration from the mass of analyte per volume of sample in the same way as grab samples obtained from a bailer or sampling pump; however, diffusion methods require that the volume within the sampler comes into mass equilibrium with the groundwater outside the sampler. It is therefore necessary to leave the samplers in the monitoring well for as long as 2 weeks.

4.2.1.1 Passive-Diffusion Bag Samplers

PDB samplers are narrow plastic sampling bags made of low-density polyethylene (LDPE), typically 18 in. long. The bags are closed at the ends after filling with pure deionized water and then lowered into a monitoring well to a specified depth for a period of 1–2 weeks. During the sampling interval, the deionized water comes into equilibrium with the surrounding well water by diffusion through the semipermeable polyethylene sampling bag, which has a pore size of less than 0.001 μm (10 Å, or 1×10^{-6} mm). Multiple PDBs can be lowered into a single monitoring well with a long screen to characterize vertical stratification of contaminants within a monitoring well screen and to reveal variations in flow rates along different portions of the screen. The primary analytes in PDB sampling programs are VOCs in groundwater monitoring wells. PDBs can also be used to monitor the discharge of contaminated groundwater to surface water (Vroblesky et al., 2000; O'Neill, 2006).

Unfortunately, 1,4-dioxane is not amenable to sampling using PDB samplers. In a field trial comparing different no-flow samplers, results for 1,4-dioxane in a PDB sampler were compared to results for samples obtained by other no-flow samplers and conventional samples. Although 1,4-dioxane was detected in other samples from the same well, it was not detected in the PDB sample. Apparently 1,4-dioxane does not diffuse well through the polyethylene bag (Parsons, 2005).

4.2.1.2 Rigid Porous Polyethylene Samplers

The RPP system is based on a rigid polyethylene tube, about 7 in. long and filled with reagent grade water. The RPP is capped at both ends and lowered into the monitoring well to equilibrate with ambient groundwater. Diffusion occurs through the rigid tube, which has pores ranging in size from 6 to 15 μm (0.006–0.015 mm). RPP samplers are used for VOCs and soluble inorganic compounds (O'Neill, 2006).

1,4-Dioxane recovery from RPP samplers was found to be comparable to recovery from samples obtained by conventional means. In a 14-day test (i.e., the RPP device remained in the monitoring well for 14 days), the RPP sampler yielded 74 parts per billion (ppb) 1,4-dioxane, whereas conventional sampling produced 80 ppb. In a 28-day comparison, the RPP device produced a result of 67 ppb; the result from conventional sampling was 64 ppb. These results are within the margin of analytical error (O'Neill, 2006). Another comparison involved ten wells completed at different depths in a saprolite aquifer formation in the South Carolina Piedmont. RPP samplers are advantageous in this setting because monitoring wells recharge very slowly when pumped or bailed for

* See the Interstate Technology & Regulatory Council Web site (http://www.itrcweb.org/) or the EPA Hazardous Waste Cleanup Information site (http://www.clu-in.org).

conventional sampling. A best-fit line to a scatter plot of RPP results versus conventional sampling results falls close to the 1:1 line, with a regression error (R^2) of 0.92 (O'Neill, 2006). A third comparison found that RPP was less reliable for 1,4-dioxane sampling than the Snap Sampler™ but was comparable to low-flow purge methods (Parsons, 2005).

4.2.1.3 Snap Sampler™

The Snap Sampler™ is a whole-water grab sampler that equilibrates with ambient natural groundwater flowing through the well screen (Britt, 2004). The Snap Sampler™ is used to retrieve a sample of unfiltered water in 40 mL volatile organic analysis (VOA) vials or 125- or 350-mL polypropylene bottles from a discrete depth (S. Britt, personal communication, 2007). The sampler is a double-ended, spring-loaded VOA vial or polypropylene bottle that seals beneath the water surface. Figure 4.2 shows a Snap Sampler™. The sample does not come into contact with the atmosphere prior to injection into a gas chromatograph. In field comparison tests, the Snap Sampler™ has consistently collected higher concentrations of VOCs in samples than are typically found in samples retrieved by using other passive sampling devices, bailers, or sampling pumps (Britt, 2004). Conventional sampling introduces air into the sample by purging monitoring wells before retrieving a sample and by decanting the sample through the air into a laboratory VOA vial.

In a field comparison conducted at a site in Los Angeles, 1,4-dioxane results in samples collected by using the Snap Sampler™ were about 80% higher than results from purging three well casing volumes (2,330 ppb Snap Sampler™ versus 1,310 ppb conventional sampling) (S. Britt, personal communication, 2007). In a comprehensive comparison of low- or no-flow sampling devices conducted at McClellan Air Force Base near Sacramento, California, the Snap Sampler™ provided higher recoveries of 1,4-dioxane than all other types of low- or no-flow samplers. The Snap Sampler™ results were comparable to but slightly lower than those of conventional sampling (Table 4.2). Comparisons of 1,4-dioxane results from different sampling approaches from the McClellan Air Force Base No-Purge Sampler Demonstration Study are summarized in Table 4.2.

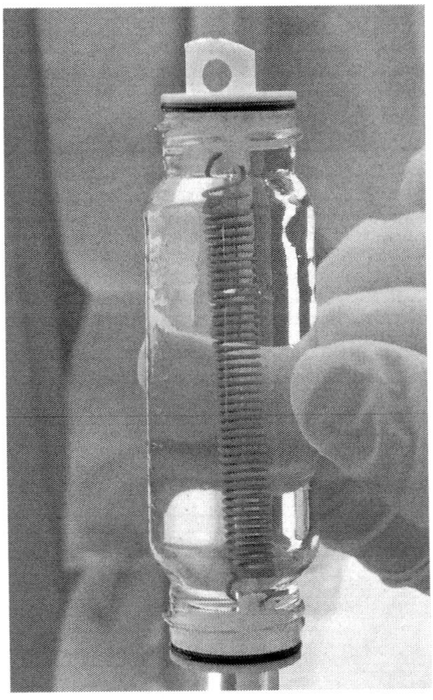

FIGURE 4.2 The Snap Sampler™. (From Britt, S., 2004, The Snap Sampler™. Paper presented to the Interstate Technology Regulatory Council Diffusion Sampler Team, Albuquerque, NM, October 28, 2004. With permission.)

TABLE 4.2
Summary of Results for 1,4-Dioxane: No-Purge Sampler Demonstration Study, McClellan Air Force Base

Sampler A	Sampler B	Number of Trials	Median RPD	Slope of *X-Y* Scatter Plot[a]	Regression Error (R^2)[b]	Sampler A		Sampler B
Low-flow	3-Volume	7	−1.5	1.41	0.97	Low flow	=	3-volume
Low-flow	Hydrasleeve®	3	0.0	1.11	1.00	Low flow	=	HSleeve
Low-flow	PSMS	4	7.7	0.94	0.99	Low flow	=	PSMS
Low-flow	RCS	3	−7.4	1.11	1.00	Low flow	<	RCS
Low-flow	RPPS	7	6.1	0.92	0.99	Low flow	=	RPPS
Low-flow	Snap Sampler™	3	20.7	1.26	1.00	Low flow	<	Snap
3-Volume	Hydrasleeve®	3	−5.4	1.05	1.00	3-Volume	=	HSleeve
3-Volume	PSMS	4	5.2	0.90	0.99	3-Volume	=	PSMS
3-Volume	RCS	3	30.3	0.74	0.99	3-Volume	=	RCS
3-Volume	RPPS	7	28.6	0.64	0.99	3-Volume	>	RPPS
3-Volume	Snap Sampler™	3	17.1	0.84	0.99	3-Volume	=	Snap
Hydrasleeve®	PSMS	6	27.6	0.83	1.00	HSleeve	>	PSMS
Hydrasleeve®	RPPS	7	25.0	0.69	0.99	HSleeve	>	RPPS
PSMS	RPPS	9	0.0	0.85	0.97	PSMS	=	RPPS
PSMS	Snap Sampler™	3	−27.3	1.20	n.a.	PSMS	<	Snap
RCS	RPPS	9	0.0	0.90	0.95	RCS	=	RPPS
RCS	Snap Sampler™	6	21.8	1.35	0.86	RCS	<	Snap
RPPS	Snap Sampler™	9	28.6	1.50	0.95	RPPS	<	Snap

Source: Parsons, 2005, Results report for the demonstration of no-purge groundwater sampling devices at former McClellan Air Force Base, CA. Army Corps of Engineers, Omaha, Nebraska. https://www.itrcweb.org/documents/mcclellan%20rpt_oct2005.pdf (accessed October 23, 2018).

Notes: RPD, relative percent difference; n.a., not applicable (too few data points were available for statistical significance). Low-flow sampling removes a small volume of water at a low flow rate from a small portion of the screened interval of a well without mixing water among vertical zones. The 3-volume sampling method rapidly removes 3–5 well-casing volumes from the well to evacuate stagnant water within the well casing and in the surrounding well filter pack. Theoretically, "stagnant" water is replaced with "fresh" groundwater from the surrounding formation with minimal mixing. Hydrasleeve® (abbreviated as HSleeve) is a flat-lying 30 × 2.75-in. polyethylene sleeve, sealed at one end, with a polyethylene reed-valve at the other end and a capacity of 2 L. The 2 × 2-in. polysulfone membrane sampler (PSMS) has 0.2 μm pore size and 108 mL capacity. The 13 × 1-in. regenerated cellulose sampler (RCS) has 0.0018 μm pore size and 400 mL capacity. The 6 × 1.5-in. rigid porous polyethylene sampler (RPPS) has 6–15 μm pore size and 150 mL capacity. The 10 × 1.6-in. Snap Sampler™ is a spring-loaded VOA holder that opens a 40 mL VOA or 125/350 mL Polypro bottle at depth.

[a] Scatter-plot slopes between 0.90 and 1.10 indicate that the populations are similar. Slopes less than 0.90 typically indicate that the concentration in sampler A was greater than the concentration in sampler B. Slopes greater than 1.10 typically indicate that the concentration in sampler B was greater than the concentration in sampler A.

[b] R^2 values less than 0.90 indicate a greater degree of scatter and a lower confidence that the slope value is meaningful. R^2 values greater than 0.90 indicate a lower degree of scatter and a higher confidence that the slope value is meaningful.

4.2.1.4 AGI Sampler (Formerly Gore Sampler)

Another approach to passive-diffusion sampling involves the semiquantitative dosimeter introduced by W.L. Gore Company and currently available from Amplified Geochemical Imaging, LLC. The AGI Sampler (formerly Gore Module or GORE-Sorber) is a mix of proprietary adsorbents encased in a tube of microporous expanded PTFE*. The sorbents are shielded from microbes, sediment, and

* PTFE = polytetrafluoroethylene.

water so that contaminants may partition into the enclosed air space according to their Henry's law constants. Contaminants adsorb by diffusion, independent of concentrations or groundwater flow rates. For low concentrations of compounds with low Henry's law constants, such as 1,4-dioxane, the AGI sampler must remain suspended in the monitoring well for a longer period of time than for more volatile compounds. The device is 6 in. long and can be used to profile vertical variation of concentrations within a well screen. Adsorbed contaminants are then thermally desorbed and analyzed by GC–MS (Anderson, 2005). More information and performance evaluation of AGI samplers can be found in the Environmental Security Technology Certification Program (ESTCP) report *Demonstration of the AGI Universal Samplers (F.K.A. the GORE® Modules) for Passive Sampling of Groundwater* (USACE, 2014).

4.2.2 Sampling for 1,4-Dioxane in Air and Landfill Gas

A variety of strategies are used to sample for gaseous air contaminants, including whole-air sampling by filling air-sampling vessels and in-field sample concentration by sorption of air contaminants to a variety of specialty sorbent tubes in the field. Air-sampling vessels include specialized bags such as Tedlar® bags or stainless steel canisters such as Summa canisters.

The objectives of air sampling vary widely. For worker health and safety, air sampling is performed to monitor personal breathing zones in the workplace. Air quality studies monitor ambient air quality, and soil vapor studies monitor for vapor intrusion into buildings from underlying soil and groundwater contamination. Air-sampling equipment is also used to monitor for the 1,4-dioxane content of landfill gas as a source of groundwater contamination at landfill sites. Air sampling for 1,4-dioxane may also be necessary for monitoring emissions from industrial facilities. As discussed in Chapter 3, 1,4-dioxane is not likely to persist in the atmosphere owing to its relatively short photodegradation half-life and low Henry's law constant. A detailed discussion of air sampling is beyond the scope of this book. This section provides a brief overview of available methods for sampling 1,4-dioxane in air.

The most common method for sampling 1,4-dioxane in air uses a stainless steel Summa canister for analysis by USEPA Method TO-15 (USEPA, 1999). The Summa canister is designed to be evacuated to a near vacuum (0.05 mm Hg) and then opened to the field atmosphere to admit the sample at a fixed rate controlled by a valve. The pressure differential draws the sample into the canister through a flow-restrictive inlet or mass-flow controller. The canister is then sealed and delivered to the laboratory for analysis by GC–MS.

Sampling gas trapped in soil is accomplished by applying a vacuum to a filtered tube and drawing the sample into a Summa canister, a Tedlar bag, a gas syringe, or a gas-tight glass bulb (CalEPA, 2003). The choice of tubing and sample container may have a bearing on 1,4-dioxane results. Some sample tubing has been shown to emit 1,4-dioxane to the sample, whereas other tubing materials retain 1,4-dioxane, on the basis of a few tests with tubing (Hayes et al., 2006). Tubing materials commonly used in soil sampling, and their potential to emit or retain 1,4-dioxane, are listed in Table 4.3.

Similar testing on the contribution of 1,4-dioxane by Tedlar bags to air samples was conducted in a Eurofins Air Toxics, formerly Air Toxics, Ltd, study (Hayes et al., 2006). Tedlar bags were flushed three times with laboratory grade, humidified "zero-air" samples and then analyzed. 1,4-Dioxane was not found. Testing with laboratory standards showed no difficulties recovering 1,4-dioxane from Tedlar bags. The average recovery was 103%, with a relative standard deviation (RSD) of 5% (Hayes et al., 2006).

4.2.3 Personnel Air Monitoring

Personnel air monitoring has typically been performed by using battery-powered portable sampling pumps and sorbent tubes. The National Institute for Occupational Safety and Health (NIOSH)

TABLE 4.3
Soil-Gas Sampling Tube Materials and 1,4-Dioxane Emissions and Recovery

Tubing Name	Tubing Material	1,4-Dioxane Emissions	Frequency of Detection (%)	1,4-Dioxane Recovery (%)	RSD[a] (%)
LM Nylaflow	Nylon	0.068 ppbv	11	100	2
PEEK	Polyetheretherketone	<DL[b]	0	79	5
Teflon	Fluoroethylene-propylene (FEP)	<DL[b]	0	77	5
Polyethylene	Low-density polyethylene (LDPE)	<DL[b]	0	113	7

Source: Hayes, H.C., Benton, D.J., and Khan, N., 2006, In: *Vapor Intrusion—The Next Great Environmental Challenge—An Update.* Los Angeles, CA: Air and Waste Management Association. With permission.

[a] Relative standard deviation – all tests were highly reproducible.

[b] Detection limit.

promulgates methods for personnel monitoring, including Method 1602 for 1,4-dioxane. NIOSH Method 1602 uses a solid sorbent tube of coconut shell charcoal. A minimum of 0.5 L to a maximum of 15 L of workplace air is drawn through the tube at a rate of 0.1–0.2 L/min. The sample is then desorbed for 30 min with carbon disulfide as the solvent, and the sample is analyzed by GC with a flame ionization detector (FID). The typical limit of detection is 0.010 mg per sample (Kurimo, 1994).

Field instruments for screening VOC vapors operate by using an FID or photoionization detector (PIDs). PIDs can detect 1,4-dioxane, whose photo-ionization potential is 9.13 electron volts (eV). PID instruments, such as those manufactured by HNU Systems, can be calibrated to detect a particular compound when it is the sole compound present within a particular ionization range; however, PIDs are not well suited for distinguishing between detectable compounds in a mixture of gases. PID readings typically provide an integrated response to the gas mixture. Field instruments based on flame ionization can detect 1,4-dioxane. The flame-ionization potential for 1,4-dioxane is 9.41 eV. Another alternative for screening 1,4-dioxane vapors is a portable gas chromatograph (Photovac, 2006).

4.3 SAMPLE PREPARATION

Sample preparation is the first step performed on a sample delivered from the field. There are many established methods for sample preparation, each dependent on the nature of the analyte and the selection of analytical method. This section offers a brief review of the major sample extraction and concentration methods that have been used for 1,4-dioxane in water and soil.

4.3.1 Purge and Trap

USEPA Method 5030C (USEPA, 2003a) describes the most common protocol for the purge-and-trap (PT) technique. The sample is purged by bubbling an inert gas, usually helium or nitrogen, through the aqueous sample at either ambient or elevated temperature. The volatile components are transferred from the aqueous phase to the vapor phase. The contaminant vapor is introduced at a specified flow rate to a sorbent column where the volatile components are adsorbed. After purging is completed, the sorbent column is heated and back-flushed with inert gas to desorb the components onto a gas chromatographic column (USEPA, 2018a). Volatile compounds with high air-water partition coefficients, such as the chlorinated solvents, are readily extracted from samples in this manner; however, hydrophilic compounds with low air-water partition coefficients, such as 1,4-dioxane, are not efficiently purged from water.

The high water solubility of 1,4-dioxane results in poor purging efficiency; however, PT systems can be modified to improve 1,4-dioxane recovery, as will be discussed later. The PT device consists of the sample purger, the trap, and the desorber. A typical purging chamber accepts 5 mL samples with a 3-cm-deep water column. The purge gas is introduced close to the sample level in the purging chamber and passes through the water column as finely divided bubbles. Trap dimensions vary by manufacturer; a typical trap may be about 1 foot long, with an inside diameter of about 0.1 in., and may contain adsorbents comprising, for example, approximately equal amounts of 2,6-diphenylene oxide polymer, silica gel, and coconut charcoal. The desorber is designed to be rapidly heated to 180°C for desorption of the target compounds for injection into the chromatographic column. Some configurations include a water trap, and most use a molecular sieve. Extensive maintenance, cleaning, and calibration protocols must be followed to produce reliable and repeatable results from PT equipment (USEPA, 2015). The carrier-gas flow rate depends on the properties of the trap. A flow rate of 15 mL/min is used for the standard silica gel trap.

4.3.2 Heated Purge and Trap

1,4-Dioxane is miscible in water and boils at higher temperatures than most GC–MS analytes; therefore, it is more efficiently purged by using a heated PT method. A 1988 USEPA publication explores the benefits of heated PT for polar, highly soluble VOCs, including 1,4-dioxane, nitriles, ketones, and alcohols (Lucas et al., 1988). Although recovery for 1,4-dioxane was poor, use of sodium sulfate to "salt out" 1,4-dioxane improved recoveries and enabled a preliminary method detection limit (MDL) of 4 ppb and limit of quantitation (LOQ) of 20 ppb. Problems with the salt clogging the columns discouraged further pursuit of this approach (Lucas et al., 1988). PT followed by GC-FID suffers from interferences that prevent attaining low detection limits for 1,4-dioxane (Song and Zhang, 1997).

Many research, water utility, and commercial laboratories have adjusted the purge times, purge and desorption temperatures, and materials used for PT to optimize extraction of the analytes of interest. A number of these innovations were targeted at fuel oxygenates, a suite of compounds chemical properties of which are similar to 1,4-dioxane and other solvent stabilizers in the ether, epoxide, and alcohol groups. Heated PT was optimized for analysis of methyl *tert*-butyl ether (MTBE), *tert*-butyl alcohol, *tert-amyl* alcohol, and other fuel oxygenates by heating the sample to 65°C and bubbling helium through the sample (Rose and Sandstron, 2003). For 1,4-dioxane, heating the sample and increasing the purge flow generally increases the recovery by a factor of 2, but poor chromatography results when the purge temperature or purge flow is increased beyond optimal levels. Erratic recoveries may be due to increased amounts of moisture introduced into the PT concentrator at higher temperatures and purge flow rates (Strout et al., 2004b).

Orange County Water District (OCWD) modified the PT extraction and instrumentation used for USEPA Method 524.2 to improve the sensitivity and reproducibility for analyzing for 1,4-dioxane in drinking water. Purge efficiency was improved by increasing the purge time from 11 to 20 min. Increasing the purge time was preferred to increasing flow rate, because purge flow rates at 40 mL/min may cause foaming in some samples. Varying the media in the trap, for example, using one that contains more Carbopack™, could improve the response and the shape of the 1,4-dioxane peak (Yoo et al., 2002).

4.3.3 Equilibrium Headspace Extraction

Equilibrium headspace (i.e., USEPA Method 5021) (USEPA, 2014b) is another approach for transferring the analyte from soil or water to the gaseous phase for analysis. Headspace methods are applicable to organic compounds with sufficiently high volatility to allow their transfer from soil samples under equilibrium headspace conditions (USEPA, 2018a). 1,4-Dioxane's hydrophilic nature and low air-water partition coefficient make headspace methods an unsuitable approach for its analysis;

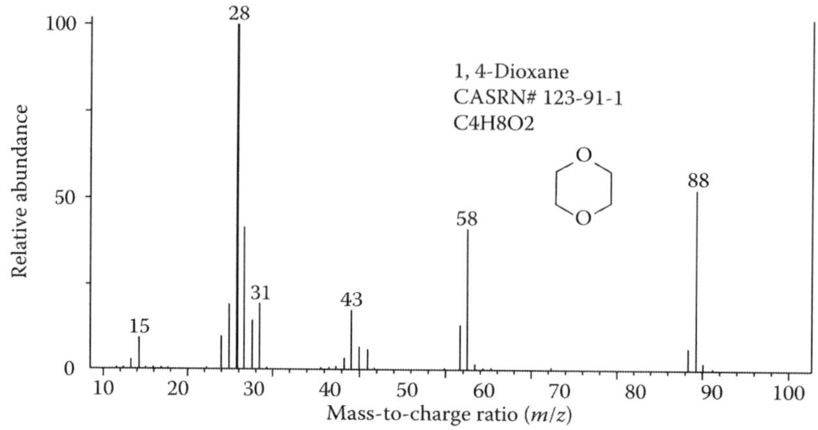

FIGURE 4.3 Mass spectrum for 1,4-dioxane-d_8 and its ion fragments. (From Yoo, L., 2002. Simultaneous determination of 1,2,3-TCP and 1,4-dioxane in drinking water by GC–MS/MS using purge/trap techniques. Orange County Water District. With permission.)

however, such methods have been used with modifications. A Danish study of 1,4-dioxane content in cosmetics and sundries adapted the equilibrium headspace method by weighing a 1 g sample in a headspace vial and adding it to 25 μL dichloromethane and 25 μL 0.1% (V/V) 1,4-dioxane-d_8 (1,4-dioxane's deuterated internal standard). The headspace vial was then heated for 16–18 hours at 80°C, and the gas was routed to a GC–MS operated in selective ion monitoring (SIM) mode for target masses 88 and 96 (for 1,4-dioxane and 1,4-dioxane-d_8 [Figure 4.3]; SIM is explained further in Section 4.5.5]. The limit of detection achieved was 300 μg/L (0.3 ppm) (Rastogi, 1990).

A Polish study also focused on the use of headspace methods for 1,4-dioxane analysis in sundries. In this study, 1 g of sample was introduced to headspace vials together with a standard of dioxane dissolved in a solvent; thereafter, the vials were heated to 130°C for 1 hour. Careful analysis of partition coefficients of dioxane from the laboratory solvent allowed quantitation to 2 μg/g (2 ppm) for dioxane in shampoos (Wala-Jerzykiewicz and Szymanowski, 1998).

4.3.4 Vacuum Distillation

Vacuum distillation is another means of transferring contaminants from samples to the carrier gas. USEPA Method 5032 (USEPA, 1996d) uses vacuum distillation and a cryogenic trapping procedure followed by GC–MS. The sample is introduced into a sample flask, which is then depressurized to the vapor pressure of water through the use of a vacuum pump. The vapor is passed over a condenser coil chilled to –10°C or less to condense water. The uncondensed distillate is cryogenically trapped on stainless steel tubing chilled with liquid nitrogen (–196°C). The condensate is then thermally desorbed and transferred to the GC by using helium carrier gas (USEPA, 2006b). The 1996 publication of USEPA Method 5032 does not identify 1,4-dioxane as suitable for analysis using vacuum distillation; however, a 2004 improvement to the method includes 1,4-dioxane as a suitable analyte (Strout et al., 2004a). USEPA Method 8261A uses a vacuum distillation unit to extract volatile and semi-VOCs, including 1,4-dioxane. The lowest MDL obtained for 1,4-dioxane was 2.5 μg/L for low-concentration water samples, but recoveries were variable (Strout et al., 2004a). See Section 4.5.4 for more discussion of vacuum distillation in USEPA Method 8261A.

4.3.5 Azeotropic Distillation

USEPA Method 5031 (USEPA, 1996c) uses azeotropic distillation to separate nonpurgeable, water-soluble, and VOCs, such as 1,4-dioxane, in aqueous samples or leachates from solid

TABLE 4.4
Results of USEPA Method 8015B for Analysis of 1,4-Dioxane at Different Concentrations and in Different Sample Types

	Low Concentration (25 μg/L)		Medium Concentration (100 μg/L)		High Concentration (750 μg/L)	
Sample Type	**Recovered (%)**	**RSD (%)**	**Recovered (%)**	**RSD (%)**	**Recovered (%)**	**RSD (%)**
Groundwater	124	16	96	10	99	8
TCLP leachate	103	20	103	16	102	7
Solid matrices	106[a]	19[a]	105[b]	10[b]	–	–

Source: U.S. Environmental Protection Agency (USEPA), 1996e, Method 8015B: Non-halogenated organics using GC/FID. U.S. Environmental Protection Agency.

Notes: Method detection limits for 1,4-dioxane in 25 μg/L spiked samples of aqueous matrices: reagent water – 12 ppb, groundwater – 15 ppb; TCLP leachate from aqueous matrix – 6 ppb. All values at each concentration of nonpurgeable volatiles are the average from seven replicate analyses by azeotropic microdistillation (USEPA Method 5031). TCLP, Toxicity Characteristic Leaching Procedure, an USEPA extraction protocol; RSD, relative standard deviation.

[a] Incinerator ash (low concentration; 0.5 mg/kg).

[b] Kaolin (medium concentration; 25 mg/kg).

matrices. An azeotrope is a liquid mixture of two or more substances that behaves like a single substance: it boils at a constant temperature, and its vapors have a constant composition. 1,4-Dioxane forms a homogeneous pressure-maximum azeotrope with water in a completely miscible system at 87.8°C where 1,4-dioxane comprises 47.2% of the vapor-phase mass (Lide, 2008). Azeotropic distillation capitalizes on the ability of selected organic compounds to form binary azeotropes with water to facilitate separation of the compounds from a complex matrix. Extraction by azeotropic distillation was used as early as 1992 for the analysis of 1,4-dioxane (Bruce et al., 1992). After 2 min of fractional distillation of the 40 mL samples, the first 100 μL of distillate was collected for analysis by GC-FID. Recoveries were determined by comparison with standards injected directly prior to distillation. An MDL of 7 ppb was achieved for 1,4-dioxane. Azeotropic microdistillation was judged superior to the PT extraction method for analysis of water-soluble VOCs (Bruce et al., 1992). USEPA Method 8015 for VOC analysis by GC–MS uses Method 5031, azeotropic distillation, as a preparatory step. Methods reporting limits for 1,4-dioxane analyzed following the standard application of USEPA Method 8015B are summarized in Table 4.4 (from USEPA, 1996e).

4.3.6 Sample Extraction

When sample matrices are incompatible with PT preparation methods or if concentration of the target analytes is desired, sample extraction procedures are needed. Sample extraction procedures isolate the target analytes from the sample matrix and concentrate the analytes into a solvent that may then be analyzed by appropriate techniques.

4.3.6.1 Solid Sample Extraction

USEPA Method 5035 (USEPA, 2002) is used in most analyses of solid environmental samples (soil and other solid media) for volatile analytes. Method 5035 uses either methanol or organic-free/reagent grade water as a solvent for extraction of organic compounds. The solvent, and surrogates used to evaluate extraction efficiency, are added to the solid sample at the same time. Solid samples with high concentrations of target analytes (>200 μg/kg) are first extracted using methanol. Following extraction, an aliquot of this methanol extract is removed, diluted into water, and internal standards are added prior to proceeding with PT. For solids with low concentrations of the water-soluble target analytes (<200 μg/kg), 5 mL of reagent grade water is added to a weighed solid sample together with surrogates and internal standards, prior to proceeding with PT.

4.3.6.2 Liquid-Liquid Extraction

Liquid environmental samples require a protocol different from that used for solid samples. Method 3510C (USEPA, 1996a), separatory funnel liquid-liquid extraction (separatory funnel LLE), and Method 3520C (USEPA 1996b), continuous liquid-liquid extraction (continuous LLE), isolate and concentrate water-insoluble and slightly water-soluble organic compounds from aqueous samples.

In the separatory funnel method (Method 3510C), 1 L of sample adjusted to a specified pH is serially extracted with dichloromethane or other solvents by using a separatory funnel. The extract is dried with anhydrous sodium sulfate, concentrated by distillation, and exchanged into a solvent compatible with the analytical method to be used. In this extraction, 1 L of sample is reduced to 10 mL for injection into the GC column.

In the continuous LLE method (Method 3520C), 1 L of sample is placed into a continuous liquid-liquid extractor and extracted with organic solvent for 18–24 hours. Then the extract is dried with anhydrous sodium sulfate, concentrated, and transferred to a solvent suitable for the analytical technique.

Because 1,4-dioxane and many of the solvent-stabilizer compounds of interest are highly soluble in water, separatory funnel LLE by Method 3510C is not generally used for their analysis. However, chemists at the OCWD were able to use an adaptation of separatory funnel LLE to analyze both 1,4-dioxane and nitrosodimethylamine (NDMA), a compound that is similarly soluble, polar, and volatile (Yoo et al., 2002). Using a separatory funnel saved 80% of the time required for continuous LLE. Three extractions were done with dichloromethane, and deuterated standards for both 1,4-dioxane and NDMA were added prior to beginning the extraction process. The MS analysis used chemical ionization with methanol to achieve an MDL of 0.2 μg/L for 1,4-dioxane and 0.3 ng/L for NDMA (nanograms per liter, i.e., parts per trillion) (Yoo et al., 2002). Figure 4.4 shows separatory funnels for LLE.

A typical separatory funnel LLE uses a 1 L sample mixed with 100 g of sodium chloride in a 2 L separatory funnel for three sequential extractions with 60 mL of dichloromethane. The extraction step involves shaking the funnel for a few minutes and then allowing the mix to separate.

FIGURE 4.4 LLE separatory funnels, OCWD Laboratory. (From Yoo, L., 2002. Simultaneous determination of 1,2,3-TCP and 1,4-dioxane in drinking water by GC–MS/MS using purge/trap techniques. Orange County Water District. With permission.)

Dichloromethane is denser than water and ideally settles to the bottom of the funnel in a separate phase. Organic compounds remain dissolved in the solvent, which is concentrated to 1 mL by bubbling nitrogen gas in a tube warmed by a 35°C water bath (Yoo et al., 2002).

In addition to the complete miscibility of 1,4-dioxane and water, LLE of 1,4-dioxane from water is made inherently difficult because of 1,4-dioxane's low octanol-water partition coefficient (K_{ow}). Various approaches have been developed to overcome these characteristics. Addition of salt has been found to improve recovery. Solvents tested for use in LLE have included dichloromethane and MTBE; however, MTBE proved to be an ineffective LLE solvent (Park et al., 2005).

4.3.6.3 Solid-Phase Extraction

Solid-phase extraction (SPE), also called liquid-solid extraction in USEPA test methods, is described by USEPA Method 3535A (USEPA, 2007). SPE involves adsorbing target analytes on a solid sorbent and then desorbing and concentrating the target analytes with a solvent before injection into the GC–MS system. The SPE protocol prescribed by USEPA Method 3535A uses a 1 L sample, spiked with internal standards and surrogates, as appropriate. pH adjustment is made depending on the target analytes, and the sample is treated with 5 mL of methanol. An SPE device, usually a coated disk, is then used to extract the organic analytes. The disks are composed of hydrophobic materials and are often prewetted with a water-miscible solvent such as acetone or acetonitrile. The sample is filtered through the SPE cartridge or disk under vacuum. The vacuum is maintained after the sample has passed through the SPE device to dry the filter. Dichloromethane or other solvents are used to elute target analytes from disks. The eluted solvent is next dried with anhydrous sodium sulfate. The sample is then concentrated using a nitrogen blow-down and hot water method, and the concentrated extract can be exchanged into a solvent as needed (USEPA, 2006a).

Earlier SPE methods used bulk graphitized carbon or powdered carbon black. Disks reduce the filtering time and cost, allow increased flow rates, improve mass transfer, and eliminate bed channeling (Markell et al., 1991). A number of researchers have developed viable methods with SPE to achieve low MDLs (Fuh et al., 2005; Isaacson et al., 2006; Shirey and Linton, 2006; USEPA, 2006a). Carbon disks work well for extraction of highly polar, water-soluble compounds, such as 1,4-dioxane and other ethers.

SPE media must be completely air-dried prior to solvent extraction, or the eluted solvent must be dehydrated using anhydrous sodium sulfate crystals. When water is present in the eluate from the SPE media, the resulting chromatographic peaks for 1,4-dioxane and 1,4-dioxane-d_8 become significantly less sharp. With increasing water content, a shoulder develops on the left side of the peak between 13.3 and 13.5 min at 7.6% water content. At 14% water content, the peak broadens dramatically between 13 and 13.5 min (Kawata et al., 2001). To eliminate water from the SPE cartridge, researchers recommend drying with clean laboratory air for 2 min using an aspirator, followed by centrifuging at 1,700 **g** (3,000 RPM) for 10 min (Kawata et al., 2001). Figure 4.5 displays the deterioration of the 1,4-dioxane peak with increasing water content.

One of the lower MDLs reported in the earlier literature for analysis of 1,4-dioxane in water, 130 parts per trillion, was achieved with SPE with solvent desorption (Isaacson et al., 2006). The method was developed at Oregon State University (OSU) and involved the following steps: 25-mm-diameter activated carbon disks with 10-μm particles and 1,100-m^2/g surface area were fitted to polypropylene disk holders and attached to a vacuum manifold. Sample volumes of 80 mL were spiked with 100 μL of 50 μg/mL dioxane-d_8. Samples were pulled through dry disks under a 67 kPa vacuum, followed by 1 hour of drying in laboratory air under vacuum. Disks were then transferred into 2-mL GC autosampler vials, to which acetone and instrumental standards were added. Acetone was found to yield greater than 90%, that is, the same recovery as dichloromethane. Acetone is preferred over dichloromethane because it is less hazardous, and wastes are more easily managed.

The parameters of GC–MS/MS analysis for the OSU SPE method are summarized in Appendix 3. The pilot study for development of this method also tested the effectiveness of SPE for extraction

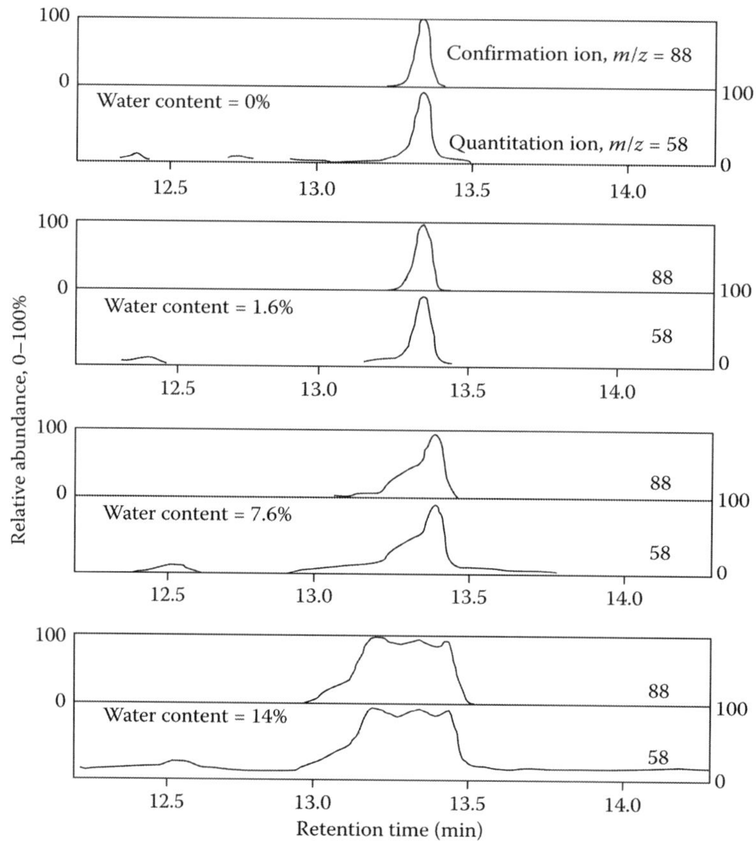

FIGURE 4.5 Deterioration of 1,4-dioxane peak with increasing water content. (After Kawata, K. et al., 2001, *Journal of Chromatography A* 911: 75–83. With permission.)

and identification of two other stabilizer compounds, tetrahydrofuran (THF) and 1,3-dioxolane. Spike and recovery experiments with THF and 1,3-dioxolane in water yielded greater than 80% recovery for THF and less than 10% for 1,3-dioxolane. Breakthrough of the 1,3-dioxolane was confirmed from the fact that 90% of the added mass was recovered from the water that passed through the disk, as determined by an in-vial LLE method. 1,3-Dioxolane is poorly suited for extraction by SPE using activated carbon. The MDL for 1,4-dioxane was 0.13 μg/L, and the LOQ was 0.31 μg/L. For THF, the MDL was 1.0 μg/L and the LOQ was 3.1 ppb (Isaacson et al., 2006). Chromatograms from the OSU study are shown in Figure 4.6.

4.3.6.4 Solid-Phase Micro Extraction

Solid-phase *micro* extraction (SPME) is a variation of SPE in which a much smaller volume of sample is sufficient to complete the analysis. The use of SPME for the analysis of 1,4-dioxane has many advantages over conventional solvent phase extraction and headspace pretreatments, including simplicity, speed, precision, lower detection limit, and minimal solvent consumption (Fuh et al., 2005). In SPME, the solid phase is coated on a fused silica fiber attached to the end of a wire plunger that is pushed through a syringe needle and immersed in the aqueous liquid phase or exposed to the headspace above the liquid. Organic compounds present in the vapor or liquid phase are absorbed into or adsorbed onto the solid phase. After a specified sampling time, the compounds are thermally desorbed from the solid phase in the heated injection port of a gas chromatograph (Black and Fine, 2001).

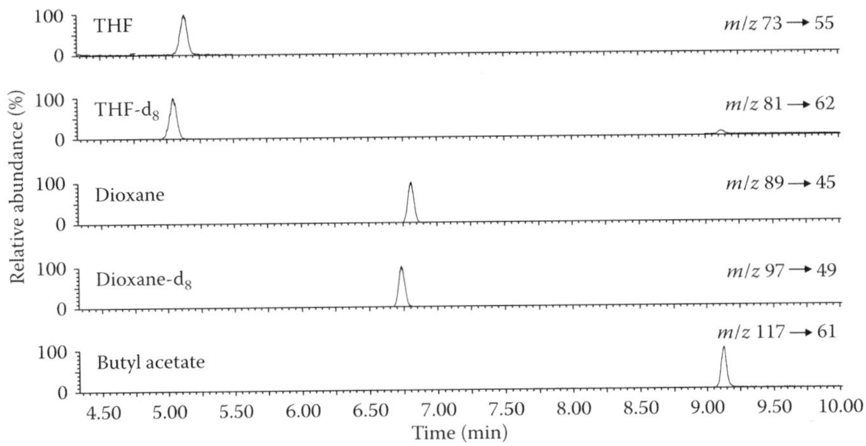

FIGURE 4.6 Chromatograms of THF, THF-d_8, dioxane, dioxane-d_8, and internal standard butyl acetate. [From Isaacson, C., Mohr, T.K.G., and Field, J.A., 2006, *Environmental Science & Technology* 40(23): 7305–7311; Chromatograms by Oregon State University. With permission.]

Variants of SPME have evolved, including stir bar sorptive extraction (SBSE), headspace sorptive extraction (HSSE), and solid-phase *dynamic* extraction (SPDE). These adaptations were developed to increase sorption capacity and to overcome some drawbacks of SPME, such as fiber fragility (Jochmann et al., 2006). As an example, SPDE uses a 2.5 mL headspace syringe equipped with a needle fabricated with an interior coating similar to a fused-silica GC column (see Section 4.4.1.1) with an immobilized extraction phase (Jochmann et al., 2006). SPDE needle coatings incorporate 4–6 times the amount of sorbent as SPME fibers. The syringe is immersed directly into a sample or the headspace above it, and the plunger is exercised several times to extract the sample, thereby adsorbing the analytes to the internal coating. After several cycles of aspirating and dispensing the sample, analytes are thermally desorbed from the coating inside the needle and purged with nitrogen gas into the GC injector (Jochmann et al., 2006).

As with LLE, the addition of sodium chloride or sodium sulfate salt to the sample improves the recovery of 1,4-dioxane in SPME extraction. The salt helps to decrease the solubility of 1,4-dioxane in water. The amount of 1,4-dioxane extracted from SPME fibers increases linearly up to 15% of sodium chloride concentration. High salt concentrations may shorten the lifetime of SPME fiber coatings. Addition of 10% (w/v) sodium chloride is considered optimal for 1,4-dioxane determination using SPME while also preserving SPME fiber coatings (Fuh et al., 2005). Increasing the ionic strength of the sample by adding sodium chloride also increases recovery of 1,4-dioxane in SPDE needles. 1,4-Dioxane recovery increased by approximately 60% when salt content was increased from 10% to 25% (w/w ratios) in samples extracted with SPDE needles (Jochmann et al., 2006). The effect of adding salt to extract more of a water-soluble organic compound is referred to as "salting out" and is quantified by the Setschenow constant, K^S,

$$\log\left(\frac{\gamma_{w,salt}}{\gamma_w}\right) = K^S\,[\text{salt}]_{total}, \tag{4.1}$$

where γ_w is the activity coefficient of 1,4-dioxane in pure water, $\gamma_{w,salt}$ is the activity coefficient of 1,4-dioxane in saline aqueous solution, and $[\text{salt}]_{total}$ is the total molar concentration of sodium chloride in the solution used to measure 1,4-dioxane activity in a sodium chloride solution. Setschenow constants for MTBE, THF, and 1,4-dioxane determined at 70°C were 0.17 L/mol ($R^2 = 0.995$), 0.16 L/mol ($R^2 = 0.988$), and 0.08 L/mol ($R^2 = 0.999$), respectively, and R^2 is the correlation coefficient from four data points (Jochmann et al., 2006). 1,4-Dioxane's Setschenow constant is considered exceptionally low, which means that the addition of sodium chloride to a sample for extraction by SPME or SPDE will have a large increase in sample recovery for 1,4-dioxane.

Selection of the adsorbent coating on the fibers is an important consideration for using SPME. A study comparing the effectiveness of different SPME media determined that carboxen-polydimethylsiloxane (CAR-PDMS) coating is most effective for extraction of 1,4-dioxane (Shirey, 2000). Another study contrasted the recovery of 1,4-dioxane from six different SPE media, each receiving a uniform quality control sample spiked at 10 μg/L. The best recovery was produced by 2OH Diol sorbent. All six media performed well* (Song and Zhang, 1997).

The quantity of carbon or other adsorbents used in SPME is also important. When cutting pieces of carbon felt for in-vial elution, at least 1.5 g are needed to produce adequate recovery. Granulated carbon has greater adsorptive power than felt but may be short-circuited by channeling of the sample through the cartridge (Kawata et al., 2001; Isaacson et al., 2006). Samples extracted onto flexible disks by using SPE may be eluted "in-vial" in a 2 mL autosampler vial and directly analyzed by GC and tandem mass spectrometry (GC–MS/MS) (Isaacson et al., 2006). Desorption of VOCs from the solid phase may use solvents such as dichloromethane or acetone. Desorption is also done in headspace extraction by heating the fiber to an elevated temperature. A comparison of these approaches showed that heated headspace extraction at 55°C is 31% more efficient than direct immersion for the extraction of 1,4-dioxane from CAR-PDMS (Shirey and Linton, 2006). Nonpolar compounds have a higher affinity for the SPME media and may displace polar compounds (Black and Fine, 2001).

Experiments were performed by chemists at Supelco (now a part of MilliporeSigma), a manufacturer of chromatographic columns and supplies, to determine whether high concentrations of methyl chloroform would compete with 1,4-dioxane for sorption onto CAR-PDMS-coated fibers. Although no displacement or competitive sorption was observed for concentrations as great as 20,000 μg/L methyl chloroform with 10 μg/L 1,4-dioxane and 25 μg/L 1,4-dioxane-d_8, an impurity of methyl chloroform was found to interfere with quantitation of 1,4-dioxane. The impurity was 1,1,2-trichloroethane (Shirey and Linton, 2006; R.E. Shirey, personal communication, 2007). A "solvent effect" on the effectiveness of SPME for 1,4-dioxane extraction was reported in a study of SPME coupled with headspace analysis that analyzed 1,4-dioxane in cosmetics (Fuh et al., 2005). Laboratory solvents were found to decrease the 1,4-dioxane yield from SPME by 30–40% after adding 1%, 5%, and 10% (w/v) of methanol and acetonitrile. The SPME media used in the study were 75 μm CAR-PDMS, 85 μm polyacrylate (PA), and 100 μm PDMS fiber assemblies made by Supelco. To avoid the solvent effect, headspace analysis was used. In another study, using hexane and dichloromethane as extraction solvents was found to interfere with quantitation of 1,4-dioxane in the mass spectrometer. Use of a lower molecular weight solvent, acetonitrile, was found to avert the interference (Song and Zhang, 1997).

The temperature at which desorption of target analytes occurs also affects the size of the peak area, that is, the recovery efficiency of the extraction. A German study of SPDE effectiveness determined that increasing the desorption temperature between 30°C and 70°C increased the efficiency of extraction at higher temperatures. A plot of chromatographic peak area versus extraction temperature creates a series of exponential curves. The curve determined for 1,4-dioxane is described by

$$y = \left(1.73 \times 10^5\right) e^{0.0376X}, \tag{4.2}$$

where y is the peak area and X is the extraction temperature in Celsius. The correlation coefficient was 0.999. For 1,4-dioxane, extraction recovery at 70°C is 4.5 times greater than at 30°C (Jochmann et al., 2006). The same study found that increases in concentrations of target ether and alcohol

* The six SPE sorbent media, sorbent manufacturers, and their performance with a 10 mg/L standard in seven replicate tests were C_{18} octadecyl (Applied Separations): 10.96 ± 0.49 mg/L; C_8 octyl (Analytichem International): 10.9 ± 0.67 mg/L; C_2 ethyl (J.T. Baker): 11.0 ± 0.37 mg/L; SI silyca (Analytichem International): 10.7 ± 0.58 mg/L; 2OH Diol (Varian): 9.94 ± 0.58 mg/L; NH_2-Aminopropyl (Varian): 9.65 ± 0.17 mg/L. Data from Kawata et al. (2001).

analytes in headspace above a liquid sample at higher temperatures are partially offset by a lower sorbent-air partition constant at higher temperatures. Because sorption is an exothermic process, the sorbent-air partition coefficient decreases as the sorbent temperature in the SPDE needle increases. The maximum extraction yields therefore occur at intermediate temperatures, with a decrease in SPDE extraction yield at higher temperatures (Jochmann et al., 2006). For SPDE, it is also important that the selected temperature not be so cool as to allow the condensation of water vapor in the needle, to prevent the 1,4-dioxane peak deterioration with increasing water content already noted (Kawata et al., 2001; Jochmann et al., 2006). The duration of extraction by SPME and the number of cycles of extraction by SPDE also have a significant bearing on the extraction recovery of 1,4-dioxane from water samples. Recovery of 1,4-dioxane approximately doubled when the number of SPDE extraction cycles was increased from 10 to 30; no appreciable change was observed when the number of SPDE extraction cycles was increased from 30 to 50 (Jochmann et al., 2006).

CASE STUDY – METHOD DEVELOPMENT FOR 1,4-DIOXANE ANALYSIS OF DRINKING WATER IN A WORKING TESTING LABORATORY

The OCWD, located in Fountain Valley, California, developed a method for time-efficient, low-level analysis of 1,4-dioxane in water. OCWD was challenged with analyzing for 1,4-dioxane because of its low-level presence in treated wastewater used as a seawater intrusion barrier through underground injection in a groundwater basin that is also used by several municipal supply wells. OCWD sought to develop a method that would reliably identify and quantify low levels of 1,4-dioxane following a protocol that would be time-efficient and cost-effective, would minimize waste generation, and would be robust for a variable sample matrix and presence of other compounds in the sample. OCWD's objective was to achieve a detection limit of 1 ppb (μg/L) with less than 20% RSD for the mean response factor (RF). The objective was achieved by pursuing a modified PT approach with GC–MS/MS. PT allows analysis to be completed with only 25 mL of sample instead of the 1 L sample required by LLE. PT generates minimal solvent wastes, whereas LLE requires 200 mL of dichloromethane per sample.

OCWD confirmed that the extraction efficiency of 1,4-dioxane is independent of pH. Extraction efficiencies ranged from 83% to 87% for spiked 1,4-dioxane samples at pH ranging from 2 to 9. Because 1,4-dioxane is a neutral compound, pH adjustment of the sample is not necessary prior to extraction and analysis.

OCWD increased the purge time from the 11 min recommended by the USEPA for drinking water to 20 min. The baking time for the trap was also increased from 15 min to 20 min. The use of a GC–MS/MS system enhanced the intensity of fragment ions for 1,4-dioxane. The extra MS function coupled with GC–MS helped to verify the sub-ppb levels of 1,4-dioxane in heavy matrix samples and improved the reproducibility and sensitivity of the system. Appendix 3 summarizes the operating parameters for the OCWD method for PT with GC–MS/MS. An MDL of 0.2 ppb (μg/L) was achieved by following this approach (Yoo et al., 2002). The time requirements for this method are comparable to LLE and SPME, but the PT method is less labor intensive.

OCWD has been running between 2,500 and 3,000 1,4-dioxane analyses annually. OCWD has also experimented with including 1,4-dioxane in a 200 mL dichloromethane LLE method for the analysis of a similarly hydrophilic compound, NDMA, using GC–MS/MS combined with chemical ionization. However, a significant portion of 1,4-dioxane was lost during concentration of the 200 mL dichloromethane extract in the sample concentration system. OCWD is now developing an extraction method to detect nitrosamines and 1,4-dioxane with GC–MS/MS/CI (chemical ionization) and SPE techniques to avoid generating large volumes of laboratory solvent wastes (Yoo, personal communication, 2007). Operating conditions and instrument settings used by OCWD for 1,4-dioxane analysis by PT GC–MS/MS are listed in Table 4.5.

TABLE 4.5
Operating Conditions for OCWD Purge and Trap, GC–MS/MS Method for Low-Level Detection of 1,4-Dioxane

	Purge and Trap Conditions[a]
Purge time	20.0 min
Bake time	15.0 min
Preheat	245°C
Desorb	250°C
Bake	260°C
	Sample Injection Conditions[b]
Settle	0.3 min
Prepurge	30 s
Sample pressurize	40–60 s
Sample transfer	75 s
Internal standard fill	5 s
Internal standard transfer	50 s
Back flush	Off
Desorb time	4 min
Transfer line rinse	Off
	GC and MS Conditions[c]
Initial column temperature	35°C
Hold time	2 min
Final temperature	220°C
Temperature rate	3.1°C/min to 100°C; 100°C; to 220°C at 30°C/min
Hold time	1 min at 30°C/min
Helium linear velocity	1 mL/min
Total run time	27.97 min
Injector temperature	125°C
Detector	Ion trap MS/MS
Transfer line	220°C
Mass range	40–100 *m/z*
Trap temperature	135°C
Background mass	40 *m/z*
EI (electron-impact ionization) current	80 μA
Scan time	0.36 s
Filament/multiply delay	14 min
Reagent ion ejection	20 V
Ion storage level	35.0 *m/z*
Ejection amplitude	20 V
High edge amplitude	15 V
Wave form	Resonance

Source: From Yoo, L.J., Fitzsimmons, S., and Wehner, M., 2002, Improved purge-trap and GC/MS/MS techniques for the trace-level determination of 1,4-dioxane in water. In: *Proceedings, Water Quality Technology Conference.* Seattle, WA: American Water Works Association, CD-ROM. With permission.

[a] Purge and trap by Tekmar ALS 3100.

[b] Sample injector by Tekmar AquaTek 50.

[c] GC – Varian Model 3800; MS – Varian Saturn 2000 (column – DB-VRX fused-silica capillary column, 60 m × 0.32 mm ID with 1.8 μm film thickness).

4.4 LABORATORY METHODS FOR THE ANALYSIS OF 1,4-DIOXANE BY GC–MS

Laboratory methods using GC–MS form the subject of numerous books and university courses. This section presents only a brief synopsis of the key operative elements of these methods, in the context of how they are modified to analyze for 1,4-dioxane.

Identification and quantification of contaminants in soil, water, air, food, and other media involve isolating the contaminants of interest from the sample by

1. Sample extraction and concentration
2. Separating the contaminants on the basis of their rate of passage through a chromatographic column designed with specific properties to retain the contaminants according to their vapor pressure, boiling point, and other properties
3. Detecting and quantifying the contaminants by ionization of the analyte followed by mass separation and detection
4. Identifying contaminants according to their mass spectra by using a computerized library of reference spectra and quantification via calibration standards

The primary approach for analysis of soil and water samples for VOCs has involved PT/GC–MS analytical methods (USEPA, 2014a, 2015; Lesage and Jackson, 1992). PT methods are designed to measure as many compounds as possible with a single procedure. Such methods are favored for those compounds that are relatively insoluble in water and that have boiling points below 200°C (USEPA, 1986, 2015) or below 150°C (Lesage and Jackson, 1992). The PT preparation techniques promote low detection limits (ppb range), and the MS detector allows positive compound identification. Compounds that contain polar functional groups such as low molecular weight ketones, alcohols, aldehydes, nitriles, and ethers (e.g., 1,4-dioxane) are generally soluble in water, do not purge well, and produce broad, tailing GC peaks that give poor quantitative estimates and are often difficult to identify by MS (Swallow, 1992). Some polar compounds are included in SW-846 Methods 8260 (USEPA, 1986, 2015), but the recovery of polar compounds is often less than 20% (Swallow, 1992). Azeotropic distillation (SW-846 Method 5031) and closed-system vacuum distillation with cryogenic condensation (SW-846 Method 5032) were introduced as preparation methods for analysis of hydrophilic analytes such as alcohols and ethers in the third update to SW-846 (Lesnik, 1993; USEPA, 1993).

4.4.1 Gas Chromatography

GC involves injecting the mixture to be analyzed into an inert gas stream that sweeps the sample into an open capillary column coated with a thin film or a column coated with a resolving stationary phase. The components in the gas stream absorb and interact with the thin film or stationary coating to varying degrees, which leads to differential separation, causing the components in the gas stream to be swept through the column in a sequence based upon their vapor pressure, boiling points, and other physical properties. As individual components of the mixture elute from the chromatographic column, they are swept by the carrier gas to a detector. The detector generates measurable electrical signals, referred to as peaks, which are proportional to the amount of analyte present. The detector response is plotted as a function of the time required for the analyte to elute from the column after it was injected. The resulting plot is called a chromatogram. The position of the peaks on the time axis may serve to identify the components, and the areas under the peaks provide a quantitative measure of the amount of each component when using a mass spectrometer. Because several compounds may possess similar retention times, the nature of the resulting peak may require further identification using MS. To identify the compound when using MS, the mass spectra corresponding to the peaks are compared to the spectra of calibration standards or to a computerized library of spectra. (McMaster and McMaster, 1998).

Mass spectrometers differentiate compounds based on the mass-to-charge ratio (*m/z*) of their ionized products. Compounds entering a mass spectrometer are ionized in a vacuum. The resulting ions are directed through a mass analyzer, which separates the ions by the *m/z* ratio. MS alone is useful for resolving the identity of individual compounds, but is less efficient for resolving compound identification in mixtures. The combination of GC and MS, GC–MS, forms a system that is effective at separating mixtures into their individual compounds and determining the amount and structure of the compounds. GC–MS has become the mainstay of environmental chemistry for volatile analytes such as solvents and related organic compounds in soil, air, and water; however, it is not without its limitations (McMaster and McMaster, 1998).

4.4.1.1 Methods in Gas Chromatography

The laboratory analyst must choose operating parameters for the analysis of each suite of analytes by GC. These include selection of carrier gas, injector type, column type, oven temperature, and temperature gradient, that is, the rate at which column temperature is increased and the duration it is held at a given temperature (Rödel and Wölm, 1987). The combinations of sample-preparation methods described in Section 4.3 with the detectors described in Section 4.4.2 differentiate the individual analytical methods for 1,4-dioxane analysis detailed in Section 4.5. Although there is a seemingly infinite number of permutations of instrumental configurations, extraction techniques, operating parameters, chromatographic columns, and detectors, seven methods for 1,4-dioxane analysis are published, and most are commercially available.

4.4.1.2 Selecting Chromatographic Columns for 1,4-Dioxane Analysis

The chromatographic column separates mixed analytes in the sample. Columns are classified according to their diameter, type, and length. Conventional columns range from 6 to 20 feet or more in length and may be of either the packed or open-tube (capillary) design. Packed columns are filled with a porous granular or beaded material, such as a specialized variety of diatomaceous earth, but more often a fused silica material. Capillary columns are longer than conventional columns, 5–100 m long with 30 m being most common, but they have a similar porous or liquid interior coating. Fused silica capillary columns are common for gas chromatographs used for environmental analysis, although a wide variety of columns are available. Capillary columns provide greater resolution of analytes than packed columns when handling small-volume samples. Most capillary columns are designed with a fused silica open tube, to which a liquid stationary phase is chemically bonded.

The liquid stationary phase is selected for its stability at high temperatures and for its low volatility. Polysiloxanes (i.e., "OV-1") and polyethylene glycol (i.e., "CarboWax 20M") are two examples of liquid stationary phases used in GC columns. For a given analyte, the thicker the liquid stationary-phase film in the column, the longer the retention time, and thicker films are used for more volatile analytes.

Oven temperature is the primary factor in determining retention time of an analyte. Temperature is used to separate analytes by programming the GC oven to increase temperature at a specified rate. Increasing the temperature and the rate of heating can cause analytes with higher boiling points to elute earlier from the column to the detector. Temperature programs are recommended for each column type by the USEPA method for particular groups of target analytes (USEPA, 2006a). Optimum temperature gradients must be independently determined for each analyte and column combination; however, lowered temperature means an increase in the elution time and the time required to complete the analysis.

The polarity of the stationary phase should be matched to the group of analytes being tested. Components of the mixture with similar polarity and therefore a high degree of affinity for the stationary phase are strongly retained, whereas components with dissimilar polarity will have low affinity for the stationary phase and migrate rapidly through the column. These differing behaviors present a challenge for analyzing samples for both solvents and stabilizers, because many solvents are nonpolar whereas many alcohol, epoxide, and ether stabilizers such as 1,4-dioxane are polar.

This challenge can be overcome with a gas chromatograph in which the sample gas flow is split between two columns of different materials to address both polar and nonpolar compounds.

Polysiloxanes, among the most common stationary phases in GC columns, contain various substituent groups to change the polarity of the stationary phase. The nonpolar end of the spectrum is polydimethyl siloxane, which is made more polar by increasing the percentage of phenyl groups on the polymer. For very polar analytes, polyethylene glycol (i.e., Carbowax) is commonly used as the stationary phase. Some gas chromatographers conclude that the column material that best separates 1,4-dioxane from the sample is DB-WAX, a fused silica column bonded with polyethylene glycol (Kawata et al., 2001). After the polymer coats the column wall or packing material, it is often cross-linked to increase the thermal stability of the stationary phase and prevent it from gradually bleeding out of the column.

The stationary phase should be similar in chemical structure to the components of the mixture targeted for analysis. To match chromatographic column selectivity to target analytes, several classification schemes of analyte polarity have been developed, including McReynold's numbers and Rohrschneider constants. McReynold's tables predict the chromatographic separation of various compounds with different stationary phases and provide the best information available for the comparison of the selectivity of the stationary phase. In the McReynold's system, 1,4-dioxane is used as an index analyte for quantifying column selectivity. 1,4-Dioxane is representative of "Class III" compounds of medium polarity. Stationary phases that will more strongly retain Class III compounds include the following:

- All polyesters
- SAIB tricresyl phosphate
- Benzyl cyanide
- Propylene carbonate
- Polyphenylether
- OV-17
- Dibutyl tetrachloro-phthalate
- STAP
- Lexan
- QF-1
- Dimethylsulfolane

The stationary phases recommended by USEPA for the analysis of ethers and other compounds of intermediate polarity ("Class III polarity") include Carbowax 400, Tricresyl phosphate, LAC, Apiezon L, and β-β'-oxydipropionitrile (USEPA, 2006c). Other listings for stationary phases suitable for the separation of ethers include Carbowax 20M, Carbowax 1500, and SE-30 (GC Grade) (Supelco/SigmaAldrich, 1997).

A variety of detectors are used in GC. The thermal conductivity detector (TCD) was the most widely used detector system during the early years of GC. The FID became more popular for common hazardous waste analyses. Specialty detectors were subsequently developed, such as the electron-capture detector (ECD) and the nitrogen-phosphorus detector. All detectors described in this section are nonspecific detectors and have largely been replaced by mass spectrometers. MS is the preferred detection technique because it is specific, sensitive, and quantitative, and good precision can be obtained at low concentrations under the right conditions.

TCDs were widely used in the early years of GC because of their simplicity, universal applicability, and low cost. Analyte detection is based on changes in the conductivity of the column effluent. The TCD is a destructive detector that can be used in series only after nondestructive detectors. The TCD detects gaseous compounds in the ppm range. TCDs are not generally used for analysis of low-concentration samples as the possibility of false identification of analytes is a large problem. Larger sample volumes are required to achieve increased sensitivity, which in turn requires using a large-diameter chromatographic column (USEPA, 2006b). TCDs are not well suited for detection of 1,4-dioxane and are no longer widely used for environmental sample analysis.

The FID is sensitive for most organic compounds containing oxidizable carbon such as aromatic and chlorinated aliphatics, petroleum compounds, semivolatile compounds, and polychlorinated biphenyls (PCBs). The FID is a destructive detector that can be used in series only after nondestructive detectors. The FID is a nonspecific detector, as most carbon-containing compounds are detectable by the FID.

An FID uses a small stainless steel jet positioned at the end of the chromatographic column. As carrier gas exits the column and flows through the jet, it mixes with hydrogen supplied in the jet and burns at the tip of the jet. Hydrocarbons and other molecules in the sample are ionized in the flame and attracted to a metal collector electrode located just to the side of the flame. The resulting electron current is amplified to convert very small currents to voltages that are recorded as the sample signal (USEPA, 2006b). An FID has been used to detect 1,4-dioxane vapors in occupational exposure testing; the detection limits are 5–190 ppmv (20–700 mg/m^3) (Cooper et al., 1971, in NICNAS, 1998).

An ECD consists of a sealed stainless steel cylinder that typically contains radioactive nickel-63. The nickel-63 emits beta particles that collide with and ionize carrier-gas molecules; in the process, a stable cloud of free electrons forms in the ECD cell. When a halogenated or other electronegative molecule enters the cell, it is immediately combined with one of the free electrons, causing a temporary but measurable reduction in the number of free electrons in the cell. The ECD is a nondestructive detector that can be used in series before other detectors (USEPA, 2006b). Compared to TCD or FID, the ECD is a more specific detector for halogenated or other electronegative compounds, but it is relatively insensitive for hydrocarbons, alcohols, ketones, and ethers such as 1,4-dioxane.

4.4.1.3 Mass Spectroscopy

A mass spectrometer is a detector used in GC. Mass spectrometers ionize gaseous molecules, separate the ions produced on the basis of the *m/z*, and then record the relative number of different ions produced. The *m/z* ratio is then plotted as the abscissa with relative intensity as the ordinate. This plot is referred to as a "mass spectrum." The mass spectrum of a compound can be considered its "fingerprint" and can be used to identify a compound through comparison with published reference spectra. MS systems interface with computers that compare experimental spectra to standard spectra and perform the identification automatically. Spectra that do not match calibration standard spectra can be compared with a library of spectra, and a tentative identification can be made.

Analytes can be fragmented by using either electron or chemical ionization. In electron-impact ionization, a 70-eV beam of electrons ionizes the analyte. Many analytes will fragment when exposed to the electron beam. The fragmentation pattern is characteristic of the ionized analyte and produces the spectrum from which the compound is identified. This experimental spectrum is then compared with spectra of compounds of the same molecular weight in a computerized spectra database or a published library of spectra (McMaster and McMaster, 1998).

Chemical ionization uses an ionization gas such as methane, butane, or carbon dioxide mixed with the sample stream. The ionization gas absorbs the initial ionizing electron and transfers the energy to the sample molecule. Chemical ionization creates a lower energy state than electron-impact ionization and thus provides less fragmentation and a larger abundance of molecular ions (McMaster and McMaster, 1998).

The major components of the MS system include a vacuum pump to maintain a vacuum in the mass spectrometer, an ionization chamber, a focusing lens to direct the stream of ions, a mass analyzer, a detector, and a data control system. Following the ionization process already described, the charged-ion stream is focused through a pinhole and a series of electrical focusing lenses to narrow and intensify the ion stream before it enters the mass analyzer. A quadrupole analyzer is a common MS configuration used in environmental analyses. The quadrupole analyzer is a set of four cylindrical metal rods clamped together in a specific configuration spaced so that their electromagnetic fields interact in a precise manner. Direct current and radio-frequency signals are applied across the rods; adjacent rods have opposite charges. The effect of the combined direct current and radio-frequency fields is to force the ion stream entering the quadrupole into a corkscrew-shaped three-dimensional sine wave. The combined electromagnetic fields establish a standing wave for just a single mass-to-charge ratio at a particular frequency, allowing it to be directed to the detector at the end of the quadrupole, while other ion fragments follow unstable paths and collide with the walls of the quadrupole rods. As the combined electromagnetic fields are varied, larger or smaller masses impinge upon the detector (Chapman, 1993; McMaster and McMaster, 1998).

Ionized analytes and their fragments leave the quadrupole and are deflected from a linear path to the ion detector. Fragment ions colliding with the conductive surface of the ion detector induce a cascade of ions in the body of the ion detector, amplifying the signal from the single fragment and producing a signal strong enough to be counted in the data system. The combination of ion detectors and data acquisition systems can record about 25,000 data points per second. The range and scan rate are instrument dependent; for example, mass spectra from 35 to 550 *m/z* can be scanned up to 10 times per second in some mass spectrometers (McMaster and McMaster, 1998).

4.4.1.4 Mass Spectroscopy and 1,4-Dioxane

To improve analytical sensitivity for detection of 1,4-dioxane and other compounds at low concentrations, MS systems can be operated in SIM. In SIM, the MS can be programmed to focus on selected narrow *m/z* range(s), refining the ability to detect ion fragments in the target range. To ensure accurate compound identification using SIM, multiple ions should be monitored for each compound (USEPA, 2006a). For 1,4-dioxane, the *m/z* of the molecular ion is 88, and the masses of fragment ions of 1,4-dioxane that are used for environmental analyses are 58, 43, and 57. For 1,4-dioxane's deuterated internal standard, 1,4-dioxane-d_8, the *m/z* of the molecular ion is 96, while the *m/z* ratios for its fragment ions are 64 and 46. Mass spectra for 1,4-dioxane and 1,4-dioxane-d_8 are shown in Figure 4.3.

4.5 ENVIRONMENTAL PROTECTION AGENCY METHODS FOR ANALYSIS OF 1,4-DIOXANE

Commercial laboratories typically offer analysis of 1,4-dioxane in water by six USEPA methods: 522 and 524.2 for drinking water and 8260 (purgeable), 8260-SIM, 8270 (extractable), and 8270-SIM for site-cleanup groundwater samples. Because of the high water solubility and poor purging efficiency of 1,4-dioxane, PT GC–MS methods without modifications can produce high detection limits, typically greater than 100 μg/L (ppb) (Strout et al., 2004b).

The development of analytical methods for 1,4-dioxane followed demand for quantifying 1,4-dioxane in the chlorinated solvents, pharmaceutical, cosmetics, and petroleum industries, as well as personnel air monitoring in these industries. Earlier test methods in these applications required quantitation in the percent range, which could be achieved by using various combinations of PT and GC with FIDs. The motivation to analyze water-borne 1,4-dioxane in the ppb range stemmed from risk analyses leading to low-level regulatory guidelines for drinking water.

Among the earlier USEPA methods for 1,4-dioxane, a heated PT method was proposed for polar, water-soluble VOCs (Lucas et al., 1988). The method involved purging samples at 90°C. To shift the aqueous phase/vapor phase equilibrium constant toward the vapor phase, salt was added to the sample. Sodium chloride was the salt selected for heated PT because it is readily available and provides the desired high ionic strength. To obtain the preferred "salting out" effect, sodium chloride and sodium or magnesium sulfate are used at concentrations that attain 80% saturation at 65°C. 1,4-Dioxane recovery with heated PT increased 21% without salt and as much as 56% with addition of sodium sulfate salt. The heated PT method provided a preliminary MDL of 4 ppb and LOQ of 20 ppb (Lucas et al., 1988).

USEPA published CLP Method OLM03.1 for 1,4-dioxane in water by SIM GC–MS in 1994 (USEPA, 1994). The method uses a heated PT at 50°C, followed by GC–MS analysis using SIM rather than full-scan analysis. The monitored ions have mass-to-charge ratios (*m/z*) of 88 ± 0.5 and 58 ± 0.5 for 1,4-dioxane and 96 ± 0.5 and 64 ± 0.5 for 1,4-dioxane-d_8 as an internal standard. Method OLM03.1 could reportedly achieve a practical quantitation limit of 5 µg/L (USEPA, 1994).

In 1997, USEPA's Central Regional Laboratory, operated by USEPA Region V in Chicago, published a method for measurement of purgeable 1,4-dioxane in water using wide-bore capillary column GC–MS (Rudinsky et al., 1997). The method uses a heated PT with sodium sulfate and 1,4-dioxane-d_8 added as the internal standard. The method was accurate for samples with 1,4-dioxane concentrations between 20 and 200 µg/L. A summary of instrumentation and operating parameters for this method is included in Appendix 3.

Researchers with the Shaw Group and USEPA in Las Vegas conducted a multilaboratory comparison study of purgeable versus extractable methods and conventional versus isotope-dilution quantitation methods (Strout et al., 2004b). The results show that quantitation by isotope dilution provides higher accuracy and precision than the conventional method of quantitation, regardless of whether the method uses purgeable or extractable sample preparation.

Carryover contamination can occur with the 1,4-dioxane purgeable method, and method blank contamination was observed in the study. High recoveries of low spiked concentrations, that is, 2 µg/L, indicate that carryover was occurring. Carryover contamination must be addressed by running an additional method or instrument blanks, which adds to the method cost and time of analysis. The study concluded that analysis of 1,4-dioxane using a 5 mL heated purge, an accelerated sample purge flow, and SIM MS detection produces the most linear calibration for a concentration range from 0.5 to 200 µg/L, as well as the highest signal-to-noise (S/N) ratio. Running a 25 mL purge volume with the same analytical conditions provided a calibration over a narrower concentration range (0.5–20 µg/L). The optimal sample temperature is approximately 50°C, and the optimal sample purge flow is approximately 50 cm^3/min for the analysis of 1,4-dioxane as a purgeable analyte on the SIM MS system (Strout et al., 2004b).

USEPA's CLP developed a draft protocol for the analysis of trace 1,4-dioxane using the SIM method. SIM requires method blanks at the beginning and end of the analytical sequence and continuing calibration verification (CCV) standards. Quantitation is performed by using the average relative response factor (RRF) from the initial calibration. USEPA Method 8260 is often used with SIM for analysis of 1,4-dioxane, as described in Section 4.5.2.

4.5.1 Environmental Protection Agency Method 8015

Method 8015, currently 8015D (revision 4, June 2013) is for the analysis of nonhalogenated volatile and semivolatile organic compounds by GC and FID. The recommended sample-preparation methods for analysis of 1,4-dioxane by USEPA Method 8015, based on the earlier version 8015B (revision 2, December 1996) are direct injection and azeotropic microdistillation. Poor purging efficiency is noted for 1,4-dioxane sample preparation by PT, and solvent extraction is not recommended for USEPA Method 8015 analysis of 1,4-dioxane (USEPA, 1996e). Four GC columns are recommended for USEPA Method 8015, including a DB-Wax column suitable for 1,4-dioxane retention. The recommended columns are summarized in Table 4.6.

For reagent grade water, groundwater, and leachate (generated according to USEPA's Toxicity Characteristic Leaching Procedure test), USEPA Method 8015 (USEPA, 2003b) can achieve MDLs of 12, 15, and 16 µg/L, respectively with sample preparation by azeotropic microdistillation following USEPA Method 5031. Detection limits for other media are shown in Table 4.4.

4.5.2 Environmental Protection Agency Method 8260

Method 8260, currently 8260D (revision 4, June 2018) (USEPA, 2018a) is a method that uses GC–MS and can therefore achieve lower detection limits for 1,4-dioxane than USEPA Method 8015 which utilizes GC-FID. For 1,4-dioxane by 8260D, the recommended sample-preparation method is

TABLE 4.6
Recommended Columns for USEPA Method 8015D

Column Length	Column ID	Column Material	Column Film Thickness (μm)
8 feet	2.54 mm (0.1 inch)	Stainless steel or glass packed, 1% SP-1000 on Carbopack-B 60/80 mesh	—
6 feet	2.54 mm (0.1 inch)	Stainless steel or glass column packed with *n*-octane on Porasil-C 100/120 mesh (Durapak)	—
30 m	0.53 mm	Fused-silica capillary column bonded with DB-Wax (or equivalent)	1
30 m	0.53 mm	Fused-silica capillary column chemically bonded with 5% methyl silicone (DB-5, SPB-5, RTx, or equivalent)	1.5

Source: U.S. Environmental Protection Agency (USEPA), 2003b, Method 8015D: Non-halogenated organics using GC/FID. U.S. Environmental Protection Agency.

USEPA Method 5031, azeotropic microdistillation. USEPA Method 8260D notes that there are several sample preparation methods for 1,4-dioxane that exhibit known difficulties that may require special treatment for proper use. These preparation methods include USEPA Methods 5030, PT for aqueous samples, 5035, PT for solids and wastes, and 5032, vacuum extraction. For many of the 128 compounds listed in the USEPA Method 8260D analyte list, historical quantitation limits for groundwater samples have been approximately 5 μg/L. If azeotropic microdistillation is used for sample preparation and if 1,4-dioxane-d_8 is used as an internal standard, the general reporting limit for 1,4-dioxane by USEPA Method 8260B is 12 μg/L (USEPA, 1996f). In practice, 1,4-dioxane reporting limits by USEPA Method 8260B are typically in the 50–250 μg/L range (based on a survey of U.S. laboratories conducted by J. Bishop in 2016).

Some laboratories provide an 8260B analysis for 1,4-dioxane with SIM, which improves the detection limit of the method. For example, commercial laboratories using 8260-SIM with a heated PT can achieve an MDL of 0.4 μg/L (Eurofins Lancaster Laboratories Environmental, 2016) and a reporting limit of 1 μg/L (Curtis and Tomkins Laboratory, 2006).

The chromatographic columns recommended for USEPA Method 8260D are listed in Table 4.7.

TABLE 4.7
Recommended Columns for USEPA Method 8260D

Column Length (m)	Column ID (mm)	Column Material	Column Film Thickness (Hm)
30	0.25	Capillary column with DB-624 (Agilent) or VOCOL (Sigma-Aldrich)	1.4
20	0.18	Capillary column with Rtx-Volatiles (Restek)	1
60	0.32	Capillary column with DB-VRX (Agilent)	1.5 or 1.8
75	0.53	Capillary column coated with DB-624 (Agilent), Rt-502.2 (RESTEK), or VOCOL (Sigma-Aldrich)	3.0
30	0.25–0.32	Capillary column coated with 95% dimethyl – 5% diphenyl polysiloxane (DB-5, Rtx-5, SPB-5, or equivalent)	1.0

Source: U.S. Environmental Protection Agency (USEPA), 2018, Method 8260D: Volatile organic compounds by gas chromatography/mass spectrometry (GC/MS). U.S. Environmental Protection Agency.

TABLE 4.8
Method 8260C Accuracy and Precision for 1,4-Dioxane by Mean Percent Recovery (*R*) and Percent Relative Standard Deviation (RSD)

	25 ppb Spike		100 ppb Spike		500 ppb Spike	
	R	RSD	*R*	RSD	*R*	RSD
1,4-Dioxane-d_8	56	34	58	11	61	18
1,4-Dioxane	63	25	55	16	54	13

Source: U.S. Environmental Protection Agency (USEPA), 2006c, Method 8260C: Volatile organic compounds by gas chromatography/mass spectrometry (GC/MS). U.S. Environmental Protection Agency.

Notes: Sample preparation by Method 5031—the microdistillation technique (single laboratory and single operator). Lower RSD values reflect increased precision. Data from analysis of seven aliquots of reagent water spiked at each concentration, using a quadrupole mass spectrometer in the selected-ion monitoring mode.

Beginning with Method 8260C (revision 3, August 2006) an option for the use of heated PT was included in Method 8260 in addition to the use of 1,4-dioxane-d_8 as a surrogate standard when using azeotropic distillation for sample preparation. The sample is heated to 90°C, which produces an approximately fivefold increase in 1,4-dioxane recovery over conventional analysis; however, the 1,4-dioxane recovery can nevertheless be as low as 1%. Using heated PT requires careful control on temperature to generate reproducible results. Because the majority of the 1,4-dioxane remains in the sparger, thorough rinsing and bake-out must be performed to prevent carryover (Fitzpatrick and Tate, 2006). The description for USEPA Method 8260C requires that the sample be heated to at least 80°C before 1,4-dioxane can be considered a viable analyte by PT. The recovery of 1,4-dioxane described in Method 8260C is low, but the recovery is indexed to the isotope standard, allowing quantitation of 1,4-dioxane. Recoveries and RSDs for 1,4-dioxane analyses by Method 8260C at a single laboratory by a single operator are listed in Table 4.8. Method 8260C does not list a lower LOQ for 1,4-dioxane.

4.5.3 Environmental Protection Agency Method 8261A

USEPA Method 8261A uses vacuum distillation to vaporize compounds and separate VOCs, some low-boiling-point semivolatile organic compounds, and polar nonpurgeable or low-purging compounds from soil, water, and other environmental sample matrices. The volatilized material passes over a chilled condenser coil to condense and remove water. The analytes are condensed by using liquid nitrogen to obtain cryogenic temperatures (down to −196°C). Volatile organic analytes are then introduced by a vacuum distiller into the gas chromatograph, which is interfaced with a mass spectrometer. USEPA Method 8261A is based on the combination of cryogenic trapping and vacuum distillation, as described by USEPA Method 5032, with the addition of the use of internal standards to correct determined concentrations based on matrix effects that may impact the sample preparation and analysis.

USEPA Method 8261A uses internal standards to measure matrix effects and to compensate analyte responses for matrix effects. This method should be considered for samples in which matrix effects are anticipated to severely affect analytical results (USEPA, 2006d). The inventor of the vacuum distillation technique, Michael Hiatt, is a USEPA scientist; USEPA was granted a patent for the method (Hiatt, 1995).

The USEPA vacuum distillation method accounts for the physical chemistry of analytes in measuring matrix effects, yielding method performance data by analyte. Results for analyses by USEPA Method 8261A are reported with confidence intervals. USEPA Method 8261A eliminates the need for matrix spike-matrix spike duplicates as well as calibration of instrumentation by matrix type (Hiatt, 2007).

In vacuum distillation, the effect of boiling point and relative volatility on the internal standards must be addressed by correcting the GC–MS response. Surrogate recovery is determined to predict

TABLE 4.9
Laboratory Performance for 1,4-Dioxane Analyses by USEPA Method 8261, Vacuum Distillation (VD/GC/MS)

Lab	Sample Size	Matrix	Number of Samples	Spike Concentration	Average Recovery (%)	Relative Standard Deviation[a]	Method Detection Limit (Estimated)
1	5 g	Soil	8	10	186	16.7	9.4 µg/kg
2	5 g	Soil	7	10	87	20.9	5.7 µg/kg
1	5 mL	Water	8	20	79	24.9	11.6 µg/L
2	5 mL	Water	7	10	113	17.9	6.3 µg/L
2	25 mL	Water	7	4	113	17.7	2.5 µg/L

Lab	Number of Samples	Sample Matrix and 1,4-Dioxane Concentration (ppb)	Average Recovery at Different Concentrations				Relative Standard Deviation[a]
			Low (%) M	Medium (%)	High (%)	Overall (%)	
1a	9	Water: 1, 10, and 50	128	103	114	116	6.4
2b	12	Water: 1, 3, and 5	327	145	178	217	13.9
1c	9	Soil: 1, 10, and 50	137	103	122	121	21.0
2d	12	Soil: 1, 3, and 5	156	127	166	150	21.0

Matrix of Five Blind Samples (5 mL)	Spike Level (µg/L)	Average Recovery (%)	Average Relative Standard Deviation[a]
Water	375	95	11.4
Water	125	111	7.8
Water	55	102	13.6
Water	7	196	24.2
Water	1	660	24.0

Source: Strout, K., et al., 2004a, Vacuum distillation unit interlaboratory study evaluation. Paper presented at the 20th Annual National Environmental Monitoring Conference, July 19–22, Independent Laboratory Institute, Washington, DC. http://nemc.us/proceedings/2004 proceedings.pdf (accessed January 2005). With permission.

[a] Lower relative standard deviation (RSD) values reflect increased precision. Recovery of volatilized 1,4-dioxane may be diminished by condensation in the apparatus (Hiatt, 2007).

the recoveries of analytes within surrogate groups on the basis of relative volatility* and boiling point. Once the recoveries of the internal standards have been corrected for boiling-point effects, recoveries then reflect only the matrix effects owing to relative volatility. Relative volatility is also corrected by using a software algorithm for processing the data generated from USEPA Method 8261A (USEPA, 2006d). Method 8261A does not list a lower LOQ for 1,4-dioxane. An estimated MDL for 1,4-dioxane on a 25 mL aliquot of water is 2.5 µg/L, whereas the routine injection of 5 mL of sample produces an estimated MDL or 6.3 µg/L (Strout et al., 2004a). Table 4.9 presents data from laboratory performance studies for USEPA Method 8261A.

4.5.4 Environmental Protection Agency Method 8270—Extracted Semivolatile Organic Compounds by Gas Chromatography-Mass Spectroscopy

USEPA Method 8270, currently 8270E (revision 6, June 2018) is a GC–MS method designed to analyze for semivolatile organic compounds on extracted samples. This method targets

* "Relative volatility" describes the distillation of one compound relative to that of another compound as a ratio of their respective partition coefficients.

compounds that are soluble in dichloromethane, or other appropriate solvents, and capable of being eluted as sharp peaks from a gas chromatographic fused-silica capillary column coated with a slightly polar grade of silicone (USEPA, 2018b). Sample-preparation methods for USEPA Method 8270 include USEPA Method 3510, LLE, and USEPA Method 3520, continuous liquid-liquid extraction (CLLE), for aqueous samples and USEPA Method 3540 or 3541, Soxhlet extraction, and USEPA Method 3545, pressurized fluid extraction, for solid samples. USEPA Method 8270 was developed using a single-column configuration, 30-m-long silicone-coated fused-silica capillary column with a 0.25 or 0.32 mm internal diameter and a 0.25, 0.5, or 1 μm film thickness, Agilent DB-5, or equivalent column (USEPA, 2018b). 1,4-Dioxane was not listed as an analyte in USEPA Method 8270 through revision D, July 2014 (USEPA, 2014c); however, 1,4-dioxane was added with revision 6, 8270E. USEPA Method 8270 notes that more volatile analytes, such as 1,4-dioxane, may be lost during sample preparation from the evaporative step for sample concentration. To minimize the potential for loss of these volatile compounds great care needs to be taken during sample preparation and the inclusion of additional surrogates including 1,4-dioxane-d_8 is advised.

In analyses using extractable sample-preparation methods, 1,4-dioxane elutes immediately after the solvent front,* which reduces interference from other compounds. In isotope dilution, samples are spiked with 1,4-dioxane-d_8 and then subjected to routine sample extraction procedures. For improved resolution, the sample extract can be reanalyzed with the mass spectrometer in SIM mode. Isotope dilution in MS analysis involves comparing the mass differences between 1,4-dioxane and its deuterated analog, 1,4-dioxane-d_8. 1,4-Dioxane-d_8 is the ideal internal standard for 1,4-dioxane because it has similar chemical properties as 1,4-dioxane, but its mass spectra are distinctly different (Linton and Alonso, 2006). Samples can be analyzed for 1,4-dioxane by using the full-scan method with other semivolatile analytes, but with higher detection limits (Strout et al., 2004b).

Isotope dilution improves precision and accuracy by compensating for the problems with calibration and sample-preparation matrix effects. Some analytes are not easily recovered during chromatography or sample extraction. Calibration problems and matrix effects are usually compensated for in part by using internal standards and surrogate analytes. In 1,4-dioxane-d_8, hydrogen atoms on the 1,4-dioxane molecule are replaced with deuterium atoms.† The 1,4-dioxane-d_8 is available as "99 atom% deuterium," meaning that 99% of the sites on the 1,4-dioxane ether ring normally occupied by ^1H are occupied by deuterium in the standard. In GC analyses, the error from methods used to inject the sample can be approximately 5%. By adding an internal standard, the error can be reduced by half to approximately 2–3%. With isotope dilution, the error should be half of that again, approximately 1% (Lindsay, 2000).

To achieve low-level reporting limits by USEPA Method 8270C, commercial laboratories have adjusted the method by adding a salting-out step that uses large-volume injection of sodium chloride or sodium sulfate and by running the mass spectrometer in single ion mode. For example, a method used by ALS (previously Columbia Analytical Services) involves separatory funnel extraction using dichloromethane on a 100 mL sample spiked with 1,4-dioxane-d_8 and prepared with sodium chloride. The extract is dried and reduced to a final volume of 5 mL. A large-volume injector is used to introduce the sample to the GC–MS, which is run in SIM mode. ALS reported that a method reporting limit of 0.1 μg/L is routinely achievable through the use of this approach (Grindstaff, 2004).

* The solvent front is the chromatographic peak corresponding to the carrier solvent, usually dichloromethane, used to extract the analyte from the sample.

† Deuterium, written as ^2H or D, is the naturally occurring stable isotope of hydrogen. It has one extra neutron and is therefore heavier than hydrogen. Deuterium has the same chemical properties as hydrogen. A "deuterated" standard is prepared so that molecular sites normally occupied by ^1H are replaced with deuterium.

CASE STUDY – COMPARISON OF 1,4-DIOXANE RESULTS FROM USEPA METHOD 8260 AND USEPA METHOD 8270

In order to better understand the potential variability of laboratory results for 1,4-dioxane in environmental samples analyzed by two of the most common analytical methods performed by commercial laboratories, USEPA Method 8260 and USEPA Method 8270, The Dow Chemical Company developed laboratory performance evaluation (PE) samples that included 1,4-dioxane at two expected concentrations (Mack and Chiang, 2016). These PE samples were developed in a real-world matrix blended from environmental material collected at multiple sites. This blended material was chosen to test the ability of the laboratories to determine the presence and concentration of 1,4-dioxane in real world samples rather than a clean matrix such as deionized water.

These PE samples were submitted to 11 laboratories with a request to analyze the samples by both USEPA Method 8260 and USEPA Method 8270 as performed by the laboratories. The performance of these methods by the laboratories including the methods as written as well as the addition of SIM and isotope dilution at some laboratories. The results for the USEPA Method 8260, including those methods that used SIM and isotope dilution, had results that were approximately twice the concentrations determined by USEPA Method 8270 and USEPA Method 8270 SIM at both of the PE sample concentrations. The only exception to the disparity between the USEPA Method 8260 and USEPA Method 8270 determined concentrations was from one laboratory that performed USEPA Method 8270 using isotope dilution. When isotope dilution was performed with USEPA Method 8270 by this one laboratory, the concentrations determined were consistent with the concentrations determined by the other laboratories following the USEPA Method 8260 procedures. Results from the PE study are shown in Figure 4.7.

The difference in the concentrations determined by the USEPA Method 8260 and USEPA Method 8270 is due to the water solubility and extraction efficiency of the sample preparation methods used by USEPA Method 8270. The calibration for USEPA Method 8260 is performed by using purge and trap for the calibration standards in the same manner as the samples are prepared for analysis and the impacts of the water solubility and air-water partitioning are minimized for this reason. However, USEPA Method 8270 is calibrated by the introduction of calibration standards without preparation into the analytical system and recovery limitations from sample preparation procedures are not incorporated into the final concentrations for the analytical technique without the use of isotope dilution.

4.5.5 Environmental Protection Agency Method 1624—Isotope-Dilution Gas Chromatography-Mass Spectroscopy

USEPA Method 1624 was similar to Method 8260 with some notable additions to improve the precision and accuracy. Method 1624 (USEPA, 1989) was developed with the inclusion of heated purging for samples with more than 1 percent solids and the addition of 1,4-dioxane-d_8 to the sample as an isotope dilution internal standard. The detection limit of USEPA Method 1624 depends more on the level of interferences than instrumental limitations (USEPA, 2001).

4.5.6 Environmental Protection Agency Method 522—Solid-Phase Extraction and Gas Chromatography-Mass Spectroscopy with Selective Ion Monitoring

USEPA's National Exposure Research Laboratory developed a method, USEPA Method 522 (USEPA, 2008), for measuring 1,4-dioxane in drinking water. The method applies SPE, using

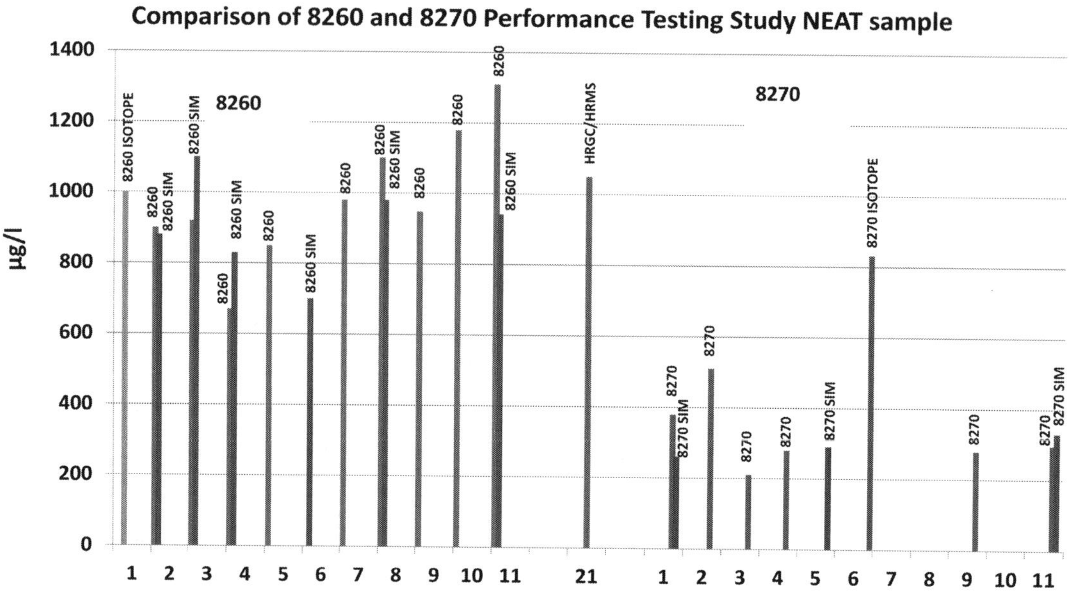

FIGURE 4.7 Dow Chemical multi-laboratory performance test illustrating major differences between EPA 8260, EPA 8270, and SIM or isotope dilution variations of 8260 and 8270.

1,4-DIOXANE METHOD COMPARISON MINI-STUDY

A groundwater investigation to delineate 1,4-dioxane occurrence was conducted at a former industrial facility located in Tallevast, Florida. Analyses for 1,4-dioxane initially used a modified Method 8270C. Beginning in 2006, samples were analyzed with both USEPA Method 8270C and USEPA Method 8260B in SIM mode. Results reported for the Method 8260B SIM analysis were consistently and significantly higher than those reported for the Method 8270C analysis, in some cases, as much as two orders of magnitude higher. Environmental scientists with the consulting firm conducting the investigation, Blasland Bouck and Lee Environmental Services (now part of Arcadis), undertook a comparison study to identify differences in analytical methods for 1,4-dioxane (Linton and Alonso, 2006).

Samples were collected in triplicate in two separate sampling events from 13 monitoring wells and two private wells by using low-flow sampling techniques. Analysis was performed by USEPA Methods 8270C, 8260B, and 8270 with isotope dilution ("8270-ID") at three different commercial laboratories. The purpose of the comparison study was to obtain a more quantitative statement of a pattern of conflicting results for analyses by different parties using different methods to analyze samples from domestic wells. The conflicting results led to controversy in the affected community in Bradenton County, Florida, and brought the subtleties of GC–MS configurations into the local newspaper. Data used in the method evaluation by Linton and Alonso were not validated by the authors, and the study was not intended to be fully validated review of these three analytical methods. Samples were analyzed by Method 8260B in standard scan mode initially and subsequently reanalyzed in SIM mode if results were below the detection limit. To check for analyte loss and potential interference, field samples were spiked at three different concentrations and used as matrix spike-matrix spike duplicate (MS/MSD) samples for all three methods to determine method accuracy (percent recovery) and method precision (relative percent difference, RPD). Recovery was

(*Continued*)

also measured on laboratory control standards, to distinguish instrument accuracy from possible matrix effects attributable to sample characteristics.

Results for samples analyzed by Method 8260B were highest among the three methods; some results were more than an order of magnitude higher than the Method 8270C results. Accuracy was consistent for the three spike concentration levels, and precision increased somewhat with spike concentration. Method 8260B in scan and SIM modes had an average recovery of 73% and an average RPD of 5%. Accuracy increased with increasing spike concentration, ranging from 49% recovery for the low spike level to 98% recovery for the high spike level. Precision was relatively constant at all spike concentration levels.

As shown in Figure 4.8, the results using Method 8260B paralleled results for the isotope-dilution method. Method 8270C had the lowest recoveries. The average recovery was 65%, and the average RPD for recoveries in MS/MSD analyses was 20%. Precision and accuracy of Method 8270C decreased as the concentration of spike addition increased. Overall accuracy for Method 8270ID averaged 100% recovery, and precision averaged 9% RPD in the first sampling event and 12% in the second sampling event. Accuracy was independent of spike concentrations, whereas precision increased somewhat with spike concentration. Figure 4.8 presents a comparison of the concentration trends for the results from the three methods.

Results for samples from each well by Methods 8260B and 8270C were compared to results from that well by Method 8270ID. In Figure 4.8, results from a specific well by Methods 8260B and 8270C (y-axis) are plotted against the concentration obtained by Method 8270ID (x-axis). Regression analysis and a paired t-test showed that the results of Method 8260B did not significantly differ from those of the isotope-dilution method, and the methods correlated well. BBL recommended that the USEPA consider performing further optimization of Method 8260B to increase the sensitivity of the method. Increasing the sample and purge volumes may decrease the MDL, and using a heated purge would increase the purging efficiency of 1,4-dioxane from the groundwater samples, thereby increasing sensitivity of the method.

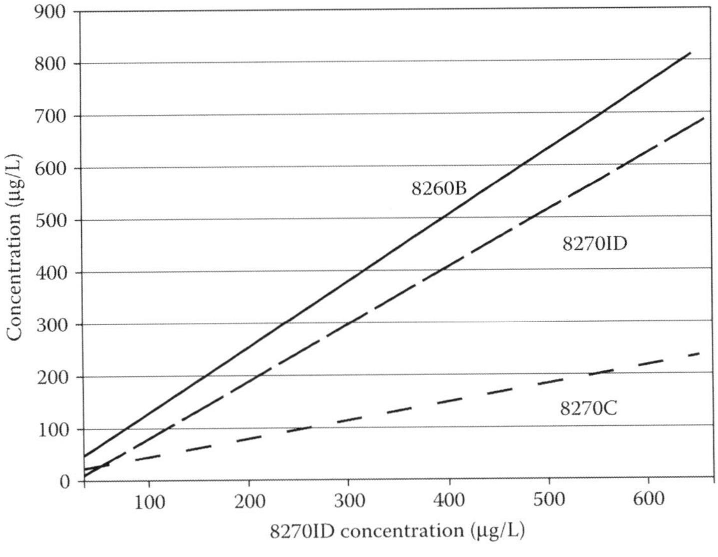

FIGURE 4.8 Paired comparisons of the results of USEPA Methods 8260B and 8270C with 8270-Isotope Dilution from Linton and Alonso Method Comparison Study. (From Linton, J. and Alonso, J.C. 2006, 1,4-Dioxane comparative analysis study—USEPA Methods 8260B, 8270C and 8270 with isotope dilution. Tampa, FL: Blasland Bouck and Lee Environmental Services [now Arcadis-US]. With permission.)

CAVEAT EMPTOR—DOUBLE-BLIND STANDARDS CAN HELP TO ENSURE GOOD DATA QUALITY

The majority of contaminated site investigations rely upon laboratories' internal quality control protocols to confirm that results are within acceptable limits of precision and accuracy. The subject of laboratory quality control is beyond the scope of this book; the reader is referred to USEPA's quality pages (http://www.epa.gov/quality/) for a thorough review of the fundamentals of laboratory quality control. Quality control tests include method blanks, initial and continuing calibration standards, matrix spike and recovery analyses, and several others. For emerging contaminants, unconventional analyses, analyses reported at the low end of the method range, and analyses performed infrequently by laboratories, it is sometimes useful to confirm laboratory performance independently by submitting double-blind standards.

Double-blind standards are usually prepared by a third-party laboratory or provider of analytical standards. The subject laboratory's sample container is shipped to the standards provider, where a whole-volume standard with a value similar to concentrations found at a field site is prepared and filled in the container and then express-shipped back to the laboratory client. The client then labels the sample with a fictitious name similar to names of other samples in the set and ships it in a cooler with other field samples from a sampling event. Laboratory results are then compared with the value of the whole-volume standard given by the standards provider. Laboratory performance can then be gauged to confirm that the laboratory's in-house quality control tests are accurate. Because there is analytical error associated with both the whole-volume standard and the subject laboratory's analysis, the combined error range can be larger than the errors of the individual analyses. Nevertheless, the laboratory client can use double-blind standards to independently verify the reasonableness of reported results.

Standards for 1,4-dioxane and 1,4-dioxane-d_8 can be purchased from laboratory standards providers. If desired, the whole-volume standard can be prepared by using the same matrix as other samples in the set—for example, groundwater from a monitoring well that has consistently tested nondetect for 1,4-dioxane—and then analyzing a second sample from a whole-volume split of that groundwater to confirm absence of 1,4-dioxane. Matrix-matched double-blind standards prepared in this manner are less obvious to laboratory analysts, who might otherwise recognize a double-blind standard by its low electrical conductivity and use extra care when conducting the analysis.

coconut charcoal, or synthetic carbon, as the solid sorbent and dichloromethane as the eluent, to extract and concentrate 1,4-dioxane from drinking water matrices. Extracts are analyzed by GC–MS using SIM. The lowest concentration minimum reporting limits for USEPA Method 522 are reported as 0.047 μg/L and 0.036 μg/L using coconut charcoal and synthetic carbon, respectively.

4.6 LABORATORY SAFETY FOR 1,4-DIOXANE HANDLING AND INSTRUMENT CLEANING

1,4-Dioxane can react with oxygen to form explosive peroxides. Discovery of old bottles of full-strength ethers can lead to expensive removal by demolition experts, or worse, accidental detonation. The peroxide crystals often form around the threads of the bottle top and explode when compressed if an unsuspecting laboratory chemist attempts to open the bottle. This problem is more likely to occur at facilities using pure 1,4-dioxane such as histology laboratories, liquid scintillation counters, and research chemistry laboratories. Several news stories have profiled incidents in which old cans or bottles of dioxane have been discovered, including a hospital in Fort Lauderdale,

TABLE 4.10
Method 522 Average Accuracy and Precision for 1,4-Dioxane by Mean Percent Recovery (R) and Percent Relative Standard Deviation (RSD)

	0.04 μg/L Spike		1 μg/L Spike		10.0 μg/L Spike	
	R	RSD	*R*	RSD	*R*	RSD
Reagent Water						
1,4-Dioxane-d_8	92	3.5	97.1	5.2	90.1	2.8
1,4-Dioxane	110	19	98.0	6.2	93.9	2.8

Source: U.S. Environmental Protection Agency (USEPA), 2008, Method 522 Version 1.0: Determination of 1,4-Dioxane in Drinkin Water by Solid Phase Extraction (SPE) and Gas Chromatography/Mass Spectrometry (GC/MS) with Selected Ion Monitoring (SIM). U.S. Environmental Protection Agency. EPA/600/R-08/101.

Notes: Average recoveries are for 7 replicates of 500 mL reagent water fortified with the listed spike concentration and extracted using 2-g of coconut charcoal.

Florida, a laser facility in Santa Clara, California, and a college laboratory in Salisbury, North Carolina. In each case, these discoveries caused facility shutdown and a bomb squad response to remove the container of 1,4-dioxane to a remote location for detonation. Once peroxides form, controlled detonation is the safest method to remove the hazard. The explosive power of a gallon of peroxidized 1,4-dioxane is probably equivalent to several hand grenades, according to the reported explosive power of other peroxidizable materials such as ethyl ether and isopropyl ether. 1,4-Dioxane is also known to form explosive mixtures with nitromethane, sulfur trioxide, silver perchlorate, and decaborane (NICNAS, 1998). Laboratory standards used for instrument calibration and quality control are generally low concentration and unlikely to form peroxides.

Unstabilized 1,4-dioxane may also form toxic and hazardous formate esters of 1,2-ethanediol over long periods of storage. The mono- and diformates of 1,2-ethanediol are formed by peroxide intermediates and have been found at concentrations as high as 1.8 M in partially consumed bottles of 1,4-dioxane. Purging containers of 1,4-dioxane with nitrogen each time they are opened, keeping containers away from light, and refrigerating 1,4-dioxane can prevent the possible health hazard caused by contamination with peroxide and formate ester (Jewett and Lawless, 1980). As described in Chapter 3, photo-oxidation of 1,4-dioxane in the atmosphere produces ethylene-1,2-diformate.

As discussed in Chapter 2, 1,4-dioxane can be stabilized to prevent peroxide and formate ester formation. Stabilizers include stannous chloride, ferrous sulfate, 2,6-tert-butyl-p-cresol, or butyl hydroxy toluene (BHT) (BASF, 1986, 2005; European Chemicals Bureau, 2002). Peroxide formation can also be prevented by filling the container headspace with nitrogen. When stabilized and stored under nitrogen in original containers, 1,4-dioxane has a shelf life of 24 months (BASF, 2005).

Absence of 1,4-dioxane to stabilize methyl chloroform has also created laboratory safety problems. A common method for cleaning parts of mass spectrometers is sonication (ultrasound cleaning) in a solvent bath. Before the use of methyl chloroform was banned by the Montreal Protocol, it was commonly used in laboratory cleaning operations. At Purdue University, a laboratory technician performed a routine sonication cleaning operation that involved placing aluminum mass spectrometer parts in a beaker with methyl chloroform. On this occasion, the technician used a new bottle of methyl chloroform. The methyl chloroform in the beaker began "boiling," became dark, and began producing a sharp odor that filled the laboratory, forcing evacuation. The mass spectrometer part was ruined. Most mass spectrometer parts are made from stainless steel. In this instance, the ion source from a Hewlett Packard (Agilent) 5970, 5992, 5995, or 5996 Series GC–MS was made of aluminum and reacted with the methyl chloroform (Purdue University, 2003). It is likely that the particular grade of methyl chloroform used was not intended for cleaning aluminum parts and was therefore not stabilized with 1,4-dioxane.

ACKNOWLEDGMENTS

The critical reviews and helpful comments of the first edition of this chapter by Carl Isaacson, with USEPA in Athens, Georgia, and Bart Simmons, retired Director of the California Department of Toxic Substances Control Hazardous Materials Laboratory, are gratefully acknowledged.

REFERENCES

Anderson, H., 2005, Memo to D. Easley, U.S. EPA region VII, from Gore survey products. EPA region VII.

BASF, 1986, *Datenblatt Dioxan [Dioxane Datasheet]*. Ludwigshafen, Germany: BASF (Badische Anilin- und Soda-Fabrik).

BASF, 2005, *Dioxane—Stab Technical Data Sheet*. Ludwigshafen, Germany: BASF (Badische Anilin- und Soda-Fabrik).

Black, L. and Fine, D., 2001, High levels of monoaromatic compounds limit the use of solid-phase micro-extraction of methyl *tert*-butyl ether and *tert*-butyl alcohol. *Environmental Science and Technology* 35(15): 3190–3192.

Britt, S., 2004, The Snap Sampler™. Paper presented to the Interstate Technology Regulatory Council Diffusion Sampler Team, Albuquerque, NM, October 28, 2004.

Bruce, M.L., Lee, R.P., and Stephens, M.W., 1992, Concentration of water soluble volatile organic compounds from aqueous samples by azeotropic microdistillation. *Environmental Science and Technology* 26(1): 160–163.

CalEPA, 2003, Advisory—Active Soil Gas Investigations. Los Angeles: Los Angeles Regional Water Quality Control Board and California Department of Toxic Substances Control.

Chapman, J.R., 1993, *Practical Organic Mass Spectrometry—A Guide for Chemical and Biochemical Analysis*. Chichester: Wiley.

Cooper, C.V., White, L.D., and Kupel, R.E., 1971, Qualitative detection limits for specific compounds utilizing gas chromatographic fractions, activated charcoal and a mass spectrometer. *American Industrial Hygiene Association Journal* 32: 383–386.

Curtis and Tomkins Laboratory, 2006, 1,4-Dioxane by GC/MS with selected ion monitoring—EPA 8270 SIM. http://www.curtisandtompkins.com (accessed February 12, 2006).

Daughton, C.G., 2003, Cradle-to-cradle stewardship of drugs for minimizing their environmental disposition while promoting human health. I. Rationale for and avenues toward a green pharmacy. *Environmental Health Perspectives* 111: 757–774.

Daughton, C.G., 2005, Emerging chemicals as pollutants in the environment: A 21st century perspective. *Renewable Resources Journal* 23(4): 6–23.

Daughton, C.G., 2006, Pharmaceuticals as environmental contaminants: An overview of the science. Paper presented at the Water Education Foundation Pacific Conference.

Donaldson, W.T., 1977, Trace organics in water: Refined methods are needed to document their occurrence and concentrations. *Environmental Science and Technology* 11(4): 348–352.

Draper, W.M., Dhoot, J.S., Remoy, J.W., and Perera, S.K., 2000, Trace-level determination for 1,4-dioxane in water by isotopic dilution GC and GC–MS. *Analyst* 125(8): 1403–1408.

Epstein, P.S., Mauer, T., Wagner, M., Chase, S., and Giles, B., 1987, Determination of parts per billion concentrations of dioxane in water and soil by purge and trap gas chromatography/mass spectrometry or charcoal tube enrichment gas chromatography. *Analytical Chemistry* 59(15): 1987–1990.

European Chemicals Bureau, 2002, European Union risk assessment report: 1,4-Dioxane. Institute for Health and Consumer Protection, Luxembourg, European Union, Report 21.

Fitzpatrick, T.W. and Tate, K., 2006, 1,4-dioxane environmental and measurement issues. In: *Proceedings, Annual Industry Workshop 2006*. Sarasota, FL: Florida Department of Environmental Protection.

Fuh, C.B., Lai, M., Tsai, H.Y., and Chang, C.M., 2005, Impurity analysis of 1,4-dioxane in nonionic surfactants and cosmetics using headspace solid-phase microextraction. *Journal of Chromatography A* 1071(1): 141–145.

Grindstaff, J., 2004, Low level analysis of 1,4-dioxane by EPA Method 8270C SIM with large volume injection. *CAS Connection* 1: 4. http://www.caslab.com/contact/cc_spring2004_email.pdf.

Guo, W. and Brodowsky, H., 2000, Determination of the trace 1,4-dioxane. *Microchemical Journal* 64: 173–179.

Hayes, H.C., Benton, D.J., and Khan, N., 2006, Impact of sampling media on soil gas measurements. In: *Vapor Intrusion—The Next Great Environmental Challenge—An Update*. Los Angeles, CA: Air and Waste Management Association.

Hiatt, M., 1995, United States Patent 5,411,707: Vacuum extractor incorporating a condenser column. Assignee: U.S. Environmental Protection Agency.

Hiatt, M., 2007, An overview of SW-846 Method 8261 chemistry: Volatile organic compounds by vacuum distillation in combination with gas chromatography/mass spectrometry (VD/GC/MS): New method for analyzing organic chemicals in a wide array of samples. http://www.epa.gov/esd/chemistry/vacuum/training/pdf/theory-rev5.pdf (accessed March 23, 2007).

Isaacson, C., Mohr, T.K.G., and Field, J.A., 2006, Quantitative determination of 1,4-dioxane and tetrahydrofuran in groundwater by solid phase extraction GC/MS/MS. *Environmental Science & Technology* 40(23): 7305–7311.

ITRC (Interstate Technology & Regulatory Council), 2007, Protocol for Use of Five Passive Sampler to Sample for a Variety of Contaminants in Groundwater. DSP-5. Washington, D.C.: Interstate Technology & Regulatory Council, Diffusion/Passive Sampler Team.

Jewett, D. and Lawless, J.G., 1980, Formate esters of 1,2-ethanediol—major decomposition products of *p*-dioxane during storage. *Bulletin of Environmental Contamination and Toxicology* 25: 118–121.

Jochmann, M.A., Kmiecik, M.P., and Schmidt, T.C., 2006, Solid-phase dynamic extraction for the enrichment of polar volatile organic compounds from water. *Journal of Chromatography A* 1115: 208–216.

Kawata, K., Ibaraki, T., Tanabe, A., et al., 2001, Gas chromatographic-mass spectrometric determination of hydrophilic compounds in environmental water by solid-phase extraction with activated carbon felt. *Journal of Chromatography A* 911: 75–83.

Kurimo, R.W., 1994, NIOSH Method 1602: Dioxane, NIOSH. http://www.cdc.gov/niosh/nmam/pdfs/1602.pdf (accessed May 2004).

Lancaster Laboratories, 2004, 1,4-Dioxane in environmental samples. http://www.lancasterlabs.com/ (accessed May 14, 2004).

Lesage, S. and Jackson, R.E., 1992, *Groundwater Contamination and Analysis at Hazardous Waste Sites*. Boca Raton, FL: CRC Press.

Lesnik, B., 1993, New SW-846 methods for the analysis of conventional and non-conventional volatile organics in solid matrices (abstract). In: R.L. Siegrist and J.J. van Ee (Eds), *Proceedings, National Symposium on Measuring and Interpreting VOCs in Soils: State of the Art and Research Needs*. Las Vegas, NV: U.S. Environmental Protection Agency, January 12–14.

Li, M., Conlon, P., Fiorenza, S., Vitale, R.J., and Alvarez, P.J.J., 2011, Rapid analysis of 1,4-dioxane in groundwater by frozen micro-extraction with gas chromatography/mass spectrometry. *Groundwater Monitoring & Remediation* 31 (4): 70–76.

Lide, D.R. (Ed.), 2008, *Handbook of Chemistry and Physics*, 88th ed. Boca Raton: Taylor & Francis Group.

Lindsay, E., 2000, Isotope dilution. *West Coast Analytical Services Digest* 8(1). http://www.wcaslab.com/tech/tech2.htm (accessed May 2004).

Linton, J. and Alonso, J.C., 2006, 1,4-Dioxane comparative analysis study—EPA Methods 8260B, 8270C and 8270 with isotope dilution. Tampa, FL: Blasland Bouck and Lee Environmental Services (now Arcadis-US). http://www.tallevast.info/pdfs/dioxane_may19.pdf (accessed September 2006).

Lucas, S.V., Burkholder, H.M., and Alford-Stevens, A., 1988, Heated purge and trap method development and testing. U.S. Environmental Protection Agency, Environmental Monitoring and Support Laboratory, EPA/600/4–88/029.

Mack, E., and Chiang, D. (2016, May). 1,4-Dioxane's Emerging Dilemma, What's a Remediation Manager Supposed to Do? Presented at the Tenth International Conference on the Remediation of Chlorinated and Recalcitrant Compounds, Palm Springs, CA.

Markell, C., Hagen, D.F., and Bunnelle, V.A., 1991, New technologies in solid-phase extraction. *LC-GC* 9: 332–337.

McMaster, M. and McMaster, C., 1998, *GC/MS: A Practical User's Guide*. New York: Wiley-VCH.

Munch, J.W., 2006, Analysis of 1,4-dioxane and other water soluble volatile organic compounds by solid phase extraction and GC/MS. Paper presented at American Water Works Association Water Quality Technology Conference, Denver, CO.

NICNAS, 1998, Priority existing chemical assessment reports—1,4-dioxane. National Industrial Chemicals Notification and Assessment Scheme (NICNAS), Australian Government.

O'Neill, D., 2006, Passive diffusion samplers: Cost effective sampling techniques for sampling groundwater for organic and inorganic analytes of interest—highlighting rigid porous polyethylene samplers (RPPS). Presented at DOD Chemist Conference, San Antonio, TX, April.

Park, Y.-M., Pyoa, H., Park, S.-J., and Park, S.-K., 2005, Development of the analytical method for 1,4-dioxane in water by liquid-liquid extraction. *Analytica Chimica Acta* 7(548): 109–115.

Parsons, 2005, Results report for the demonstration of no-purge groundwater sampling devices at former McClellan Air Force Base, CA. Army Corps of Engineers, Omaha, Nebraska. http://www.geoinsight-online.com/resources/mcclellan_report_cover.html (accessed October 12, 2006).

Photovac, 2006, VOC detection instruments. http://www.photovac.com/White_Papers/VOC_Compounds_Detected_by_Photovac_Instruments.htm (accessed July 14, 2006).

Purdue University, 2003, The Lilly incident. http://web.ics.purdue.edu/~swihart/safety/LillyIncidentOct2003.htm (accessed August 28, 2005).

Rankin, T., 1996, Residual solvents in drugs by static headspace analysis. In: *Pharmaceutical Analysis: GC Technical Report*. Austin, TX: ThermoQuest CE. www.tmqaustin.com (accessed February 2005).

Rastogi, S.C., 1990, Headspace analysis of 1,4-dioxane in products containing polyethoxylated surfactants by GC–MS. *Chromatographia* 29(9–10): 441–445.

Rödel, W. and Wölm, G., 1987, *A Guide to Gas Chromatography*. Berlin, German Democratic Republic: VEBDeutscher Verlag der Wissenschaften.

Rose, D.L. and Sandstron, M.W., 2003, Determination of gasoline oxygenates, selected degradates, and BTEXin water by heated purge and trap/gas chromatography/mass spectrometry—methods of analysis by the U.S. Geological Survey National Water Quality Laboratory. USGS Method O-4024–03:39. http://web1.er.usgs.gov (accessed February 2007).

Rudinsky, R., Tang, C.M., and Elly, C.T., 1997, CRL Method 624 VOC-dioxane: Standard operating procedure for measurement of purgeable 1,4-dioxane in water by wide-bore capillary column gas chromatography/mass spectrometry. U.S. Environmental Protection Agency, U.S. EPA Region V.

Shirey, R.E., 2000, Optimization of extraction conditions for low molecular weight analytes using solid-phase microextraction. *Journal of Chromatographic Science* 38(3): 109–115.

Shirey, R.E. and Linton, C.M., 2006, The extraction and analysis of 1,4-dioxane from water using solid phase microextraction coupled with gas chromatography and gas chromatography-mass spectrometry. *Journal of Chromatographic Science* 44(7): 444–450.

Simmons, B., 1997, Non-target compounds—you can't always get what you want. *Hydrovisions* 6(2): 1. Groundwater Resources Association of California. http://www.grac.org/summer97/non-target.htm (accessed May 2004).

Song, D. and Zhang, S., 1997, Rapid determination of 1,4-dioxane in water by solid-phase extraction and gas chromatography-mass spectrometry. *Journal of Chromatography A* 787: 283–287.

Strout, K., Hedin, C., Zimmerman, M., and Smith, T., 2004a, Vacuum distillation unit interlaboratory study evaluation. Paper presented at the 20th Annual National Environmental Monitoring Conference, July 19–22, Independent Laboratory Institute, Washington, DC. http://nemc.us/proceedings/2004proceedings.pdf (accessed January 2005).

Strout, K., Zimmerman, M., Hedin, C., and Smith, T., 2004b, Comparison of 1,4-dioxane as a volatile and semivolatile analyte in single and multi-laboratory studies. Paper presented at the 20th Annual National Environmental Monitoring Conference, Book of Abstracts, No. 29. July 19–22, Independent Laboratory Institute, Washington, DC. http://nemc.us/proceedings/2004proceedings.pdf (accessed January 2005).

Supelco/SigmaAldrich, 1997, *The Retention Index System in Gas Chromatography: McReynolds Constants*. Bellefonte, PA. http://www.sial.com (accessed January 2005).

Swallow, K.C., 1992, Nonpriority pollutant analysis and interpretation. In: S. Lesage and R.E. Jackson (Eds), *Groundwater Contamination and Analysis at Hazardous Waste Sites*. New York: Marcel Dekker.

U.S. Army Corps of Engineers (USACE), 2014, *Demonstration of the AGI Universal Samplers (F.K.A. the GORE® Modules) for Passive Sampling of Groundwater*, Environmental Security Technology Certification Program (ESTCP) Project ER-200921, ERDC/CRREL TR-14-4, Alexandria, Virginia, March 2014, 250 p.

U.S. Environmental Protection Agency (USEPA), 1989, Method 1624, Revision C: Volatile Organic Compounds by Isotope Dilution GCMS. U.S. Environmental Protection Agency.

U.S. Environmental Protection Agency (USEPA), 1993, Behavior and determination of volatile organic compounds in soil: A literature review. U.S. Environmental Protection Agency, Office of Research and Development. EPA/600/R-93/140.

U.S. Environmental Protection Agency (USEPA), 1994, 1,4-Dioxane in water by selective ion monitoring (SIM) gas chromatography/mass spectrometry (GC/MS)—EPA CLP Method OLM03.1. U.S. Environmental Protection Agency.

U.S. Environmental Protection Agency (USEPA), 1996a, Method 3510C: Separatory Funnel Liquid-Liquid Extraction. U.S. Environmental Protection Agency.

U.S. Environmental Protection Agency (USEPA), 1996b, Method 3520C: Continuous Liquid-Liquid Extraction. U.S. Environmental Protection Agency.

U.S. Environmental Protection Agency (USEPA), 1996c, Method 5031: Volatile, Nonpurgeable, Water-Soluble Compounds by Azeotropic Distillation. U.S. Environmental Protection Agency.

U.S. Environmental Protection Agency (USEPA), 1996d, Method 5032: Volatile Organic Compounds by Vacuum Distillation. U.S. Environmental Protection Agency.

U.S. Environmental Protection Agency (USEPA), 1996e, Method 8015B: Non-halogenated organics using GC/FID. U.S. Environmental Protection Agency.

U.S. Environmental Protection Agency (USEPA), 1996f, Method 8260B: Volatile organic compounds by gas chromatography/mass spectrometry (GC/MS).

U.S. Environmental Protection Agency (USEPA), 1996g, Method 8270C semivolatile organic compounds by gas chromatography/mass spectrometry (GC/MS). U.S. Environmental Protection Agency.

U.S. Environmental Protection Agency (USEPA), 1999, Compendium Method TO-15: Determination of volatile organic compounds (VOCs) in air collected in specially prepared canisters and analyzed by gas chromatography/mass spectrometry (GC/MS). U.S. Environmental Protection Agency.

U.S. Environmental Protection Agency (USEPA), 2001, Method 1624 revision B—Volatile Organic Compounds by Isotope Dilution GC/MS. 40 CFR Chapter I, Part 136, Appendix A, Method 1624, July 1, Edition.

U.S. Environmental Protection Agency (USEPA), 2002, Method 5035A: Closed-System Purge-and-Trap and Extraction for Volatile Organics in Soil and Waste Samples. U.S. Environmental Protection Agency.

U.S. Environmental Protection Agency (USEPA), 2003a, Method 5030C: Purge-and-Trap for Aqueous Samples. U.S. Environmental Protection Agency.

U.S. Environmental Protection Agency (USEPA), 2003b, Method 8015D: Nonhalogenated Organics Using GC/FID. U.S. Environmental Protection Agency.

U.S. Environmental Protection Agency (USEPA), 2006a, Test Methods SW-846 on line. http://www.epa.gov/epaoswer/hazwaste/test/main.htm (accessed May 4, 2005).

U.S. Environmental Protection Agency (USEPA), 2006b, Characterization methods web site. http://www.clu-in.org/char/technologies/gc.cfm (accessed December 2006).

U.S. Environmental Protection Agency (USEPA), 2006c, Method 8260C: Volatile organic compounds by gas chromatography/mass spectrometry (GC/MS). U.S. Environmental Protection Agency.

U.S. Environmental Protection Agency (USEPA), 2006d, Method 8261A: Volatile organic compounds by vacuum distillation in combination with gas chromatography/mass spectrometry (VD/GC/MS). U.S. Environmental Protection Agency.

U.S. Environmental Protection Agency (USEPA), 2007, Method 3535A: Solid-Phase Extraction (SPE). U.S. Environmental Protection Agency.

U.S. Environmental Protection Agency (USEPA), 2008, Method 522 Version 1.0: Determination of 1,4-Dioxane in Drinking Water by Solid Phase Extraction (SPE) and Gas Chromatography/Mass Spectrometry (GC/MS) with Selected Ion Monitoring (SIM). U.S. Environmental Protection Agency. EPA/600/R-08/101

U.S. Environmental Protection Agency (USEPA), 2014a, *Test Methods For Evaluating Solid Waste, SW-846*, 3rd ed. Washington, DC: U.S. Environmental Protection Agency, Office of Solid Waste and Emergency Response.

U.S. Environmental Protection Agency (USEPA), 2014b, Method 5021A: Volatile Organic Compounds in Various Sample Matrices Using Equilibrium Headspace Analysis. U.S. Environmental Protection Agency.

U.S. Environmental Protection Agency (USEPA), 2014c, Method 8270E: Semivolatile organic compounds by gas chromatography/mass spectrometry (GC/MS). U.S. Environmental Protection Agency.

U.S. Environmental Protection Agency (USEPA), 2015, Test Methods for Evaluating Solid Waste, Physical/Chemical Methods, EPA publication SW-846, Third Edition, Final Updates I (1993), II (1995) IIA (1994), IIB (1995), III (1997), IIIA (1999), IIIB (2005), IV (2008), and V (2015).

U.S. Environmental Protection Agency (USEPA), 2018a, Method 8260D: Volatile organic compounds by gas chromatography/mass spectrometry (GC/MS). U.S. Environmental Protection Agency.

U.S. Environmental Protection Agency (USEPA), 2018b, Method 8270E: Semivolatile organic compounds by gas chromatography/mass spectrometry (GC/MS). U.S. Environmental Protection Agency.

Vainberg, S., McClay, K., Masuda, H., et al., 2006, Biodegradation of ether pollutants by *Pseudonocardia* sp. strain ENV478. *Applied and Environmental Microbiology* 72(8): 5218–5224.

Vroblesky, D.A., Borchers, J.W., Campbell, T.R., and Kinsey, W., 2000, Investigation of polyethylene passive diffusion samplers for sampling volatile organic compounds in ground water at Davis Global Communication Station, Sacramento, CA, August 1998 to February 1999. U.S. Geological Survey Open-File Report 00–307.

Wala-Jerzykiewicz, A. and Szymanowski, J., 1998, Headspace gas chromatography analysis of toxic contaminants in ethoxylated alcohols and alkylamines. *Chromatographia* 48(3–4): 299–304.

Weck Laboratories, 2006, Method reporting limits. http://www.wecklabs.com/Resources/MethodReportingLimits/SemivolatileOrganics/tabid/130/Default.aspx (accessed February 12, 2006).

Wilken, M., 2016. 1,4-Dioxane's Emerging Dilemma What's a Remediation Manager Supposed to Do? (Panel Discussion Presentation) Remediation of Chlorinated and Recalcitrant Compounds—2016. Tenth International Conference on Remediation of Chlorinated and Recalcitrant Compounds (Palm Springs, CA; May 2016).

Yasuhara, A., Shiraishi, H., Nishikawa, M., et al., 1999, Organic components in leachates from hazardous waste disposal sites. *Waste Management and Research* 17(3): 186–197.

Yasuhara, A., Tanaka, Y., Tanabe, A., and Katami, T. 2003, Elution of 1,4-dioxane from waste landfill sites. *Bulletin of Environmental Contamination and Toxicology* 71: 641–646.

Yoo, L., 2002. Simultaneous determination of 1,2,3-TCP and 1,4-dioxane in drinking water by GC–MS/MS using purge/trap techniques. Orange County Water District.

Yoo, L.J., Fitzsimmons, S., and Wehner, M., 2002, Improved purge-trap and GC/MS/MS techniques for the trace-level determination of 1,4-dioxane in water. In: *Proceedings, Water Quality Technology Conference*. Seattle, WA: American Water Works Association, CD-ROM.

5 Review of the Human Health Toxicology and Ecotoxicology of 1,4-Dioxane

Janet K. Anderson and William B. Kappleman

CONTENTS

International regulations related to 1,4-dioxane as an environmental contaminant have been largely based on the concerns for human health risk. Within the past decade, several regulatory and/or public health agencies have conducted reviews of the available 1,4-dioxane toxicity data with varying scope and depth (Health Canada, 2010, 2018; ATSDR, 2012; USEPA, 2013; NTP, 2016; USEPA, 2018a). The data available as of early 2019, including epidemiology studies, laboratory animal bioassays, and *in vitro* and *in silico* studies, are reviewed herein, with a focus on data that are relevant to inform assessment of potential human health and ecological risk from 1,4-dioxane as an environmental contaminant (i.e., environmental exposures).

1,4-Dioxane exposure to humans and ecological receptors may occur via inhalation, dermal, and/or oral routes; however, ingestion is the most relevant route of exposure to 1,4-dioxane as an environmental contaminant for humans. Metabolism does not appear to be route- or species-specific, with the majority of 1,4-dioxane being metabolized by mammalian cytochrome P-450 enzymes and excreted as 2-hydroxyethoxyacetic acid (aka β-hydroxyethoxyacetic acid) (HEAA) in the urine of both rodents and humans. Liver and kidney toxicities are the primary health effects associated with exposure to 1,4-dioxane in humans and laboratory animals. Short-term exposures to high levels of 1,4-dioxane have been reported in occupational settings, resulting in human fatalities from kidney failure and liver necrosis and central nervous system nerve fiber damage has also been reported. Similar results have been observed in animal studies wherein severe kidney and liver damage occurred following short-term high dose exposures. Effects from chronic exposure to 1,4-dioxane in humans have not been well characterized, as available studies are inadequate to evaluate the causal relationship between 1,4-dioxane exposure and human disease outcome. Chronic effects seen in rodent bioassays predominately include neoplastic and non-neoplastic lesions in the kidney, liver, and nasal tissue/respiratory tract. In experimental animal bioassays, cancer is the most severe health effect associated with exposure to 1,4-dioxane. Therefore, the most relevant critical effects associated with exposure to 1,4-dioxane in humans are tumorigenesis and liver and kidney damage, and the cancer risk is the predominant concern for environmental human health risk assessments. The most recent collective evidence indicates that 1,4-dioxane is not likely to be genotoxic; however, different regulatory and public health opinions currently exist as to how 1,4-dioxane may cause cancer and therefore, whether non-linear threshold doses for carcinogenicity should be utilized for quantitative human health risk assessment approaches. Specifically, recent analysis related to the 1,4-dioxane carcinogenic mode of action (MoA) have suggested that default assumptions used by some regulatory and public health agencies (e.g., the USEPA) are overly stringent (Health Canada, 2010, 2018); however, other agencies have determined that sufficient uncertainty remains to warrant these more protective default public health assumptions (USEPA, 2013; MDEQ, 2015).

This chapter summarizes the potential concerns related to human and ecological exposure to 1,4-dioxane in the environment and highlights the different options for assessing environmental risk.

5.1 MAMMALIAN TOXICOKINETICS

The data specific to the absorption, distribution, metabolism, or excretion of 1,4-dioxane in humans are limited to a few volunteer studies; however, extensive data in rodents are available that allow for adequate toxicokinetic (TK) characterization. There does not appear to be significant species-specific TK processes, such that rodent TK data are assumed relevant to humans. Based on these data, it is apparent that 1,4-dioxane is rapidly absorbed and metabolized via all routes of exposure. The primary metabolites are well characterized and include HEAA, and metabolism and excretion occur via nonlinear and saturable kinetics.

5.1.1 Absorption

1,4-Dioxane is readily absorbed in mammals following both oral and inhalation exposures. Absorption of 1,4-dioxane in humans has been investigated following short-term (between 6 and

8 hours) inhalation exposures (Young et al., 1976, 1977; Göen et al., 2016). The human studies demonstrate rapid absorption via time-course measurements of HEAA in blood and urine. In an occupational worker study, five volunteers were exposed to a time-weighted average of 1.6 parts per million (ppm) 1,4-dioxane in air for 7.5 hours and an absorbed dose of 0.37 mg/kg (for a 70-kg person) was calculated assuming 100% lung absorption (Young et al., 1976). However, no breath measurements were taken to confirm that 100% of the inhaled 1,4-dioxane was absorbed. In another study, four volunteers were exposed to 50 ppm 1,4-dioxane in air chambers for 6 hours and plasma and urine concentrations of 1,4-dioxane and HEAA were measured hourly (Young et al., 1977). The authors determined that uptake of 1,4-dioxane into plasma was rapid within the first 2 hours and approached steady-state concentrations by the 6-hour time point. The concentration of HEAA peaked at 1 hour postexposure and was undetectable by 4 hours postexposure. The authors calculated a mean absorbed dose of 5.4 mg/kg at a mean rate of 76.1 mg/h; however, exhaled breath was not analyzed and the absorbed fraction not determined. Göen et al. (2016) exposed 18 volunteers to 20 ppm 1,4-dioxane for 8 hours, some of whom performed exercises on a bicycle ergometer. 1,4-Dioxane and HEAA levels in blood were determined at 4- and 8.75-hour time points and urine samples were collected and analyzed over 24 hours. The authors reported that 1,4-dioxane levels were barely detectable in both the blood and urine; however, HEAA was detected in urine at a maximum of 9.8 ± .9 hours after the beginning of exposure and levels increased with increasing exercise. The authors did not calculate an absorbed dose of 1,4-dioxane.

Rodent studies demonstrate rapid and nearly complete absorption of 1,4-dioxane by both inhalation and oral exposure routes. Early studies conducted in male Sprague-Dawley rats exposed to a single dose or 17 consecutive daily doses of ^{14}C-1,4-dioxane via oral gavage show that less than 2% of the administered 1,4-dioxane was recovered in feces regardless of dose level or frequency of exposure, demonstrating high levels of gastrointestinal absorption (Young et al., 1978a, 1978b). The same study authors also exposed male Sprague-Dawley rats to 50 ppm 1,4-dioxane vapors (head only) for 6 hours and measured plasma and urine concentrations of 1,4-dioxane and HEAA. 1,4-Dioxane was rapidly absorbed, with plasma concentrations peaking at the 6-hour time point and decreasing to non-detect by 5 hours postexposure. The mean total absorbed dose was calculated as 71.9 mg/kg (Young et al., 1978a, 1978b). In a more recent study, male F344/DuCrj specific pathogen-free rats were administered a single oral dose of 1,4-dioxane or were exposed via inhalation to 250 ppm in whole-body chambers for 6 hours (Take et al., 2012). The study authors noted that in the rats exposed orally, the concentrations of 1,4-dioxane in blood rose rapidly, peaked 1 hour after exposure, then decreased to undetectable levels at 7 hours. For the inhalation exposed rats, blood concentrations of 1,4-dioxane increased during the first 3 hours of exposure, then held constant throughout the remaining exposure duration. The amount absorbed was not calculated by the study authors for either exposure scenario.

Data describing the dermal absorption of 1,4-dioxane are limited. Little dermal absorption (2–3% of the applied dose) was demonstrated in a study in rhesus monkeys using either a methanol or skin lotion vehicle applied to unoccluded clipped skin for 24 hours (Marzulli et al., 1981). Low levels of 1,4-dioxane dermal absorption (3.2% of the applied dose) were also indicated in an *in vitro* experiment using excised occluded human abdominal skin and either water, lotion, or isopropyl myristate vehicle preparations (Bronaugh, 1982).

5.1.2 Distribution

There are no data describing the tissue distribution of 1,4-dioxane in humans. Following a single oral exposure to male F344/DuCrj specific pathogen-free rats, Take et al. (2012) detected 1,4-dioxane in all tested tissues, with peak concentrations at 60 min postexposure in the lung (215 μg/g), liver (185 μg/g), kidney (180 μg/g), brain (175 μg/g), and abdominal fat (85 μg/g). The authors also tested the distribution of 1,4-dioxane following combined exposures to a single gavage dose to deuterated 1,4-dioxane (1,4-dioxane-d_8) followed by whole-body exposure to unlabeled 1,4-dioxane vapors; 1,4-dioxane and 1,4-dioxane-d_8 were detected in all tissues with a similar pattern as the single oral exposure assay indicating that route of exposure had no effect on tissue distribution patterns. Early studies using 6.97 mg/kg 3H-1,4-dioxane via intraperitoneal injections into the abdominal cavity of male Sprague-Dawley rats also indicate a generally uniform tissue distribution across multiple tissues including liver, kidney, spleen, lung, colon, and skeletal muscle (Woo et al., 1977a). Mikheev et al. (1990) exposed rats to ^{14}C-1,4-dioxane by intraperitoneal injection and evaluated the levels of 1,4-dioxane distribution to the blood, brain, testes, liver, and kidney, and calculated the tissue:blood concentration ratios for each. For all tissues evaluated across several time points (minutes to up to 6 hours post-injection), the tissue:blood ratio was between 0.5 and 1.5, demonstrating that 1,4-dioxane is generally evenly distributed and does not accumulate in any specific tissue.

5.1.3 Metabolism

Metabolism of 1,4-dioxane is depicted in Figure 5.1. The proposed metabolic scheme for 1,4-dioxane includes the production of HEAA as the primary urinary metabolite (Young et al., 1976; Braun and Young, 1977; ATSDR, 2012), and physiologically based pharmacokinetic (PBPK) models suggest that this metabolic scheme is similar across species, including humans, rats, and mice (Sweeney et al., 2008). HEAA can be generated from oxidation of 1,4-dioxane via diethylene glycol (DEG), 1,4-dioxane-2-ol, or 1,4-dioxane-2-one by cytochrome P450 enzymes in the liver (Woo et al., 1977b, 1977c, 1978; Nannelli et al., 2005). Young et al. (1976) calculated a HEAA:1,4-dioxane ratio in urine from humans exposed to 1,4-dioxane via inhalation for 7.5 hours as 118:1, suggesting that nearly all 1,4-dioxane is metabolized to HEAA. Göen et al. (2016) estimated that the urinary excretion of HEAA is approximately 53% of the inhaled dose of 1,4-dioxane in human male volunteers exposed to 20 ppm for 8 hours. Similarly, in male volunteers exposed to 50 ppm 1,4-dioxane for 6 hours, more than 99% of the inhaled 1,4-dioxane was detected in the urine as HEAA (Young et al., 1977). Additional evidence for HEAA as the primary metabolite is available from structure-activity relationship analyses of the genotoxicity of the putative 1,4-dioxane metabolites; this analysis suggests that HEAA is only weakly genotoxic compared to 1,4-dioxane-2-one and the observed results

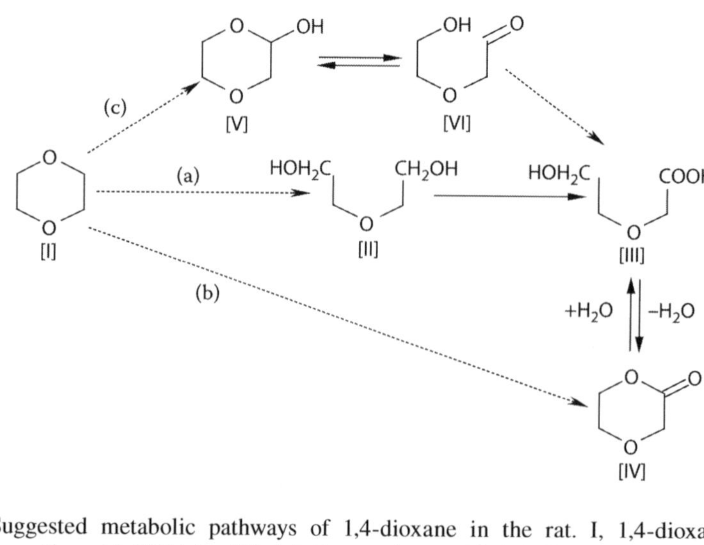

FIGURE 5.1 Suggested metabolic pathways of 1,4-dioxane in the rat. I, 1,4-dioxane; II, DEG; III, β-hydroxyethoxy acetic acid (HEAA); IV, 1,4-dioxane-2-one; V, 1,4-dioxane-2-ol; VI, β-hydroxyethoxy acetaldehyde. (Adapted from Woo, Y., Argus, M.F., and Arcos, J.C., 1977c, *Biochemical Pharmacology* 25: 1539–1542. With permission.)

of genotoxicity assays for 1,4-dioxane with metabolic activation correlate with the predicted results from HEAA (ATSDR, 2012).

The exact cytochrome P450 enzymes involved in 1,4-dioxane metabolism have not been fully characterized; however, several have been shown to be involved. Pretreatment with known inducers of cytochrome P450 enzymes, such as phenobarbital and Arochlor 1254, increased the total amount of HEAA urinary excretion, whereas pretreatment with an inducer of cytochrome P488 enzymes (3-methycholanthrene) had no effect (Woo et al., 1977c). Additionally, suppression of cytochrome P450 (CYP) activity with 2,4-dichloro-6-phenylphenoxy ethylamine or cobaltous chloride significantly decreased the HEAA levels found in the rat urine (Woo et al., 1977c). Increased enzyme activity of CYP2B1, CYP2B2, CYP2C11, and CYP2E1 in the liver microsomes were demonstrated following treatment of male Sprague-Dawley rats exposed to 1,4-dioxane in drinking water or via gavage (Nannelli et al., 2005). Gavage treatment also resulted in increased CYP3A activity in the liver (Nannelli et al., 2005). Liver microsome content from male Sprague-Dawley rats treated with 500 mg 1,4-dioxane per kilogram body weight (bw) per day (mg/kg/day) via intraperitoneal injection for 3 days shows increased CYP2B and CYP2E activities, however decreased CYP2C activity (Takano et al., 2010). At high doses, 1,4-dioxane can induce increased expression of CYP enzymes. For example, 1,4-dioxane administration in male and female mice at 500 and 1,000 mg/kg/day for 3 days resulted in increased levels of several CYPs, including CYP2B1, CYP2B2, CYP2C11, and CYP2E1, in hepatic microsomes (Nannelli et al., 2005) and increased the CYP-mediated metabolism of several barbiturates (Mungikar and Pawar, 1978).

1,4-Dioxane metabolism has been shown to be a saturable process, as demonstrated by various rodent assays of oral and intravenous injection exposure routes. Sprague-Dawley rats administered 1,4-dioxane by intravenous injection or oral dosing showed saturable metabolism occurred at plasma concentrations of approximately 100 μg/mL (Young et al., 1978a, 1978b). These same authors demonstrated that, as the oral doses of 1,4-dioxane increased, the percentage of urinary excretion of 1,4-dioxane decreased with concomitant increase in 1,4-dioxane in exhalation, suggesting that as the metabolism of 1,4-dioxane reached saturation, more 1,4-dioxane was able to circulate in the blood and be eliminated by evaporation through the lungs. One mouse study suggests that mice may have a higher rate of metabolism than rats, as the metabolic saturation in B6C3F1 mice was suggested to occur between 245 and 2,230 mg/kg/day for mice administered a single oral dose; however, the metabolic profile from isolated human hepatocytes were very similar to those from isolated rat and mice hepatocytes (Sweeney et al., 2008).

5.1.4 Excretion

Urinary excretion is the primary excretory pathway following both oral and inhalation exposure; 1,4-dioxane and HEAA are detected predominantly (>90%) in the urine in both human and rodent studies. However, as mentioned above, circulating 1,4-dioxane can also be eliminated via expired air through the lungs if metabolism becomes saturated (Young et al., 1978a, 1978b). In workers exposed to a time-weighted average of 1.6 ppm 1,4-dioxane for 7.5 hours, 99% of 1,4-dioxane eliminated in urine was in the form of HEAA (Young et al., 1976). The elimination half-life was approximately 1 hour in adult male volunteers exposed to 50 ppm 1,4-dioxane for 6 hours, with 90% of urinary 1,4-dioxane and 47% of urinary HEAA excreted within 6 hours (Young et al., 1977). The elimination half-life in rats exposed to 50 ppm 1,4-dioxane for 6 hours was calculated to also be approximately 1 hour (Young et al., 1978a, 1978b).

5.1.5 Physiologically Based Pharmacokinetic (PBPK) Models

PBPK models are quantitative and mathematical descriptions of the absorption, distribution, metabolism, and elimination of chemical substances within the body. PBPK models are used in human health risk assessments to predict the concentrations of a chemical or its metabolite in any given target tissue following various exposure scenarios (route, dose level, across species). PBPK models have been developed for 1,4-dioxane in rats, mice, humans, and lactating women (Leung and Paustenbach, 1990; Reitz et al., 1990; Fisher et al., 1997; Sweeney et al., 2008; Takano et al., 2010). Given that 1,4-dioxane absorption and metabolism are accomplished through common mechanisms in both rats and humans (Young et al. 1977, 1978a, 1978b), identical model structures are used for both rodents (rats and mice) and humans. In all three models, physiologically relevant, species-specific parameter values for tissue volume, blood flow, and metabolism and elimination are used. Sweeney et al. (2008) evaluated the available PBPK models (Leung and Paustenbach, 1990; Reitz et al., 1990) for 1,4-dioxane. To address uncertainties and deficiencies in these models, the authors conducted studies to fill data gaps and reduce pharmacokinetic uncertainties pertaining to 1,4-dioxane and HEAA in rats, mice, and humans. They demonstrated that the metabolic profile and rate constants in primary human hepatocytes and primary hepatocytes from rats and mice were all similar (Sweeney et al., 2008). USEPA (2013) recently conducted a comprehensive review of the available models to assess their utility for use in human health risk assessments, and readers are encouraged to refer to the USEPA assessment for more details on each model. Specifically, the USEPA evaluated available 1,4-dioxane PBPK models for their ability to predict observations made in experimental studies of rat and human exposures to 1,4-dioxane (see Appendix B; USEPA, 2013). The USEPA ultimately determined that the use of these models did not provide improved estimates of 1,4-dioxane's cross-species TK over default assumptions and use of body weight allometric scaling (USEPA, 2013).

5.2 EPIDEMIOLOGY STUDIES AND CASE REPORTS

The available human data for 1,4-dioxane are limited to studies primarily on the health risks following inhalation exposure in occupational exposure settings and volunteer studies. There are no oral 1,4-dioxane exposure studies in humans. There have been two case reports of acute occupational poisonings with 1,4-dioxane, showing that exposure to high concentrations resulted in liver, kidney, and central nervous system toxicity (Barber, 1934; Johnstone, 1959). Barber (1934) described a small number of deaths that occurred among factory workers exposed to high (unspecified) concentrations of 1,4-dioxane via inhalation, and possibly dermal exposures. Post-mortem analyses revealed extensive lesions (hemorrhagic nephritis) in the kidneys, which was attributable as the cause of death, and centrilobular necrosis of the liver. Johnstone (1959) discussed one fatality following a worker exposed for 1 week to 208–650 ppm 1,4-dioxane, with a mean air concentration of 470 ppm. Post-mortem evaluation showed significant histopathological changes in the liver (centrilobular necrosis), kidney (hemorrhagic necrosis), and brain (perivascular widening with small foci of demyelination in the cortex and basal nuclei). Acute inhalation studies in human volunteers show that acute exposure (a few minutes) to concentrations of 1,4-dioxane >200 – 5,500 ppm results in irritation of the eyes, nose, and throat, as well as vertigo (Yant et al., 1930; Wirth and Klimmer, 1936; Silverman et al., 1946; Young et al., 1977). Two studies reported no symptoms after exposure to 1,000–2,000 ppm 1,4-dioxane vapors for 3–5 min (Fairley et al., 1934) or to 20 ppm 1,4-dioxane for 2 hours (Ernstgård et al., 2006).

There are only two epidemiology studies available for occupational cohorts exposed to 1,4-dioxane and results are predominantly negative (Thiess et al., 1976; Buffler et al., 1978). Both studies had relatively small cohort sizes and small number of human health effects available for assessment, which limits the conclusions that can be drawn. Thiess et al. (1976) conducted a cross-sectional study of 74 male workers in a 1,4-dioxane production plant in Germany. Workers were employed at the plant for 5–41 years. The air concentrations were measured and modeled to be between 0.06 and 7.2 ppm. Compared to the general German population, no statistically significant effects were found for any of the studied health effects, including hematological and serum biochemistry parameters, renal function tests and urinalysis, liver size assessment, mortality, and cancer. Chromosome analysis was also performed on six of the workers still actively employed and six control persons;

no differences were noted. Buffler et al. (1978) evaluated a mortality study on 165 workers (100 manufacturing employees and 65 process employees) at a chemical manufacturing and processing plant in Texas. 1,4-Dioxane exposure level was estimated to be below 25 ppm for all individuals in both cohorts. There were no significant increases in cancer-related or all-cause mortality between the workers and the general population in Texas; however, the mean exposure duration of the study was only 5 years (and less than 2 years for 43% of the workers), and the latency period for evaluation was less than 10 years for 59% of the employees.

5.3 ANIMAL BIOASSAYS TO INFORM HUMAN HEALTH RISK

5.3.1 Acute and Short-Term Studies

There have been many acute and short-term toxicity studies of 1,4-dioxane in laboratory animals (reviewed in ECJRC, 2002; ATSDR, 2012; USEPA, 2013; Health Canada 2010, 2018). The lowest oral LD_{50} values are between 1,270 and 2,000 mg/kg for guinea pigs, rabbits, and cats, and 4,500 and 5,170 mg/kg for mice and rats. In general, experimental animals exposed orally to doses between 2,500 and 11,000 mg/kg exhibited increased central nervous system depression and hepatic and renal degeneration and necrosis, and histopathological lesions in the nasal cavity and brain.

Given that the primary concern regarding environmental exposure to media contaminated with 1,4-dioxane is long-term exposure and low-level concentrations from the oral route (i.e., ingestion of contaminated drinking water or soil), only subchronic and chronic oral animal bioassays and data relevant to the interpretation of potential environmental health effects are discussed further below. As outlined below, the most severe long-term health effect associated with exposure to 1,4-dioxane in experimental animals is cancer, and the liver, kidney, and respiratory tract are the major and most sensitive target organs.

5.3.2 Subchronic and Chronic Non-cancer Toxicity

Long-term exposures are of most relevance to the assessment of potential risk from 1,4-dioxane environmental exposure, and subchronic and chronic animal toxicity studies are useful for identification of potential health effects associated with longer term and environmentally relevant exposures. An evaluation of consistent patterns of toxicity across multiple animal bioassays helps determine the most probable adverse health effects following exposure to 1,4-dioxane, and mechanistic studies help inform biological plausibility and MoA for which 1,4-dioxane may cause those identified adverse health effects to occur. The summary provided below gives a brief overview of findings from the most relevant studies for assessment of human health risk from environmental exposure to 1,4-dioxane. Each study's original publication and/or more robust comprehensive reviews should be consulted for additional detailed information on methodology and data analyses.

Most of the available published subchronic and chronic 1,4-dioxane animal bioassays are oral drinking-water studies, summarized in Table 5.1. The liver, kidney, and respiratory tract have been identified as the major sensitive target organs affected by 1,4-dioxane exposure.

5.3.2.1 Hepatic Effects

As discussed above, acute lethal doses of 1,4-dioxane exposure in humans caused serious liver damage (Barber, 1934; Johnstone, 1959); however, a cross-sectional occupational study found no conclusive evidence of serious liver damage (Thiess et al., 1976). In oral (drinking water) subchronic and chronic repeat exposure studies conducted in experimental animals, the collective evidence shows that 1,4-dioxane is toxic to the liver; adverse effects include cytotoxicity, hepatocellular degeneration, pre-neoplastic cellular changes, increased liver weight and increases in liver clinical pathology parameters.

TABLE 5.1
Subchronic and Chronic Oral Toxicity Studies for 1,4-Dioxane

Species	Dose (mg/kg/day)	Duration	NOAEL/ LOAEL (mg/kg/ day)	Noncancer Effects at the LOAEL	Target Organ and Tumor Type (if Applicable)	Lowest Dose with Significant Increase in Tumor Incidence (mg/kg/day)[a]	Reference
Subchronic Studies							
Rat ($N = 6$); unknown strain	1,900[b]	67 days	NA/1,900	Renal cortical degeneration and necrosis, hemorrhage; hepatocellular degeneration	NA	NA	Fairley et al. (1934)
Mouse ($N = 6$); unknown strain	3,300[b]	67 days	NA/3,300	Renal cortical degeneration and necrosis, hemorrhage; hepatocellular degeneration	NA	NA	Fairley et al. (1934)
Male Sprague-Dawley rat ($N = 4–6$)	0, 10, or 1,000[c]	11 weeks	10/1,000	Minimal centrilobular hepatocyte swelling; increased DNA synthesis	NA	NA	Stott et al. (1981)
F344/DuCrj rat ($N = 10$ per sex per group)	Males: 0, 52, 126, 274, 657, or 1,554; Females: 0, 100, 83, 185, 427, 756, or 1,614	13 weeks	52/150	Nuclear enlargement of nasal respiratory epithelium; hepatocyte swelling	NA	NA	Kano (2008)
Crj:BDF$_1$ mouse ($N = 10$ per sex per group)	Males: 0, 86, 231, 585, 882, or 1,570; Females: 0, 0, 170, 387, 898, 1,620, or 2,669	13 weeks	170/410	Nuclear enlargement of bronchial epithelium	NA	NA	Kano (2008)
Chronic Studies							
Male Wistar rat ($N = 26$ treated, 9 controls)	0 or 640	63 weeks	NA/640	Hepatocytes with enlarged hyperchromic nuclei; glomerulonephritis	Liver (hepatocellular carcinoma)	640	Argus et al. (1965)
Male Sprague-Dawley rats ($N = 30$)	0, 430, 574, 803, or 1,032	13 months	NA/430	Hepatocytomegaly; glomerulonephritis	Nasal cavity (squamous cell carcinoma), liver (hepatocellular carcinoma)	NA[d]	Argus et al. (1973), Hoch-Ligeti et al. (1970)

(Continued)

TABLE 5.1 ***(Continued)***
Subchronic and Chronic Oral Toxicity Studies for 1,4-Dioxane

Species	Dose (mg/kg/day)	Duration	NOAEL/ LOAEL (mg/kg/ day)	Noncancer Effects at the LOAEL	Target Organ and Tumor Type (if Applicable)	Lowest Dose with Significant Increase in Tumor Incidence (mg/kg/day)[a]	Reference
Sherman rat (*N* = 60 per sex per group)	Males: 0, 9.6, 94, or 1,015; Females: 0, 19, 148, or 1,599	2 years	9.6/94	Degeneration and necrosis of renal tubular cells and hepatocytes	Nasal cavity (squamous cell carcinoma), liver (hepatocellular carcinoma)	1,307 (average of male and female dose), 1,307 (average of male and female dose)	Kociba et al. (1974)
Osborne–Mendel rat (*N* = 35 per sex per group)	Males: 0, 240, or 530; Females: 0, 350, or 640	110 weeks	NA/240	Pneumonia, gastric ulcers, and cortical tubular degeneration in the kidney	Nasal cavity (squamous cell carcinoma), liver (hepatocellular adenoma)	530 (males), 350 (females);, 350 (females)	NCI (1978)
B6C3F1 mouse (50 per sex per group)	Males: 0, 720, or 830; Females: 0, 380, or 860	90 weeks	NA/380	Pneumonia and rhinitis	Liver (hepatocellular adenoma or carcinoma)	720 (males), 380 (females)	NCI (1978)
F344/DuCrj rat (*N* = 50 per sex per group)	Males: 0, 16, 81, or 398; Females: 0, 21, 103, or 514	2 years	81/398	Atrophy of nasal olfactory epithelium; nasal adhesion and inflammation	Nasal cavity (squamous cell carcinoma, sarcoma, rhabdomyo-sarcoma, and esthesio-neuro-epithelioma), liver (hepatocell-ular adenoma or carcinoma)	398 (males), 514 (females); 398 (males), 514 (females)	Kano (2009)
Crj:BDF$_1$ mouse (*N* = 50 per sex per group)	Males: 0, 66, 251, or 768; Females: 0, 77, 323, or 1,066	2 years	77/323	Nasal inflammation	Liver (hepatocell-ular adenoma or carcinoma)	768 (males), 77 (females)	Kano (2009)
Developmental and Reproductive Studies							
Sprague-Dawley rat (*N* = 18–20 per group)	Pregnant dams: 0, 250, 500, or 1,000	GD 6–15	500/1,000	Delayed ossification of the sternebrae and reduced fetal body weights	NA	NA	Giavini et al. (1985)

Note: NA = not applicable.
[a] $p \leq 0.05$ by Fisher's exact test pair-wise comparison with controls.
[b] as calculated and reported in USEPA (2013).
[c] The methods section of Stott et al. (1981) describes the high dose as 100 mg/kg/day; however, the rest of the article describes the high dose as 1,000 mg/kg/day.
[d] Precise incidences cannot be calculated because the number of rats per group was reported as a range (28–32); incidence in control rats was not reported; no statistical analysis of the results was conducted.

Shorter duration (9.5–13 weeks of exposure) studies include Fairley et al. (1934), Stott et al. (1981), and Kano et al. (2008), and generally show that non-neoplastic lesions and increases in liver weight occur at doses ≥126 mg/kg/day for males and ≥756 mg/kg/day for female rats. These are the lowest reported lowest-observed-adverse-effect-levels (LOAELs) for liver effects, and are for hepatic histopathological changes including cellular necrosis, hypertrophy, and vacuolitic changes in F344/DuCrj rats exposed to 0, 52, 126, 274, 657, or 1,554 mg/kg/day (males) or 0, 83, 185, 427, 756, or 1,614 mg/kg/day (females) average 1,4-dioxane in drinking water per day for 13 weeks (Kano et al., 2008). Centrilobular swelling, with a statistically significant increase in relative liver weight in male Sprague-Dawley rats was observed following exposure to 1,000 mg/kg/day for 11 weeks (Stott et al., 1981). Hepatocellular degeneration in rats (strain unspecified) exposed to 1,900 mg/kg/day (as estimated by the USEPA, 2013) for 67 days (approximately 9.5 weeks) has also been observed as a sensitive hepatic endpoint in rats (Fairley et al., 1934). Studies in mice have also reported liver toxicity following subchronic 1,4-dioxane exposure, including histopathological alterations (Fairley et al., 1934; Kano et al., 2008). Mice (strain unspecified) exposed to 1,4-dioxane at 3,300 mg/kg/day in drinking water for 67 days (approximately 9.5 weeks) showed signs of hepatocellular degeneration (Fairley et al., 1934). Single cell necrosis and centrilobular swelling was observed in Crj:BDF1 mice exposed to ≥585 mg/kg/day (male) and ≥898 mg/kg/day (female) for 13 weeks (Kano et al., 2008). Kano et al. (2008) also evaluated changes in serum liver enzyme levels and activity following 1,4-dioxane exposure for 13 weeks; high-dose male and female rats showed increases in alanine aminotransferase (males only), aspartate aminotransferase, leucine aminopeptidase, and alkaline phosphatase.

Hepatic effects have also been observed in rodents following long-term exposure to 1,4-dioxane; effects include increased liver weight and concomitant histopathological changes at doses ≥55 mg/kg/day (Argus et al., 1965; Kociba et al., 1974; NCI, 1978; Yamazaki et al., 1994; Kano et al., 2009). Argus et al. (1965) dosed 26 adult male Wistar rats with 1,4-dioxane in drinking water for 64.5 weeks at approximately 640 mg/kg/day (calculated by USEPA, 2013). Although no statistical analysis of the results was provided, observed histological changes in the liver included hepatocytes with enlarged hyperchromic nuclei and large cells with reduced cytoplasmic basophilia. Kociba et al. (1974) dosed groups of 6- to 8-week-old Sherman rats with 1,4-dioxane in drinking water for up to 716 days (just over 102 weeks) with mean daily doses of 9.6, 94, and 1,015 mg/kg/day for male rats and 19, 148, and 1,599 mg/kg/day for female rats. At study termination, the authors noted a significant increase in absolute and relative liver weights in male and female high-dose rats. Histopathological liver lesions were restricted to the mid- and high-dose groups and consisted of variable degrees of hepatocellular degeneration and necrosis (no quantitative incidence data were provided in the original publication, data provided by Dow Chemical Company were published in Dourson et al., 2014). The authors determined a LOAEL of 94 mg/kg/day and corresponding NOAEL of 9.6 mg/kg/day based on the liver and kidney effects (discussed below) in male rats.

The National Cancer Institute (NCI) in collaboration with the U.S. National Toxicology Program (NTP) exposed groups of male and female Osborne-Mendel rats and B6C3F1 mice to 1,4-dioxane in drinking water for 110 or 90 weeks, respectively, at dose levels of approximating 0, 240, or 530 mg/kg/day in male rats; 0, 350, or 640 mg/kg/day in female rats; 0, 720, and 830 mg/kg/day in male mice; and 0, 380, and 860 mg/kg/day in female mice (NCI, 1978). During the 1978 study, it was the practice of NTP and NCI to typically only record the worst diagnosis for given sample (i.e., adenoma or carcinoma), and thus non-neoplastic lesions were frequently not recorded. Since that time, terminology and best practices for reporting liver lesions have changed. Given these changes, Dourson et al. (2014) had the mouse liver slides from the NCI study (1978) reevaluated for non-neoplastic lesions at an independent contract pathology laboratory. Several non-neoplastic lesions were identified, including hepatocellular hypertrophy, necrosis, inflammation, and development of hyperplastic hepatocellular foci at doses greater than 720 and 860 mg/kg/day in the livers of male and female mice, respectively.

A 2-year drinking water study with 1,4-dioxane conducted by the Japan Bioassay Research Center (JBRC) provides the lowest LOAEL for histopathological changes in rats exposed to 1,4-dioxane (Kano et al., 2009).* Male and female F344/DuCrj rats were exposed to 0, 11, 55, or 274 mg/kg/day (males) or 0, 18, 83, 429 mg/kg/day (females) 1,4-dioxane in drinking water for 104 weeks. Spongiosis and hyperplasia were observed in the livers at doses ≥55 and ≥83 mg/kg/day for males and females, respectively. Dourson et al. (2017) conducted a reanalysis of the original JBRC laboratory reports (translated to English) and also identified hepatocellular hypertrophy, necrosis, and foci development in these same dose levels. Kano et al. (2009) also reported a significant increase in relative liver weight in male and female rats exposed at ≥55 and ≥429 mg/kg/day, respectively. This same study also dosed Crj:BDC1 mice with 1,4-dioxane in drinking water at 0, 49, 191, or 677 mg/kg/day for males and 0, 66, 278, or 964 mg/kg/day for females for 104 weeks (Kano et al., 2009). The reevaluation of the original laboratory report by Dourson et al. (2017) noted hypertrophy and necrosis in male mice exposed to ≥191 mg/kg/day 1,4-dioxane. The JBRC study also evaluated clinical pathology parameters following long-term 1,4-dioxane exposure and reported a significant increase in serum alanine aminotransferase, alkaline phosphate, and aspartate aminotransferase and serum gamma-glutamyltransferase activity in both male and female rats at 274 and 428 mg/kg/day, respectively, and in male and female mice at ≥191 and ≥278 mg/kg/day, respectively.

5.3.2.2 Kidney Effects

Kidney histopathological changes and damage, including degeneration of cortical tubule cells, papillary necrosis with hemorrhage and glomerulonephritis, and increases in kidney weight, have been observed in subchronic and chronic repeat dose rodent studies with 1,4-dioxane administered via drinking water (Fairley et al., 1934; Argus et al., 1965; Kociba et al., 1974; NCI, 1978; Kano et al., 2008).

In subchronic studies, doses of ≥185 mg/kg/day 1,4-dioxane administered to rats for 13 weeks in drinking water have been shown to cause adverse kidney effects. Male and female F344/DuCrj rats from the Kano et al. (2008) study had a significant increase in kidney histopathological alterations at doses of ≥657 mg/kg/day and ≥756 mg/kg/day for male and females respectively. Observed changes included reversible cellular swelling and vacuolar changes and nuclear enlargement of the proximal tubules (Kano et al., 2008). Significant increases in relative kidney weight were reported at lower doses of ≥274 and ≥185 mg/kg/day, respectively, in the male and female F344/DuCrj rats. Rats dosed for up to 34 days with the lethal dose of 1,428 mg/kg/day had vascular congestion in the kidneys and cell degeneration in the cortical epithelium (Fairley et al., 1934). Mice exposed to 2,916 mg/kg/day for up to 67 days had similar lesions (Fairley et al., 1934). The rats and mice exposed to 1,4-dioxane at 1,900 and 3,300 mg/kg/day, respectively, also exhibited non-neoplastic lesions in the renal cortex, including degeneration, necrosis, hemorrhages and vascular congestion, and kidney weight and size were increased (Fairley et al., 1934).

The lowest LOAEL for chronic exposure to 1,4-dioxane reported to produce kidney damage were 55 and 94 mg/kg/day in mice and rats, respectively (Argus et al., 1973; Kociba et al., 1974; NCI, 1978; Kano, et al., 2009). Chronic (2-year) exposure to 1,4-dioxane in drinking water resulted in kidney damage indicated by renal tubular epithelial degeneration and necrosis, in male and female Sherman rats at doses of ≥94 and ≥148 mg/kg/day, respectively (Kociba et al., 1974). The NOAELs in males and females were 9.6 and 19 mg/kg/day, respectively, which were also the NOAELs for liver effects in the study (Kociba et al., 1974). The JBRC studies of F344/

* Kano et al. (2009) is the peer-reviewed publication associated with the Japan Bioassay Research Center (JBRC) chronic 1,4-dioxane study reported in a conference proceedings (Yamazaki et al., 1994) and laboratory report (JBRC, 1998). Given the improved quantitative detail, updated use of more current pathological diagnostic criteria, and peer-review, the data as presented in Kano et al. (2009) is relied upon herein.

DuCrj rats and Crj:BDF1 mice reported nuclear enlargement of the kidney proximal tubules at doses >55 mg/kg/day in male and 83 mg/kg/day in female rats, and in male mice dosed with 677 mg/kg/day (Kano et al., 2009). The NCI (1978) study reported kidney lesions consisting of vacuolar degeneration and focal tubular epithelial regeneration in proximal cortical tubules in male and female Osborne-Mendel rats exposed to 1,4-dioxane in drinking water for 90 weeks at doses ≥240 and at 640 mg/kg/day, respectively. Finally, glomerulonephritis and thickening of the glomerular capsule was observed in the kidneys of male Wistar rats (number not specified) exposed to 640 mg/kg/day (estimated by USEPA, 2013) of 1,4-dioxane in drinking water for 64.5 weeks (Argus et al., 1965). Conversely, no significant kidney histological changes were reported in the 2-year drinking water study with F344/DuCrj rats or Crj:BDF1 mice dosed with up to 429 or 964 mg/kg/day, respectively (Kano et al., 2009) and no kidney lesions associated with 1,4-dioxane exposure were reported in the B6C3F1 mice exposed up to 860 mg/kg/day via drinking water for 90 weeks (NCI, 1978).

5.3.2.3 Nasal and Respiratory Tract Effects

Both subchronic and chronic oral exposure to 1,4-dioxane have resulted in several different effects observed in the lung, respiratory track, and nasal epithelium of experimental animals. For subchronic exposures, the most notable respiratory and nasal effects were reported by Kano et al. (2008). In the male and female F344/DuCrj rats, non-neoplastic lesions, including nuclear enlargement of respiratory and olfactory epithelium in the nasal cavity, tracheae, and bronchial epithelium, were observed at doses ≥126 and ≥185 mg/kg/day, respectively. The Crj:BDF1 mice exposed to 1,4-dioxane from the same study also had non-neoplastic lesions including nuclear enlargement of respiratory and olfactory epithelia in the nasal cavity, vacuolar changes in the olfactory nerve, nuclear enlargement of tracheal epithelium, and nuclear enlargement and degeneration of bronchial epithelium, at doses ≥585 and ≥387 mg/kg/day in males and females, respectively (Kano et al., 2008).

Respiratory and nasal effects following chronic exposure to 1,4-dioxane have been observed in rats, mice, and guinea pigs (Argus et al., 1965, 1973; Hoch-Ligeti and Argus, 1970; Hoch-Ligeti et al., 1970; NCI, 1978; Yamazaki et al., 1994; Kano et al., 2009). The lowest LOAEL for respiratory tract effects was reported at 55 mg/kg/day in male F344/DuCrj rats, with a corresponding NOAEL of 11 mg/kg/day (Kano et al., 2009). At all higher doses within this study, additional non-neoplastic lesions were observed, including nuclear enlargement and atrophy of the olfactory epithelium, squamous cell metaplasia of the respiratory epithelium, hydropic change and sclerosis in the lamina propria, adhesion, inflammation, and/or proliferation of the nasal gland. Argus et al. (1965) reported bronchitis with epithelial hyperplasia, marked prebronchial infiltration, and abscesses in the lungs of male Wistar rats exposed to 640 mg/kg/day (reported by USEPA, 2013). NCI (1978) reported increased incidence of pneumonia in female Osborn-Mendel rats exposed to 640 mg/kg/day 1,4-dioxane in drinking water for 110 weeks.

Chronic studies in mice and guinea pigs report consistent adverse effects in the respiratory tract following oral 1,4-dioxane exposure. The JBRC studies reported non-neoplastic lesions in the respiratory tract epithelium including nuclear enlargement of the nasal olfactory epithelium in male and female mice and nuclear enlargement of the bronchial epithelium in female mice at 191 and 278 mg/kg/day, respectively. Nuclear enlargement of nasal respiratory epithelium, atrophy of nasal olfactory epithelium and tracheal epithelium, atrophy of the bronchial epithelium, inflammation of the nasal cavity, and accumulation of foamy cells in the lungs were observed at the next highest exposure level for male (677 mg/kg/day) and female (964 mg/kg/day) Crj:BDF1 mice (Kano et al., 2009). Relative and absolute lung weights were increased in these male and female mice, as well (Kano et al., 2009). The NCI (1978) study of B6C3 F1 mice reported pneumonia and non-neoplastic lesions in male and female mice at doses ≥720 and ≥380 mg/kg/day, respectively, and rhinitis in the females at this same dose level. Guinea pigs from the Hoch-Ligeti and Argus (1970) study were reported to have bronchial epithelial hyperplasia and mononuclear infiltration in the lung following exposure to 0.5–2% 1,4-dioxane (reported by the authors to be 588–635 grams of 1,4-dioxane over the study period) in drinking water for 23–28 months.

5.3.2.4 Other Effects

Although hepatic, renal, and respiratory/nasal effects are the most sensitive non-cancer endpoints identified in animal studies, other effects have been reported following long-term exposure of rodents to 1,4-dioxane in drinking water. These include gastrointestinal effects, hematological effects, reductions in body weight, increased mortality, and potential neurological effects. All of these effect endpoints occurred at higher 1,4-dioxane doses than liver, kidney, and respiratory effects. 1,4-Dioxane does not appear to be an endocrine disruptor, as no gross or histological changes have been observed in the thyroid, parathyroid, adrenal, pituitary, pancreas, or salivary glands, testes and epididymis, prostate and seminal vesicle, uterus, ovaries, thymus from rats or mice dosed with 1,4-dioxane in drinking water for 13 weeks (Kano et al., 2008) or 2 years (Kociba et al., 1974; NCI 1978; Kano et al., 2009). 1,4-Dioxane is reported as inactive in all of the endocrine-related binding and activation assays utilized by the USEPA in their endocrine disruptor screening program and Tox21 analysis (USEPA, 2018a). However, a recent oral dosing (via food) study in rabbits shows that 30 days of exposure to 1,000, 1,500, and 2,000 mg/kg resulted in statistically significant decreases in testosterone in male rabbits, decreases in progesterone in pregnant female rabbits, and decreases in progesterone and estradiol in non-pregnant female rabbits (Hashim et al., 2018); however, the published study report lacks critical information to assess the methodology and validity of these findings. Standard rodent reproductive toxicity studies for 1,4-dioxane have not been conducted; however, as stated above, no gross or histopathological changes were observed in any reproductive organ following high-level exposure for 13 weeks or 2 years in rats and mice (Kano et al., 2008, 2009). There is only one study that evaluated the potential developmental effects of 1,4-dioxane exposure. Giavini et al. (1985) exposed groups of pregnant Sprague-Dawley rats to 1,4-dioxane via gavage (0, 258, 516, or 1,033 mg) on gestation days 6–15 and sacrifice was conducted on gestational day 21. High-dose dams gained less weight than controls, and fetal weight in this group was reduced by 5.3% relative to controls. A higher incidence of reduced sternum ossification was also noted in the high-dose fetuses. No other developmental endpoints were found to be different between exposed groups and controls.

5.3.3 Cancer

Cancer is the primary risk driver regarding long-term environmental exposures. Although human occupational studies assessing the association between 1,4-dioxane exposure and increased risk of cancer are inconclusive, they are limited by small cohort sizes and a small number of reported cancer cases. 1,4-Dioxane has been identified as "reasonably anticipated to be a human carcinogen" by the NTP (NTP, 2016) based on sufficient evidence of carcinogenicity in experimental animals. Likewise, 1,4-dioxane is classified as "likely to be carcinogenic to humans" by the USEPA (2013) based on evidence of carcinogenicity in several rodent bioassays. The International Agency for Research on Cancer (IARC, 1999) has classified 1,4-dioxane as "possibly carcinogenic to humans" (group 2B) based on sufficient evidence in experimental animals and inadequate evidence in humans. Exposure of experimental animals (rats, mice, and guinea pigs) to 1,4-dioxane is associated with increased incidence of tumors in the liver, kidney, nasal cavity, peritoneum, mammary gland, and Zymbal gland. However, some of the experimental cancer bioassays suffer from study limitations such as a small number of animals, lack of statistical analyses, only one dose tested prohibiting dose-response analyses, and low tumor incidences (Argus et al., 1965, 1973; Hoch-Ligeti and Argus, 1970; Hoch-Ligeti et al., 1970).

Studies adhering more closely to standard cancer bioassay study designs (Kociba et al., 1974; NCI, 1978; Kano et al., 2009) are discussed in the most detail herein and include hepatic and nasal cavity tumor incidences.

5.3.3.1 Hepatocellular Adenomas, and Carcinomas

Hepatocellular adenomas and carcinomas have been observed in mice, rats, and guinea pigs chronically exposed to 1,4-dioxane in drinking water from multiple studies (Argus et al., 1965, 1973; Hoch-Ligeti and Argus, 1970; Hoch-Ligeti et al., 1970; Kociba et al., 1974; NCI, 1978; Yamazaki et al., 1994; Kano et al., 2009). In the Kociba et al. (1974) study, 6- to 8-week-old Sherman rats (60/sex/dose level) were exposed to 1,4-dioxane in drinking water for 716 days at levels corresponding to doses of 0, 9.6, 94, and 1,015 mg/kg/day in males and 0, 19, 148, and 1,599 mg/kg/day in females based on body weight and water consumption data. Exposure to 1,4-dioxane significantly increased mortality in the high-dose males and females beginning at about 2–4 months of treatment. Rats showed degenerative changes in both the liver and kidney, as discussed above. Body weight gain was also significantly reduced in high-dose animals from the beginning of the study. Carcinogenic effects were identified in the liver and nasal turbinates of the high-dose (1,015 and 1,599 mg/kg/day males and females, respectively) animals. The study authors combined the incidence of tumors in males and females and expressed them as the effective incidences in the number of rats that survived for 12 months. The incidence of all hepatic tumors (adenoma and carcinoma) was 2/106 (1.9%), 0/110 (0%), 1/106 (0.9%), and 12/66 (18.2%, statistical significance of $p = 0.0022$) in control, low-, mid-, and high-dose rats, respectively. The corresponding incidences of hepatocellular carcinomas alone were 1/106 (0.9%), 0/110 (0%), 1/106 (0.9%), and 10/66 (15.2%, $p = 0.00033$). These data show that the high-dose exposure group was the only exposure that increased the formation of hepatic carcinomas and also caused liver toxicity. As discussed above, the mid-dose group exhibited hepatic degeneration and necrosis, and regenerative proliferation in hepatocytes; however, no increase in tumor incidence was noted (Kociba et al., 1974).

The NCI/NTP (1978) cancer bioassay also identified hepatocellular adenomas and carcinomas in female rats and in male and female mice. In this study, groups of Osborne-Mendel rats (35/sex/dose level) were administered 1,4-dioxane in the drinking water for 110 weeks at estimated doses of 0, 240, and 530 mg/kg/day in males and 0, 350, and 640 mg/kg/day in females. Hepatocellular adenomas were observed in the female rats at incidences of 0/31, 10/33 (30%), and 11/32 (34%) in controls, low-, and high-dose groups, respectively. Non-neoplastic lesions and hepatocytomegaly associated with exposure to 1,4-dioxane were also observed in the female rat livers (described above). In this same study, groups of B6C3F1 mice (50/sex/dose level) were administered 1,4-dioxane in drinking water for 90 weeks at doses of 0, 720, and 830 mg/kg/day in males, and 0, 380, and 860 mg/kg/day in females (NCI, 1978). A significant increase in the incidence of hepatocellular carcinomas or adenomas was observed in male and female mice exposed to ≥720 and ≥380 mg/kg/day, respectively. The incidences of hepatocellular carcinoma were 2/49 (4%), 18/50 (36%), and 24/47 (51%), in controls, low-, and high-dose males, respectively; the incidences in females were 0/50, 12/48 (25%), and 29/37 (78%) in controls, low-, and high-dose males, respectively. It should also be noted that mortality was significantly increased in the female mice in a dose-dependent manner, with survival rates of 90%, 78%, and only 56% in the control, low-, and high-dose groups, respectively.

JBRC researchers exposed groups of Crj:BDF1 mice (50/sex/dose level) 1,4-dioxane in the drinking water for 2 years at mean doses of 0, 49, 191, 677 mg/kg/day for males and 0, 66, 278, and 964 mg/kg/day for females (Kano et al., 2009). Early mortality attributed to liver tumors occurred in the female mice; survival rates were 29/50, 29/50, 17/50, and 5/50 in control, low-, mid-, and high-dose groups, respectively (as reported in USEPA, 2013). Liver angiectasis, an abnormal dilatation and/or lengthening of a blood or lymphatic vessel, was seen in male mice administered 1,4-dioxane at a dose of 677 mg/kg/day; however, this finding was not reported to be statistically significant

(Kano et al., 2009). A significant and dose-related increase in the incidence of hepatic adenomas and carcinomas was found in male and female mice. The combined adenomas or carcinomas incidence rates in male mice were 23/50, 31/50, 37/50 ($p < 0.01$) and 40/50 ($p < 0.05$). The incidences of combined adenomas or carcinomas in control, low-, mid-, and high-dose females were 5/50, 35/50, 41/50, and 46/50 ($p < 0.01$ for all treated groups).

F344/DuCrj rats (50/sex/dose level) were also exposed to 1,4-dioxane at mean concentrations of 0, 11, 55, and 274 mg/kg/day for male, and 0, 18, 83, or 429 mg/kg/day for female rats (Kano et al., 2009). As discussed above, non-neoplastic lesions in the liver of the high- and mid-dose rats were observed, as well as spongiosis hepatis, and clear and mixed cell foci at ≥55 mg/kg/day in male rats and spongiosis hepatis, cyst formation, and mixed cell foci at 429 mg/kg/day in female rats. These lesions are likely pre-neoplastic changes (Bannasch, 2003; Dourson et al., 2014; Dourson et al., 2017). Significantly increased incidences of liver tumors (adenomas and carcinomas) occurred in high-dose male (39/50) and female (48/50) rats ($p < 0.01$).

5.3.3.2 Nasal Cavity Tumors

Tumors in the respiratory tract, including nasal cavity, have been reported in rats chronically exposed to 1,4-dioxane in drinking water (Argus et al., 1965; Kociba et al., 1974; NCI, 1978; Kano et al., 2009). Kociba et al. (1974) reported an increased incidence of cholangiomas and squamous cell carcinomas of the nasal turbinates in male (n =1) and female (n = 2) Sherman rats exposed to the high dose of 1,015 and 1,500 mg/kg/day, respectively ($p < 0.05$ for combined male and female incidence). NCI (1978) reported a significant increase in nasal cavity squamous cell carcinoma and adenocarcinomas of the nasal cavity in male and female Osborne-Mendel rats exposed to 530 (male) and ≥350 mg/kg/day (female); however, a reexamination of the nasal tissue sections from this study determined that the tumors likely resulted from inhalation of water droplets, based on the location of the tumors (Goldsworthy et al., 1991). There were no nasal or respiratory tumors reported in the mouse portion of this study that were attributable to 1,4-dioxane exposure by the study authors (NCI, 1978).

Finally, JBRC researchers attributed the significant decrease in survival of male and female F344/DuCrj rats exposed to 274 and 429 mg/kg/day, respectively, in part, to the nasal squamous cell carcinomas observed (incidence rates were 3/50 for males and 7/50 for females) (Kano et al., 2009). Microscopic examination of the tissues showed non-neoplastic alterations in the nasal cavity mainly in high-dose rats. Alterations in high-dose (274 mg/kg/day) male rats consisted of nuclear enlargement and metaplasia of the olfactory and respiratory epithelia, atrophy of the olfactory epithelium, hydropic changes and sclerosis of the lamina propria, adhesion, and inflammation. In female rats, nuclear enlargement of the olfactory epithelium occurred at doses ≥83 mg/kg/day, and nuclear enlargement and metaplasia of the respiratory epithelium, squamous cell hyperplasia, respiratory metaplasia of the olfactory epithelium, hydropic changes and sclerosis of the lamina propria, adhesion, inflammation, and proliferation of the nasal gland occurred at 429 mg/kg/day. Metaplasia and hyperplasia of the nasal epithelium were also observed in high-dose male and female rats. USEPA (2013) identified the NOAEL and LOAEL values for rats in this study as 55 and 274 mg/kg/day, respectively, based on toxicity observed in nasal tissue of male rats. USEPA determined that these effects are preneoplastic and likely to be associated with the formation of nasal cavity tumors in these dose groups. No nasal or respiratory tumors were reported in the Crj:BDF1 mice component of this study.

5.3.4 Cancer Mode of Action

A chemical's cancer MoA is a postulated sequence of key events, starting with a chemical's interaction with a cell, that leads to chemical-mediated tumor induction (Sonich-Mullin et al. 2001; Boobis et al., 2006). Elucidating the MoA often requires combining *in vitro* mechanistic data with animal and/or epidemiologic studies to establish a biologically plausible explanation of the "key events" in chemical-induced tumor development (USEPA, 2005). Key events are the "measurable events that are critical to the induction of tumors (Sonich-Mullin et al., 2001), based on experimental and empirical results. The MoA is different than the mechanism of action, which is a more detailed understanding of each molecular step along the pathway from internal exposure to tumor development. Identifying the MoA for tumor development in an animal is a key component in the process of human health risk assessment as it contributes to the overall conclusion of human relevancy of the animal findings to humans (Sonich-Mullin et al., 2001; Boobis et al., 2006). For chemical carcinogens, the MoA is also a critical component informing the quantitative assessment of human health risk (USEPA, 2005). In the absence of "sufficiently, scientifically justifiable MoA information" the USEPA uses default assumptions regarding the interpretation of toxicological data for dose-response, which most often result in more stringent risk assessments (USEPA, 2005).

The MoA for 1,4-dioxane-induced tumors is of critical importance for the quantitative assessment of human health risk and is a debated topic of research. 1,4-Dioxane exposure has been shown to cause tumors in multiple sites; however, the two tumor types of primary importance identified from the comprehensive rodent cancer bioassays are nasal and hepatic tumors. Carcinogenic effects appear to be the most sever in the liver.

The nasal tumor formation in rats exposed to 1,4-dioxane via drinking water may be due to the breathing of water into the nasal cavity during drinking, resulting in exposure directly to nasal tissue and subsequent cytotoxicity (Reitz et al., 1990; Goldsworthy et al., 1991; Stickney et al., 2003; Sweeney et al., 2008). Additionally, Dourson et al. (2014) note that all but one of the tumor types for nasal tumors occur at sites with a high historical background incidence. As such, nasal tumors could be related to the rat nasal anatomy and experimental methodology and not relevant to environmental human health risk. However, in a 2-year inhalation study in rats (Kasai et al., 2009), preneoplastic and neoplastic lesions in rat nasal cavity were observed; therefore, the human relevance of nasal tumor induced by 1,4-dioxane is uncertain. The USEPA (2013) concluded that "there is insufficient biological support for potential key events and to have reasonable confidence in the sequence of events and how they related to the development of nasal tumors following exposure to 1,4-dioxane." Therefore, the agency concluded that nasal tumors may be of relevance to human health. However, there are insufficient data to support the development of a MoA for 1,4-dioxane-mediated nasal tumors in rodents.

The hepatic tumors identified in numerous chronic animal bioassays have two proposed modes of action (Figure 5.2): (1) genotoxicity leading to tumor initiation and promotion or (2) hepatic cytotoxicity leading to mitosis and regenerative hyperplasia. Both proposed modes of action would be relevant to human health; however, they result in different dose-response analysis methodologies, which significantly impacts the quantitative assessment of 1,4-dioxane's cancer risk. Each proposed MoA is discussed further below.

5.3.4.1 Genotoxicity

Genotoxic compounds are those that damage the genetic information within a cell, often causing mutations or other direct DNA insult. The potential genotoxicity of 1,4-dioxane has been investigated in many *in vitro* and *in vivo* assays (Table 5.2). Weight-of-evidence evaluations of the genotoxicity database have concluded that 1,4-dioxane is not genotoxic (NICNAS, 1998; Netherlands Organization for Applied Scientific Research, 1999; ECJRC, 2002; Health Canada, 2010, 2018), is only weakly genotoxic (Kitchin and Brown, 1990; Ashby, 1994), or that both options may be possible and data are inconclusive (ATSDR, 2012; USEPA, 2013).

Three *in vivo* studies have evaluated whether 1,4-dioxane is a tumor promotor or initiator (King et al., 1973; Bull et al., 1986; Lundberg et al., 1987). In the earliest analysis, King et al. (1973) evaluated 1,4-dioxane as a tumor initiator and promoter in mouse skin. To test for tumor initiation, 1,4-dioxane was applied in acetone to shaved skin on the back of Swiss Webster mice three times a week for 78 weeks. To test if 1,4-dioxane can promote tumor growth, each animal was also treated with dimethylbenzanthracene. 1,4-Dioxane was unable to initiate or promote tumor development on its own via dermal exposure. 1,4-Dioxane treatment in the presence of the

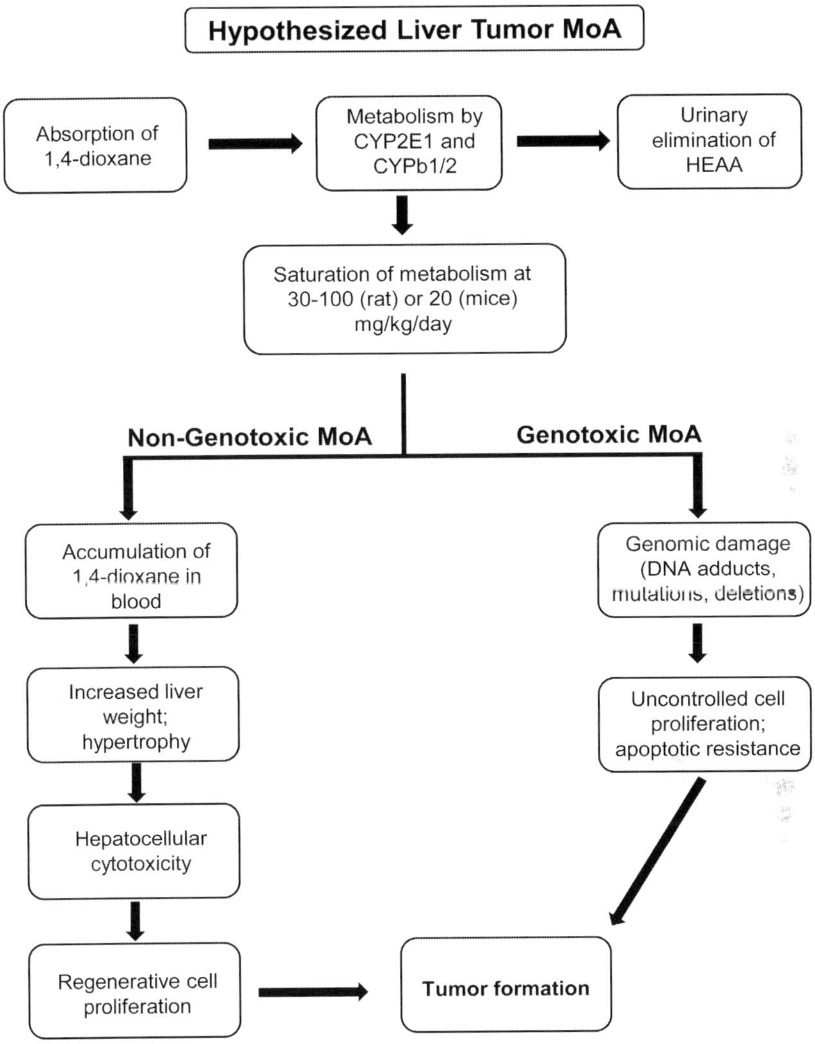

FIGURE 5.2 Proposed modes of action for 1,4-dioxane-mediated hepatic tumors (Original artwork, J. Anderson, 2019).

mutagen increased the percentage of mice with skin tumors after 10 weeks of exposure, indicating evidence of tumor promotion ability following dermal exposure. Bull et al. (1986) also tested the ability of 1,4-dioxane to promote tumor development in mice using oral, subcutaneous, and topical routes of exposure and obtained the opposite finding. Female SENCAR mice were administered a single dose of 1,4-dioxane by gavage, subcutaneous injection, or topical administration. Two weeks later, 12-O-tetradecanoylphorbol-13-acetate (TPA) was applied to the shaved back of the mice three times a week for 20 weeks. In this study, 1,4-dioxane did not increase the incidence of papillomas compared to mice with TPA alone, by all three exposure routes, including dermal application. Finally, Lundberg et al. (1987) also tested the ability of 1,4-dioxane to promote tumor development in rat livers, specifically. Male Sprague-Dawley rats underwent a partial hepatectomy followed by an intraperitoneal injection of diethylnitrosamine (DEN) as an initiator. 1,4-Dioxane was then administered daily by gavage at 0, 100, or 1,000 mg/kg/day for 5 days/week for 7 weeks. Only the high dose of 1,4-dioxane administered as a promoting agent produced an increase in the number of γ-glutamyl trans-peptidase (GGT)-foci and total foci volume; however, there were also

TABLE 5.2
1,4-Dioxane Genotoxicity Studies

Test System	Endpoint	Test Conditions	Activation: Without	Activation: With	References
Bacteria					
Salmonella typhimurium strains TA98, TA100, TA1535, and TA1537	Reverse mutation	Plate incorporation assay	–	–	Haworth et al. 1983
Salmonella typhimurium strains TA98, TAIOO, TA 1530, TA1535, and TA1537	Reverse mutation	Plate incorporation assay	–	–	Khudoley et al. 1987
Salmonella typhimurium strains TA98, TAIOO, TA1535, and TA1537	Reverse mutation	Plate incorporation and preincubation assays	–	–	Morita and Hayashi, 1998
Salmonella typhimurium strains TAIOO, TAI535	Reverse mutation	Preincubation assay	–	–	Nestmann et al. 1984
Salmonella typhimurium strains TA98, TA100, TA1535, TA1537, and TA1538	Reverse mutation	Plate incorporation assay	–	–	Stott et al. 1981
E. coli K-12uvrB/recA	DNA repair	Host-mediated assay	–	–	Hellmer and Bolcsfoldi, 1992
E. coli WP2/WP2uvrA	Reverse mutation	Plate incorporation and preincubation assays	–	–	Morita and Hayashi, 1998
P. phosphoreum M169	Mutagenicity, DNA damage	Mutatox assay	–	ND	Kwan et al. 1990
Yeast/Drosophila					
S. cerevisiae D61.M	Aneuploidy	Standard 16-hour incubation or cold-interruption regimen	– T	ND	Zimmermann et al. 1985
D. melanogaster	Meiotic nondisjunction	Oocytes were obtained for evaluation 24 and 48 hour after mating	+ T	ND	Muñoz and Barnett, 2002
D. melanogaster	Sex-linked recessive lethal test	Exposure by feeding and ingestion	–	ND	Yoon et al. 1985
Mammalian Cells					
L5178Y mouse lymphoma cells	Forward mutation assay	Thymidine kinase mutagenicity assay (trifluorothymidine resistance)	–	–	McGregor et al. 1991
L5178Y mouse lymphoma cells	Forward mutation assay	Thymidine kinase mutagenicity assay (trifluorothymidine resistance)	–	–T	Morita and Hayashi, 1998
Rat hepatocytes	DNA damage; single-strand breaks measured by alkaline elution	3-hour exposure to isolated primary hepatocytes	+ T	ND	Sina et al. 1983
Primary hepatocyte culture from male F344 rats	DNA repair	Autoradiography	–	ND	Goldsworthy et al. 1991

(*Continued*)

TABLE 5.2 ***(Continued)***
1,4-Dioxane Genotoxicity Studies

Test System	Endpoint	Test Conditions	Activation		References
			Without	With	
Calf thymus DNA	Covalent binding to DNA	Incubation with microsomes from 3-methylcholanthrene-treated rats	–	–	Woo et al. 1977a
B ALB/3 T3 cells	Cell transformation	48-hour exposure followed by 4-week incubation; 13-day exposure followed by 2.5-week incubation	+ T	ND	Sheu et al. 1988
CHO cells	Sister chromatid exchange	5-bromo-2'-deoxyuridine (BrdU) was added 2 hour after 1,4-dioxane addition; chemical treatment was 2 hour with S9 and 25 hour without S9	±	–	Galloway et al. 1987
CHO cells	Chromosomal aberration	Cells were harvested 8–12 hour or 18–26 hour after treatment (time of first mitosis)	–	–	Galloway et al. 1987
CHO cells	Sister chromatid exchange	3-hour pulse treatment; followed by continuous treatment of BrdU for 23 or 26 hour	–	–	Morita and Hayashi, 1998
CHO cells	Chromosomal aberration	5-hour pulse treatment, 20-hour pulse and continuous treatments, or 44-hour continuous treatment; cells were harvested 20 or 44 hours following exposure	–	–	Morita and Hayashi, 1998
CHO cells	Micronucleus formation	5-hour pulse treatment or 44-hour continuous treatment; cells were harvested 42 hour following exposure	–	–	Morita and Hayashi, 1998
Laboratory Animal Studies					
Male F344 rat	DNA repair in hepatocytes (autoradiography)	Gavage and drinking-water exposure; thymidine incorporation	–	ND	Goldsworthy et al. 1991
Male F344 rat	Replicative DNA synthesis (i.e., cell proliferation) in hepatocytes	Gavage and drinking-water exposure; thymidine incorporation	+(a)	ND	Goldsworthy et al. 1991
Male F344 rat	DNA repair in nasal epithelial cells from the nasoturbinate or maxilloturbinate	Gavage and drinking-water exposure; thymidine incorporation	–	ND	Goldsworthy et al. 1991

(Continued)

TABLE 5.2 *(Continued)*
1,4-Dioxane Genotoxicity Studies

Test System	Endpoint	Test Conditions	Activation Without	Activation With	References
Male F344 rat	Replicative DNA synthesis (i.e., cell proliferation) in nasal epithelial cells	Drinking-water exposure; thymidine incorporation	–	ND	Goldsworthy et al. 1991
Female Sprague-Dawley rat	DNA damage; single-strand breaks measured by alkaline elution	Two gavage doses given 21 and 4 hour prior to sacrifice	+	ND	Kitchin and Brown 1990
Male Sprague-Dawley rat	RNA synthesis; inhibition of RNA polymerase A and B	i.v. injection; activity measured in isolated hepatocytes	+	ND	Kurl et al. 1981
Male F344 rat	DNA synthesis in hepatocytes	Gavage; thymidine and BrdU incorporation	+	ND	Miyagawa et al. 1999
Male F344 rat	DNA synthesis in hepatocytes	Thymidine incorporation	±	ND	Uno et al. 1994
Male Sprague-Dawley rat	DNA synthesis in hepatocytes	Drinking water; thymidine incorporation	+	ND	Stott et al. 1981
Male Sprague-Dawley rat	DNA repair in hepatocytes	Drinking water; thymidine incorporation with hydroxyurea to repress normal DNA synthesis	–	ND	Stott et al. 1981
Male Sprague-Dawley rat	DNA alkylation in hepatocytes	Gavage; DNA isolation and HPLC analysis 4 hour after dosing	–	ND	Stott et al. 1981
Male B6C3F! mouse	Micronucleus formation in bone marrow	i.p. injection; analysis of polychromatic erythrocytes 24 or 48 hour after dosing	–	ND	McFee et al. 1994
Male and female C57BL6 mice; male BALB/c mouse	Micronucleus formation in bone marrow	Gavage; analysis of polychromatic erythrocytes 24 or 48 hour after dosing	+(b)	ND	Mirkova, 1994
Male CD-I mouse	Micronucleus formation in peripheral blood	Two i.p. injections (one per day); micronucleated reticulocytes measured 24, 48, and 72 hour after the second dose	–	ND	Morita, 1994
Male CD-I mouse	Micronucleus formation in hepatocytes	Gavage; partial hepatectomy 24 hour after dosing; hepatocytes analyzed five days after hepatectomy	+	ND	Morita and Hayashi, 1998
Male CD-I mouse	Micronucleus formation in peripheral blood	Gavage; partial hepatectomy 24 hour after dosing; peripheral blood obtained from tail vein 24 hour after hepatectomy	–	ND	Morita and Hayashi, 1998

(Continued)

TABLE 5.2 *(Continued)*
1,4-Dioxane Genotoxicity Studies

Test System	Endpoint	Test Conditions	Activation Without	Activation With	References
Male CD-I mouse	Micronucleus formation in hepatocytes and bone marrow	Gavage; analysis of bone marrow erythrocytes and proliferating BrdU-labeled hepatocytes	+	ND	Roy et al. 2005
Male CBA and C57BL6 mice	Micronucleus formation in bone marrow	Gavage; analysis of polychromatic erythrocytes from specimens prepared 24 hour after dosing	–	ND	Tinwell and Ashby, 1994
gpt delta transgenic F344 rats	*gpt* assay for mutagenesis	Drinking-water exposure; hepatocytes analyzed after 16 weeks of exposure	+(c)	ND	Gi et al. 2018
Male F344/DuCrlCrlj	Micronucleus formation in hepatocytes	Juvenile rat method; gavage dosing 2 days in a row; hepatocytes analyzed 4 days after dosing	+	ND	Itoh and Hattori, 2019
Male F344/DuCrlCrlj	Micronucleus formation in hepatocytes	Gavage; dosing 24 hour after partial hepatectomy; hepatocytes analyzed 3 days after dosing	+	ND	Itoh and Hattori, 2019
Male F344/DuCrlCrlj	Micronucleus formation in hepatocytes	Gavage; hepatectomy 24 hour after dosing; hepatocytes analyzed 5 days after dosing	+	ND	Itoh and Hattori, 2019
Male F344/DuCrlCrlj	Micronucleus formation in bone marrow	Gavage; bone marrow analyzed 24 and 48 hour after dosing	–	ND	Itoh and Hattori, 2019
Male F344/DuCrlCrlj	Pig-α assay for mutagenicity in peripheral blood	Gavage; peripheral blood analyzed 15 and 30 days after dosing	–	ND	Itoh and Hattori, 2019

Notes: +, positive; ±, weak positive; –,negative; T, toxicity; ND, no data; (a) 1–2-week exposure, (b) (C57BL6)-(BALB/c), (c) highest dose only; Table modified and updated based on Stickney (2010) Table 5.4.

increased histopathological changes within the treated rat livers at this high dose, compared to controls, suggesting that the 1,4-dioxane was cytotoxic at this level.

There is a plethora of available *in vitro* and *in vivo* assays that test for genotoxic potential of a chemical. As summarized in Table 5.2 and discussed in detail in ATSDR (2012) and USEPA (2013), all standard *in vitro* tests for mutagenicity with 1,4-dioxane were negative, including bacterial, yeast, and mammalian cell assays with or without metabolic activation. An *in vitro* study in *Saccharomyces cerevisiae* also gave negative results for chromosomal aneuploidy without activation (Zimmermann et al., 1985), but was not tested in the presence of metabolic activation. All standard studies (assays for induction of micronuclei, sister chromatid exchanges, and chromosomal aberrations in Chinese hamster ovary cells (CHO)) for chromosomal stability in isolated mammalian cells exposed to 1,4-dioxane have also been negative, except for a slight increase in the incidence of sister chromatic exchange in the absence of activation in one study of chromosomal aberrations in CHO cells (Galloway et al., 1987). 1,4-Dioxane was also found to increase the number of cellular transformations in BALB/3t3 cells; however, this occurred at cytotoxic concentrations

(Sheu et al., 1988). Similarly, 1,4-dioxane was positive for the induction of DNA single strand breaks in rat hepatocytes; however, only at cytotoxic concentrations (Sina et al., 1983).

Studies of the genotoxicity of 1,4-dioxane *in vivo* also have been mostly negative, although some positive results have been reported (see Table 5.2). Micronuclei formation in mouse bone marrow cells following exposure to a chemical is a standard assay for genotoxicity; however, several studies conducted for 1,4-dioxane reveal inconsistent results. C57BL6 mice exposed to a single gavage dose of 1,4-dioxane exhibited a dose-dependent increased in micronuclei; however, exposed BALB/c mice were negative in the same study (Mirkova, 1994). A separate study of similar design also reported negative results for 1,4-dioxane-induced micronuclei in CBA mice and C57BL6 mice (Tinwell and Ashby, 1994); however, equivocal results were found by McFee et al. (1994). Morita and Hayashi (1998) treated CD-1 mice with a single gavage dose of 1,4-dioxane and showed an increase in micronuclei in hepatocytes, but not in peripheral blood reticulocytes. In a repeat exposure study looking at micronuclei formation, Roy et al. (2005) reported an increased incidence of micronuclei following 5 days of oral (gavage) exposure to 1,4-dioxane in both hepatocytes and marrow in male CD-1 mice. Itoh and Hattori (2019) evaluated liver and bone marrow micronuclei formation in male F344/DuCrlCrlj rats exposed to 1,4-dioxane at 1,000, 2,000, and 3,000 mg/kg for several (4 to 6) days, with and without partial hepatectomy. Pig-α gene mutations in peripheral blood were also evaluated. The authors report an increase in the frequency of micronuclear hepatocytes only, consistent with structural chromosome aberrations occurring in the liver, but not in bone marrow (i.e., no bone marrow micronuclear cells or Pig-α gene mutations were observed at any dose level) (Itoh and Hattori, 2019). Gi et al. (2018) exposed *gpt* delta transgenic F344 rats to 0, 2, 20, 200, 2,000, or 5,000 ppm 1,4-dioxane in drinking water for 16 weeks and examined the overall mutation frequency, A:T to G:C transition and A:T to T:A transversions. Signs of mutations, transitions, and/or transversions were seen at the high doses and authors conclude that 1,4-dioxane exposure caused these effects via genetic damage (Gi et al., 2018).

Other *in vivo* studies have looked for signs of DNA repair or alkylation following 1,4-dioxane exposure in rats, and mixed results are also evident. In summary, although all standard *in vitro* tests for genotoxicity are negative without cytotoxicity and most *in vivo* tests have been negative, a few positive results have been reported in animals treated with high levels of 1,4-dioxane. The available data present mixed results.

5.3.4.2 Cytotoxicity Followed by Regenerative Hyperplasia

Recent data and reevaluations of previous cancer bioassays have led to an alternative proposed carcinogenic MoA for 1,4-dioxane, involving hepatic cytotoxicity leading to enhanced cell cycle progression and regenerative hyperplasia, which leads to the promotion of endogenously formed mutagenic cells (Dourson et al., 2014, 2017). This is a non-mutagenic MoA based on a threshold dose that would support a nonlinear dose-response assessment as opposed to the linear default dose response risk assessment used by the USEPA. The nonlinear and threshold dose response determines a dose of a chemical below which adverse effects are not occurring due to protective cellular mechanisms. At or above this threshold concentration, cells are not able to compensate for the cellular injury and adverse effects are expected. This is in contrast to the low-dose linear dose-response modeling, which assumes that any incremental increase in dose above zero would increase an effect.

As discussed above, Dourson et al., 2014 had the liver slides from the NCI (1978) study re-reviewed using today's more refined and updated understanding of cancer biology and pathological practices/terminology. The purpose of the reevaluation was to account for differences in histologic approaches when quantifying and reporting non-neoplastic changes in animal bioassays, namely that when adenomas or carcinomas were observed in the tissue histopathology, no additional non-neoplastic evaluations were conducted or recorded. In addition to the peer-reviewed publication, this work was presented and reviewed by an expert panel of toxicologist and risk assessors through the Alliance for Risk Assessment. The re-review of the 1978 slides revealed hepatic non-cancer toxicity at all doses, as evident by an increase incidence and severity in many non-neoplastic endpoints, including depletion of hepatocellular glycogen, hepatocellular hypertrophy, necrosis, inflammation, Kupffer cell hyperplasia, bile duct hyperplasia, basophilic, eosinophilic, clear cell and mixed cell foci, and increased incidence of various types of neoplasms. In follow-up work, Dourson and colleagues conducted a detailed evaluation of the findings reported in the second long-term oral mouse bioassay, and the original Japanese rat and mouse bioassays (JBRC, 1998; Kano, 2008, 2009). Dourson et al. then incorporated multiple lines of evidence to determine whether a regenerative hyperplasia MoA for hepatic tumor formation is supported. The authors conclude that the study reports from JBRC (1998) confirm information found in the publications of Kano et al. (2008, 2009), and add additional information relevant to the hypothesized MoA of cytotoxicity and regenerative hyperplasia. Dourson et al. (2017) conclude that the sequence of key MoA events from the subchronic and chronic mouse studies reporting hepatic tumors "generally support" the regenerative hyperplasia MoA; extensive toxicity is seen within the liver of dosed animals, suggesting a regenerative cell proliferation MoA. This proposed MoA is also supported by positive findings in promotion bioassays and DNA replication bioassays. Furthermore, dose response and temporal concordance for non-cancer precursors to liver tumors are evident for rats and generally supportive for mice (Dourson et al., 2017). Utilizing these recent data and updated analyses, Health Canada evaluated the regenerative proliferation MoA based on the modified Bradford Hill criteria (Meek et al., 2014) and concluded that the weight of evidence supports the conclusions that there is a dose-related non-neoplastic response in the liver following 1,4-dioxane exposure, consisting of a hypertrophic response of hepatocytes, followed by necrosis, inflammation, and hyperplastic hepatocellular foci, all of which, per USEPA guidelines (USEPA, 2005) and standard best risk assessment practices support a nonlinear approach (Health Canada, 2018).

5.3.4.3 Differing Mode of Action Conclusions Still Exist, Resulting in Different Human Health Threshold Levels

USEPA (2013) concluded that the available literature indicates that 1,4-dioxane may induce cancer via several modes of action including regenerative hyperplasia or genotoxicity, but available evidence was inadequate to establish a MoA. Following the USEPA Cancer Guidelines (USEPA, 2005), USEPA determined that for the liver specifically, "The available evidence in support of any hypothesized MoA for liver tumors is not conclusive" (USEPA, 2013). For nasal tumors, they state that "there is insufficient biological support for potential key events and to have reasonable confidence in the sequence of events and how they related to the development of nasal tumors following exposure to 1,4-dioxane." Therefore, they concluded that no single MoA can account for the formation of liver, nasal, or other tumors seen in laboratory animals exposed to 1,4-dioxane, and the default and most stringent assumption of low-dose linearity was used to quantify cancer risk (USEPA, 2013). The resulting human health cancer toxicity value, the oral cancer slope factor, of 0.10 $(mg/kg/day)^{-1}$ was based on liver tumors in female mice from Kano et al. (2009) was derived, which when applied to a default drinking water threshold equation, results in a drinking water health advisory of 0.35 μg/L to 35 μg /L for 10^{-6} to 10^{-4} risk, respectively (USEPA, 2013). Conversely, using the same available data, the Government of Australia, Health Canada, and the European Commission concluded that the total weight of evidence indicates that 1,4-dioxane is a non-genotoxic compound (NICNAS, 1998; ECJRC, 2002; Health Canada, 2010).

As described above, Dourson et al. (2014, 2017) have provided new data to support the proposed cytotoxicity following by regenerative hyperplasia MoA. These new data were not previously available for USEPA or ATSDR when they conducted their analysis; however, Health Canada considered these new data in their most recent reassessment (Health Canada, 2018) and uses these data to strengthen their assertion that 1,4-dioxane exposure leads to tumor development via a MoA involving cytotoxicity and regenerative hyperplasia with a threshold dose-response. This conclusion results in an oral cancer toxicity value (tolerable daily intake) of 0.0054 mg/kg/day and a health-based threshold level for human consumption of drinking water of 50 μg/L. Conversely, a review of the initial Dourson et al. (2014) publication by Michigan Department of Environmental Quality (MDEQ) reveals that some agencies remain uncertain about the hypothesized non-genotoxic MoA for 1,4-dioxane and are unwilling to move away from the USEPA cancer assessment and slope factor. Specifically, concerns include: (a) whether the observed foci of phenotypically altered hepatocytes are pre-neoplastic and therefore, should not be classified as regenerative hyperplasia, (b) whether spongiosis hepatis is pre-neoplastic, and (c) concerns regarding the conflicting information from the NCI (1978) female mice. MoAs for other observed rodent tumors have not been developed.

5.4 CONCLUSIONS – HUMAN HEALTH TOXICITY

Epidemiological studies for 1,4-dioxane are available only for acute and high-dose occupational studies and/or accidental poisonings and are therefore of limited value for the assessment of human environmental exposure to 1,4-dioxane contaminated media. In experimental animals, 1,4-dioxane from long-term oral exposures predominately causes tumors in the liver and nasal cavity, and non-cancer adverse effects occur at higher doses in the liver, kidney, and nasal/respiratory tract. Human environmental exposures to 1,4-dioxane are generally unlikely to reach the high doses necessary to produce the tumors evident in rodent studies.

5.5 ECOTOXICOLOGY

This section summarizes available information on the toxicity of 1,4-dioxane to aquatic and terrestrial ecological receptors. An ecological conceptual site model (CSM) that evaluates the exposure potential of 1,4-dioxane to ecological receptors is discussed in Chapter 7.

5.5.1 Toxicity of 1,4-Dioxane to Aquatic Organisms

1,4-Dioxane aquatic toxicity data are available for aquatic plants (algae), bacteria and protozoa, aquatic invertebrates, and fish. Table 5.3 lists the abbreviations of common toxicity endpoints discussed in this section. Aquatic toxicity data tabulated in this section were obtained primarily from secondary sources (BUA, 1991; NICNAS, 1998; ECJRC, 2002; CCME, 2008; Health Canada, 2010; USEPA, 2015a, 2018d). Citations of original studies are provided for the reader's convenience but were generally not researched for this chapter.

TABLE 5.3
Abbreviations of Common Toxicity Endpoints/Parameters for Ecological Receptors

EC_0	Concentration causing an effect in 0% of the test organisms
EC_{50}	Concentration causing an effect in 50% of the test organisms
EC_{100}	Concentration causing an effect in 100% of the test organisms
ESL	Ecological screening level
ESV	Ecological screening value
LC_0	Concentration lethal to 0% of the test organisms
LC_{50}	Concentration lethal to 50% of the test organisms
LC_{100}	Concentration lethal to 100% of the test organisms
LD_{50}	Lethal dose for 50% of the test population
LOAEL	Lowest observed adverse effect level
LOEC	Lowest observed effect concentration
MATC	Maximum acceptable toxicant concentration
NOAEL	No observed adverse effect level
NOEC	No observed effect concentration
PNEC	Predicted no effect concentration
TRV	Toxicity reference value
UF	Uncertainty (safety) factor

TABLE 5.4
Toxicity of 1,4-Dioxane to Freshwater Aquatic Plants

Test Species	Duration/ Endpoint	Concentration (mg/L)	Effect	References
Subchronic Studies				
Pseudokirchnerella subcapitata (green algae)	72-hour NOEC	580	Growth rate	USEPA (2015a)
Pseudokirchnerella subcapitata	72-hour NOEC	1,000	Biomass	USEPA (2015a)
Chronic Studies				
Microcystis aeruginosa (blue-green algae)	8-day toxic threshold	575	Population growth rate	Bringmann and Kühn (1976, 1978)
Scenedesmus quadricauda (green algae)	8-day toxic threshold	5,600	Population growth rate	Bringmann and Kühn (1977b, 1978)

1,4-Dioxane is not highly toxic to aquatic organisms. The aquatic bioassay results tabulated in this section show that 1,4-dioxane is acutely toxic to bacteria, algae, invertebrates, and fish at concentrations ranging from hundreds to tens of thousands of milligrams per liter (mg/L; ppm).

5.5.1.1 Aquatic Plants

Four subchronic or chronic values are available for aquatic plants (algae); all are no observed effect concentrations (NOECs) for freshwater species. These NOECs range from 575 to 5,600 mg/L (Table 5.4). These data indicate that the toxicity of 1,4-dioxane from water-based exposures to aquatic plants is low.

5.5.1.2 Microorganisms and Protozoa

Eight values are available for microorganisms and protozoa. The five NOECs range from 2,700 to >10,000 mg/L, while the three EC_{50} values range from 6,000 to 16,000 mg/L (Table 5.5). These data indicate that the toxicity of 1,4-dioxane from water-based exposures to microorganisms and protozoa is very low.

TABLE 5.5
Toxicity of 1,4-Dioxane to Microorganisms/Protozoa

Test Species	Duration/ Endpoint	Concentration (mg/L)	Effect	References
Pseudomonas fluorescens (bacterium)	16-hour NOEC	2,700	glucose assimilation	Bringmann (1973)
Pseudomonas putida (bacterium)	16-hour NOEC	2,700	cell multiplication	Bringmann and Kühn (1977b)
Entosiphon sulcatum (protozoan)	72-hour NOEC	5,340	cell multiplication	Bringmann (1978)
Uronema parduczi (ciliate)	20-hour NOEC	5,620	cell multiplication	Bringmann and Kühn (1980)
Photobacterium phosphoreum (bacterium)	EC_{50}	6,000	luminescence reduction	Thomulka and McGee (1992)
Vibrio harveyi (bacterium)	EC_{50}	6,500	luminescence reduction	Thomulka and McGee (1992)
Chilomonas paramecium (cryptomonad)	48-hour NOEC	>10,000	cell multiplication	Bringmann et al. (1980)
Vibrio harveyi	EC_{50}	16,000	luminescence reduction	Thomulka and McGee (1992)

TABLE 5.6
Toxicity of 1,4-Dioxane to Freshwater Aquatic Invertebrates

Test Species	Duration/ Endpoint	Concentration (mg/L)	Effect	References
Acute Studies				
Ceriodaphnia dubia (water flea)	48-hour EC_{50}	163	immobilization	Dow (1989a, 1989b, 1995)
Ceriodaphnia dubia	24-hour EC_{50}	299	not specified	Dow (1989a, 1989b, 1995)
Daphnia magna (water flea)	24-hour EC_0	2,070	intoxication	Bringmann and Kühn (1977a)
Gammarus pseudolimnaeus (amphipod)	48-hour EC_{50}	2,274	not specified	Dow (1989b, 1995)
Daphnia magna	24-hour EC_{50}	4,700	immobilization	Bringmann and Kühn (1977a)
Daphnia magna	48-hour EC_{50}	5,500	not specified	USEPA (1996a)
Daphnia magna	24-hour EC_0	6,210	behavior	Bringmann and Kühn (1982)
Daphnia magna	24-hour EC_{50}	8,450	behavior/ immobilization	Bringmann and Kühn (1982)
Daphnia magna	24-hour EC_{100}	10,000	intoxication	Bringmann and Kühn (1977a)
Daphnia magna	24-hour EC_{100}	10,000	behavior	Bringmann and Kühn (1982)
Aedes agyptii (mosquito larvae)	4-hour LC_{50}	40,000	mortality	Kramer et al. (1983)
Subchronic Studies				
Gammarus pseudolimnaeus	96-hour LC_{50}	2,274	mortality	Brooke (1987)
Chronic Studies				
Ceriodaphnia dubia	7-day NOEC	625	not specified	Dow (1995); Springborn (1989)
Daphnia magna	21-day NOEC	>1,000	survival, growth, reproduction	USEPA (2015a)
Ceriodaphnia dubia	7-day LOEC	1,250	not specified	Dow (1995)

5.5.1.3 Aquatic Invertebrates

Eleven acute and four subchronic or chronic values are available for aquatic invertebrates (Table 5.6), all of them for freshwater species. Of the acute values, EC_0 values (NOECs) range from 2,070 to 6,210 mg/L, while EC_{50} values range from 163 to 8,450 mg/L; there are two EC_{100} values of 10,000 mg/L and a single LC_{50} value of 40,000 mg/L. Of the subchronic/chronic values, chronic NOECs range from 625 to >1,000 mg/L; there is one chronic lowest observed effect concentration (LOEC) of 1,250 mg/L and one subchronic LC_{50} of 2,274 mg/L.

The overlap in the acute EC_{50} data and the chronic NOEC data is due to two very low acute EC_{50} values for the water flea (*Ceriodaphnia dubia*) of 163 and 299 mg/L, which are inconsistent with the remainder of the data set. The next lowest acute EC_{50} is 2,274 mg/L, which is in line with the other available acute and chronic data. Since the original test report for these two values could not be obtained, most of the researchers who cite this study (e.g., ECJRC, 2002; Health Canada, 2010) consider these values to be anomalous and do not use them to derive a predicted no effect concentration (PNEC) or comparable value. Excluding these two values, the available data indicate that the acute toxicity of 1,4-dioxane from water-based exposures to aquatic invertebrates is very low and the chronic toxicity is low.

5.5.1.4 Fish

The largest database for the aquatic toxicity of 1,4-dioxane to aquatic organisms is for fish; there are 23 acute values (22 for freshwater species and 1 for a saltwater species) and 8 chronic values (all for freshwater species) (Table 5.7). The single saltwater value is a 96-hour LC_{50} for the inland silverside (*Menidia beryllina*) of 6,700 mg/L. Of the freshwater acute values (all but two are based on mortality), LC_0 values (NOECs) range from 6,180 to 6,700 mg/L, LC_{50} values range from 4,269 to 13,000 mg/L, and LC_{100} values range from 9,790 to 15,450 mg/L.

TABLE 5.7
Toxicity of 1,4-Dioxane to Fish

Test Species[a]	Duration/ Endpoint	Concentration (mg/L)	Effect	References
Acute Studies				
Lepomis macrochirus (bluegill sunfish)	48-hour LC_{50} (neonate)	4,269	mortality	Brooke (1987)
Ictalurus punctatus (channel catfish)	96-hour LC_{50}	6,155	mortality	USEPA (1996a)
Leuciscus idus (carp or golden orfe)	48-hour LC_0	6,180	mortality	Juhnke and Ludemann (1978)
Leuciscus idus	48-hour LC_0	6,700	mortality	Juhnke and Ludemann (1978)
Menidia beryllina (inland silverside)[b]	96-hour LC_{50}	6,700	mortality	Dawson et al. (1977)
Oncorhyncus mykiss (rainbow trout)	96-hour LC_{50}	7,961	mortality	USEPA (1996a)
Leuciscus idus	48-hour LC_{50}	8,450	mortality	Juhnke and Ludemann (1978)
Pimephales promelas (fathead minnow)	96 hour EC_{50}	9,340	equilibrium	Geiger et al. (1990)
Pimephales promelas	96-hour EC_{50}	9,550	behavior	Geiger et al. (1990)
Leuciscus idus	48-hour LC_{50}	9,630	mortality	Juhnke and Ludemann (1978)
Leuciscus idus	48-hour LC_{100}	9,790	mortality	Juhnke and Ludemann (1978)
Pimephales promelas	96-hour LC_{50}	9,850	mortality	Geiger et al. (1990)
Pimephales promelas	96-hour LC_{50}	9,872	mortality	Brooke (1987)
Lepomis macrochirus	96-hour LC_{50}	>10,000	mortality	Dawson et al. (1977)
Oryzias latipes (Japanese medaka)	48-hour LC_{50}	10,500	mortality	CITI-Japan (1992)
Pimephales promelas	96-hour LC_{50}	10,800	mortality	Geiger et al. (1990)
Pimephales promelas	96-hour LC_{50}	10,820	mortality	Geiger et al. (1990)
Pimephales promelas	96-hour LC_{50}	11,080	mortality	Geiger et al. (1990)
Pimephales promelas	96-hour LC_{50}	11,260	mortality	Geiger et al. (1990)
Cyprinus carpio (common carp)	48-hour LC_{50}	12,000	mortality	Nishiushi (1981)
Pimephales promelas	96-hour LC_{50}	12,326	mortality	Brooke (1987)
Pimephales promelas	96-hour LC_{50}	13,000	mortality	Dow (1989a)
Leuciscus idus	48-hour LC_{100}	15,450	mortality	Juhnke and Ludemann (1978)
Chronic Studies				
Pimephales promelas (embryo-larval)	32-day NOEC	>103	survival, growth, reproduction	USEPA (2015a); ECJRC (2002)
Pimephales promelas (embryo-larval)	32-day MATC	>145	survival, growth, reproduction	USEPA (2015a); ECJRC (2002)
Oryzias latipes (fry)	28-day NOEC	994	histopathology (kidney)	Johnson et al. (1993)
Oryzias latipes (fry)	28-day LOEC	1,870	histopathology (kidney)	Johnson et al. (1993)
Oryzias latipes (fry)	28-day NOEC	1,870	growth	Johnson et al. (1993)
Oryzias latipes (fry)	28-day LOEC	3,536	growth	Johnson et al. (1993)
Oryzias latipes (fry)	28-day NOEC	3,536	survival	Johnson et al. (1993)
Oryzias latipes (fry)	28-day LOEC	6,933	survival	Johnson et al. (1993)

[a] Freshwater species, except where noted.
[b] Saltwater species.

Of the chronic values, NOECs range from >103 to 3,536 mg/L, while LOECs range from 1,870 to 6,933 mg/L. Overall, the acute toxicity of 1,4-dioxane to fish from water-based exposures is very low and the chronic toxicity is low.

5.5.2 Toxicity of 1,4-Dioxane to Benthic Organisms

The toxicity of 1,4-dioxane to benthic invertebrates (and other sediment-dwelling organisms) from sediment exposures could not be directly evaluated because no toxicity studies evaluating this medium were found in the literature for 1,4-dioxane. Since 1,4-dioxane has a low potential to adsorb to sediments, it is unlikely to accumulate in this medium; therefore, ecological exposures are expected to be low. However, the water-based toxicity data can be used to develop useable toxicity thresholds for sediment pore water and, using equilibrium partitioning models, for bulk sediment. This is addressed in Section 5.5.4.

5.5.3 Toxicity of 1,4-Dioxane to Terrestrial Organisms

Very little data were found evaluating direct exposures of terrestrial organisms to 1,4-dioxane (USEPA, 2018d). What little data that are available (for plants, fungi, and insects) are from studies that used agar or filter paper, not soil, as the exposure medium. Thus, none of these studies are considered relevant for evaluating real-world exposures.

USEPA (2015b) indicates that, in general, the bulk of exposures for lower trophic level terrestrial ecological receptors (such as plants and soil invertebrates) are expected to occur from the soil surface to a depth of 25 to 30 cm (10 to 12 inches). 1,4-Dioxane, if released to soil, is not expected to persist in the surface horizons due to its propensity to volatilize or, more commonly, leach to deeper horizons and groundwater. Thus, ecological exposures for this medium are likely to be very low.

Exposures to terrestrial wildlife through the food chain are not significant with respect to the levels of 1,4-dioxane typically observed in contaminated soils (Huntley et al., 2004). However, several literature sources (e.g., Sample et al., 1996; USEPA, 1999) have developed ingestion-based toxicity reference values (TRVs) for various terrestrial mammalian wildlife species (e.g., meadow vole, short-tailed shrew) using data from laboratory studies with rodents (e.g., laboratory rats and mice). Such data are compiled in sources such as ATSDR (2012) for ingestion exposures; there are no known data for birds, reptiles, or amphibians. While such TRVs, or the laboratory data to derive them, are not compiled as part of this section (since such exposures are not likely to be ecologically significant), soil screening values developed using these TRVs and food chain models are provided in Section 5.5.4

5.5.4 Summary of Available Ecotoxicity Thresholds

Medium-specific ecotoxicity thresholds, often called ecological screening values (ESVs), ecological screening levels (ESLs), or ecological screening benchmarks, have been derived by various researchers using the available ecotoxicology data (in the past [e.g., USEPA, 1999], these values were sometimes called toxicity reference values [TRVs], although this term is now typically used to refer only to dietary-based [not medium-based] toxicity values). These values are generally intended to provide a conservative chronic threshold for the toxicity of a contaminant for a specific exposure medium (such as soil or marine surface water) and all receptors, or for only a specific receptor group (e.g., terrestrial plants).

Available ecotoxicity thresholds for aquatic systems are summarized in Table 5.8, and include values for fresh surface water, freshwater sediment, marine surface water, and marine sediment, as well as for microorganisms. Available ecotoxicity thresholds for terrestrial systems are summarized in Table 5.9 and include values for air and soil.

TABLE 5.8
Chronic Ecological Toxicity Thresholds for 1,4-Dioxane – Aquatic Habitats

Medium	Threshold/ESV	Basis	References
Fresh surface water	22,000 μg/L	Michigan Water Quality Standard (Rule 57); USEPA Regions 4 and 5 ESV/ESL	USEPA (2003, 2018c); MDEQ (2018)
Fresh surface water	36,000 μg/L	Illinois Aquatic Life Criterion (chronic "advisory" value)	Illinois EPA (2013)
Fresh surface water	57,500 μg/L	PNEC: 8-day algae NOEC; UF of 10	ECJRC (2002); Health Canada (2010)
Fresh surface water	58,000 μg/L	Interim Provincial Water Quality Guideline for Ontario – fish acute EC_{50}; UF of 160	CCME (2008)
Fresh surface water	62,100 μg/L	Invertebrate acute EC_0; UF of 100	USEPA (1999)
Freshwater sediment[a]	91.1 μg/kg	Fresh surface water ESV (22,000 μg/L); K_{oc} calculated using log K_{ow} of –0.39 (USEPA, 1995)[b]	–
Freshwater sediment[a]	119 μg/kg	USEPA Region 5 ESL	USEPA (2003)
Freshwater sediment[a]	544 μg/kg	Fresh surface water ESV (62,100 μg/L); K_{oc} of 0.876	USEPA (1999)
Freshwater sediment	43.3 mg/kg (wet weight)	PNEC using equilibrium partitioning	ECJRC (2002)
Marine-Estuarine surface water	67,000 μg/L	Fish LC_{50}; UF of 100	USEPA (1999)
Marine sediment[a]	277 μg/kg	Marine surface water ESV (67,000 μg/L); K_{oc} calculated using log K_{ow} of –0.39 (USEPA, 1995)[b]	–
Marine sediment[a]	587 μg/kg	Marine surface water ESV (67,000 μg/L); K_{oc} of 0.876	USEPA (1999)
Microorganisms	2,700 mg/L	PNEC	ECJRC (2002)

[a] Sediment ESVs are calculated using equilibrium partitioning (EqP), an assumed fraction of organic carbon in sediment (f_{oc}) of 1%, and a K_{oc} estimated from a measured log K_{ow} as follows: $ESV_{sed} = K_{oc} \times f_{oc} \times ESV_{sw}$ (USEPA, 1996b).

[b] K_{oc} calculated as in USEPA (1996b): $K_{oc} = 10^{(0.00028 + 0.983 * \log Kow)}$

TABLE 5.9
Chronic Ecological Toxicity Thresholds for 1,4-Dioxane – Terrestrial Habitats

Medium	Threshold/ESV	Basis	References
Air	367 mg/m^3	USEPA Region 5 ESL	USEPA (2003)
Soil	174 μg/kg	Mammal (domestic sheep) exposures via food chain; used as soil protection value for agricultural land uses	OMOE (2007)
Soil	1,820 μg/kg	Mammal (meadow vole) exposures via food chain; used as soil protection value for residential/parkland and commercial/industrial land uses	OMOE (2007)
Soil	2,050 μg/kg	USEPA Region 5 ESL (mammal [meadow vole] exposures via food chain)	USEPA (2003)
Soil	14,000 μg/kg (wet weight)	PNEC calculated using equilibrium partitioning	ECJRC (2002)
Soil	176,000 μg/kg	Mammal (short-tailed shrew) exposures via food chain	OMOE (2007)
Soil	625,000 μg/kg	Mammal (red fox) exposures via food chain	OMOE (2007)

Various researchers from around the world, including the United States (e.g., USEPA, 2015a), Canada (e.g., Health Canada, 2010), Europe (e.g., ECJRC, 2002), and Australia (e.g., NICNAS, 1998), have concluded that, based on the available ecotoxicological data and measured or predicted environmental concentrations of 1,4-dioxane, this chemical does not pose an unacceptable risk to ecological receptors.

REFERENCES

Agency for Toxic Substances and Disease Registry (ATSDR), 2012, Toxicological profile for 1,4-dioxane. Centers for Disease Control and Prevention (CDC), Agency for Toxic Substance Disease Registry. Atlanta, GA: U.S. Department of Health and Human Services, Public Health Service. April. http://www.atsdr.cdc.gov/toxprofiles/tp.asp?id=955&tid=199.

Argus, M.F., Arcos, J.C., and Hoch-Ligeti, C., 1965, Studies on the carcinogenic activity of protein-denaturing agents: Hepatocarcinogenicity of dioxane. *Journal of the National Cancer Institute* 35: 949–958.

Argus, M.F., Sohal, R.S., Bryant, G.M., Hoch-Ligeti, C., and Arcos, J.C., 1973, Dose-response and ultrastructural alterations in dioxane carcinogenesis: Influence of methylcholanthrene on acute toxicity. *European Journal of Cancer* 9: 237–243.

Ashby, J., 1994, Series: Current issues in mutagenesis and carcinogenesis, No. 45 – The genotoxicity of 1,4-dioxane. *Mutation Research* 322(2): 141–150.

Bannasch, P., 2003, Comments on R. Karbe and R.L. Kerlin (2002). Cystic degeneration/spongiosis hepatis. *Toxicologic Pathology* 30(2): 216–227. *Toxicologic Pathology* 31: 566–570.

Barber, H., 1934, Haemorrhagic nephritis and necrosis of the liver from dioxane poisoning. *Guys Hospital Report* 84: 267–280.

Beratergremium für Umweltrelevante Altstoffe (BUA), 1991, 1,4-Dioxane, BUA report 80. German Chemical Society, Committee on Existing Chemicals of Environmental Relevance, Beratergremium für Umweltrelevante Altstoffe (BUA) [Advisory Committee for Environmentally Relevant Materials] S. Hirzel Wissenschaftliche Verlagsgesellschaft.

Boobis, A.R., Cohen, S.M., Dellarco, V., McGregor, D., Meek, M.E., Vickers, C., and Farland, W. (2006), IPCS framework for analyzing the relevance of a cancer mode of action for humans. *Critical Reviews in Toxicology* 36(10): 781–792.

Braun, W.H. and Young, J.D., 1977, Identification of β-hydroxyethoxyacetic acid as the major urinary metabolite of 1,4-dioxane in the rat. *Toxicology and Applied Pharmacology* 39: 33–38.

Bringmann, G., 1973, Determination of the adverse biological effects of water contaminants from the inhibition of glucose assimilation in Pseudomonas fluorescens [in German]. *Gesundheits-Ingenieur [Sanitation Engineer]* 94(12): 366–369.

Bringmann, G., 1978, Determination of the harmful biological action of water contaminants on protozoa: I. Bacteria fed flagellates. *Zeitschrift für Wasser- und Abwasser-Forschung [Journal of Water and Wastewater Research]* 11: 210–215.

Bringmann, G. and Kühn, R., 1976, Vergleichende Befunde der Schadwirkung wassergefahrdender Stoffe gegen Bakterien (Pseudomonas putida) und Blaualgen (Microcystis aeruginosa) [Comparative results of the adverse effects of aquatic toxins on bacteria (Pseudomonas putida) and blue algae (Microcystis aeruginosa)]. *GWF-Wasser/Abwasser* 117: 410–413.

Bringmann, G. and Kühn, R., 1977a, Findings of the adverse effects of water contaminants on Daphnia magna [in German]. *Zeitschrift für Wasser- und Abwasser-Forschung [Journal of Water and Wastewater Research]* 10: 161–166.

Bringmann, G. and Kühn, R., 1977b, Limiting values of the harmful action of water contaminants on bacteria (Pseudomonas putida) and green algae (Scenedesmus quadricauda) in the cell multiplication inhibition test. *Zeitschrift für Wasser- und Abwasser-Forschung [Journal of Water and Wastewater Research]* 10: 87–98.

Bringmann, G. and Kühn, R., 1978, Testing of substances for their toxicity threshold: Model organisms Microcystis (Diplocystis) aeruginosa and Scenedesmus quadricauda. *Int. Ver. Theor. Angew. Limnol.* 21: 275–284.

Bringmann, G. and Kühn, R., 1980, Determination of the harmful biological action of water contaminants on protozoa: II bacteria fed ciliates. *Zeitschrift für Wasser- und Abwasser-Forschung [Journal of Water and Wastewater Research]* 13: 26–31.

Bringmann, G. and Kühn, R., 1982, Results of the adverse effects of water contaminants against Daphnia magna in an improved standardized test procedure [in German]. *Zeitschrift für Wasser- und Abwasser-Forschung [Journal of Water and Wastewater Research]* 15: 1–6.

Bringmann, G., Kühn, R., and Winter, A., 1980, Determination of the harmful biological action of water contaminants on protozoa: III Saprozoic flagellates. *Zeitschrift für Wasser- und Abwasser-Forschung [Journal of Water and Wastewater Research]* 13: 170–173.

Bronaugh, R.L., 1982, Percutaneous absorption of cosmetic ingredients. In: P. Frost (Ed.), *Principles of Cosmetics for the Dermatologist*, pp. 277–284. St. Louis, MO: C.V. Mosby.

Brooke, L., 1987, Report of the flow-through and static acute test comparisons with fathead minnows and acute tests with an amphipod and a cladoceran. *Center for Lake Superior Environmental Studies*, p. 24. Superior, Wisconsin: University of Wisconsin-Superior.

Buffler, P.A., Wood, S.M., Suarez, L., and Kilian, D.J., 1978, Mortality follow-up of workers exposed to 1,4-dioxane. *Journal of Occupational and Environmental Medicine* 20(4): 255–259.

Bull, R.J., Robinson, M., and Laurie, R.D., 1986, Association of carcinoma yield with early papilloma development in SENCAR mice. *Environmental Health Perspectives* 68: 11–17.

Canadian Council of Ministers of the Environment (CCME), 2008, Canadian water quality guidelines for the protection of aquatic life: 1,4-dioxane. 5 pp.

CITI-Japan, 1992, Biodegradation and bioaccumulation data of existing chemicals based on CSCL Japan. Chemicals Inspection and Testing Institute [cited in ECJRC, 2002].

Dawson, G.W., Jennings, A.L., Drozdowski, D., and Rider, E., 1977, The acute toxicity of 47 industrial chemicals to fresh and saltwater fishes. *Journal of Hazardous Materials* 1: 303–318.

Dourson, M., Reichard, J., Nance, P., Burleigh-Flayer, H., Parker, A., Vincent, M., and McConnell, E.E., 2014, Mode of action analysis for liver tumors from oral 1,4-dioxane exposures and evidence-based dose response assessment. *Regulatory Toxicology and Pharmacology* 68(3). 387–401.

Dourson, M., Higginbotham, J., Crum, J., Burleigh-Flayer, H., Nance, P., Forsberg, N.D., Lafranconi, M., and Reichard, J., 2017, Update: Mode of action (MOA) for liver tumors induced by oral exposure to 1,4-dioxane. *Regulatory Toxicology and Pharmacology* 88: 45–55.

Dow Chemical Company (Dow), 1989a, Toxic Substances Control Act test submission, October 17th, 1989. OTS0000719, Dow Chemical Company, D004057, 0158–0179. (cited in Health Canada, 2010 and ECJRC, 2002).

Dow Chemical Company (Dow), 1989b, Summaries of environmental data for phenol, 1,4-dioxane and acrylonitrile with attached reports and cover letter. U.S. Environmental Protection Agency, Office of Science and Technology.

Dow Chemical Company (Dow), 1995, Toxicity and environmental references for 1,4-dioxane. Personal communication. (cited in NICNAS, 1998; Health Canada, 2010; CCME 2008).

ECJRC (European Commission, Joint Research Centre), 2002, European Union risk assessment report: 1,4-dioxane. (EUR 19833 EN). Luxembourg: Office for Official Publications of the European Communities. Available from: https://echa.europa.eu/documents/10162/a4e83a6a-c421-4243-a8df-3e84893082aa.

Ernstgård, L., Iregren, A., Sjögren, B., and Johanson, G., 2006, Acute effects of exposure to vapours of dioxane in humans. *Human & Experimental Toxicology* 25: 723–729.

Fairley, A., Linton, E.C., and Ford-Moore, A.H., 1934, The toxicity to animals of 1,4-dioxane. *The Journal of Hygiene* 34: 486–501.

Fisher, J., Mahle, D., Bankston, L., Greene, R., and Gearhart, J., 1997, Lactational transfer of volatile chemicals in breast milk. *American Industrial Hygiene Association Journal* 58: 425–431.

Galloway, S.M., Armstrong, M.J., Reuben, C., et al., 1987, Chromosome aberrations and sister chromatid exchanges in Chinese hamster ovary cells: Evaluations of 108 chemicals. *Environmental and Molecular Mutagenesis* 10(Suppl 10): 1—175.

Geiger, D.L., Brooke, L.T., and Call, D.J. (Eds), 1990, *Acute Toxicities of Organic Chemicals to Fathead Minnows (Pimephales Promelas)*, Vol. V, p. 332. Superior, WI: University of Wisconsin, Center for Lake Superior Environmental Studies.

Gi, M., Fujioka, M., Kakehashi, A., et al., 2018, In vivo positive mutagenicity of 1, 4-dioxane and quantitative analysis of its mutagenicity and carcinogenicity in rats. *Archives of Toxicology* 92(10): 3207–3221.

Giavini, E., Vismara, C., and Broccia, M.L., 1985, Teratogenesis study of dioxane in rats. *Toxicology Letters* 26: 85–88.

Göen, T., von Helden, F., Eckert, E., Knecht, U., Drexler, H., and Walter, D., 2016, Metabolism and toxicokinetics of 1, 4-dioxane in humans after inhalational exposure at rest and under physical stress. *Archives of Toxicology* 90(6): 1315–1324.

Goldsworthy, T.L., Monticello, T.M., Morgan, K.T., et al., 1991, Examination of potential mechanisms of carcinogenicity of 1,4-dioxane in rat nasal epithelial cells and hepatocytes. *Archives of Toxicology Links – Archives of Toxicology* 65: 1–9.

Hashim, M., Iqbal, R., Afsheen, S., Fahad, A.U.M., Asghar, U., Daman, M., and Ghazanfar, M., 2018, Effect of 1,4-dioxane on rabbit (Oryctolagus cuniculus) reproductive hormonal level and histopathology of testes and ovaries. *Pure and Applied Biology (PAB)* 7(2): 466–469.

Haworth, S., Lawlor, T., Mortelmans, K., Speck, W., and Zeiger, E., 1983, Salmonella mutagenicity test results for 250 chemicals. *Environmental Mutagenesis* 1(Supp 1): 1, 10.

Health Canada, 2010, Screening assessment for the challenge: 1,4-Dioxane. Environment Canada, Health Canada. Available from: http://www.ec.gc.ca/ese-ees/789BC96E-F970-44A7-B306-3E32419255A6/batch7_123-91-1_en.pdf

Health Canada, September 2018, 1,4-Dioxane in drinking water: Guideline technical document for public consultation. Available from: https://www.canada.ca/en/health-canada/programs/consultation-1-4-dioxane-drinking-water/document.html.

Hellmer, L. and Bolcsfoldi, G., 1992, An evaluation of the E. coli K-12 uvrB/recA DNA repair host-mediated assay I: In vitro sensitivity of the bacteria to 61 compounds. *Mutation Research* 272: 145–160.

Hoch-Ligeti, C. and Argus, M.F., 1970, Effect of carcinogens on the lung of guinea pigs. In: P. Nettesheim, M.G. Hanna, and J.W. Deatherage (Eds), *Morphology of Experimental Respiratory Carcinogenesis*, pp. 267–279. AEC Symposium Series 21, National Cancer Institute and U.S. Atomic Energy Commission. CONF700501.

Hoch-Ligeti, C., Argus, M.F., and Arcos, J.C., 1970, Induction of carcinomas in the nasal cavity of rats by dioxane. *British Journal of Cancer* 24(1): 164–167.

Huntley, S., Amarai, M., and Schell, J., 2004, Toxicokinetics of 1,4-dioxane and implications for assessing hazards to terrestrial wildlife. Paper presented at the Society of Environmental Toxicology and Chemistry, 25th Annual Meeting, Portland, OR, November 14–18.

IARC (International Agency for Research on Cancer), 1999, Some chemicals that cause tumours of the kidney or urinary bladder in rodents, and some other substances – 1,4-dioxane (group 2B). *IARC Monographs on the Evaluation of Carcinogenic Risks to Humans*, Vol. 71. Lyon, France: World Health Organization. Available at http://www-cie.iarc.fr/htdocs/monographs/vol71/019-dioxane.html.

Illinois Environmental Protection Agency (EPA), 2013, Derived water quality criteria list. April 4. https://www2.illinois.gov/epa/topics/water-quality/standards/Pages/derived-criteria.aspx.

Itoh, S. and Hattori, C., 2019, In vivo genotoxicity of 1,4-dioxane evaluated by liver and bone marrow micronucleus tests and Pig-a assay in rats. *Mutation Research/Genetic Toxicology and Environmental Mutagenesis* 837: 8–14.

JBRC (Japan Bioassay Research Center), 1998, Laboratory Report: Two-year studies of 1,4-dioxane in F344 rats and BDF1 mice (drinking water). Kanagawa, Japan: Japan Bioassay Research Center.

Johnson, R., Tietge, J., Stokes, G., and Lothenbach, D., 1993, The medaka carcinogenesis model. In: Compendium of the FY1988 & FY1989 research reviews for the Research Methods Branch, pp. 147–172. U.S. Army Biomedical Research & Development Laboratory, Fort Detrick, Frederick, Maryland. Technical Report 9306.

Johnstone, R.T., 1959, Death due to dioxane? *A.M.A. Archives of Industrial Health* 20: 445–447.

Juhnke, I. and Ludemann, D., 1978, Ergebnisse der Untersuchung von 200 chemischen Verbindungen auf akute Fischtoxizität mit dem Goldorfentest [Results of the investigation of 200 chemical compounds on acute fish toxicity with the Golden Orfe test]. *Zeitschrift für Wasser- und Abwasser-Forschung [Journal of Water and Wastewater Research]* 11: 161–164.

Kano, H., Umeda, Y., Kasai, T., Sasaki, T., Matsumoto, M., Yamazaki, K., Nagano, K., Arito, H., and Fukushima, S., 2009, Carcinogenicity studies of 1,4-dioxane administered in drinking-water to rats and mice for 2 years. *Food and Chemical Toxicology* 47(11): 2776–2784.

Kano, H., Umeda, Y., Saito, M., et al., 2008, Thirteen-week oral toxicity of 1,4-dioxane in rats and mice. *Journal of Toxicological Sciences* 33: 141–153.

Kasai, T., Kano, H., Umeda, Y., et al., 2009, Two-year inhalation study of carcinogenicity and chronic toxicity of 1,4-dioxane in male rats. *Inhalation Toxicology* 21(11): 889–897.

Khudoley, V.V., Mizgireuv, I., and Pliss, G.B., 1987, The study of mutagenic activity of carcinogens and other chemical agents with Salmonella typhimurium assays: Testing of 126 compounds. *Archiv für Geschwulstforschung* 57: 453–462.

King, M.E., Shefner, A.M., and Bates, R.R., 1973, Carcinogenesis bioassay of chlorinated dibenzodioxins and related chemicals. *Environmental Health Perspectives* 5: 163–170.

Kitchin, K.T. and Brown, J.L., 1990, Is 1,4-dioxane a genotoxic carcinogen? *Cancer Letters* 53(1): 67–71.

Kociba, R.J., McCollister, S.B., Park, C., Torkelson, T.R., and Gehring, P.J., 1974, 1,4-dioxane. I. Results of a 2-year ingestion study in rats. *Toxicology and Applied Pharmacology* 30: 275–286.

Kramer, V.C., Schnell, D.J., and Nickerson, K.W., 1983, Relative toxicity of organic solvents to aedes aegypti larvae. *Journal of Invertebrate Pathology* 42: 285–287.

Kurl, R.N., Poellinger, L., Lund, J., and Gustafsson, J.A., 1981, Effects of dioxane on RNA synthesis in the rat liver. *Archives of Toxicology* 49(1): 29–33.

Kwan, K.K., Dutka, B.J., Rao, S.S., and Liu, D., 1990, Mutatox test: A new test for monitoring environmental genotoxic agents. *Environmental Pollution* 65: 323–332.

Leung, H. and Paustenbach, D.J., 1990, Cancer risk assessment for dioxane based upon a physiologically based pharmacokinetic approach. *Toxicology Letters* 51(2): 147–162.

Lundberg, I., Hogberg, J., Kronevi, T., and Holmberg, B., 1987, Three industrial solvents investigated for tumor promoting activity in the rat liver. *Cancer Letters* 36: 29–33.

Marzulli, F., Anjo, D.M., and Maibach, H.I., 1981, In vivo skin penetration studies of 2,4-toluenediamine, 2,4-diaminoanisole, 2-nitro-p-phenylenediamine, p-dioxane and N-nitrosodiethanolamine in cosmetics. *Food and Cosmetics Toxicology* 19: 743–747.

McFee, A.F., Abbott, M.G., Gulati, D.K., and Shelby, M.D., 1994, Results of mouse bone marrow micronucleus studies on 1,4-dioxane. *Mutation Research* 322: 141–150.

McGregor, D.B., Brown, A.G., Howgate, S., McBride, D., Riach, C., and Caspary, W.J., 1991, Responses of the L5178Y mouse lymphoma cell forward mutation assay. V. 27 coded chemicals. *Environmental and Molecular Mutagenesis* 17(3): 196–219.

Michigan Department of Environmental Quality (MDEQ), 2015, Review of a 1,4-Dioxane Presentation by Michael Dourson, Ph.D., on October 8, 2013.

Michigan Department of Environmental Quality (MDEQ), 2018, Rule 57 water quality values, surface water assessment section.15 March.

Mikheev, M.I., Gorlinskaya, Y.E.P., and Solovyova, T.V., 1990, The body distribution and biological action of xenobiotics. *Journal of Hygiene, Epidemiology, Microbiology, and Immunology* 34: 329–336.

Mirkova, E.T., 1994, Activity of the rodent carcinogen 1,4-dioxane in the mouse bone marrow micronucleus assay. *Mutation Research* 322: 142–144.

Miyagawa, M., Shirotori, T., Tsuchitani, M., and Yoshikawa, K., 1999, Repeat-assessment of 1,4-dioxane in a rat-hepatocyte replicative DNA synthesis (RDS) test: Evidence for stimulus of hepatocyte proliferation. *Experimental and Toxicologic Pathology* 51: 555–558.

Morita, T., 1994, No clastogenicity of 1,4-dioxane as examined in the mouse peripheral blood micronucleus test. *Mammalian Mutagenesis Study Group Communication* 2: 7–8.

Morita, T. and Hayashi, M., 1998, 1,4-Dioxane is not mutagenic in five in vitro assays and mouse peripheral blood micronucleus assay, but is in mouse liver micronucleus assay. *Environmental and Molecular Mutagenesis* 32: 269–280.

Mungikar, A.M. and Pawar, S.S., 1978, Induction of the hepatic microsomal mixed function oxidase system in mice by p-dioxane. *Bulletin of Environmental Contamination and Toxicology* 20: 797–804.

Muñoz, E.R. and Barnett, B.M., 2002, The rodent carcinogens 1,4-dioxane and thiourea induce meiotic nondisjunction in drosophila melanogaster females. *Mutation Research* 517(1–2): 231–238.

Nannelli, A., Rubertis, A., Longo, V., and Gervasi, P.G., 2005, Effects of dioxane on cytochrome P450 enzymes in liver, kidney, lung and nasal mucosa of rat. *Archives of Toxicology* 79(2): 74–82.

National Cancer Institute (NCI), 1978, Bioassay of 1,4-dioxane for possible carcinogenicity. Bethesda, MD: National Cancer Institute. National Institutes of Health Publication No. 78–1330 NCICGTR-80.

National Toxicology Program (NTP), 2016, 14th Report on Carcinogens. Research Triangle Park, NC: U.S. Department of Health and Human Services, Public Health Service. https://ntp.niehs.nih.gov/pubhealth/roc/index-1.html

Nestmann, E.R., Otson, R., Kowbel, D.J., Bothwell, P.D., and Harrington, T.R., 1984, Mutagenicity in a modified Salmonella assay of fabric-protecting products containing 1,1,1-trichloroethane. *Environmental Mutagenesis* 6: 71–80.

Netherlands Organization for Applied Scientific Research, 1999, Risk assessment: 1,4-Dioxane. Netherlands Organization for Applied Scientific Research (TNO) and the National Institute of Public Health and the Environment (RIVM). Chemical Substances Bureau, Ministry of Housing, Spatial Planning and the Environment (VROM), Netherlands, Final Version, November 5, EINECS-No. 204–661–8.

NICNAS (National Industrial Chemicals Notification and Assessment Scheme), 1998, Priority existing chemical assessment no. 7 – 1,4-dioxane. Department of Health and Aging, Australian National Industrial Chemicals Notification and Assessment Scheme (NICNAS). Commonwealth of Australia. June.

Nishiushi, Y., 1981, Evaluation of the effect of pesticides on some aquatic organisms. *Setai Kagaku* 4(3): 45–47 [cited in ECJRC, 2002].

Ontario Ministry of the Environment (OMOE), 2007, Rationale for the development of generic soil and groundwater standards for use at contaminated sites in Ontario. Ontario Ministry of the Environment, Standards Development Branch. March.

Reitz, R.H., McCroskey, P.S., Park, C.N., Andersen, M.E., and Gargas, M.L., 1990, Development of a physiologically based pharmacokinetic model for risk assessment with 1,4-dioxane. *Toxicology and Applied Pharmacology* 105(1): 37–54.

Roy, S.K., Thilagar, A.K., and Eastmond, D.A., 2005, Chromosome breakage is primarily responsible for the micronuclei induced by 1,4-dioxane in the bone marrow and liver of young CD-1 mice. *Mutation Research* 586: 28–37.

Sample, B.E., Opresko, D.M., and Suter II, G.W., 1996, Toxicological benchmarks for wildlife: 1996 revision. Environmental restoration division ORNL Environmental Restoration Program. ES/ER/TM-86/R3.

Sheu, C.W., Moreland, F.M., Lee, J.K., and Dunkel, V.C., 1988, In vitro BALB/3T3 cell transformation assay of nonoxynol-9 and 1,4-dioxane. *Environmental and Molecular Mutagenesis* 11: 41–48.

Silverman, L., Schulte, H.F., and First, M.W., 1946, Further studies of sensory response to certain industrial solvent vapors. *The Journal of Industrial Hygiene and Toxicology Journal of Industrial Hygiene* 28: 262–266.

Sina, J.F., Bean, C.L., Dysart, G.R., Taylor, V.I., and Bradley, M.O., 1983, Evaluation of the alkaline elution/rat hepatocyte assay as a predictor of carcinogenic/mutagenic potential. *Mutation Research* 113: 357–391.

Sonich-Mullin, C., Fielder, R., Wiltse, J., Baetcke, K., Dempsey, J., Fenner-Crisp, P., and Mangelsdorf, I., 2001, IPCS conceptual framework for evaluating a mode of action for chemical carcinogenesis. *Regulatory Toxicology and Pharmacology* 34(2): 146–152.

Springborn Laboratories, Inc. (Springborn), 1989, 1,4-Dioxane – chronic toxicity to Ceriodaphnia dubia under static renewal conditions. Wareham, MA: Springborn Laboratories Inc., SLI Report 89–9–3089.

Stickney, J.A., Sager, S.L., Clarkson, J.R., et al., 2003, An updated evaluation of the carcinogenic potential of 1,4-dioxane. *Regulatory Toxicology and Pharmacology* 38: 183–195.

Stott, W.T., Quast, J.F., and Watanabe, P.G., 1981, Differentiation of the mechanisms of oncogenicity of 1,4-dioxane and 1,3-hexachlorobutadiene in the rat. *Toxicology and Applied Pharmacology* 60: 287–300.

Sweeney, L.M., Thrall, K.D., Poet, T.S., et al., 2008, Physiologically based pharmacokinetic modeling of 1,4-dioxane in rats, mice, and humans. *Toxicological Sciences* 101(1): 32–50.

Takano, R., Murayama, N., Horiuchi, K., Kitajima, M., Shono, F. and Yamazaki, H., 2010, Blood concentrations of 1,4-dioxane in humans after oral administration extrapolated from in vivo rat pharmacokinetics, in vitro human metabolism, and physiologically based pharmacokinetic modeling. *Journal of Health Sciences* 56(5): 557–565.

Take, M., Ohnishi, M., Yamamoto, S., Matsumoto, M., Nagano, K., and Fukushima, S., 2012, Distribution of 1,4-dioxane by combined inhalation plus oral exposure routes in rats. *International Journal of Environmental Analytical Chemistry* 92(15): 1715–1728.

Thiess, A.M., Tress, E., and Fleig, I., 1976, Industriell-medizinische Untersuchungsentdeckungen von der Arbeiter Einatmung des Dioxans [Industrial-medical investigation results in the case of workers exposed to dioxane]. *Arbeitsmedizin, Sozialmedizin, Präventivmedizin* 11: 35–46.

Thomulka, K.W. and McGee, D.J., 1992, Evaluation of organic compounds in water using Photobacterium phosphoreum and Vibrio harveyi bioassays. *Fresenius Environmental Bulletin* 1: 815–820.

Tinwell, H. and Ashby, J., 1994, Activity of 1,4-dioxane in mouse bone marrow micronucleus assays. *Mutation Research* 322: 148–150.

U.S. Environmental Protection Agency (USEPA), 1995, Internal report on summary of measured, calculated and recommended log Kow values. Environmental Research Laboratory, Athens, GA. 10 April.

U.S. Environmental Protection Agency (USEPA), 1996a, ASTER database. U.S. Environmental Protection Agency, National Health and Environmental Effects Research Laboratory. (cited in NICNAS, 1998)

U.S. Environmental Protection Agency (USEPA), 1996b, Ecotox thresholds. Eco Update, Volume 3, Number 2. EPA/540/F-95/038. 12 pp.

U.S. Environmental Protection Agency (USEPA), 1999, Screening level ecological risk assessment protocol for hazardous waste combustion facilities. EPA/530/D-99/001A. Peer Review Draft. August.

U.S. Environmental Protection Agency (USEPA), 2003, Region 5 RCRA ecological screening levels. 22 August.

U.S. Environmental Protection Agency (USEPA), 2005, Guidelines for carcinogen risk assessment. Risk Assessment Forum, Washington, DC; EPA/630/P-03/001B. Available at http://www2.epa.gov/osa/guidelines-carcinogen-risk-assessment.

U.S. Environmental Protection Agency (USEPA), 2013. Toxicological review of 1,4-Dioxane (CAS No. 123-91-1) with Inhalation Update. In: Integrated Risk Information System (IRIS). (EPA-635/R-09-005-F). Washington, DC: Environmental Protection Agency. Available from: https://cfpub.epa.gov/ncea/iris/iris_documents/documents/toxreviews/0326tr.pdf.

U.S. Environmental Protection Agency (USEPA), 2015a, TSCA work plan chemical problem formulation and initial assessment, 1,4-dioxane. 740-R1-5003. April.

U.S. Environmental Protection Agency (USEPA), 2015b, Determination of the biologically relevant sampling depth for terrestrial and aquatic ecological risk assessments. EPA/600/R-15/176. ERASC-015F. October.

U.S. Environmental Protection Agency (USEPA), May 2018a, Problem formulation of the risk evaluation for 1,4-Dioxane, CASRN: 123-91-1. EPA-740-RI-7012. Office of Chemical Safety and Pollution Prevention.

U.S. Environmental Protection Agency (USEPA). 2018b, Chemistry dashboard, 1,4-dioxane chemical activity summary. Accessed November 2018. https://comptox.epa.gov/dashboard/dsstoxdb/results?search=1,4-dioxane#invitrodb.

U.S. Environmental Protection Agency (USEPA), 2018c, Region 4 ecological risk assessment supplemental guidance, March 2018 update.

U.S. Environmental Protection Agency (USEPA), 2018d, ECOTOX database. Accessed September 2018. https://cfpub.epa.gov/ecotox/.

Uno, Y., Takasawa, H., Miyagawa, M., Inoue, Y., Murata, T., and Yoshikawa, K., 1994, An in vivo-in vitro replications DNA synthesis (RDS) test using rat hepatocytes as an early prediction assay for nongenotoxic hepatocarcinogens: Screening of 22 known positives and 25 noncarcinogens. *Mutation Research* 320(3): 189–205.

Wirth, W. and Klimmer, O., 1936, Zur Toxikologie der Lösungsmittel: 1,4-dioxan [On the Toxicology of Organic Solvents: 1,4-Dioxane (diethylene dioxide).] *Archiv für Gewerbepathologie und Gewerbehygiene* 17: 192–206.

Woo, Y.-T., Argus, M.F., and Arcos, J.C., 1977a, Tissue and subcellular distribution of 3H-dioxane in the rat and apparent lack of microsome-catalyzed covalent binding in the target tissue. *Life Sciences* 21(10): 1447–1456.

Woo, Y., Arcos, J.C., Argus, M.F., Griffin, G.W., and Nishiyama, K., 1977b, Structural identification of p-dioxane-2-one as the major urinary metabolite of p-dioxane. *Naunyn-Schmiedeberg's Archives of Pharmacology* 299: 283–287.

Woo, Y., Argus, M.F., and Arcos, J.C., 1977c, Metabolism in vivo of dioxane: Effect of inducers and inhibitors of hepatic mixed-function oxidases. *Biochemical Pharmacology* 25: 1539–1542.

Woo, Y.T., Argus, M.F., and Arcos, J.C., 1978, Effect of mixed-function oxidase modifiers on metabolism and toxicity of the oncogen dioxane. *Cancer Research* 38: 1621–1625.

Yamazaki, K., Ohno, H., Asakura, M., Narumi, A., Ohbayashi, H., Fujita, H., Ohnishi, M., Katagiri, T., Senoh, H., Yamanouchi, K., Nakayama, E., Yamamoto, S., Noguchi, T., Nagano, K., Enomoto, M. and Sakabe, H., 1994, Two- year toxicological and carcinogenesis studies of 1,4-dioxane in F344 rats and BDF1 mice. In: *Proceedings of the Second Asia-Pacific Symposium on Environmental and Occupational Health 22-24 July, 1993*, Sumino, K., Sato, S. and Shinkokai, N.G. (eds.). Kobe University School of Medicine, International Center for Medical Research, Kobe, Japan, pp. 193–198.

Yant, W.P., Schrenk, H.H., Waite, C.P., and Patty, F.A., 1930, Acute response of guinea pigs to vapors of some new commercial organic compounds – VI dioxane. *Public Health Reports* 45: 2023–2032.

Yoon, J.S., Mason, J.M., Nalencia, R., Woodruff, R.C., and Zimmering, S., 1985, Chemical mutagenesis testing in drosophila. IV. Results of 45 coded compounds tested for the national toxicology program. *Environmental Mutagenesis* 7: 349–367.

Young, J.D., Braun, W.H., Gehring, P.J., Horvath, B.S., and Daniel, R.L., 1976, Short communication. 1,4-Dioxane and beta-hydroxyethoxyacetic acid excretion in urine of humans exposed to dioxane vapors. *Toxicology and Applied Pharmacology* 38: 643–646.

Young, J.D., Braun, W.H., Rampy, L.W., Chenowith, M.B., and Blau, G.E., 1977, Pharmacokinetics of 1,4-dioxane in humans. *Journal of Toxicology and Environmental Health* 3: 507–520.

Young, J.D., Braun, W.H., and Gehring, P.J., 1978a, The dose-dependent fate of 1,4-dioxane in rats. *Journal of Environmental Pathology and Toxicology* 2: 263–282.

Young, J.D., Braun, W.H., and Gehring, P.J., 1978b, Dose-dependent fate of 1,4-dioxane in rats. *Journal of Toxicology and Environmental Health* 4(5–6): 709–726.

Zimmermann, F.K., Mayer, V.W., Scheel, I., and Resnick, M.A., 1985, Acetone, methyl ethyl ketone, ethyl acetate, acetonitrile and other polar aprotic solvents are strong inducers of aneuploidy in saccharomyces cerevisiae. *Mutation Research* 149: 339–351.

6 Regulation of 1,4-Dioxane

Thomas K.G. Mohr and William H. DiGuiseppi

CONTENTS

A fundamental maxim of toxicology is "The dose makes the poison."* Regulating chemical use, handling, and disposal protects us against risk from exposure to harmful doses of toxic chemicals used widely in industry, consumer products, crop protection, and other applications. Regulations establishing exposure thresholds protect human health by providing regulatory agencies with a framework for evaluating and enforcing the safety of chemical use and associated exposure. The many ways by which chemical exposure may occur are mirrored in the complex web of regulations that have evolved to govern:

- Worker health and safety
- Drinking water quality
- Contaminated environmental media (groundwater, soil, air)
- Chemical residues in food
- Pharmaceuticals and sundries
- Ecological effects of wastewater discharges
- Ambient air quality
- Transportation safety
- Consumer right-to-know of the presence of carcinogens or reproductive toxins in consumer products
- Community right-to-know of large quantities of dangerous chemicals stored and used by local industries

* This adage is attributed to the Swiss scientist "Paracelsus," aka Phillip von Hohenheim, 1493–1541, who said, "Alle Ding' sind Gift und nichts ohn' Gift; all ein die Dosis macht, dass ein Ding kein Gift ist." ("All things are poison and nothing is without poison, only the dose permits something not to be poisonous.") The axiom is sometimes wrongly attributed to Pericles, a Greek statesman.

This chapter reviews the full complement of regulations and regulatory guidance thresholds for 1,4-dioxane, with an emphasis on the derivation of water quality standards. Regulation of 1,4-dioxane is rapidly evolving and it is likely that some of these standards will have been updated since the publication of this book. The reader is advised to seek guidance from regulatory agencies before making decisions based on the following compilation.

6.1 DRINKING WATER REGULATIONS FOR 1,4-DIOXANE

There are several federal statutes that regulate chemicals in surface, ground, and drinking water. Water quality standards are mandated by the Clean Water Act* and preserve the beneficial uses of a surface water body by designating criteria to protect it against degradation from pollutants. A water quality standard accounts for

- The designated beneficial uses of the water body (i.e., recreation, drinking water supply, aquatic life, and agriculture)
- The water quality criteria to protect beneficial uses (numeric pollutant concentrations and narrative requirements)
- The maintenance and protection of existing beneficial uses and high-quality waters (anti-degradation policy)
- Practical implementation issues (i.e., low flows, variances, mixing zones) (USEPA, 2006)

The National Pollution Discharge Elimination System (NPDES), a permit program that controls water pollution by regulating point-source discharges into waters of the United States, is the primary framework for water quality standards from the Clean Water Act.

The National Oil and Hazardous Substances Pollution Contingency Plan (NCP; 53 FR 51394) provides the policies and authorities for the cleanup of environmental contaminants. The Comprehensive Environmental Response, Compensation, and Liability Act (CERCLA, or "Superfund") of 1980, and as amended, and the Resource Conservation and Recovery Act (RCRA) establish a national program for addressing releases of hazardous substances and hazardous wastes into the environment, including soil, air, groundwater and surface water, and requires that actions selected be protective of human health and the environment. The Safe Drinking Water Act of 1974 (SDWA) authorizes the U.S. Environmental Protection Agency (USEPA) to establish national health-based standards for drinking water to protect against both naturally occurring and anthropogenic contaminants in drinking water. The SDWA was amended in 1986 to include specific health goals and defined approaches for risk management. The SDWA also outlines criteria for adding chemicals to the list of contaminants regulated by Maximum Contaminant Levels (MCLs). Currently, USEPA has the flexibility to determine whether to regulate a contaminant after completing a required review of at least five contaminants every 5 years.

The 1996 SDWA amendments require that once every 5 years EPA issue a new list of no more than 30 unregulated contaminants to be monitored by public water systems (PWSs) under the Unregulated Contaminant Monitoring Rule (UCMR), which helps assess the nationwide occurrence frequency and magnitude of an unregulated chemical that is being considered for MCL regulation. The Revisions to the Unregulated Contaminant Monitoring Rule for Public Water Systems; 77 FR 26072, May 2, 2012 (USEPA, 2012) required monitoring of Public Water Supplies (PWSs) serving more than 10,000 people and 800 representative PWSs serving 10,000 or fewer people within a 12-month period between January 2013 through December 2015 for 21 contaminants using six EPA-approved analytical methods and four equivalent consensus methods. 1,4-Dioxane was included in this Assessment Monitoring program. The data are stored in the National Contaminant Occurrence Database (NCOD) (USEPA, 2017a) and discussed in Chapter 2.

* The Clean Water Act or CWA is the Federal Water Pollution Control Act Amendments of 1972 and 1977.

Pursuant to the SDWA, the determination to set a drinking water MCL is at the sole discretion of the USEPA Administrator. The USEPA Administrator uses three review criteria to determine whether or not to regulate a contaminant in public drinking water systems under the SDWA:

1. The contaminant adversely affects human health
2. The contaminant is likely to occur in public water systems at sufficient frequency and concentrations to create a public health concern
3. Regulation of the contaminant presents a meaningful opportunity for health risk reduction (USEPA, 2006)

A number of events would have to transpire before 1,4-dioxane could be regulated with a federal MCL. Firstly, identification of 1,4-dioxane as a possible health concern in U.S. drinking water has been completed with the data collected in UCMR3. Secondly, the USEPA must determine whether an opportunity exists to reduce public health risks by regulation using the agency's human health risk assessments, which determine potential drinking water exposure risks and standards. An additional element to support regulation of 1,4-dioxane in drinking water, the availability of viable analytical methods for analyzing 1,4-dioxane in drinking water at the levels of interest for reducing risk, has also been completed with the introduction of USEPA Method 522 in 2008 (see Chapter 4).* Review of 1,4-dioxane treatment methods required to analyze costs for compliance and enforcement would also be needed to adopt an MCL. The draft MCL and other regulatory elements would then be reviewed by USEPA and the President's Office of Management and Budget (OMB). Thereafter, public comments are solicited and addressed, the final National Primary Drinking Water Regulations (NPDWR) document is reviewed by USEPA and OMB again, and then the regulation is promulgated. Implementation and enforcement would follow a schedule outlined in the NPDWR (Macler, 2006).

As described in Chapter 4, of the more than 10 million commercially available organic and inorganic chemicals, only about a quarter million are subject to any form of regulation, far fewer are routinely analyzed, and fewer still have been profiled and studied for health effects. Reviewing more chemicals for addition to the list of regulated contaminants is essential to ensure public health protection. At the same time, some compounds on the list of regulated contaminants are being revisited to determine whether to lower, raise, or eliminate their MCLs. For example, California in 2006 revised the draft Public Health Goals for several chlorinated solvent compounds including cis- and trans-1,2-dichloroethylene, suggesting that these compounds may not be as toxic as was first thought when they were originally listed. Subsequently, based on new data from Camp LeJeune, NC and elsewhere, California lowered the Public Health Goals for cis- and trans-1,2-dichloroethylene from 100 and 60 to 13 and 50 μg/L, respectively (OEHHA, 2018). The California MCLs for cis- and trans-1,2-dichloroethylene remain 6 and 10 μg/L. Similarly, USEPA is reviewing whether to lower the toxicity threshold used to establish the Preliminary Remediation Goal (PRG) for trichloroethylene (TCE). California has established a Public Health Goal for perchloroethylene at 0.06 μg/L, whereas the MCL remains at 5 μg/L. Setting drinking water standards can be described as a fluid process that is by no means smooth or certain. It is in this uncertain and sometimes controversial regulatory environment that the evaluation of regulating 1,4-dioxane is being made.

The multiyear process of evaluating, listing, and approving additions to the list of regulated contaminants includes many tiers of review and can be subjected to legal challenge and political initiatives. For example, although methyl *tert*-butyl ether (MTBE) and perchlorate have been investigated and remediated for many years, the fact that these compounds are not yet regulated by a federal MCL is due in part to a protracted and often politicized review process. States may independently adopt and implement regulation of drinking water contaminants on the basis of the prevalence and magnitude of drinking water contaminants within their jurisdictions. To date, 1,4-dioxane and other solvent stabilizer compounds are not regulated by USEPA, except for epichlorohydrin (formerly used to stabilize TCE), which has an MCL goal of zero (USEPA, 2007a).

* Method 522 (2008) uses solid-phase extraction and gas chromatography-mass spectrometry with selected-ion monitoring.

6.1.1 Drinking Water Guidelines

As discussed above, there is currently* no federal MCL or MCL Goal for 1,4-dioxane, however, there are several federal programs with health-based screening levels for 1,4-dioxane in drinking water. USEPA's Office of Water derived a 1-day health advisory of 4.0 milligrams per liter (mg/L) and a 10-day health advisory of 400 μg/L in drinking water for a 10-kg child, and a noncancer lifetime health advisory of 200 μg/L in drinking water for adult exposure (EPA, 2012, 2018a). The agency also derived a cancer risk level of 35 μg/L, based on 1×10^{-4} excess cancer risk level (USEPA, 2012, 2018a).

USEPA regional offices have derived screening levels for 1,4-dioxane in soil, air, and potable groundwater (called "tapwater") for both residential and industrial exposure settings at contaminated sites. The USEPA regions originally used a groundwater screening level (Regional Screening Level [RSL]) of 6.1 μg/L for 1,4-dioxane found in potentially potable groundwater for residential use. The RSLs have been recently updated to incorporate new toxicological data and the tapwater RSL for 1,4-dioxane is currently 0.46 μg/L based on a 1×10^{-6} excess cancer risk level (USEPA, 2018b).

Internationally and within the United States, advisory levels asserting acceptable levels of 1,4-dioxane in drinking water and cleanup levels for 1,4-dioxane in soil and groundwater vary widely. The absence of a federal MCL for 1,4-dioxane has left the states to determine whether to promulgate regulatory standards or guidance levels. Representative state, federal, and international regulations and guidelines for drinking water or potable groundwater, surface water, soil, and vapor intrusion, as of the third quarter of 2019, are presented in Tables 6.1 through 6.5. Even recently established guidance values range from 0.3 to 50 μg/L, demonstrating that there are more than two orders of magnitude divergence in health-based values for the oral ingestion of water containing 1,4-dioxane.

6.1.2 Basis for Variation in Standards for 1,4-Dioxane in Groundwater

Chapter 5 provides a comprehensive review of the 1,4-dioxane toxicology knowledge base. The primary exposure routes leading to 1,4-dioxane toxicity are ingestion of contaminated drinking water and inhalation (which is likely to occur only in occupational settings). Liver and kidney toxicity are the primary noncancer health effects associated with exposure to 1,4-dioxane in humans and laboratory animals, and cancer risk following exposure to 1,4-dioxane by any route is the primary driver for public health and regulatory action. In recent years, federal, state, and international authorities have established a number of health-based regulatory values and evaluation criteria (see Tables 6.1 through 6.5). As is the case for most emerging contaminants, the regulatory process for 1,4-dioxane is in various stages of development and there remain differences in science interpretation and policy decisions, resulting in values and criteria that vary over two orders of magnitude between states, the U.S. federal government, and other international agencies. Differences are largely due to the interpretation of key toxicity studies and the approaches used for modeling cancer risk. (See Chapter 5 for a discussion of the different scientific interpretations related to how 1,4-dioxane may cause carcinogenicity and how that impacts the cancer toxicity values used by regulatory and public health agencies.) The choice of exposure assumptions may also differ between agencies. Thus, both differences in scientific conclusions and public health policies affect the available regulatory and guidance values for 1,4-dioxane.

All health-based values (promulgated standards or guidance values) for a chemical in drinking water combine information about a chemical's toxicity with exposure assumptions that are applicable to the specific regulatory paradigm. In the United States, the current federal

* As of late 2019.

TABLE 6.1
Representative Drinking Water and Groundwater Regulations and Guidance

State	Regulatory Agency	State Guidance	Year	Standard	Concentration (µg/L)
		Drinking Water			
California	SWRCB-DDW	Notification level	2019	Drinking water	1.0
Connecticut	CTDEP	Action level	2013	Drinking water	3.0
Delaware	DNREC	Screening level table	2018	Groundwater (for drinking)	0.46
Florida	FL DOH	Health advisory level	2016	Drinking water	0.35
Hawaii	HDOH	Environmental action level	2017	Groundwater IS current or potential source of drinking water	0.46
Indiana	IDEM	Screening level	2019	Groundwater (for tap water)	4.6
Maine	MDEP	Remedial action guidelines	2018	Residential water	4.6
Massachusetts	MADEP	Drinking water standard	2012	Groundwater (drinking water aquifer)	0.3
Michigan	MI DEQ	Cleanup criteria or screening levels	2017	Drinking water (residential)	7.2
				Drinking water (industrial)	350
Missouri	MO DNR	Risk-based target concentrations	2006	Domestic use of groundwater	61
Minnesota	Minnesota Department of Health	Guidance value	2015	Drinking water	1
Nebraska	NE DEQ	Remediation goal	2018	Residential groundwater	0.46
Nevada	NE DEP	Basic comparison level	2017	Residential water	0.672
New Hampshire	NHDES	Ambient groundwater quality standard	2018	Drinking water	0.32
New Mexico	NMED	Screening levels	2017	Drinking water (Cancer)	4.59
				Drinking water (Noncancer)	56.7
New York	NYSDOH	Recommended maximum contaminant level	2019	Drinking water	1
North Carolina	NCDEQ	North Carolina protective values for surface waters	2017	Surface water supply	0.35
Oregon	ODEQ	Ingestion and inhalation from tap water risk-based concentrations	2018	Groundwater (residential)	0.46
				Groundwater (occupational)	2.4
Vermont	VTDEC	Preventative Action Level	2019	Drinking water	0.3
		Groundwater			
Alaska	ADEC	Human health risk-based concentration	2017	Groundwater	4.6
California	SWRCB-DDW	Response Level		Groundwater	35
Colorado	CDPHE	Interim water quality standard	2017a	Groundwater	0.35
Connecticut	CTDEP	Groundwater Protection Criteria	2018	Groundwater	3.0
Delaware	DNREC	Screening level table	2018	Groundwater (for drinking)	0.46
Florida	FDEP	Clean-up Target Level	2005a	Groundwater	3.2
Georgia	GDNR	Media Target Concentration	2019	Groundwater	4.6
Hawaii	HDOH	Environmental Action Level	2017	Groundwater IS NOT a current or potential source of Drinking Water	5,000

(Continued)

TABLE 6.1 (Continued)
Representative Drinking Water and Groundwater Regulations and Guidance

State	Regulatory Agency	State Guidance	Year	Standard	Concentration (µg/L)
Illinois	Illinois EPA	Groundwater remediation objective	2016	Groundwater (Class I)	7.7
Iowa	IDNR	Statewide standards	2019	Groundwater (protected source)	200
				Groundwater (non-protected source)	1,000
Kansas	KDHE	Tier 2 risk-based standards	2015	Residential groundwater	8.49
				Non-residential groundwater	28.5
Maine	MDEP	Remedial action guidelines	2018	Construction worker direct contact water	8,600
Massachusetts	MADEP	Contingency plan standards	2012	Groundwater (potential drinking water)	0.3
				Groundwater (shallow aquifer with vapor-intrusion potential)	6,000
				Groundwater (ecological risk)	50,000
Mississippi	Mississippi-DEQ	Target remedial goal	2002	Groundwater	6.09
Montana	MT DEQ	Human health standard	2019	Groundwater	3
New Jersey	NJDEP	Groundwater quality standard	2018	Groundwater	0.4
North Carolina	NCDEQ	Groundwater quality standard	-	Groundwater	3
Pennsylvania	PADEP	Media-specific concentrations for organic-regulated substances in groundwater	2018	Groundwater <2500 (residential—used aquifer)	6.4
				Groundwater (nonresidential—used aquifer)	32
				Groundwater (residential—nonuse aquifer)	64
				Groundwater (nonresidential—nonuse aquifer)	320
Texas	TCEQ	Protected concentration levels	2018	Groundwater (residential)	9.1
				Groundwater (commercial or industrial)	20
Vermont	VTDEC	Groundwater Enforcement Standard	2019	Groundwater	0.3
		Preventative action level	2019	Groundwater	0.3
Washington	WA ECY	Groundwater cleanup levels	2019	Groundwater – Method B (unrestricted) non-cancer	240
				Groundwater – Method B (unrestricted) cancer	0.440
				Groundwater – Method C (industrial) non-cancer	530
				Groundwater – Method C (industrial) cancer	4.40
West Virginia	WVDEP	Risk-based concentrations	2018	Groundwater	0.46

(Continued)

TABLE 6.1 (Continued)
Representative Drinking Water and Groundwater Regulations and Guidance

State	Regulatory Agency	State Guidance	Year	Standard	Concentration (μg/L)
Wisconsin	Wisconsin DNR	Groundwater quality standards	2017	Groundwater (enforcement standard)	3
				Groundwater (preventative action limit)	0.3
Wyoming	WYDEQ	Cleanup level lookup table	2018	Water clean-up level	0.897

Sources: AK DEC, 2018, 18 AAC 75 Oil and Other Hazardous Substances Pollution Control, As amended through October 27, 2018, https://dec.alaska.gov/media/1055/18-aac-75.pdf (accessed September 22, 2019);

CDPHE, 2017a, *Regulation No. 41—The Basic Standards for Ground Water.* Department of Public Health and Environment;

CTDEP, 2013, *Action Level List for Private Wells.* Connecticut Department of Public Health. www.ct.gov/dph/lib/dph/environmental_health/eoha/groundwater_well_contamination/110916_ct_action_levels_list_nov_2016_update.pdf (accessed September 20, 2019);

CTDEP, 2018, Technical Support Document: Recommended Criteria Values for Common Additional Polluting Substances and Alternative Criteria Requests. http://www.ct.gov/deep/lib/deep/site_clean_up/remediation_regulations/Technical_Support_Document_Additional_Polluting_Substances__Alternative_Criteria.pdf (accessed September 22, 2019);

DNREC, 2018, Screening Level Table, January 1, 2013, Last Updated February 2018, http://www.dnrec.delaware.gov/dwhs/SIRB/Documents/Screening%20Level%20Table.pdf (accessed September 22, 2019);

FDEP, 2005a, Technical Report, Table 1, Groundwater Clean-up Target Levels, https://floridadep.gov/sites/default/files/62-777-TechRprt-Table1-Groundwater-CTLs_Feb2005.xlsx (accessed September 22, 2019);

FDOH, 2016, *Chemicals in Private Drinking Water Wells Fact Sheet.* Florida Department of Health, Bureau of Environmental Health. http://www.floridahealth.gov/environmental-health/drinking-water/_documents/dioxanefs2016updated.pdf (accessed September 20, 2019);

GDNR, 2019, Georgia Administrative Code, Subject 391-3-19 Hazardous Site Response. Department of Natural Resources, September 6, 2019 (accessed September 22, 2019);

HDOH, 2017, Evaluation of environmental hazards at sites with contaminated soil and groundwater. Hawaii Department of Health Environmental Management Division. Fall 2017 (accessed September 22, 2019);

IDEM, 2019, 2019 Screening Level Table. Effective beginning March 4, 2019. https://www.in.gov/idem/cleanups/2392.htm (accessed September 2019);

IDNR, 2019, Chapter 137 Statewide Standards, Iowa Department of Natural Resources. https://programs.iowadnr.gov/riskcalc/Home/statewidestandards (accessed September 22, 2019);

IEPA, 2012, Groundwater Quality Standards for Class I: potable resource groundwater. http://www.ilga.gov/commission/jcar/admincode/035/035006200D04100R.html (accessed September 22, 2019);

KDHE, 2015, Risk-Based Standards for Kansas, Kansas Department of Health and Environment – Bureau of Environmental Remediation, October 2010, Revised September 2015. http://www.kdheks.gov/remedial/download/RSK_Manual_15.pdf (accessed September 22, 2019);

MADEP, 2014, 310 CMR 40.0000: Massachusetts Contingency Plan. April 25, 2014. Massachusetts Department of Environmental Protection. https://www.mass.gov/files/documents/2017/10/17/310cmr40.pdf (accessed September 22, 2019);

MADEP, 2019, FAQs: 1,4-Dioxane: What you need to know about 1,4-dioxane in drinking water. Massachusetts Department of Environmental Protection. https://www.mass.gov/service-details/faqs-14-dioxane (accessed September 22, 2019);

MDH, 2019, Comparison of State Water Guidance and Federal Drinking Water Standards. Minnesota Department of Health. August 2019. https://www.health.state.mn.us/communities/environment/risk/guidance/waterguidance.html (accessed September 22, 2019);

MEDEP, 2018, "Maine Remedial Action Guidelines (RAGs) for Sites Contaminated with Hazardous Substances." Maine Department of Environmental Protection. Effective Date October 19, 2018. https://www.maine.gov/dep/spills/publications/guidance/rags/ME-Remedial-Action-Guidelines-10-19-18cc.pdf (accessed September 20, 2019);

MI DEQ, 2017, Cleanup Criteria Requirements for Response Activity. Michigan Department of Environmental Quality. October 20, 2017. https://www.michigan.gov/documents/deq/deq-rrd-2017-041EQFinalEnvironmentalContaminatioResponseActivityDioxane_604807_7.pdf (accessed September 23, 2019);

(*Continued*)

TABLE 6.1 ***(Continued)***

Sources: (*continued*)

MO DNR, 2006, Missouri Risk-Based Corrective Action (MRBCA), Appendix A, Updates and Revisions to the MRBCA Technical Guidance. April 2006. https://dnr.mo.gov/env/hwp/mrbca/docs/mrbca-append6-06.pdf (accessed September 23, 2019);

MS DEQ, Risk Evaluation Procedures for Voluntary Cleanup and Redevelopment of Brownfield Sites, Appendix A, Tier 1 Target Remedial Goal Table. https://www.mdeq.ms.gov/wp-content/uploads/2017/05/Proced.pdf (accessed September 22, 2019);

MT DEQ, 2019, Montana Numeric Water Quality Standards, Circular DEQ-7, June 2019. http://deq.mt.gov/Portals/112/Water/WQPB/Standards/PDF/DEQ7/DEQ-7_June2019_Final.pdf (accessed September 23, 2019);

NCDEQ, 15A NCAC 02L .0202 Groundwater Standards. https://files.nc.gov/ncdeq/documents/files/02L%20Groundwater%20Standards%20Table%205-21%202013_0.pdf (accessed September 26, 2019);

NCDWR (North Carolina Department of Water Resources), Surface Water Quality Standards, Criteria & In-Stream Target Values. https://files.nc.gov/ncdeq/documents/files/StandardsTable_06102019.xlsx (accessed September 26, 2019);

NEDEP, 2017, User's Guide and Background Technical Document for the Nevada Division of Environmental Protection Basic Comparison Levels for Human Health for the BMI Complex and Common Areas. Revised July 2017. https://ndep.nv.gov/uploads/documents/july-2017-bcl-guidance-doc.pdf (accessed September 23, 2019);

NEDEQ, 2018, Nebraska Voluntary Clean-up Program Guidance Document. http://deq.ne.gov/Publica.nsf/xsp/.ibmmodres/domino/OpenAttachment/Publica.nsf/D243C2B56E34EA8486256F2700698997/Attach/VCP 2018 Revised VCP Guidance Document Final.pdf (accessed September 23, 2019);

NHDES, 2018, Risk Characterization and Management Policy, Method 1 Groundwater Standards. https://www.des.nh.gov/organization/divisions/waste/hwrb/documents/rcmp.pdf (accessed September 20, 2019);

NJDEP, 2018, "Ground Water Quality Standards. New Jersey Department of Environmental Protection. August 9, 2018. https://www.nj.gov/dep/rules/rules/njac7_9c.pdf (accessed September 20, 2019);

NMED, 2017, "Risk Assessment Guidance for Site Investigations and Remediation. Volume I. Soil Screening Guidance for Human Health Risk Assessments. https://www.env.nm.gov/wp-content/uploads/2016/11/NMED_SSG_VOL-I_-March_2017_revised.pdf (accessed September 26, 2019);

NYSDOH, 2019, "Governor Cuomo Announces Availability of $350 Million for Water System Upgrades Statewide and Directs Health Department to Begin Adopting Maximum Contaminant Levels for PFOA, PFOS and 1,4-Dioxane." July 8, 2019. https://www.governor.ny.gov/news/governor-cuomo-announces-availability-350-million-water-system-upgrades-statewide-and-directs (accessed September 26, 2019);

ODEQ, 2018, Risk-Based Concentration for Individual Chemicals. https://www.oregon.gov/deq/FilterDocs/RBDMTable.pdf (accessed September 26, 2019);

PADEP, 2018, Chapter 250, Administration of Land Recycling Program, Appendix A. http://files.dep.state.pa.us/EnvironmentalCleanupBrownfields/LandRecyclingProgram/LandRecyclingProgramPortalFiles/SWHTables-2016/Table%201.pdf (accessed September 26, 2019);

SWRCB, 2019, *Groundwater Information Sheet.* California State Water Resources Control Board – Division of Water Quality. Revised May, 2019, Sacramento, California;

TCEQ, 2018, Table 3: Tier 1 Groundwater PCLs – Residential and Commercial/Industrial. http://www.tceq.texas.gov/assets/public/remediation/trrp/2018-PCL-Tables.xlsx (accessed September 26, 2019);

VTDEC, 2019, Chapter 12 of the Environmental Protection Rules: Groundwater Protection Rule and Strategy. https://dec.vermont.gov/sites/dec/files/dwgwp/DW/2019.07.06%20-%20GWPRS.pdf (accessed September 26, 2019);

Washington Department of Ecology (ECY), 2019, Clean-up Levels and Risk Calculation (CLARC) tables. https://www.ezview.wa.gov/Portals/_1987/Documents/Documents/CLARCInterimUpdate.xlsx (accessed September 26, 2019); WVDEP, 2018, Series 3: Voluntary Remediation and Redevelopment Rule. http://apps.sos.wv.gov/adlaw/csr/readfile.aspx?DocId=50235&Format=PDF (accessed September 26, 2019);

WIDNR, 2017, Chapter NR 140: Groundwater quality. https://docs.legis.wisconsin.gov/code/admin_code/nr/100/140.pdf (accessed September 26, 2019);

WYDEQ, 2018, Cleanup Level Look-up Table. http://deq.wyoming.gov/media/attachments/Solid%20%26%20Hazardous%20Waste/Voluntary%20Remediation%20Program/Fact%20Sheets/October%202018%20Cleanup%20Level%20Lookup%20Table.pdf (accessed September 26, 2019).

TABLE 6.2
Representative International Water Regulations and Guidance

Country/Regulatory Agency	Guidance	Standard	Concentration (μg/L)
Health Canada	Proposed maximum acceptable concentration (MAC)	Drinking water	50
German Federal Environmental Agency	Precautionary guideline limit for weak or non-genotoxic compounds	Drinking water	0.1
World Health Organization	Drinking water threshold value	Drinking water	50
Japan Ministry of Health, Labour, and Welfare	Drinking water quality standard	Drinking water	50
New Zealand Ministry of Health	Drinking water standards	Drinking water	50

Sources: Cheremisinoff, N.P., 2017, Groundwater Remediation: A Practical Guide for Environmental Engineers and Scientists. JT Wiley and Sons, June 13, 2017. 416 pp.

German Federal Environmental Agency, 2019, https://www.umweltbundesamt.de/sites/default/files/medien/421/dokumente/02_oasen_the_benefits_of_pmt_criteria_under_reach.pdf (September 20, 2019).

Health Canada, 2018, Guideline Technical Document for Public Consultation, 1,4 dioxane in Drinking Water. November 9, 2018. https://www.canada.ca/en/health-canada/programs/consultation-1-4-dioxane-drinking-water/document.html#a11 (accessed September 26, 2019);

Japan Ministry of Health, Labour, and Welfare, 2019, Drinking Water Quality Standards in Japan (April 2015). https://www.mhlw.go.jp/english/policy/health/water_supply/dl/4a.pdf (September 20, 2019);

New Zealand Ministry of Health, 2008, Drinking Water Standards for New Zealand, 2005 (Revised 2008). https://www.health.govt.nz/system/files/documents/publications/drinking-water-standards-2008-jun14.pdf (accessed September 20, 2019);

World Health Organization, 2005, 1,4-Dioxane in Drinking Water. https://www.who.int/water_sanitation_health/dwq/chemicals/14dioxane0505.pdf (accessed September 20, 2019);

TABLE 6.3
Representative Surface Water Regulations and Guidance

State	Regulatory Agency	State Guidance	Year	Standard	Concentration (μg/L)
Colorado	CDPHE	Water quality standards	2017b	Surface water	0.35
Connecticut	CTDEP	Recommended Protection Level	2018	Surface water	960
Florida	FDEP	Cleanup target levels	2005a	Surface water	120
Hawaii	HDOH	Environmental Action levels	2017	Surface water (freshwater)	0.46
				Surface water (marine or estuary)	50,000
Michigan	MI DEQ	Human health values	2019	Surface water (non-cancer, drink value)	890
				Surface water (non-cancer, non-drink value)	72,000
				Surface water (cancer, drink value)	3.5
				Surface water (cancer, non-drink value)	280

Sources: CDPHE, 2017b, Regulation No. 31—The Basic Standards and Methodologies for Surface Water. Colorado Department of Public Health and Environment;

CTDEP, 2018, Technical Support Document: Recommended Criteria Values for Common Additional Polluting Substances and Alternative Criteria Requests. September 2018. http://www.ct.gov/deep/lib/deep/site_clean_up/remediation_regulations/Technical_Support_Document_Additional_Polluting_Substances__Alternative_Criteria.pdf (accessed September 22, 2019);

FDEP, 2005a, Technical Report, Table 1, Groundwater Clean-up Target Levels, https://floridadep.gov/sites/default/files/62-777-TechRprt-Table1-Groundwater-CTLs_Feb2005.xlsx (accessed September 22, 2019);

HDOH, 2017, Evaluation of Environmental Hazards at Sites with Contaminated Soil and Groundwater. Hawai'i Department of Health Environmental Management Division. Fall 2017 (accessed September 22, 2019);

MI DEQ, 2019, Rule 57 Surface Water Quality Values. https://www.michigan.gov/documents/egle/egle-wrd-swas-rule57_662210_7.xlsx (accessed September 23, 2019).

TABLE 6.4
Representative Soil Regulations and Guidance

State	Regulatory Agency	State Guidance	Year	Standard	Concentration (mg/Kg)
Alaska	ADEC	Human health risk-based concentration	2017	Soil in direct contact	100
				Soil in direct contact (>40 ft to groundwater)	73
				Soil in direct contact (<40 ft to groundwater)	58
				Migration to groundwater	0.012
Arizona	DEC	Remediation level	2009	Residential soil (10E-6 Risk)	0.050
				Nonresidential soil	1.6
Colorado	CDPHE	Groundwater protection value	2014	Leaching to groundwater	0.0016
Connecticut	CTDEP	Recommended Exposure Criteria	2018	Residential direct exposure	6.1
				Industrial/commercial direct exposure	57
Delaware	DNREC	Screening level table	2018	Soil	5.3
Florida	FDEP	Cleanup target levels	2005b	Soil (with direct exposure—commercial or industrial)	38
				Soil (with direct exposure - residential)	23
				Soil (leachable by groundwater)	0.01
				Soil (leachable by surface water)	0.5
Georgia	GDNR	Notification level	2019	Soil	0.13
Hawaii	HDOH	Action levels	2017	Soil (migration to groundwater used for drinking water)	0.00021
				Soil (migration to groundwater NOT used for drinking water)	5.3
Indiana	IDEM	Screening levels	2019	Residential soil	74
				Commercial/industrial soil	240
				Excavation soil (contact during)	13,000
				Soil migration to groundwater	0.019
Iowa	IDNR	Statewide standard	2019	Soil standard	31
Kansas	KDHE	Tier 2 risk-based standards	2010	Residential soil	79.6
				Residential soil to groundwater	0.0384
				Nonresidential soil	247
				Nonresidential soil to groundwater	0.129
Maine	ME DEP	Remedial action guidelines	2018	Leaching to groundwater	0.052
				Residential	74
				Commercial worker	340
				Park user	260
				Recreator sediment	310
				Construction worker	1,700
Massachusetts	MA DEP	Contingency plan standards	2012	Soil (direct contact)	20
Michigan	MI DEQ	Cleanup criteria or screening levels	2018	Soil (industrial Drinking water protection)	7
				Soil (residential drinking water protection)	1.7
Minnesota	MPCA	Soil Reference Values	2009	Soil residential	150
				Soil industrial	250

(Continued)

TABLE 6.4 (Continued)
Representative Soil Regulations and Guidance Values

State	Regulatory Agency	State Guidance	Year	Standard	Concentration (mg/Kg)
Mississippi	MS DEQ	Target remediation goal	2002	Soil (restricted)	520
				Soil (unrestricted)	58.1
Missouri	MO DNR	Lowest Default Target Level	2006	Soil	0.235
Nebraska	NE DEQ	Remediation goals	2018	Soil (residential)	5.3
				Soil (industrial)	2,400
				Soil (protection of groundwater)	0.0019
Nevada	NE DEP	Basic comparison levels	2017	Soil (residential)	6.4
				Soil (indoor industrial/commercial w/o dermal)	65.4
				Soil (outdoor industrial/commercial)	36.3
New Hampshire	NHDES	Soil remediation standards	2007	Soil	5
New Mexico	NMED	Soil screening levels	2017	Soil (residential – cancer)	53.3
				Soil (residential – noncancer)	1850
				Soil (Industrial/occupational – cancer)	257
				Soil (Industrial/occupational – noncancer)	27,500
				Soil (Risk-based screening level Dilution Factor 1 – soil migration to groundwater)	0.000814
				Soil (Risk-based screening level Dilution Factor 20 – soil migration to groundwater)	0.0163
New York	NYSDEC (New York Department of Environmental Conservation)	Soil cleanup objectives	2019	Soil (unrestricted)	0.1
				Soil (residential)	9.8
				Soil (restricted residential)	13
				Soil (commercial)	130
				Soil (industrial)	250
				Soil (protection of groundwater)	0.1
North Carolina	NCDEQ	Preliminary Soil Remediation Goals	2018	Soil (residential health based)	5.4
				Soil (industrial/commercial health based)	25
				Soil (protection of groundwater)	0.012
Oregon	ODEQ	Risk-based concentrations – soil ingestion, dermal contact, and inhalation	2018	Soil (residential)	5.4
				Soil (residential – leaching to groundwater)	0.0023
				Soil (occupational)	24
				Soil (occupational – leaching to groundwater)	0.012
				Soil (construction worker)	210
Pennsylvania	PA DEP	Medium-Specific Concentrations for	2016	Soil (residential)	58
				Surface soil (non-residential)	290
				Subsurface soil (non-residential)	330
		Draft statewide health standard screening values for VI assessment	2016	Soil (residential)	0.23
				Soil (nonresidential)	5.9

(*Continued*)

TABLE 6.4 ***(Continued)***
Representative Soil Regulations and Guidance Values

State	Regulatory Agency	State Guidance	Year	Standard	Concentration (mg/Kg)
Texas	TCEQ	Protected concentration levels	2018	Soil (groundwater protection—residential)	0.018
				Soil (groundwater protection—industrial)	0.04
				Soil (total combined pathways—residential)	45
				Soil (total combined pathways—industrial)	150
Vermont	VTDEC	Vermont Soil Standards		Soil (resident)	2.8
				Soil (non-resident)	17
Washington	WA ECY	Soil clean-up levels	2019	Soil – Method B (unrestricted) non-cancer	2,400
				Soil – Method B (unrestricted) cancer	10.0
				Soil – Method C (industrial) non-cancer	110,000
				Soil – Method C (industrial) cancer	1,300
West Virginia	WVDEP	Risk-based concentrations	2018	Soil (residential)	5.4
				Soil (industrial)	400
				Soil (migration to groundwater)	0.000094
Wyoming	WYDEQ	Cleanup level lookup table	2018	Soil (residential)	5.30
				Soil (migration to groundwater)	0.000180

Sources: AK DEC, 2018, 18 AAC 75 Oil and Other Hazardous Substances Pollution Control, As amended through October 27, 2018, https://dec.alaska.gov/media/1055/18-aac-75.pdf (accessed September 22, 2019);

AZDEQ, 2009, Arizona Administrative Code Title 18, Ch. 7, Department of Environmental Quality – Remedial Action, March 31, 2009, https://apps.azsos.gov/public_services/Title_18/18-07.pdf (accessed September 22, 2019);

CDPHE, 2014, Groundwater Protection Values Soil Cleanup Table, https://environmentalrecords.colorado.gov/HPRMWebDrawerHM/RecordView/413311. March 2014 (accessed September 22, 2019);

CTDEP, 2018, Technical Support Document: Recommended Criteria Values for Common Additional Polluting Substances and Alternative Criteria Requests. September 2018. http://www.ct.gov/deep/lib/deep/site_clean_up/remediation_regulations/Technical_Support_Document_Additional_Polluting_Substances__Alternative_Criteria.pdf (accessed September 22, 2019);

DNREC, 2018, Screening Level Table, January 1, 2013, Last Updated February 2018, http://www.dnrec.delaware.gov/dwhs/SIRB/Documents/Screening%20Level%20Table.pdf (accessed September 22, 2019);

FDEP, 2005b, Technical Report, Table 2, Soil Clean-up Target Levels, https://floridadep.gov/sites/default/files/62-777TechReportApril2005Table2SoilCTLs-CAS_24Oct13.xlsx (accessed September 22, 2019);

GDNR, 2019, Georgia Administrative Code, Subject 391-3-19 Hazardous Site Response. Department of Natural Resources, September 6, 2019 (accessed September 22, 2019);

HDOH, 2017, Evaluation of Environmental Hazards at Sites with Contaminated Soil and Groundwater. Hawai'i Department of Health Environmental Management Division. Fall 2017. file:///C:/Users/wdiguise/Downloads/Volume%20%202%20App%201%20(HDOH%202017)%20(1).pdf (accessed September 22, 2019);

IDEM, 2019, 2019 Screening Level Table. Effective beginning March 4, 2019. https://www.in.gov/idem/cleanups/2392.htm (accessed September 2019);

IDNR, 2019, Chapter 137 Statewide Standards, Iowa Department of Natural Resources. https://programs.iowadnr.gov/riskcalc/Home/statewidestandards (accessed September 22, 2019);

KDHE, 2015, Risk-Based Standards for Kansas, Kansas Department of Health and Environment – Bureau of Environmental Remediation, October 2010, Revised September 2015. http://www.kdheks.gov/remedial/download/RSK_Manual_15.pdf (accessed September 22, 2019);

(*Continued*)

TABLE 6.4 *(Continued)*

Sources: (continued)

MADEP, 2014, 310 CMR 40.0000: Massachusetts Contingency Plan. April 25, 2014. Massachusetts Department of Environmental Protection. https://www.mass.gov/files/documents/2017/10/17/310cmr40.pdf (accessed September 22, 2019);

MEDEP, 2018, Maine Remedial Action Guidelines (RAGs) for Sites Contaminated with Hazardous Substances. Maine Department of Environmental Protection. Effective Date October 19, 2018. https://www.maine.gov/dep/spills/publications/guidance/rags/ME-Remedial-Action-Guidelines-10-19-18cc.pdf (accessed September 20, 2019);

MI DEQ, 2018, Part 201 Generic Cleanup Criteria and Screening Levels. Michigan Department of Environmental Quality. Updated June 25, 2018. https://www.michigan.gov/documents/deq/deq-rrd-Rules-Table2SoilResidential_447072_7.pdf (accessed September 24, 2019);

MO DNR, 2006, Missouri Risk-Based Corrective Action (MRBCA), Appendix A, Updates and Revisions to the MRBCA Technical Guidance. April 2006. https://dnr.mo.gov/env/hwp/mrbca/docs/mrbca-append6-06.pdf (accessed September 23, 2019);

MPCA, 2006, Risk-Based Site Evaluation Guidance for the Soil - Human Health Pathway, Tier 1 Table. Minnesota Pollution Control Agency. https://www.pca.state.mn.us/sites/default/files/c-r1-01.xls (accessed September 22, 2019);

MS DEQ, Risk Evaluation Procedures for Voluntary Cleanup and Redevelopment of Brownfield Sites, Appendix A, Tier 1 Target Remedial Goal Table. https://www.mdeq.ms.gov/wp-content/uploads/2017/05/Proced.pdf (accessed September 22, 2019);

NCDEQ, 2018, Preliminary Soil Remediation Goals (PSRG). https://files.nc.gov/ncdeq/Waste%20Management/DWM/SF/IHS/guidance/Feb2018_PSRGs.pdf (accessed September 26, 2019);

NEDEP, 2017, User's Guide and Background Technical Document for the Nevada Division of Environmental Protection Basic Comparison Levels for Human Health for the BMI Complex and Common Areas. Revised July 2017. https://ndep.nv.gov/uploads/documents/july-2017-bcl-guidance-doc.pdf (accessed September 23, 2019);

NEDEQ, 2018, Nebraska Voluntary Clean-up Program Guidance Document. http://deq.ne.gov/Publica.nsf/xsp/.ibmmodres/domino/OpenAttachment/Publica.nsf/D243C2B56E34EA8486256F2700698997/Attach/VCP 2018 Revised VCP Guidance Document Final.pdf (accessed September 23, 2019);

NHDES, 2007, New Hampshire Code of Administrative Rules, Soil Remediation Standards. February 1, 2007. https://www.des.nh.gov/organization/commissioner/legal/rules/documents/env-or600.pdf (accessed September 24, 2019);

NMED, 2017, "Risk Assessment Guidance for Site Investigations and Remediation. Volume I. Soil Screening Guidance for Human Health Risk Assessments. https://www.env.nm.gov/wp-content/uploads/2016/11/NMED_SSG_VOL-I_-March_2017_revised.pdf (accessed September 26, 2019);

NYSDEC, 2019, 6 CRR-NY 375-6.8. July 15, 2019. https://govt.westlaw.com/nycrr/Document/I4eadfca8cd1711dda432a117e6e0f345?viewType=FullText&originationContext=documenttoc&transitionType=CategoryPageItem&contextData=%28sc.Default%29 (accessed September 26, 2019;

ODEQ, 2018, Risk-Based Concentration for Individual Chemicals. https://www.oregon.gov/deq/FilterDocs/RBDMTable.pdf (accessed September 26, 2019);

PA DEP, 2016, Draft Land Recycling Program Technical Guidance Manual for Vapor Intrusion into Buildings from Groundwater and Soil under Act 2. June 13, 2016. http://files.dep.state.pa.us/EnvironmentalCleanupBrownfields/LandRecyclingProgram/LandRecyclingProgramPortalFiles/CSSAB/2016/July13/VI_guidance_final_draft_20160613.pdf (accessed September 27, 2019).

TCEQ, 2018, Table 3: Tier 1 Groundwater PCLs – Residential and Commercial/Industrial. http://www.tceq.texas.gov/assets/public/remediation/trrp/2018-PCL-Tables.xlsx (accessed September 26, 2019);

VTDEC, 2019, Environmental Protection Rules: Chapter 35: Investigation and Remediation of Contaminated Properties Rule. https://dec.vermont.gov/sites/dec/files/dwgwp/DW/2019.07.06%20-%20GWPRS.pdf (accessed September 26, 2019);

Washington Department of Ecology (ECY), 2019, Clean-up Levels and Risk Calculation (CLARC) tables. https://www.ezview.wa.gov/Portals/_1987/Documents/Documents/CLARCInterimUpdate.xlsx (accessed September 26, 2019);

WVDEP, 2018, Series 3: Voluntary Remediation and Redevelopment Rule. http://apps.sos.wv.gov/adlaw/csr/readfile.aspx?DocId=50235&Format=PDF (accessed September 26, 2019);

WYDEQ, 2018, Cleanup Level Look-up Table. http://deq.wyoming.gov/media/attachments/Solid%20%26%20Hazardous%20Waste/Voluntary%20Remediation%20Program/Fact%20Sheets/October%202018%20Cleanup%20Level%20Lookup%20Table.pdf (accessed September 26, 2019);

TABLE 6.5
Representative Vapor Intrusion Related Regulations and Guidelines

State	Regulatory Agency	State Guidance	Year	Standard	Concentration (μg/m³)
Connecticut	CTDEP	Recommended Indoor Air Concentration	2018	Indoor Air (residential)	0.24
Delaware	DEDNR	Screening level table	2018	Sub-slab and soil gas	18
Hawaii	HDOH	Environmental Action Level	2017	Indoor air (residential)	0.56
				Indoor Air (Commercial/ Industrial)	2.5
				Shallow Soil Vapor (Residential)	1,100
				Shallow Soil Vapor (Commercial/Industrial)	9,800
Illinois	Illinois EPA	Tier 1 soil gas remediation objectives for outdoor inhalation exposure	2013	Soil gas (residential)	16
				Soil gas (industrial/commercial)	30
				Soil gas (construction worker)	42
		Tier 1 soil gas and groundwater remediation objectives for indoor inhalation exposure	2013	Soil gas (residential)	0.22
				Soil gas (industrial/commercial)	2.3
Indiana	IDEM	Screening level	2019	Indoor air (residential)	5.6
				Indoor air (commercial/ industrial)	25
Maine	MDEP	Remedial action guidelines	2018	Indoor air (residential)	5.6
				Indoor air (commercial)	25
Michigan	MI DEQ	Initial risk screening level	2014	Inhalation (annual)	0.2
Montana	MT DEQ	Risk screening level	2012	Indoor air	320
Nebraska	NE DEQ	Remediation goals	2018	Indoor air (residential)	0.56
				Indoor Air (industrial)	25
				Sub-slab/soil gas (residential)	19
				Sub-slab/soil gas (industrial)	8,200
New Hampshire	NHDES	Ambient Air Limits	2018	24 hr ambient air	258
				Annual ambient air	30
New Mexico	NMED	Vapor intrusion (soil gas) and indoor air screening levels	2017	Indoor air (residential)	5.62
				Indoor Air (industrial)	27.5
				Vapor intrusion (residential)	187
				Vapor Intrusion (industrial)	918
Nevada	NE DEP	Basic comparison level	2017	Ambient air	0.562
Oregon	ODEQ	Soil vapor intrusion into buildings – risk-based concentrations	2018	Soil (residential)	28
				Soil (occupational)	370
		Soil gas vapor intrusion into buildings – risk-based concentrations	2018	Soil gas (residential)	110
				Soil gas (occupational)	2,500
		Groundwater vapor intrusion into buildings – risk-based concentrations	2018	Groundwater (residential)	340,000 ug/L
				Groundwater (occupational)	4,500,000 ug/L
		Air inhalation – risk-based concentrations	2018	Indoor air (residential)	0.56
				Indoor air (occupational)	2.5

(Continued)

TABLE 6.5 *(Continued)*
Representative Vapor Intrusion Related Regulations and Guidelines

State	Regulatory Agency	State Guidance	Year	Standard	Concentration (μg/m³)
Pennsylvania	PA DEP	Statewide health screening values	2016	Near-source soil gas (residential)	630
				Near-source soil gas (non-residential)	16,000
				Sub-slab (residential)	120
				Sub-slab (non-residential)	2,000
				Indoor air (residential)	3.2
				Indoor air (non-residential)	16
Rhode Island	RIDEM	Proposed acceptable ambient levels (AALs)	2008	One-hour AAL	3,000
				Annual AAL – for processes that achieved the lowest achievable emissions rate	1
				Annual AAL – For all other processes	0.1
Washington	WA ECY	Air clean-up levels	2019	Air – Method B (unrestricted) non-cancer	14.0
				Air – Method B (unrestricted) cancer	0.500
				Air – Method C (industrial) non-cancer	30.0
				Air – Method C (industrial) cancer	5.00

Sources: CTDEP, 2018, Technical Support Document: Recommended Criteria Values for Common Additional Polluting Substances and Alternative Criteria Requests. September 2018. http://www.ct.gov/deep/lib/deep/site_clean_up/remediation_regulations/Technical_Support_Document_Additional_Polluting_Substances__Alternative_Criteria.pdf (accessed September 22, 2019);

DNREC, 2018, Screening Level Table, January 1, 2013, Last Updated February 2018, http://www.dnrec.delaware.gov/dwhs/SIRB/Documents/Screening%20Level%20Table.pdf (accessed September 22, 2019);

HDOH, 2017, Evaluation of Environmental Hazards at Sites with Contaminated Soil and Groundwater. Hawai'i Department of Health Environmental Management Division. Fall 2017 (accessed September 22, 2019);

IDEM, 2019, 2019 Screening Level Table. Effective beginning March 4, 2019. https://www.in.gov/idem/cleanups/2392.htm (accessed September 2019);

IEPA, 2013, Part 742 Tiered Approach to Corrective Action Objectives, Table G - Tier 1 Soil Gas Remediation Objectives for the Outdoor Inhalation Exposure Route. May 15, 2013. http://www.ilga.gov/commission/jcar/admincode/035/03500742ZZ9996bgR.html (accessed September 24, 2019);

MEDEP, 2018, "Maine Remedial Action Guidelines (RAGs) for Sites Contaminated with Hazardous Substances." Maine Department of Environmental Protection. Effective Date October 19, 2018. https://www.maine.gov/dep/spills/publications/guidance/rags/ME-Remedial-Action-Guidelines-10-19-18cc.pdf (accessed September 20, 2019);

MI DEQ, 2014, Updated Screening Levels for 1,4-Dioxane. January 29, 2014. http://www.deq.state.mi.us/aps/downloads/ATSL/123-91-1/123-91-1_1hr_annual_ITSL_IRSL.pdf (accessed September 23, 2019);

MT DEQ, 2012, Typical Indoor Air Concentrations of Volatile Organic Compounds in Non-Smoking Montana Residences Not Impacted by Vapor Intrusion - A Montana Indoor Air Quality Investigation. August 2012. https://deq.mt.gov/Portals/112/Land/StateSuperfund/Documents/VI_guide/CompleteIndoorVOCReport.pdf (accessed September 23, 2019);

NEDEP, 2017, User's Guide and Background Technical Document for the Nevada Division of Environmental Protection Basic Comparison Levels for Human Health for the BMI Complex and Common Areas. Revised July 2017. https://ndep.nv.gov/uploads/documents/july-2017-bcl-guidance-doc.pdf (accessed September 23, 2019);

NEDEQ, 2018, Nebraska Voluntary Clean-up Program Guidance Document. http://deq.ne.gov/Publica.nsf/xsp/.ibmmodres/domino/OpenAttachment/Publica.nsf/D243C2B56E34EA8486256F2700698997/Attach/VCP 2018 Revised VCP Guidance Document Final.pdf (accessed September 23, 2019);

(Continued)

TABLE 6.5 *(Continued)*

Sources: (*continued*)

NHDES, 2018, Part Env-A 1450 Table of all Regulated Toxic Air Pollutants. January 5, 2018. https://www.des.nh.gov/organization/commissioner/legal/rules/documents/env-a1400.pdf (accessed September 24, 2019);

NMED, 2017, "Risk Assessment Guidance for Site Investigations and Remediation. Volume I. Soil Screening Guidance for Human Health Risk Assessments. https://www.env.nm.gov/wp-content/uploads/2016/11/NMED_SSG_VOL-I_-March_2017_revised.pdf (accessed September 26, 2019);

ODEQ, 2018, Risk-Based Concentration for Individual Chemicals. https://www.oregon.gov/deq/FilterDocs/RBDMTable.pdf (accessed September 26, 2019);

PA DEP, 2016, Draft Land Recycling Program Technical Guidance Manual for Vapor Intrusion into Buildings from Groundwater and Soil under Act 2. June 13, 2016. http://files.dep.state.pa.us/EnvironmentalCleanupBrownfields/LandRecyclingProgram/LandRecyclingProgramPortalFiles/CSSAB/2016/July13/VI_guidance_final_draft_20160613.pdf (accessed September 27, 2019).

RIDEM, 2008, Rhode Island Air Toxics: Guideline. http://www.dem.ri.gov/programs/benviron/air/pdf/airtoxgl.pdf (accessed September 26, 2019);

Washington Department of Ecology (ECY), 2019, Clean-up Levels and Risk Calculation (CLARC) tables. https://www.ezview.wa.gov/Portals/_1987/Documents/Documents/CLARCInterimUpdate.xlsx (accessed September 26, 2019;

and state drinking water and groundwater guidance levels are based in some way on a human health toxicity assessment conducted by the USEPA. In 2004, USEPA's National Center for Environmental Assessment in the Office of Research and Development began a review of the Integrated Risk Information System (IRIS) health assessment for 1,4-dioxane (USEPA, 2004). IRIS is a database of scientific positions on potential adverse human health effects that may result from chronic (or lifetime) exposure to chemicals in the environment. IRIS contains chemical-specific summaries of qualitative and quantitative health information in support of chemical-specific toxicity values. The IRIS database and toxicity values therein is used to support the development of risk assessments, site-specific environmental decisions, and federal and state rule making.

The IRIS Program completed the full toxicological review of 1,4-dioxane in September 2013 (USEPA, 2013). The assessment presents reference values for the noncancer effects of 1,4-dioxane (the reference dose [RfD], and the reference concentration [RfC]), and a cancer assessment and corresponding cancer toxicity value (oral cancer slope factor). The Toxicological Review and IRIS Summary were subjected to internal peer review, federal expert review, and external scientific peer review.

The USEPA IRIS cancer toxicity value of 0.10 (mg/kg-day^{-1}), the oral slope factor, was based on low-dose linear extrapolation of liver tumors in mice (see discussion in Chapter 5). Although this toxicity value currently serves as the toxicological basis for USEPA health-based values, more recent data and analyses suggest that a less stringent toxicity assessment may be warranted (Dourson et al., 2017), which has led some state agencies to adopt a less stringent application of the cancer toxicity value (i.e., a higher excess cancer risk level), and some international agencies to adopt completely different toxicity values (Health Canada, 2018).

Exposure parameters used in setting human health-based standards or guidance are often prescribed by the regulatory agency statutes or may also be subject to professional opinion and discretion. For example, the USEPA Office of Water and the USEPA Regionals Screening Level use different default exposure frequency and duration assumptions and different water ingestion rate assumptions. Thus, even when the same cancer slope factor and cancer risk level is utilized, resulting screening levels may be different. Table 6.6 presents examples of differing use of standard parameters for creating regulatory criteria amongst different agencies.

TABLE 6.6
Parameters Used in Some of the 1,4-Dioxane Drinking Water Standards or Guidance Levels

Parameter	USEPA OW	USEPA RSL	Alaska	Michigan	Canada
Cancer slope factor ((mg/kg-day)$^{-1}$)	0.1	0.1	0.1	0.1	NA
Cancer risk range	10^{-4}	10^{-6}	10^{-5}	10^{-5}	NA
Tolerable daily intake (mg/kg-day)[a]	NA	NA	NA	NA	0.0054
Body weight (kg)	70	80	80	80	70
Drinking water ingestion rate (L/day)	2	2.5	2.5	2.5	1.5
Relative source contribution (unitless)	NA	NA	NA	NA	0.2
Level (µg/L)	**35**	**0.46**	**4.6**	**7.2**	**50**

NA = Not applicable
[a] Tolerable daily intake (TDI) is the same things as the USEPA reference dose (RfD).

6.2 REGULATING 1,4-DIOXANE IN INDUSTRIAL WASTEWATER DISCHARGES AND RECYCLED WATER

1,4-Dioxane is not easily removed through aerobic wastewater treatment, the primary means used to treat domestic sewage (Abe, 1999). Discharge of 1,4-dioxane to sewers should be restricted where the discharge would cause a detectable concentration in the plant influent or effluent. In addition to water quality considerations, discharge of 1,4-dioxane to sanitary sewers is potentially hazardous because 1,4-dioxane is flammable.

1,4-Dioxane is not routinely analyzed in industrial pretreatment programs.* Title 33 of the Code of Federal Regulations governs navigable waters and includes Chapter 26, which addresses water pollution prevention and control. The National Pretreatment Program for wastewater is derived from these and other regulations. The program is a cooperative effort of federal, state, and local regulatory environmental agencies established to protect water quality by reducing the level of pollutants discharged by industry and other nondomestic wastewater sources into municipal sewer systems. Pretreatment of industrial wastewater prior to sewer discharge is the primary mechanism used to limit the amount of pollutants released to the sewer. The National Pretreatment Program protects POTWs from pollutants that could disrupt the chemical and biological balance necessary to operate the POTW and prevent introduction of pollutants to the POTW that may pass through the treatment works untreated. Limiting the chemical load arriving at the POTW improves opportunities to reuse wastewater and sludges generated by the POTW. Concentration limits for discharge to sewers are established by USEPA, state, or regional and local sewer districts and municipalities.

6.2.1 Industrial Wastewater Discharge Regulations

Pharmaceutical production is among the industries generating wastewater containing 1,4-dioxane. In 1995, revisions to the Clean Water Act introduced guidelines for effluent limitations and pretreatment standards for pollutants discharged to POTWs from pharmaceutical manufacturing plants (USEPA, 1997). The guidelines addressed 1,4-dioxane discharges by establishing pretreatment standards for existing sources, and best available technology effluent limitations, as listed in Table 6.7.

* Mohr et al., (2007) reviewed two decades of organic compound analyses at five POTWs in the San Francisco Bay area as part of a study on historical dry cleaner discharges.

TABLE 6.7
1,4-Dioxane Pretreatment Standards for Pharmaceutical Manufacturing Effluent

Standard	Long-Term Mean Concentration (mg/L)	Maximum for Any One Day (mg/L)	Monthly Average (mg/L)
Best available technology	0.8	8.4	2.6
Pretreatment standards for existing sources	1240	3160	1760

Source: U.S. Environmental Protection Agency (USEPA), 1997, *Federal Register* 62(153): 42,720–42,732.

USEPA's review of chemicals in pharmaceutical manufacturing effluent noted the occurrence of 1,4-dioxane. Although all six plants surveyed discharged 1,4-dioxane, USEPA determined that the average loading (0.061% of total loading, about 24,000 pounds annually) did not warrant regulation. In making this determination, USEPA acknowledged that 1,4-dioxane might pass untreated through POTWs and therefore solicited data from dischargers regarding the fate of 1,4-dioxane in biological treatment processes. As a placeholder, USEPA assumed that 1,4-dioxane should be at least as biodegradable as tetrahydrofuran on the basis of structural similarities (USEPA, 1998a). Subsequent research has shown that the first-order aerobic biodegradation decay constant for tetrahydrofuran is approximately twice that of 1,4-dioxane; therefore, 1,4-dioxane is unlikely to degrade as quickly as tetrahydrofuran (Zenker et al., 2000).

USEPA groups 1,4-dioxane in pharmaceutical wastewater with "alcohols and related pollutants," including methanol, ethanol, *n*-propanol, isopropanol, *n*-butyl alcohol, *tert*-butyl alcohol, amyl alcohol, formamide, *N,N*-dimethylaniline, pyridine, aniline, and petroleum naphtha. The economic feasibility of removing this entire group of compounds was evaluated by USEPA, but the cost, $40 million/year, was deemed excessive compared to the benefit, because USEPA assumed all compounds in the group were easily biodegradable.

Substantial efforts have been made to eliminate 1,4-dioxane discharge in treated wastewater effluent from the production of polyester fibers. In Spartanburg, North Carolina, the Hoechst Celanese polyester plant invested millions of dollars in a large-scale treatment system including construction of a 54,000-gallon holding tank, an azeotropic distillation column for separation of dioxane from the process wastewater, and a gas-fired thermal oxidizer for the destruction of 1,4-dioxane (O'Neal Inc., 2007).

1,4-Dioxane testing is required for effluent from leachate treatment at several landfills, for example, Laraway Waste Management facility in Elwood, Illinois (IEPA, 2007). At the Lowry Landfill Superfund Site in Aurora, Colorado, wastewater containing 1,4-dioxane is permitted for discharge to the City of Aurora's wastewater treatment plant.

Permissible discharges of 1,4-dioxane to the sewer decrease with increasing flow rates. Discharges of 10, 20, and 30 gallons/min are allowed to contain 1,4-dioxane at 5930, 3950, and 2970 μg/L, respectively (MWRD, 1999). As discussed in Chapter 8, aerobic wastewater treatment is not expected to effectively remove 1,4-dioxane, as demonstrated by the persistent detections of 1,4-dioxane in tertiary or secondary treated recycled water and described in the next section.

A challenge for groundwater treatment systems approved for chlorinated solvents includes discharging effluent to surface water bodies when 1,4-dioxane remains present at detectable concentrations in the treatment system effluent. In EPA Region 1, the General NPDES Permit issued for Remediation Activities allows 1,4-dioxane to remain in groundwater treatment system discharge at concentrations up to 200 μg/L (USEPA, 2016). In Japan and Korea, effluent discharge criteria for 1,4-dioxane are linked to the World Health Organization's 50 μg/L 1,4-dioxane drinking water standard, and set at 50 μg/L (JMOE, 2009; An et al., 2014). Discharge of 1,4-dioxane in industrial effluent has also been a problem for Barcelona, Spain, where discharges affect water quality in the Llobregat River, whose flows are often low or irregular. Mean 1,4-dioxane detections in

TABLE 6.8
1,4-Dioxane Detections in U.S. Municipal Treatment Plant Effluents

Summary Statistic	Dioxane Concentration in WWTP Effluent (μg/L)	In-stream Dilution Factors at 1323 Drinking Water Intakes Downstream
Mean ± Std. Dev.	1.11 ± 0.60	875 ± 2470
Minimum	<0.30	2.6
10th Percentile	0.4	44.5
Median	1.13	245
90th Percentile	1.75	1700
Maximum	3.30	48,113

Source: Simonich, S.M., Sun, P., Casteel, K., Dyer, S., Wernery, D., Garber, K., Carr, G., and Federle, T., 2013, Probabilistic analysis of risks to US drinking water intakes from 1,4-dioxane in domestic wastewater treatment plant effluents. *Integrated Environmental Assessment and Management* 9(4): 554–559.

Barcelona groundwater recharged by the Llobregat River and used to supply drinking water, ranged from 1.7 to 50.5 μg/L, with maximum concentration in one area measured at 241 μg/L (Carrera et al., 2017).

A comparatively unregulated discharge of wastewater containing detectable concentrations of 1,4-dioxane is wastewater treatment plant effluent. 1,4-Dioxane is not effectively removed during aerobic or anaerobic wastewater treatment; influent concentrations have been widely reported in the United States, Germany, and Japan as approximately equal to effluent concentrations. In Germany, where average wastewater influent concentrations in four wastewater plants are 834 ± 480 ng/L(nanograms per liter), no reduction of 1,4-dioxane during sewage treatment was observed (Stepien et al., 2014). A survey of 40 U.S. wastewater treatment plants for 1,4-dioxane in treated effluent showed the results summarized in Table 6.8 (Simonich et al., 2013).

Researchers at Oregon State University conducted a probabilistic analysis of the possible combination of high 1,4-dioxane in wastewater effluent and low dilution factors for downstream drinking water treatment intakes, and concluded that the probability of exceeding the USEPA 2012 Health Advisory Level for 1,4-dioxane in drinking water, 0.35 μg/L, was 0.0031 (Simonich et al., 2013). The subsequent publication of UCMR3 testing for 1,4-dioxane in surface water sources of drinking water had the surprising result that 10% of surface water sources had detections of 1,4-dioxane (>0.07 μg/L) and 2% of public water systems supplied by surface water exceeded the 0.35 μg/L reference concentration. The implication is that sources other than municipal wastewater treatment effluent, for example, large industrial sources upstream of drinking water intakes, are responsible for 1,4-dioxane detections in surface water sources of drinking water. It is also possible that in sewer systems with sewer mains below the water table in industrial areas, infiltration of 1,4-dioxane-contaminated groundwater is contributing to 1,4-dioxane in municipal wastewater treatment effluent.

While 1,4-dioxane was not routinely monitored in municipal wastewater treatment effluent in the 1970s and 1980s, measured concentrations of methyl chloroform provide an indication of the presence and magnitude of possible 1,4-dioxane in effluent discharges. In 1978, measurements of methyl chloroform on the effluent from the Joint Water Pollution Control Plant (JWPCP) of Los Angeles County Sanitation Districts were 130 ± 15 μg/L methyl chloroform on primary treated effluent, 180 ± 15 μg/L methyl chloroform on secondary treated effluent, and 190 ± 5 μg/L methyl chloroform on final effluent. Los Angeles City effluents in 1978 had 340 ± 250 μg/L methyl chloroform on primary treated effluent, and <10 μg/L on secondary treated effluent, while Orange County had 4000 ± 2700 μg/L methyl chloroform on primary treated effluent, and <10 μg/L on secondary treated effluent (Young, 1978). Knowledge of past wastewater effluent methyl chloroform concentrations can inform site investigations for potential sources of 1,4-dioxane, provided other wastewater indicators (e.g., caffeine, hormones, pharmaceuticals) are also detected.

The UCMR3 surface water testing for 1,4-dioxane shows that discharge of effluents containing 1,4-dioxane, even if at low enough concentrations to prevent aquatic toxicity, presents problems for drinking water quality, with 2% of public water systems supplied by surface water exceeding the 0.35 μg/L reference concentration, and 10% of public water systems supplied by surface water showing detectable concentrations of 1,4-dioxane (Adamson et al., 2017).

6.2.2 Recycled Water Regulations

Groundwater recharge projects in California using recycled water from POTWs in which 1,4-dioxane is present in the sewer influent must use reverse osmosis and advanced oxidation to achieve at least 0.5 log reduction in concentration. Post-treatment concentrations can be no greater than California's notification level (NL) for 1,4-dioxane, 1 μg/L (CDPH, 2007). The secondary/tertiary treated wastewater measured environmental concentration (MEC) 90th percentile concentration of 1,4-dioxane in California was determined to be 7.2 μg/L (SWRCB, 2018). The 7.2 μg/L 90th percentile MEC is based on measurements collected since the "Blue Ribbon Panel" report was first issued in 2010. In that report, the 90th percentile MEC for 1,4-dioxane was reported as 2.6 μg/L. The monitoring trigger level for 1,4-dioxane in recycled water destined for potable reuse is 1 μg/L, and 10 μg/L for recycled water designated for nonpotable reuse.

Recycled water projects designed for potable reuse, for example, indirect potable reuse, such as groundwater replenishment with recycled water, must be subjected to advanced water purification to achieve at least a 0.5-log (69%) reduction in 1,4-dioxane concentration (CDPH, 2014). Recycled water system operators must include challenge or spiking tests, using 1,4-dioxane, to demonstrate the proposed oxidation process will achieve the minimum 0.5-log reduction of influent 1,4-dioxane concentrations under the proposed oxidation process's normal full-scale operating conditions. Advanced water purification may include UV-oxidation such as UV-peroxidation, ozone peroxidation, or UV and sodium hypochlorite. California's Title 22 regulations governing recycled water are not prescriptive regarding treatment technology. Instead, California uses performance-based metrics to confirm that a utility's adopted treatment technology achieves treatment performance goals, including 0.5-log reduction of 1,4-dioxane.

6.3 AIR QUALITY REGULATIONS: OCCUPATIONAL HEALTH AND SAFETY AND AMBIENT AIR QUALITY

As described in Chapter 2, ambient air sampling at 45 locations in 12 cities during the early 1980s confirmed the presence of 1,4-dioxane concentrations ranging from less than 1–30 (parts per billion by volume; ppbv). Because 90% of the 1,4-dioxane produced for the United States was for stabilizing methyl chloroform, it is likely that the source of 1,4-dioxane in ambient air was from vapor degreasing with methyl chloroform, as well as the other direct industrial uses of 1,4-dioxane and the industries that produce 1,4-dioxane as a by-product.

The explosion hazard of 1,4-dioxane should not be underestimated. 1,4-Dioxane is a hazardous material that presents fire and explosion dangers from its direct use as a solvent. 1,4-Dioxane is a highly volatile and highly flammable liquid with an explosive/flammability limit of 2% in air at standard temperature and pressure. In addition, the vapor density of 1,4-dioxane is about three times denser than air, causing its vapors to sink. The combination of 1,4-dioxane's volatility, flammability, and vapor density account for its potential to travel considerable distances to an ignition source with a consequent risk of "flash back." The accident that killed the welder working on the space shuttle solid rocket booster casings (described in Chapter 1) serves as a grim reminder of the hazard that 1,4-dioxane, nitromethane, and other flammable stabilizers can pose to workers who handle these chemicals.

As described in Chapter 4, 1,4-dioxane can form explosive peroxide crystals on the threads of screw cap bottles. 1,4-Dioxane may also react vigorously with oxidizing agents. Risks from products containing 1,4-dioxane as an ingredient depend on the physicochemical properties of other ingredients present in the product and should be determined for the mixture as a whole (NICNAS, 1998).

Regulation of 1,4-dioxane as a workplace inhalation hazard and as a toxic air contaminant is reviewed in the remainder of this section.

6.3.1 Occupational Health and Safety Regulations for 1,4-Dioxane

Occupational use of 1,4-dioxane or work at facilities that produce 1,4-dioxane as a by-product creates the greatest potential for inhalation exposure to 1,4-dioxane. Beginning in the late 1970s, major advances were made in worker health and safety regulations and practices in the United States and other developed nations. Today, the likelihood of exposure is greatly diminished, if not eliminated, because methyl chloroform was banned in the 1990 Montreal Protocol Amendments to the Clean Air Act.

A variety of regulations have been promulgated by health and labor agencies to ensure that exposure to airborne contaminants in the workplace is controlled to prevent adverse health effects. Regulations establish safe exposure levels. The wide variety in established exposure levels reflects differences in the time period for the measurement, the underlying toxicity values upon which the exposure limits are based, and the safety factor applied. Regulatory bodies throughout the world use exposure standards including the Threshold Limit Value (TLV), Occupational Exposure Standard (OES), Maximum Exposure Limit (MEL), maximum allowable concentration (MAK), and Short-Term Exposure Limit (STEL). The common convention for an exposure period is either an 8-hour workday for chronic toxins or a 15-minute exposure period for acute toxins. Table 6.9 provides definitions for the common exposure standards, and Table 6.10 lists inhalation exposure limits developed for 1,4-dioxane. Figure 6.1 presents a graphical comparison of health limits and regulatory advisory thresholds for inhalation of 1,4-dioxane.

6.3.2 Emissions Standards and Air Quality Regulations for 1,4-Dioxane

1,4-Dioxane is designated as a hazardous air pollutant in Title 42, Public Health and Welfare, Chapter 85, on Air Pollution Prevention and Control. As early as 1978, 1,4-dioxane was recognized in New Jersey as among the top eight air pollutants for which emission reductions through regulation were necessary, together with benzene, carbon tetrachloride, chloroform, ethylene amine, ethylene dibromide, tetrachloroethylene, and TCE (Chemical Week, 1978).

Most ambient air quality monitoring programs do not monitor for 1,4-dioxane; however, there are many indications that it was likely to be present in solvent-polluted air in air basins with significant industrial activity. In 1999, 1,4-dioxane was ranked as 14th among 40 priority pollutants by the California Air Resources Board (Cal EPA, 1999; Kyle et al., 2001).

Regulation of 1,4-dioxane emissions varied by industry and by state. For example, vapor degreasing operations, a major source of 1,4-dioxane emissions, were regulated by the National Emission Standards for Halogenated Solvent Cleaning (40 CFR, 63.460 through 63.469). Local air pollution control districts inspect and enforce standards on degreasing operations to control solvent emissions. After 1998, wood furniture-manufacturing facilities releasing more than 1200 pounds of 1,4-dioxane per year were required to report emissions and track annual usage (USEPA, 1998b). California's Toxic Hot Spots rule requires all facilities emitting more than 85 pounds of 1,4-dioxane per year to prepare emission inventory plans and reports (CARB, 2008). Connecticut's emission regulation stipulates that 1,4-dioxane emissions must not exceed an 8-h hazard limiting value of 450 μg/m^3 or a half-hour maximum of 2250 μg/m^3 (CDEP, 2002).

TABLE 6.9
Common Occupational Exposure Standards (OESs) in the United States and Europe

Standard	Acronym	Definition
Threshold Limit Value (United States, EU)	TLV	The time-weighted average concentration of airborne substances to which nearly all workers may be repeatedly exposed, without adverse effect
Time-Weighted Average (United States, EU)	TWA	The time-weighted average concentration for a conventional 8-h day and 40-h week (ACGIH and EU)
Short-Term Exposure Limit (United States, EU)	STEL	The concentration to which workers can be exposed for short periods of time, usually 15 min, 4 times per day, without suffering adverse effects (ACGIH and EU)
Occupational Exposure Standard (UK)	OES	The concentration of an airborne substance averaged over a reference period, at which, according to current knowledge, there is no evidence that it is likely to be injurious to employees, if they are exposed, by inhalation, day after day, to that concentration (EH/40 Occupational Exposure Limits, Health & Safety Executive, UK)
Maximum Exposure Limit (UK)	MEL	The *maximum* concentration of an airborne substance, averaged over a reference period, to which employees may be exposed by inhalation under any circumstances. Unlike the OES, there may be a residual risk to health at this level of exposure.
Maximum Workplace Concentration (Germany)	MAK	The *maximum* allowable concentration of a working substance in the workplace atmosphere as a gas, steam, or aerosol that according to current knowledge does not impair the health of employees exposed during 8-h working days over the long term (German Commission for the Investigation of Health Hazards of Chemical Compounds in the Work Area)
Maximum Workplace Concentration (Netherlands)	MAC	The maximum concentration of a gas, steam, or aerosol of a substance in workplace air, which when inhaled during a workday has no adverse impact on the health of the employees (Dutch Expert Committee for Occupational Standards, Netherlands)
Level Limit Value and Ceiling Limit Value	LLV and CLV	LLV: An occupational exposure limit value for exposure during one working day; CLV: An occupational exposure limit value for exposure during a reference period of 15 min (Swedish Work Environment Authority)
Permissible Exposure Limits	PEL	The legal limit for exposure of an employee to a substance, usually given as a time-weighted average (TWA) (U.S. Occupational Safety and Health Administration)
Occupational Exposure Limit	OEL	The maximum concentration of an airborne substance, averaged over a reference period, to which employees may be exposed by inhalation under any circumstances (South African Variation of UK OES)
Immediately Dangerous to Life and Health	IDLH	The limit beyond which an individual will not be capable of escaping death or permanent injury without help in less than 30 min (U.S. NIOSH)
Minimal Risk Level	MRL	An estimate of the daily human exposure to a hazardous substance that is likely to be without appreciable risk of adverse noncancer health effects over a specified duration of exposure (U.S. ATSDR)

Notes: ACGIH, American Conference of Governmental Industrial Hygienists; EU, European Union; and UK, United Kingdom.

TABLE 6.10
Recommended or Regulated Inhalation Exposure Limits for 1,4-Dioxane

Agency	Level (ppm)/(mg/m³)	Context	References
NIOSH	500/1800	IDLH	NIOSH (2005)
OSHA	100/360	PEL; 8-h TWA	OSHA (2007)
United Kingdom	25/91	Long-term exposure limit: 8-h TWA	UKHSE (2007)
	100/366	STEL: 15-min reference period	
South Africa	25/90	OEL: 8 h	SAIOH (2007)
	100/360	OEL: 15-min peak	
Finland	25/91	HTP: 8-h exposure level	STM (2005)
	40/150	HTP: 15-min exposure level	
Germany	20/72	MAK: 8 h	Greim (2000)
	40/144	MAK: 15 min—based on eye irritation	
The Netherlands	12/43	MAC: 8 h	ECB (1999)
	24/66	MAC: 15 min	
Argentina	20/72	CMP	ILO (2007)
Hungary	10/36	Occupational exposure level: 8 h	ILO (2007)
National Authority for Occupational Safety and Health (Ireland)	20/72	OEL value: 8 h (reference period)	NAOSH (2002)
France	10/35	VME: 8-h exposure level	INRS (2006)
	40/150	VCLT: Level 15-min exposure level	
Japan	10/36	Occupational exposure level: 8 h	JSOH (2007)
Sweden	10/35	LLV: 8 h	SWEA (2005)
	25/90	CLV: 15 min	
United States	2/7.2	MRL: 14-day inhalation	ATSDR (2012)
	0.2/0.7	MRL: 15-365-day inhalation	

Sources: ATSDR, 2012, Toxicological Profile for 1,4-Dioxane. Centers for Disease Control and Prevention (CDC), Agency for Toxic Substances and Disease Registry.

European Chemicals Bureau (ECB), 1999, Risk assessment report. 1,4-Dioxane. Joint Research Center, European Chemicals Bureau;

Greim, H., 2000, 1,4-Dioxan. Gesundheitsschädliche Arbeitsstoffe, Toxikologisch-arbeitsmedizinische Begründungen von MAK-Werten [1,4-Dioxane. Workplace Health Hazards from Chemicals: Toxicological and Occupational Health Rationale for Maximum Workplace Concentration Values]. Weinheim, Germany: VCH Verlag;

ILO, 2007, Chemical exposure limits. International Occupational Safety and Health Information Centre. http://www.ilo.org/public/english/protection/safework/cis/products/explim.htm#swe (accessed November 11, 2007);

INRS, 2006, Valeurs limites d'exposition professionnelle aux agents chimiques en France, Institut National de Recherche et de Securite [Exposure limit values for chemical agents in the workplace (in French), National Institute of Research and Safety]. http://www.inrs.fr/htm/valeurs_limites_exposition_professionnelle_agents.html (accessed March 11, 2007);

JSOH, 2007, Recommendation of occupational exposure limits. Journal of Occupational Health 49: 328–344;

NAOSH, 2002, Code of Practice for the Safety, Health and Welfare at Work. [Irish] National Authority for Occupational Safety and Health;

NIOSH, 2005, NIOSH pocket guide to chemical hazards. U.S. Department of Health and Human Services, Centers for Disease Control and Prevention and National Institute for Occupational Safety and Health;

OSHA, 2007, Occupational safety and health standards. U.S. Department of Labor, Occupational Safety and Health Administration, 1910.000, Table Z-1 (Limits for Air Contaminants);

SAIOH, 2007, Occupational Exposure Levels. Parktown, South Africa: South African Institute for Occupational Hygiene;

STM, 2005, HTP-arvot 2005 (Finnish Occupational Exposure Levels). Helsinki, Finland: Ministry of Social Affairs and Health;

SWEA, 2005, Occupational exposure limit values and measures against air contaminants. Swedish Work Environment Authority, Solna, Sweden. www.av.se/dokument/inenglish/legislations/eng0517.pdf (accessed March 18, 2007);

UKHSE, 2007, EH/40 Occupational Exposure Limits, United Kingdom Health & Safety Executive. http://www.hse.gov.uk/coshh/table1.pdf (accessed November 17, 2007);

Note: CMP, Concentración Máxima Permisible; HTP, Haitallisiksi Tunnetut Pitoisuudet (concentrations known to be hazardous); NIOSH, U.S. National Institute for Occupational Safety and Health; OSHA, U.S. Occupational Safety and Health Administration; VCLT, Valeurs Limites Court Term (short-term exposure limits, i.e., STELs); VME, Valeur Limite de Moyenne d'Exposition (average exposure limit values).

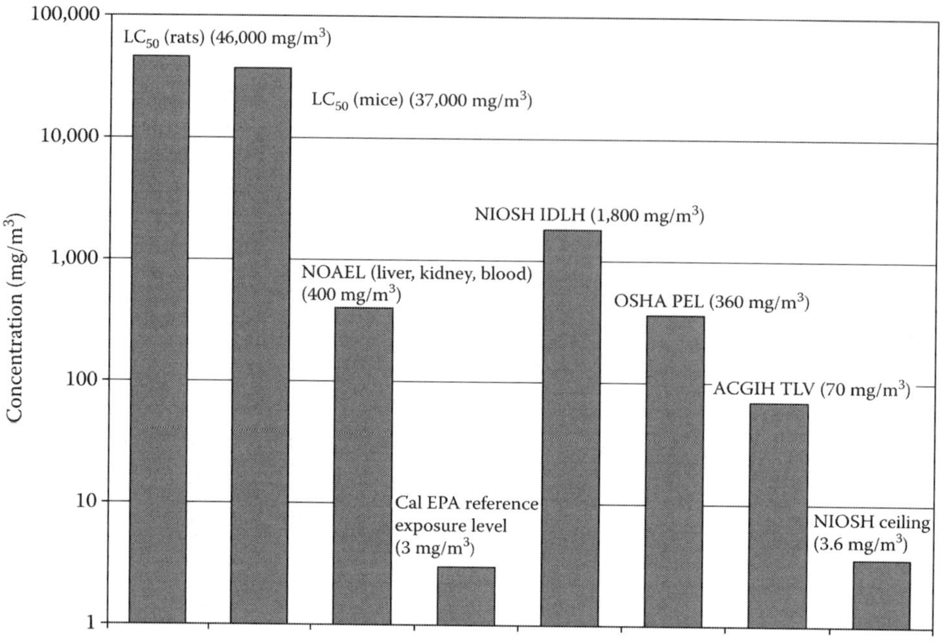

FIGURE 6.1 Comparison of health guidance and regulatory levels of 1,4-dioxane in 1 m^3 of air at standard temperature and pressure for inhalation of 1,4-dioxane. [After U.S. Environmental Protection Agency (USEPA), 2007b, 1,4-Dioxane (1,4-diethyleneoxide) hazard summary. U.S. Environmental Protection Agency, Technology Transfer Network, Office of Air and Radiation. http://www.epa.gov/ttn/atw/hlthef/dioxane.html (accessed November 11, 2007)].

Notes:

ACGIH TLV: American conference of governmental and industrial hygienists' threshold limit value expressed as a time-weighted average; the concentration of a substance to which most workers can be exposed without adverse effects.

LC_{50} Lethal concentration (50% mortality); a calculated concentration of a chemical in air to which exposure for a specific length of time is expected to cause death in 50% of a defined experimental animal population.

NIOSH IDLH: National Institute of Occupational Safety and Health's immediately-dangerous-to-life-or-health limit; NIOSH recommended exposure limit to ensure that a worker can escape from an exposure condition that is likely to cause death or immediate or delayed permanent adverse health effects or prevent escape from a hazardous atmosphere.

NIOSH: NIOSH's recommended exposure limit ceiling; the concentration that should not be exceeded ceiling at any time.

NOAEL: No observed adverse effect level.

OSHA PEL: Occupational safety and health administration's permissible exposure limit expressed as a time-weighted average; the concentration of a substance to which most workers can be exposed without adverse effect averaged over a normal 8-h workday or a 40-h workweek.

Cal EPA: California Environmental Protection Agency; the reference exposure level is based on nasal reference and eye irritation in healthy human volunteers where exposure averaging time is 1 h. The exposure chronic exposure REL is based on the liver, kidney, and hematological changes in rats and is level an 8-h exposure level. Both the acute and chronic RELs are 3 mg/m^3.

Verschueren (2001) offered the following equation to convert the mass concentration of an airborne contaminant (*W*/*V*) to its volume concentration (C_{vol}):

$$C_{vol} = \frac{W}{V}\frac{RT}{Pm}, \tag{6.1}$$

where C_{vol} is the volumetric concentration of the contaminant (in ppb), *W* is the mass of the compound present in volume *V* (in grams), *V* is the volume of contaminated air (in m^3; $V = 1$ m^3), *T* is the temperature (in kelvins; $T = 298.16$ K [25°C]), P is the pressure (in atmospheres; $P = 1$ atm = 101.325 kPa), m is the molecular weight (in g/mol), $R = 8.205 \times 10^{-5}$ (m^3 atm)/(mol K), and $W/m = n$, the number of moles.

For 1,4-dioxane at 25°C (molecular weight, 88.20), 1 μg/m^3 is equivalent to 0.278 ppb, and 1 ppb is equivalent to 3.6 μg/m^3.

6.4 REGULATING 1,4-DIOXANE IN FOOD, PHARMACEUTICALS, COSMETICS, AND SUNDRIES

As stated by EPA, "1,4-Dioxane is currently used in industrial processes and for industrial and commercial uses." (USEPA, 2018c). However, as discussed in Chapter 2, 1,4-dioxane may be present in food, personal care products, and pharmaceuticals. Although 1,4-dioxane has been measured in residual concentrations in food, pharmaceuticals, personal care products including cosmetics, shampoos, and lotions, and in dish-washing detergents and soaps (ATSDR, 2012), a study conducted by the U.S. Centers for Disease Control and Prevention collected human blood samples in 2007–2008 from a geographically diverse population of U.S. residents ≥12 years old and found no detectable 1,4-dioxane (Wang et al., 2009), suggesting that widespread or significant exposure to the U.S. generable population is not likely occurring. 1,4-Dioxane produced as a by-product in the production of other chemicals is currently excluded from the scope of the EPA upcoming Toxic Substances Control Act (TSCA) evaluation (USEPA, 2017b, 2018c). However, 1,4-dioxane by-product and residual contaminant issues may be considered in an evaluation of ethoxylated chemicals, if the EPA conducts such an assessment.

The only current federal U.S. regulation pertaining to 1,4-dioxane in consumer products or food is the U.S. FDA regulation of 1,4-dioxane as an indirect food additive, which also includes its presence in food contact materials, at a limit of 10 ppm (21 CFR 172.736). Based on a review of analyses done by international scientific agencies, the FDA concluded that the levels of 1,4-dioxane observed in cosmetics do not present a hazard to consumers and the agency does not have a current recommendation or required limit for 1,4-dioxane in cosmetics (see https://www.fda.gov/cosmetics/potential-contaminants-cosmetics/14-dioxane-cosmetics-manufacturing-byproduct). The FDA continues to recommend that cosmetic and personal care product manufactures use vacuum stripping to ensure negligible 1,4-dioxane levels in final products.

Finally, the allowable limit for 1,4-dioxane present as a residual solvent in veterinary and human medicines in the United States and Europe is 3.8 mg/d (the "permitted daily exposure," or PDE), and the permissible concentration limit in veterinary and human pharmaceuticals is 380 ppm (FDA, 2001; EMEA, 2006). There are no other applicable federal regulations for 1,4-dioxane in personal care products, foods, or pharmaceuticals.

6.4.1 California Proposition 65

In 1986, California voters approved an initiative that requires the state to publish a list of toxic chemicals known to the state to cause cancer, birth defects, or other reproductive harm. The initiative, called the Safe Drinking Water and Toxic Enforcement Act of 1986, came to be known as Proposition 65. California businesses are required to notify the public about significant amounts of chemicals in the products for sale or used in their homes or workplaces, so that they may make informed decisions about protecting themselves from exposure to the listed chemicals. Proposition 65 also prohibits discharge of the listed chemicals to sources of drinking water. Proposition 65 is enforced by the California Attorney General's Office.

The California Environmental Protection Agency (Cal EPA) lists chemicals under Proposition 65 if a panel of independent scientists and health professionals finds that the chemical causes cancer, birth defects, or other reproductive harm. Chemicals may also be listed if other organizations, including USEPA, FDA, National Institute for Occupational Safety and Health (NIOSH), National Toxicology Program (NTP), and International Agency for Research on Cancer (IARC), classifies a compound in any of these categories. If a chemical on the list is present in a product, businesses must post a warning unless exposure is low enough to pose no significant risk of cancer or is significantly below levels observed to cause birth defects or other reproductive harm. The list of chemicals must be updated at least annually; it now includes approximately 900 chemicals (https://www.p65warnings.ca.gov/).

To help businesses determine when they are subject to the rules of Proposition 65, Cal EPA's OEHHA publishes "safe harbor numbers," below which businesses are exempt from the Proposition 65 warning requirements. California's safe harbor numbers consist of No Significant Risk Level (NSRL) for chemicals listed as causing cancer and Maximum Allowable Dose Levels (MADLs) for chemicals listed as causing birth defects or other reproductive harm. For carcinogens such as 1,4-dioxane, the NSRL is the level of exposure that would result in not more than one excess risk of cancer in 100,000 individuals exposed to the chemical over a 70-year lifetime. 1,4-Dioxane first appeared on the California Proposition 65 list of chemicals "known to the state to cause cancer" on January 1, 1988. The NSRL for 1,4-dioxane, is 30 μg/d (Cal EPA, 2007). Because 1,4-dioxane causes neither birth defects nor reproductive toxicity, an MADL has not been established.

6.4.2 Washington Children's Safe Products Act

The Washington State Department of Ecology is responsible for executing the state's 2011 Children's Safe Product Act, which requires manufacturers of children's products to annually notify the department when a chemical of high concern to children (CHCC) is present in items sold in the state (CSPA Reporting Rule, Chapter 173-334 Washington Administrative Code, authorized by the Children's Safe Products Act, Chapter 70.240 Revised Code of Washington). The presence of a CHCC in a product does not mean that the product is necessarily harmful, but the reported information is geared to provide transparent information to consumers and agencies. Similar to California Proposition 65, the Washington Children's Safe Products Reporting Rule prioritizes carcinogenicity, reproductive/developmental toxicity and endocrine disruption, and also prioritizes potential for child exposure, specifically. 1,4-Dioxane was added to the state's initial CHCC list in 2011 after evaluations by the state departments of ecology and health.

ACKNOWLEDGMENT

The critical review, questions, corrections, and comments from Janet Anderson, GSI Environmental Inc., are greatly appreciated. The author retains responsibility for the interpretation and accuracy of the information provided in this chapter.

REFERENCES

Abe, A., 1999, Distribution of 1,4-dioxane in relation to possible sources in the water environment. *The Science of the Total Environment* 227: 41–48.

Adamson, D.T., Piña, E.A., Cartwright, A.E., Rauch, S.R., Anderson, R.H., Mohr, T.K., and Connor, J.A., 2017, 1,4-Dioxane drinking water occurrence data from the third unregulated contaminant monitoring rule. *Science of the Total Environment* 596–597: 236–245.

AFCEE, 2007, 1,4-Dioxane—a primer for air force remedial program managers and risk assessors. Air Force Center for Environmental Excellence.

An, Y-J., Kwak, J., Nam, S-H., and Jung, M.S., 2014, Development and implementation of surface water quality standards for protection of human health in Korea. *Environmental Science and Pollution Research* 21: 77–85.

ATSDR, 2012, Toxicological Profile for 1,4-Dioxane. Centers for Disease Control and Prevention (CDC), Agency for Toxic Substances and Disease Registry.

Cal EPA, 1999, Air toxics hot spots program risk assessment guidelines, part I: The determination of acute reference exposure levels for airborne toxicants. California Environmental Protection Agency, Office of Environmental Health Hazard Assessment.

Cal EPA, 2007, Proposition 65 safe harbor levels. California Environmental Protection Agency, Office of Environmental Health Hazard Assessment.

Cal EPA, 2011, "1,4-Dioxane." Drinking Water Systems. California Department of Public Health. www.waterboards.ca.gov/drinking_water/certlic/drinkingwater/14-Dioxane.shtml (accessed September 20, 2019).

CARB, 2008, California toxics inventory database. California Air Resources Board. http://www.arb.ca.gov/toxics/cti/cti.htm (accessed January 11, 2008).

Carrera, G., Vegué, L., Rosa Boleda, R., and Ventura, F. 2017. Simultaneous determination of the potential carcinogen 1,4-dioxane and malodorous alkyl-1,3-dioxanes and alkyl-1,3-dioxolanes in environmental waters by solid-phase extraction and gas chromatography tandem mass spectrometry. *Journal of Chromatography A* 1487: 1–13.

CDPH, 2007, Groundwater recharge reuse draft regulation. Title 22, California Code of Regulations, Division 4, Environmental Health, Chapter 3, Recycling Criteria. California Department of Public Health.

CDPH, 2014, DPH-14-003E GW Replenishment Using Recycled Water. May 30, 2014. Title -22, California Code Of Regulations Division 4, Chapter 3.

Chemical Week, 1978, Jersey fights smog. *Chemical Week* (15): 24.

Dourson, M., Higginbotham, J., Crum, J., Burleigh-Flayer, H., Nance, P., Forsberg, N., Lafranconi, M., and Reichard, J., 2017, Update: Mode of action (MOA) for liver tumors induced by oral exposure to 1,4-dioxane. *Regulatory Toxicology and Pharmacology* 88: 45–55.

EMEA, 2006, Note for guidance on impurities: Residual solvents. *International Conference on Harmonisation (ICH) of Technical Requirements for Registration of Pharmaceuticals for Human Use. 24*. London, UK: European Medicines Agency.

Food and Drug Administration (FDA), 2001, Impurities: Residual solvents in new veterinary medicinal products, active substances and excipients. Rockville, MD: Center for Veterinary Medicine, U.S. Food and Drug Administration, http://www.fda.gov/cvm/Guidance/guide100.doc (accessed June 8, 2006).

German Federal Environmental Agency, 2019, https://www.umweltbundesamt.de/sites/default/files/medien/421/dokumente/02_oasen_the_benefits_of_pmt_criteria_under_reach.pdf (September 20, 2019).

IEPA, 2007, NPDES permit no. IL0063479 draft reissued NPDES permit to discharge into waters of the state. Illinois Environmental Protection Agency.

JMOE, 2009, Review of environmental standards on the protection of human health related to water pollution, second report. Japanese Ministry of Environment, Committee on Health Items Environmental Standards.

Kyle, A.D., Wright, C.C., Caldwell, J.C., Buffler, P.A., and Woodruff, T.J., 2001, Evaluating the health significance of hazardous air pollutants using monitoring data. *Public Health Reports* 116: 32–44.

Macler, B., 2006, Regulating emerging contaminants. … What can you expect? Paper presented at Groundwater Resources Association of California symposium Emerging Contaminants in Groundwater—A Continually Moving Target, Concord, CA, June 7–8. http://www.grac.org (accessed August 30, 2006).

Mohr, T.K.G., Tulloch, C., Chan, S., Cook, G., Giri, S., Broughton, P., and Crowley, J., 2007, *Study of Potential for Groundwater Contamination from Past Dry Cleaner Operations in Santa Clara County*. San Jose: Santa Clara Valley Water District. https://www.waterboards.ca.gov/sanfranciscobay/publications_forms/documents/SCVWD_dry_cleaner_study_2007.pdf

MWRD, 1999, Industrial Wastewater Discharge Permit 1–118. City of Aurora Utilities Department, Metro Wastewater Reclamation District.

NCDEQ, 2018, "Groundwater Classification and Standards." North Carolina Department of Environmental Quality. https://deq.nc.gov/about/divisions/water-resources/water-resources-rules/nc-administrative-code-statutes (accessed September 20, 2019).

NICNAS, 1998, Priority existing chemical assessment no. 7—1,4-dioxane. Department of Health and Aging, Australian National Industrial Chemicals Notification and Assessment Scheme (NICNAS).

O'Neal Inc., 2007, Dioxane reduction system. Hoechst Celanese. http://www.onealinc.com/process.cfm (accessed October 31, 2007).

OEHHA, 2018, Public Health Goals: cis- and trans-1,2-Dichloroethylene in Drinking Water July 2018. Office of Environmental Health Hazard Assessment.

PADEP, 2016, Statewide Health Standards: Revised MCLs and HALs. Pennsylvania Department of Environmental Protection. http://www.dep.pa.gov/Business/Land/LandRecycling/Standards-Guidance-Procedures/Pages/Statewide-Health-Standards.aspx (accessed September 20, 2019).

Simonich, S.M., Sun, P., Casteel, K., Dyer, S., Wernery, D., Garber, K., Carr, G., and Federle, T., 2013, Probabilistic analysis of risks to US drinking water intakes from 1,4-dioxane in domestic wastewater treatment plant effluents. *Integrated Environmental Assessment and Management* 9(4): 554–559.

Stepien, D., Diehl, P., Helm, J., Thomas, A., and Püttmann, W., 2014, Fate of 1,4-dioxane in the aquatic environment: From sewage to drinking water. *Water Research* January 1 (48): 406–419.

SWRCB, 2018, Draft Final Report Monitoring Strategies for Constituents of Emerging Concern (CECs) in Recycled Water Recommendations of a Science Advisory Panel. State Water Resources Control Board January 31, 2018, Sacramento, California.

TCEQ, 2017, March 2017 Tier 1 PCL Tier and supporting tables. Texas Commission on Environmental Quality. https://www.tceq.texas.gov/remediation/trrp/trrppcls.html (accessed September 20, 2019).

U.S. Environmental Protection Agency (USEPA), 1997, Notice of availability; effluent limitations guidelines, pretreatment standards, and new source performance standards: Pharmaceutical manufacturing category—40 CFR part 439. *Federal Register* 62(153): 42,720–42,732.

U.S. Environmental Protection Agency (USEPA), 1998a, Development document for final effluent limitations guidelines and standards for the pharmaceutical manufacturing point source category. *U.S. Environmental Protection Agency*, Office of Science and Technology, Engineering and Analysis Division.

U.S. Environmental Protection Agency (USEPA), 1998b, National emission standards for hazardous air pollutants: Wood furniture manufacturing operations. *Federal Register* 63(248): 71,376–71,385.

U.S. Environmental Protection Agency (USEPA), 2004, Integrated risk information system (IRIS) announcement of 2004 program request for information. *Federal Register* 69(26): 5971–5976. http://www.wais.access.gpo.gov (accessed June 17, 2006).

U.S. Environmental Protection Agency (USEPA), 2006, Water quality standards. http://www.epa.gov/water-science/standards/(accessed November 11, 2007).

U.S. Environmental Protection Agency (USEPA), 2007a, Options for discharging treated water from pump and treat systems. U.S. Environmental Protection Agency, Office of Solid Waste and Emergency Response.

U.S. Environmental Protection Agency (USEPA), 2007b, 1,4-Dioxane (1,4-diethyleneoxide) hazard summary. U.S. Environmental Protection Agency, Technology Transfer Network, Office of Air and Radiation. https://www.epa.gov/sites/production/files/2016-09/documents/1-4-dioxane.pdf (accessed September 20, 2019).

U.S. Environmental Protection Agency (USEPA), 2012, Revisions to the Unregulated Contaminant Monitoring Regulation (UCMR 3) for Public Water Systems, pages 43523–43524 [FR DOC #C2-2012-9978].

U.S. Environmental Protection Agency (USEPA), 2013, Toxicological Review of 1,4-Dioxane, September 2013.

U.S. Environmental Protection Agency (USEPA), 2016, Re-Issuance of the National Pollutant Discharge Elimination System (NPDES) General Permit for Remediation Activity Discharges – The Remediation General Permit (RGP); NPDES Permit MAG910000 and NHG910000. December 28, 2016.

U.S. Environmental Protection Agency (USEPA), 2017a, The Third Unregulated Contaminant Monitoring Rule (UCMR 3): Data Summary, January 2017.

U.S. Environmental Protection Agency (USEPA), 2017b, Scope of the Risk Evaluation for 1,4-Dioxane, Office of Chemical Safety and Pollution Prevention, Washington, DC, June 2017, EPA-740-R1-7003.

U.S. Environmental Protection Agency (USEPA), 2018a, Drinking Water Standards and Health Advisories, Office of Water, Washington, DC, March 2018, EPA 822-F-18-001.

U.S. Environmental Protection Agency (USEPA), 2018b, Regional Screening Level (RSL) Summary Table (TR=1E-06, HQ=0.1) November 2018. https://www.epa.gov/risk/regional-screening-levels-rsls-whats-new (accessed November 20, 2018).

U.S. Environmental Protection Agency (USEPA), 2018c, Problem Formulation of the Risk Evaluation for 1,4-Dioxane, Office of Chemical Safety and Pollution Prevention, Washington, DC, May 2018, EPA-740-R1-7012.

Verschueren, K., 2001, *Handbook of Environmental Data on Organic Chemicals*, 4th ed. New York, NY: John Wiley.

Wang, R.Y., Blount, B.C., Chambers, D.M., et al., 2009, Measuring 1,4-dioxane in blood as a biomarker of exposure [abstract]. *Toxicologist* 108: 336.

Washington Department of Ecology (ECY), 2015, https://ecology.wa.gov/Regulations-Permits/Guidance-technical-assistance/Contamination-clean-up-tools/CLARC (accessed September 20, 2019).

Water Research Foundation, 2014, 1,4-Dioxane White Paper https://www.savmn.com/DocumentCenter/View/290/14-Dioxane-White-Paper-from-the-Water-Research-Foundation-PDF (accessed 9/27/2019).

West Virginia Department of Environmental Protection (WV DEP), 2009, "Voluntary Remediation and Redevelopment Rule." www.dep.wv.gov/dlr/oer/voluntarymain/Documents/60CSR3%20VRRA%20rule%206-5-09.pdf. (accessed September 26, 2019).

Young, D.R., 1978, Priority Pollutants in Municipal Wastewaters. Annual Report. South California Coastal Water Project. ftp://ftp.sccwrp.org/pub/download/DOCUMENTS/AnnualReports/1978AnnualReport/ar06.pdf

Zenker, M.J., Borden, R.C., and Barlaz, M.A., 2000, Mineralization of 1,4-dioxane in the presence of a structural analog. *Biodegradation* 11: 239–246.

7 Risk Assessment

Barrie Selcoe and William B. Kappleman

CONTENTS

Risk assessments evaluate the likelihood and severity of harm to human health and environmental receptors that may result from exposure to a chemical substance. Risk assessments include four components: source or release assessment, exposure assessment, dose-response assessment, and risk characterization. Source assessment includes monitoring and modeling to estimate the magnitude of the release and data screening to identify chemicals of potential concern. Exposure assessment evaluates the chemical concentrations to which populations may be exposed by determining the routes of exposure and quantifying the magnitude of exposure. Dose-response assessment is the quantitative evaluation of toxicity as a function of exposure. Risk characterization integrates source assessment, exposure assessment, and dose-response assessment into a statement of the types and magnitudes of adverse effects that the contaminant may cause, qualified by probability of occurrence and uncertainty associated with the estimates (Cohrssen and Covello, 1989).

7.1 HUMAN HEALTH RISK ASSESSMENT OF 1,4-DIOXANE

A number of hazard assessments for 1,4-dioxane have been published by various government agencies (see Chapter 5). The key risk summaries reviewed for this book are listed in Table 7.1.

7.1.1 Hazard Identification and Data Screening

1,4-Dioxane has been detected in air, groundwater, surface water, soil, sediment, fish, and consumer products (primarily detergents and personal care products). Through environmental fate processes, it may migrate from one impacted environmental medium to another (see Chapter 3).

Risk-based concentrations are available from various sources for comparison to site data. Table 7.2 presents a compilation of screening levels (risk-based concentrations) for various media from a variety of sources and indicates the risk basis for the value.

The equations and default assumptions incorporated into the USEPA RSLs (USEPA, 2019a) for air, groundwater, and soil are provided in the *Regional Screening Levels (RSLs) – User's*

TABLE 7.1
Key Reports on 1,4-Dioxane Exposure

Year	Title	Publishing Agency
2018	1,4-Dioxane in Drinking Water – Guideline Technical Document for Public Consultation	Health Canada (HC)
2016	Report on Carcinogens, Fourteenth Edition, 1,4-Dioxane	National Toxicology Program
2015	1,4-Dioxane: Reevaluation of the Carcinogenicity and Genotoxicity	Health Council of the Netherlands (HCN)
2013	Toxicological Review of 1,4-Dioxane (with Inhalation Update)	USEPA
2012	Toxicological Profile for 1,4-Dioxane	ATSDR
2010	Screening Assessment for the Challenge, 1,4-Dioxane	Environment Canada, HC
2005	1,4-Dioxane in Drinking Water	World Health Organization (WHO)
2002	European Union Risk Assessment Report: 1,4-Dioxane	European Chemicals Bureau (ECB)
1998	1,4-Dioxane: Priority Existing Chemical No. 7	National Industrial Chemicals Notification and Assessment Scheme (NICNAS)

Guide (USEPA, 2019b). However, because there are no default tabulated RSLs for surface water or sediment, screening levels were calculated using USEPA's RSL Calculator, an assumed exposure frequency of 52 days/year (once per week), assumed exposure time of 2 hours/day (on the days exposed), and all other exposure factor values set at default values embedded in the calculator.

Derivation of the ACGIH threshold limit value (TLV) and time-weighted average (TWA) value is provided in "1,4-Dioxane: Threshold Limit Values for Chemical Substances and Physical Agents and Biological Exposure Indices" (ACGIH, 2011).

The source for the Health Council of the Netherlands (HCN) value is provided in "1,4-Dioxane – Reevaluation of the Carcinogenicity and Genotoxicity" (HCN, 2015).

TABLE 7.2
Screening Levels for 1,4-Dioxane

Medium	Value	Source	Basis	Type
Air – Residential	0.56 μg/m^3	USEPA RSL	TR = 10^{-6}, THQ = 1	Screening level
Air – Industrial	2.5 μg/m^3	USEPA RSL	TR = 10^{-6}, THQ = 1	Screening level
Air – Occupational	20 mg/m^3	ACGIH		TLV TWA
	20 mg/m^3	HCN	LOAEL	8-hr TWA
Groundwater	0.46 μg/L	USEPA RSL	TR = 10^{-6}, THQ = 1	Screening level
	50 μg/L	WHO	TR = 10^{-5}, THQ = 1	Guideline value
	50 μg/L	MOECC	TR = 10^{-6}, THQ = 0.2	Standard
Surface water – Recreational	35.6 μg/L	USEPA RSL	TR = 10^{-6}, THQ = 1	Screening level
Soil – Agricultural	0.2 mg/kg	MOECC	TR = 10^{-6}, THQ = 0.2	Standard
Soil – Residential	5.3 mg/kg	USEPA RSL	TR = 10^{-6}, THQ = 1	Screening level
	1.8 mg/kg	MOECC	TR = 10^{-6}, THQ = 0.2	Standard
Soil – Industrial	24 mg/kg	USEPA RSL	TR = 10^{-6}, THQ = 1	Screening level
	1.8 mg/kg	MOECC	TR = 10^{-6}, THQ = 0.2	Standard
Sediment – Recreational	45.6 mg/kg	USEPA RSL	TR = 10^{-6}, THQ = 1	Screening level

Notes: ACGIH, American Conference of Governmental Industrial Hygienists; HCN, Health Council of the Netherlands; MOECC, Ontario Ministry of the Environment and Climate Change; RSL, Regional Screening Level; THQ, target hazard quotient; TLV, threshold limit value; TR, target cancer risk; TWA, time-weighted average; USEPA, United States Environmental Protection Agency; WHO, World Health Organization.

The toxicity values and assumptions incorporated into the World Health Organization (WHO) guideline value are provided in "1,4-Dioxane in Drinking-water" (WHO, 2005).

The Ontario MOECC standards were obtained from Table 7.2 (Full Depth Generic Site Condition Standards in a Potable Ground Water Condition) of "Soil, Ground Water and Sediment Standards for Use under Part XV.1 of the Environmental Protection Act" (MOECC, 2011). The exposure assumptions used to derive the standards are presented in the "1,4-Dioxane in Drinking Water – Guideline Technical Document for Public Consultation" (HC, 2018).

7.1.2 Exposure Assessment

Potential exposure routes to 1,4-dioxane present in environmental media or released from consumer products include inhalation (of particulates, vapors, or aerosols), dermal contact (from direct application to skin or contact with residues on surfaces), and ingestion (in drinking water or from hand-to-mouth or object-to-mouth contact). When determining the potential exposure routes to 1,4-dioxane that are specific to a site or consumer-product interaction, a site-specific conceptual exposure model (CEM), sometimes referred to as a "conceptual model" (USEPA, 1989) or a "conceptual site model" (CSM), should be developed so that the applicable exposure scenarios can be identified and addressed in the human health risk assessment. Guidance for developing site-specific CEMs is provided in various sources including USEPA, 1989; USEPA, 2011; ASTM, 2014; and ITRC, 2015.

A compilation of potential exposure scenarios resulting from use of consumer products is provided on USEPA's website: https://www.epa.gov/expobox/exposure-assessment-tools-media-consumer-products.

The website includes links to guidance for European Union human health and environmental risk assessment for exposure from household use of cleaning products, USEPA guidance on methods for assessing consumer exposure to chemicals, and Australian risk assessment guidance. Links to calculation tools from the USEPA and Dutch National Institute for Health and Environment (RIVM) are also provided (USEPA, 2019c).

7.1.2.1 Oral Exposure

Oral exposure is the primary exposure route of concern for 1,4-dioxane, and most of the laboratory animal studies have evaluated this exposure route. Oral exposure to 1,4-dioxane may occur through consumption of impacted drinking water, pharmaceutical products, or food. In addition, oral exposure may occur through incidental ingestion of small quantities of impacted soil, surface water, or sediment during outdoor activities. Incidental ingestion of consumer products containing varying amounts of 1,4-dioxane may also occur, as well as exposures in laboratories or industrial facilities where 1,4-dioxane is produced or used. However, it should be noted that 1,4-dioxane has a low potential for bioaccumulation (NTP, 2016).

The information needed to estimate the dose by oral exposure to 1,4-dioxane in water, food, products, soil, surface water, or sediments includes the following:

- 1,4-Dioxane concentration in the exposure medium
- Ingestion rate of water, food, consumer product, soil, surface water, or sediments
- Fraction of 1,4-dioxane absorbed in the gastrointestinal tract
- Frequency of exposure
- Duration of exposure
- Body weight of the exposed person

7.1.2.2 Dermal Exposure

Dermal exposure to 1,4-dioxane may occur while using potable water contaminated by 1,4-dioxane. In addition, dermal exposure to 1,4-dioxane may occur from contact with soil, surface water, or sediment. Dermal exposure may also occur from applying consumer products containing 1,4-dioxane (e.g., lotions or cosmetics) directly to the skin or by contacting consumer products (e.g., shampoo, laundry detergents, or surface cleaners) during use. In the workplace, dermal exposure to 1,4-dioxane may occur by workers handling 1,4-dioxane if appropriate occupational exposure training and use of personal protective equipment is not implemented.

Skin contact with consumer products containing 1,4-dioxane may result in absorption by diffusion through the outermost layer of skin (the dead stratum corneum) and through the living epidermis, where metabolic processes may be involved. Unlike the stratum corneum, the cutaneous tissue is primarily an aqueous medium into which hydrophilic compounds like 1,4-dioxane will partition. Some of the 1,4-dioxane in contact with the skin may be absorbed, while some may be subject to evaporation. Many of the measured values of dermal permeability for chemicals are obtained by percutaneous absorption studies conducted by covering the site of application with an occlusive wrap to prevent loss of the compound. Values obtained by this approach (occlusion) may overestimate skin permeability for volatile compounds such as 1,4-dioxane. Up to 3.2% of applied 1,4-dioxane (dissolved in lotion) was absorbed under occlusion for 3.5 hours, whereas 0.3% absorption occurred under nonoccluded conditions (Bronaugh, 1982). The difference is due to the relative volatility of 1,4-dioxane (ATSDR, 2012).

Rates of chemical permeation of the skin vary in humans by gender, skin type, skin condition, skin hydration, skin temperature, circulation of blood to the skin, capacity of the compound to bind to keratin, and the anatomic region exposed. Dermal permeability in humans is relatively invariant by skin age (Wester et al., 1985). Much of the experimental data for dermal exposure factors are obtained from tests on laboratory animals such as rabbits and guinea pigs. Intraspecies variability therefore becomes an important factor for estimating human dermal exposure.

Skin penetration may also occur via skin structures such as hair follicles and sweat glands, which may allow 1,4-dioxane to diffuse through otherwise rate-limiting barriers. Less than 1% of the human skin surface is composed of these structures; therefore, this route of dermal exposure is negligible for lipophilic compounds such as chlorinated solvents. 1,4-Dioxane is expected to diffuse slowly through the stratum corneum; therefore, skin structures may play a more important role than diffusion for 1,4-dioxane passage through the skin (USEPA, 2004). A different view is presented by the European Union, which notes that 1,4-dioxane is a "fat solvent" and can cause eczema upon prolonged or repeated contact (ECB, 2002).

The predicted permeability coefficient (K_p)* for 1,4-dioxane in aqueous media is 3.3×10^{-4} cm/h, and the 95% upper confidence limit K_p value is 8.6×10^{-3} cm/h. For comparison, the measured K_p of water is 5.0×10^{-4} cm/h (USEPA, 2004).

Available approaches to estimating dermal exposures to 1,4-dioxane in water, soil, and sediment are provided in USEPA (2004). The information needed to estimate the dose by dermal exposure to 1,4-dioxane in water includes the following:

- 1,4-Dioxane concentration in water
- Period of time skin is in contact with water per event
- Area of skin to which water is applied
- Fraction of 1,4-dioxane absorbed through skin during water contact
- Dermal permeability coefficient in water
- Frequency of exposure
- Exposure duration (years) over which events occur
- Body weight of the exposed person

Tables 2.11 and 2.12 present 1,4-dioxane concentrations measured in some common consumer products in 2006 and 2018, illustrating possible sources of dermal exposure to 1,4-dioxane.

---

* Permeability coefficient (K_p) is a flux value, normalized for concentration, that represents the rate at which the chemical penetrates the skin (cm/h) (USEPA, 2004).

7.1.2.3 Inhalation Exposure

Inhalation exposure to 1,4-dioxane may result from a variety of occupational and domestic activities involving use of groundwater containing 1,4-dioxane or products containing 1,4-dioxane as an additive or impurity.

7.1.2.3.1 Occupational Settings

Inhalation is the most common route of exposure to chemicals used in the workplace. Chapter 2 provides a detailed description of 1,4-dioxane occurrence in various manufacturing settings, and Table 2.7 lists the industrial processes in which 1,4-dioxane emissions may occur. The greatest inhalation risk from 1,4-dioxane exposure is with its direct use without proper personal protective equipment and ventilation measures. Unfortunately, 1,4-dioxane does not provide workers with warning properties such as a distinct and low-threshold odor or an effect such as watery eyes or respiratory irritation. The odor threshold for 1,4-dioxane in air is 24 ppm (Amoore and Hautala, 1983).

The greatest occupational exposure to 1,4-dioxane by inhalation is likely to have occurred during degreasing operations in earlier decades. Degreasing with methyl chloroform containing up to 3% 1,4-dioxane will cause inhalation exposure from 3 to 15 ppm (12–60 mg/m^3). The scenario under which such exposure would occur is an open-top vapor degreaser without supplied ventilation (VROM, 1999). Since use of methyl chloroform was banned by the Montreal Protocol, this exposure scenario no longer occurs in most developed countries. However, some occupational exposure to 1,4-dioxane may remain where it is used in laboratories, as a solvent for producing cellulose acetate membranes, or where it occurs as a byproduct from the production of ethoxylated surfactants.

Table 7.3 lists the leading occupations with potential exposure to 1,4-dioxane; actual exposure is increasingly unlikely in the more developed countries because of improvements in worker health and safety training, use of personal protective equipment, and regulation and monitoring of worker exposure. However, occupational exposure to chemicals such as 1,4-dioxane is a continuing issue in manufacturing facilities located in countries with less developed health and safety regulations.

7.1.2.3.2 Domestic Settings

A sampling survey of 12 homes in Woodland, California, found a geometric mean concentration of 0.03 ppb of 1,4-dioxane in indoor air, an arithmetic mean of 0.39 ppb, and a maximum concentration of 39 ppb (Sheldon et al., 1991; Hodgson and Levin, 2003). An earlier review of surveys estimated that the geometric mean 1,4-dioxane concentration in indoor air was 0.02 ppb (Pellizzari et al., 1986). A National VOC Indoor Air Database, with 585 measurements for 1,4-dioxane, listed the average 1,4-dioxane concentration at 1.03 ppbv (Shah and Heyerdahl, 1988; Shah and Singh, 1988). A survey of indoor air in Los Angeles and Contra Costa County, California, found 1,4-dioxane present in 12–55% of the homes sampled. In Los Angeles, 112 homes sampled in the summer of 1984 had an average 1,4-dioxane concentration of 0.15 ppbv; the maximum detected value was 1.2 ppbv. A winter 1984 survey of 51 homes in Los Angeles detected 1,4-dioxane at an average concentration of 0.044 ppbv. The sampling of indoor air in Contra Costa homes occurred in the summer of 1984 and found an average concentration of 0.042 ppbv and a maximum of 0.81 ppbv (Wallace, 1986).

1,4-Dioxane may be present in indoor air from adhesives and coatings used in construction and interior finishes (see Chapter 2), from showering with contaminated water, and from indoor use of consumer products. These uses are discussed in the following sections.

7.1.2.3.3 Materials and Finishes

A NASA survey of indoor air sources determined that a minor percentage of commonly encountered home construction practices can introduce 1,4-dioxane. By testing materials in closed containers at elevated temperatures not ordinarily encountered in the home, researchers were able to measure 1,4-dioxane present in 16 construction adhesives, coatings, and rubber materials. Adhesives were the major source of 1,4-dioxane; their median emission rate was 4.3 mg/g of adhesive, and their maximum emission rate was 21 mg/g of adhesive (McDonnell Douglas Corporation, 1986). Draperies and

TABLE 7.3
Occupational Settings with Potential Exposure to 1,4-Dioxane

Occupational Setting	1,4-Dioxane Occurrence	Type of Exposure	Exposure Level	Reference
Research and analytical laboratories	Solvent and reagent for research, development, and analysis	Inhalation, dermal	TWA—1.8 ppm	1–3
Celluloid film processing	Producing, splicing, repairing, or cementing movie film	Inhalation, dermal	TWA—1 ppm 7–14 ppm	4, 5
Optical lens manufacture	Solvent fumes from coating ovens	Inhalation	Modeled: 10–50 ppm (high estimate)	1
Production of ethoxylated surfactants[a]	Closed-loop process; potential for exposure is during drumming off of finished product	Inhalation	Alkyl ether sulfates (100–200 ppm before vacuum stripping); current potential for worker exposure, <1 ppm	1
Production of microwave popcorn bags	In heat-resistant water-based epoxy adhesives	Dermal, ocular, inhalation	Less than hazardous levels but irritation observed	6
Automotive parts manufacturing	Vapor degreasing and parts washing	Inhalation, dermal	1.5–13.3 ppm in 15-min PBZ samples; 2.5–51 ppm at breathing level near sink and degreaser	7
Assembling and cleaning shirts from finished textiles	Dry cleaning	Inhalation	<0.1–0.5 ppm	8
Painting steel	Paint solvents (methyl chloroform)	Inhalation, dermal	Up to 1.7 mg/m^3	9
Food packaging	Solvent printing and coating materials for individual serving packages of condiments	Inhalation	Not investigated	10
Resin production	Unsaturated polyester resins	Inhalation	Modeled: Maximum annual average, 0.121 $\mu g/m^3$; maximum 1-hour average, 3.26 $\mu g/m^3$	11
Histology using electron microscopy	Direct use of 1,4-dioxane to dehydrate samples	Inhalation, dermal	Not investigated	12
Pharmaceutical industry	Used in chemical synthesis of pharmaceuticals	Inhalation	<3.6 mg/m^3 6.5 mg/m^3	13
Medical radioisotope production	Used in ion-exchange process to recover radioisotopes	Inhalation	Not investigated	14
Wet textile processing	Wool scouring and finishing, woven and knit fabric finishing, carpet finishing, yarn finishing, felted fabric processing	Inhalation	Total emissions can be up to 0.65 lb/h; exposure not investigated	15
Production of dyes	Solvent in colorants and adhesives in felt tip pens and fabric dyes	Inhalation, dermal (also oral in the case of children)	Felt tip pen dye, 11 mg/kg; fabric dye, 4.7 mg/kg; worker exposure not measured; modeled child exposure, 0.303 µg/kg-day for felt tip pens used 0.5 h/day	16

(Continued)

TABLE 7.3 *(Continued)*
Occupational Settings with Potential Exposure to 1,4-Dioxane

Occupational Setting	1,4-Dioxane Occurrence	Type of Exposure	Exposure Level	Reference
Production of cellulose membranes	Direct use in solution deposition of cellulose acetate and cellulose triacetate membranes for production of filters used in water treatment and kidney dialysis	Inhalation, oral if used in reverse osmosis of drinking water, renal if used in kidney dialysis	Risk is absent unless filters are not adequately flushed in final production stage; worker exposure not investigated	17
Production of magnetic tape	Solvent for magnetic tape pigment, up to 40% 1,4-dioxane	Inhalation	37–75 mg/m³	18
Production of 1,4-dioxane at chemical plants[b]	Synthesis (closed loop)	Inhalation	0.18 mg/m³ (59)/ 0.9 mg/m³ 0.08 mg/m³ (18)/ 1.1 mg/m³	18
	Storage/drumming	Inhalation	0.07 mg/m³ (37)/ 40 mg/m³ 0.10 mg/m³ (8)/ 10 mg/m³	
	Waste disposal	Inhalation	0.07 mg/m³ (35)/ 0.4 mg/m³ 0.15–4.7 mg/m³ (2)	
	Laboratory Production and drumming Around storage tanks	Inhalation Inhalation Inhalation	0.11 mg/m³ (305)/ 0.6 mg/m³ 43 mg/m³ (30) 360–2,880 mg/m³ (2)	

Sources: [1] NICNAS (1998); [2] Hertlein (1980, cited in NICNAS, 1998); [3] Rimatori et al. (1994, cited in NICNAS, 1998); [4] Okawa and Coye (1982, cited in NICNAS, 1998); [5] Salisbury and Arnold (1984); [6] Crandall and Wilcox (1987); [7] Krake and Herrera-Moreno (1995); [8] Kullman (1993); [9] Hills et al. (1989); [10] MDH (2000); [11] ATSDR (2005); [12] Shearer and Hunsicker (1980); [13] USEPA (1997a); [14] Taylor et al. (1995); [15] Smith (1986); [16] Hansen (2005); [17] Sirkar et al. (1978); and [18] VROM (1999).

Notes: PBZ, personal breathing zone; TWA, time-weighted average.

[a] Ethoxylated surfactants are comprised of alkyl and alkyl phenol ether sulfates, PEGs, and ethoxylates of alcohols, alkyl phenols, sorbitan esters, amides, and amines.

[b] First value is median, numbers in parentheses are the number of samples; value following slash "/" is 90th percentile; multiple rows are for data from different companies or different periods of record.

drapery linings emitted 1.3 mg/m²/h in one study (Bayer and Papanicolopoulos, 1990). A German study identified 1,4-dioxane at less than 50 mg/m³ (14 ppbv) among volatile organic compounds emitted in subcarpet construction coatings (Fischer, 1998).

7.1.2.3.4 Contaminated Water

Showering or bathing in water contaminated with 1,4-dioxane may result in inhalation exposure (USEPA, 2018). Although 1,4-dioxane is hydrophilic and less susceptible to partitioning from the liquid to vapor phase, the rate of volatilization is increased when water contaminated with 1,4-dioxane is heated. The total surface area available for mass transfer is a key determinant of the rate of volatilization. Shower nozzles break up the water stream into fine droplets, maximizing the opportunity for volatilization and production of aerosols. The relatively confined space of a shower in a bathroom will exacerbate the potential for exposure to 1,4-dioxane in shower water.

7.1.2.3.5 Consumer Products

Incidental low-level inhalation exposure could result from the presence of residual 1,4-dioxane in personal care products such as soaps, shampoos, lotions, cosmetics, and related products. Japanese domestic wastewater was evaluated for 1,4-dioxane content derived from shampoos and soaps; the potential exposure was determined to be approximately 0.25 mg/person/day (Abe, 1999). The highest cumulative worst-case uptake in the ECB (2002) risk assessment would amount to only 4% of the total mass of 1,4-dioxane ascribed to soaps and sundries in Japan. The majority of 1,4-dioxane present in shampoos and sundries ends up in the sewer. The 1,4-dioxane concentrations that someone might breathe while shampooing, coupled with dermal exposure, were concluded to be inconsequential and to not constitute a health risk (ECB, 2002).

7.1.2.3.6 Estimating Inhalation Exposures

The information needed to estimate the dose by inhalation exposure to 1,4-dioxane in water or consumer products includes the following:

- Modeled/measured concentration in air
- Period of time spent inhaling impacted air
- Frequency of exposure
- Duration of exposure

7.1.2.4 Breast Milk Exposure

Infants nursed by mothers exposed to elevated levels of 1,4-dioxane in the workplace (via oral or inhalation pathways) may be exposed to 1,4-dioxane through breast milk. A PBPK* modeling study simulated maternal exposure to 1,4-dioxane at an indoor air concentration of 25 ppm in the workplace. The PBPK model used estimated blood/air and milk/air partitioning coefficients to calculate blood/milk partitioning. The model predicted that perchloroethylene, bromochloroethane, and 1,4-dioxane would be present in breast milk at elevated concentrations. The PBPK simulation of lactational transfer is not well developed for volatile organic compounds; the study necessarily applies assumptions regarding the kinetics and properties of lactational transfer, but nonetheless suggests that this route of exposure could be significant for infants nursed by mothers exposed to 1,4-dioxane by breathing vapors in the workplace (Fisher et al., 1997).

Analysis of breast milk has shown that the inhalation-lactational transfer route of exposure was completed in a woman who visited her husband at his dry-cleaning establishment for the lunch hour each day. Her breast milk contained elevated levels of perchloroethylene because daily exposure saturated her body's ability to eliminate the chemical (Bagnell and Ellenberger, 1977). Methyl chloroform, carbon tetrachloride, dichloromethane, TCE, and perchloroethylene have all been detected in breast milk (Pellizzari et al., 1982; Labrèche and Goldberg, 1997; Solomon and Weiss, 2002).

7.1.3 Toxicity Assessment

A key decision in preparing a risk assessment is the selection of suitable toxicity data for a given route of exposure. Risk assessments review available toxicological studies and evaluate the relevance of the derived toxicity values to the environmental exposure scenario. From among the of available toxicity sources, the specific toxicity values selected for the risk assessment should be the most recent and credible peer-reviewed values. USEPA guidance and policy (USEPA, 2003) provides a hierarchy of toxicity value sources and endorses selecting toxicity values with consideration of values that are the most recent and have undergone credible peer-review.

The relationship between the dose of a chemical and the incidence of adverse health effects forms the basis for the quantitative dose response. Toxicity values are derived from these relationships and can be used to estimate the incidence or potential for adverse effects in an exposed population. A detailed discussion of the human health toxicology of 1,4-dioxane is presented in Chapter 5.

* PBPK, physiologically based pharmacokinetic model; see Chapter 5.

7.1.3.1 Non-Carcinogenic Effects

For non-carcinogenic effects, the point of departure may be the no-observed-adverse-effect-level (NOAEL) or a benchmark dose level that is determined by modeling. Uncertainty factors (UFs) are applied to these values to derive reference doses (RfDs), reference concentrations (RfCs), and acceptable or tolerable daily intakes (TDIs).

Oral noncancer health effects are generally evaluated by using an RfD value, which is defined as an estimate (with uncertainty spanning perhaps an order of magnitude) of a daily oral exposure to the human population (including sensitive subgroups) that is likely to be without an appreciable risk of deleterious effects during a lifetime. An oral RfD of 3×10^{-2} mg/kg-day is available for 1,4-dioxane in USEPA's Integrated Risk Information System (IRIS) database (USEPA, 2019d) and is based on liver and kidney toxicity.

Inhalation noncancer health effects are generally evaluated by using an RfC value, which is defined as an estimate (with uncertainty spanning perhaps an order of magnitude) of a continuous inhalation exposure to the human population (including sensitive subgroups) that is likely to be without an appreciable risk of deleterious effects during a lifetime (USEPA, 2019e). An inhalation RfC of 3×10^{-2} mg/m^3 is available for 1,4-dioxane in USEPA's IRIS database (USEPA, 2019d) and is based on atrophy and respiratory metaplasia of the olfactory epithelium.

The WHO identified a TDI of 96 μg/kg of body weight per day for a noncancer end-point (including renal tubular epithelial and hepatocellular degeneration and necrosis) (WHO, 2005).

ATSDR's Toxicological Profile for 1,4-dioxane derives oral and inhalation minimal risk levels (MRLs) for various durations of exposure (ATSDR, 2012). The MRL is defined as the daily human exposure to a substance that is likely to be without an appreciable risk of adverse noncarcinogenic effects over a specified duration and exposure route. MRLs are based only on noncarcinogenic health effects and do not consider carcinogenic effects. MRLs for dermal exposure were not derived due to the lack of appropriate methodology to develop dermal MRLs. ATSDR (2012) derived the following MRLs for noncancer effects:

- An acute-duration inhalation MRL of 2 ppm based on the NOAEL of 20 ppm from a human study by Ernstgård et al. (2006).
- An acute-duration oral MRL of 5 mg/kg/day based on a 14-day gavage study in rats (Giavini et al., 1985).
- An intermediate-duration inhalation MRL of 0.2 ppm based on a 13-week inhalation study in rats (Kasai et al., 2008).
- An intermediate-duration oral MRL of 0.5 mg/kg/day based on a 13-week drinking water study in rats (Kasai et al., 2008).
- A chronic-duration inhalation MRL of 0.03 ppm based on a 2-year inhalation study in rats (Kasai et al., 2009).
- A chronic-duration oral MRL of 0.1 mg/kg/day based on a 2-year drinking water study in rats (Kociba et al., 1974).

Health Canada identified a draft TDI of 0.0054 mg/kg of body weight per day for a noncancer end-point (based on liver toxicity) (HC, 2018).

7.1.3.2 Carcinogenic Effects

For chemicals that are considered to be known, probable, or possible carcinogens to humans, the cancer risk assessment is an important component of environmental risk assessment. The USEPA has concluded that 1,4-dioxane is likely to be carcinogenic to humans by all routes of exposure (USEPA, 2013). The oral cancer slope factor (CSF) (or inhalation unit risk [IUR]) is used to estimate an upper-bound probability of an individual developing cancer as a result of a lifetime of exposure to a specific level of a potential carcinogen (USEPA, 1989). The CSF (or IUR) is a plausible upper-bound estimate of the probability of a response per unit intake of a chemical over a lifetime. The CSFs and IURs are developed from laboratory animal studies for use in cancer risk assessment.

An oral CSF of 1×10^{-1} per mg/kg-day is available for 1,4-dioxane in USEPA's IRIS database (USEPA, 2019d). The CSF is based on liver tumors and derived from the Kano et al. (2009) study.

An IUR of 5×10^{-6} per μg/m^3 is available for 1,4-dioxane in USEPA's IRIS database (USEPA, 2019d). The IUR is based on reproductive, liver, urinary, gastrointestinal, respiratory, and other tumors, and derived from the Kasai et al. (2009) study.

The WHO identified a TDI of 16 μg/kg of body weight per day for a cancer end-point (liver tumors) (WHO, 2005).

Health Canada identified a draft TDI of 0.0054 mg/kg of body weight per day for a cancer end-point (liver tumors) (HC, 2018).

7.2 ECOLOGICAL RISK ASSESSMENT OF 1,4-DIOXANE

The first step of an ecological risk assessment (ERA) is typically termed "problem formulation" (e.g., USEPA, 1992, 1997b, 1998). One key component of problem formulation is the development of an ecological CSM that identifies and evaluates sources, transport pathways, fate and transport mechanisms, exposure media, exposure pathways and routes, and receptors. In this section, an ecological CSM is developed for 1,4-dioxane and is used to evaluate the exposure and risk potential of this chemical to ecological receptors.

7.2.1 Ecological Conceptual Site Model for 1,4-Dioxane

A transport pathway describes the mechanisms whereby a chemical (such as 1,4-dioxane), once released, might be transported from a source to ecologically relevant media (such as surface soils) where exposures might occur. An exposure pathway links a source with one or more ecological receptors (such as terrestrial plants or fish) through exposure via one or more environmental media (such as surface water) and exposure routes (such as direct contact, inhalation, or ingestion). Exposure, and thus potential risk, can occur only if complete exposure pathways exist.

Certain chemical properties influence environmental fate and transport, and thus ecological exposure and toxicity (risk) potential. Data on several key chemical and physical properties of 1,4-dioxane, and its fate and transport potential in the environment, are discussed in detail in Chapter 3 and found in various review documents (e.g., NICNAS, 1998; USEPA, 1999; ECB, 2002; CCME, 2008; Health Canada, 2010; ATSDR, 2012; USEPA, 2015a; USEPA, 2017; and NAVFAC, 2018). Based on its physical and chemical properties, 1,4-dioxane is characterized by high water solubility (>800,000 mg/L; totally miscible in water), moderate to high vapor pressure (38.1 mm Hg at 25°C), low to moderate Henry's law constant (4.29 to 4.80×10^{-6} atm-m^3/mole at 25°C), very low log K_{ow} (-0.10 to -0.49 with a recommended value of −0.39 [USEPA, 1995]), and a very low estimated log K_{oc} of 1.07–1.23.

If released to the atmosphere (e.g., through volatilization from soil or water), 1,4-dioxane would be present in vapor form but has low persistence in air, with an estimated half-life of <1–3 days due primarily to photooxidation. Thus, the potential for chronic exposures to ecological receptors from inhalation and foliar contact are expected to be negligible under typical environmental conditions, and this potential exposure pathway would rarely, if ever, be ecologically significant.

The typical zone of highest biological activity for soil-dwelling ecological receptors commonly utilized in ERAs (such as soil invertebrates and terrestrial plants) is estimated to extend from the soil surface down to a depth of approximately 25–30 cm (10–12 inches) (USEPA, 2015b). If released to surface soil, 1,4-dioxane is not expected to persist in this zone of the soil column, significantly reducing the exposure potential for terrestrial plants (through root uptake) and soil invertebrates (through direct contact and ingestion), since some of it will volatilize to the air and, due to its very low log K_{oc} and very high water solubility, most of the remainder will rapidly leach to deeper soil horizons and groundwater, where it is expected to persist for relatively long periods of time (months to years) due to slow rates of biodegradation. Thus, the predominant fate of 1,4-dioxane released to soils is expected to be relatively rapid transport to underlying groundwater, with little potential for exposure in surface soils.

Bioconcentration, bioaccumulation, and biomagnification through terrestrial food chains are not considered important environmental fate processes for 1,4-dioxane. Based on bioaccumulation modeling (no empirical data were found), 1,4-dioxane is not bioconcentrated in terrestrial invertebrates, animals, or plants; however, it may accumulate in plants due to its high water solubility during transpiration. In mammals, any 1,4-dioxane that is absorbed or ingested is rapidly metabolized and excreted, primarily through the urine. Information was not available on the fate of 1,4-dioxane in birds, soil invertebrates, or terrestrial plants. However, based on its low bioaccumulation potential and rapid metabolism, exposures to terrestrial wildlife through the food chain are not significant with respect to the levels of 1,4-dioxane typically observed in contaminated soils (Huntley et al., 2004).

Groundwater is the medium most associated with the presence of 1,4-dioxane in the environment. However, ecological receptors (other than microbes) have little, if any, direct exposure to groundwater. Exposures become possible when groundwater surfaces as a seep or discharges (through sediment pore water) to a surface water body or wetland and enters the water column. This groundwater to surface water transport pathway, with subsequent exposures in the sediment pore water and water column of the receiving water body (if one is present in relatively close proximity to a source), will typically be the most ecologically relevant exposure pathway for evaluation in an ERA.

1,4-Dioxane is highly soluble in water. Volatilization from water surfaces occurs at low to moderate rates. Section 3.1.3.3 lists volatilization half-lives from a model river and lake. 1,4-Dioxane's modeled $t_{1/2}$ in rivers was estimated to be 5 days, while its model lake $t_{1/2}$ was estimated to be 56 days. 1,4-Dioxane biodegrades slowly in water (half-live in surface water of 1–6 months). However, as discussed in Chapter 5, 1,4-dioxane exhibits low to very low toxicity to aquatic biota from water-based exposures. 1,4-Dioxane is not expected to adsorb to suspended sediments, bed sediments, or detritus due to its low K_{oc} value. Thus, partitioning to sediments is expected to be very low, reducing potential exposures for sediment-dwelling organisms.

Based on its volatility from water and low adsorption to sediments, bioaccumulation is not expected to be a significant fate process for 1,4-dioxane in water and sediment. Thus, bioconcentration, bioaccumulation, and biomagnification through aquatic food chains are not considered important environmental fate processes for 1,4-dioxane. The only empirical aquatic bioaccumulation data are for freshwater fish from water exposures. In a bioaccumulation test using the common carp (*Cyprinus carpio*), measured bioconcentration factor (BCF) values of 0.3–0.7 L/kg (wet weight, at an exposure concentration of 10 mg/L) and 0.2–0.6 L/kg (wet weight, at an exposure concentration of 1 mg/L) have been reported (USEPA, 2015a). Based on bioaccumulation modeling (no empirical data were found), estimated BCFs for other biota from water exposures, and from sediment exposures, are expected to be very low for 1,4-dioxane based on its very low log K_{ow}. In mammals, any 1,4-dioxane that is absorbed or ingested is rapidly metabolized and excreted, primarily through the urine. Information was not available on the fate of 1,4-dioxane in birds, fish, aquatic or benthic invertebrates, or aquatic plants.

7.2.2 Ecological Exposure and Risk Potential for 1,4-Dioxane

The exposure potential to 1,4-dioxane for ecological receptors is expected to be very low for air, soil, sediment, and food chain pathways (both terrestrial and aquatic). Currently, exposure to 1,4-dioxane in surface water and sediment pore water from discharging groundwater represents the most likely ecological exposure pathway. However, 1,4-dioxane exhibits low to very low toxicity to aquatic biota from water-based exposures. Thus, the risk potential is correspondingly low. In addition, various researchers from around the world, including the United States (e.g., USEPA, 2015a), Canada (e.g., Health Canada, 2010), Europe (e.g., ECB, 2002), and Australia (e.g., NICNAS, 1998), have concluded that, based on the available ecotoxicological data and measured or predicted environmental concentrations of 1,4-dioxane, this chemical does not pose an unacceptable risk to ecological receptors except possibly under the most extreme, and unlikely, scenarios.

REFERENCES

Abe, A., 1999, Distribution of 1,4-dioxane in relation to possible sources in the water environment. *The Science of the Total Environment* 227: 41–48.

Agency for Toxic Substances and Disease Registry (ATSDR), 2005, Health consultation—Interplastic Corporation: Review of emission testing data, City of Minneapolis, Hennepin County, Minnesota. Centers for Disease Control and Prevention (CDC), Agency for Toxic Substances and Disease Registry.

Agency for Toxic Substances and Disease Registry (ATSDR), 2012, Toxicological profile for 1,4-dioxane. Centers for Disease Control and Prevention (CDC), Agency for Toxic Substance Disease Registry. April.

American Conference of Governmental Industrial Hygienists (ACGIH), 2011, 1,4-Dioxane: Threshold limit values for chemical substances and physical agents and biological exposure indices.

American Society for Testing and Materials (ASTM), 2014, Standard Guide for Developing Conceptual Site Models for Contaminated Sites. ASTM E1689-95(2014). ASTM International, West Conshohocken, PA, 2014.

Amoore, J.E. and Hautala, E., 1983, Odor as an aid to chemical safety: Odor thresholds compared with threshold limit values and volatilities for 214 industrial chemicals in air and water dilution. *Journal of Applied Toxicology* 3(6): 272–290.

Bagnell, P.C. and Ellenberger, H.A., 1977, Obstructive jaundice due to a chlorinated hydrocarbon in breast milk. *Canadian Medical Association Journal* 117(9): 1047–1048.

Bayer, C.W. and Papanicolopoulos, C.D., 1990, Exposure assessments to volatile organic compound emissions from textile products. *In* Proceedings of the 5th International Conference on Indoor Air Quality and Climate, Toronto, Canada, July 29 to August 3, Vol. 3, pp. 725–730.

Bronaugh, R.L., 1982, Percutaneous absorption of cosmetic ingredients. In: P. Frost, S.N. Horwitz (Eds), *Principles of Cosmetics for the Dermatologist*, pp. 277–284. Minneapolis, MN: University of Minnesota Press.

Canadian Council of Ministers of the Environment (CCME), 2008, Canadian water quality guidelines for the protection of aquatic life: 1,4-Dioxane. 5 pp.

Cohrssen, J.J. and Covello, V.T., 1989, *Risk Analysis: A Guide to Principles and Methods for Analyzing Health and Environmental Risks*. Washington, DC: Council on Environmental Quality.

Crandall, M.S. and Wilcox, T.G., 1987, Graphic Packaging Corporation, Paoli, Pennsylvania. Atlanta, GA: National Institute for Occupational Safety and Health Report No. HETA-87–181–0000.

Ernstgård, L., Iregren, A., Sjögren, B., and Johanson, G., 2006, Acute effects of exposure to vapours of dioxane in humans. *Human & Experimental Toxicology* 25: 723–729.

European Chemicals Bureau (ECB), 2002, European Union risk assessment report: 1,4-Dioxane. *Institute for Health and Consumer Protection, European Chemicals Bureau.*

Fischer, M., 1998, Teppichböden müssen nicht schuld an Innenraumbelastungen sein [Carpeted floors are not always the reason for indoor air contamination]. *Umwelt & Gesundheit [Environment & Health]* 1998(1): 6–13. https://www.iug-umwelt-gesundheit.de/htdocs/publikationen/pdf/ug981.pdf.

Fisher, J., Mahle, D., Bankston, L., Greene, R., and Gearhart, J., 1997, Lactational transfer of volatile chemicals in breast milk. *American Industrial Hygiene Association Journal* 58(6): 425–431.

Giavini, E., Vismara, C., and Broccia, M.L., 1985, Teratogenesis study of dioxane in rats. *Toxicology Letters* 26: 85–88.

Hansen, O.C., 2005, Screening for health effects from chemical substances in textile colorants. In Survey of Chemical Substances in Consumer Products. Danish Ministry of the Environment, Danish Technological Institute, Report No. 57. https://www2.mst.dk/udgiv/publications/2005/87-7614-672-3/pdf/87-7614-673-1.pdf.

Health Canada (HC), 2010, Screening Assessment for the Challenge: 1,4-Dioxane. Environment Canada, Health Canada. http://www.ec.gc.ca/ese-ees/789BC96E-F970-44A7-B306-3E32419255A6/batch7_123-91-1_en.pdf.

Health Canada (HC), 2018, 1,4-Dioxane in Drinking Water – Guideline Technical Document for Public Consultation. August.

Health Council of the Netherlands (HCN), 2015, 1,4-Dioxane – Re-evaluation of the carcinogenicity and genotoxicity. The Hague: Health Council of the Netherlands; publication no. 2015/26. November.

Hertlein, F., 1980, Monitoring airborne contaminants in chemical laboratories. *Analytical Techniques in Occupational Health Chemistry*, ACS Symposium Series No. 120: pp. 215–230 [cited in NICNAS, 1998].

Hills, B., Klincewicz, S., Blade, L.M., and Sack, D., 1989, Health Hazard Evaluation (HHE) report for BMY Corporation, a Division of Harsco Corporation, York, Pennsylvania. Atlanta, GA: National Institute for Occupational Safety and Health Report No. HETA 87–367–1987.

Hodgson, A.T. and Levin, H., 2003, *Volatile Organic Compounds in Indoor Air: A Review of Concentrations Measured in North America Since 1990*. Berkeley, CA: Lawrence Berkeley National Laboratory Report LBNL-51715.

Huntley, S., Amarai, M., and Schell, J., 2004, Toxicokinetics of 1,4-dioxane and implications for assessing hazards to terrestrial wildlife. Paper presented at the Society of Environmental Toxicology and Chemistry, 25th Annual Meeting, Portland, OR, November 14–18.

Interstate Technology and Regulatory Council, 2015, Decision Making at Contaminated Sites: Issues and Options in Human Health Risk Assessment (RISK-3). January.

Kano, H., Umeda, Y., Kasai, T., Sasaki, T., Matsumoto, M., Yamazaki, K., Nagano, K., Arito, H., and Fukushima, S., 2009, Carcinogenicity studies of 1,4-dioxane administered in drinking-water to rats and mice for 2 years. *Food and Chemical Toxicology* 47: 2776–2784. http://dx.doi.org/10.1016/j.fct.2009.08.012.

Kasai, T., Kano, H., Umeda, Y., Sasaki, T., Ikawa, N., Nishizawa, T., Nagano, K., Arito, H., Nagashima, H., and Fukushima, S., 2009, Two-year inhalation study of carcinogenicity and chronic toxicity of 1,4-dioxane in male rats. *Inhalation Toxicology* 21: 889–897.

Kasai, T., Saito, M., Senoh, H., Umeda, Y., Aiso, S., Ohbayashi, H., Nishizawa, T., Nagano, K., and Fukushima, S., 2008, Thirteen-week inhalation toxicity of 1,4-dioxane in rats. *Inhalation Toxicology* 20: 961–971.

Kociba, R.J., McCollister, S.B., Park, C., Torkelson, T.R., and Gehring, P.J., 1974, 1,4-Dioxane. I. Results of a 2-year ingestion study in rats. *Toxicology Applied Pharmacology* 30: 275–286.

Krake, A.M. and Herrera-Moreno, V., 1995, *Automotive Controls Corporation, Independence, Kansas*. Atlanta, GA: National Institute for Occupational Safety and Health Report No. HETA-95–0296–2547.

Kullman, G.J., 1993, *Meyersdale Manufacturing Co., Meyersdale, Pennsylvania*. Atlanta, GA: National Institute for Occupational Safety and Health Report No. MHETA-88–269–1993.

Labrèche, F.P. and Goldberg, M.S., 1997, Exposure to organic solvents and breast cancer in women: A hypothesis. *American Journal of Industrial Medicine* 32(1): 1–14.

McDonnell Douglas Corporation, 1986, Materials Testing Data Base. Software Technology Development Laboratory, Houston, Texas. [Data reported in A.T. Hodgson and J.D. Wooley, 1991, Assessment of indoor concentrations, indoor sources and source emissions of selected volatile organic compounds. Prepared for Research Division, California Air Resources Board by Indoor Environment Program, Lawrence Berkeley Laboratory, University of California, Berkeley.]

Ministry of Housing and the Environment (VROM), 1999, *Risk Assessment: 1,4-Dioxane*. Bilthoven, Netherlands: S. P. Netherlands Ministry of Housing and the Environment (VROM), Chemical Substances Bureau.

Ministry of the Environment and Climate Change (MOECC) Ontario, 2011, Soil, Ground Water and Sediment Standards for Use under Part XV.1 of the Environmental Protection Act. April 15.

Minnesota Department of Health (MDH), 2000, Health consultation: Pechiney Plastic Packaging Inc., Minneapolis, Hennepin County, Minnesota.

National Industrial Chemicals Notification and Assessment Scheme (NICNAS), 1998, Priority existing chemical assessment no. 7—1,4-dioxane. Department of Health and Aging, Australian National Industrial Chemicals Notification and Assessment Scheme (NICNAS). June.

National Toxicology Program (NTP), 2016, 14th Report On Carcinogens. Research Triangle Park, NC: U.S. Department of Health and Human Services, Public Health Service. https://ntp.niehs.nih.gov/pubhealth/roc/index-1.html.

Naval Facilities Engineering Command (NAVFAC), Department of the Navy, 2018, 1,4-Dioxane guidance/frequently asked questions. Ser EV/007. 6 September.

Okawa, M. and Coye, M.J., 1982, *HETA* 80–144–1109, *Film Processing Industry, Hollywood, California*. Atlanta, GA: National Institute for Occupational Safety and Health.

Pellizzari, E.D., Hartwell, T.D., Harris, B.S.H., Waddell, R.D., Whitaker, D.A., and Erickson, M.D., 1982, Purgeable organic compounds in mother's milk. *Bulletin of Environmental Contamination and Toxicology* 28: 322–328.

Pellizzari, E.D., Hartwell, T.D., Perritt, R.L., Sparacino, C.M., Sheldon, L.S., Zelon, H.S., Whitmore, R.W., Breen, J.J., and Wallace, L., 1986, Comparison of indoor and outdoor residential levels of volatile organic chemicals in five U.S. geographical areas. *Environment International* 12(6): 619–623.

Rimatori, V., Fronduto, M., Staiti, D., and Niu, Q., 1994, Controllo e valutazione dell'esposizione lavorativa di addetti di laboratorio (Monitoring and evaluation of laboratory worker exposure [in Italian]). *Archivio di Scienze del Lavoro* 10(3–4): 481–484 [cited in NICNAS, 1998, Priority existing chemical assessment No. 7—1,4-Dioxane. Department of Health and Aging, Australian National Industrial Chemicals Notification and Assessment Scheme (NICNAS).].

Salisbury, S.A. and Arnold, S., 1984, Niemand Industries, Inc., Statesville, North Carolina. Atlanta, GA: National Institute for Occupational Safety and Health Report No. HETA-84–108–1821.

Shah, J. and Heyerdahl, E.K., 1988, National ambient volatile organic compounds (VOCs) database update. *Research Triangle Park, NC: U.S. Environmental Protection Agency, EPA* 600/3–88/010.

Shah, J. and Singh, H.B., 1988, Distribution of volatile organic chemicals in outdoor and indoor air: A national VOCs data base. *Environmental Science and Technology* 22(12): 1381–1388.

Shearer, T.P. and Hunsicker, L.G., 1980, Rapid method for embedding tissues for electron microscopy using 1,4-dioxane and polybed 812. *Journal of Histochemistry and Cytochemistry* 28(5): 465–467.

Sheldon, L.S., Clayton, A., and Jones, B., 1991, Indoor pollutant concentrations and exposures. *Final Report.* Sacramento: California Air Resources Board.

Sirkar, K.K., Agarwal, N.K., and Rangaiah, G.P., 1978, The effect of short air exposure periods on the performance of cellulose acetate membranes from casting solutions with high cellulose acetate content. *Journal of Applied Polymer Science* 22(7): 1919–1944.

Smith, B., 1986, Identification and reduction of pollution sources in textile wet processing. Raleigh, NC, North Carolina State University, School of Textiles.

Solomon, G. and Weiss, P., 2002, Chemical contaminants in breast milk: Time trends and regional variability. *Environmental Health Perspectives* 110(6): A339–A348.

Taylor, W.A., Jamriska, D.J., Hamilton, V.T., Heaton, C., Phillips, D.R., Staroski, R.C., Garcia, J.B., Garcia, J.G., and Ott, M.A., 1995, Waste minimization in the Los Alamos medical radioisotope program. *Journal of Radioanalytical and Nuclear Chemistry* 195(2): 287–295.

U.S. Environmental Protection Agency (USEPA), 1989, Risk Assessment Guidance for Superfund Volume I Human Health Evaluation Manual (Part A). Interim Final. EPA/540/1-89/002. December.

U.S. Environmental Protection Agency (USEPA), 1992, Framework for ecological risk assessment. EPA/630/R-92/001.

U.S. Environmental Protection Agency (USEPA), 1995, Internal report on summary of measured, calculated and recommended log K_{ow} values. Environmental Research Laboratory, Athens, GA. April 10.

U.S. Environmental Protection Agency (USEPA), 1997a, Profile of the pharmaceutical manufacturing industry. U.S. Environmental Protection Agency, Office of Enforcement and Compliance Assurance.

U.S. Environmental Protection Agency (USEPA), 1997b, Ecological risk assessment guidance for Superfund: Process for designing and conducting ecological risk assessments. Interim Final. EPA/540/R-97/006.

U.S. Environmental Protection Agency (USEPA), 1998, Guidelines for ecological risk assessment. EPA/630/R-95/002F. April. Final.

U.S. Environmental Protection Agency (USEPA), 1999, Screening level ecological risk assessment protocol for hazardous waste combustion facilities. EPA/530/D-99/001A. Peer Review Draft. August.

U.S. Environmental Protection Agency (USEPA), 2003, Human Health Toxicity Values in Superfund Risk Assessments. OSWER Directive 9285.7-53. December 5.

U.S. Environmental Protection Agency (USEPA), 2004, Risk Assessment Guidance for Superfund Volume I Human Health Evaluation Manual (Part E, Supplemental Guidance for Dermal Risk Assessment). Final. EPA/540/R/99/005. OSWER 9285.7-02EP. July.

U.S. Environmental Protection Agency (USEPA), 2011, Environmental Cleanup Best Management Practices: Effective Use of the Project Life Cycle Conceptual Site Model. Office of Solid Waste and Emergency Response (5102G). EPA 542-F-11-011. July.

U.S. Environmental Protection Agency (USEPA), 2013, Toxicological Review of 1,4-Dioxane (With Inhalation Update). EPA/635/R-11/003F. September.

U.S. Environmental Protection Agency (USEPA), 2015a, TSCA work plan chemical problem formulation and initial assessment, 1-4-dioxane. 740-R1-5003. April.

U.S. Environmental Protection Agency (USEPA), 2015b, Determination of the biologically relevant sampling depth for terrestrial and aquatic ecological risk assessments. EPA/600/R-15/176. ERASC-015F. October.

U.S. Environmental Protection Agency (USEPA), 2017, Technical Fact Sheet – 1,4-Dioxane. EPA 505-F-17-011. November.

U.S. Environmental Protection Agency (USEPA), 2018, Problem Formulation of the Risk Evaluation for 1,4-Dioxane. United States Office of Chemical Safety and Pollution Prevention. EPA-740-R1-7012. May. https://www.epa.gov/sites/production/files/2018-06/documents/14-dioxane_problem_formulation_5-31-18.pdf.

U.S. Environmental Protection Agency (USEPA), 2019a, Regional Screening Levels (RSLs).

U.S. Environmental Protection Agency (USEPA), 2019b, Regional Screening Levels (RSLs) – User's Guide.

U.S. Environmental Protection Agency (USEPA), 2019c, EPA ExpoBox. Exposure Assessment Tools by Media – Consumer Products. https://www.epa.gov/expobox/exposure-assessment-tools-media-consumer-products.

U.S. Environmental Protection Agency (USEPA), 2019d, Integrated Risk Information System. 1,4-Dioxane. https://cfpub.epa.gov/ncea/iris2/chemicalLanding.cfm?&substance_nmbr=326.

U.S. Environmental Protection Agency (USEPA), 2019e, Vocabulary Catalog. Integrated Risk Information System (IRIS) Glossary. https://ofmpub.epa.gov/sor_internet/registry/termreg/searchandretrieve/glossariesandkeywordlists/search.do;jsessionid=qDdfJlcBQGbxzBQWLc4K3xMVWf-g4Z6J1EZ7_MQijWgx0jvcrcqU!84008612?details=&vocabName=IRIS%20Glossary&filterTerm=reference%20concentration&checkedAcronym=false&checkedTerm=false&hasDefinitions=false&filterTerm=reference%20concentration&filterMatchCriteria=Contains.

Wallace, L.A., 1986, Personal exposure, indoor and outdoor air concentrations, and exhaled breath concentrations of selected volatile organic compounds measured for 600 residents of New Jersey, North Dakota, North Carolina and California. *Toxicological and Environmental Chemistry* 12(3): 215–236.

Wester, R.C., Maibach, H.I., Surinchak, J., and Bucks, D.A.W., 1985, Predictability of *in vitro* diffusion systems: Effect of skin types and ages on percutaneous absorption of triclocarban. *Dermatology* 1985(6): 223–226.

World Health Organization, 2005, 1,4-Dioxane in Drinking-water. Background document for development of WHO Guidelines for Drinking-water Quality. WHO/SDE/WSH/05.08/120.

8 Treatment Technologies

William H. DiGuiseppi and James W. Hatton

CONTENTS

As discussed in Chapter 3, 1,4-dioxane has a low K_{oc} (soil adsorption coefficient for soil organic matter) and a high solubility and is therefore not highly sorbed to soils. Because of these characteristics, 1,4-dioxane is dominantly found in groundwater. Therefore, the following discussion of remedial technologies is focused primarily on groundwater remediation methods, both *ex situ* and *in situ*, with limited soil treatment approaches discussed as appropriate. In this chapter, we first describe each technology, and then we present literature studies to demonstrate the effectiveness of the technology at 1,4-dioxane removal or destruction. The literature studies include laboratory bench-scale studies and field-scale tests, documenting the removal of 1,4-dioxane from groundwater or effluent. Finally, we assess the technology's likelihood of success in the remediation of actual sites contaminated with 1,4-dioxane.

Remediation technologies presented are:

a. Vapor-phase transfer
b. Sorption
c. Natural attenuation
d. Ultraviolet photolysis
e. Phytotechnology
f. Bioremediation
g. Chemical oxidation

The assessment of treatment technologies for 1,4-dioxane presented here is comprehensive; however, it is not intended to capture each and every study of 1,4-dioxane treatment. The selected citations provide an assessment of the viability of a given technology to remediate groundwater and soil contaminated with 1,4-dioxane. Preference is given to citations that provide insight into the fate of 1,4-dioxane or the treatment mechanism. Summary compilations of 1,4-dioxane treatment technologies can be found in DiGuiseppi et al. (2016) and Bell et al. (2019). An assessment of sustainability considerations related to available 1,4-dioxane treatment technologies was presented in Favara et al. (2016).

8.1 VAPOR-PHASE TRANSFER

This section discusses technologies for transfer of 1,4-dioxane from the dissolved or sorbed phase, in the case of soil, to the vapor phase for removal or further treatment.

8.1.1 Air Stripping

Air stripping is a common method of removing volatile organic compounds (VOCs) from groundwater because the relationship between solubility and volatility is optimal in this group of contaminants. This method is typically conducted in a packed tower or a low-profile aeration system. Water is introduced to the top of the stripper and flows down through the system under gravity. Air is pumped into the bottom of the stripper under pressure and exits the top. This arrangement, called countercurrent flow, optimizes the exchange of contaminants from water to air. Air stripping is a nondestructive transfer of contaminants from the liquid phase to the vapor phase. The resulting gaseous phase contaminants often require treatment. Granular activated carbon (GAC) adsorption and thermal destruction are commonly used to treat contaminated air before it is discharged to the atmosphere. Although 1,4-dioxane is highly volatile when in pure form, its infinite solubility in water is generally considered to prevent efficient air-stripping removal from groundwater to concentrations suitable for discharge. Nevertheless, several studies have been performed to assess the ability of air stripping to remove 1,4-dioxane from groundwater.

An air-stripping optimization test was performed at the U.S. Air Force Plant 44 (AFP 44), located in Tucson, Arizona, to determine whether adjustments to the operating parameters for the large-scale treatment system could effectively reduce the influent 1,4-dioxane concentrations (10–15 μg/L) to meet a target level of 6.1 μg/L (Earth Tech, 2004). The AFP 44 groundwater treatment plant utilizes three parallel trains of two-stage air-stripping towers (primary and secondary stripping towers) with a design capacity of ~5000 gallons/min. At an air:water ratio of seven, the 25-feet-tall primary towers, in combination with GAC treatment of the off-gas, are capable of removing as much as 85% of the trichloroethene (TCE) and 1,1-dichloroethylene. At an air:water ratio of 25, the 42-feet-tall secondary towers are designed to remove the remaining TCE and 1,1-dichloroethylene from the water; the stripped vapors are then vented to the atmosphere. This system had no measurable effect on 1,4-dioxane removal. Optimization of the existing treatment system to remove 1,4-dioxane was accomplished by modifying the system to allow higher airflow rates, thereby increasing the air:water ratios. The air:water ratio in the primary tower was set at 69, the maximum air:water ratio possible, and the secondary tower's air:water ratio was varied from 183 to as high as 291. The study indicated that the maximum removal rate for 1,4-dioxane was ~10%, even with air:water ratios as high as 291, which is more than 10 times the designed air:water ratio for the removal of TCE and dichloroethene. Chlorinated VOC removal was complete regardless of the air:water ratio. The results of the optimization test indicated that utilizing the air-stripping system as the primary means to remove 1,4-dioxane would not achieve the target level of 6.1 μg/L.

Uppal et al. (2014) performed an air-stripping/sparging analysis, case study review, and bench-scale treatability testing to further evaluate the potential applicability of *in situ* air sparging for 1,4-dioxane removal from groundwater at low concentrations. The modeling analysis performed using Henry's Law Constant and mass transfer coefficients suggested that *in situ* air sparging using multiple sparge curtains (e.g., rows of sparge wells) would theoretically be capable of removing low concentrations of 1,4-dioxane from groundwater with a removal efficiency of approximately 70%. Sensitivity analysis was performed to evaluate air injection flow rate, groundwater velocity, ratio of volumetric air flow to volumetric water flow (air:water ratio), mass transfer coefficient, treatment residence time, number of air sparge curtains, and subsurface temperature. Bench tests were performed to optimize air:water ratios, subsurface temperature conditions, and air injection techniques (e.g., pressurized microbubble sparging, diffused bubble aeration, steam injection, etc.) for maximum removal from groundwater. The air:water ratios over the experimental runs ranged from 0 to 9,600:1, substantially higher than typical air stripping or air sparging operations. The treatability testing results demonstrated 1,4-dioxane concentration reductions from 100 micrograms per liter (μg/L) to less than 27 μg/L, representing a 73% removal. These data suggest that at low levels, and with limited removal targets, air sparging may have application in specific circumstances (Uppal et al., 2014).

A variation on *ex situ* air stripping has recently been patented (Oberle, 2018), which involves heating the water to be treated to the boiling point of water and then introducing air into the vapor stream. The air:steam ratio must be sufficient to facilitate transfer of the 1,4-dioxane from the steam to the air. The steam-air-contaminant vapor mixture is then cooled rapidly to condense the steam, leaving the miscible liquid in the air. The condensed steam and air are separated in a vapor-liquid separating device, for instance gravity knock-out tank or centrifugal separator, where the contaminated air is then directed to a vapor phase GAC system or thermal oxidizer. A field pilot test of this concept involved air:steam separation of emissions from an electrical resistance heating (ERH) soil treatment application. In that test, 95% of the 1,4-dioxane mass was contained in the air after separation and only 5% in the steam, which was removed with carbon before atmospheric discharge (Oberle, 2018).

8.1.2 Accelerated Remediation Technologies, LLC In-Well System

A more focused and comprehensive mass-transfer approach – combining air stripping, air sparging, soil-vapor extraction, enhanced bioremediation, and underground circulation – has been demonstrated to be effective for 1,4-dioxane removal under specific circumstances (Odah et al., 2005). The Accelerated Remediation Technologies, LLC (ART) In-Well System is designed to operate in a 4-inch or larger well, screened both above and below the water table. The system uses several concurrent physical, chemical, and biological processes to reduce contaminant levels in groundwater (Figure 8.1). Air is forced into the lower part of the well to remove some levels of contamination in the groundwater through sparging. Additionally, the aeration of the

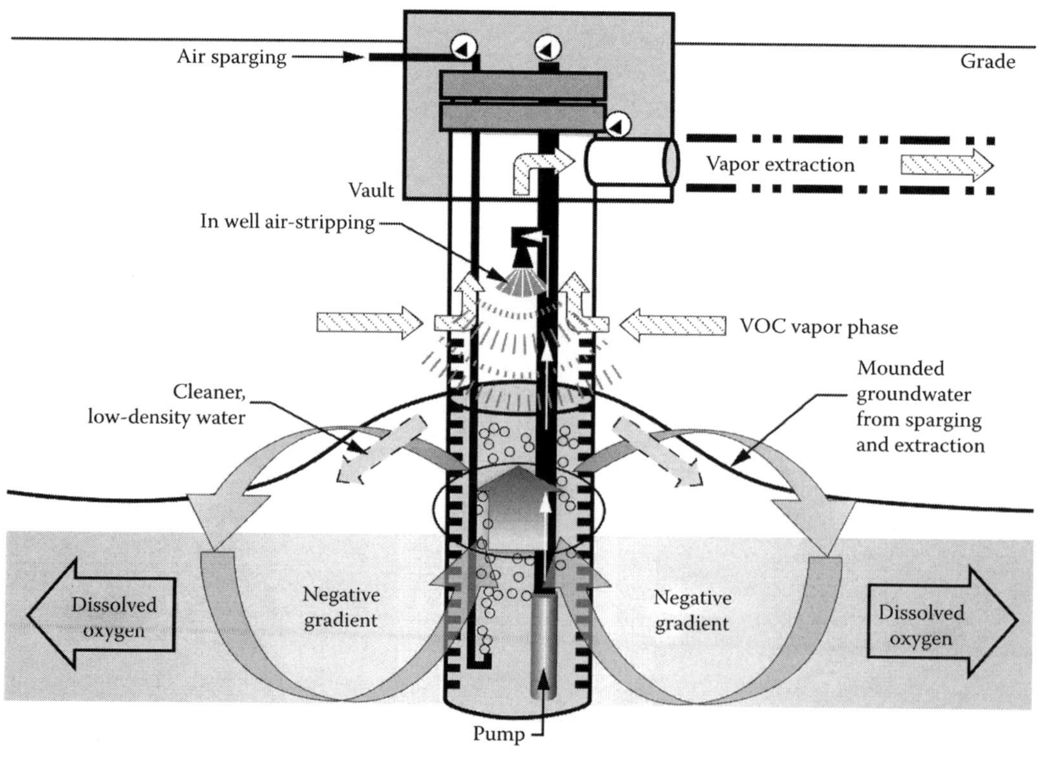

FIGURE 8.1 Various removal processes simultaneously applied in the ART In-Well System. (From Odah, M.M., Powell, R., and Riddle, D.J., 2005, ART in-well technology proves effective in treating 1,4-dioxane contamination. *Remediation Journal* 15(3): 51–64.)

groundwater reduces its density, which causes a drop in hydraulic head in the well. The lower head causes cleaner, low-density groundwater to be pushed out from the upper part of the well and contaminated, high-density groundwater to be drawn into the lower part of the well to create what the inventor refers to as dynamic subsurface circulation. The circulated water effectively flushes soluble contaminants from the lower vadose zone. Water in the well bore is also pumped to the top of the well and sprayed downward through the rising air column, creating an air-stripping effect without using a packing medium to increase the surface area available for phase transfer. The In-Well stripping requires multiple stripping passes, to compensate for 1,4-dioxane's general resistance to stripping from groundwater. Vapors and air are extracted from the wellhead, creating a soil-vapor extraction effect for the screened section of the well above the water table. Lastly, the increased movement of air from all this sparging and extraction leads to higher levels of dissolved oxygen (DO) in the groundwater and more oxygenation of the vadose zone, potentially stimulating aerobic biologic processes that may otherwise be absent under natural conditions. The radius of influence of the ART In-Well System has been documented at ~10 times the water-column height, both up gradient and down gradient from the well (Odah et al., 2005). The ART In-Well technology was applied at a major aerospace manufacturing facility in North Carolina, where the subsurface was saprolitic soil over fractured bedrock. A single ART well was installed, and monitor wells were positioned 10 and 20 ft down gradient (MW-1 and MW-2, respectively) to gauge the system's effectiveness. 1,4-Dioxane concentrations in that part of the aquifer were as high as 43,000 μg/L, and high levels of chlorinated VOCs were also present. Within 90 days, levels in MW-1 declined from 25,000 to 7,400 μg/L, and those in MW-2 declined from 28,000 to 2,400 μg/L. These reductions represented more than 90% removal. The fact that 1,4-dioxane levels in extracted vapors were as high as mg/m^3 demonstrated that removal in the vapor phase was a significant mechanism. Chlorinated VOC removal rates were predictably high as well. The results of this pilot study with a single well led the state to approve the ART In-Well technology as the primary remediation method for the site.

In 2018, Oberle received a patent for "miscible solvent treatment" which involves heating water containing the solvent to its boiling point and introducing air to the steam to allow the solvent to transfer to the air phase (Oberle, 2018).

8.1.3 Pervaporation

An additional vaporization technique for the separation of contaminants from water, generally in industrial applications, is referred to as pervaporation. This process is a combination of evaporation and membrane permeation, which uses relatively low temperatures and pressures for cost-effective *ex situ* separation of liquids with similar boiling points. In the pervaporation process, the membrane is gas permeable, which allows VOCs in the gas phase to pass through the membrane, while water is prevented from passing (Figure 8.2). Chemicals in a mixture can be separated from each other because their partial pressures and membrane permeabilities differ. Pervaporation effectively separates azeotropic mixtures, which have similar boiling points and evaporation rates.

Specific studies related to separation of 1,4-dioxane from water have been performed by using a variety of innovative pervaporation membranes to increase flux rates, separation efficiency, or resistance to chemical and thermal degradation. Pintauro and Jian (1995) assessed the capability of separation of 1,4-dioxane from water with a polyvinylidene fluoride resin membrane. Their results indicated that pervaporation separation of 1,4-dioxane from an aqueous solution did occur, but at significantly lower rates than for other contaminants, such as benzene, toluene, or chloroform. Flux rates through the membrane, at percent-level concentrations of 1,4-dioxane, were on the order of 5 $g/(m^2 h)$.

Much higher 1,4-dioxane flux rates, on the order of 30–100 $g/(m^2 h)$ and efficiencies, with separation factors two to three orders of magnitude higher, were achieved by using blended membranes

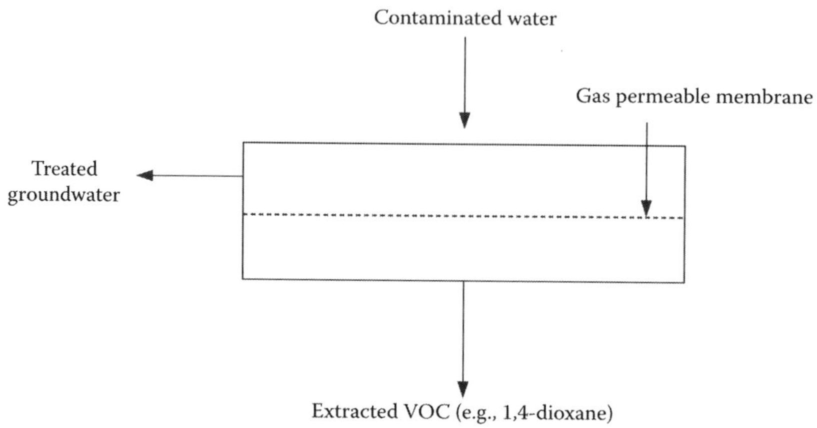

FIGURE 8.2 Overview of pervaporation process. [After Pintauro, P.N. and Jian, K., 1995, United States Patent 5,387,378: Integral asymmetric fluoropolymer pervaporation membranes and method of making the same. Assignee: Tulane University, New Orleans. http://www.patentstorm.us/patents/5387378-description.html (accessed October 12, 2007).]

composed of chitosan and nylon 66 (Smitha et al., 2006). Proportions of these two components were varied, and their tensile, mechanical, thermal, and molecular characteristics were evaluated with a variety of 1,4-dioxane feed mixtures. Because this testing was performed at percent-level concentrations and the pervaporation method is generally used for separation of liquids in industrial manufacturing scenarios, it is not clear whether these materials would be effective at the part per billion (ppb) concentration levels and gallons per minute flow rates typically being evaluated in groundwater remediation programs.

8.1.4 Enhanced/Extreme Soil Vapor Extraction

While 1,4-dioxane physical properties do not favor remediation of 1,4-dioxane in the vadose zone by soil-vapor extraction, successful treatment of 1,4-dioxane has been achieved through application of "enhanced" or "extreme" SVE (XSVE) (Hinchee, 2017; Hinchee et al., 2018). Treatment of 1,4-dioxane by conventional SVE methods is limited by 1,4-dioxane's miscibility in water and low Henry's Law Constant at ambient temperature, which favor partitioning of 1,4-dioxane into the aqueous phase (vadose zone pore water). XSVE enhances conventional SVE methods by applying heated air injection and more focused SVE extraction (Hinchee et al., 2018). The heat influences the Henry's Law Constant, which is temperature-dependent and increases with temperature and helps dry out the soil. In the absence of moisture, 1,4-dioxane readily volatilizes. A pilot-scale demonstration of an XSVE system at Former McClellan Air Force Base, CA resulted in a 94% decrease in 1,4-dioxane soil concentrations following 14 months of operation (Hinchee et al., 2018). Soil moisture decreased by ~45%, which was especially relevant because of 1,4-dioxane's tendency to be present in pore water (Hinchee et al., 2018). Sampling of 1,4-dioxane in soil gas during the demonstration was made with a sampling apparatus designed to capture both vapor and condensate (Burris et al., 2018a). Sampling of the vapor phase alone during XSVE can yield inaccurate results for 1,4-dioxane because of partitioning of 1,4-dioxane into water that condenses in ambient temperature sampling vessels (Burris et al., 2018a). An Excel-based screening-level tool (HyperVent XSVE) has been developed to aid professionals in assessing the feasibility of treating 1,4-dioxane with XSVE and as a design tool (Hinchee, 2017; Burris et al., 2018b).

8.1.5 Thermal Desorption

1,4-dioxane has a boiling point of 101.320°C (Smith and Wojciechowski, 1937), which is slightly higher than that of water. This would suggest that 1,4-dioxane would not be volatilized from water until all the water was boiled off. However, 1,4-dioxane is azeotropic in water and therefore will volatilize from water at a lower temperature (87.7°C) than the boiling point. The Henry's Law Constant also increases 50-fold between the standard temperature of 25°C and the boiling point of water (Oberle et al., 2015), resulting in higher concentrations in the off-gas than released from the initial solution. Because of these physical characteristics, 1,4-dioxane can be removed from water at lower temperature than the boiling point of water.

Oberle et al. (2015) observed 1,4-dioxane reductions in groundwater during a field application of thermal treatment for volatile co-contaminants, which occurred well below the boiling point of water. 1,4-Dioxane concentration declined from 140 μg/L to 1.4 μg/L during the field project. To validate the observations and document the mechanism for these reductions, Oberle et al. (2015) performed bench tests with spiked deionized (DI) water with a high concentration (1,000 mg/L) of 1,4-dioxane, which was then heated slowly to the boiling point of water. The condensate and residual DI water were sampled at regular intervals during the heating. Results indicated that 1,4-dioxane was reduced by 69.2% with only 13.2% water removal. More dramatically, 1,4-dioxane was reduced by 98.9% with merely 39.5% water removal. This confirms the field observation that one need not completely remove the water in an aquifer or vadose zone to reduce 1,4-dioxane mass. Additional experimentation with solids involved heating spiked soil samples. Soil test results yielded conclusions comparable to the water testing, with 1,4-dioxane reductions in soil of 87.6% after soil moisture levels were reduced by 20% (Oberle et al., 2015).

Another set of bench tests was performed on both sandy and clay-rich soil using three different heating mechanisms: thermal conductive heating (TCH), steam enhanced extraction (SEE), and hot water flushing (Fortune, 2015). Results ranged from 91.55% 1,4-dioxane removal in the clay-rich soil using TCH to 99.996% removal in the sandy soil using SEE.

8.1.6 Summary

Some level of 1,4-dioxane removal from conventional *ex situ* air-stripping systems is achievable under ideal circumstances, but in most cases, it is unlikely that this technology will achieve the removal levels required. In general, increasing 1,4-dioxane concentrations and temperature improves performance. The ART In-Well technology – which maximizes air stripping – along with several other physical, chemical, and biological processes has been shown to reduce 1,4-dioxane levels from high concentrations, but the exact mechanism for the removal is not known. Concentrations in the demonstration well were very high, which likely aided mass transfer. Pervaporation has been demonstrated to be capable of separating 1,4-dioxane from water at high percent levels at rates that may be applicable in some *ex situ* situations. It is not clear how effective these latter two technologies would be at the lower concentration levels typically found at most 1,4-dioxane-contaminated sites. XSVE may be a promising option for vadose zone pore water treatment but is still essentially experimental. Field application of thermal desorption from vadose zone soils has demonstrated effectiveness at a limited number of sites where 1,4-dioxane soil contamination was problematic.

8.2 SORPTION

The "sorption" process refers to a combination of chemical adsorption and physical absorption of a contaminant to a "sorptive" medium. Some materials, like engineered resins, are specifically used for one process or the other (i.e., they are either "absorptive" or "adsorptive"), whereas other materials, such as clays and activated carbons, can be used for both processes. Different chemicals often have an affinity for one process over another. Many types of sorptive media are used to remove contaminants from groundwater, most commonly in tandem with a "pump-and-treat" system, although some technologies can be applied *in situ*. For the typical *ex situ* system, the sorptive medium is contained in a series of vessels connected to the extracted groundwater stream. The groundwater passes through the vessels containing the medium, and contaminants are transferred to the medium before the treated groundwater exits the vessel. The contaminants removed are trapped within or form a coating on the medium's sorptive surfaces until they reach the point of saturation, after which no additional contaminants can be removed and breakthrough occurs. The saturation point typically varies in proportion to the applied concentration and flow rates. As with air stripping, sorption technologies constitute a transference of contamination from one medium to another, and the contaminated sorptive media must be disposed of properly and/or treated. Typical treatment and disposal options include stream regeneration for GAC, regeneration of resins, or disposal in landfills. As discussed in the following sections, research into the efficiency of several types of sorptive media has been performed to assess their ability to remove 1,4-dioxane from groundwater.

8.2.1 Adsorptive Media

A bench-scale treatability test was performed for AFP 44 to assess the removal efficiency of several adsorptive media that were identified as potentially applicable for 1,4-dioxane treatment (Earth Tech, 2004). These included a surfactant-modified zeolite (SMZ) (referred to as 14–40 SMZ) and an SMZ with zero-valent iron (ZVI), both provided by Robert Bowman, New Mexico Institute of Mining and Technology, and activated Tri-Base Pelletized Carbon™ (Tri-Base Carbon™) (Hiatt Distributors Limited, Long Beach, California). Zeolites are natural or synthetic aluminum silicates that form a regular crystal lattice and release water at high temperature. Zeolites are polar in nature and have been used for the filtration of a number of organic and inorganic contaminants (Bowman et al., 2000). The results of the partitioning phase of the study indicate that the SMZ with ZVI media adsorbed measurable concentrations of 1,4-dioxane at media:water ratios of 1:20 through 1:2, indicating that sorption is possible at low 1,4-dioxane concentrations and low media:water ratios. No measurable sorption of the 1,4-dioxane to the standard SMZ medium took place. Tri-Base Carbon adsorbed 1,4-dioxane measurably. However, a high media:water ratio was required in order to significantly reduce 1,4-dioxane concentrations. An additional potential filtration technology using macroporous polymer (MPP) extraction was also tested by the manufacturer (Akzo Nobel, Incorporated), but was ineffective at removing 1,4-dioxane from the tested water. In conclusion, the Tri-Base Carbon exhibited a measurable sorption rate for groundwater samples from the site. However, results of the testing and calculations to scale-up the test led to the estimate that ~1,500 tons of Tri-Base Carbon would be required to treat AFP 44 regional groundwater per year. This volume of material is prohibitive from a logistical and cost standpoint.

AMBERSORB™ 560 (AMBERSORB) is a carbonaceous synthetic media manufactured by the Dow Chemical Company. It is produced by the partial pyrolysis of a sulfonated styrene–divinylbenzene copolymer. Carbonaceous synthetic media are high surface area, high porosity, hard, non-dusting spherical adsorbents that have been used for over 30 years in chemical processing and surface water and groundwater remediation. Unlike GAC, which is a broad-spectrum adsorbent, AMBERSORB has more specificity and selectivity for many polar compounds, such as 1,4-dioxane. With GAC, there is less control over the characteristics of the end product, because the starting materials are natural products, such as bituminous coal and coconut shells. This can result in batch-to-batch variability in adsorption capacity and regeneration efficiency. With AMBERSORB, the starting material is an engineered product, resulting in more predictable adsorption and regeneration properties, because pore size distribution and other properties are more easily controlled (Rohm and Haas, 1999). The exact mechanism of 1,4-dioxane removal by AMBERSORB has not been established, but it is believed to be an adsorption process, which is generally classified as interaction of weak van der Waals forces and/or electrostatic attraction

(Neely and Isacoff, 1982). AMBERSORB is a relative newcomer to the array of available 1,4-dioxane treatment technologies that overcomes many of the operating challenges faced by existing technologies. Under typical operating EBCTs of 5 to 20 min, AMBERSORB has 5 to 10 times the sorption capacity as GAC for most compounds of concern, including 1,4-dioxane (Woodard et al., 2014). This is largely due to its unique pore structure, particle size distribution, and hydrophobicity. AMBERSORB is typically regenerated in place, using low-pressure steam. Regeneration liquids require further treatment or disposal. There are several full-scale AMBERSORB treatment systems operating throughout the United States, most for the remediation of contaminated groundwater. The majority of these systems are designed to process flowrates of 100 gpm or less, with influent 1,4-dioxane concentrations that vary from 3 μg/L to as high as 30,000 μg/L. Some of the systems are installed inside intermodal shipping containers, while others are housed in new buildings or have been adapted to fit inside existing structures. AMBERSORB can consistently and reliably deliver 1,4-dioxane treatment down to sub-0.3 μg/L levels, differentiating synthetic media technology from other 1,4-dioxane sorption technologies (S. Woodard, personal communication, 2019).

Multiple synthetic polymeric resins were screened by Martineau et al. (2014) for 1,4-dioxane sorption, including Dowex Optipore L-493, Amberlite XAD4, Dowex Marathon C, and Amberlite XAD7HP. The results of the isotherm testing indicated that Dowex Optipore L-493 was the most effective for 1,4-dioxane removal from water.

8.2.2 Activated Carbon

Johns et al. (1998) performed a series of laboratory studies to determine the sorption characteristics of activated carbons derived from a variety of waste agricultural by-products. Two groups of materials were evaluated: soft materials (made from rice straw, soybean hulls, sugarcane waste, and peanut shells) and hard materials (made from pecan and walnut shells). Materials were commingled with molasses as a binder and then pressed into briquettes or pellets. All materials were carbonized and then crushed and sieved for a consistent size fraction, prior to activating with carbon dioxide or steam. The study specifically utilized lower carbonization temperatures (750°C) than normally used in the industry (1000°C) to demonstrate potential reductions in energy demands and overall manufacturing costs. The sorption effectiveness of the agricultural products was evaluated for a variety of common organic contaminants, including benzene, toluene, 1,4-dioxane, acetonitrile, acetone, and methanol. Several types of commercially available GAC were also tested for comparison. Of the chemicals evaluated, 1,4-dioxane had the lowest overall adsorption rates, as well as the most variability in sorption rates during the tests. The only GAC capable of effective removal of 1,4-dioxane was made from hard nutshells, which achieved a little more than 50% removal. Johns et al. (1998) concluded that the low removal rates could have been due to competition for sorption sites with other adsorbates, steric hindrance, reduced intraparticle diffusion, or lower GAC binding affinity for 1,4-dioxane. More recent column testing of 10 different types of GAC indicated none of them removed 1,4-dioxane suitably to suggest full-scale implementation would be cost-effective (Fotta, 2012).

GAC has been reported to effectively remove 1,4-dioxane from extracted groundwater at the Stanford Linear Accelerator Center (SLAC) in California (Sabba and Witebsky, 2003; USEPA, 2006). On the basis of the evaluation of several different kinds of GAC by Johns et al. (1998), it is unlikely that the reductions noted at SLAC are due solely to sorption onto the GAC. The GAC unit at the SLAC site has moderate influent concentrations of 1,4-dioxane (average 725 μg/L) and very low flow rates (~0.5 gpm). The observed reduction in 1,4-dioxane concentration (<10 μg/L in effluent) may result from cometabolic biodegradation, in which a co-contaminant, in this case tetrahydrofuran (THF), supports biological activity while a by-product of the biological activity (i.e., an enzyme) effectively destroys the target contaminant. Cometabolic degradation of 1,4-dioxane in the presence of THF is well documented in the literature (Zenker et al., 2000; Parales et al., 1994) and is discussed further in Section 8.6.

Recent research suggests GAC does remove 1,4-dioxane, it just does so inefficiently under most circumstances. In general, 1,4-dioxane sorption to GAC increases as concentrations increase. At higher concentrations, it may be possible to sorb significant amounts of 1,4-dioxane to GAC, although it would be difficult to achieve drinking water concentrations. At low 1,4-dioxane concentrations, GAC can remove 1,4-dioxane, but the capacity of the GAC is very small. A study of 1,4-dioxane sorption to GAC estimated the sorptive capacity of three different types of GAC was 14–15 μg 1,4-dioxane per gram of GAC at the low concentrations expected in household systems (Curry, 2009). Another study found that a particular GAC had a relatively high affinity for 1,4-dioxane (Martins et al., 2014) and could adsorb as much as 40 μg 1,4-dioxane per gram of GAC at an initial concentration of 20 μg/L 1,4-dioxane. Myers (2016) found that several different types of GAC, especially hard nutshell GAC, had a higher affinity for 1,4-dioxane than other GACs. At roughly 20 μg/L initial 1,4-dioxane concentration, adsorption to these types of GAC was determined to be ~20 μg/g. These low adsorption rates are enough to allow treatment of a small flow if a large volume of GAC is used and changed out frequently.

Myers (2016) and Myers et al. (2018) published studies on enhancing GAC with the 1,4-dioxane-degrading microbes *Pseudonocardia dioxanivorans* CB1190 and *Mycobacterium austroafricanum* JOB5. These studies demonstrated that 1,4-dioxane-degrading microbes could be grown on GAC and doing so improved overall removal performance. The studies indicated the process would extend GAC longevity but the authors did not quantify the effect.

Biochar, a pyrolyzed organic product produced from biomass, is currently being used for soil remediation and is also being developed for groundwater remediation of other contaminants (Xie et al., 2015). One remediation construction company has developed Bioavailable Absorbent Media (BAM), an injectable biochar for groundwater and soil remediation (Orin Technologies, 2016). Unlike GAC, whose surface is typically chemically palliated during manufacturing to render it nonreactive, biochar's surface is left reactive from the combustion process. This allows biochar to be used as a mild oxidizer for recalcitrant compounds (like 1,4-dioxane) (Kinsman and Mohs, 2018) and as an activator for oxidizers (Fang et al., 2014; Fang et al., 2015; Ouyang et al., 2017; Pignatello et al. 2017) that can be used to treat the recalcitrant compounds (Ouyang et al., 2017).

8.2.3 Organoclays

Organoclays have been used as adsorbents at wood-treating sites for polynuclear aromatic hydrocarbon and creosote removal. They are typically bentonite clays (primarily montmorillonite), chemically modified to make them hydrophobic (water repelling) and organophilic (oil attracting). One application uses quaternary amines to produce an ion exchange capacity of ~70–90 milliequivalents/g (Alther, 2006, 2010). Amine chains bind to the clay particles and attach to organic compounds in the influent water, binding contaminants to the organoclay material. Organoclays are sometimes mixed with GAC, or a secondary carbon unit may be used on the effluent to achieve higher removal efficiencies. Alther (2006, 2010) performed a number of laboratory tests to evaluate the removal efficiency of organoclays on a variety of contaminants, including 1,4-dioxane. Laboratory batch tests were run by combining water containing 1,4-dioxane with nonionic organoclay in glass vials and by shaking for 1 day. Aqueous concentration of 1,4-dioxane was reduced by only 2.8% from levels as high as 958,000 μg/L, which are higher than those that are typically found on contaminated sites.

A proprietary modification technology used by Aqua Technologies of Wyoming, Inc. (A.B. Brown, personal communication, 2007), resulted in an organoclay product (ET-1) capable of removing up to 83% of 1,4-dioxane, from 726,500 to 126,000 μg/L. Test waters were provided by a plastics manufacturer in South Carolina from a contaminated site. The results were verified by an independent laboratory using Environmental Protection Agency Method 8260 with selected-ion monitoring (SIM). Although this removal rate is impressive, the very high concentrations in the extracted groundwater are not typical of most hazardous waste sites contaminated with 1,4-dioxane, and the treatment end points are well above most 1,4-dioxane standards.

8.2.4 Experimental Sorbents

Azad et al. (2000) evaluated the sorption characteristics of 1,4-dioxane on palladium-111 (^{111}Pd). Their research indicated that sorption occurred at very low temperatures (–193°C) in a multilayer that desorbs to leave an overlayer at a somewhat higher temperature (–113°C). The 1,4-dioxane overlayer desorbs and thermally decomposes at ~27°C. The temperatures involved in this process are not likely to be achieved cost-effectively in environmental remediation projects.

Cortés-Arriagada et al. (2017) performed a chemistry study to understand the sorption properties of oxidized graphene (GO) and silicon-doped graphene (SiG) nanosheets for use in 1,4-dioxane removal from water. Theoretical calculations show that GO and SiG sorb through non-covalent and covalent interactions, more effectively than unmodified graphene. With oxidized graphene, hydroxyl and carboxyl functional groups appear to enhance removal efficiency and are responsible for the relatively high adsorption stability in water and at room temperature. The evaluation suggests that these modified graphene sheets should be considered as bench testing alternatives for removal of 1,4-dioxane in liquids.

8.2.5 Summary

The following sorptive media have been tested to evaluate their capacity to remove 1,4-dioxane from water: SMZs, SMZ with ZVI, Tri-Base Carbon, MPPs, GACs made from agricultural by-products and modified with bacteria, Ambersorb and other polymeric synthetic resins, organoclays, palladium-111, and modified graphene sheets. Although some of the media tested demonstrated somewhat effective removal rates, the test conditions often required high media:water ratios that may not be feasible in many field applications. Specific sorbents may be a consideration in situations with low levels of 1,4-dioxane and low groundwater extraction rates.

8.3 NATURAL ATTENUATION

Natural attenuation, or monitored natural attenuation (MNA), has been used to demonstrate passive remediation at sites contaminated with a variety of organic and inorganic contaminants. MNA is most commonly applied to sites contaminated with chlorinated ethenes and petroleum compounds, primarily because these are among the most common contaminants at hazardous waste sites and there are published protocols for leveraging natural attenuation of these compounds for site remediation. The U.S. Environmental Protection Agency's (USEPA's) policy directive on natural attenuation describes MNA processes as those that "act without human intervention to reduce the mass, toxicity, volume or concentrations of contaminants in soil or groundwater" (USEPA, 1999). Natural attenuation of contaminant mass is attributable to a variety of physical, chemical, or biological processes that can include biodegradation, dilution, dispersion, advection, sorption, volatilization, radioactive decay, and chemical and biological stabilization, transformation, and destruction of contaminants; destructive processes are preferred over nondestructive processes (USEPA, 1999).

8.3.1 Evidence of the Effectiveness of Natural Attenuation

Demonstrating that MNA will reliably achieve the desired passive remediation typically requires identifying the active attenuation process and then establishing its effectiveness through several different lines of evidence. USEPA (1999) has established the following evaluation elements, or lines of evidence, that should be assessed:

1. Establish historical trends
2. Indirectly verify the attenuation process by using hydrogeologic and chemical data
3. Directly confirm the attenuation process; for instance, microbial activity can be proven through microcosm studies

Implementing an MNA solution typically involves demonstrating any or all available lines of evidence, the availability of MNA protocols for the contaminant of interest, or other factors. Because MNA has not been commonly used for 1,4-dioxane, implementing MNA for 1,4-dioxane will require a more rigorous weight of evidence to satisfy regulatory authorities. Furthermore, many of the MNA processes commonly invoked for chlorinated solvent sites may be less effective with 1,4-dioxane because of its chemical characteristics, especially its miscibility and low soil partitioning coefficient. Demonstrating a natural attenuation mechanism for 1,4-dioxane is more difficult than for other more well-documented contaminants, but is worth consideration, depending on site conditions and remedial objectives.

A robust statistical evaluation of trends in 1,4-dioxane groundwater monitoring data using California state (GeoTracker) and U.S. Air Force records was performed to estimate natural attenuation of 1,4-dioxane at a variety of field sites (Adamson et al., 2015). Source area attenuation rates were derived from GeoTracker sites using changes over time (average 6.8 years monitoring duration) in site-wide maximum concentrations. Positive attenuation rates were determined to be statistically significant at 22 of the sites assessed, with an estimated equivalent half-life of 20 years. In looking at recent maximum concentrations in comparison to historical maximum concentrations, 82% of the sites exhibited an apparent concentration reduction, generally of less than one order of magnitude, although 23% of the sites demonstrated greater than one order of magnitude reductions over the monitoring period. In assessing the monitoring data from 441 U.S. Air Force wells, it was observed that 269 (61%) exhibited no significant trend, 41 (9%) exhibited significantly increasing trends and 131 (30%) exhibited significantly decreasing trends. Factors that may have an influence over 1,4-dioxane attenuation include the presence of metals and chlorinated solvents, which were negatively correlated with attenuation rates, and DO, which was positively correlated with 1,4-dioxane attenuation rates in individual wells. These data taken as whole, suggest that natural attenuation of 1,4-dioxane is occurring at some sites, but was not universally observed in either the GeoTracker or the U.S. Air Force data sets (Adamson et al., 2015).

8.3.2 Diffusion into Immobile Porosity

One potentially applicable attenuation mechanism relates to the hydraulic characteristics of fine-grained materials in the subsurface. Finer-grained, less permeable aquifer materials such as clays and silts in contact with more transmissive, coarser-grained aquifer materials may trap contaminants migrating through the aquifer. Fine-grained sediments reduce the effective porosity of the aquifer to the point that molecules of the contaminant can diffuse from the transmissive zones into less transmissive zones (aka dead-end pore spaces) within the finer-grained aquifer material. Diffusion into dead-end pores will cause a reduction in monitored aqueous concentrations similar to the reduction caused by chemical sorption of typical organic contaminants to organic materials in the aquifer. Fine-grained clay and silt are present in most unconsolidated aquifers. Clays, though highly porous, are relatively impermeable and hence restrict the advective flow of groundwater. Dead-end pores exist that are connected to the more permeable materials of the aquifer but are not subject to through-flow. Because water is not actually flowing into or through these dead-end pore spaces, the contaminant migration process is governed by molecular diffusion. The transfer of 1,4-dioxane into dead-end pores is a slow process driven by the concentration gradient between the contaminated free-flowing groundwater and the relatively clean water in the dead-end pore spaces. Unfortunately, unlike chemical adsorption, where the contaminant may be permanently bound to the organic material in the aquifer, this attenuation process involving matrix diffusion into low permeability zones is a reversible process. Back-diffusion from the low permeability zones into the high permeability zones occurs after the concentration gradient reverses to reflect cleaner water in the aquifer and more contaminated water in the dead-end pore spaces. This back-diffusion is a slower process than the initial diffusion into low transmissivity zones because the gradients are typically less pronounced, resulting in long-duration, low concentration contributions to the high transmissivity zones.

8.3.3 Modeling Natural Attenuation

Statistical trend analysis and a numerical flow model were used to gain the approval of a natural attenuation remedial approach for an industrial site in North Carolina (Chiang et al., 2006, 2007b, 2008). Active sources of 1,4-dioxane had been removed and dissolved-phase plumes were present in four distinct hydrogeologic layers beneath the site. Data collected from 60 wells at the site since the early 1990s were evaluated yearly from 2003 through 2005 by using Mann-Kendall trend analysis. The evaluation in 2005 noted that 12 wells (20%) exhibited a downward trend and 21 wells (35%) exhibited a stable trend. Twenty-six wells (43%) exhibited no discernable trend. An upward trend was noted in only one well (2%). To demonstrate a continued attenuation trend into the future, a Modular Three-Dimensional Groundwater Flow and Transport Model (MODFLOWT) was constructed by using four hydrogeologic layers representative of the varying hydraulic conductivities in subsurface sand versus clay or silt units. The 1,4-dioxane plume was assumed to be more sensitive to advection than to hydrodynamic dispersion. (Advection refers to the process of contaminant migration due to movement of groundwater. Hydrodynamic dispersion is the sum of mechanical dispersion and diffusion.) The modeling effort simulated advective-dispersive transport but considered diffusion to be relatively insignificant and therefore omitted this term from the model.

The flow model was calibrated by using measured heads from monitoring and production wells at the site. Particle tracking and transport modeling were performed to estimate future concentrations of 1,4-dioxane at a local surface-water body to the east. The modeling under an assumption of no degradation underestimated the observed decline in 1,4-dioxane concentrations. To assess the degradation rate, historical field measurements were evaluated at several monitoring wells within different hydrogeologic layers. Time-series plots were prepared under the assumption of no degradation versus degradation half-lives of 3, 5, and 7 years. The optimal correlation between observed and predicted values occurred with a 7-year degradation half-life. For each of the four layers, Figure 8.3 presents the 10-year predictions, which show significant reductions in the distribution of 1,4-dioxane as a result of using a 7-year half-life. The modeling results predict that 1,4-dioxane migration at levels above North Carolina's surface-water standards would not reach the surface-water body via any of the modeled hydrogeologic layers. Among the natural attenuation mechanisms, Chiang et al. surmised that 1,4-dioxane is more highly influenced by advection because it is not readily sorbed to the aquifer matrix. Because no active sources of contamination existed at the site and the combination of modeling and field observations demonstrated that the plume was stable or shrinking, the proposed MNA approach was approved by the regulatory agency for implementation as the final remedy (Chiang et al., 2008).

Adamson et al. (2016) performed modeling to better understand the significance of matrix diffusion as an attenuation mechanism, as well as a predictor of long-term 1,4-dioxane concentrations in groundwater plumes. Two simple two-layer model simulations were performed using the U.S. Department of Defense Matrix Diffusion Toolkit, a publicly available free software tool. In the first case, the 1,4-dioxane release was assumed to involve high concentrations (e.g., 28,900 mg/L) in conjunction with a methyl chloroform DNAPL release within a well-delineated on-site source zone, as would be expected for a typical historical chlorinated solvent release site. Because of the miscibility of 1,4-dioxane, the loading period for the release to the aquifer was assumed to be only 2.7 years, with the 1,4-dioxane migrating away from the source area rapidly compared to an ongoing (39.9 years) chlorinated solvent release from the NAPL to the aquifer. In the second model simulation, the 1,4-dioxane source was assumed to be upsteam of the study area, contributing lower concentrations (1 mg/L) for a shorter duration (1 year) as the "slug" migrates through the aquifer. The results of the modeling effort indicate that for the first case, the source area 1,4-dioxane concentrations declined by three orders of magnitude within a mere 3 years, suggesting that recent monitoring is measuring concentrations substantially lower than historic (e.g., 1970's era) source area concentrations. Additionally, the 1,4-dioxane mass remaining at the end of the model simulation in 2050 was almost twice the amount of methyl chloroform mass because of the greater amount trapped in low

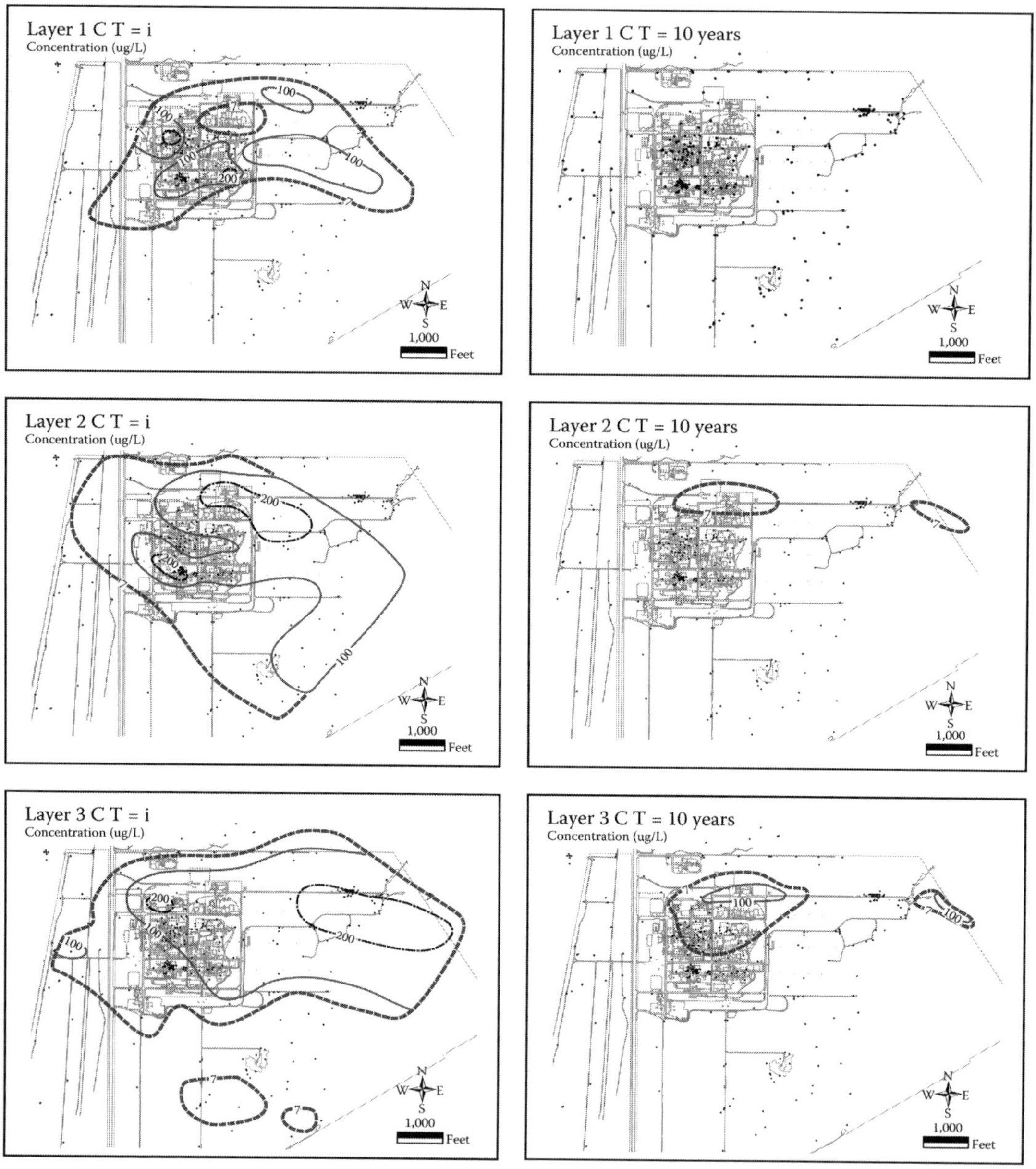

FIGURE 8.3 Ten-year predictions of 1,4-dioxane simulations in three hydrogeologic layers under pump-off scenario with 1,4-dioxane half-life of 7 years. (After Chiang, D., Zhang, Y., Glover, E., Harrigan, J., and Woodward, D., 2006, In-situ oxidation of 1,4-dioxane. In: B.M. Sass (Ed.), *Proceedings of the Fifth International Conference on Remediation of Chlorinated and Recalcitrant Compounds*, Paper G-70. Columbus, OH: Battelle Press.)

transmissivity zones. This additional mass results in a longer "tail" at higher concentrations for 1,4-dioxane relative to methyl chloroform. In the second case, the overall mass loading to the low transmissivity zone was less due to the lower initial concentration, but the resulting diffusion-fed groundwater concentrations were still above potential standards well into the future. The simulations predict that back diffusion from the low-transmissivity zone will act as a secondary source of 1,4-dioxane to the high transmissivity zone long after source material in the transmissive zone has been removed through remedial actions (Adamson et al., 2016). This does not bode well for achieving aquifer cleanup, even after substantial effort is invested in removing source area mass or attempting to clean up the high transmissivity zones within an aquifer.

8.3.4 Biological Mechanisms for Natural Attenuation

Numerous papers have been published documenting 1,4-dioxane biodegradation in environmental and laboratory studies, and biological degradation of this compound is widely recognized (see Chapter 3). However, robust assessment of biological mechanisms responsible for natural attenuation in aerobic environments is generally lacking in the literature. Several recent studies accomplishing this evaluation are discussed below.

Li et al. (2015) assessed the potential for 1,4-dioxane biodegradation at several California sites through microcosm studies using soil and groundwater samples from the presumed source zone, middle and leading edge of the plume at each of the three sites. 1,4-Dioxane biodegradation was noted in 12 of the 16 microcosms, which simulated natural attenuation over a 28-week period. The fate of 1,4-dioxane in the microcosms was determined using ^{14}C-labeled dioxane, which was determined to assimilate into cell material only in microcosms with significant dioxane biodegradation occurring. Confirmation of biological activity was determined through quantifying the abundance of tetrahydrofuran monooxygenase (*thm*) A and dioxane monooxygenase (*dxm aka DXMO*) A, both of which had been previously determined to play a crucial role in initiating dioxane metabolism. Degradation decreased significantly with increasing distance from the source area, possibly due to less acclimatization of the biological community to site conditions. The results demonstrate that indigenous microorganisms capable of degrading dioxane were present at all three sites and suggest that MNA could be a viable remedial approach (Li et al., 2013; Li et al., 2015).

Possible biological mechanisms for natural attenuation of 1,4-dioxane were evaluated at Air Force Plant 44, in Tucson, Arizona, using a parallel approach to that typically undertaken for the evaluation of natural TCE biodegradation (Chiang et al., 2012). Baseline geochemical conditions were defined, including analysis of cations, anions, dissolved gases, total organic carbon (TOC), and alkalinity. BioTrap™ sampling for quantitative polymerase chain reaction (qPCR) was performed to verify the presence and abundance of bacteria and enzymes capable of aerobically degrading TCE and 1,4-dioxane. The targeted microorganisms and enzymes include methanotrophs (MOB), soluble methane monooxygenase (sMMO), phenol hydroxylase (PHE), ring hydroxylating toluene monooxygenase (RMO), and toluene dioxygenase (TOD). This phase of study revealed that MOB and TOD are ubiquitous with the density of MOB ranging from 10^3 to 10^7 cell/mL and TOD ranging from 10^4 to 10^8 cell/mL. PHE and sMMO are also present at significant levels. Based on the presence of key bacteria and enzymes in all samples collected, further testing for naturally occurring biodegradation was performed. Phospholipid fatty acid analysis with stable isotope probes (PLFA-SIP) was conducted using BioTraps™ baited with 1,4-dioxane tagged with the ^{13}C isotope. The ^{13}C-1,4-dioxane-baited BioTraps™ were evaluated for ^{13}C fractionation (indicating biodegradation of 1,4-dioxane), incorporation of ^{13}C into cell phospholipid fatty acids (i.e., PLFA), and generation of ^{13}C carbon dioxide. Proof that mineralization of 1,4-dioxane was occurring at low levels during the study period was provided through the quantification of ^{13}C enriched dissolved inorganic carbon. Elevated ^{13}C in biomass (max. +5885‰) confirm that *in situ* 1,4-dioxane intrinsic biodegradation was occurring (Chiang et al., 2012).

Another well documented study of natural attenuation of 1,4-dioxane is discussed further in Section 9 – Case Studies. In summary, Gedalanga et al. (2016) presented results of an in-depth assessment of the long-term declining 1,4-dioxane concentration trends and microbiological testing results across a groundwater plume in the Midwest. A thorough Conceptual Site Model was developed, including assessment of the complex hydrogeological and geochemical characteristics of the site. Temporal and spatial trends in 1,4-dioxane and colocated THF, as well as iron, sulfate, and methane were evaluated, and linear regression results calculated for each constituent. Compound-specific isotope analysis

(CSIA) analysis was performed along with analysis of 1,4-dioxane biomarkers and sMMO gene targets to confirm biological activity. Distinct downward concentration trends were well documented in the >10 years of groundwater monitoring data in source area and mid-plume locations. Additionally, 1,4-dioxane mass declined over time suggested degradation rather than dispersion. A weak ^{13}C enrichment was noted in CSIA analysis as 1,4-dioxane was depleted with distance downgradient, suggesting evidence that a biochemical degradation process is active. The most compelling evidence of biodegradation was the correlation between locations with the presence of 1,4-dioxane degradation biomarkers: dioxane monooxygenase (DXMO) and aldehyde dehydrogenase (ALDH). Detections of DXMO/ALDH matched the 1,4-dioxane plume geography with reasonable accuracy, strongly suggesting these markers are stimulated by the presence of 1,4-dioxane and confirming biological degradation of 1,4-dioxane (Gedalanga et al., 2016). More details are provided in Chapter 9. In another recent weight of evidence study (Jackson and Lemke, 2019), mass balance and spatial moment analyses suggest potentially significant rates of natural attenuation, although other indicators were less conclusive.

8.3.5 Summary

Natural attenuation processes may reduce concentrations of 1,4-dioxane in the groundwater under specific site conditions. The most likely processes are advection, diffusion (especially into dead-end pores), and dispersion. Intrinsic biodegradation has been demonstrated to be a viable mechanism in several studies, but additional field study is necessary to verify occurrence and effectiveness. Volatilization and sorption to conventional soil materials are not likely to be significant attenuation mechanisms for 1,4-dioxane.

8.4 ULTRAVIOLET PHOTOLYSIS

The energy from an ultraviolet lamp can often directly break apart the chemical bonds of an organic contaminant, ultimately resulting in harmless constituent compounds, such as carbon dioxide and water. When photons are absorbed by a compound, the chemical bonds holding the elements together may be broken if their energy is less than the energy contained in the photon. On the basis of historical research into quantum physics by Albert Einstein and Max Planck, the energy E of a "light quanta" is described by the equation $E = hc/x$, where h is Planck's constant (4.14×10^{-15} eV s), c is the speed of light (3.0×10^{8} m/s), and x is the wavelength (Tailored Lighting, 2007). Thus, shorter-wavelength light, including ultraviolet light, contains more energy than longer-wavelength light.

Hentz and Parrish (1971) demonstrated that the decomposition of 1,4-dioxane occurred at a "far" (ISO, 2005) ultraviolet wavelength of 147 nm, which carries 6.2–12.4 eV of energy per photon. Ninety-eight percent of the total decomposition yielded ethane and formaldehyde. In general, this application is restricted to ultrapure, laboratory-grade water, and restricted low wavelengths of ultraviolet light.

A. Festger (personal communication, 2007) of Trojan Technologies has stated that manufacturers of ultraviolet-peroxide water treatment systems use 1,4-dioxane as a reagent in bench-scale tests quantifying the effects of ultraviolet energy on oxidation systems because 1,4-dioxane does not react by strict ultraviolet photolysis at middle ultraviolet wavelengths commonly applied to water treatment (230–300 nm), without the addition of oxidation chemicals. The middle ultraviolet wavelengths contain less energy (4.13–6.20 eV), which likely limits the destructive effect on 1,4-dioxane. The use of 1,4-dioxane in bench-scale tests allows Trojan Technologies to demonstrate the treatment efficacy of ultraviolet plus oxidation, as opposed to limited ultraviolet photolysis, by quantifying the degradation of 1,4-dioxane.

Although there is some evidence of direct ultraviolet photolysis of 1,4-dioxane, the demonstrated applications are at wavelengths not typically applied to groundwater treatment and were tested on extremely pure water. Therefore, direct ultraviolet photolysis is not considered as a viable remediation technology for 1,4-dioxane.

8.5 PHYTOTECHNOLOGY

Phytotechnology refers to contaminant removal, remediation, or containment through one or more of several plant-mediated processes. Phytotechnology processes have been summarized previously as follows (Interstate Technology and Regulatory Council [ITRC], 2009):

- *Phytosequestration:* The ability of plants to sequester certain contaminants in the rhizosphere through exudation of phytochemicals and on the root through transport proteins and cellular processes (resulting in containment).
- *Rhizodegradation:* Exuded phytochemicals can enhance microbial biodegradation of contaminants in the rhizosphere (resulting in remediation by destruction).
- *Phytohydraulics:* The ability of plants to capture and evaporate water off the plant and take up and transpire water through the plant (resulting in containment by hydraulic control).
- *Phytoextraction:* The ability of plants to take up contaminants into the plant with the transpiration stream (resulting in remediation by removal of the plants).
- *Phytodegradation:* The ability of plants to take up and break down contaminants in the transpiration stream through internal enzymatic activity and photosynthetic oxidation/reduction (resulting in remediation by destruction).
- *Phytovolatilization:* The ability of plants to take up, translocate, and subsequently transpire volatile contaminants in the transpiration stream (resulting in remediation by removal through the plants).

Limitations of phytoremediation include depth to groundwater, growing season, and soil conditions.

8.5.1 Bench-Scale Tests

Imperial Carolina hybrid poplar tree cuttings were rooted hydroponically in a series of experiments by Aitchison et al. (2000). Cuttings were tested in individual flasks containing an aqueous solution of 20,000 μg/L 1,4-dioxane. Several variables were evaluated in this study, including plant uptake, transpiration, incorporation into the stem material, sorption to roots, and degradation by microorganisms associated with the root zone. The researchers used a carbon isotope (^{14}C) to label 1,4-dioxane and assess its distribution among the plant roots, stems, and leaves. Between 30% and 79% (average 54%) of the 1,4-dioxane was removed from the solution after 8 days. However, the fact that only 10% was removed from the solution in experiments in which the leaves had been detached from the stem suggests that transpiration was an important process. Eight percent removal occurred in the control flask that did not contain a poplar cutting, confirming that the 1,4-dioxane was not degrading or volatilizing from the flasks by some other mechanism. The data indicate that transpiration was the dominant removal mechanism and that volatilization of the 1,4-dioxane from the plant's leaves accounted for 77% of the 1,4-dioxane removed by the plant. Degradation within the root zone was considered to be minimal.

Researchers at the University of Iowa experimented with adding *Amycolata sp.* CB1190 to planted and unplanted soil (Kelley et al., 2001). CB1190 has been demonstrated to effectively degrade 1,4-dioxane in water (Parales et al., 1994). The effectiveness of bioaugmentation of poplar (*Populus sp.*) root zones for the destruction of 1,4-dioxane was evaluated by using a known 1,4-dioxane-degrading bacteria to enhance phytoremediation. Soil and basal salt medium (BSM) microcosms were prepared and spiked with 100,000 μg/L of 1,4-dioxane prior to the introduction of CB1190 bacteria, which were grown on 1,000,000 μg/L of THF. Several controls were also used, including one in an aqueous solution with CB1190 and no soil, to demonstrate the viability of the bacteria to degrade 1,4-dioxane. Additional controls demonstrated that neither sterile conditions nor the native soil bacteria were effective in 1,4-dioxane degradation. Complete removal of the 1,4-dioxane occurred in all CB1190-amended microcosms within 30 days in the BSM and within 45 days in a soil amended with poplar root extract.

Adding the THF apparently led to enhanced enzyme creation. Poplar root extract and 1-butanol stimulated additional growth of CB1190. Planting hybrid poplar trees in 1,4-dioxane-contaminated soils caused some of the dioxane to be evapotranspired through the trees. Adding CB1190 to planted soils was less effective because evapotranspiration removed the dioxane, making it less available for cometabolic biodegradation. 1,4-Dioxane was not directly degraded by poplar root exudates; the root exudates only enhanced the growth of CB1190. Only THF produced a dioxane-specific activity.

Unplanted soils amended with CB1190 at 10^7 cells/g of soil completely destroyed ~25% ± 5% of the ^{14}C-labeled 1,4-dioxane added to the soil. CB1190 can survive in the soil substrate and compete with indigenous soil and rhizosphere microflora. CB1190 could colonize the poplar rhizosphere, which suggests that bioaugmenting poplar roots with CB1190 might enhance the remediation process. However, cells grow slowly and the cell yield is relatively low. To bioaugment 1 acre of contaminated soil over a 1-foot thickness at 10 mg of dry cell mass/kg of soil (i.e., ~10^7 cells/g), ~15,000 gallons of cell suspension would be needed (Kelley et al., 2001).

The effects of co-contaminants on 1,4-dioxane phytoremediation have also been explored. Edwards et al. (2011) evaluated the effects of ethylene glycol by measuring changes in plant growth, water transpiration, and the amount of 1,4-dioxane phytotranspired in closed bench-top experiments. Ethylene glycol was found to decrease the rate of plant growth, in both hybrid poplar and hybrid willow species as well as in the model organism *Arabidopis thaliana* (thale cress). The effects of ethylene glycol on hydroponically-grown poplars and willows grown in the presence of 1,4-dioxane was also investigated. Ethylene glycol was found to lower water transpiration and to disproportionally lower the rate of 1,4-dioxane (i.e., the decrease in 1,4-dioxane phytovolatilization could not be explained by decreases in transpiration alone).

8.5.2 Mathematical Modeling

Ouyang (2002) modified an existing one-dimensional mathematical model (CTSPAC—Coupled Transport of Water, Solutes, and Heat in the Soil-Plant-Atmosphere Continuum [Cawfield et al., 1990]) to predict the removal of 1,4-dioxane from groundwater and soil. Calibration of the model was performed by using the hydroponic and soil experimental results of Aitchison et al. (2000). Model simulations indicated that uptake from soil would be relatively rapid and that 30% of the 1,4-dioxane mass would be removed within 7 days. The modeled distribution of 1,4-dioxane in the various parts of the plant correlated well with research by others. The findings, shown in Figure 8.4, indicate that 1,4-dioxane was concentrated early into the leaves, and, after 7 days, the leaves still contained the greatest mass, relative to the proportional mass of the various plant parts. The stem contained some 1,4-dioxane, but the root system held little, if any, at the conclusion of the simulations. Ouyang (2002) concluded that the model effectively simulated the removal rates and distribution of 1,4-dioxane in the plant material that have been documented by others in laboratory experiments. The CTSPAC model could therefore be used to estimate removal rates from contaminated sites.

In a second modeling study, Ouyang (2008) developed a dynamic model for the uptake and translocation of contaminants from a soil-plant ecosystem (UTCSP) using the STELLA® modeling tool (isee systems, inc., Lebanon, New Hampshire) UTCSP assesses simultaneous transport, accumulation, and transpiration of water and contaminants in plant systems. The model was calibrated with experimental data and then used to predict 1,4-dioxane phytoremediation from a sandy soil by a poplar tree. Results indicated ~20% 1,4-dioxane removal from the soil in 90 days. Ouyang (2008) identified a steadily increasing mass of 1,4-dioxane in the plant material over time, as well as a typical diurnal distribution pattern in the uptake and translocation in the plant system, resulting from daily variations of plant leaf transpiration. The study concluded that the UTCSP model could be useful for estimating 1,4-dioxane phytoremediation in soil-plant systems.

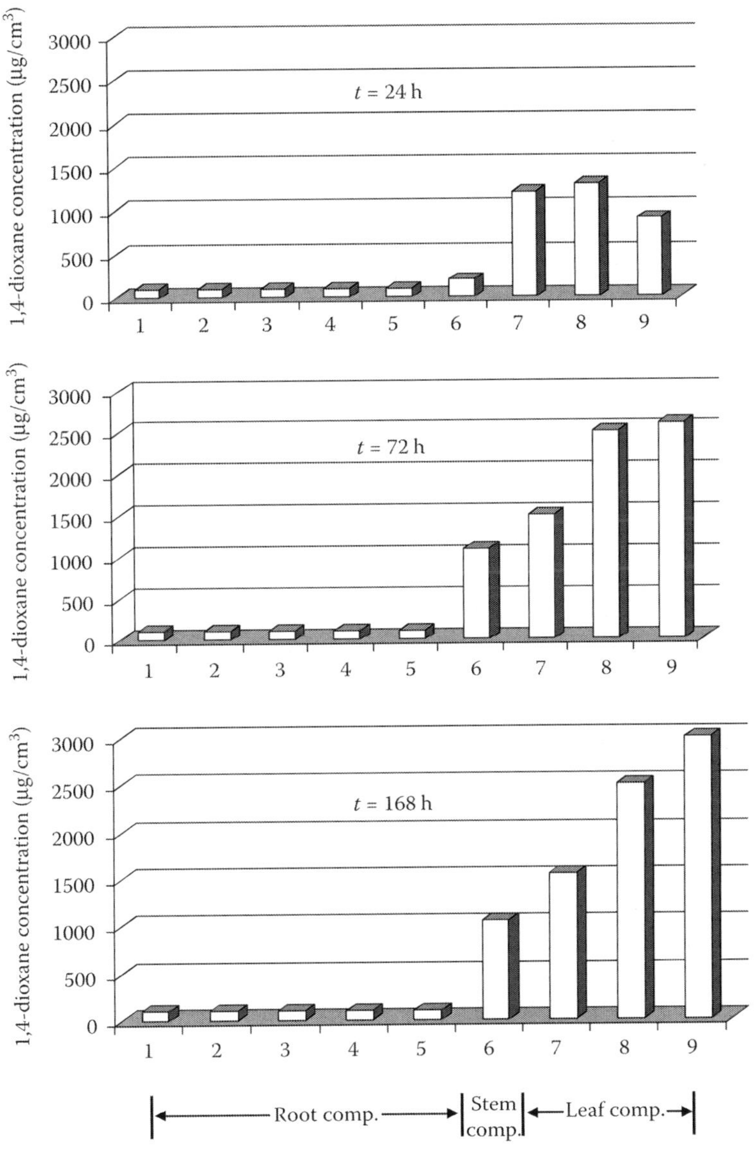

FIGURE 8.4 1,4-Dioxane concentration in plant compartments ("Comp.") at times of 24, 72, and 168 hours. (From Ouyang, Y., 2002, Phytoremediation: Modeling plant uptake and contaminant transport in the soil-plant-atmosphere continuum. *Journal of Hydrology* 266(1–2): 66–82.)

8.5.3 Greenhouse and Field Studies

Ferro et al. (2005) performed greenhouse and field pilot studies toward the design of a full-scale 1,4-dioxane phytoremediation project at an industrial site in North Carolina. Greenhouse studies were utilized to identify candidate species of both deciduous and coniferous trees to employ in the full-scale design. The study concluded that the full-scale project should go forward with both types of trees in use for continuous groundwater phytoremediation. In this scenario, the coniferous trees address the contaminants during the winter when the deciduous trees, which are the major water-extraction plants, are dormant.

A field pilot study of phytoremediation of 1,4-dioxane has been documented by Chiang et al. (2007a). This study was performed in an area of approximately 8,000 ft^2 to address a groundwater seep with the potential to impact surface water. Over 100 poplar trees were planted in 12 rows perpendicular to the groundwater flow direction. Later, an additional 100 hybrid poplar cuttings were planted between the trees to increase the immediate water uptake capacity. Seep sample locations were not able to be sampled the following summer because they were dry, which was interpreted to be a direct result of the dewatering (i.e., water uptake) capacity of the trees. Thus, although the specific removal of 1,4-dioxane could not be documented, the removal of impacted groundwater was deemed a successful outcome.

Another pilot study was conducted in 2010 to test 1,4-dioxane removal by trees grown in plots irrigated with recovered groundwater extracted from downgradient of a former solvent recycling facility and nearby closed landfill (Ferro et al., 2013). In this study, four plots of mature trees were drip-irrigated with recovered groundwater containing 2.5 mg/L 1,4-dioxane and 17 mg/L total VOCs (including chlorinated ethanes and ethenes, mono- and dichloromethane, and BTEX compounds). The study plots contained either: hybrid poplar, Japanese pine, mixed coniferous species, or mixed hardwood species. The pilot was conducted over two study periods lasting approximately 1 month each. Irrigation was controlled to avoid leaching of 1,4-dioxane below the root zone (monitored with a bromide tracer). After the first study period, during which irrigation was well-controlled, an average of only 14% of the applied 1,4-dioxane remained in the soil. The tentative conclusion was that the 1,4-dioxane had been phytovolatilized, although other processes could not be ruled out. During the second study period, excessive rainfall occurred and poor bromide tracer recovery suggested that leaching below the root zone had occurred. A planned full-scale treatment will locate trees upgradient of the extraction wells so that contaminants that leach below the root zone can be recaptured. Coniferous trees were selected for this planned full-scale application to maximize water uptake and phytovolatilization during the winter.

Gatliff et al. (2013) described multiple applications of engineered phytoremediation systems (Applied Natural Sciences' patented TreeWell) including a manufacturing site in the Netherlands where 1,4-dioxane concentrations were as high as 822,000 μg/L in the groundwater. More than 200 hybrid poplar (*Populus deltoides* x *nigra*) trees were planted, including approximately 50 using a "straw" method where a water extraction tube was placed adjacent to the root structure to allow access to water as deep as 7 m to 8 m below ground surface. After a little over a year, the groundwater gradient had been altered toward the planting area, suggesting mitigation of off-site plume migration (Gestler, 2016). Lack of tree mortality suggests high tolerance of 1,4-dioxane present at high concentrations, and rhizodegradation was documented. At a site in Sarasota, Florida, 1,4-dioxane-impacted groundwater is migrating in fractured bedrock, which was inaccessible to existing vegetation, and was migrating into a residential area. More than 150 slash pine (*Pinus elliottii*), pond cypress (*Taxodium ascendens*), and laurel oak (*Quercus laurifolia*) trees were planted on 20-ft centers. These trees were also planted utilizing the TreeWell® system to draw deeper water into the root zone for removal and degradation. The phytoremediation system reversed the groundwater gradient at the site in under 2 years' time, thereby limiting off-site migration. 1,4-Dioxane concentrations were reduced two orders of magnitude to nondetect levels in a well 10 m downgradient of the planting area (Gestler, 2016).

8.5.4 Summary

Phytoremediation to remove 1,4-dioxane from soil and groundwater has been demonstrated in bench-, pilot-, and full-scale studies using various tree species. Removal mechanisms include rhizodegradation and uptake into the plant and transpiration of the 1,4-dioxane out of the leaves to the atmosphere. Because of the short half-life of 1,4-dioxane in the atmosphere, this process could be considered an effective long-term method for the removal and destruction of 1,4-dioxane. Climatic (i.e., growing season) and hydrogeologic (i.e., depth to groundwater) limitations are important factors when considering phytoremediation of 1,4-dioxane.

8.6 BIOREMEDIATION

Bioremediation of common organic contaminants such as chlorinated solvents and petroleum hydrocarbons has been demonstrated to be effective at many project sites, such that it is now considered a presumptive remedy for certain contaminants. In general, bioremediation is implemented by providing amendments to the existing biocommunity in the form of nutrients, food, or oxygen. In some cases, the native bacterial community is not robust enough to effectively address the contaminants present. In these cases, bioaugmentation with specialized bacteria is often performed. 1,4-Dioxane is a relatively stable molecule with strong internal chemical bonding. Therefore, it has been commonly considered to be nonbiodegradable. However, a substantial amount of research has been performed to identify specific organisms that can degrade 1,4-dioxane, the mechanisms of degradation, and the geochemical conditions necessary for these organisms to thrive.

Information presented in this chapter will focus on studies related to furthering the possibility of field application of bioremediation as a technology to carry sites toward closure. Additional information has been presented in Chapter 3 that addresses the identification of organisms capable of degrading 1,4-dioxane, the definition of degradation pathways and new molecular-biological tools (MBTs) that can be used to identify the presence of degraders and confirm degradation. Although there is some overlap between these two chapters, the theoretical approaches are primarily addressed in Chapter 3; Chapter 8 deals with potential applied approaches. Note that there has been some work using microorganisms in conjunction with adsorbents. That work has been discussed in Section 8.2.2, Activated Carbon.

8.6.1 Mixed Bacterial Cultures

8.6.1.1 Actinomycete Culture

Adamus et al. (1995) patented a culture comprising an actinomycete, genus *Amycolata*, identified as ATCC 55486, which they claimed was capable of utilizing dioxane as a sole carbon growth source under growth or resting aerobic conditions. The actinomycete strain was tested between 15°C and 35°C on 1,4-dioxane concentrations as high as 10,000,000 μg/L. Adamus et al. (1995) noted in their patent application that, among the cyclic ethers, 1,4-dioxane was considered to be remarkably refractory to biological degradation and generally considered to be not readily biodegradable.

The *Amycolata* strain can be stored for at least 1 month at 4°C with little loss of activity. Its activity is maintained indefinitely at −70°C without cryoprotectants such as skim milk or glycerol, provided that cultures are stored in BSM.

Adamus et al. (1995) incubated bacteria in the presence of 100,000–300,000 μg/L of THF in a BSM containing 0.1% weight per volume yeast extract. Once THF degradation is confirmed, the culture is supplied with both THF and 100,000–500,000 μg/L of 1,4-dioxane. Transfer cultures degraded the THF and dioxane to less than 1,000 μg/L within 78 hours.

The culture – designated as "CB1184" – represents a mixed population including irregularly branched, filamentous organisms with clumps or mats of organisms formed within these filaments. The specific rate of 1,4-dioxane degradation, normalized to 1 mg of the protein of these organisms, was ~0.2–1.2 mg/min when the 1,4-dioxane concentration was maintained at 300,000 μg/L (i.e., 2–3 nmol/min per 1 mg of the organisms' protein).

Because the mixed population includes both eukaryotic microorganisms (protozoa, fungi, and yeast) and prokaryotic organisms (bacteria), tests were conducted to determine which degraded 1,4-dioxane. Antibiotics, which terminate activity in specific bacteria, were added to the culture; treatment with 50 mg/mL tetracycline, streptomycin, and penicillin inhibited the degradation of 1,4-dioxane, whereas treatment with cycloheximide (inhibitor of eukaryotic protein synthesis) did not inhibit the degradation of 1,4-dioxane. Therefore, the degradation of dioxane was traceable to the prokaryotic organisms.

Adamus et al. (1995) built a large-scale (40 L capacity) packed-bed reactor to treat groundwater contaminated with 1,4-dioxane, biphenyl, biphenyl ether, ethylene glycol, chlorobenzene, acetone, chloroform, 1,1-dichloroethane, 1,1-dichloroethylene, 1,2-dichloroethene, methylene chloride, methyl chloroform, and toluene. At an effluent temperature of ~15°C, influent and effluent 1,4-dioxane concentrations were 255,000 and 40,000 μg/L, respectively, and the hourly volumetric degradation rate was calculated to be 0.016 g/L. When the bioreactor temperature was 27.4°C, influent and effluent concentrations were 250,000 and less than 1,000 μg/L, respectively, and the rate was calculated to be 0.019 g/L.

8.6.1.2 Indigenous Soil Microbes

Taylor et al. (1997) investigated the ability of indigenous soil microbes to degrade a series of contaminants related to the production of plastic films in Rochester, New York. Aerobic and anaerobic tests were conducted on representative groundwater samples. Quantification and isolation of the aerobic and anaerobic bacterial populations and fungi were conducted for growth on the site contaminants acetone, isopropanol, methanol, toluene, dichloroethane, dichloromethane, cyclohexane, and 1,4-dioxane as the sole source of carbon and energy. All the major site contaminants, except for 1,4-dioxane, were utilized by the aerobic and anaerobic bacteria and fungi.

Zenker et al. (1999) sought to determine whether cyclic ethers and alkyl ether oxygenates such as MTBE biodegrade under ambient conditions in soil microcosms. The microcosms used 1,4-dioxane concentrations of 1,000 and 200,000 μg/L as well as mixtures of 1,4-dioxane and THF at 200,000 μg/L (Zenker et al., 1999). The researchers noted the complete biodegradation of 1,4-dioxane and THF at 35°C and high concentrations (200,000 μg/L) with the addition of nitrogen, phosphorous, and trace minerals. Biodegradation of 1,4-dioxane was dependent upon the presence of THF. Microcosms using activated sludge and incubated under ambient conditions exhibited no biodegradation of 1,4-dioxane or THF, which suggests that breakdown of 1,4-dioxane is unlikely to succeed by intrinsic bioremediation. The latter involves a demonstration that indigenous subsurface microflora do not have the capability to biodegrade the target contaminants without artificial augmentation.

More recent work suggests 1,4-dioxane degraders are more diverse and widespread than previously thought. Soluble di-iron monooxygenase (SDIMO) genes, which are associated with known 1,4-dioxane degraders including CB1190 and various propanotrophs, were used as a biomarker for 1,4-dioxane degradation in four groundwater samples collected at a petroleum industry site on the north slope of Alaska, above the arctic circle (Li et al., 2013). SDIMO genes were detected in all four samples, including samples not thought to have been impacted by 1,4-dioxane. Six propane monooxygenase genes were also detected. Genes were enriched near the source, but detected at the source of a plume, middle, edge and in background locations, although the location of the samples appeared to impact the extent to which the microbial community was impacted. Microcosm's prepared from site groundwater amended with ^{14}C-tagged dioxane degraded 1,4-dioxane with little lag time evident in microcosms prepared with groundwater from the source zone, and less degradation and more lag time in microcosms prepared using groundwater from the middle of the plume. After a few weeks when degradation ceased, adding acetate to the microcosm revived the organisms and degradation resumed, suggesting the substrate had been depleted. Degradation slowed by 40% when acetylene was added to the microcosms, confirming degradation was related to monooxygenase activity, but suggesting other enzymes were involved. Propane monooxygenases were suspected of causing the continued degradation activity (Li et al., 2013).

As was previously discussed, in Chapter 3, several 1,4-dioxane-degrading organisms have been identified recently. Huang et al. (2014) isolated a gram-negative bacteria *Acinetobacter baumannii* DD1 that could directly degrade 1,4-dioxane. Sei et al. (2013) found four bacteria in soil from a drainage ditch at a chemical factory that could degrade 1,4-dioxane (*Mycobacterium sp.* D6, *Mycobacterium sp.* D11, *Pseudonocardia sp.* D17 and *Afipia sp.* D1). Aoyagi et al. (2018) used SIP-rRNA to identify ten organisms likely to degrade 1,4-dioxane in a sludge sample from an industrial wastewater treatment plant in Japan. Deng et al. (2018) identified a bacterium (*Azoarcus sp.* strain DD4) capable of cooxidizing 1,4-dioxane and 1,1-DCE using propane as the primary substrate.

In addition to those previously discussed, He et al. (2018) identified two bacterial cultures in uncontaminated soils capable of catabolically degrading 1,4-dioxane. While inhibition of degradation by acetylene suggested degradation involved monooxygenases, qPCR analysis did not identify the THF/dioxane monooxygenase (thmA/dxmA) gene, suggesting degradation by an unknown pathway. Yabuki et al. (2018) found evidence of biodegradation in two landfill wastewaters but did not identify the organisms present. Treatment appears to be enhanced by treatment of other organic compounds in the wastewaters.

8.6.1.3 Degradation of Tetrahydrofuran and 1,4-Dioxane Together

In the second important paper on the biodegradation of 1,4-dioxane from North Carolina State University, "Mineralization of 1,4-Dioxane in the Presence of a Structural Analog," a mixed bacterial culture (a "consortium") capable of aerobically biodegrading 1,4-dioxane in the obligate presence of THF was enriched from a 1,4-dioxane-contaminated aquifer sediment. Zenker et al. (2000) used a continuous-flow rotating biological contactor (RBC) to inoculate samples from each soil microcosm bottle at 35°C. The RBC was supplied with 6 L/day of a 25% BSM containing 20 mg/L of THF and 30 mg/L uniformly ^{14}C-labeled 1,4-dioxane. Radio-labeled 1,4-dioxane was utilized to monitor destruction and assimilation (labeling molecules allows the tracing and quantitation of by-products).

THF and 1,4-dioxane disappeared after 3 months in one microcosm and after 10 months in the remaining two. Microcosms incubated without THF did not exhibit 1,4-dioxane biodegradation, even when respiked with additional 1,4-dioxane. When more THF was added, both compounds were degraded within 75 days. Zenker explained that these data show that the enrichment culture can completely oxidize 1,4-dioxane. A small amount (16.1%) of ^{14}C-labeled 1,4-dioxane was observed in the particulate fraction, which the researchers considered to be cell mass. The majority of the 1,4-dioxane (78.1%) was converted to CO_2, whereas 5.8% remained in the liquid.

Zenker et al. (2000) noted that the biodegradation of 1,4-dioxane was dependent on the presence of THF. Repeated addition of 1,4-dioxane in the absence of THF failed to stimulate biodegradation. During the 400 days in which the RBC was operated, samples of the bioreactor culture were periodically placed in BSM containing only 1,4-dioxane. No disappearance of 1,4-dioxane or microbial growth was observed during the course of these experiments.

The toxicity to by-products is due to the formation of HEAA that can denature enzymes or other cellular components. Therefore, the continued biodegradation of the nongrowth substrate is limited by the effects of product toxicity. In contrast, 1,4-dioxane biodegradation did not decrease the observed THF biodegradation rate.

The study showed that 1,4-dioxane does not begin to disappear until THF reaches relatively low levels. This result suggests competitive inhibition, a commonly observed phenomenon in cometabolic processes. Even though Zenker et al.'s 2000 study shows that 1,4-dioxane is completely destroyed, produces no toxic by-products, and is partially assimilated into biomass, the ability of the bacterial culture to biodegrade 1,4-dioxane is lost in the absence of THF.

The North Carolina State University research suggests that the loss of biodegradation activity may be due to the depletion of enzymatic cofactors, or coenzymes. The biodegradation of 1,4-dioxane may exert an abnormally high demand on the resources of the consortium. For example, the consortium may invest more energy and/or coenzymes in the biodegradation of 1,4-dioxane than is ultimately yielded. This greater required bacterial effort is consistent with the high energy demand required for scission of the ether bond, in contrast to the small yield of assimilable carbon. The loss of biodegrading activity could also be due to a rapid turnover of the active enzyme once the THF is depleted.

8.6.1.4 Kinetics of Cometabolism

In Zenker et al.'s (2002) third noteworthy contribution to the body of 1,4-dioxane biodegradation knowledge, the kinetics of cometabolism of cyclic ethers by a mixed culture was modeled by using data from their 2000 study to estimate the maximum specific utilization rate of 1,4-dioxane. Note that in biodegradation studies, total suspended solids (TSS) is often used as a surrogate for biomass.

The maximum specific utilization rates were determined to be 1.09 mg of THF per milligram of TSS per day and 0.45 mg of 1,4-dioxane per milligram of TSS per day. The half-saturation coefficients (K_S) were measured as 10.8 and 12.6 mg/L for THF and 1,4-dioxane, respectively. No evidence of toxic by-products was observed, and Zenker et al. (2002) noted that THF biodegradation was not inhibited by the presence of 1,4-dioxane. However, 1,4-dioxane biodegradation may be competitively inhibited by the presence of THF. The model was capable of predicting the biodegradation of 1,4-dioxane and THF at 1,4-dioxane:THF molar ratios of 0.9:3.3.

The presence of the nongrowth substrate, 1,4-dioxane, did not significantly affect the biodegradation of the growth substrate, THF. If the same enzyme is responsible for the biotransformation of both ethers, then THF apparently binds much more aggressively to the enzyme than does 1,4-dioxane. Zenker et al. (2002) also noted that in systems treating dilute industrial effluents, 1,4-dioxane concentrations may be present below the threshold necessary to maintain growth.

Mahendra and Alvarez-Cohen (2006) tested 16 bacterial cultures for their ability to degrade 1,4-dioxane and determined kinetic constants for *Pseudonocardia dioxanivorans* CB1190 and *Pseudonocardia benzenivorans* B5. For CB1190, the maximum substrate degradation rate was 1.1 mg 1,4-dioxane/hour/mg protein; the substrate half-saturation constant was 160 mg/L, the cell yield was 0.09 mg protein/equivalent 1,4-dioxane and the maximum specific growth rate was 2.4/day. For B5, the maximum substrate degradation rate was 0.1 mg 1,4-dioxane/h/mg protein; the substrate half-saturation constant was 330 mg/L, the cell yield was 0.03 mg protein/equivalent 1,4-dioxane and the maximum specific growth rate was 0.07/day. Degradation rates were also presented for other 1,4-dioxane-degrading cultures that utilize Monooxygenase-expressing bacteria.

Mahendra et al. (2013) investigated the impact of methyl chloroform and 1,1-DCE on 1,4-dioxane degradation kinetics under metabolic [*Pseudonocardia dioxanivorans* strain CB1190 (CB1190)] and co-metabolic [*Pseudomonas mendocina* strain KR1 (KR1) and *Escherichia coli* TG1 (e. coli T4MO)] conditions. CB1190 was unable to degrade either methyl chloroform or 1,1-DCE and the presence of either compound inhibited degradation of 1,4-dioxane, but degradation recovered when the methyl chloroform or 1,1-DCE were removed. The inhibition constant (K_I) was 1.2 for methyl chloroform and 3.3 for 1,1-DCE. KR1 and e. coli T4MO were able to degrade 1,4-dioxane, methyl chloroform, and 1,1-DCE simultaneously, but 1,4-dioxane degradation rates were reduced. Analysis of the data suggested the inhibition was caused by competition. KR1 was more inhibited by 1,1-DCE (KI was 3.8 for methyl chloroform and 0.52 for 1,1-DCE), while e. coli T4MO was more inhibited by methyl chloroform (KI was 0.23 for methyl chloroform and 22 for 1,1-DCE). Degradation rates did not recover when the methyl chloroform or 1,1-DCE were removed, suggesting the inhibition was permanent.

8.6.2 Bacteria Genus: *Rhodococcus*

Additional work by Cowan et al. (1994) identified a bacterium from the genus *Rhodococcus* that could degrade 1,4-dioxane as the sole carbon and energy source. They found that the bacterial growth rate and the associated biodegradation rate were optimal at 35°C and decreased dramatically above and below that temperature. Cowan et al. evaluated the performance of a fluidized-bed reactor (FBR) inoculated with the *Rhodococcus*, which effectively reduced an initial 1,4-dioxane concentration of 100,000 to less than 1,000 µg/L. However, the growth rates were very slow; adequate removal was only achievable with long residence times.

8.6.3 Bacteria Genus: *Pseudonocardia*

8.6.3.1 Degradation of Multiple Ether Pollutants

Vainberg et al. (2006) published a study of the biodegradation of THF, 1,4-dioxane, 1,3-dioxolane, bis-2-chloroethylether (BCEE), and methyl tert-butyl ether (MTBE) by *Pseudonocardia sp.* strain ENV478. Strain ENV478 degraded 1,4-dioxane after growth on sucrose, sodium lactate, yeast extract, 2-propanol, and propane, which showed that there was some level of constitutive degradation activity. The highest rates of 1,4-dioxane degradation occurred after growth on THF. Degradation of 1,4-dioxane caused the accumulation of 2-hydroxyethoxyacetic acid (2HEAA). Although ENV478 does not grow directly on 1,4-dioxane, ENV478 degraded 1,4-dioxane for more than 80 days in aquifer microcosms in the presence of THF. The inability of strain ENV478 to grow solely on 1,4-dioxane is related to its inability to efficiently metabolize the 1,4-dioxane degradation product 2HEAA. Vainberg et al. (2006) proposed that strain ENV478 may nonetheless be useful as a biocatalyst for remediating 1,4-dioxane-contaminated aquifers.

The Vainberg et al. (2006) paper discusses the identification of urinary metabolites of 1,4-dioxane from toxicology studies using rats as indicators of possible bacterial 1,4-dioxane biodegradation pathways. Two urinary metabolites of 1,4-dioxane are 2HEAA and 1,4-dioxane-2-one (PDX). δ-Hydroxy acids are generally unstable in aqueous solutions except as salts; they more commonly occur in the lactone form. The lactonization of 2HEAA produces PDX, as identified in the study of Woo et al. (1977). The initial monooxidation of 1,4-dioxane likely results in the production of 1,4-dioxane-2-ol, the hemiacetal of 2-hydroxyethoxy-2-acetaldehyde. A proposed degradation pathway for the biodegradation of 1,4-dioxane by Strain ENV478 is shown in Figure 8.5.

Strain ENV478 degradation rates for bacteria grown on THF were as follows (TSS is a surrogate for biomass):

Organic Contaminant	Rate per Gram of TSS (mg/h)
1,4-Dioxane	21
1,3-Dioxolane	19
THF	63
BCEE	12
MTBE	9.1

The BCEE degradation rates were about threefold higher after ENV478 growth on propane (32 mg/h/g of TSS) than after ENV478 growth on THF, and MTBE degradation resulted in the accumulation of *tert*-butyl alcohol.

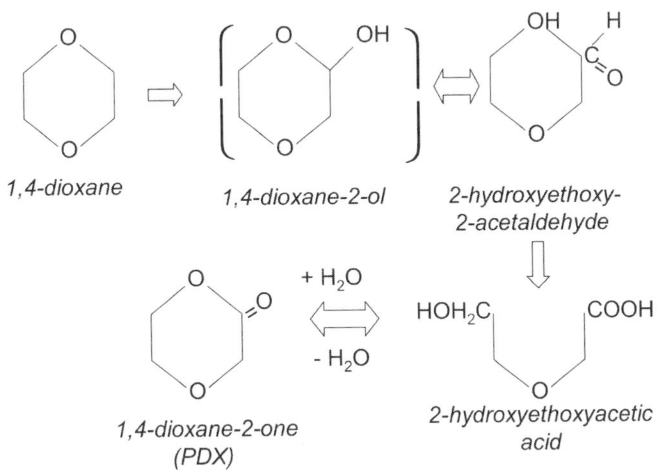

FIGURE 8.5 Proposed partial pathway for biodegradation of 1,4-dioxane by strain ENV478. (From Vainberg, S., McClay, K., Masuda, H., et al., 2006, Biodegradation of ether pollutants by *Pseudonocardia sp.* strain ENV478. *Applied and Environmental Microbiology* 72(8): 5218–5224. With permission.)

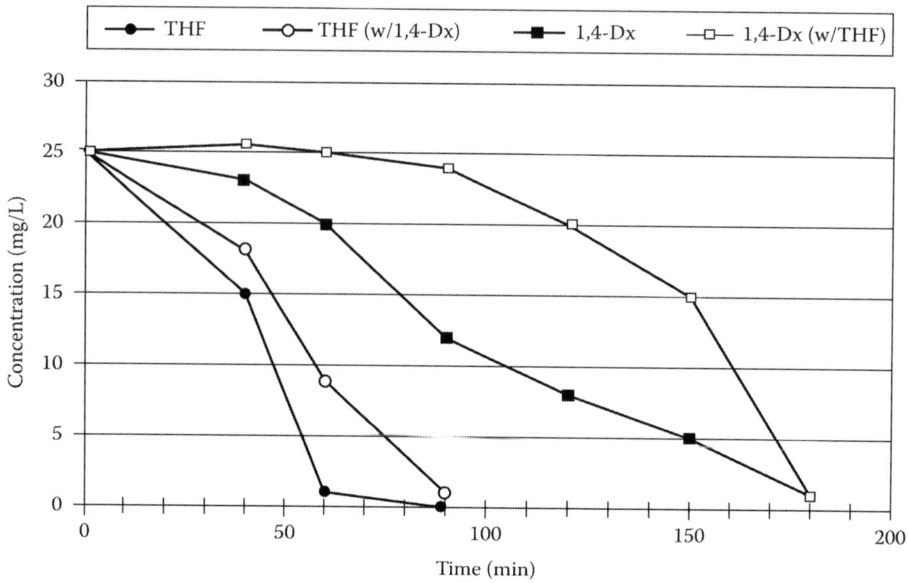

FIGURE 8.6 Biodegradation of 1,4-dioxane and THF by THF-grown strain ENV478 when the compounds were added alone or as a 50:50 mixture. The concentration of volatile suspended solids was 0.38 mg/L. (From Vainberg, S., McClay, K., Masuda, H., et al., 2006, Biodegradation of ether pollutants by *Pseudonocardia sp.* strain ENV478. *Applied and Environmental Microbiology* 72(8): 5218–5224.)

Strain ENV478, when first grown on THF, degraded 1,4-dioxane at an initial rate of ~21 mg/h/g of TSS, which was approximately one-third the rate at which ENV478 degraded THF. When the THF and 1,4-dioxane were added together in equal amounts, 1,4-dioxane degradation did not occur until all the THF in the medium had been degraded (Figure 8.6). When Strain ENV478 cells grown on THF were incubated with both 1,4-dioxane and 1,3-dioxolane, they degraded both compounds simultaneously at approximately the same rate (Vainberg et al., 2006).

Much slower cell growth and lower 1,4-dioxane degradation rates (per gram of TSS) were observed when Strain ENV478 was grown on sucrose (0.71 mg/h), yeast extract (1.1 mg/h), sodium lactate (0.6 mg/h), 2-propanol (1.5 mg/h), and propane (3.2 mg/h). After growing on lactate, the strain also degraded 1,3-dioxolane (1.0 mg/h). The team concluded that Strain ENV478 has a low level of constitutive activity with these ethers. The team also found that Strain ENV478 does not grow on 2HEAA as a sole carbon source. To identify the biodegradation products, radio-labeled 1,4-dioxane was used. Analysis of the culture liquor showed that the [^{14}C] dioxane converted to a product that coeluted with authentic 2HEAA. The apparent 2HEAA product was not degraded further, even after 402 hours of incubation at 25°C. Addition of THF to cultures that had depleted the 1,4-dioxane and accumulated 2HEAA also did not result in the further degradation of 2HEAA. At the end of the experiment, the radioactivity in the supernatant was not less than the radioactivity in the control. This observation indicated that no carbon dioxide had been produced and no 1,4-dioxane was converted to the Strain ENV478 biomass.

Vainberg et al. (2006) incubated aerobic aquifer microcosms for 80 days at 15°C to simulate ambient aquifer temperatures. 1,4-Dioxane at ~700 μg/L was rapidly degraded in the augmented microcosms before the first sampling time (day 20); when repeatedly added at ~1000 μg/L, the 1,4-dioxane was also rapidly degraded. Fresh groundwater was added to the microcosms after

53 days of incubation to replace the volume that had been removed for sampling. This addition created an ~30% dilution of the resident microbial population in the microcosms, with a commensurate decrease in 1,4-dioxane degradation. In the study by Zenker et al. (2000), the system required augmentation with relatively large amounts of THF on a continuous basis to maintain 1,4-dioxane degradation activity. The ability of Strain ENV478 to maintain 1,4-dioxane degradation activity in microcosms for 80 days without THF was, therefore, surprising and encouraging; however, additional work is needed to confirm whether inoculating contaminated aquifers with Strain ENV478 will be effective for economical remediation.

Strain ENV478 is phylogenetically related to Strains CB1190 and K1 but is not able to grow on 1,4-dioxane as a sole carbon source even though it appears that the same enzyme or enzymes in the strain are required to oxidize both substrates. The inability of Strain ENV478 and other THF-degrading bacteria to grow on 1,4-dioxane alone could be related to a lack of induction of the requisite enzyme(s) by 1,4-dioxane or to an inability to efficiently process metabolites for energy production. The research team focused on evaluating why this strain cannot grow on 1,4-dioxane and on whether this strain can be used to degrade 1,4-dioxane in contaminated environments.

Degradation of 1,4-dioxane by Strain ENV478 was greatest after its growth on THF; the presence of THF inhibited the degradation of 1,4-dioxane by this strain. During 1,4-dioxane degradation, Strain ENV478 appeared to accumulate 2HEAA, an expected product of a monooxygenation reaction with 1,4-dioxane. The research team therefore concluded that the degradation of 1,4-dioxane and THF by Strain ENV478 involves an initial oxidation of the cyclic ether by a monooxygenase that presumably is homologous to the THF monooxygenase of Strain K1 (Vainberg et al., 2006).

8.6.3.2 Degradation of 1,4-Dioxane

Parales et al. (1994) isolated CB1190, a pure culture of a *Nocardioform actinomycete*, likely of the family *Pseudonocardia*, in mixed cultures of bacteria from industrial sludge generated at a 1,4-dioxane-contaminated site in Darlington, South Carolina. Bacteria were then grown aerobically on BSM, yeast extract, and THF. CB1190 was demonstrated to have a specific activity for dioxane degradation at a rate of 0.33 mg dioxane/min/mg of protein. Initial concentrations of 5.5 mM 1,4-dioxane (48.4 μg/L) were degraded to less than 0.55 mM (4.84 μg/L).

Parales et al. (1994) also confirmed the complete destruction of 1,4-dioxane by CB1190, when 4,000 mM 1,4-dioxane and CB1190 were incubated at 30°C for 18 hours and degraded to less than 10 mM. About 60% of the carbon in the dioxane molecule was converted to carbon dioxide. CB1190 was also found to degrade the ethers THF, tetrahydropyran, 1,3-dioxane, diethyl ether, and 2-methyl-1,3-dioxolane, as well as the alcohols ethanol, isopropanol, 1-butanol, and 2-butanol. CB1190 would not degrade 1,4-dioxane at temperatures higher than 42°C. Moderately increased salinity did not affect the growth of CB1190 on 1,4-dioxane; 2% NaCl did not slow the growth at all, and some growth was observed at up to 4% NaCl. No growth was observed at 8% NaCl.

Attempts to grow the organism directly on 1,4-dioxane were unsuccessful. Early experiments apparently cometabolized 1,4-dioxane in the presence of THF. Long-term attempts eventually yielded a culture capable of growth on 1,4-dioxane alone, but THF remains the preferred substrate for growth of CB1190. Once the THF was exhausted and only 1,4-dioxane was provided, the cultures were capable of degrading 10 mM 1,4-dioxane daily (Parales et al., 1994).

In a study by Polasko et al. (2019), a mixed culture of KB-1 (a culture designed to degrade chlorinated organic compounds like TCE under anaerobic conditions, containing *Dehalococcoides sp.* and other organisms) and known dioxane degrader CB1190 was used to sequentially degrade TCE to cis-1,2-DCE under anaerobic conditions, and then degrade the cis-1,2-DCE and 1,4-dioxane aerobically. The results of the study indicated CB1190 was resilient enough to remain viable during the anaerobic phase and was capable of degrading both the cis-1,2-DCE and the 1,4-dioxane when oxygen was added to the culture (Polasko et al., 2019).

8.6.4 Gasotrophic Bacteria: Cometabolic Biodegradation

Propane-utilizing aerobic bacteria (propanotrophs) capable of cometabolic 1,4-dioxane degradation in laboratory studies have been cultured from contaminated groundwater at several sites (Fam et al., 2005). Natural enzyme systems within these cultures, which are designed to oxidize propane, are capable of cometabolically converting 1,4-dioxane to carbon dioxide. Propane degradation occurs initially, and the destruction of 1,4-dioxane occurs only after most of the propane has been destroyed. The bacterial cultures are able to degrade up to 10,000 μg/L of 1,4-dioxane, but higher concentrations inhibit degradation. The aerobic propanotrophs compete for oxygen with other aerobically degraded contaminants such as acetone, Freon, and chlorinated solvents; therefore field applications with competing contaminants may inhibit propanotroph growth on 1,4-dioxane.

Innovative Engineering Solutions, Inc. (IESI) (2008) performed microcosm studies using groundwater from a 1,4-dioxane-contaminated industrial facility in Michigan and designed a field pilot study as a follow-on. Baseline conditions indicated that the site groundwater was only slightly anaerobic, in the range of nitrate and iron reduction, and therefore was conducive to bio-enhancement to stimulate aerobic propanotrophs. Aerobic microcosms were constructed and filled with site groundwater and nutrients. The presence of propane-utilizing bacteria was confirmed through the addition of propane and observation of species growth. Subsequently, 1,4-dioxane was added to the reactors at the rate of approximately 3 mg/L, and 1–4% propane was sparged through the test solution. Two of the well samples were confirmed to have the appropriate propanotrophic bacteria in sufficient quantities and were observed to reduce 1,4-dioxane concentrations. Bioaugmentation with SL-D, IESI's proprietary propanotrophic bacteria consortium, resulted in greater microbial growth and more rapid 1,4-dioxane degradation. IESI did not observe any evidence that the 1,4-dioxane was directly degraded as the sole carbon and energy source during the 88-day test period.

Global BioSciences, Inc. (2007) evaluated the capability of butane biostimulation to enhance the degradation of 1,4-dioxane in groundwater from a brownfield site in Massachusetts. Samples were enriched by using standard microbiological subculturing techniques and incubated at ~10°C in a BSM for 4 weeks. Microcosm studies were performed by using static headspace methods over a 5-day period to evaluate butane consumption and 1,4-dioxane degradation. Duplicate serum bottles of the biostimulated and control microcosms were sacrificed each day for analysis. Samples demonstrated 1,4-dioxane losses from 6.73 to 7.66 mg/L over the 5-day period, whereas the control bottles exhibited losses from 1.35 to 2.07 mg/L, which were attributed to abiotic losses, specifically leaks through a serum bottle septum. The difference between the total loss and abiotic loss was attributed to undefined microbial activity, either direct metabolism or cometabolism. After 2 days, the 1,4-dioxane concentrations in the serum bottles were below the laboratory detection limit of 0.1 mg/L.

He et al. (2017) identified a novel group-6 propane monooxygenase gene cluster (prmABCD) in *Mycobacterium* dioxanotrophicus PH-06, which is associated with degradation of 1,4-dioxane. The organism lacks the thmA/dxmA biomarker associated with 1,4-dioxane degradation related to other organisms. A new biomarker prmA is suggested to enhance the industry's ability to detect 1,4-dioxane degrading organisms.

Chu et al. (2018) conducted a field trial of *in situ* aerobic cometabolic biodegradation using propane as the stimulant and simultaneously targeting 1,4-dioxane (60 μg/L) and chlorinated volatile organic compounds (CVOCs) (1.4 to 10 μg/L). After 265 days, the stimulated indigenous microbial population degraded 1,4-dioxane to below 3 μg/L, TCE to less than 1 μg/L, and 1,2-DCA to less than 0.18 μg/L. 1,4-Dioxane and 1,2-DCA removal was 95% and TCE removal was 90%.

8.6.5 Bioreactors and *In Situ* Biological Treatment

Building on the laboratory work related to the bacterial degradation of 1,4-dioxane, several authors have constructed and tested bioreactors or have proposed the use of bacteria in a bio-barrier concept.

Zenker et al. (2004) demonstrated the effectiveness of a trickling filter inoculated with bacteria derived from a 1,4-dioxane-contaminated site. The bacteria were grown in an RBC in the presence of THF and 1,4-dioxane at concentrations of 20,000 and 30,000 μg/L, respectively. The bacteria were then inserted into the trickling filter system, which was maintained at 35°C. Various THF and 1,4-dioxane concentration regimes were evaluated; 1,4-dioxane levels were tested at 1,000 and 200 μg/L. Removal efficiencies for 1,4-dioxane were 93% at both concentrations. Zenker et al. (2004) confirmed the obligate presence of THF because biodegradation of 1,4-dioxane slowed and then disappeared once THF was depleted.

A bioreactor for *ex situ* cometabolic biotreatment of 1,4-dioxane was designed and implemented at the Lowry Landfill, in suburban Denver, Colorado, following a series of bench-scale studies described in Stanfill et al. (2004). An existing multiprocess treatment plant, including ultraviolet oxidation and GAC, was effective at removing other organic contaminants, but did not effectively remove 1,4-dioxane because of low light transmittance of the influent groundwater and potential interference by inorganic compounds. Groundwater at the landfill contained high levels of 1,4-dioxane (up to 60,000 μg/L) and other organic contaminants. The extracted groundwater was fortuitously contaminated with THF (up to 200,000 μg/L), which had been shown by Zenker et al. (2000, 2004) and others to foster cometabolic degradation of 1,4-dioxane. The bioreactors were constructed to operate in a batch processing mode, with an anoxic initial stage, which created activated sludge. The sludge flowed to a multichamber aerobic reactor, the first stage of which was aerated and agitated; the second stage was designed as a clarifier. The reactors were seeded with biomass derived from a polyester manufacturing facility that treats 1,4-dioxane. Two distinctly different feed waters were used: One was from the toe of the contamination plume and contained high levels of 1,4-dioxane and THF, as well as other organic compounds, and the other was a combination of the high-organic-content water and water from an extraction trench with much lower organic contamination levels. This combination was used to evaluate 1,4-dioxane degradation in the potential combined influent that would be used in a full-scale operating system. A significant proportion of the organic content in both tested waters was attributable to the high concentrations of 1,4-dioxane and THF, which were demonstrated to be degraded in the bioreactor tests. In the highly organic water, the removal rates were lower than in the blended water, suggesting that toxic chemicals in the more contaminated water may have had an inhibitory effect on the biodegradation of both 1,4-dioxane and THF. The removal rate was as high as 97.5% for 1,4-dioxane and 99.3% for THF in the blended water tests, but only 87.3% and 97.9%, respectively, in the highly organic water. Although the field pilot-testing indicated successful reduction of 1,4-dioxane, below the detection limit in some cases, the detection limit was 360 μg/L, well above health-based and regulatory cleanup criteria in many states and EPA regions.

Evans et al. (2006) measured the specific activity of 1,4-dioxane metabolized by CB1190 at 0.33 μg/min per milligram of protein. A variety of bioreactor configurations was tested to measure maximum achievable biodegradation rates under optimized conditions. The bioreactors used the microbe strain CB1185, a mixed culture that includes CB1190, grown in batch on BSM with initial 1,4-dioxane concentrations of 5,000 mg/L and fed additional 1,4-dioxane as needed to maintain the required biomass and growth rate. Reactors supported by silica sand were inoculated with 100 L cultures containing 2,500 mg of protein per liter. The dimensions of the packed-bed reactor were 0.5 ft in diameter and 10 ft in height; the groundwater flow rate was 0.5–0.8 L/min, and the airflow rate was 2 L/min. Evans et al. (2006) measured the best performance in the sequence batch reactor with biomass carriers. Biodegradation rates were slow, and reactor efficiency declined over time, possibly because of bacteria washout, low temperatures, toxic contaminants or by-products, and

TABLE 8.1
Laboratory Bioreactor Results Summary

Bioreactor	Run Time (h)	1,4-Dioxane In (mg/L)	1,4-Dioxane Out (mg/L)	Biodegradation Rate[a]
Activated sludge with recycle	60	100	0.42	1.8
Activated sludge with recycle	43	1,200	270	24
RBC	55	91	1.0	1.8
RBC	55	5,000	1,800	54
Packed-bed reactor	<9	100	<0.04	12
Packed-bed reactor	<9	2	<0.05	0.24
Sequencing batch reactor	24	100	<1.0	1.8
Sequencing batch reactor with biomass carriers	12	1,200	<1.0	250
Fluidized-bed reactor	15	10	<1.0	117

Source: Evans, P.J., Parales, R.E., and Parales, J.V., 2006, Bioreactors for treatment of 1,4-dioxane. In: *18th Symposium in Groundwater Resources Association Series on Groundwater Contaminants: Emerging Contaminants in Groundwater: A Continually Moving Target*, June 7–8, 2006. Concord, CA. Housed at the University of California Water Resources Archives, Berkeley, CA.

[a] Units are milligrams of protein per liter per hour.

imbalance between the rates of bacteria growth and die-off (Evans et al., 2006). Table 8.1 presents a summary of the laboratory testing results of various bioreactor designs and 1,4-dioxane concentrations for influent and effluent.

Microvi Biotech LLC has developed a process that removes 1,4-dioxane from groundwater in an *in situ* biobarrier or an *ex situ* bioreactor (F. Shirazi, personal communication, 2007). This treatment uses *Rhodococcus sp.*, which under aerobic conditions can degrade 1,4-dioxane as the sole carbon or energy source, without the addition of a secondary substrate (Cowan et al., 1994, as referenced in Zenker et al., 2000). The bacteria are isolated, grown on glucose, and encapsulated in a porous polymeric matrix to increase survivability, reduce bacterial washout, and provide an improved environment for the stability and longevity of the bacteria. Figure 8.7 shows a scanning electron microscope photograph of the encapsulated microorganisms within the matrix. Bench testing demonstrated that the isolated *Rhodococcus sp.* was capable of degrading greater than 10,000 μg/L 1,4-dioxane in less than 50 hours (Figure 8.8); however, the effectiveness at lower 1,4-dioxane concentrations has not been demonstrated. The rate of 1,4-dioxane degradation per day (46,000–56,000 μg/L) was maintained for more than 300 days in the laboratory reactor without preconditioning the bacteria to 1,4-dioxane or adding any substrate to support cometabolic biodegradation.

Yamamoto et al. (2018) used a small bioreactor to test *ex situ* bioremediation using *Pseudonocardia sp.* D17. Bacteria for the reactor was grown in an MGY media, harvested and used to maintain the cell count in the reactor. The reactor was able to reduce 1,4-dioxane concentrations from 0.17 mg/L to 0.050 mg/L (70.5% reduction in concentration) at a 15-hour retention time.

Lippincott et al. (2015) report the installation and successful operation of a field-pilot scale cometabolic propane sparging system at Vandenburg Air Force Base, California. This system, which was augmented using *Rhodococcus ruber* ENV425 propanotrophic bacteria, was the first field pilot-scale system to report *in situ* 1,4-dioxane biological treatment. The system included a single sparging well and four monitoring wells. 1,4-Dioxane concentrations were reduced from as high as 1,090 μg/L to less than 2 μg/L, although concentrations and removal rates varied by monitoring well. 1,4-Dioxane degradation was not observed in a control well installed in the deep aquifer which received propane and oxygen but not microbes, suggesting the treatment required augmentation with the proper culture. A full-scale propane biosparging system is being implemented at the base in 2020 (C. Bell, personal communication, 2019).

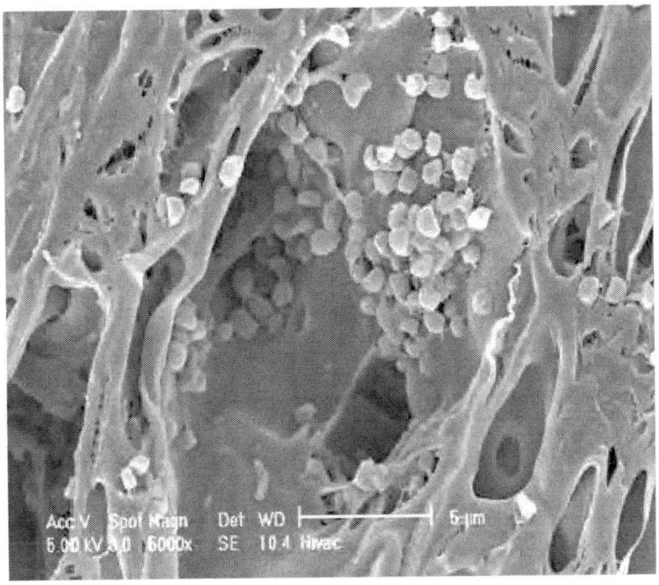

FIGURE 8.7 Encapsulated microorganisms shown inside MicroBead. (From F. Shirazi, personal communication, 2007.)

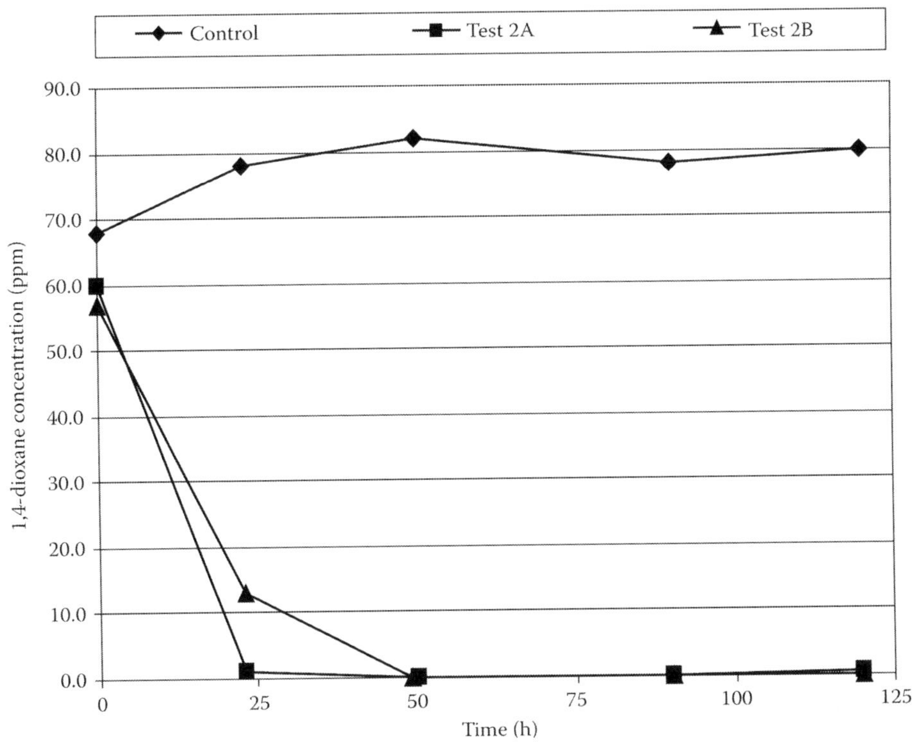

FIGURE 8.8 Reduction in 1,4-dioxane in duplicate *Rhodococcus sp.* bioreactor tests (2A and 2B), as well as a sterile control. (From F. Shirazi, personal communication, 2007.)

Sadeghi et al. (2016) reported on *in situ* enhanced biostimulation of an aquifer in Tucson, Arizona using methane to treat 1,4-dioxane by aerobic cometabolism. Two studies are included: in one test area, the aquifer was stimulated with an emulsified vegetable oil solution (EVOS), with nutrients and a consortium of anaerobic bacteria designed to degrade CVOCs. In this area, methane generated by the EVOS treatment was oxidizing to promote cometabolic treatment of the 1,4-dioxane. In the second study area, wells were stimulated with methane, air, and nutrients. Methane and air were added cyclically. In the EVOS area, the chemical data and CSIA analysis suggest 1,4-dioxane was not being degraded by the methane produced from EVOS. In the methane stimulation, concentrations in one well were reduced by 71% and CSIA revealed deuterium enrichment, suggesting biodegradation occurred. However, evidence of 1,4-dioxane treatment was not observed in the three other methane-stimulated wells. The study is tantalizing; there is strong evidence that under the "right" circumstances methane can be used to stimulate biodegradation, but the process appears to have serious limitations. Another study by Hatzinger et al. (2017), discussed in Section 3.3, presents data from another attempt to use methane generated during anaerobic bioremediation, with similar results. In another study, direct methane stimulation was evaluated for 1,4-dioxane treatment at a confidential commercial site (Alvarez et al., 2014). In this case, testing suggested that the methane stimulated particulate methane monooxygenase (pMMO) rather than sMMO. Since pMMO does not degrade 1,4-dioxane, the test failed, and the research team moved on to other methods of treating the site. Methane stimulation, while theoretically possible for 1,4-dioxane treatment, is apparently more difficult to implement and other stimulants are more likely to be successful.

8.6.6 Anaerobic Bacteria

Anaerobic degradation of 1,4-dioxane has only been reported rarely and the ability of microorganisms to degrade 1,4-dioxane is a matter of debate in the environmental remediation community. The anaerobic degradation of 1,4-dioxane was evaluated under iron-reducing conditions involving anaerobic sludges by Pan and Chen (2006) of Zhejiang University in Hangzhou, China. They determined that the iron-reducing bacteria present in the anaerobic sludge could intrinsically degrade 40% of 1,4-dioxane in the microcosm without the need for amendments. Adding humic substances, which may have acted as a carbon source, stimulated the anaerobic degradation of up to 70% of the 1,4-dioxane present. The degradation was caused by iron-reducing bacteria in which Fe(III) was available as an electron acceptor. 1,4-Dioxane biodegradation was greatest in the presence of added humic substances and Fe(III) under all the redox conditions. Complete degradation of 1,4-dioxane was observed (Pan and Chen, 2006).

Shen et al. (2008) presented additional information related to their studies of iron-reducing bacteria for 1,4-dioxane degradation. The degradation observed was determined to be entirely from biological activity, because sterile conditions resulted in no statistically significant declines in 40 days. Degradation of 1,4-dioxane in reactors incubated with iron-reducing bacteria derived from the anaerobic sludge resulted in a decline of 25% (from 13 mg/L) over 40 days with no amendments. Introduction of Fe(III) oxide caused a decline of 37% and adding Fe(III) oxides and humic acids led to a reduction of 62% over the 40-day period. The humic acids are thought to act as a catalyst, promoting biodegradation by the iron-reducing bacteria, rather than creating reducing conditions, as is common in biodegradation of other contaminants, such as chlorinated VOCs. A second set of experiments conducted after the initial 40-day period, using the same inoculated bottles but replenishing 1,4-dioxane up to 50 mg/L, resulted in even higher daily degradation rates: 0.7 mg/L versus the initial rate of 0.2 mg/L. Shen et al. (2008) attributed this higher rate to the presence of an established and robust bacterial population. They concluded that the iron-reducing bacteria were using the 1,4-dioxane as the sole source of carbon; however, depending on the source of humic material added, other organic compounds may have favored cometabolism of 1,4-dioxane.

An alternate explanation of the observed anaerobic degradation is that it relates to microbial cycling of iron from iron (II) to iron (III). Work by Sekar et al. (2016) suggests an alternative role for iron in the degradation of 1,4-dioxane. Son et al. (2009) found that oxidizing reduced or ZVI could degrade 1,4-dioxane because of the production of hydrogen peroxide and hydroxyl radicals during the odixation of iron (a Fenton's reagent-like reaction). The hydroxyl radicals then oxidized 1,4-dioxane. Sekar et al. (2016) took this one step further and created an iron-rich biological system which alternately oxidized and then reduced the iron in the system, producing hydrogen peroxide and hydroxyl radicals as a by-product, and subsequently degrading the 1,4-dioxane. In addition to providing a potential remediation technology for a mixed 1,4-dioxane plume, this work also suggests that the presence of iron and shifting redox conditions can boost the degradation of 1,4-dioxane by abiotic degradation and might explain degradation of 1,4-dioxane in high-iron anaerobic systems.

8.6.7 Fungi

A patented remediation process to reduce concentrations of 1,4-dioxane in groundwater has been developed by Patt and Abebe (1995). The remediation requires the addition of a pure culture of a fungal microorganism that degrades cyclic ethers, a cometabolite (ideally THF), and oxygen. The process involves fermentation of the chemical pollutant as the substrate – in this case 1,4-dioxane – in the presence of THF and oxygen by a nonbacterial fungus (*Aureobasidium pullulans*), identified by the Northern Regional Research Laboratories as NRRL 21064. Compared to bacterial cultures, the fungus has a broader range of pH tolerance and therefore would be suited to a wider range of site conditions. Additionally, sufficient DO, as is normally found in natural groundwater, is a prerequisite for fungal survival. The fungus is also not as sensitive to temperature, being functional between 5°C and 37°C, although lower temperatures result in slower metabolism and higher temperatures kill the fungus. The process is proposed as being applicable in batch mode (lakes, ponds, and reactors) or in continuous mode (reactors, ponds with inflow and outflow, and *in situ* groundwater). The information provided in the patent does not elaborate on the relationship between the THF and the 1,4-dioxane except to clearly define the THF as an obligate cometabolite.

Patt and Abebe's patent cites two applied examples of 1,4-dioxane degradation using *Aureobasidium pullulans*. In the first application, a 2-million-gallon aeration lagoon containing 1,4-dioxane at a concentration of 8,000 μg/L and THF at concentrations between 5,000 and 8,000 μg/L was treated with a 300-gallon volume of *Aureobasidium pullulans* inoculum containing about a million cells per milliliter; hence, trillion cells were added to the 2-million-gallon lagoon. 1,4-Dioxane concentrations decreased to below an unspecified detection limit within 2 months.

In the second example, a FBR was inoculated with 10 billion cells of *Aureobasidium pullulans* and supplied with groundwater contaminated with 1,4-dioxane at a concentration of 8,000 μg/L and supplemented with THF. The FBR retention time was 1 hour, and the pH was between 6 and 7. Effluent contained 1,4-dioxane in concentrations from just a few micrograms per liter to 2,000 μg/L and less THF than the 500 μg/L detection limit (Patt and Abebe, 1995).

Kim et al. (2006) made a comparison between the destruction of 1,4-dioxane by advanced oxidation technology (AOT) and biodegradation using BAC-TERRA, a filamentous fungi. AOT was demonstrated with ultraviolet + hydrogen peroxide and a modified Fenton's chemistry, both of which demonstrated pseudo first-order kinetics, with a rate coefficient of 5×10^{-4} s^{-1}. The fungal degradation also exhibited pseudo first-order kinetics, but with a rate coefficient that was two orders of magnitude lower (2.38×10^{-6} s^{-1}) than that shown for AOT.

8.6.8 Enhancing Biodegradability with *In Situ* Chemical Oxidation

Several authors have performed various studies to determine the effect of *in situ* chemical oxidation (ISCO) on the biodegradation of 1,4-dioxane.

Adams et al. (1994) evaluated the use of hydrogen peroxide and ozone to enhance the biodegradation of 1,4-dioxane using a mixed microbial culture obtained from a municipal wastewater treatment plant. Upon subjecting the test cells to hydrogen peroxide/ozone ratios of 0.0 (ozone only), 0.5, and 1.0, the biological oxygen demand (BOD_5) increased, and, following a lag period, the chemical oxygen demand (COD) decreased. This rise in BOD_5 and the limited drop in COD suggests that biodegradability of the initial oxidation by-products occurred more readily than the biodegradation of the original 1,4-dioxane. Biodegradability was also assessed by using shaker-table bioassays to measure COD removal rates at different hydrogen peroxide/ozone molar ratios. Minimal biodegradation was observed in the samples with no oxidant, but up to 85% reduction in COD was observed in 57 hour in oxidized samples. Higher residual COD (up to 37%) was present in the samples with lower oxidant input because of residual untreated 1,4-dioxane. The ratio of oxidant to 1,4-dioxane in these experiments was lower than the ratio required for direct and complete oxidation of the 1,4-dioxane. Adams et al. (1994) concluded that, for the treatment of 1,4-dioxane in wastewater, significant reduction in oxidant usage could be obtained by optimizing an oxidation system to enhance biodegradability of the 1,4-dioxane, rather than full oxidation of the contaminant.

Suh and Mohseni (2004) looked at the kinetics of 1,4-dioxane oxidation with ozone and peroxide and the relationship of these oxidants to enhanced biodegradability. They also looked at the effects of 1,4-dioxane concentration, pH, and hydrogen peroxide concentration on 1,4-dioxane biodegradability. The fact that initial seeded solutions contained minimal BOD_5 indicated that 1,4-dioxane was not biodegradable if conventional bacterial methods were used. After treatment with hydrogen peroxide and ozone, 1,4-dioxane and COD decreased and BOD_5 increased, indicating oxidation of 1,4-dioxane and creation of biodegradable intermediates. Oxidation of 1,4-dioxane was found to follow a Langmuir-Hinshelwood-type model following first-order kinetics and transitioning to zero-order kinetics at higher concentrations (>60,000 µg/L). The overall kinetics, as well as the transition from first order to zero order, is affected by pH, hydrogen peroxide, and presence of other intermediate species. Increases in pH, up to ~9, increase biodegradability. Increasing hydrogen peroxide levels enhances biodegradability up to a point (0.4–0.45 mol:mol hydrogen peroxide:ozone), after which the hydrogen peroxide has a negative effect, reduces 1,4-dioxane destruction, and thereby decreases biodegradability.

Miao et al. (2018) developed a novel catalyzed hydrogen peroxide (CHP)-based ISCO process designed to degrade CVOCs and 1,4-dioxane and applied the process to site soil and groundwater samples in microcosms. They used the MiSeq System by Illumina and bioinformatics to track microbial changes during the test. During the test, more than 80% of the 1,4-dioxane was degraded over 12 hours. A second, catalyzed oxidation test removed an additional 20% of the residual 1,4-dioxane. Tests were conducted with "high concentration" (6,800 µg/L 1,4-dioxane and 7,400 µg/L CVOCs) and "low concentration" (1,100 µg/L 1,4-dioxane and 1,000 µg/L CVOCs) conditions. Results indicated that in general, ISCO reduced the overall numbers of microorganisms, but more microorganisms survived under low concentration conditions. Diversity was only adversely impacted under the high 1,4-dioxane and high CVOCs scenario. This is thought to be related to the different catalysts used to treat the higher concentrations.

8.6.9 Other Processes affecting Biodegradability

Inhibition, which is quantifiable through process kinetics, is discussed in Section 8.6.1.4 Kinetics of Cometabolism. Here it is important to note that work by leading researchers (Mahendra and Alvarez-Cohen, 2006 and Mahendra et al., 2013) found that the presence of some CVOCs, especially 1,1-DCE, can inhibit degradation of 1,4-dioxane, to the point where degradation ceases permanently on one hand, or by slowing the process down until the CVOCs have been depleted. Conducting bioremediation of 1,4-dioxane in the presence of CVOCs, which is commonly the circumstance, therefore must consider potential inhibitory effects.

Arulazhagan et al. (2013) studied the impact of nutrients on the 1,4-dioxane bioremediation process, using a bacterial consortium enriched from 1,4-dioxane contaminated industrial sludge. They tested the impact of various enhancers on that process including glucose, yeast extract, and THF. Without the nutrient, the consortium degraded 74% of 1,4-dioxane in 72 hours. THF is a known 1,4-dioxane cometabolite and adding it to the tests resulted in complete degradation of 1,4-dioxane in 72 hours. Yeast extract also substantially boosted the removal rate to 92% in 72 hours, while glucose slowed degradation to 59% in 72 hours.

Li et al. (2010) looked at degradation of 1,4-dioxane under artic/low temperature conditions. The study found that biodegradation was active at 4°C. Enhanced biodegradation using CB1190 and 1-butanol proceeded at a higher rate at 14°C than at 4°C but proceeded nonetheless. Native microorganisms degraded 1,4-dioxane in microcosm when stimulated with added oxygen to simulate aerobic natural attenuation conditions and when stimulated with 1-butanol and nutrients. When the microcosms were augmented with DVS 5a1, a significant lag was observed before 1,4-dioxane degradation began, suggesting a need for the culture to acclimatize to the site conditions including temperature, however, once acclimatized, DVS 5a1 outperformed CB1190 under low concentration conditions.

8.6.10 Bacteria Summary

Laboratory and field pilot studies have identified many different types of bacteria that are effective at either utilizing 1,4-dioxane as an energy source or cometabolically degrading 1,4-dioxane while consuming some other energy source. Chapter 3 summarized the laboratory studies aimed at identifying the requisite bacteria and optimal conditions, defining degradation pathways, and determining biodegradation rates. The available literature related to the potential field application for soil or groundwater remediation of 1,4-dioxane indicates that the bacteria are dominantly aerobic and fairly common. However, the natural conditions generally encountered at hazardous waste sites are not conducive to high degradation rates, either because groundwater is anaerobic and the effective processes all appear to be aerobic, or because an appropriate co-substrate is missing, or because the natural process is slow. Because of this, some sort of stimulation is required to achieve more effective removal rates. Biodegradation of 1,4-dioxane has been applied *ex situ* in bioreactors at the field scale. There have been several attempts to apply *in situ* bioremediation for 1,4-dioxane by biostimulation and bioaugmentation, and some of them have succeeded. In general, propane appears to be the most robust and available co-substrate; but ethane also appears to be a viable co-substrate. Methane is a known 1,4-dioxane degrading co-substrate, well demonstrated under laboratory conditions, but applying it in the field has not been successful. Other co-substrates have been tested on a limited basis. When considering cometabolic treatment of 1,4-dioxane, testing and selecting the proper co-substrate is a critical step.

8.7 CHEMICAL OXIDATION

Oxidation is a contaminant-destruction technology in which oxidation-reduction (redox) chemical reactions convert hazardous contaminants to nonhazardous or less toxic compounds. Redox reactions involve the transfer of electrons from one compound to another. During chemical oxidation, the redox reaction oxidizes one reactant (the contaminant, which loses electrons), whereas the other reactant (the oxidizer, which gains electrons) is reduced. Oxidizing agents that may be used in the treatment of drinking water, wastewater, and groundwater include ozone, hydrogen peroxide, persulfate, hypochlorite, chlorine, potassium permanganate, Fenton's reagent (hydrogen peroxide and iron), and UV-catalyzed solid phase reactants. Applications of chlorine-based methods for the treatment of groundwater are not common. The strength of the oxidant – measured as oxidation potential – varies, and oxidation potentials of more than 2.0 V are generally considered to effectively address 1,4-dioxane. The strongest oxidant is fluoride, which is impractical for water treatment. The hydroxyl radical is nearly as strong as fluoride and has been effectively employed for both *in situ* and

TABLE 8.2
Oxidation Potentials for Some Common Oxidizers

Oxidant	Oxidation Potential (V)	Relative Strength (Chlorine = 1)
Hydroxyl radical (OH•)	2.7	2
Sulfate radical (SO_4)	2.6	1.8
Ozone (O_3)	2.2	1.5
Persulfate anion ($S_2O_8^{2-}$)	2.1	1.5
Hydrogen peroxide (H_2O_2)	1.8	1.3
Permanganate ion (MnO_4^-)	1.7	1.2
Chlorine (Cl^-)	1.4	1
Oxygen (O_2)	1.2	0.9

Source: DiGuiseppi, W.H. and Whitesides, C., 2007, *Environmental Engineer* 43(2): 36–41.

ex situ remediation. Table 8.2 presents a compilation of common oxidants, their oxidation potential, and their strength relative to chlorine. Chemical oxidation for the destruction of 1,4-dioxane *ex situ* is a proven technology demonstrated by numerous full-scale applications.

Figure 8.9 presents a comparison of laboratory simulations of various *ex situ* chemical oxidation treatment technologies for 1,4-dioxane removal, provided by ITT-WEDECO (H. Davis, personal communication, 2007). Tap water samples from Herford, Germany, were spiked with 1,4-dioxane

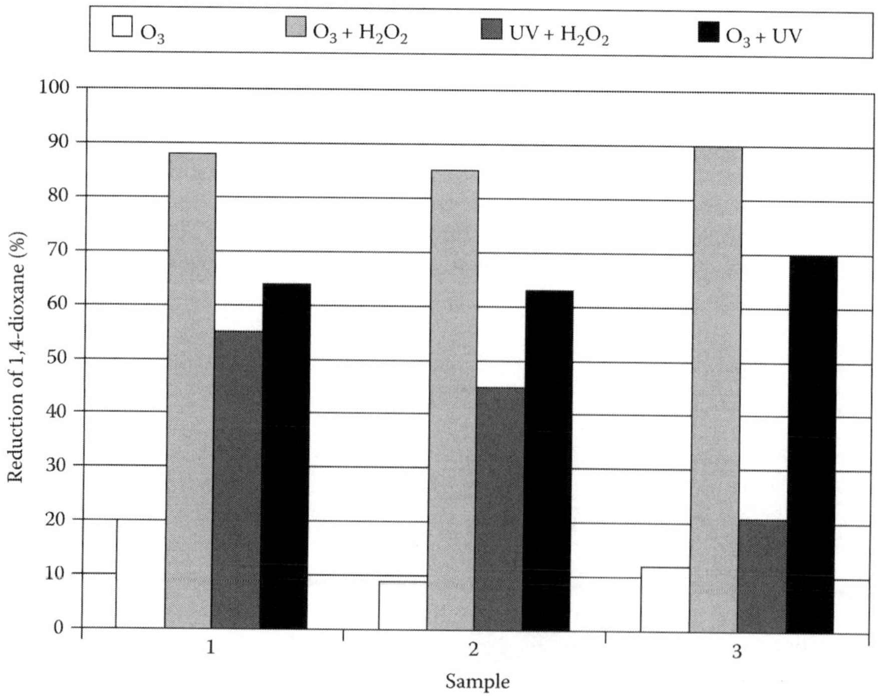

FIGURE 8.9 Reduction in 1,4-dioxane in three samples of drinking water (start value: 230 μg/L) through the use of four different remediation treatments. (H. Davis, personal communication, 2007; Courtesy of ITT – WEDECO.)

at a concentration of 230 μg/L and treated with ozone (O_3), ozone plus hydrogen peroxide (H_2O_2), hydrogen peroxide plus ultraviolet light, and ozone plus ultraviolet light. These specific technologies are discussed in further detail in this section, but with 1,4-dioxane removal rates of up to 90%, these are clearly demonstrated methods to be considered for 1,4-dioxane treatment. Site heterogeneity and undiscovered source areas often hinder successful applications of pump-and-treat technologies. When applying oxidation technologies, the effect of oxidizing conditions on organic and inorganic elements present in extracted groundwater or *in situ* materials must be evaluated and addressed. For inorganics, bromate formation from naturally occurring bromide may occur in chemical oxidation applications involving hydroxyl radicals. Additionally, hexavalent chromium, which is often a co-contaminant at solvent sites, is not addressed by chemical oxidation, and its concentration may be increased by creating strong oxidizing conditions. Arsenic, selenium, and other metals may also exhibit increased solubility, causing elevated concentrations under oxidizing conditions.

8.7.1 Hydrogen Peroxide and Ozone—*Ex Situ*

The combination of ozone (O_3) and hydrogen peroxide (H_2O_2) is known in chemistry as peroxone and has been used for specific contaminant remediation applications, such as TCE and explosives (Fleming et al., 1997). At equilibrium in water, hydrogen peroxide is present as $HO_2^- + H^+$. The hydroperoxide ion (HO_2^-) reacts with ozone to form hydroxyl radicals (OH•). *Ex situ* peroxone applications have been demonstrated in the treatment of drinking water and in the remediation of contaminated groundwater.

Bowman et al. (2007) demonstrated the effectiveness of injecting hydrogen peroxide and ozone under high pressure in a continuous in-line process to treat feed water containing an aqueous mixture with 1,4-dioxane and chlorinated compounds. In the HiPOx™ (Applied Process Technology, Inc., Pleasant Hill, CA) system, hydrogen peroxide is injected into the water stream first, followed by high-pressure ozone. Ozone is produced at 40–50 psig on site from a solid-state ozone generator supplied with liquid oxygen or from oxygen from the atmosphere, generated with a pressure swing absorption system. The ozone is injected at specific locations along the in-line reactor flow path to limit the occurrence of excess ozone and thereby avoid the creation of undesirable by-products, such as bromate (Figure 8.10).

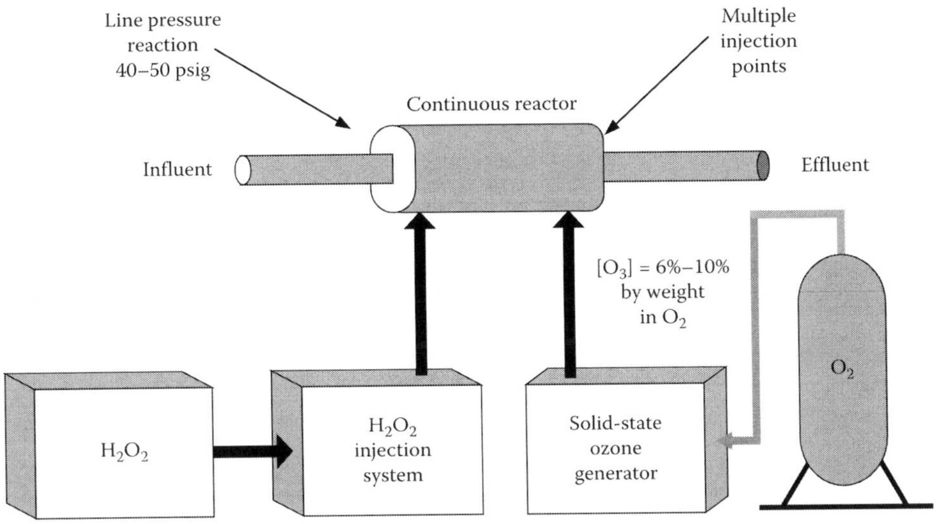

FIGURE 8.10 Schematic of HiPOx™ advanced oxidation system. (From Bowman, R.H., Miller, P., Purchase, M., and Schoellerman, R., 2007, Ozone-peroxide advanced oxidation water treatment system for treatment of chlorinated solvents and 1,4-dioxane. http://www.aptwater.com/assets/tech_papers/Paper-1.4Dioxane.pdf [accessed August 1, 2007].)

The HiPOx system was tested with water from three groundwater extraction and treatment systems operating in California with elevated chlorinated VOC and 1,4-dioxane levels. A system at South El Monte, California, was pilot-tested at 10 gpm for perchloroethylene levels of 150 μg/L and 1,4-dioxane of 20.2 μg/L. Effluent levels in initial testing achieved perchloroethylene reductions to 2.1 μg/L and 1,4-dioxane reductions to below detection limit of 2 μg/L. Because the existing water treatment for the contaminated drinking-water production well included liquid-phase GAC for effective treatment of chlorinated VOCs, the scaled-up design for the permanent production well solution was optimized for the removal of 1,4-dioxane. The final 490-gpm system was installed pre-GAC and effectively lowered the 1,4-dioxane concentration from 4.6 μg/L to below the detection limit at 2 μg/L (Bowman et al., 2007), but allowed low levels of chlorinated VOCs to exit the system. This approach addressed the 1,4-dioxane issue, while keeping the inexpensive GAC system operational and extending the life of the existing GAC media by lowering influent chlorinated VOC levels.

A second system in the City of Industry, California, was tested as pre- and post-treatment to an existing air stripper that was effectively treating chlorinated ethenes and ethanes as high as 730 μg/L (for methyl chloroform) to below California's regulatory threshold. Pilot-testing was optimized for 1,4-dioxane removal and not for the removal of chlorinated ethanes (e.g., methyl chloroform, 1,1-dichloroethane) because these compounds are approximately one-tenth as susceptible to hydroxyl radical oxidation as the corresponding chlorinated ethene (Bowman et al., 2007). Increasing hydrogen peroxide and ozone injections to eliminate the methyl chloroform would have increased costs dramatically for the air-stripper system that was already fully capable of removing chlorinated ethanes. Pilot-testing of the HiPOx advanced oxidation process (AOP) system as a post-treatment option revealed that the elevated pH of the stripper effluent, caused by the addition of an antifouling agent, reduced the efficiency of the AOP. Several different injection rates for ozone were tested to maximize 1,4-dioxane removal and assess the optimal order of treatment for the system (i.e., pre- or post-air stripper) (Figure 8.11). On the basis of the results, the system was designed for

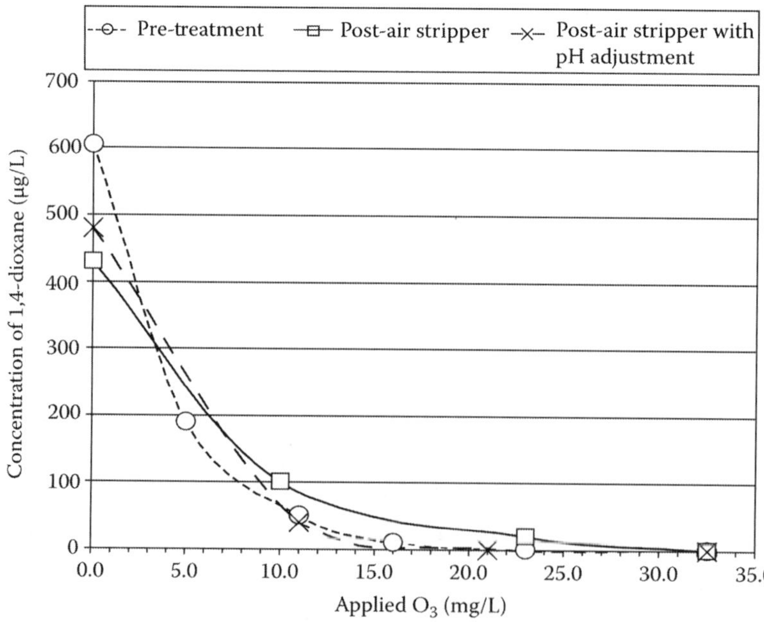

FIGURE 8.11 Plot of 1,4-dioxane concentration as a function of applied ozone at the City of Industry AOP Pilot Study. (From Bowman, R.H., Miller, P., Purchase, M., and Schoellerman, R., 2007, Ozone-peroxide advanced oxidation water treatment system for treatment of chlorinated solvents and 1,4-dioxane. http://www.aptwater.com/assets/tech_papers/Paper-1.4Dioxane.pdf [accessed August 1, 2007].)

use before the existing air stripper and was optimized to achieve the removal of 1,4-dioxane to levels below the detection limit of 2 µg/L. Chlorinated VOCs leaving the AOP system were effectively treated in the air stripper.

The third test site, in Mountain View, California, was treating chlorinated VOCs (as high as 8,800 µg/L for TCE). 1,4-Dioxane levels as high as 16 µg/L were untreated by the existing system (Boarer and Milne, 2004). Testing confirmed that treatment goals for all the contaminants of concern, including 1,4-dioxane, could be achieved with the HiPOx as a replacement for the air stripper while retaining the GAC system as a polishing unit to address compounds resistant to complete removal by oxidation (e.g., 1,1-dichloroethane) (Bowman et al., 2007). Overall, the HiPOx technology has been demonstrated in a variety of applications and conditions and is capable of complete destruction of 1,4-dioxane and associated chlorinated ethenes. The HiPOx system can also be used to destroy 1,4-dioxane and decrease the chlorinated VOCs to sufficiently low levels for enhancing or prolonging the effectiveness of existing treatment systems.

A key variable of applying perozone is the ratio of H_2O_2 to O_3. At low H_2O_2/O_3 ratios, the formation of OH• is limited. However, excess amounts of H_2O_2 act as an OH• scavenger, which has a negative effect on removal efficiency. A H_2O_2/O_3 ratio of 0.25 to 0.75 has been reported for >90% removal of 1,4-dioxane from the water tested (100 mg/L) (Kwon et al., 2012).

Andaluri and Suri (2017) tested ozone and ozone peroxide treatment of 1,4-dioxane and chlorinated compounds in laboratory reactors using milli-Q water and groundwater to assess the impact of solution pH and ozone dose on the oxidation process and 1,4-dioxane treatment. This testing included quantifying the relationship between gas flow, ozone concentration and ozone transfer, and quantifying 1,4-dioxane removal by ozone at varying pHs. In milli-Q water, ozone was capable of removing 1,4-dioxane, with the highest removal at the highest pH (10.2). It was suggested that conversion of ozone to hydroxyl radicals improved as pH increased and that creation of hydroxy radicals improved treatment. Another means of producing hydroxyl radicals from ozone is adding hydrogen peroxide. The study found the optimal ratio of ozone to hydrogen peroxide as 1.4:1 in milli-Q water. In groundwater systems, tests were run on influent (1,900 µg/L TCE) and effluent samples (360 µg/L TCE), with traces of other CVOCs in both samples. Hydrogen peroxide was added at 5 mg/L, ozone was varied up to 65 mg/L. TCE was removed within 15 min contact time, with ozone doses of 22 to 30 mg/L. However, chloroethanes present at trace concentrations (up to 9 µg/L) were present even after 30 min contact time at 65.3 mg/L ozone. This study illustrates the difficulty in treating the chloroethanes often associated with 1,4-dioxane in a hydroxyl radical-based process.

8.7.2 Hydrogen Peroxide and Ozone—*In Situ*

Schreier et al. (2006) performed a series of bench-scale studies assessing the viability of using ozone alone for ISCO, as opposed to mixing ozone and hydrogen peroxide *in situ*, which can be difficult in heterogeneous aquifer materials. The work was performed in support of remedial action at the Cooper Drum Superfund site in South Gate, Los Angeles County, California, where 1,4-dioxane concentrations targeted for treatment by ISCO were as high as 750 µg/L (Yunker, 2007). Peroxone, the mixture of ozone and hydrogen peroxide, is a demonstrated oxidizer, as already described, but ozone alone had been commonly regarded as not being of sufficient oxidation potential to destroy 1,4-dioxane. Ozone sparging of site groundwater resulted in the complete removal of 1,4-dioxane. To assess whether the removal was due to volatilization or destruction, vapor samples were collected from the off-gas and a sparge with inert gas (nitrogen) was evaluated. The data indicate that 1,4-dioxane removal was due to oxidation. Schreier et al. (2006) postulated that naturally occurring inorganic materials in the site soil or groundwater were enhancing the capability of ozone, which could eliminate or reduce the need for hydrogen peroxide. Ferrous iron, chelated iron, bicarbonate, and co-contaminants such as TCE were evaluated as potential oxidation enhancers. A test of spiked DI water indicated that ozone alone was capable of reducing

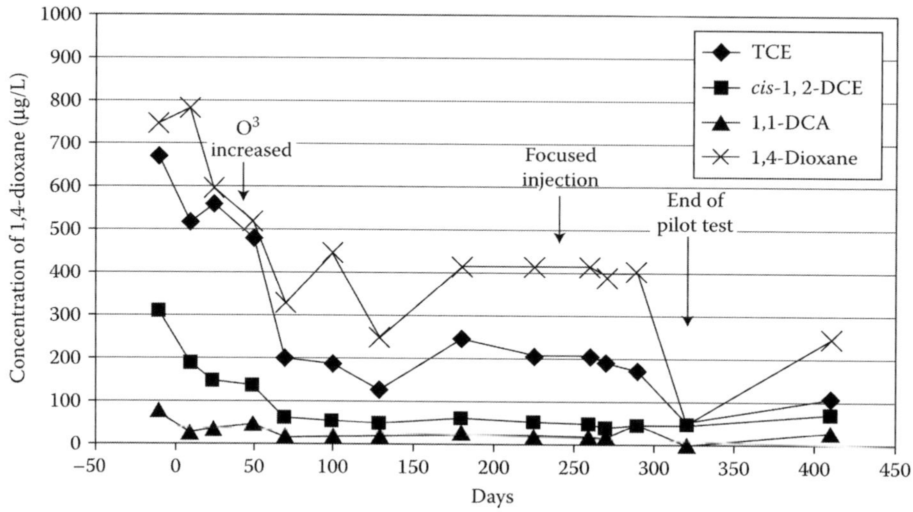

FIGURE 8.12 ISCO field pilot study results. (From Sadeghi, V.M. and Gruber, D.J., 2007, In situ oxidation of 1,4-dioxane with ozone and hydrogen peroxide. Poster presentation at URS's Environmental Technology and Management Seminar, Oakland, CA.)

concentrations from 400 to 98 μg/L, but not to the detection limit of 3 μg/L. Additional tests run with ferrous iron, chelated iron, and bicarbonate resulted in reductions down to the detection limit. 1,4-Dioxane removal in the presence of TCE did not significantly differ from the removal observed with ozone alone. Schreier et al. (2006) concluded that natural waters may contain elements or compounds that allow ozone to succeed at complete 1,4-dioxane removal.

A follow-on field pilot study at the same site (Sadeghi et al., 2006; Sadeghi and Gruber, 2007; USEPA, 2007; Yunker, 2007) involved the use of specially constructed wells with an ozone diffuser in the bottom of the well and a hydrogen peroxide diffuser higher in the well, such that the two amendments interact in the subsurface to form the hydroxyl radical. This system is patented by Applied Process Technology as the Pulse-OX™ (Applied Process Technology, Inc., Pleasant Hill, CA) system. Initial injections were of ozone alone, which resulted in significant reductions in 1,4-dioxane (and TCE) concentrations (Figure 8.12). Five months later, hydrogen peroxide injections were initiated, and it was determined that optimal results were achieved with the combination of 16% hydrogen peroxide and 2 pounds per day per injection well or 1 pound per day per injection interval of ozone. The injection interval was best situated not more than 10 ft below the target interval for maximum removal. Reductions as high as 88% were achieved in some of the monitoring wells (Table 8.3), and all the target contaminant concentrations decreased during the pilot test period. However, none of the concentrations were reduced to the detection limits during the field pilot test. The recommendation of this study was for the design and implementation of a full-scale system with more intense oxidation in the source zone and lesser efforts in the dissolved portion of the plume.

Kerfoot and Brolowski (2006) and Kerfoot (2008) described bench- and field-scale pilot-testing for 1,4 dioxane oxidation using ozone alone and ozone with hydrogen peroxide. Ozone reactivity exhibited a 1,4-dioxane removal rate of 0.15% per minute and total removal rates of 69–78% in soil slurries contaminated with 1,4-dioxane and chloroethene. Overall, the reaction rates were computed to be in the range of values previously defined for TCE and perchloroethylene, indicating that ISCO application of this technology could be as effective for 1,4-dioxane removal as for TCE and perchloroethylene removal. A field application was performed with the Perozone® system (Kerfoot Technologies, Inc., Mashpee, MA), which uses hydrogen peroxide-coated ozone microbubbles (<50 μm) delivered with a proprietary sparging system, referred to as the C-Sparger®

TABLE 8.3
ISCO Field Pilot Study Results

Well	COC	Initial Concentration in July 2005 (μg/L)	Concentration in June 2006 (μg/L)	Change Since Start of Pilot Study (%)	Concentration in August 2006 (μg/L)	Change Since Start of Pilot Study (%)
EW-1 (63 ft)	TCE	660	65	−90	120	−82
EW-1 (63 ft)	1,4-Dioxane	750	47	−94	250	−67
MW-33A	TCE	940	180	−81	130	−86
MW-33A	1,4-Dioxane	630	99	−84	74	−88
MW-20	TCE	520	110	−79	140	−73
MW-20	1,4-Dioxane	140	79	−44	71	−49

Source: Sadeghi, V.M. and Gruber, D.J., 2007, In situ oxidation of 1,4-dioxane with ozone and hydrogen peroxide. Poster presentation at URS's Environmental Technology and Management Seminar, Oakland, CA.

(Kerfoot Technologies, Inc., Mashpee, MA) (Kessel et al., 2005). The 2-month-long study evaluated the technology application in a barrier configuration designed to protect an open-water channel affected by groundwater discharges. Results indicated decreases in 1,4-dioxane concentrations within the ~25-ft radius of influence of the injection wells. Kerfoot and Brolowski (2006) noted that the removal rates depended upon contaminant concentrations and competitive destruction, because of the presence of other organic contaminants. Later bench-scale testing of nanoscale (250 nm) ozone sparge bubbles, produced by a miniature Laminar Spargepoint® (Kerfoot Technologies, Inc., Mashpee, MA), indicated equivalent effectiveness between ozone alone and ozone plus hydrogen peroxide. The nanobubbles exhibited a greater resistance to collapse and a longer half-life (more than 20 hours) than microbubbles (Kerfoot, 2008).

A field pilot study performed by Pall Corporation (2004) involved injecting hydrogen peroxide alone (to be combined with naturally occurring ferrous iron as a Fenton's reagent reaction), ozone alone, and ozone plus hydrogen peroxide. Results indicated that all three methods were effective to varying degrees. Ozone injection caused significant decreases in 1,4-dioxane concentrations, but also caused the formation of bromate above the maximum contaminant level of 10 ppb. The contaminated aquifer exhibited high levels of bromide, which exacerbated the bromate formation issue. Above-ground treatment using ozone and peroxide can be adjusted to control the reactions, whereas subsurface reactions cannot be adequately controlled to ensure that bromate levels do not become a problem.

8.7.3 Hydrogen Peroxide and Ultraviolet Light

Ultraviolet + peroxide oxidation systems cleave the hydrogen peroxide molecule (H_2O_2) into two hydroxyl radicals (OH•) by direct photolysis caused by the energy absorbed from an ultraviolet lamp.

Experiments performed by Stefan and Bolton (1998) defined the degradation pathways for 1,4-dioxane by hydroxyl radicals generated by direct ultraviolet photolysis of hydrogen peroxide. They determined that the destruction of 1,4-dioxane by ultraviolet + peroxide follows pseudo first-order kinetics and was more rapid than direct photolytic destruction of 1,4-dioxane using ultraviolet light. Within the first 5 min of irradiation, almost 90% of the initial concentration of 1,4-dioxane was depleted.

Ultraviolet + peroxide is a reliable, demonstrated *ex situ* technology, that is commercially available from a number of vendors. A case study presented in *Pollution Engineering* in July 1999 described a system installed to remove 1,4-dioxane following an air stripper that removes chlorinated compounds. The system demonstrated 1,4-dioxane removal from an initial concentration of 600 μg/L to below unspecified detection limits at a rate of 100 gpm.

An improvement to the typical ultraviolet + peroxide system was proposed and tested by Safarzadeh-Amiri et al. (1996), who used an ultraviolet + ferrioxalate + hydrogen peroxide system for treatment of highly contaminated waste streams. The ferrioxalate complex $[Fe(C_2O_4)_3^{3-}]$ is widely studied and is used to measure light intensity. The photolysis of ferrioxalate yields ferrous iron, which then reacts with the hydrogen peroxide to generate the hydroxyl radicals and provide a constant source of Fenton's reagent (discussed further in the following section). Safarzadeh-Amiri et al. (1996) demonstrated removal of 1,4-dioxane, among other contaminants, to below unspecified detection limits at substantially higher efficiency than that of ultraviolet + peroxide systems. This greater efficiency would result in fewer ultraviolet bulbs being needed and therefore lower maintenance and energy costs.

In a German application, bench testing for 5,000 gallons per day of process wastewater containing 15,000,000–20,000,000 μg/L 1,4-dioxane achieved a 99.95% removal efficiency by using ultraviolet light and peroxide (Sörensen and Weckenmann, 2006).

Chitra et al., (2012) compared UV/H_2O_2, UV/Fenton's Reagent, and natural sunlight/Fenton's Reagent. The study demonstrated that UV/H_2O_2 was effectively achieving almost complete removal. This result was similar to the UV/Fenton's Reagent processes, however, the UV/H_2O_2 was two to four times slower (4 hours at an initial concentration of 8,800 mg/L). However, results from the comparative study on the kinetics of degradation of 0.1 M 1,4-dioxane at pH 3.0 under stoichiometric addition of H_2O_2 revealed that the reaction using sunlight was faster than that obtained from UV.

A study of chloramines in UV-peroxide systems by Patton et al. (2018) looked at the impact of these compounds on the removal of 1,4-dioxane. The study found that monochloramine significantly decreased the removal efficiency of 1,4-dioxane relative to the impact of dichloramine. This is thought to be because the extra chlorine in dichloramine contributes to the generation of chlorine free-radicals, which boosts the treatment of 1,4-dioxane. Monochloramine, on the other hand, tends to act as a radical sink in the process.

8.7.4 Hydrogen Peroxide and Ferrous Iron

Fenton's chemistry involves catalyzing hydrogen peroxide, typically with a transition metal such as ferrous iron, to produce hydroxyl radicals, which are strong, nonspecific oxidizers. For the reaction to work properly, water must be acidified to a pH between 3 and 5. The iron catalyst is typically added as iron sulfate ($FeSO_4$), and then hydrogen peroxide is slowly added to control generation of heat from the reaction.

Pall Corporation (2004) performed multiple phases of field testing to evaluate Fenton's chemistry for the treatment of 1,4-dioxane *in situ*. Initial testing of hydrogen peroxide injections, which would have relied on naturally occurring ferrous iron, indicated that the 1,4-dioxane was completely broken down in the test area, without the production of bromate. However, a follow-up field study concluded that the 1,4-dioxane could not be destroyed *in situ* to a significant extent by peroxide injections alone. Potential injections of ferrous iron and peroxide were evaluated but are not documented in the literature.

Rushing et al. (2006) performed large-scale batch experiments using contaminated soil and groundwater from a coastal-plain pharmaceutical site. Naturally occurring ferrous iron was sufficient to catalyze the hydrogen peroxide injected at a concentration of 10 g/L into the reactors that were maintained at a pH of 4. Samples were collected before adding peroxide and after 24 hours of reaction time. The concentration of 1,4-dioxane, which was not the focus of the study, was reduced 99% from 15,000 to 110 μg/L. Rushing et al. (2006) also noted a relatively small decline in COD and TOC compared to the large reductions in the targeted contaminants of concern. They concluded that the parent compounds were degraded, but complete destruction of the organic compounds was not achieved.

An additional Fenton's reagent pilot study was performed to address 1,4-dioxane in saprolitic soils and groundwater at an active printing facility in North Carolina (Masten, 2004). Analytical data, which were collected 45 days postinjection, demonstrated that 1,4-dioxane was decreased by 98% in the pilot-testing area and similar concentration reductions were confirmed down gradient.

Laboratory bench-scale tests using modified Fenton's chemistry (Isotec, 2008) resulted in greater than 99% reduction in 1,4-dioxane levels in groundwater from 26,500 μg/L to ND at 55.1 μg/L. Although this detection level is well above many states' promulgated groundwater standards, this level of reduction, with one application, of a demonstrated aggressive oxidant suggests that achievement of lower concentration standards is likely. Soil remediation testing was also performed with comparably positive results, with 1,4-dioxane declining from 330 to less than the 1.1 mg/kg detection limit. This removal from soil took two doses of the modified Fenton's oxidant to achieve.

As referenced in Section 8.7.3, Chitra et al. (2012) demonstrated that Fenton's Reagent performance could be improved by incorporating UV into the Fenton's Reagent chemistry. Barndõk et al. (2016) studied photo-Fenton's processes using ZVI. In UV photolysis-Fenton treatment of synthetic waters, complete removal of 1,4-dioxane was achieved (initial 1,4-dioxane at 7,300 mg/L) and it was reported that UV was more efficient than sunlight (Barndõk et al., 2016). Ratios of H_2O_2/COD_0 were evaluated and the order of organics destruction was found to occur in the following ratios: 2.625 > 2.125 > 4 > 1.625. Complete removal of 1,4-dioxane was reached in as little as 5 min, while an almost total removal of COD and up to 96% removal of TOC were reached in 30 min (Barndõk et al., 2016).

Work has been done to investigate the oxidation kinetics of 1,4-dioxane (Ghosh et al., 2011). The researchers concluded that pH reduction during the reaction corroborates formation of organic acids when 1,4-dioxane is oxidized. Also, it was established that the change in concentration of Fe^{2+} and H_2O_2 during the reaction shows that Fe^{2+} is the limiting reactant. As referenced in Section 8.7.3, Chitra et al. (2012) demonstrated that Fenton's Reagent performance could be improved by addition of UV or sunlight, with sunlight having the greater increase in the reaction rate constant.

Zeng et al. (2017) demonstrated a novel way to use iron-rich clay minerals to degrade 1,4-dioxane through production of OH• without the need of H_2O_2 or UV radiation. In this study, nontronite (iron-rich smectite) was used to produce OH• under neutral pH and aerobic conditions such as aeration (*ex situ*) or air sparging (*in situ*). The initial 1,4-dioxane concentration of 35 mg/L was reduced by about 75%. Clay mineral regeneration via sodium dithionite was further demonstrated but the authors indicated that alternative reductants such as ascorbic acid or iron bacteria could also be used (Zeng et al., 2017).

8.7.5 Persulfate

Persulfate is a strong oxidizer, with a standard oxidation potential of 2.1 V, which is comparable to ozone (2.1 V) but higher than that of hydrogen peroxide (1.8 V) and the permanganate anion (1.7 V) (Block et al., 2004). Persulfate can be "activated" in a number of ways to create sulfate radicals (SO_4^-) that have an oxidation potential (2.6 V) almost as high as that of the hydroxyl radical (2.7 V). Activation methods for persulfate include heat, high pH, transitional metals, and hydrogen peroxide. Reactions involving the sulfate radical are kinetically fast, and the sulfate radical is more stable, and therefore longer lived, than the hydroxyl radical. The hydroxyl radical has a half-life of minutes to hours, whereas the sulfate radical has a half-life of hours to weeks, depending on aquifer geochemistry (Huling and Pivetz, 2006), allowing greater migration distances in the subsurface when injected. Persulfate also has a lower affinity for natural soil organic matter than some other oxidizers, most notably permanganate, and would therefore be more effective in highly organic soils. Note that the sulfate formed through activation has a secondary drinking water standard of 250,000 μg/L for taste.

Evaluation of the effectiveness of alkaline activation of sodium persulfate ($Na_2S_2O_8$) for 1,4-dioxane degradation was performed by Block et al. (2004); in this study, potassium hydroxide (KOH) was

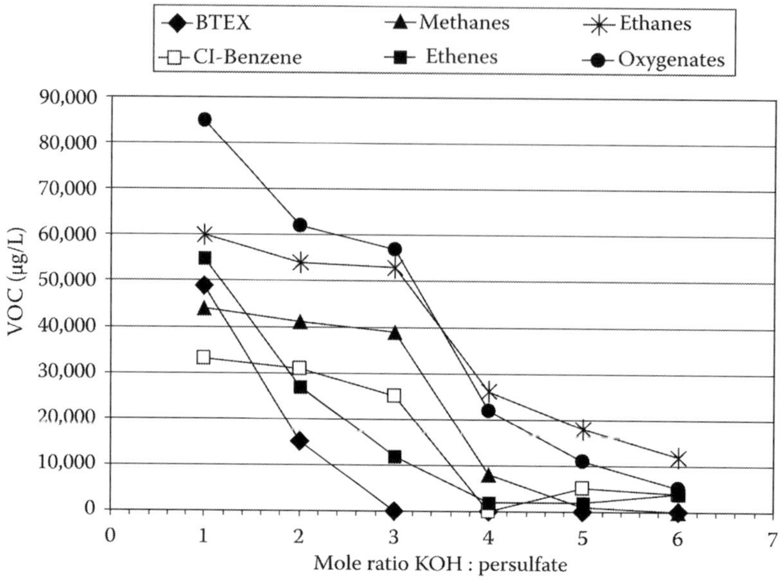

FIGURE 8.13 Effect of KOH ratio on persulfate reactivity. (From Block, P.A., Brown, R.A., and Robinson, D., 2004, Novel activation technologies for sodium persulfate in situ chemical oxidation. In: A.R. Gavaskar and A.S.C. Chen [Eds], *Proceedings of the Fourth International Conference on Remediation of Chlorinated and Recalcitrant Compounds*, Paper 2A-05. Columbus, OH: Battelle Press.)

used as the base to increase pH to >10. Several contaminants were tested, including chlorinated VOCs and oxygenated compounds (e.g., MTBE, TBA, and 1,4-dioxane), at a range of molar ratios of KOH:persulfate. The laboratory-scale studies demonstrated that alkaline activation of persulfate, in addition to the buffering effect of the KOH, resulted in significant declines in target compound concentrations (Figure 8.13). Block et al. (2004) additionally reviewed activation with inorganic compounds and found that Fe^{2+} was an effective activator; however, natural conditions would not favor using ferrous iron for activation because of solubility issues. Evaluation of various chelating agents and inorganic compounds identified a combination of iron and ethylenediaminetetraacetic acid (EDTA) to be an effective activator at typical pH ranges found in natural waters.

Félix-Navarro et al. (2007) studied the sensitivity of sodium persulfate oxidation of 1,4-dioxane at different persulfate concentrations, temperatures, and pH levels. At temperatures ranging from 25°C to 50°C, a 25 mM persulfate solution at pH 7 oxidized 1.13 mM 1,4-dioxane at half-lives ranging from 5 to 64 min; higher temperatures yielded more rapid destruction. Persulfate solutions ranging from 12.5 to 100 mM, at pH 7 and 25°C, oxidized 1.13 mM 1,4-dioxane within half-lives of 38–122 min; the higher-concentration solution oxidized the fastest. Dependency on pH was also evaluated at pH values of 3, 5, 7, 9, and 11, for a 25 mM persulfate solution at 25°C containing 1.13 mM 1,4-dioxane. The reaction rate was found to decrease with increasing pH values. Apparently, sulfate and hydroxyl radicals decay rapidly because of reactions with hydroxyl ions in high-pH solutions. Additionally, carbon dioxide from oxidation leads to the formation of bicarbonate and carbonate ions, which may hinder the oxidation of organic compounds.

FMC (2007) discussed an industrial site with a methyl chloroform rail car loading/unloading operation where contaminant concentrations were elevated as follows: methyl chloroform, 203,000 µg/L; 1,1-dichloroethylene, 82,000 µg/L; and 1,4-dioxane, 50,000 µg/L. Initial testing at the site using injected calcium hydroxide (hydrated lime) and steam encountered difficulties with steam daylighting – that is, steam escape through to the ground surface – resulting in an inability to provide the required heat to the treatment area. Sodium hydroxide was then used to catalyze the persulfate with high alkalinity. Postinjection concentrations of 1,4-dioxane were below the local cleanup standard and detection limit of 5 μg/L. Chlorinated VOCs were reduced dramatically within the first year and continued to decline during the following 9-month monitoring period.

Zhao et al. (2014) evaluated the reaction of 1,4-Dioxane (at 100 mg/L) and heat activated persulfate, with and without an iron activator, at acidic pH values (2.3 to 2.9). The results demonstrated that addition of Fe^{2+} improved the reaction time at the lower temperature (40°C) but was observed to slow the reaction time at the higher temperatures (50°C and 60°C). Zhao et al. (2014) theorized that the higher temperature can accelerate the reaction between Fe^{2+} and SO_4^-, limiting the availability of SO_4^- to react with 1,4-dioxane.

Eberle et al. (2016) evaluated perozone activated persulfate (PAP) to treat 1,4-dioxane with chlorinated solvent co-contaminants at ambient temperatures. The testing demonstrated that PAP was capable of degrading 1,4-dioxane, TCE, and 1,1,1-TCA at 25°C, which sets it apart from high-heat activated persulfate approaches. The findings of this study support PAP as a potential *in situ* treatment alternative (Eberle et al., 2016).

Kambhu et al. (2017) reported on ZVI-activated persulfate. ZVI has the advantage of being a solid that can serve as electron donor by itself as well as provide a longer-term supply of Fe^{2+} for persulfate activation (Kambhu et al., 2017). The ZVI activated persulfate reactions were evaluated at room temperature (20°C). The results suggest efficient degradation of 1,4-dioxane from the test solutions, however, consistent with the research by Zhao et al. (2014), care must be taken to avoid excess production of Fe^{2+}, as excess iron competes with the sulfate radical for oxidation of 1,4-dioxane (Kambhu et al., 2017).

In a similar concept of iron activation of a sulfur-based oxidant, Feng et al. (2018) reported a novel oxidation approach using pyrite as a slow-release source of iron and peroxymonosulfate (PMS) (HSO_5^-). Near-100 percent degradation of 1,4-dioxane (from 50 mg/L) was achieved after 40 min using pyrite with PMS at room temperature (24°C). The high efficiency of pyrite-PMS oxidation suggests that it may be an economical and effective approach for the treatment of groundwater (Feng et al., 2018).

Removal of 1,4-dioxane with persulfate has been successfully demonstrated in numerous field pilot studies with a variety of activation methods. Following a successful bench-scale study in which 100% of 1,4-dioxane and 85–99% of chlorinated VOCs were removed, Cronk (2006) executed a source-area removal at a site in Huntington Beach, California, with 1,4-dioxane levels as high as 260,000 μg/L, as well as TCE and dichloromethane as contaminants of concern. Approximately 5,000 pounds of persulfate activated with CHP was applied to the tight clayey soil through eight application wells with a radius of influence of 10–12 ft each. 1,4-Dioxane levels were reduced by 88–95% in the source zone, down to 21,000 μg/L, during the source-removal project. In La Mirada, California, another field pilot study used chelated iron (EDTA) activation of sodium persulfate. 1,4-Dioxane concentrations in three wells located down gradient of the injection area declined from between 100 and 225 μg/L to nondetectable levels over the 9-month test period. Levels have remained below detection limits for several months, and the groundwater will be monitored periodically to ensure that rebound does not occur (Cronk, 2008).

Considerable work has been accomplished using persulfate (including sodium and potassium persulfate) for ISCO in the past 10 years. Persulfate has emerged as a powerful tool for *in situ* remediation, and while it is beyond the scope of this book to detail its varied activations, suffice it to say that a wide variety of reactive species can be produced from the persulfate ion under engineered conditions and these species can be applied to a variety of compounds (ITRC, 2005). Persulfate readily forms the hydroxyl radical by reacting in water (and boosted by elevated pH), and the hydroxyl radical is sufficient to treat 1,4-dioxane.

Alkaline activation has been the industry standard for many years, but other activations are possible and allow management of the treatment chemistry. Activations applied to 1,4-dioxane in the literature include heat (Zhao et al., 2014), sonolysis (Li and Zhu, 2016), soluble iron (Zhao et al., 2014), ZVI (Zhong et al. 2015), iron minerals (Yan, 2018), peroxide mixtures (Eberle et al., 2016; Yan et al., 2018), and natural minerals (Berggren and Hatton, 2016; Brubaker et al., 2016; Yan et al., 2019), all of these publications documented some degree of success in using these activations for treatment.

Documentation of full scale *in situ* treatment using base-activated persulfate (pH 10.5) at a co-contaminated solvent site has been reported (Dombrowski et al., 2010). These results demonstrated that reduction of PCE and TCE was generally good (70% to 98% removal) whereas 1,4-dioxane was <40% removal, suggesting that base activation at ambient temperature was not as effective with respect to 1,4-dioxane compared to other activation methods reported (Dombrowski et al., 2010).

Borchert et al. (2014) performed bench-testing of activated sodium persulfate and CHP prior to embarking on full-scale design for an ISCO solution in low permeability source area soils at a site impacted with mixed CVOCs (TCE at 596,000 µg/L), petroleum contaminants, and 1,4-dioxane (45,200 µg/L). Laboratory studies were performed to evaluate the effectiveness of various stabilizers (citrate, sodium phytate, and diethylenetriaminepenta acetic acid) to enhance subsurface longevity of CHP. Phytate extended the CHP effectiveness to 4 days, which was 1 day longer than the activated persulfate longevity. The CHP achieved 1,4-dioxane reductions of 87% versus 29% reduction for activated persulfate in the bench tests, which led to the conclusion that CHP would be the optimal for full-scale implementation. However, during predesign drilling late in the planning and design process, petroleum light nonaqueous-phase liquid (LNAPL) was identified at the site, which had the potential to generate exothermic reactions during oxidation. To avoid this possibility, activated persulfate was chosen for field implementation because although it was less effective on 1,4-dioxane, it was suitable for the other high concentration contaminants present at the site. Following two persulfate applications, groundwater monitoring indicated 1,4-dioxane reductions between 78% and 91% (S. Borchert, personal communication, 2016).

Natural mineral activation (NMA) of sodium persulfate is a novel activation approach which has been evaluated in bench testing, field pilot testing at four sites, and full-scale field implementation at two sites (Berggren and Hatton, 2016; Brubaker et al., 2016). NMA activation avoids the cost and difficulties with specific activators. It also has the potential to last longer than more aggressive, quick-acting activators. It is presumed that site soils often contain iron and manganese on the soil surfaces, which act as activators for the sodium persulfate. Other possible mechanisms for activation could include naturally occurring dissolved iron and manganese, reactive organic compounds generated in the aquifer from anaerobic processes, and the presence of other possible activating minerals (Berggren and Hatton, 2016). Brubaker et al. (2016) demonstrated proof of concept in bench testing which demonstrated effective sodium persulfate activation and complete 1,4-dioxane removal from levels as high as 250 µg/L within 21 days. Follow-on field pilot testing occurred at a site in Texas, where 1,4-dioxane concentrations decreased from 205 µg/L to less than 0.465 µg/L (99.8% reduction). Unactivated persulfate was noted 650 ft downgradient of the injection wells 4 year after injection, indicating the persulfate longevity was extended by relying on naturally occurring minerals rather than rapid activation which limits effective radius from an injection well (Brubaker, personal communication, 2019). A second field pilot test was conducted with 65,000 pounds of unactivated persulfate at a site in Texas with 2,750 µg/L 1,4-dioxane in a 150-ft-thick aquifer. Two months after injection, the treatment zone concentrations had declined 90% initially, followed by a brief rebound and then stabilized at 65% reduction. A well 50 ft downgradient realized a decrease of 50% at 2 months and geochemical parameters suggested the injection impacts at that location were minimal. The two case studies differed in that the oxidation-reduction potential (ORP) readings of 0 to 200 mV and DO readings of 1 mg/L to 4 mg/L at the first site indicated an aerobic aquifer, whereas the ORP readings of −50 to −150 mV and DO readings of 0 mg/L in the second aquifer suggested it was mildly anaerobic (Brubaker et al., 2016). The anaerobic conditions may have represented an increased oxidant demand that the persulfate injection struggled to overcome.

Li et al. (2018) looked at sodium persulfate as a source of radicals in a UV/AOP process and found the process effective for 1,4-dioxane removal. Similar to the results for ozone-peroxide discussed above, chloramines present released chlorine as they broke down and the chloride participated in the radical-forming process. At an optimal ratio of dichloramine to persulfate (0.1 to 1, molar basis), treatment of 1,4-dioxane was enhanced, but outside this range, treatment was hindered. Control of the dichloramine-persulfate ratio is critical for controlling this process.

A means of treating 1,4-dioxane and CVOCs using slow-release, unactivated persulfate was presented by Evans et al. (2014). Originally developed as permanganate candles, unactivated persulfate candles are slow release cylinders made of an oxidant and wax, designed to slowly dissolve and release the oxidant. Persulfate candles were shown to treat 1,4-dioxane at concentrations up to 10,000 μg/L, removals as high as 99% were reported in bench scale tests. Unactivated persulfate candle 1,4-dioxane degradation rates were about 30 times higher than those determined for permanganate. An Environmental Security Technology Certification Program (ESTCP) project reported on the results of implementing slow release candle technology at Naval Air Station North Island Operable Unit 11, in San Diego, California (Evans et al., 2018). Unactivated persulfate slow release cylinders were tested in microcosms and flow-through columns. Both 1,4-dioxane and CVOCs were removed by persulfate in the microcosms and column studies. Persulfate slow release cylinders were selected for field-scale testing. Field removals exceeded 90% for 1,4-dioxane and total CVOCs. Upgradient 1,4-dioxane and CVOC concentrations exceeded 20,000 μg/L. The downgradient 1,4-dioxane concentration was as low as 140 μg/L (99.3% removal) and the total CVOC concentrations were as low as 200 μg/L (90% removal).

8.7.6 Permanganate

Potassium permanganate ($KMnO_4$) is a common oxidizer used in water treatment. A survey of surface-water treatment systems serving more than 10,000 people reveals that 36.8% of the treatment systems use potassium permanganate for the removal of organic compounds (Waldemer and Tratnyek, 2004). Although ISCO with potassium permanganate is an accepted and widely used remedial method for chlorinated VOCs, it has been generally thought that the oxidation potential (1.7 V) of potassium permanganate is too low to have an impact on 1,4-dioxane. However, Waldemer and Tratnyek (2006) determined the kinetics of permanganate oxidation for a number of environmental contaminants and found that the secondary rate constant for permanganate oxidation of 1,4-dioxane was four orders of magnitude lower than that found for TCE or perchloroethylene. Although at a very slow rate, oxidation of 1,4-dioxane was nonetheless observed. Given the very rapid destruction of chlorinated VOCs (on the order of minutes), an oxidation rate for 1,4-dioxane four orders of magnitude slower would be on the order of months, which may be suitable for some applications.

Evidence from an ongoing ISCO pilot study at Air Force Plant 44 (AFP 44), in Tucson, Arizona – a site contaminated with chlorinated VOCs and 1,4-dioxane – indicates long-term decline in 1,4-dioxane concentrations following injection of potassium permanganate (Earth Tech, 2008). 1,4-Dioxane levels in two monitoring wells located near the center of the $KMnO_4$ injection area (M-96 and M-98) declined from 250 to 100 mg/L and from 650 to 170 mg/L, respectively, between May 2004 and May 2007. Although the permanganate oxidation kinetics for 1,4-dioxane are slow, the reaction is likely occurring and, given enough time, could reduce concentrations sufficiently to meet regulatory standards. $KMnO_4$ oxidation of 1,4-dioxane was further studied at AFP 44 through a bench-scale test using site groundwater contaminated with 1,4-dioxane in the range of 40 μg/L. Triplicate samples analyzed quarterly, where one sample was an untreated control, one sample was acidified to sterilize and eliminate possible bioactivity, and one sample was treated with approximately 4 mg/L of $KMnO_4$, intended to approximate the $KMnO_4$ concentrations maintained in site groundwater since 2004. Results from the first quarter indicate 1,4-dioxane reductions of ~20% in the control and the acidified control, likely because of interactions with dissolved minerals in the site groundwater. The $KMnO_4$-treated replicate indicated ~30% reduction over the same time period, which suggested that oxidation was occurring. Results from subsequent quarters were ambiguous, indicating that 1,4-dioxane destruction was not unequivocal. Therefore, the test was terminated.

Oxidation reactions, while often modeled as "pseudo first order" reactions to cover site-specific conditions, are better generalized as second-order reactions, where the "rate" is dependent on the concentration of the reactants rather than a constant (Tratnyek et al., 2007). Therefore, 1,4-dioxane's slow reaction with permanganate can be improved by either increasing the concentration of permanganate or only applying it to high concentrations of 1,4-dioxane. The problem is how to do so economically. A means of treating 1,4-dioxane and CVOCs using permanganate was presented by Evans et al. (2014). Permanganate candles are slow release cylinders made of an oxidant and wax, designed to slowly dissolve and release the oxidant. Permanganate candles were shown to treat 1,4-dioxane at concentrations up to 10,000 μg/L, removals as high as 90% were reported in bench scale tests. An ESTCP project reported on the results of implementing slow release candle technology at Naval Air Station North Island Operable Unit 11, in San Diego, California (Evans et al., 2018). Permanganate slow release cylinders were tested in microcosms and then flow-through columns. While both 1,4-dioxane and CVOCs were removed by permanganate in the microcosms, 1,4-dioxane was not removed in the column study. The lack of removal of 1,4-dioxane in the column experiments was thought to be due to the development of a coating on the section of slow-release cylinder, which hindered release of the permanganate.

8.7.7 Photocatalytic Oxidation

In aqueous solutions, ultraviolet light activates a catalyst, such as titanium dioxide (TiO_2), to produce hydroxyl radicals on its surface from oxygen in air or dissolved oxygen in water.

Hill et al. (1996) determined that 1,4-dioxane in water can be completely broken down by exposure to oxygen and ultraviolet light, in the presence of TiO_2 as a catalyst. No reaction occurred in any experiment conducted that omitted any one of the three requisite elements. The study focused on degradation intermediates, particularly ethylene diformate, which appears to be created more readily than it is destroyed. Formate esters, which are likely present even in "pure" 1,4-dioxane, should therefore be considered in toxicological studies of the effects of 1,4-dioxane.

Mehrvar et al. (2000) further evaluated the photocatalytic destruction of 1,4-dioxane, alone and in an aqueous solution combined with THF, which is associated with 1,4-dioxane at many sites. They defined the hypothetical reaction mechanism pathways for both THF and 1,4-dioxane and developed kinetic models to predict degradation behavior. They concluded that 1,4-dioxane is converted to hydroxylated 1,4-dioxane; this process leads to ring opening and formation of organic acids, which ultimately convert to CO_2 and water.

Yanagida et al. (2006) evaluated the effect on an applied voltage to a TiO_2-impregnated screen exposed to ultraviolet light for the enhanced treatment of 1,4-dioxane. They determined that the photocatalytic decomposition rate of 1,4-dioxane was increased when a voltage swing of ±0.4 V was applied, and formation of toxic intermediates was suppressed.

Vescovi et al. (2010) studied the effects of pH and TiO_2 configuration types with 1,4-dioxane at 36 mg/L. The results demonstrated little difference in reaction rate constants between the pH 7 and pH 11 tests, and a slower reaction rate constant at pH 3 (Vescovi et al., 2010). Using the TiO_2 sol-gel reactor, a robust reaction rate constant was observed at pH 7, achieving complete removal in 45 min versus 60 min at pH 3 (Figure 8.14).

Ex situ photocatalytic treatment of 1,4-dioxane in extracted groundwater is a demonstrated technology, as described in the trade literature (www.purifics.com). Several case studies are presented, with influent 1,4-dioxane levels as high as 1,500 μg/L and post-treatment levels of below 3 μg/L.

An *ex situ* advanced oxidation system evaluation was performed by Wannamaker (2005), where cost and anticipated performance were compared between systems utilizing hydrogen peroxide + ultraviolet, hydrogen peroxide + ozone, and ultraviolet + titanium dioxide (photocatalytic oxidation). Additionally, bromate production was evaluated for each of the technologies. Field tests of the ultraviolet + titanium dioxide system were performed that yielded a 1,4-dioxane decrease from 150 to less than 1.9 μg/L (the detection limit). Bromate conversion, from naturally occurring bromide at 600 μg/L, was minimal in the effluent from the ultraviolet + titanium dioxide system.

A novel approach to UV catalyzed reactions for water treatment systems has been published by Patton et al. (2016). This study looked at photocatalyzation of monochloramine, a disinfectant

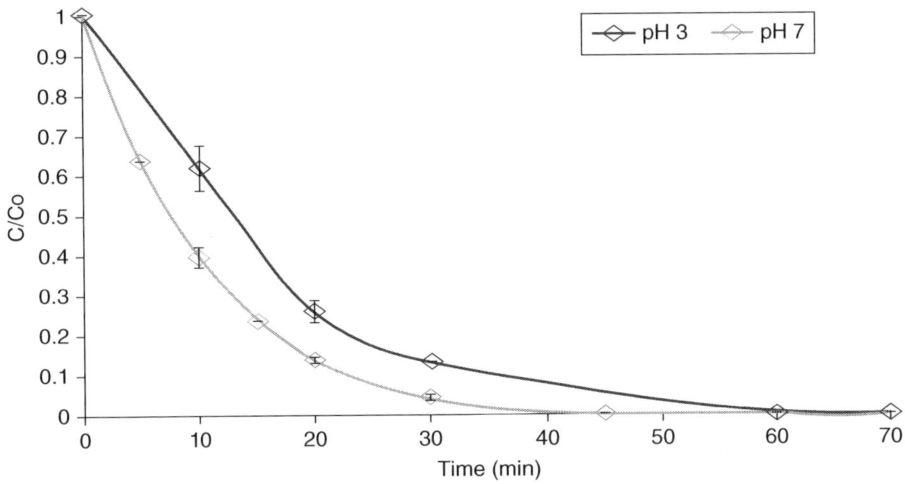

FIGURE 8.14 1,4-Dioxane degradation at pH 3 and pH 7 using a TiO_2 sol-gel reactor system and UVA light. Reprinted (adapted) with permission from Vescovi, T., Coleman, H., and Amal, R., 2010, The effect of pH on UV-based advanced oxidation technologies- 1,4-Dioxane degradation. *Journal of Hazardous Materials* 182:75–79. Copyright 2010 Elsevier.

commonly used in water treatment. As both 1,4-dioxane and monochloramine are poorly rejected by reverse osmosis (RO) membranes, demonstrating this reaction has application for water reuse systems (Patton et al., 2016). The results strongly suggested that both Cl_2^- and OH• significantly contributed to degradation of 1,4-dioxane, whereas the NH_2 played a less critical role (Patton et al., 2016).

8.7.8 Ozonation

Ozone alone had been commonly regarded as not being of sufficient oxidation potential to destroy 1,4-dioxane. However, ozone sparging has resulted in the complete removal of 1,4-dioxane (Schreier et al., 2006; Barndõk et al., 2014a). A series of bench-scale studies were performed to assess the viability of using ozone alone for ISCO. The work was performed in support of remedial action at the Cooper Drum Superfund site in South Gate, Los Angeles County, California, where 1,4-dioxane concentrations targeted for treatment by ISCO were as high as 750 mg/L (Yunker, 2007). To assess whether the removal was due to volatilization or destruction, vapor samples were collected from the off-gas and a sparge with inert gas (nitrogen) was evaluated. The data indicate that 1,4-dioxane removal was due to oxidation. Schreier et al. (2006) postulated that naturally occurring inorganic materials in the site soil or groundwater were enhancing the capability of ozone, which could eliminate or reduce the need for hydrogen peroxide. Ferrous iron, chelated iron, bicarbonate, and co-contaminants such as TCE were evaluated as potential oxidation enhancers. A test of spiked DI water indicated that ozone alone was capable of reducing concentrations from 400 to 98 mg/L, but not to the detection limit of 3 mg/L. Additional tests run with ferrous iron, chelated iron, and bicarbonate resulted in reductions down to the detection limit. 1,4-Dioxane removal in the presence of TCE did not significantly differ from the removal observed with ozone alone. The research team concluded that natural waters may contain elements or compounds that allow ozone to succeed at complete 1,4-dioxane removal.

Barndõk et al. (2014a) showed that ozone alone is a viable treatment for the degradation of 1,4-dioxane under appropriate operation conditions. The testing demonstrated removal of 90% of COD and almost all of the 1,4-dioxane was achieved. The degradation at $pH > 9$ was shown to occur faster through the formation of ethylene glycol as a primary intermediate; whereas the decomposition in acidic conditions ($pH < 5.7$) results in the formation and slower degradation of ethylene glycol diformate (EGD).

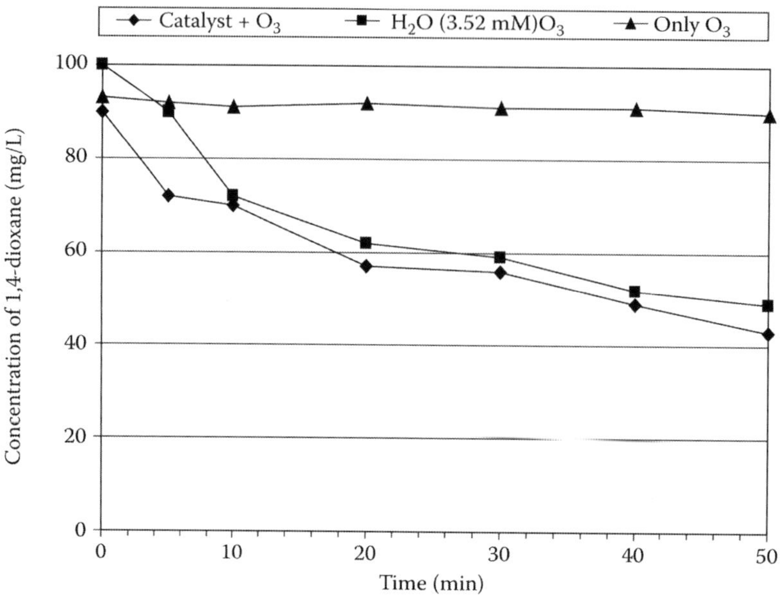

FIGURE 8.15 Destruction of dioxane during oxidation experiments for three different sources of oxidant; pH = 10; ozone dosage = 10 mg/min. (From Suh, J.H., Kang, D.J., Park, J.D., and Lee, H.S., 2005, A study on the catalytic ozonation of 1,4-dioxane. In: *Proceedings from the 9th Russian-Korean International Symposium on Science and Technology* [KORUS], Novosibirsk, Russia.)

Kishimoto et al. (2005) studied ozonation combined with electrolysis for COD removal from 1,4-dioxane solution. They defined the destruction pathway as starting with hydroxyl radicals produced through ozonation near the cathode, in a high-pH environment. Additional cathodic ozone reduction occurred, which helps destroy the 1,4-dioxane. Ozone was noted to then destroy the 1,4-dioxane degradation products produced in the initial stages. Carbon dioxide produced by the oxidation of organics typically forms bicarbonate, which inhibits 1,4-dioxane oxidation; however, this process was mitigated by using a two-compartment flow cell to strip off the carbon dioxide in the anodic cell. Chlorine, produced from the oxidation of chloride at the anode, was found to enhance oxidation of the degradation by-products in a one-compartment flow-cell test.

Suh et al. (2005) used a palladium (Pd) catalyst deposited on activated carbon to evaluate 1,4-dioxane oxidation using ozone, ozone + peroxide, and ozone + Pd catalyst. No significant concentration decreases were observed with ozone alone (Figure 8.15), and there were no significant differences between the 1,4-dioxane concentration decrease from ozone + peroxide or ozone + Pd catalyst. Suh et al. (2005) noted that oxidation of 1,4-dioxane to more easily biodegraded intermediates would support the integration of advanced oxidation and more conventional biological treatment methods for wastewater. Cyclodextrin has found use as a stabilizer for ozone in the subsurface environment. Dettmer et al. (2017) reported that cyclodextrin could complex with and stabilize ozone, extending its reactivity and mobility in the subsurface. Khan et al. (2019) determined reaction rate constants for unstabilized and stabilized ozone and found the reaction of stabilized cyclodextrin was much slower and that reactivity with 1,4-dioxane, in the presence of other contaminants, was improved.

8.7.9 Sonochemical Oxidation

Sonochemical oxidation of organic contaminants is caused by the production of hydroxyl radicals and other oxidizing species during the collapse of cavitation bubbles formed during high-frequency acoustic stimulation of the aqueous media. High temperatures (up to 4,000°C) and pressures (100,000 kilopascals) can be present in the cavitation bubbles when they collapse. VOCs are degraded through combustion, high-temperature chemical reactions, supercritical water oxidation, and the production of free radicals. Hydroxyl and other free radicals formed during this process are capable of oxidizing organic compounds, such as 1,4-dioxane.

Beckett and Hua (2000) and Hua (2000) performed an evaluation of the viability of using sonochemical destruction for 1,4-dioxane and defined the decomposition products and pathways for destruction of 1,4-dioxane during sonolysis at discrete ultrasonic frequencies. Ninety-six percent of the 1,4-dioxane was completely broken down in the first 120 min, and Beckett and Hua (2000) and Hua (2000) identified five major reaction intermediates: ethylene glycol diformate (EGD), methoxyacetic acid, formic acid, glycolic acid, and formaldehyde (Figure 8.16). In general, these breakdown products are less toxic and less recalcitrant than 1,4-dioxane. The frequency proved to be an important factor because the bubble size, and resultant collapse energy, is inversely correlated to the sonic frequency.

Beckett and Hua (2003) assessed how the addition of ferrous iron affected 1,4-dioxane sonochemical degradation rates. Ferrous iron addition increased the degradation rate and efficiency of 1,4-dioxane decomposition in all tested ultrasonic frequencies. The combination of ferrous iron and ultrasound yielded a higher concentration of the hydroxyl radical, thereby causing more rapid destruction than either sonication or Fenton's reagent alone. Nakajima et al. (2007) further assessed the effect on sonochemical degradation rates by addition of reduced TiO_2 powder, as compared to an inert powder (SiO_2). They found that the TiO_2 increased both the temperature and cavitation in the aqueous solutions. These changes led to an increase in the ultrasonication energy consumed in the water and the destruction efficiency of 1,4-dioxane. Nakajima et al. (2004) coupled photocatalytic treatment of 1,4-dioxane using ultraviolet and TiO_2 with sonication in the presence of TiO_2. They also assessed hydrofluoric acid treatment of TiO_2, which had been shown elsewhere to enhance sorption of organic contaminants to the TiO_2 surface. The research findings indicate that the synergistic effect of these combined methods improves the decomposition rate of 1,4-dioxane. The reaction constant for the decomposition of 1,4-dioxane using the combination of ultrasound, ultraviolet, and a TiO_2 catalyst is three times higher than that derived by using only ultraviolet and the TiO_2 catalyst.

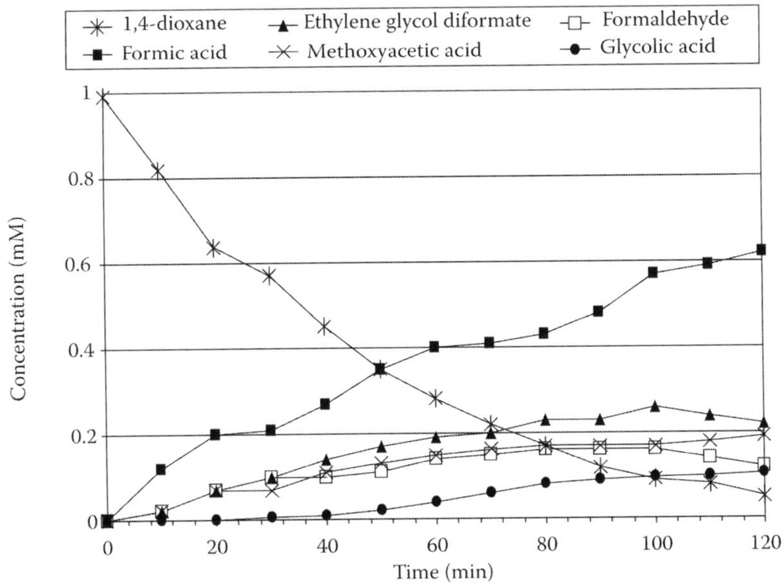

FIGURE 8.16 Formation of sonolytic by-products of 1,4-dioxane over time. (From Beckett, M.A. and Hua, I., 2000, Elucidation of the 1,4-dioxane decomposition pathway at discrete ultrasonic frequencies. *Environmental Science and Technology* 34(19): 3944–3953.)

Dietrich (2016) tested the effect of ozone activation by sonolysis. The study looked at ozone doses, operating pressures, ultrasonic intensity, pH effects and radical scavenging, and then attempted to model the combined process. Results indicated there was a synergistic effect to combining ozone and ultrasonic treatment: in the combined technology tests removal rates exceeded the sum of removals that would be expected by applying the technologies separately. Increasing the pressure proportionally increased the influent ozone concentration and boosted consumption of ozone. Increasing pH improved ozone transfer but appears to compete with sonolysis effects and only resulted in a small improvement in performance. Increasing sonolytic intensity in the presence of ozone increased performance significantly. The technology was found to hold great promise as an effective means of treating recalcitrant contaminants.

8.7.10 Electrochemical Oxidation

Electrochemical oxidation is treatment of organic contaminants by passing an electrical current through the water via an anode and a cathode. At the anode, oxidizing conditions are created, and oxidizers may also be created. Electrochemical treatment proceeds via direct and indirect anodic oxidation. In direct electrolysis, contaminants are adsorbed onto and degraded directly at the electrode while in indirect electrolysis, contaminants are degraded in the bulk liquid in reactions with oxidizing agents (i.e., hydroxyl radicals) formed at the electrode (Radjenovic and Sedlak, 2015). Hydroxyl radicals formed during this process are capable of oxidizing organic compounds, such as 1,4-dioxane.

A variety of materials have been used as anodes for electrochemical oxidation. Most research on 1,4-dioxane has been conducted using a boron-doped diamond (BDD) (Choi et al., 2010; DeClercq et al. 2010; Barndõk et al. 2014b). Wang et al. (2015) tested platinum/titanium anodes, which alone show limited destruction of 1,4-dioxane (Figure 8.17) (Choi et al., 2010; Wang et al., 2015). In this

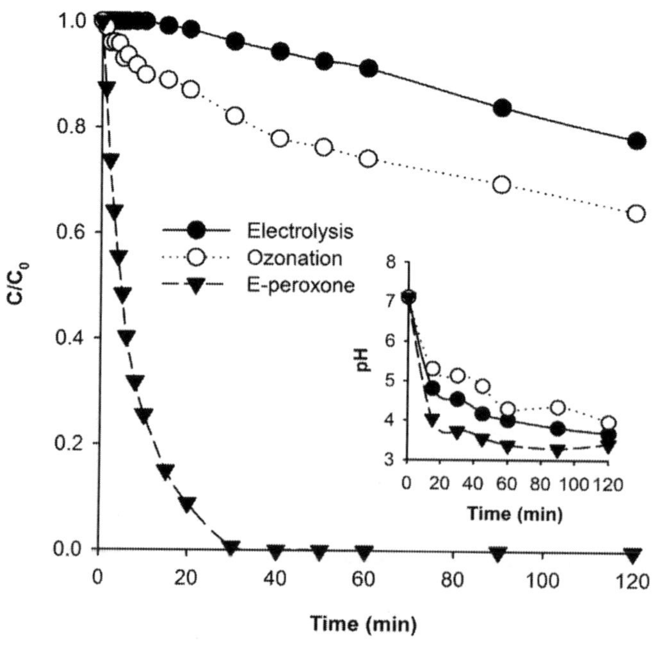

FIGURE 8.17 Degradation of 1,4-dioxane by ozonation electrolysis and E-peroxone processes. Reprinted (adapted) with permission from Wang, H., Bakheet, B., Yuan, S., Li, X., and Yu, G., Murayama, S., and Wang, Y., 2015, Kinetics and energy efficiency for the degradation of 1,4-dioxane by electro-peroxone process. *Journal of Hazardous Materials* 294: 90–98. Copyright 2015 Elsevier.

research, perozone chemistry (see Section 8.7.1) was incorporated with the platinum/titanium electrode, demonstrating significant improvement of the 1,4-dioxane destruction (Wang et al., 2015).

According to Wang et al. (2015), during the E-peroxone process, pollutants are primarily degraded by aqueous OH• generated from the cathode induced peroxone reaction, while anodic oxidation plays a minor role in pollutant degradation and may provide a cost-effective and energy-efficient alternative to 1,4-dioxane wastewater treatment compared to BDD-based systems.

BDD electrodes are effective without the need for additional chemicals. Choi et al. (2010) provides a comparison of platinum and BDD electrodes for removal of 1,4-dioxane (Figure 8.18).

Because of the high electrochemical activity of the BDD electrode, which generates strong oxidants on its surface, it showed a higher current efficiency than Pt-coated steel electrode in removal of 1,4-dioxane, with a removal efficiency as high as 97% (Choi et al., 2010). Similar research supports the use of BDD electrodes. At a current density of about 32 mA/cm^2 the oxidation process is completely controlled by mass transfer and no intermediates are formed, and removal of 92% can be realized with an electrical energy consumption of 7 Ah/L (DeClercq et al., 2010). Similarly, using BDD electrodes, Barndõk et al. (2014b) showed that up to 98% of the COD was removed in the electrooxidation of the industrial wastewater effluent containing 1,4-dioxane, and required 114 kWh/kg COD removed.

A series of publications from the Colorado State University (CSU) further explores electrochemical treatment of 1,4-dioxane. Jasmann et al. (2016) describe electrolytic oxidation of 1,4-dioxane using a pelletized titanium oxide catalyst in flow-through reactors. The paper describes the pellet-making process and reactor design, including how light was excluded from the experiment. Experiments included varying the sodium sulfate electrode concentration and the applied voltage. Reactors were designed to remove from 50% to 70% of the 1,4-dioxane. Removal rates were generally in the target range, with removal rates as high as 85%. Increasing the electrolyte concentration improved removal, but adding the titanium catalyst also improved performance, especially at the lower electrolyte concentrations. Results also helped elucidate the electrochemical degradation pathways for 1,4-dioxane and quantify mineralization of 1,4-dioxane in the process.

In a subsequent study, the CSU group studied the addition of a biological process to the electrochemical process (Jasmann et al., 2017). The electrochemical reactors were bioaugmented with *Pseudonocardia dioxanivorans* CB1190, a known 1,4-dioxane degrader. Biological degradation

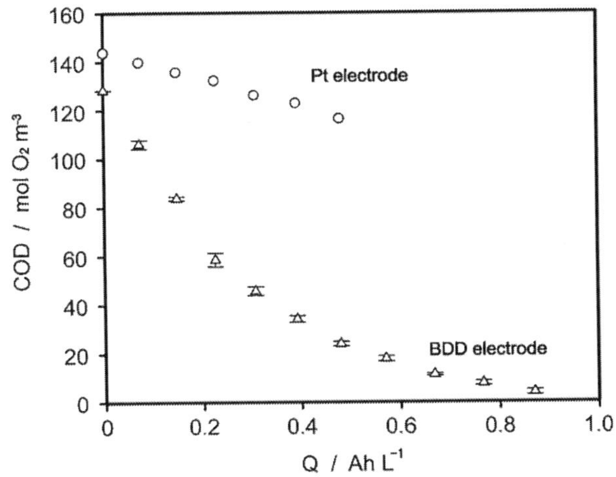

FIGURE 8.18 Comparison of COD reduction at BDD or Pt-coated electrodes as a function of specific charge. Reprinted (adapted) from Choi, J.Y., Lee, Y., Shin, J., and Yang, J., 2010, Anodic oxidation of 1,4-dioxane on boron-doped diamond electrodes for wastewater treatment. *Journal of Hazardous Materials* 179:762–768. Copyright 2010 Elsevier.

rates were 2.5 times greater in the electrochemical process compared to a control, apparently because of the creation of oxygen at the anode. The microbial population was adversely impacted at higher voltages.

Currently, electrochemical processes are becoming more commonly implemented in the field and systems are now commercially available. Provectus is now marketing a down-hole system (Provectus, 2019) that can be powered by a solar panel as both a creator of hydroxyl radicals and free oxygen. Provectus also holds a recent patent for a BDD electrode process (Provectus, 2018).

8.7.11 Chemical Oxidation Summary

Chemical oxidation effectively destroys 1,4-dioxane, and the degradation mechanisms and kinetics are well understood. *Ex situ* treatment systems utilizing several of the technologies described in this chapter are in place and functioning as designed, decreasing 1,4-dioxane concentrations to below clean-up or discharge standards at several sites. *In situ* applications of a variety of chemical oxidation technologies have been applied full scale and researchers are developing improvements to the processes. Rapid evolution of unconventional approaches, such as electrochemical treatment, holds great promise in improving treatment effectiveness and reducing costs.

8.8 MUNICIPAL WELL TREATMENT AND PURIFICATION OF RECYCLED WATER

Municipal well treatment solutions for reliable removal of 1,4-dioxane are similar to *ex situ* remediation options for site cleanup. Scaling treatment technologies to high flow rates (i.e., 500 to 3,000 gpm) eliminates some options, leaving AOP as the primary method for municipal well treatment. For example, synthetic media (Ambersorb 560) is an effective solution at moderate flow rates encountered for site remediation (i.e., 50–500 gpm), but hydraulic limitations render it noncompetitive at higher flow rates. Similarly, while GAC may be useful for domestic wells at low flow rates, scaling up to high flow rates is not feasible, owing to the high cost of frequent change-outs of carbon vessels.

AOP generates hydroxyl radicals that are effective at disassembling the 1,4-dioxane molecule, as described in Section 8.7. The primary variations of advanced oxidation employed for municipal well treatment include UV-oxidation (e.g., Trojan UV-Phox, Calgon RayOx and Sentinel), catalyzed photooxidation (Purifics PhotoCAT), and ozone-peroxidation (APT Water's HiPOx).

Upon discovering the presence of 1,4-dioxane in a municipal well and determining that alternative supply or blending are not viable options, water utilities must assess a number of logistical considerations for planning a treatment system capable of removing 1,4-dioxane, including:

- Land acquisition to secure space required for installing AOP system;
- Secure sufficient electrical supply for this potentially high-energy process;
- Integration of AOP technology with existing treatment, if present, for other contaminants (e.g., VOCs, nitrate, iron, etc.);
- Space and access for installation and future change-out of GAC vessels, which may be needed to remove residual peroxide and oxidation by-products;
- Secure storage of peroxide and/or ozone generator (peroxide is regulated by the Department of Homeland Security*), and
- Additional plumbing and discharge permits for filter backwash and other system requirements.

* The Department of Homeland Security regulates or restricts storage of hydrogen peroxide in quantities greater than 400 lbs and when hydrogen peroxide concentrations are greater than 34.5%. [6 CFR 27.203(c)]

Logistical challenges for installing AOP for 1,4-dioxane where other contaminants are already undergoing treatment may include:

- Retaining air stripping for VOC removal ahead of AOP;
- Removing dissolved organic carbon and other oxidant demand to improve AOP effectiveness;
- Conditioning water to improve UV-transmittance (for UV-Ox and catalyzed photooxidation);
- Scaling GAC for both peroxide quenching and removal of PFAS compounds (AOP is generally not effective for PFAS removal); and
- Accounting for presence of compounds whose oxidation by-products may further impair drinking water quality, such as bromide (oxidizes to bromate, a carcinogen), chromium, or arsenic.

Requirements and costs for installing AOP on municipal systems vary widely with water chemistry and site conditions. It is therefore critical to complete a phase of thorough pilot testing to obtain data needed for successful treatment system design. Pilot testing may be limited to laboratory column tests for where conditions are well understood and data from nearby treatment systems are available. For a new project with limited or no prior data, full field pilot testing using a mobile treatment system is recommended.

Designing AOP municipal well treatment systems for 1,4-dioxane removal must include a detailed assessment of water chemistry, which is an important determinant in the capital and operating costs. Certain compounds interfere with the transmittance of ultraviolet light. When present, these compounds must be addressed with filtration or pretreatment to ensure sufficient ultraviolet light transmittance (UVT) for efficient removal of target contaminants. Other compounds and water quality characteristics will limit treatment effectiveness. For example, bicarbonate alkalinity and natural organic matter (NOM) (measured as TOC) will also be oxidized by hydroxyl radicals produced in AOP, thereby limiting the destruction of the target contaminant. Influent water with a TOC content of 2.5 mg/L can cost in the range of \$0.38/1,000 gallons treated, whereas a slightly higher TOC, such as 3.3 mg/L, could increase costs to \$0.44/1,000 gallons treated. Adding a coagulation or other treatment step to reduce TOC can save \$75,000 per year in reduction of AOP operating cost, while only adding \$0.01 to \$0.02/1,000 gallons (Nowack et al., 2017).

Other water quality features that compete for hydroxyl radicals include sulfate, chloride, and nitrate. Where high nitrate concentrations are encountered (>22 mg/L), nitrate may accumulate on the GAC used for post-AOP peroxide quenching, and then slough off in unacceptably high concentrations following system shut-off and start-up.

While AOP has been proven effective at removing 1,4-dioxane, ozone peroxidation AOP (e.g., HiPOx) can also convert bromide to bromate, another carcinogen. When bromide is present, HiPOx and other ozone peroxidation systems can be optimized to mitigate bromate formation. Accordingly, water chemistry is an important determinant of the appropriate AOP technology design for treating 1,4-dioxane.

8.8.1 Comparison of 1,4-Dioxane Wellhead Treatment for Three Advanced Oxidation Technologies

AOP technologies were discussed in Section 8.7, and are relatively common in municipal systems. The advantages and limitations for two types of AOP are compared in Table 8.4, below. Each system generates hydroxyl radicals (OH•), which have a strong oxidation potential of 2.7 eV*, and destroy the 1,4-dioxane molecule in a ring-opening reaction.†

A 1-month treatability study conducted in 2016 for the City of New Brighton, Minnesota, compared Trojan UV-Phox with APT Water HiPOx for 1,4-dioxane removal from the same source

---

* The electron volt (eV) is a unit of energy equal to approximately 1.602×10^{-19} joule, the amount of energy gained by the charge of a single electron moved across an electric potential difference of one volt.

† See Figure 3.2 for the successive reactions leading to 1,4-dioxane destruction via UV-peroxidation.

TABLE 8.4
Comparison of Leading Advanced Oxidation Treatment Technologies for 1,4-Dioxane

Technology/Vendor	Principle	Advantages	Limitations
UV-Peroxide (e.g., Trojan UV-Phox, Calgon RayOx, and Calgon Sentinel)	UV + peroxide generates hydroxyl radicals (OH•)	Does not form bromate. Remote monitoring of automated peroxide dosing; automated lamp cleaning; stackable modular design minimizes footprint; 12,000 hour lamp life (~16 months); Low-pressure UV lamps with wipers generally resistant to fouling from natural minerals.*	Water may require pretreatment to improve UV-transmittance and remove natural constituents that compete for OH•‡; GAC post-treatment required to quench high levels of residual peroxide. High concentrations of iron may be problematic for lamp fouling, requiring pretreatment for iron removal. Requires delivery and on-site storage of hydrogen peroxide, which is regulated.
Ozone Peroxidation/ (e.g., APT Water HiPOx)	Ozone + peroxide (without UV) generates OH•	Does not require replacement of expensive UV lamps. Efficient mixing for optimal production of OH• (>90% MTE†); Reaction times adjustable to target contaminants; fully automated; built-in redundancy.	Generates bromate if not controlled‡; Requires delivery and on-site storage of hydrogen peroxide, which is regulated. Potential formation of nitrosamines or perfluorochemicals.§ May require very high ozone dosing for water with elevated total organic carbon.¶ HiPOx often requires higher peroxide dosing than Trojan UV-Phox, and produces higher peroxide residuals than Trojan UV-Phox.**

water, and evaluated important performance features (Ling and Macejkovic, 2018). The findings of the study are summarized in Table 8.5. Apparently, both systems will require post-treatment GAC polishing to remove residual peroxide and VOCs that are not destroyed by either system. The higher peroxide dosing and peroxide residual found for the HiPOx system will entail higher costs for peroxide, and more frequent GAC replacement for peroxide quenching. However, the lower operating cost per thousand gallons quoted for HiPOx may compensate for the increased peroxide demand.

8.8.2 Municipal Well Treatment Cost Engineering

A survey of treatment vendors' equipment costs can provide a partial picture of the overall cost to address 1,4-dioxane in municipal wells. Costs for treatment equipment among the three leading variations of AOP for 1,4-dioxane are similar, and in the range of \$300,000 to \$500,000 for

* Drinking Water UV Disinfection, Carollo Engineers. https://www.carollo.com/sites/default/files/docs/UVDisinfectionWater.pdf

† MTE = Mass Transfer Efficiency

‡ Most HiPOx installations have successfully limited bromate formation using APT Water's patented bromate control technology.

§ PFAS precursors can be oxidized to perfluoroalkyl acids (i.e., PFOA). This occurs during the ozonation of wastewater (Pisarenko, et al. 2015, Environmental Science: Water Research and Technology, 1:668–678).

¶ Craig, K., 2018. Breakthroughs in 1,4-Dioxane Treatment: Cost Effectively Treating 1,4-Dioxane in a Chemically Complex Water Stream. 14th Annual Gatekeeper Regulatory Roundup, March 22nd, 2018. Geosyntec.

** Barr Engineering, 2016. 1,4-Dioxane Response: Minneapolis Interconnection Status AOP Technology Evaluation Results and Recommendations. Presentation to New Brighton City Council Meeting, August 23, 2016.

TABLE 8.5
Performance Comparison of Trojan UV-Phox vs APT Water HiPOx for 1,4-Dioxane

Performance Attribute	Trojan UV-Phox	APT Water HiPOx
1,4-DX removal to ≤0.1 ug/L	100%	100%
Peroxide dosing	8 mg/L	23 mg/L
Peroxide residual in effluent	4 mg/L	16 – 18 mg/L
Bromate formation	None	Some > threshold
Removal of other VOCs	Incomplete	Incomplete

1,000 gpm flow and 1,4-dioxane concentrations less than 10 μg/L. However, equipment purchase is a minor portion of overall capital costs, which are dominated by construction costs and land acquisition. Municipal wells are often located in residential areas, where the aesthetic impact of constructing a treatment system may be unwelcome. Water utilities therefore find it necessary to design equipment buildings that fit into the neighborhood, sometimes developing structures that resemble homes. The overall construction cost, including land acquisition, chemical storage, and backwash or truck and crane access for GAC replacement, can equal or exceed ten times the cost of the treatment equipment.

Operating expenses are often measured by vendor quotes for electrical power and costs for consumables, presented as an annual cost or as $/1,000 gallons. However, comparison of these fixed operating costs alone do not reveal the total operating costs, which will also include the variable cost categories, such as labor, maintenance (including lamp replacement), laboratory analysis costs, and infrequent change-out of the GAC unit used for peroxide quenching. Variable operating costs generally exceed fixed operating costs, making comparison of fixed operating costs less meaningful. A few site-specific cost comparisons, inclusive of full construction costs and variable operating costs, can be found online, for example as presentations to city councils considering staff recommendations for contract award. Benchmarking costs is useful for budget preparation, but otherwise water utilities should expect that costs paid by other utilities may not accurately reflect the costs particular to their circumstances, for which there is no substitute for a well-specific technical assessment.

The reader will be most interested in cost comparisons among the leading municipal well treatment technologies. As described above, the site-specific costs comprise the largest portion of the overall costs, and publishing a cost comparison will only be useful for the timeframe for which the data were collected. Moreover, the initial capital and operating costs are only the beginning. A forecast is needed to estimate the duration of operations, so that a Net Present Value calculation can be made for the life cycle costs (LCC). Some larger water utilities have staff engineers who possess the knowledge and experience needed to develop an effective program for assessing and acquiring municipal well treatment for 1,4-dioxane removal. However, most utilities will prefer to retain the services of an experienced engineering consultant to complete the technology selection, develop and oversee the pilot testing, and manage the procurement, construction, testing, and permitting phases to ensure a successful project.

8.8.3 1,4-Dioxane Removal for Purification of Recycled Water

A growing number of water utilities, particularly in the southwestern United States, Australia, and island nations, have been investing in projects to purify recycled wastewater. Recycled water is viewed as a "drought-proof" source of water that can be leveraged to improve water supply reliability. Many sanitation districts and water districts have already developed recycled water programs to produce pathogen-free water suitable for irrigating parks, golf courses, street trees and landscaping, cemeteries, etc. Adding a purification step can produce potable water that meets all of the California Title 22 drinking water quality criteria; however, in the United States, the regulatory framework is still evolving and has not yet allowed direct distribution of purified recycled wastewater to be served as drinking water.

The strategy employed by some southwestern U.S. water districts is to develop the treatment capacity and infrastructure for "indirect potable reuse", aka IPR. Another term used to describe IPR projects is "groundwater replenishment and recharge projects" (GRRP). California Water Code §13561 requires multiple barrier treatment processes be used to purify recycled water for replenishment of a groundwater basin or an aquifer that has been designated as a source of water supply for a public water system. Additional treatment steps to render tertiary treated recycled water suitable for IPR, i.e., in compliance with California Title 22 drinking water quality criteria, include microfiltration (MF), RO, and UV-oxidation.

Recycled wastewater contains trace organic contaminants including nitrosamine compounds, pharmaceutical compounds and 1,4-dioxane. 1,4-Dioxane was present in tertiary treated recycled water in the range of 1 to 12.6 ppb, with a median concentration of 1.8 ppb, and a 90th percentile concentration of 2.6 ppb based on 75 California recycled water sample analyses between 2007 and 2009 (SWRCB, 2010), as illustrated in Figure 8.19.

Most commonly, the presence of 1,4-dioxane in recycled water is attributable to the discharge of soaps, detergents, shampoos, and sundries in which 1,4-dioxane is present as an impurity of alcohol- and alkyl-phenol ethoxylated surfactants. Because 1,4-dioxane is ubiquitous in tertiary treated recycled water, the state of California's regulations for IPR require that water purification systems

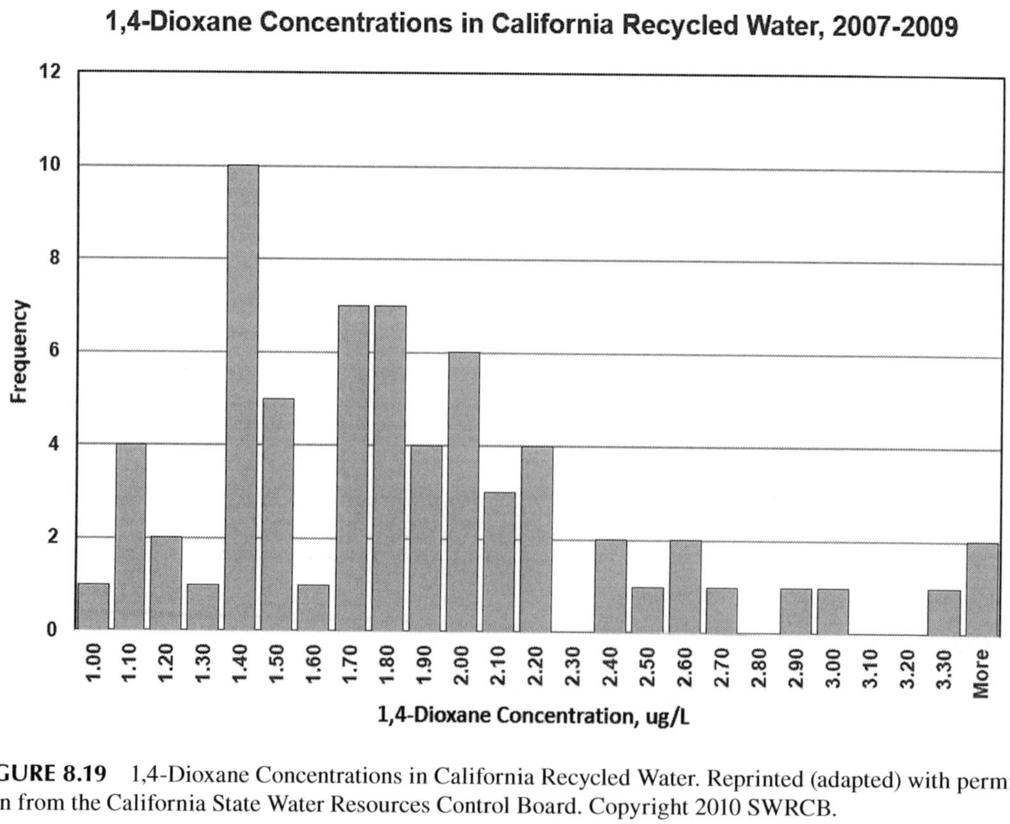

FIGURE 8.19 1,4-Dioxane Concentrations in California Recycled Water. Reprinted (adapted) with permission from the California State Water Resources Control Board. Copyright 2010 SWRCB.

achieve a 0.5-log reduction of 1,4-dioxane (regardless of whether 1,4-dioxane is present in wastewater). The 0.5-log reduction requirement ensures that 1,4-dioxane removal capacity is included in the design of water purification systems. Those water purification systems whose recycled water source does not include detectable 1,4-dioxane use 1,4-dioxane as a performance indicator to demonstrate compliance by spiking the recycled water to complete the test (TWDB, 2016). Similarly, water purification systems must demonstrate 1.2-log removal of nitrosodimethylamine (NDMA). Surrogate indicators can be used to gauge AOP performance. For example, 0.38-log removal of chloramines (mono-, di-, and trichloroamine, formed when HOCl reacts with ammonia) is a reliable indication that 1.2-log removal of NDMA has been achieved. Chloramines can be analyzed on-site with a spectrophotometer or colorimetric methods, allowing real-time feedback on system performance. Chloramine removal or similar surrogates may also be useful in reducing the cost of 1,4-dioxane monitoring, where a reliable relationship between removal of the surrogate and removal of 1,4-dioxane can be established.

Recycled wastewater presents a challenging matrix in which to treat trace organics with UV-oxidation. It generally has elevated total dissolved solids, in the range of 500 to 1,000 mg/L, and includes NOM. UV-absorbing compounds (nitrate, NOM) cause competition for UV light, while other water constituents (dissolved organic carbon [DOC], nitrite, carbonate/bicarbonate) act as OH•-radical scavengers (Martijn et al., 2010). The extent to which UV-absorbing and OH•-scavenging compounds remain present in recycled water treatment trains will determine the efficiency of UV-oxidation treatment for 1,4-dioxane removal and compliance with the 0.5-log reduction requirement.

RO is effective at removing bacteria, colloids, most salts and some polar organic compounds with molecular weights greater than 200 Daltons; however, non-ionic compounds and low molecular weight compounds such as NDMA and 1,4-dioxane exhibit only partial removal by RO (Drewes et al., 2011). RO membranes filter ionic and particulate contaminants through an engineered semi-permeable membrane, but it operates on the basis of osmotic pressure, which is absent for the dominant non-polar conformation of 1,4-dioxane. 1,4-Dioxane's molecular weight is 88.11 g/mol, which is well below the operating range of RO membranes (often cited as >200 g/mol; more reliably >300 g/mol) (Sedlak and Kavanaugh, 2006). While 1,4-dioxane may exist in a polar conformation, it is not ionically charged, and is therefore not amenable to complete removal by RO. Similarly, micro-filtration (MF) is ineffective at removing 1,4-dioxane because it is designed to remove particulate matter, while 1,4-dioxane is infinitely soluble and cannot be filtered out by MF. The MF and RO steps are nonetheless useful to support 1,4-dioxane treatment by UV-oxidation, as they remove UV-absorbing and OH•-scavenging compounds that are present in the mg/L range, allowing more efficient removal of 1,4-dioxane and other trace organic compounds in recycled water in the µg/L range.

AOP based on hydroxyl radicals may be more susceptible to OH•-scavengers than other radicals. For example, sulfate radicals (SO_4^-) generated in the UV-peroxodisulfate process react with Cl^- and $HCO3^-$ at rates three orders of magnitude lower than OH• (Velo-Gala et al., 2019). AOP for water purification continues to evolve with ongoing research into the multiple challenges of optimizing removal of trace organics from this complex matrix. Unlike UV-oxidation for municipal supply wells, purified recycled water used in IPR projects is not distributed directly as drinking water. Therefore, residual peroxide will be consumed as purified water is percolated or reinjected back into the aquifer by mineral and organic reactants.

Water utilities have been challenged to fine-tune their AOP treatment technologies to address multiple target contaminants. For example, the optimal dosage of H_2O_2 to remove NDMA may not be sufficient to achieve 0.5-log removal of 1,4-dioxane. Some utilities have found that sodium hypochlorite (NaOCl) is more effective than H_2O_2 for generating sufficient hydroxyl radicals to effectively remove 1,4-dioxane. For example, Figure 8.20 compares UV-NaOCl with UV-H_2O_2 for 1,4-dioxane removal from recycled water which has undergone MF and RO.

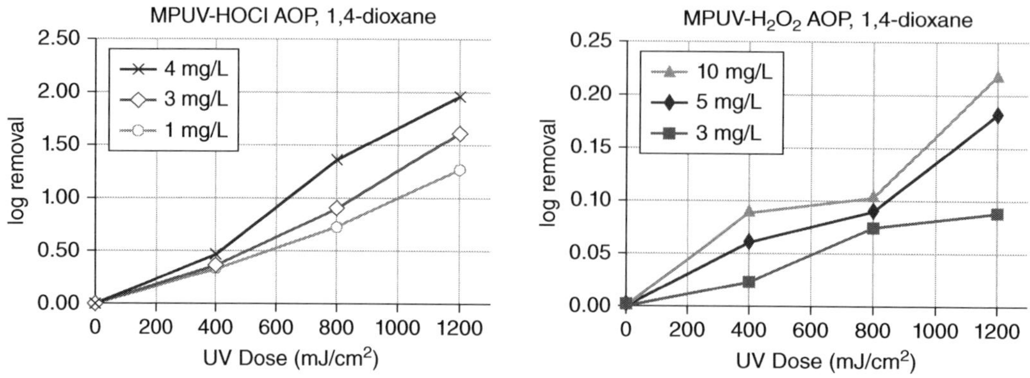

FIGURE 8.20 Comparison of UV-H_2O_2 with UV-HOCl for 1,4-Dioxane Removal from MF-RO Recycled Water. Reprinted (adapted) with permission from the Santa Clara Valley Water District. Copyright 2014 SCVWD.

REFERENCES

Adams, C.D., Scanlan, P.A., and Secrist, N.D., 1994, Oxidation and biodegradability enhancement of 1,4-dioxane using hydrogen peroxide and ozone. *Environmental Science and Technology* 28: 1812–1818.

Adamson, D.T., de Blanc, P.C., Farhat, S.K., and Newell, C.J., 2016, Implications of matrix diffusion on 1,4-dioxane persistence at contaminated groundwater sites. *Science of the Total Environment* 562: 98–107.

Adamson, D.T. Anderson, R.H., Mahendra, S., and Newell, C.J., 2015, Evidence of 1, 4-dioxane attenuation at groundwater sites contaminated with chlorinated solvents and 1, 4-dioxane. *Environmental Science and Technology* 49(11): 6510–6518.

Adamus, J.E., May, H.D., Paone, D.A., Evans, P.J., and Parales, R.E., 1995, United States Patent 5,474,934: Biodegradation of ethers. Assignee: Celgene Corporation, Warren, NJ.

Aitchison, E.W., Kelley, S.L., Alvarez, P.J.J., and Schnoor, J.L., 2000, Phytoremediation of 1,4-dioxane by hybrid poplar trees. *Water Environment Research* 72: 313–321.

Alther, G.R., 2006, Organoclays trap PNAHs and creosote in permeable barriers. In: B.M. Sass (Ed.), *Proceedings of the Fifth International Conference on Remediation of Chlorinated and Recalcitrant Compounds*, Paper L-71. Columbus, OH: Battelle Press.

Alther, G.R., 2010, The Versatility of Organoclays for Permeable Barriers. In: Proceedings of the Annual International Conference on Soils, Sediments, Water and Energy: Vol. 12, Article 22.

Alvarez, P., Li M., and Liu, Y., 2014, Final Report: Bench-scale 1,4-dioxane biodegradation assessment at a western Texas site. Rice University – Department of Civil and Environmental Engineering. November 5.

Andaluri, G. and Suri, R., 2017, Removal of 1,4-dioxane and volatile organic compounds from groundwater using ozone-based advanced oxidation process. Ozone: Science & Engineering, 39:6, 423-434, DOI: 10.1080/01919512.2017.1327802

Aoyagi, T., Morishita, F., Sugiyama, Y., Ichikawa, D., Mayumi, D., Kikuchi, Y., Ogata, A., Muraoka, K., Habe, H. and Hori, T., 2018, Identification of active and taxonomically diverse 1,4-dioxane degraders in a full-scale activated sludge system by high-sensitivity stable isotope probing. *The ISME Journal* 12: 2376–2388.

Arulazhagan, P., Yeom, I.T., Sivaraman, C., Srikanth, M., and Banu, J.R., 2013, Role of nutrients on biodegradation of 1,4-dioxane by a bacterial consortium enriched from industrial sludge. *Advances in Environmental Biology* 7(9): 2081–2090.

Azad, S., Bennett, D.W., and Tysoe, W.T., 2000, Investigation of the surface chemistry of crown ethers: The adsorption and reaction of 1,4-dioxane on palladium (111). *Surface Science* 464: 183–192.

Barndõk, H., Cortijo, L., Hermosilla, D., Negro, C., and Blanco, Á., 2014a, Removal of 1,4-dioxane from industrial wastewaters: Routes of decomposition under different operation conditions to determine the ozone oxidation capacity. *Journal of Hazardous Materials* 280: 340–347.

Barndõk, H., Hermosilla, D., Cortijo, L., Torres E., and Blanco, Á., 2014b, Electrooxidation of industrial wastewater containing 1,4-dioxane in the presence of different salts. *Environmental Science Pollution Research* 21: 5701–5712.

Barndõk, H., Blanco, L., Hermosilla, D., and Blanco, Á., 2016, Heterogeneous photo-Fenton processes using zero valent iron microspheres for the treatment of wastewaters contaminated with 1,4-dioxane. *Chemical Engineering Journal* 284: 112–121.

Beckett, M.A. and Hua, I., 2000, Elucidation of the 1,4-dioxane decomposition pathway at discrete ultrasonic frequencies. *Environmental Science and Technology* 34(19): 3944–3953.

Beckett, M.A. and Hua, I., 2003, Enhanced sonochemical decomposition of 1,4-dioxane by ferrous iron. *Water Research* 37(10): 2372–2376.

Bell, C.H., Gentile, M., Kalve, E., Ross, I., Horst, J. and Suthersan, S., 2019, *Emerging Contaminants Handbook*, p. 469. Taylor and Francis/CRC Press.

Berggren, D. and Hatton, J., 2016, Testing to evaluate the effectiveness of natural mineral activated sodium persulfate for treating chlorinated contaminants. Poster Presented at Remediation of Chlorinated and Recalcitrant Compounds, The Tenth International Conference, Monterey, California, May 21.

Block, P.A., Brown, R.A., and Robinson, D., 2004, Novel activation technologies for sodium persulfate in situ chemical oxidation. In: A.R. Gavaskar and A.S.C. Chen (Eds), *Proceedings of the Fourth International Conference on Remediation of Chlorinated and Recalcitrant Compounds*, Paper 2A-05. Columbus, OH: Battelle Press.

Boarer, J.E. and Milne, R., 2004, 1,4-Dioxane treatment in Mountain View, California. *Pollution Engineering* 36(7): 68–69.

Borchert, S., Depoy, A., and Watts, R., 2014, Treatability Study and Full-Scale ISCO for Mixed VOCs and 1,4-Dioxane in Low Permeable Formation, Ninth International Conference on Remediation of Chlorinated and Recalcitrant Compounds, sponsored by Battelle, Monterey, California. May.

Borchert, S., 2016, Personal communication.

Bowman, R.H., Miller, P., Purchase, M., and Schoellerman, R., 2007, Ozone-peroxide advanced oxidation water treatment system for treatment of chlorinated solvents and 1,4-dioxane. http://www.aptwater.com/assets/tech_papers/Paper-1.4Dioxane.pdf (accessed August 1, 2007).

Bowman, R.S., Sullivan, E.J., and Li, Z., 2000, Mechanisms of cationic, anionic, and nonpolar organic solute sorption by surfactant-modified zeolite. In: C. Colella and F.A. Mumpton (Eds), *Natural Zeolites '97: Occurrence, Properties, and Use*, pp. 287–297. Brockport, NY: International Committee on Natural Zeolites.

Brown, A.B., 2007, Personal communication.

Brubaker, S., Hatton, J., and Strunk, J., 2016, Field implementation of natural mineral activated sodium persulfate for 1,4-dioxane treatment. Tenth International Conference on remediation of Chlorinated and Recalcitrant Compounds, sponsored by Battelle, Palm Springs, California. May.

Burris, D.R., Dahlen, P.R., and Hinchee, R.E., 2018a, Soil gas sampling for 1,4-dioxane during heated soil vapor extraction. *Groundwater Monitoring and Remediation* 38: 85–89. doi: 0.1111/gwmr.12255.

Burris, D.R., Johnson, P.C., Hinchee, R.E., and Dahlen, P.R., 2018b, 1,4-Dioxane soil remediation using enhanced soil vapor extraction (XSVE): II. Modeling. *Groundwater Monitoring and Remediation* 38: 49–56. doi: 10.1111/gwmr.12277.

Cawfield, D.E., Lindstrom, F.T., and Boersma, L., 1990, *User's Guide to CTSPAC: Mathematical Model for Coupled Transport of Water, Solutes, and Heat in the Soil-Plant-Atmosphere Continuum*. Corvallis, OR: Oregon State University, Agricultural Experiment Station.

Chiang, D., Glover, E., Dahlgren, B., Peterman, J., Harrigan, J., and Woodward, D., 2007a, Approach, assessment and implementation of monitored natural attenuation at a 1,4-dioxane-contaminated site. In: A.R. Gavaskar and C.F. Silver (Eds), *Proceedings of the Ninth International In Situ and On-Site Bioremediation Symposium*, Paper M-03. Columbus, OH: Battelle Press.

Chiang, D., Glover, E., Dahlgren, B., Woodward, D., and Aitchison, E.W., 2007b, Phytoremediation of 1,4-dioxane for groundwater seep remediation. In: A.R. Gavaskar and C.F. Silver (Eds), *Proceedings of the Ninth International In Situ and On-Site Bioremediation Symposium*, Paper M-04. Columbus, OH: Battelle Press.

Chiang, D., Glover, E., Peterman, J., Harrigan, J., DiGuiseppi, W., and Woodward, D., 2008, Evaluation of natural attenuation at a 1,4-dioxane contaminated site. *Remediation* 19(1): 19–37.

Chiang, D., Zhang, Y., Glover, E., Harrigan, J., and Woodward, D., 2006, 1,4-Dioxane solute transport modeling in support of natural attenuation determination. In: B.M. Sass (Ed.), *Proceedings of the Fifth International Conference on Remediation of Chlorinated and Recalcitrant Compounds*, Paper G-70. Columbus, OH: Battelle Press.

Chiang, D.S., Mora, R., DiGuiseppi, W.H., Davis, G., Sublette, K., Gedalanga, P., and Mahendra, S., 2012, Characterizing the intrinsic bioremediation potential of 1,4-dioxane and trichloroethene using innovative environmental diagnostic tools. *Journal of Environmental Monitoring* 14: 2117–2326.

Chitra, S., Paramasivan, K., Cheralathan, M., and Kumar Sinha, P., 2012, Degradation of 1,4-dioxane using advanced oxidation processes. *Environmental Science Pollution Research* 19: 871–878.

Choi, J.Y., Lee, Y., Shin, J., and Yang, J., 2010, Anodic oxidation of 1,4-dioxane on boron-doped diamond electrodes for wastewater treatment. *Journal of Hazardous Materials* 179: 762–768.

Chu, M.Y.J., Bennett, P.J., Dolan, M.E., Hyman, M.R., Peacock, A.D., Bodour, A., Anderson, R.H., Mackay, D.M., and Goltz, M.N., 2018, Concurrent treatment of 1,4-dioxane and chlorinated aliphatics in a groundwater recirculation system via aerobic cometabolism. *Groundwater Monitoring and Remediation* 38(3): 53–64. Summer.

Cortés-Arriagada, D., Miranda-Rojas, S., Ortega, D.E., and Toro-Labbé, A., 2017. Oxidized and Si-doped graphene: emerging adsorbents for removal of dioxane. *Physical Chemistry Chemical Physics* 2017 Jul 21;19(27): 17587–17597. Epub Jun 28.

Cowan, R.M., Morin, M.D., Sock, S.M., Grady, C.P.L., Jr., and Hughes, T.A., 1994, Isolation and identification of microorganisms responsible for 1,4-dioxane mineralization. In: *Abstracts of the 94th General Meeting of the American Society of Microbiology*, Abstract Q-444, Washington, D.C.: American Society of Microbiology.

Cronk, G., 2006, Effective treatment of recalcitrant compounds using activated sodium persulfate. In: B.M. Sass (Ed.), *Proceedings of the Fifth International Conference on Remediation of Chlorinated and Recalcitrant Compounds*, Paper D-81. Columbus, OH: Battelle Press.

Cronk, G., 2008, Case study: Comparison of multiple activation methods for sodium persulfate ISCO treatment. In: B.M. Sass (Ed.), *Proceedings of the Sixth International Conference on Remediation of Chlorinated and Recalcitrant Compounds*. Columbus, OH: Battelle Press.

Curry, M.A., 2009, 1,4-Dioxane removal from groundwater using point-of-entry water treatment techniques. *Master's Thesis*. University of New Hampshire. 2009.

Davis, H., 2007, Personal communication.

DeClercq, J., Van de Steene, E., Verbeken, K., and Verhaege, M., 2010, Electrochemical oxidation of 1,4-dioxane at boron-doped diamond electrode. *Journal of Chemical Technology and Biotechnology* 85: 1162–1167.

Deng, D., Li, F., Wu, C., and Li, M., 2018, Synchronic biotransformation of 1,4-dioxane and 1,1-dichloroethylene by a gram-negative propanotroph *azoarcus sp.* DD4. *Environmental Science and Technology Letters* 5 (8): 526–532.

Dettmer, A., Ball, R., Boving, T.B., Khan, N.A., Schaub, T., Sudasinghe, N., Fernandez, C.A., and Carroll, K.C., 2017, Stabilization and prolonged reactivity of aqueous-phase ozone with cyclodextrin. *Journal of Contaminant Hydrology* 196: 1–9.

Dietrich, M.T., 2016, Combined ozone and ultrasound process for the destruction of 1,4-dioxane in continuous flow reactor. Dissertation, Temple University.

DiGuiseppi, W.H. and Whitesides, C., 2007, Treatment options for remediation of 1,4-dioxane in groundwater. *Environmental Engineer* 43(2): 36–41.

DiGuiseppi, W.H., Walecka-Huchison, C., and Hatton, J.W., 2016, 1,4-Dioxane treatment technologies. *Remediation* 27(1): 71–92. http://onlinelibrary.wiley.com/doi/10.1002/rem.21498/full

Dombrowski, P.M., Weir, B.A., Kelly, K.M., and Brown, J., 2010, Fast-track remedial design of full-scale ISCO application using pilot scale testing and field screening parameters. *Proceedings of the Annual International Conference on Soils, Sediments, Water and Energy*: Vol. 15, Article 16.

Drewes, J., Dickenson, E., and Snyder, S., 2011, Development of surrogates to determine the efficacy of groundwater recharge systems for the removal of trace organic chemicals. *WateReuse Research Foundation*. Alexandria, Virginia, 137 pages.

Earth Tech, Inc., 2004, Technology evaluation for treatment of 1,4-dioxane in groundwater. Prepared for the Air Force Center for Environmental Excellence, December 2004.

Earth Tech, Inc., 2008, Air Force Plant 44 semi-annual report. Prepared for the Air Force Center for Engineering and the Environment, May 2008.

Eberle, D., Ball, R., and Boving, T.B., 2016, Peroxone activated persulfate treatment of 1,4-dioxane in the presence of chlorinated solvent co-contaminants. *Chemosphere* 144, 728–735.

Edwards, M.R.A., Hetu, M., Columbus, A.S., and Lefebvre, D.D., 2011, The effect of ethylene glycol on the phytovolatilization of 1,4-dioxane. *International Journal of Phytoremediation* 13: 702–716.

Elgressy, N.E., 2018, US Patent US 9,975,156 B2. Breakdown of fuel components and solvents in groundwater and contaminated soil, May 22.

Evans, P., Dugan, P., and Crimi, M., 2014, In situ chemical oxidation of dioxane using slow-release chemical oxidant candles. Presented at the California the Emerging Contaminants Symposium. Groundwater Resources Association of California, Concord, California, February 5.

Evans, P., Hooper, J., Lamar, M., Nguyen, D., Dugan, P., Crimi, M., and Ruiz, N., 2018, Sustained In situ Chemical Oxidation (ISCO) of 1,4-Dioxane and Chlorinated VOCs Using Slow-release Chemical Oxidant Cylinders. Environmental Security Technology Certification Program (ESTCP) Cost and Performance Report Project ER-201324, July 2018. US Department of Defense.

Evans, P.J., Parales, R.E., and Parales, J.V., 2006, Bioreactors for treatment of 1,4-dioxane. In: *18th Symposium in Groundwater Resources Association Series on Groundwater Contaminants: Emerging Contaminants in Groundwater: A Continually Moving Target*, June 7–8, 2006. Concord, CA. (Housed at the University of California Water Resources Archives, Berkeley, CA.)

Fam, S.A., Fogel, S., and Findlay, M., 2005, Rapid degradation of 1,4-dioxane using a cultured propanotroph. In: B.C. Alleman and M.E. Kelley (Eds), *Proceedings of the Eighth International In Situ and On-Site Bioremediation Symposium*, Paper I-10. Columbus, OH: Battelle Press.

Fang, G., Gao, J., Liu, C., Dionysiou, D.D., Wang, Y., and Zhou, D., 2014, Key role of persistent free radicals in hydrogen peroxide activation by biochar: Implications to organic contaminant degradation. *Environmental Science and Technology* 48(3): 1902–1910.

Fang, G., Liu, C., Gao, J., Dionysiou, D.D., and Zhou, D., 2015, Manipulation of persistent free radicals in biochar to activate persulfate for contaminant degradation. *Environmental Science and Technology* 49(9): 5645–5653.

Favara, P., Tunks, J., Hatton, J., and DiGuiseppi, W.H., 2016, Sustainable remediation considerations for treatment of 1,4-dioxane treatment in groundwater. *Remediation* 27(1): 133–158. http://onlinelibrary.wiley.com/doi/10.1002/rem.21501/full

Félix-Navarro, R.M., Lin-Ho, S.W., Barrera-Díaz, N., and Pérez-Sicairos, S., 2007, Kinetics of the degradation of 1,4-dioxane using persulfate. *Journal of the Mexican Chemical Society* 1(2): 67–71.

Feng, Y., Li, H., Lin, L., Kong, L., Li, X., Wu, D., Zhao, H., and Shih, K., 2018, Degradation of 1,4-dioxane via controlled generation of radicals by pyrite activated oxidants: Synergistic effects, roles of disulfides, and activation sites. *Chemical Engineering Journal* 336: 416–426.

Ferro, A., Tammi, C.E., Hodgen, J., and LaRue, J., 2005, Treatment of recovered groundwater containing 1,4-dioxane: Year-round phytovolatilization by irrigating stands of deciduous and coniferous trees. In: *The Annual Conference on Soils*, Sediments and Water. Amherst, Massachusetts: University of Massachusetts. http://www.onlybusiness.com/Member/AEHSFoundation/Images/ImageGallery/ECC/Proceedings/2005-Proceedings.pdf

Ferro, A.M., Kennedy, J., and LaRue, J.C., 2013, Phytoremediation of 1,4-dioxane-containing recovered groundwater. *International Journal of Phytoremediation* 15: 911–923.

Festger, A., 2007. Personal communication.

Fleming, E.C., Zappi, M.E., Toro, E., Hernandez, R., and Myers, K., 1997, Laboratory Assessment of Advanced Oxidation Processes for Treatment of Explosives and Chlorinated Solvents in Groundwater from the Former Nebraska Ordnance Plant. Vicksburg, MS: U.S. Army Engineer Waterways Experiment Station.

FMC, 2007, In-situ chemical oxidation with Klozur™ activated persulfate: Co-mingled plume of chlorinated solvents and 1,4-dioxane. http://www.envsolutions.fmc.com/Klozur/ResourceCenter/tabid/356/Default.aspx (accessed August 14, 2007).

Fortune, A., 2015, Bench Scale Thermal Treatment of 1,4-Dioxane, Blog posted on June 18, 2015. http://terratherm.com/blog/?p=1715

Fotta, M.E., 2012, Effect of granular activated carbon type on adsorber performance and scale-up approaches for volatile organic compound removal. *MS Thesis*. North Carolina State University.

Gatliff, E., Linton, P.J., Riddle, D.J., and Thomas, P., 2013, Engineered phytoremediation of recalcitrant VOCs in deep groundwater horizons. 2013 20th Annual International Petroleum Environmental Conference, San Antonio, TX. November 12–14.

Gedalanga, P., Madison, A., Miao, Y., Richards, T., Hatton, J., DiGuiseppi, W.H., Wilson, J., and Mahendra, S., 2016, A multiple lines of evidence framework to evaluate intrinsic biodegradation of 1,4-dioxane. *Remediation* 27(1): 93–114.

Gestler, R., 2016, Case studies for phytoremediation of 1,4-dioxane contaminated aquifers. *Emerging Contaminants Summit, Westminster, Colorado*, March 1–3. Troy, MI: BNP Media.

Ghosh, P., Samanta, A., and Ray, S., 2011, Oxidation kinetics of degradation of 1,4-dioxane in aqueous solution by H_2O_2/Fe (II) system. *Journal of Environmental Science and Health* 45: 395–399.

Global BioSciences, Inc., 2007, Enhancing natural attenuation: 1,4-Dioxane degradation using butane biostimulation. http://www.globalbiosciences.com/downloads/Natural_Attenuation.pdf (accessed October 26, 2007).

Hatzinger, P.B., Banerjee, R., Rezes, R., Streger, S.H., McClay, K., and Schaefer, C.E., 2017, Potential for cometabolic biodegradation of 1,4-dioxane in aquifers with methane or ethane as primary substrates. Biodegradation; published online October 11, 2017.

He, Ya, Mathieu, J., da Silva, M.L.B., Li, M., and Alvarez, P.J.J., 2018, 1,4-Dioxane-degrading consortia can be enriched from uncontaminated soils: Prevalence of Mycobacterium and soluble di-iron monooxygenase genes. *Microbial Biotechnology* 11(1): 189–198.

He, Y., Mathieu, J., Yang, Y., Yu, P., da Silva, M.L.B., and Alvarez, P.J.J., 2017, 1,4-Dioxane biodegradation by mycobacterium dioxanotrophicus PH-06 is associated with a group-6 soluble di-iron monooxygenase. *Environmental Science and Technology Letters* 4: 494–499.

Hentz, R.R. and Parrish, C.F., 1971, Photolysis of gaseous 1,4-dioxane at 1470 A. *Journal of Physical Chemistry* 75(25): 3899–3901.

Hill, R.R., Jeffs, G.E., and Roberts, D.R., 1996, *Photocatalytic Degradation of 1,4-Dioxane in Aqueous Solution*. Milton Keynes, UK: The Open University, Department of Chemistry, Walton Hall.

Hinchee, R., 2017, 1,4-Dioxane Remediation by Extreme Soil Vapor Extraction (XSVE) Screening-Level Feasibility Assessment and Design Tool in Support of 1,4-Dioxane Remediation by Extreme Soil Vapor Extraction (XSVE). ESTCP Project ER-201326. ESTCP.

Hinchee, R.E., Dahlen, P.R., Johnson, P.C., Burris, D.R., 2018, 1,4-Dioxane soil remediation using enhanced soil vapor extraction: I. Field demonstration. *Groundwater Monitoring and Remediation* 38: 40–48. doi:10.1111/gwmr.12264.

Hua, I., 2000, An investigation of homogeneous and heterogeneous sonochemistry for destruction of hazardous waste. Final Report 09/16/1996–09/14/2000. Department of Energy, Office of Scientific and Technical Information. http://www.osti.gov/bridge/servlets/purl/13487-CVTUXk/webviewable/13487.PDF (accessed September 4, 2007).

Huang, H., Shen, D., Li, N., Shan, D., Shentu, J., and Zhou, Y., 2014, Biodegradation of 1,4-dioxane by a novel strain and its biodegradation pathway. *Water Air Soil Pollution* 225: 2135.

Huling, S.G. and Pivetz, B.G., 2006, In-situ chemical oxidation. EPA Engineering Issue Paper EPA/600/R-06/072.

Innovative Engineering Solutions, Inc. (IESI), 2008, Microcosm testing and design of a bioenhancement remediation system, Michigan (1,4 dioxane). http://www.iesionline.com/projects/Michigan.pdf (accessed August 24, 2008).

International Organization for Standardization (ISO), 2005, Space environment (natural and artificial)—process for determining solar irradiances. ISO TC 20/SC 14, December 1, 2005. http://www.spacewx.com/pdf/ISO_DIS_21348_E_revB.pdf (accessed September 21, 2008).

ITRC, 2005, *Technical and Regulatory Guidance for In Situ Chemical Oxidation of Contaminated Soil and Groundwater*, 2nd ed. ISCO-2. January 2005. Washington, D.C.: Interstate Technology & Regulatory Council, In Situ Chemical Oxidation Team. Available on the Internet at http://www.itrcweb.org.

ITRC, 2009, *Phytotechnology Technical and Regulatory Guidance and Decision Trees*, Revised. PHYTO-3. February 2009. Washington, D.C.: Interstate Technology & Regulatory Council, Phytotechnologies Team, Tech Reg Update. www.itrcweb.org

Isotec, 2008, ISOTEC technology effectively treats 1,4-dioxane. https://isotec-inc.com/pdfs/ISOTEC_Treats_1-4Dioxane.pdf

Jackson, L.E. and Lemke, L.D., 2019, Evidence for natural attenuation of 1,4-dioxane in a glacial aquifer system. *Hydrogeology Journal*, https://doi.org/10.1007/s10040-019-02028-6, published online October 2, 2019 (accessed November 1, 2019).

Jasmann, J.R., Borch, T., Sale T.C., and Blotevogel, J., 2016, Advanced electrochemical oxidation of 1,4-dioxane via dark catalysis by novel titanium dioxide (TiO2) pellets. *Environmental Science & Technology* 50: 8817–8826.

Jasmann, J.R., Gedalanga, P.B., Borch, T., Mahendra, S., and Blotevogel, J., 2017, Synergistic treatment of mixed 1,4-dioxane and chlorinated solvent contaminations by coupling electrochemical oxidation with aerobic biodegradation. *Environmental Science and Technology* 51: 12619–12629.

Johns, M.M., Marshall, W.E., and Toles, C.A., 1998, Agricultural by-products as granular activated carbons for absorbing dissolved metals and organics. *Journal of Chemical Technology and Biotechnology* 71: 131–140.

Kambhu, A., Gren, M., Tang, W., Comfort, S., and Harris, C., 2017, Remediating 1,4-dioxane-contaminated water with slow-release persulfate and zerovalent iron. *Chemosphere* 175: 170–177.

Kelley, S.L., Aitchison, E.W., Deshpande, M., Schnoor, J.L., and Alvarez, P.J.J., 2001, Biodegradation of 1,4-dioxane in planted and unplanted soil: Effect of bioaugmentation with *Amycolata* sp. CB1190. *Water Research* 35(16): 3791–3800.

Kerfoot, W.B., 2008, In situ 1,4-dioxane remediation at HVOC sites. In: B.M. Sass (Ed.), *Proceedings of the Sixth International Conference on Remediation of Chlorinated and Recalcitrant Compounds*. Columbus, OH: Battelle Press.

Kerfoot, W.B. and Brolowski, A., 2006, In-situ 1,4 dioxane remediation in HVOC sites. Paper presented at The Annual Conference on Soils, Sediments and Water, University of Massachusetts, Amherst. http://www.umasssoils.com/abstracts2005/Tuesday/ozone.htm (accessed August 15, 2007).

Kessel, L., Hajali, P., Nielsen, B., and Olivo, K., 2005, Competitive contaminant destruction during ozone-peroxide chemical oxidation: Implications for in-situ remediation of 1,4-dioxane. In: 15th Annual West Coast Conference on Soil, Sediment and Water of the Association for Environmental Health and Sciences (AEHS), San Diego, CA.

Khan, N.A., Johnson, M.D., Kubicki, J.D., Holguin, F.O., Dungan, B., and Carroll, K.C., 2019, Cyclodextrin-enhanced 1,4-dioxane treatment kinetics with TCE and 1,1,1-TCA using aqueous ozone. *Chemosphere* 219: 335–344.

Kim, C.G., Seo, H.J., and Lee, B.R., 2006, Decomposition of 1,4-dioxane by advanced oxidation and biochemical process. *Journal of Environmental Science and Health*, Part A 41: 599–611.

Kinsman, K. and Mohs, A., 2018, US Patent US9561530B1. Method for the in situ remediation of contaminants. Current Assignee: Orin Technologies LLC. January 30.

Kishimoto, N., Yasuda, Y., Oomura, T., and Ono, Y., 2005, Ozonation combined with electrolysis as an alternative technology for removing 1,4-dioxane from landfill leachates. In: R. Cossu and R. Stegmann (Eds), *Sardinia 2005, 10th International Waste Management and Landfill Symposium*, Cagliari, Italy. CISA, Environmental Sanitary Engineering Centre, Italy.

Kwon, S.C., Kim, J.Y., Yoon, S.M., Bae, W., Kang, K.S., and Rhee, Y.W., 2012, Treatment characteristic of 1,4-dioxane by ozone-based advanced oxidation processes. *Journal of Industrial and Engineering Chemistry* 18: 1951–1955.

Li, B. and Zhu, J., 2016, Simultaneous degradation of 1,1,1-trichloroethane and solvent stabilizer 1,4-dioxane by a sono-activated persulfate process. *Chemical Engineering Journal* 284: 750–763.

Li, M., Fiorenza, S., Chatham, J.R., Mahendra, S., and Alvarez, P.J.J., 2010, 1,4-Dioxane biodegradation at low temperatures in Arctic groundwater samples. *Water Research* 44: 2894–2900.

Li, M., Mathieu, J., Liu, Y., Van Orden, E.T., Yang, Y., Fiorenza, S., and Alvarez, P.J.J., 2013, The abundance of tetrahydrofuran/ dioxane monooxygenase genes (thmA/dxmA) and 1,4-dioxane degradation activity are significantly correlated at various impacted aquifers. *Environmental Science and Technology Letters* 1(1): 122–127.

Li, M., Van Orden, E.T., DeVries, D.J., Xiong, Z., Hinchee, R., and Alvarez, P.J., 2015, Bench-scale biodegradation tests to assess natural attenuation potential of 1, 4-dioxane at three sites in California. *Biodegradation* 26(1): 39–50.

Li, W., Patton, S., Gleason, J. M., Mezyk, S. P., Ishida, K. P., and Liu, H. 2018. UV photolysis of chloramine and persulfate for 1,4-1 dioxane removal in reverse osmosis permeate for potable water reuse. *Environmental Science and Technology 2018* 52 (11): 6417–6425.

Ling, A., and Macejkovic, J., 2018, New Brighton: Advanced Oxidation Process Pilot and Final Design for 1,4-Dioxane Removal, World Environmental and Water Resources Congress 2018, June 3–7, 2018, Minneapolis, Minnesota.

Lippincott, D., Streger, S.H., Schaefer, C.E., Hinkle, J., Stormo J., and Steffan, R.J., 2015, Bioaugmentation and propane biosparging for in situ biodegradation of 1,4-dioxane. *Groundwater Monitoring and Remediation* 35(2): 81–92. Spring.

Mahendra, S. and Alvarez Cohen, L., 2006, Kinetics of 1,4-dioxane biodegradation by monooxygenase-expressing bacteria. *Environmental Science & Technology* 40: 5435–5442.

Mahendra, S., Grostern, A., and Alvarez-Cohen, L., 2013, The impact of chlorinated solvent co-contaminants on the biodegradation kinetics of 1,4-dioxane. *Chemosphere* 91(1): 88–92.

Martijn, B., Fuller, A., Malley, J., and Kruithof, J., 2010. Impact of IX-UF pretreatment on the feasibility of UV/H2O2 treatment for degradation of NDMA and 1,4-dioxane. *Ozone: Science and Engineering* 32: 383–390.

Martineau, C., Prasad, H., and Roll, P., 2014, *Removal of 1,4-Dioxane from Water by Adsorption to Resins*. Worcester, MA: Worcester Polytechnic Institute.

Martins, K., DiGuiseppi, W.H., Hatton, J., and Suedbeck, B. 2014. Treatability testing outcome for 1,4-dioxane removal using resin adsorption. Internal Research Project. CH2M, Corvallis, Oregon. June 5.

Masten, S.J., 2004, Chemical oxidation treatment technologies: Gelman Sciences Dioxane Project, Ann Arbor, MI. Technical Outreach Services for Communities (TOSC). http://www.egr.msu.edu/tosc/gelman/Masten_Community_Meeting.ppt (accessed October 3, 2007).

Mehrvar, M., Anderson, W.A., and Moo-Young, M., 2000, Photocatalytic degradation of aqueous tetrahydrofuran, 1,4-dioxane, and their mixture with TiO_2. *International Journal of Photoenergy* 2: 67–80.

Miao, Y., Johnson, N.W., Heck, K., Guo, S., Powell, C.D., Phan, T., Gedalanga, P.B., Adamson, D.T., Newell, C.J., Wong, M.S., and Mahendra, S., 2018, Microbial responses to combined oxidation and catalysis treatment of 1,4-dioxane and co-contaminants in groundwater and soil. *Frontiers of Environmental Science and Engineering* 12: 2. https://doi.org/10.1007/s11783-018-1071-6

Myers, M.A., 2016, 1,4-Dioxane Biodegradation Using Bioaugmented Granular Activated Carbon. Thesis. UCLA.

Myers, M.A., Johnson, N.W., Marin, E.Z., Pornwongthong, P., Liu, Y., Gedalanga, P.B., and Mahendra, S., 2018, Abiotic and bioaugmented granular activated carbon for the treatment of 1,4-dioxane-contaminated water. *Environmental Pollution* 240: 916–924.

Nakajima, A., Sasaki, H., Kameshima, Y., Okada, K., and Harada, H., 2007, Effect of TiO_2 powder addition on sonochemical destruction of 1,4-dioxane in aqueous systems. *Ultrasonics Sonochemistry* 14(2): 197–200.

Nakajima, A., Tanaka, M., Kameshima, Y., and Okada, K., 2004, Sonophotocatalytic destruction of 1,4-dioxane in aqueous systems by HF-treated TiO_2 powder. *Journal of Photochemistry and Photobiology A: Chemistry* 167: 75–79.

Neely, J.W. and Isacoff, E.G., 1982, *Carbonaceous Adsorbents for the Treatment of Ground or Surface Waters*. New York, NY: Marcel Dekker, Inc.

Nowack, K., Rung, B., Ostapczuk, R., and Bryant, D., 2017. Surface Water Treatment of Poly- And Perfluoroalkyl Substances. NYS AWWA Annual Meeting, April 2017.

Oberle, D., Crownover, E., and Kluger, M., 2015, In situ remediation of 1,4-dioxane using electrical resistance heating. *Remediation* 25(2): 35–42.

Oberle, D., 2018, United States Patent 9,926,208: Miscible Solvent Treatment. Assignee: TRS China, LLC., Kalamazoo, MI. https://patents.google.com/patent/US9926208B1/en?oq=patent+9926208 (accessed November 11, 2018).

Odah, M.M., Powell, R., and Riddle, D.J., 2005, ART in-well technology proves effective in treating 1,4-dioxane contamination. *Remediation* 15(3): 51–64.

Orin Technologies, 2016, *BAM – Bioavailable Absorbent Media. Product Information Sheet*. Verona, WI: Orin Technologies, September 16.

Ouyang, Y., 2002, Phytoremediation: Modeling plant uptake and contaminant transport in the soil-plant-atmosphere continuum. *Journal of Hydrology* 266(1–2): 66–82.

Ouyang, Y., 2008, Modeling the mechanisms for uptake and translocation of dioxane in a soil-plant ecosystem with STELLA. *Journal of Contaminant Hydrology* 95(1–2): 17–29.

Ouyang, D., Yan, J., Qian, L., Chen, Y., Han, L., Su, A., Zhang, W., Ni, H. and Chen, M., 2017, Degradation of 1,4-dioxane by biochar supported nano magnetite particles activating persulfate. *Chemosphere* 2017 184: 609–617. May 30.

Pall Corporation, 2004, *Final Feasibility Study and Proposed Interim Response Plan for the Unit E Plume*. Ann Arbor, MI: Pall Life Sciences Corporation.

Pan, S. and Chen, H., 2006, Anaerobic degradation of 1,4-dioxane under humic-reducing or Fe(III)-reducing conditions. The 231stAmerican Chemical Society National Meeting, Abstract ENVR 144.

Parales, R.E., Adamus, J.E., White, N., and May, H.D., 1994, Degradation of 1,4-dioxane by an *actinomycete* in pure culture. *Applied and Environmental Microbiology* 60(12): 4527–4530.

Patt, T.E. and Abebe, H.M., 1995, United States Patent 5,399,495: Microbial degradation of chemical pollutants. Assignee: The Upjohn Company, Kalamazoo, MI. http://www.patentstorm.us/patents/5399495-fulltext.html (accessed September 12, 2007).

Patton, S., Li, W., Couch, K., Mezyk, S., Ishida, K., and Liu, H., 2016, Impact of the ultraviolet photolysis of monochloramine on 1,4-dioxane removal: New insights into potable water reuse. *Environmental Science and Technology Letters* 4: 26–30.

Patton, S., Romano, M., Naddeo, V., Ishida, K.P., and Liu, H., 2018, Photolysis of mono- and dichloramines in UV/hydrogen peroxide: Effects on 1,4-dioxane removal and relevance in water reuse. *Environmental Science and Technology* 52: 11720–11727.

Pignatello, J.J., Mitch, W.A., and Xu, W., 2017, Activity and reactivity of pyrogenic carbonaceous matter toward organic compounds. *Environmental Science and Technology* 51(16): 8893–8908.

Pintauro, P.N. and Jian, K., 1995, United States Patent 5,387,378: Integral asymmetric fluoropolymer pervaporation membranes and method of making the same. Assignee: Tulane University, New Orleans. http://www.patentstorm.us/patents/5387378-description.html (accessed October 12, 2007).

Polasko, A., Zulli, A., Gedalanga, P.B., Pornwongthong, P., and Mahendra, S., 2019, A mixed microbial community for the biodegradation of chlorinated ethenes and 1,4-dioxane. *Environmental Science and Technology Letters* 6(1): 49–54.

Pollution Engineering, 1999, UV oxidation provides the missing link. *Pollution Engineering* 31(7): 2.

Provectus, 2018, US Patent US 9,975,156 B2 Breakdown of Fuel Components and Solvents in Groundwater and Contaminated Soil. May 22, 2018.

Provectus, 2019, Provect-EBR® ISCO Generator Technical Data Sheet.

Radjenovic, J., and Sedlak, D.L., 2015, Challenges and Opportunities for Electrochemical Processes as Next-Generation Technologies for the Treatment of Contaminated Water, *Environ. Sci. Technol.* 2015, 49, 11292–11302.

Rohm and Haas, 1999, Product data sheet for AMBERSORBTM 563. Philadelphia, PA. June.

Rushing, J., Vikesland, R., Love, N., Mutuc, M., Chan, K., Casselberry, R., and Cichy, P., 2006, Evaluating in-situ chemical and biological treatment approaches for two chlorinated aliphatic ethers, BCEE and BCEM. In: B.M. Sass (Ed.), *Proceedings of the Fifth International Conference on Remediation of Chlorinated and Recalcitrant Compounds*, Paper D-62. Columbus, OH: Battelle Press.

Sabba, D. and Witebsky, S., 2003, Documenting the nature and extent of 1,4-dioxane at a solvent plume site. In: Symposium Resource Binder: 1,4-Dioxane and Other Solvent Stabilizer Compounds in the Environment, December 10, 2003. San Jose, CA, Groundwater Resources Association of California. (Housed at the University of California Water Resources Archives, Berkeley, CA.)

Sadeghi, V.M. and Gruber, D.J., 2007, In situ oxidation of 1,4-dioxane with ozone and hydrogen peroxide. Poster presentation at internal Environmental Technology and Management Seminar, Oakland, CA.

Sadeghi, V.M., Gruber, D.J., Yunker, E., Simon, M., and Gustafson, D., 2006, In situ oxidation of 1,4-dioxane with ozone and hydrogen peroxide. In: B.M. Sass (Ed.), *Proceedings of the Fifth International Conference on Remediation of Chlorinated and Recalcitrant Compounds*, Paper D-31. Columbus, OH: Battelle Press.

Sadeghi, V., Mora, R., Jacob, P., and Chiang, S.Y., 2016, Characterizing in situ methane-enhanced biostimulation potential for 1,4-dioxane biodegradation in groundwater. *Remediation* 27(1): 1–173. Winter.

Safarzadeh-Amiri, A., Bolton, J.R., and Cater, S.R., 1996, *Calgon Carbon Oxidation Technologies*. Ontario, Canada: Markham.

Santa Clara Valley Water District, 2014. Silicon Valley Advanced Water Purification Center Test Protocol Review. February 13, 2014.

Schreier, C.G., Sadeghi, V.M., Gruber, D.J., Brackin, J., Simon, M., and Yunker, E., 2006, In-situ oxidation of 1,4-dioxane. In: B.M. Sass (Ed.), *Proceedings of the Fifth International Conference on Remediation of Chlorinated and Recalcitrant Compounds*, Paper D-21. Columbus, OH: Battelle Press.

Sedlak, D. and Kavanaugh, M., 2006, Removal and destruction of NDMA and NDMA precursors during wastewater treatment. *WateReuse Foundation*, Alexandria, VA.

Sei, K., Miyagaki, K., Kakinoki, T., Fukugasako, K., Inoue, D., and Ike, M., 2013, Isolation and characterization of bacterial strains that have high ability to degrade 1,4-dioxane as a sole carbon and energy source. *Biodegradation* 24: 665–674.

Sekar, R., Taillefert, M., and DiChristina, T.J., 2016, Simultaneous transformation of commingled trichloroethylene, tetrachloroethylene, and 1,4-dioxane by a microbially driven fenton reaction in batch liquid cultures. *Applied and Environmental Microbiology* 82: 21. November.

Shen, W., Chen, H., and Pan, S., 2008, Anaerobic biodegradation of 1,4-dioxane by sludge enriched with iron-reducing microorganisms. *BioResources Technology* 99(7): 2483–2487.

Shirazi, F., 2007. Personal communication.

Smith, E.R. and Wojciechowski, M., 1937, Boiling point-composition diagram of the system dioxane-water. Research paper RP985. *Part of Journal of Research of the National Bureau of Standards* 18: 461–465.

Smitha, B., Dhanuja, G., and Sridha, S., 2006, Dehydration of 1,4-dioxane by pervaporation using modified blend membranes of chitosan and nylon 66. *Carbohydrate Polymers* 66(4): 463–472.

Son, H.S., Im, J.K., and Zoh, K.D., 2009, A fenton-like degradation mechanism for 1,4-dioxane using zero-valent iron (Fe0) and UV light. *Water Research* 43(5): 1457–1463. March.

Sörensen, M. and Weckenmann, J., 2006, 1,4-Dioxan: Was tun? [1,4-Dioxane: What to do?]. *Process*. http://www.process.vogel.de/index.cfm?pid=2511&pk=58794 (accessed October 10, 2007).

Stanfill, J.C., Koon, J., Shangraw, T., and Bollmann, D., 2004, 1,4-Dioxane bio-degradation bench study at the lowry landfill superfund site. In: *Proceedings of the Water Environment Federation, 10th Annual Industrial Wastes Technical and Regulatory Conference*, (14), pp. 420–433. WEF/A&WMA Industrial Wastes.

State Water Resources Control Board (SWRCB), 2010. Monitoring Strategies for Chemicals of Emerging Concern (CECs) in Recycled Water – Recommendations of a Science Advisory Panel.

Stefan, M.I. and Bolton, J.R., 1998, Mechanism of the degradation of 1,4-dioxane in dilute aqueous solutions using the UV/hydrogen peroxide process. *Environmental Science and Technology* 32: 1588–1595.

Suh, J.H., Kang, D.J., Park, J.D., and Lee, H.S., 2005, A study on the catalytic ozonation of 1,4-dioxane. In: *Proceedings from the 9th Russian-Korean International Symposium on Science and Technology* (KORUS), Novosibirsk, Russia.

Suh, J.H. and Mohseni, M., 2004, A study on the relationship between biodegradability enhancement and oxidation of 1,4-dioxane using ozone and hydrogen peroxide. *Water Research* 38(10): 2596–2604.

Tailored Lighting, Inc., 2007, Solux ultraviolet radiation primer. http://spectralfidelity.com/cgi-bin/tlistore/infopages/uv.html (accessed September 20, 2008).

Taylor, S.W., Lange, C.R., and Lesold, E.A., 1997, Biofouling of contaminated ground-water recovery wells: Characterization of microorganisms. *Ground Water* 35(6): 973–980.

Texas Water Development Board (TWDB), 2016, Direct Potable Reuse Monitoring: Testing Water Quality in a Municipal Wastewater Effluent Treated to Drinking Water Standards. Volume 1 of 2. Water Environment and Reuse Foundation.

Tratnyek, P., Powell, J., and Waldemer, R., 2007, FINAL REPORT – OBJECTIVE I: Improved Understanding of In Situ Chemical Oxidation Contaminant Oxidation Kinetics. SERDP Project ER-1289. December.

U.S. Environmental Protection Agency (USEPA), 1999, Use of Monitored Natural Attenuation at Superfund, RCRA Corrective Action, and Underground Storage Tank Sites. Washington, DC: U.S. Environmental Protection Agency Office of Solid Waste and Emergency Response Directive 9200.4–17P.

U.S. Environmental Protection Agency (USEPA), 2006, Treatment Technologies for 1,4-Dioxane: Fundamentals and Field Applications. Washington, DC: U.S. Environmental Protection Agency Office of Solid Waste and Emergency Response, EPA-542-R-06–009.

U.S. Environmental Protection Agency (USEPA), 2007, Superfund and Technology Liaison Region 9 Newsletter, 39th Edition. Spring 2007. San Francisco, CA: Office of Research and Development. http://www.epa.gov/osp/hstl/sn_spring2007.pdf (accessed October 23, 2007).

Uppal, O.J., Kaur, K., Najib, N., Abrams, S.H., and Marley, M.C., 2014, In-Situ Air Stripping Bench-Scale Testing for Treatment of 1,4-Dioxane in Groundwater. Battelle's Ninth International Conference on Remediation of Chlorinated and Recalcitrant Compounds, Monterey, California, May 19–22, 2014.

Vainberg, S., McClay, K., Masuda, H., Root, D., Condee, C., Zylstra, G.J., and Steffan, R.J., 2006, Biodegradation of ether pollutants by *Pseudonocardia* sp. strain ENV478. *Applied and Environmental Microbiology* 72(8): 5218–5224.

Velo-Gala, I., Farré, M.J., Radjenovic, J., and Gernjak, W., 2019, N-Nitrosodimethylamine (NDMA) degradation by the UV/peroxodisulfate process. *Environmental Science and Technology Letters* 6(2): 106–111. January 2019. doi: 10.1021/acs.estlett.8b00667.

Vescovi, T., Coleman, H., and Amal, R., 2010, The effect of pH on UV-based advanced oxidation technologies – 1,4-Dioxane degradation. *Journal of Hazardous Materials* 182: 75–79.

Waldemer, R.H. and Tratnyek, P.G., 2004, The efficient determination of rate constants for oxidations by permanganate. In: A.R. Gavaskar and A.S.C. Chen (Eds), *Proceedings of the Fourth International Conference on Remediation of Chlorinated and Recalcitrant Compounds*, Paper 2A-09. Columbus, OH: Battelle Press.

Waldemer, R.H. and Tratnyek, P.G., 2006, Kinetics of contaminant degradation by permanganate. *Environmental Science and Technology* 40: 1055–1061.

Wang, H., Bakheet, B., Yuan, S., Li, X., and Yu, G., Murayama, S., and Wang, Y., 2015, Kinetics and energy efficiency for the degradation of 1,4-dioxane by electro-peroxone process. *Journal of Hazardous Materials* 294: 90–98.

Wannamaker, E., 2005, Treatment of 1,4-dioxane in groundwater. Presented at Colorado Hazardous Waste Management Society (CHWMS) luncheon, Denver, CO.

Woo, Y.T., Argus, M.F., and Arcos, J.C., 1977, Metabolism *in vivo* of dioxane: Effect of inducers and inhibitors of hepatic mixed function oxidases. *Biochemical Pharmacology* 26: 1539–1542.

Woodard, S., 2019. Personal Communication.

Woodard, S.W., Mohr, T., and Nickelsen, M.G., 2014, Synthetic media: A promising new treatment technology for 1,4-dioxane. *Remediation* 24(4): 27–40. doi: 10.1002/rem, Wiley Periodicals, Inc.

Xie, T., Reddy, K.R., Wang, C., Yargicoglu, E., and Spokas, K., 2015, Characteristics and applications of biochar for environmental remediation: A review. *Critical Reviews in Environmental Science and Technology* 45(9): 939–969. January 13.

Yabuki, Y., Yoshida, G., Daifuku, T., Ono, J., and Banno, A., 2018, Biological treatment of 1,4-dioxane in wastewater from landfill by indigenous microbes attached to flowing carriers. *Journal of Water and Environment Technology* 16(6): 245–255.

Yamamoto, N., Inoue, D., Sei, K., Saito, Y., and Ike, M., 2018, Field test of on-site treatment of 1,4-dioxane-contaminated groundwater using *Pseudonocardia sp.* D17. *Journal of Water and Environment Technology* 16(6): 256–268.

Yan, N., 2018, In situ oxidation of 1,4-dioxane with persulfate-based oxidation activated by iron minerals. *Dissertation, Department of Hydrology and Atmospheric Sciences*. Tucson, Arizona: The University of Arizona.

Yan, N., Liu, F., Liu, B., and Brusseau, M.L., 2018, Treatment of 1,4-dioxane and trichloroethene co-contamination by an activated binary persulfate-peroxide oxidation process. *Environmental Science and Pollution Research* 25: 32088–32095.

Yan, N., Zhong, H., and Brusseau, M.B., 2019, The natural activation ability of subsurface media to promote in-situ chemical oxidation of 1,4-dioxane. *Water Research* 149: 386–393.

Yanagida, S., Nakajima, A., Kameshina, Y., and Okada, K., 2006, Effect of applying voltage on photocatalytic destruction of 1,4-dioxane in aqueous system. *Catalysis Communications* 7: 1042–1046.

Yunker, E., 2007, In Situ Chemical Oxidation of 1,4-Dioxane and VOCs with Ozone and Hydrogen Peroxide. Pittsburgh, PA: The National Association of Remedial Project Managers (NARPM) 17th Annual Training Conference.

Zeng, Q., Dong, H., Wang, X., Yu, T., and Cui, W., 2017, Degradation of 1,4-dioxane by hydroxyl radicals produced from clay minerals. *Journal of Hazardous Materials* 331: 88–98.

Zenker, M.J., Borden, R.C., and Barlaz, M.A., 1999, Investigation of the intrinsic biodegradation of alkyl and cyclic ethers. In: B.C. Alleman and A. Leeson (Eds), *Natural Attenuation of Chlorinated Solvents, Petroleum Hydrocarbons, and Other Organic Compounds, The Fifth International In Situ and On-Site Bioremediation Symposium*, pp. 165–170. San Diego, CA: Battelle Press.

Zenker, M.J., Borden, R.C., and Barlaz, M.A., 2000, Mineralization of 1,4-dioxane in the presence of a structural analog. *Biodegradation* 11: 239–246.

Zenker, M.J., Borden, R.C., and Barlaz, M.A., 2002, Modeling cometabolism of cyclic ethers. *Environmental Engineering Science* 19(4): 14.

Zenker, M. J., Borden, R.C., and Barlaz, M.A., 2004, Biodegradation of 1,4-dioxane using trickling filter. *Journal of Environmental Engineering* 130(9): 926–931.

Zhao, L., Hou, H., Fujii, A., Hosomi, M., and Li, F., 2014, Degradation of 1,4-dioxane in water with heat- and Fe2+-activated persulfate oxidation. *Environmental Science Pollution Research* 21: 7457–7465.

Zhong, H., Brusseau, M.L., Wang, Y., Yan, N., Quig, L., and Johnson, G.R., 2015, In-situ activation of persulfate by iron filings and degradation of 1,4-dioxane. *Water Research* 83: 104–11. October 15.

9 Case Studies of 1,4-Dioxane Releases, Treatment, and Drinking Water Contamination

Thomas K.G. Mohr and William H. DiGuiseppi

CONTENTS

Groundwater contamination case studies provide an appreciation for the range of release modes and contaminant behavior one might expect to encounter at different contamination sites. The rate of migration of 1,4-dioxane relative to that of chlorinated solvents is among the more interesting attributes of this contaminant. Migration rates reflect both the higher mobility of 1,4-dioxane and its refractory nature relative to methyl chloroform because 1,4-dioxane is miscible and resistant to biodegradation and abiotic degradation. Will 1,4-dioxane be flushed out of the aquifer by flowing groundwater and diluted to inconsequential concentrations? Should one expect to find a substantial mass of residual 1,4-dioxane near the point of release, withheld in the fine-grained pores of silts and clays? What conditions support persistence of 1,4-dioxane such that the quality of drinking water is degraded? The answers to these questions are helpful for planning site characterization at 1,4-dioxane release sites.

Case studies provide useful benchmarks for the success or failure and cost of remedies selected to treat water contaminated with 1,4-dioxane. The experience of the water utilities and the drinking water regulators in places where 1,4-dioxane has contaminated the drinking water is helpful to other utilities that unfortunately discover 1,4-dioxane in their drinking water supplies.

This chapter provides several case studies profiling migration patterns, treatment solutions, and responses to drinking water contamination.* Benchmarking previously studied sites to determine solutions for another site is useful to bracket the range of possible conditions one might encounter, but there is no substitute for field data from the site in question to make informed site-characterization decisions. Additional case studies can be found in the 1st edition of this book.

9.1 SUPERFUND SITE AT SEYMOUR, INDIANA

In Jackson County, Indiana, Seymour Recycling Corporation (SRC) operated a waste chemical storage and processing facility on a 14-acre property next to the former Freeman Army Air Field about 2 miles southwest of the town of Seymour, at the junction of U.S. Highway 50 and State Route 11. The town of Seymour's population in the mid-1980s was approximately 15,300; about 100 homes were located within a 1-mile radius of the SRC site (USEPA, 1986). SRC processed, stored, and incinerated chemical wastes at the site from about 1970 to early 1980. Most of the area immediately surrounding the site is used for agriculture. Approximately 100 residences and businesses relied on private wells that used groundwater from the shallow aquifer near the site before these homes were connected to a municipal water system. The neighboring Freeman Municipal Airport obtains its water supply from a well completed in the deep aquifer (USEPA, 1987).

* The sites in these case studies were all characterized by professional geologists and engineers who have a great deal more familiarity with site conditions than the authors; any unintended misrepresentation of site conditions is the responsibility of the authors alone.

FIGURE 9.1 Photograph of the Seymour Superfund Site in 1980, before the drums were removed.

The Seymour community experienced problems with the SRC site beginning in 1976. Residents complained about air and surface-water discharges migrating from the site, which they blamed for health problems. In March 1980, about 100 homes were evacuated when a reaction among incompatible chemicals at the site released toxic fumes that drifted into a residential neighborhood. Following that incident, USEPA began enforcement actions at the site (USEPA, 1987). SRC had accumulated about 60,000 55-gallon drums and 98 bulk tanks that contained waste solvents, metal finishing wastes, phenols, cyanides, acids, and C-56*, as well as hundreds of small containers of hazardous materials, primarily from laboratory operations. Some wastes were highly explosive (USEPA, 1983). These wastes leaked and spilled from their containers, leading to fire and odor problems, as well as substantial soil, groundwater, surface water, and sediment contamination (USEPA, 1986). The site was ordered to be closed in 1978 after violating a State of Indiana order to stop receiving wastes; the property was placed in receivership by court order in 1980, and USEPA entered into a consent decree with PRPs (potentially responsible parties) in 1982. USEPA began removal and emergency response actions in 1980. These actions included fencing the site and constructing runoff control dikes. Two PRPs removed several thousand drums in 1980, and in 1981, USEPA removed chemicals from the bulk tanks and transferred those wastes to authorized disposal sites (USEPA, 1987). By 1984, all of the drums and bulk storage tanks were removed, and the top foot of contaminated soil was removed from about 75% of the site and also transported to authorized disposal sites; clean fill replaced the contaminated soil (USEPA, 1987). Figure 9.1 shows a photograph of the site in 1980, before the drums were removed.

9.1.1 Contaminant Distribution and Hydrogeology

Chemicals from the SRC site contaminated a shallow aquifer, 6–8 ft below ground, and a "deep" aquifer, about 55–70 ft below ground, beneath a leaky silt-clay aquitard. Groundwater in the shallow aquifer discharges into a local creek and a drainage ditch during the wet season; in the dry season, the creek is dry and groundwater flows beneath the creek bed toward the nearby private wells. Although flows in the shallow aquifer are to the northwest, flows in the deep aquifer are primarily to the south, toward supply wells at the airport east of the site.

Sampling via monitoring wells showed that the shallow aquifer beneath the 14-acre site was highly contaminated with more than 35 different hazardous organic chemicals. In 1985, the volatile organic compound (VOC) plume maps of the shallow aquifer showed high-concentration

* C-56 is Hexachlorocyclopentadiene (HCCPD), a toxic percursor to pesticides, formula C_5Cl_6.

contamination extending about 400 ft beyond the site boundaries and trace amounts of organic chemicals as far as 1,100 ft down-gradient of the site boundary. Shallow groundwater flow was estimated to be 400 ft/year (USEPA, 1987; Feldman, 2000).

1,4-Dioxane is associated with soil, groundwater, and sediment at the SRC site (USEPA, 2007a). The soils were contaminated by 54 organic chemicals, including high concentrations of 1,1,2-trichloroethane, carbon tetrachloride, 1,1,2,2-tetrachloroethane, and trichloroethylene (TCE). The presence of 1,4-dioxane in the groundwater was discovered by placement and sampling of monitoring wells during the remedial design phase of the project (USEPA, 1997). Subsequent sampling revealed that 1,4-dioxane had migrated about 3,900 ft north-northwest of the site boundary, as shown in Figure 9.2. Although diverse chemicals, including methyl chloroform and tetrahydrofuran (THF) (which extends northwest of the site boundary by about 1,300 ft), have contaminated the groundwater near the SRC site, the plume itself is defined by the 1,4-dioxane. In the space of 6 years, between 1984 and 1990, the 1,4-dioxane plume advanced approximately 2,000 ft (Feldman, 2000).

In mid-1990, at a monitoring well located on the plume centerline about 1,500 ft downgradient of the site boundary (where the plume turns to the left in Figure 9.2), the 1,4-dioxane concentrations had decreased from about 2,300 ppb to less than 100 ppb. Other shallow monitoring wells along the plume core generally decreased from about 500 ppb to less than 100 ppb (Feldman, 2000).

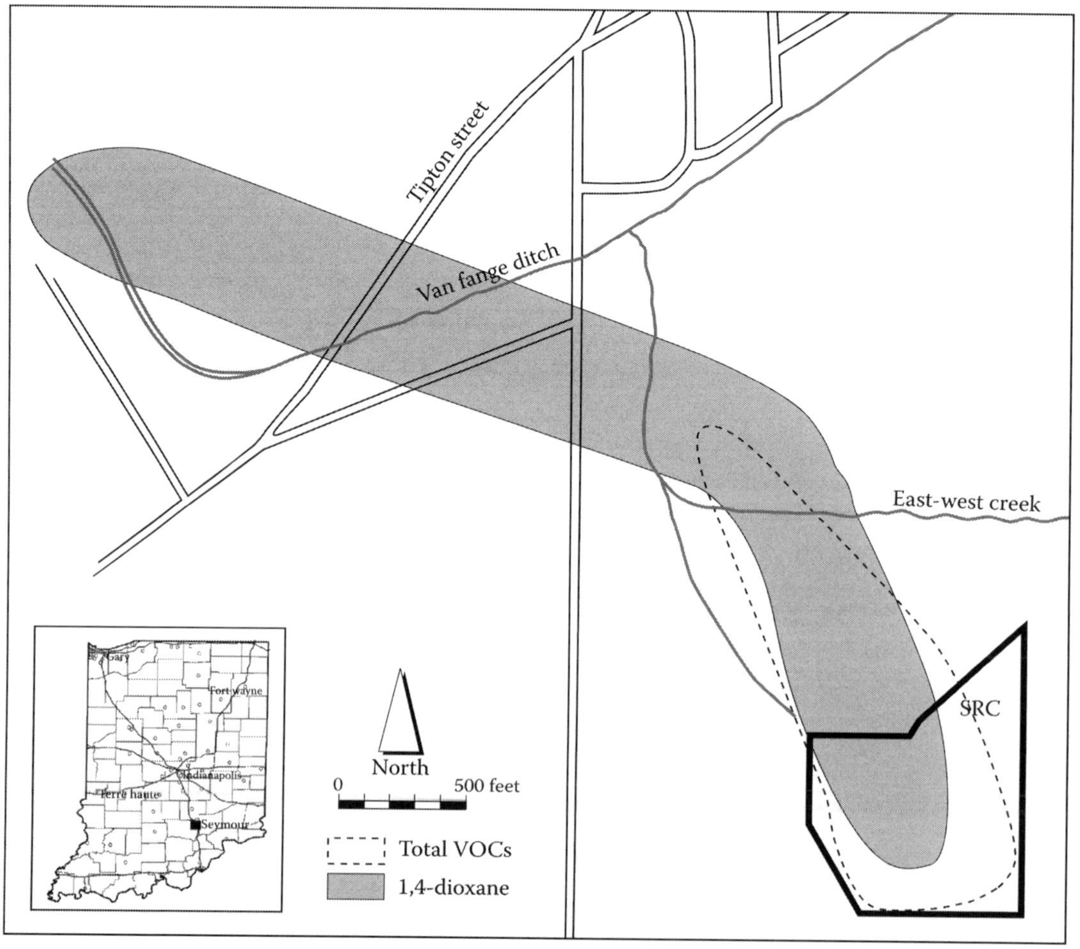

FIGURE 9.2 July 1999 distribution of 1,4-dioxane (10 ppb contour) relative to total VOCs at Seymour Indiana Superfund site, where the SRC operated (Feldman, 2000).

9.1.2 Risk Management

The foremost risk at the SRC site included off-site migration of highly mobile organic contaminants that could arrive at the nearest private well, one-quarter mile northwest of the site, at concentrations greater than those equivalent to the 1 in 1,000,000 excess lifetime cancer risk. 1,4-Dioxane was specifically identified as potentially exceeding the recommended risk levels at the nearest private well in less than 5 years (USEPA, 1987). USEPA analyzed the cumulative excess cancer risk from all carcinogenic compounds emanating from the site, including benzene, methylene chloride, chloroform, perchloroethylene, 1,2-dichloroethane, 1,1,2-trichloroethane, 1,1-dichloroethylene (DCE), TCE, 1,4-dioxane, and vinyl chloride. To prevent exposure, the airport well and the nearest operational downgradient private supply well were monitored for volatile organic contaminants (USEPA, 1987).

9.1.3 Treatment

A remedial investigation and feasibility study concluded that pump and treat prior to discharge to the local publicly-owned treatment works (POTWs) would achieve City of Seymour pretreatment requirements (total toxic organics <2.13 ppm).* Pretreatment consisted of mixed media filtration (a sand filter to remove iron to prevent formation of iron oxide in the air stripper), followed by an air stripper and carbon adsorption. During remedial investigations, USEPA discovered that bacteria were naturally aiding in soil remediation, which allowed construction of a smaller groundwater treatment facility than originally planned, leading to substantial savings (USEPA, 2007c). USEPA capitalized on contaminant biodegradability by applying nutrients to contaminated soil to stimulate biodegradation (U.S. Congress, 1988). Soil-vapor extraction (SVE) and a multimedia cap over the most contaminated soils were also included in the remedy. SVE was preferred over soil excavation because of the dangers of worker exposure associated with excavating soil with high concentrations of volatile compounds. The selected remedial alternative was projected to cost $18 million (U.S. Congress, 1988).

The discharge of treated groundwater to the POTW was projected at 150 gallons per minute (gpm), which was approximately 20% of the excess capacity of the Seymour POTW and approximately 15% of the average daily wastewater flow. An evaluation showed that there would be no detectable levels of contaminants in the discharge of the POTW to the South Fork of the White River (USEPA, 1987).

The SVE system removed the bulk of the contaminant mass during its first year of operation; more than 20,000 pounds of VOCs were vented. The system reached asymptotic levels after approximately 34,000 pounds were removed, and operations were discontinued. About 85% of the total mass of VOCs extracted by the SVE system over a 4-year period consisted of methyl chloroform (31.8%), TCE (23.2%), perchloroethylene (9.7%), *cis*-1,2-DCE (8.7%), 1,1,2-trichloro-1,2,2-trifluoroethane (Freon 113) (7.0%), and toluene (4.8%).

After 12 years of operating the groundwater pump and treatment system, more than 90% of the contamination had been removed from the groundwater (USEPA, 2002). USEPA concluded that the plume was stable and shrinking, because the areal extent of the 1–4 dioxane and THF plumes had decreased when compared to data from 4 years earlier. The pump and treat system was shutdown in October 2001 because it was deemed no longer efficient in removing the lower levels of groundwater contaminants that were still above cleanup standards. The approval for the shutdown required (1) increased groundwater monitoring and (2) a contingency to restart the groundwater treatment system if the groundwater plume expanded from its position at system shutdown (USEPA, 2002). The groundwater treatment was originally estimated to take 28–42 years to complete groundwater remediation, but system shutdown was approved within less than 20 years (U.S. Congress, 1988).

* Most total toxic organics analyses are not performed in a manner that is likely to detect 1,4-dioxane.

The 5-year review report of the groundwater monitoring network found that the plume remained stable, but two additional monitoring wells were needed to sample for 1,4-dioxane (USEPA, 2007b). The overall extent and concentration distribution of 1,4-dioxane in the shallow aquifer has not changed significantly since 2001. 1,4-Dioxane concentrations in one deep-aquifer monitoring well near the capped area showed a significant increase to 690 μg/L as of the 5-year review; however, other monitoring wells continued to show low or nondetectable levels of 1,4-dioxane.

9.1.4 Discussion

Figure 9.2 shows that by 1999, conditions at the SRC site caused 1,4-dioxane to migrate about 2.3 times farther than total VOCs. Contaminant transport modeling accounting only for retardation and dispersion, but discounting the biotransformation of contaminants, accurately predicted 1,4-dioxane migration. The model overestimated the migration of chloroethane and THF because these contaminants are biodegradable (Nyer et al., 1991).

The remedies selected for this site—soil bioremediation, SVE, air stripping, and carbon adsorption—are effective at removing the bulk of the chlorinated VOCs at the site, but are relatively ineffective at removing 1,4-dioxane. The areal extent of 1,4-dioxane distribution changed little between 1999 and 2008, which suggests that natural attenuation is also not actively removing 1,4-dioxane at this site. The primary means of attenuating 1,4-dioxane have apparently been by pumping 1,4-dioxane out of the shallow aquifer and discharging to the POTW, where it is substantially diluted prior to discharge to the receiving water, where it is presumably diluted further. The part of the plume outside the influence of pumping wells will eventually attenuate through dispersion and dilution from recharge due to rainfall, irrigation, and losing streams. Risk of exposure at this site was eliminated by connecting private well owners to the municipal water supply, which was not contaminated by this site, and by frequent monitoring of the nearest actively used supply wells. Maximum concentrations of 1,4-dioxane measured along the plume centerline in 1990, ~2,300 ppb, were relatively low compared to the much higher concentrations seen at other solvent-waste handling and recycling facilities. The 2,300 ppb maximum was measured 10 years after the cessation of active discharges, raising the possibility that the bulk of the 1,4-dioxane mass has dispersed much farther downgradient.

9.2 PALL/GELMAN SCIENCES INC., ANN ARBOR, MICHIGAN

Gelman Sciences Inc., acquired by Pall Life Sciences in 1997 (hereafter P/GSI), produced cellulose triacetate filters used in medical and other applications. As described in Chapter 2, 1,4-dioxane is used as a rheological agent to produce microporous cellulose filters for scientific and medical applications. In 1965, company scientists began experimenting with a polymer called cellulose triacetate that had higher thermal stability and better chemical compatibility than any other membrane on the market. Because this polymer was more stable, traditional solvents, like acetone, could not dissolve it to create pores. In 1966, the company began using 1,4-dioxane to create a stronger solvent system (Fotouhi et al., 2006). In 1986, the company discontinued the use of 1,4-dioxane and phased out the line of filters that required its use in production. At the time that P/GSI stopped using 1,4-dioxane, it was discovered that 30 private wells in the Scio Township area were contaminated with 1,4-dioxane. Approximately 850,000 pounds of 1,4-dioxane was used to form triacetate filters during the 20 years before contamination was discovered (Fotouhi et al., 2006).

The P/GSI site has a long and complicated history that includes protracted legal disputes, complex hydrogeology, a vociferous citizenry, and an evolution in regulatory requirements. Pall Corporation (Pall) acquired GSI in 1997, long after the discovery and initial litigation of 1,4-dioxane contamination. P/GSI acquired GSI's environmental liability and has been engaged in an epic effort to contain a multidirectional set of plumes that have migrated beneath developed neighborhoods. Pall Corporation was acquired by Danaher Corporation in May of 2015.

The engineers at P/GSI, and the regulators with the State of Michigan and Washtenaw County, together with scientists at the University of Michigan, were pioneers who developed the earliest 1,4-dioxane analysis, treatment technologies, toxicological profiles, and regulatory solutions to a massive 1,4-dioxane contamination problem. The P/GSI site is the oldest and largest single 1,4-dioxane contamination case in the United States.

9.2.1 Wastewater Disposal

When Gelman Instruments moved to Scio Township, there were no sanitary sewers available on-site. P/GSI explains that the Michigan Department of Natural Resources (DNR) recommended discharge of wastewater into infiltration lagoons and removal of natural geologic barriers between the pond bottom and the water table 15–40 ft below, to facilitate slow infiltration to groundwater (Kellogg, 2005). The state issued a permit in 1964 to install a three-million-gallon wastewater lagoon to hold and infiltrate process wastes, which may have contained 1,4-dioxane in concentrations as high as 25 mg/L (City of Ann Arbor, 2006; SRSW, 2006); other sources familiar with the case state that the 1,4-dioxane concentration in wastewater may have been higher, in the range of 200–300 mg/L.* After installing two additional lagoons in 1966 and 1967, the company still required additional wastewater capacity. In 1976, DNR granted P/GSI temporary permission to try managing treated process wastes by using spray irrigation. Seven months later, in May 1977, the Michigan Water Resources Commission issued a groundwater discharge permit authorizing effluent spraying onto open fields on the company's 40-acre property. Later, USEPA recommended that Gelman drill an injection well, which they did in 1981. The injection well was drilled at a cost of approximately $1,000,000. The well was drilled more than 1 mile into sandstone (the Mt. Simon Formation), which is geologically isolated from shallower production aquifers (Fotouhi et al., 2006).

In 1984, a University of Michigan graduate student from the School of Public Health, Dan Bicknell, discovered 1,4-dioxane in Third Sister Lake near the P/GSI property in a section of the Saginaw Forest owned by the university. The following year, 1,4-dioxane contamination was discovered in nearby private wells by an investigation conducted by the Washtenaw County Health Board. After learning of 1,4-dioxane contamination in late 1985, P/GSI used a water supply well to remove more than 25,000 pounds of 1,4-dioxane from groundwater near the plant between 1987 and 1994. The contaminated water was not treated; instead, it was discharged into the injection well under a USEPA permit. In 1994, P/GSI decided not to renew the injection well permit because of the cost of upgrading and operating the well, and the well was sealed (Kellogg, 2005). The permitted practice of treating 1,4-dioxane-laden wastewater in unlined lagoons and spray irrigating wastewater on lawns and fields caused widespread contamination of groundwater with 1,4-dioxane.

9.2.2 Geologic Setting

The advance and retreat of glaciers defines the highly heterogeneous hydrostratigraphy of the Ann Arbor/Scio Township area. The subsurface consists of unconsolidated glacial outwash sediments, tills, and lacustrine deposits.† Multiple glacial outwash channels provide high-permeability aquifers facilitating rapid contaminant migration in different directions. The glacial deposits are up to 300 ft thick and overlie the Coldwater Formation of Mississippian age (i.e., older than 300 million years), which is predominantly shale (Brode et al., 2005; Fotouhi et al., 2006). Groundwater is generally shallow, averaging 15 ft below the ground surface. Three major aquifers were initially identified; they are designated as the Unit C3 (Core Area), Unit D0 (Western System), and Unit D2 (Evergreen System) aquifers. A Unit E aquifer that is deeper was later determined to be present.

* Jim Brode (of Fishbeck, Thompson, Carr and Huber, Kalamazoo, Michigan, February 2008, personal communication).

† Lacustrine deposits are lake bed deposits.

A confining layer within the glacial sediments was thought to isolate deeper aquifers from contamination. Subsurface characterization performed in 2001 revealed a geologic window in the confining layer that allowed contamination to migrate into deeper aquifers. The complex geology in the vicinity of the P/GSI property contributes to the widespread nature of the contamination (MDEQ, 2005).

9.2.3 Delineating 1,4-Dioxane Contamination

Beginning in 1986, investigations by the company identified soil contamination on the P/GSI property and groundwater contamination extending off the property. Groundwater concentrations of 1,4-dioxane were as high as 221,000 μg/L. The plumes initially encompassed an area of approximately 0.6 square miles, as defined by the Michigan drinking water action level of 85 μg/L (Brode et al., 2005). Citizen groups define the plume to a 1 μg/L contour, which in 2005 had an area of 0.65 square miles (Kellogg, 2005). 1,4-Dioxane method detection levels have ranged from 100 to 1 ppb during the course of the cleanup (Fotouhi et al., 2006). As of 2005, groundwater was monitored routinely at 50–100 locations (Brode et al., 2005).

In 2016, the Michigan DEQ adopted an emergency Rule that set the 1,4-dioxane residential drinking water criteria to 7.2 μg/L (MDEQ, 2016). While most residents overlying the plume do not rely on private wells for water supply, lowering the drinking water criteria from 85 to 7.2 μg/L had the effect of increasing the mapped plume area and elevating residents' concerns regarding 1,4-dioxane groundwater contamination. Figure 9.3 displays the extent of the shallow and deep plumes delineated to 85 ppb, the former Michigan 1,4-Dioxane Drinking Water Criteria.

Deeper contamination was discovered during an investigation in spring 2001, following a detection of 1,4-dioxane at 2 ppb in the City of Ann Arbor's Northwest Supply well. The city turned off the well immediately following the detection, and investigations of deep aquifer contamination began. After an intensive effort to characterize deep contamination, concentrations as high as approximately 5,000 ppb were found in the deep Unit E aquifer. A clay layer separating the two shallower aquifers from the deeper Unit E aquifer in areas to the east was found to be absent in areas to the west of the site. The Unit E aquifer generally flows to the east toward the Huron River, a major regional hydraulic feature (MDEQ, 2005). State regulators have required vigilant

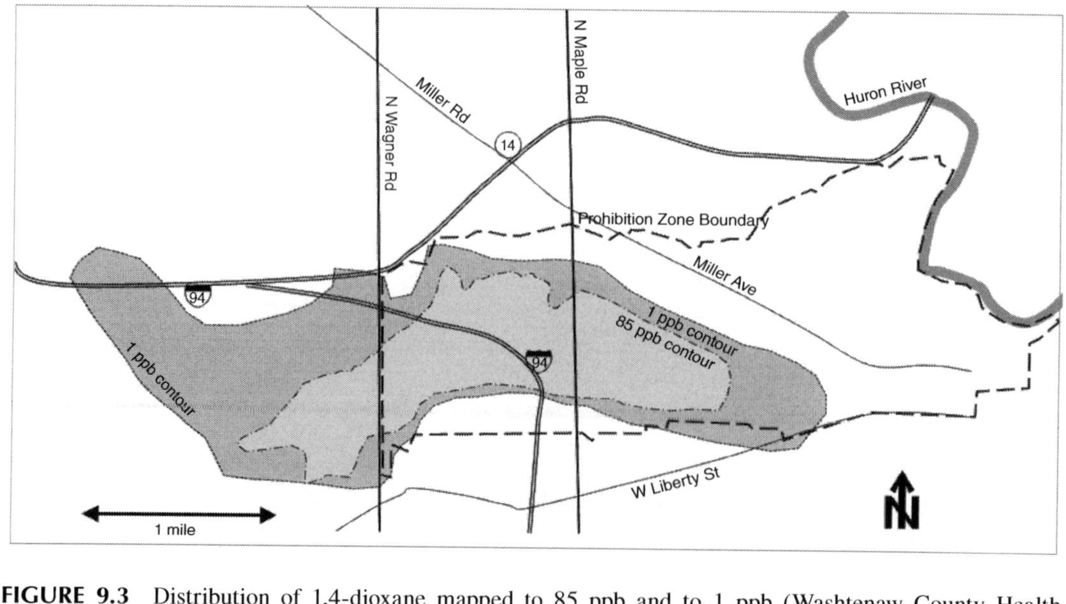

FIGURE 9.3 Distribution of 1,4-dioxane mapped to 85 ppb and to 1 ppb (Washtenaw County Health Department, 2019).

monitoring to ensure that spreading plumes of 1,4-dioxane do not pose a threat to private wells outside of the city. The plume is headed in the direction of the Huron River, which is the source of about 80% of the City of Ann Arbor's water supply; however, Ann Arbor's drinking water supply is not directly threatened by 1,4-dioxane because the intake location is upstream of the groundwater discharge area. In 2019, citizen groups and the City of Ann Arbor raised concerns that 1,4-dioxane in groundwater may eventually migrate to the Barton Ponds drinking water intake (Stanton, 2019). The City of Ann Arbor in 2019 was considering installation of sentry wells to provide advance warning of the possible arrival of the plume at Barton Ponds. 1,4-Dioxane has already been detected at Barton Ponds, sporadically and at low concentrations, due to the discharge of treated groundwater from which 1,4-dioxane has not been completely eliminated to creeks that flow to the ponds.

While the first private well contamination was discovered in 1986, full cleanup of off-site groundwater began the late 1990s. The delay was partly due to legal disputes between the State of Michigan and the city and to countersuits by P/GSI, as described in Section 9.2.6. For more than 10 years, off-site cleanup was limited to some extraction of contaminated groundwater from a former supply well, and 1,4-dioxane continued to migrate away from the site. The plumes have expanded at rates very close to the linear groundwater velocities (Fotouhi et al., 2006).

9.2.4 Regulation

Upon discovery of private well contamination by 1,4-dioxane, P/GSI provided bottled water for residents and businesses, and agreed to contribute toward the cost of extending municipal water lines to affected subdivisions (City of Ann Arbor, 2006). By 2000, more than 120 homes with contaminated private water supply wells had been connected to the municipal water supply system (MDEQ, 2000). Washtenaw County staff collects regular samples from private wells near the edges of the plumes. 1,4-Dioxane detections in private wells have ranged from 1 to 24 ppb; with nondetect results in most private wells. In 2018, 67 supply wells were sampled, none exceeded the 7.2 ppb limit. P/GSI pays the costs for sampling and connecting to the municipal supply (MDEQ, 2000). P/GSI conducts its own sampling of an extensive network of monitoring wells. In 2018, samples were collected at 431 monitoring well locations.

Among the more interesting regulatory actions in the P/GSI case is the prohibition zone (PZ) (depicted in Figure 9.3). In order to protect the public, the Washtenaw County Circuit Court issued an order on May 17, 2005, to restrict the use of groundwater in portions of the City of Ann Arbor and in Ann Arbor and Scio Townships. The order prohibits the use of groundwater that is or may become contaminated with unacceptable levels of 1,4-dioxane in a mapped area called the PZ (MDEQ, 2007). At the time the PZ was created, the threshold concentration was 85 µg/L. Since the residential drinking water criteria were lowered to 7.2 µg/L in 2016, the Attorney General and Danaher Corporation, the successor to Pall Life Sciences, have been negotiating a revised Consent Judgment (MDAG, 2019). A few of the properties in this area receive their drinking water from private wells; however, most areas are already connected to the city water supply. In the PZ area, the order prohibits the consumption or use of well water, the installation of new water supply wells, and requires the proper destruction of all private wells at P/GSI's expense. P/GSI must also fund connecting properties within the PZ that are on private wells to the municipal water supply. Exceptions are available for ground-source heat-pump systems that operate with closed circulation loops and for dewatering wells used in construction and maintenance. Property owners must disclose the order restrictions upon selling or leasing a property within the PZ, as well as disclosing the presence of 1,4-dioxane contamination beneath the property. Because the depth to groundwater in most areas of the PZ is about 40 ft, typical construction activities are unlikely to encounter contaminated groundwater. The order requires that P/GSI continue monitoring the deep plume beneath the PZ and install additional monitoring wells if concentrations change in a manner suggesting that the plume is heading in a different direction than anticipated in the delineated PZ. The restriction on new wells is enforced by the Washtenaw County Department of Planning and Environment, which issues well permits in the PZ.

The PZ order is an institutional control mechanism that is effective to the extent property owners within the designated area are aware of it and adhere to its requirements. During 2006, a small lake in west Ann Arbor showed detections of 1,4-dioxane at 6 ppb. The lake had not previously shown any 1,4-dioxane detections. Investigation of the contamination revealed that the owners of a local inn and restaurant were using a well to cool the buildings and water the lawns. The owners did not realize that they were in the PZ or that there was a 1,4-dioxane contamination problem in the area. 1,4-Dioxane levels in their cooling system water, which discharges into a storm drain that drained to the lake, were several times higher than the 85 ppb action level (Davis, 2006b).

In 2016, MDEQ published draft regulations updating Part 201 Vapor Intrusion Criteria and proposing to establish a groundwater cleanup standard of 29 ppb to prevent vapor intrusion of 1,4-dioxane where basements are flooded with groundwater. The draft regulation relied upon USEPA's vapor intrusion protocol, but did not adequately consider the hydrophilic nature of 1,4-dioxane. In responses to questions raised in a Town Hall meeting on 1,4-dioxane contamination on October 27, 2016, MDEQ published a Q&A in which they acknowledge that 1,4-dioxane does not easily volatilize from groundwater. The Q&A explained that re-evaluation of the basis for calculating vapor intrusion risk factors led to a revised flooded basement scenario accounting for homeowners' use of sump pumps. More appropriate physical parameter values were employed as well, resulting in a revised draft vapor intrusion threshold of 1,900 ppb 1,4-dioxane in groundwater (MDEQ, 2018). The 29 ppb vapor intrusion threshold was set aside in April 2017 to allow adoption of the revised residential drinking water criterion of 7.2 μg/L 1,4-dioxane in an emergency rule. The 1,900 ppb 1,4-dioxane vapor intrusion threshold will be considered in the full Part 201 rulemaking package (MDEQ, 2018).

9.2.5 Treatment

P/GSI embarked on a series of treatability studies in the decades following the discovery of 1,4-dioxane contamination in private wells. Early treatability studies focused on separation methods using gas-phase separation, such as air stripping and steam stripping, and on solid-phase separation using GAC. P/GSI's treatability studies also tested destructive treatment methods including chemical destruction using hydrogen peroxide, ozone, and chlorine; physical destruction using UV light; and combinations of physical and chemical destruction methods. Biological treatability studies included microbial methods such as activated sludge and bioactivated carbon, enzymes, and inoculation with engineered organisms (Fotouhi et al., 2006).

Site remediation began with contaminated soil removal. The most highly contaminated soils were removed in 1979 and 1987 (MDEQ, 2000). Groundwater remediation near the source zone began in 1987 with the reinjection of contaminated water from a nearby supply well. This remediation continued until 1994, when P/GSI found it too expensive and impractical to renew their injection well permit. The injection well also had limited capacity; therefore, it could not accommodate the purge volumes produced at the site. Between 1994 and 1997, there was a hiatus in off-site groundwater remediation while P/GSI responded to a lawsuit filed by the Michigan attorney general on behalf of the DNR to force P/GSI to cleanup groundwater (City of Ann Arbor, 2006).

Remediation of the two shallow aquifers has been under way since 1997. Treated groundwater is discharged to Honey Creek, a small ephemeral tributary to the Huron River. This remediation has significantly decreased the concentration and mass of 1,4-dioxane contamination in the shallow aquifers.

The selected remedy for full-scale groundwater treatment was UV light + hydrogen peroxide (H_2O_2). P/GSI operated a full-scale UV + H_2O_2 system at its facility to remediate groundwater containing high concentrations of 1,4-dioxane. As of 2004, more than 60,000 pounds of 1,4-dioxane had been extracted from the groundwater and destroyed by UV + H_2O_2 since treatment began in 1997 (Brode et al., 2005).

Installation of a centralized treatment system with high-cost equipment and hazardous chemicals creates a logistical challenge. It is not practical or safe to install many treatment centers throughout the wide area of the plumes, yet it is difficult to pipe extracted groundwater back to a centralized treatment center. Residents in developed neighborhoods overwhelmingly oppose putting extraction wells and pipelines conveying contaminated water in their residential streets. In order to overcome property owner and local government objections, P/GSI was forced to install a 4,479-ft-long horizontal infiltration well to extract and divert groundwater to the treatment system (Brode et al., 2005). This well may be the longest trenchless horizontal well yet installed for environmental remediation purposes anywhere in the world.

The UV + H_2O_2 remediation treated 1,4-dioxane-contaminated water produced at up to 1,300 gpm from 18 extraction wells. Influent 1,4-dioxane concentrations were 5,000–10,000 ppb, and effluent concentration was 5 ppb or less (Fotouhi et al., 2006). Treated water was discharged to Honey Creek, a tributary to the Huron River that enters the river upstream of the City of Ann Arbor's water supply intake. Prior to treatment, extracted water entered a lined pretreatment pond where it was mixed with a 93% solution of sulfuric acid to lower the pH to 3.8, which P/GSI determined to be optimal for UV + H_2O_2 reactions (Brode et al., 2005). Following pH reduction, a 50% H_2O_2 solution was injected into the treatment line and mixed into groundwater by a static mixer. The mixed solution was then exposed to UV when it passed through multiple chambers containing 22 UV lamps. Following UV exposure, the pH was raised back to 6.9 with a 40% solution of sodium hydroxide. Sodium bisulfite was added in order to comply with surface discharge requirements and to remove excess H_2O_2 (Brode et al., 2005). Although the areal extent of the plume did not decrease significantly as a result of this treatment, 1,4-dioxane concentrations in the plume core decreased 100-fold to below 10,000 ppb.

P/GSI is permitted to discharge up to 1,300 gpm to Honey Creek. The system was the largest known 1,4-dioxane treatment operation anywhere in the world, operating 24/7/365, excluding brief maintenance periods (Fotouhi et al., 2006). P/GSI's UV + H_2O_2 treatment system operated from October 1997 through March 2005, prior to changing the treatment technology to ozone-peroxide.

The successful operation of the UV + H_2O_2 treatment system required large volumes of reactive chemicals (hydrogen peroxide, sulfuric acid, sodium hydroxide, and sodium bisulfite), and the UV bulbs consumed a lot of electricity, averaging 600 kW h/day. The overall cost of the UV + H_2O_2 treatment system was $3.50/1,000 gallons. Following careful engineering analysis, P/GSI determined that an ozone-peroxide system would use considerably less energy and would result in a significant reduction in chemical use, making the technology safer, more environmentally friendly, and less expensive to operate (Brode et al., 2005).

The ozone-peroxide system requires 50% less H_2O_2 consumption and eliminates the need for sulfuric acid and sodium hydroxide (Brode et al., 2005). The new system transfers water from extraction wells to a pretreatment pond where insoluble iron settles out. Peroxide is introduced via injector quills and then ozone is separately administered through a series of Venturi injectors. Following the addition of post-treatment chemicals, the water again enters a settling pond prior to surface discharge. The flow capacity for the ozone-peroxide system is 1,200 gpm. In April 2019, 867 gpm were treated in all areas, with influent concentrations ranging from 2,000 to 7,000 ppb, whereas effluent 1,4-dioxane concentrations are below 5 ppb. Treatment costs using the ozone-peroxide technology are approximately $1.50/1,000 gallons, yielding a savings of approximately $450K compared to the UV + H_2O_2 costs of approximately $1.4 million/year for a 1,200 gpm flow (Brode et al., 2005; Fotouhi et al., 2006). In addition to the reduction in chemical use, the electrical demand for the ozone-peroxide treatment system is approximately one-third the demand of the UV + H_2O_2 treatment system. P/GSI also deployed a remotely operated mobile ozone-peroxide treatment unit to treat the Unit E plume (Fotouhi et al., 2006).

An important consideration for operating ozone-peroxide treatment is the formation of bromate, as described in Chapter 8. Bromate, a probable human carcinogen, forms as a by-product of ozone treatment of groundwater containing bromide. Michigan allows discharge to surface water of bromate in concentrations of up to 50 ppb; P/GSI's treatment system attains bromate concentrations of less than 10 ppb in the effluent of its ozone-peroxide treatment system for 1,4-dioxane (Kellogg, 2005). Residual 1,4-dioxane and bromate in the effluent are a concern because Honey Creek runs past several private wells that could be affected, and Honey Creek discharges to the Huron River upstream of the City of Ann Arbor's drinking water intake (MDEQ, 2004).

In 2003, P/GSI submitted a work plan to evaluate the feasibility of an *in situ* oxidation system using hydrogen peroxide and ozone to degrade 1,4-dioxane in the aquifers without removing the groundwater. ISCO was pilot-tested by injecting H_2O_2 as a catalyst and Fenton's reagent (see Chapter 8) into one of the confined aquifers, but only a minor decrease in 1,4-dioxane concentrations was observed. Bromate formation exceeded the 10 μg/L MCL. The *in situ* oxidation approach was therefore considered infeasible for large-scale subsurface remediation.

Altogether, P/GSI has pumped about 8.4 billion gallons of water and removed more than 100,000 pounds of 1,4-dioxane (Fotouhi et al., 2006). Since 1997, the company has been spending $5.5 million/year to cleanup 1,4-dioxane pollution (Kellogg, 2005). The value of the affected groundwater resources supplying residents of Ann Arbor, Scio Township, and Ann Arbor Township has not been established.

9.2.6 Legal Actions

In addition to major technical and logistical challenges, the P/GSI case has been punctuated by many lawsuits. Litigants against P/GSI have included the State of Michigan, the City of Ann Arbor, several citizen groups, local businesses, a class-action suit, and an insurance company. In addition, P/GSI filed a product liability suit against 1,4-dioxane manufacturers (Fotouhi et al., 2006). In 1988, the Michigan attorney general, on behalf of the DNR, sued GSI* to force a cleanup of the groundwater (City of Ann Arbor, 2006). The attorney general sought penalties of more than $400,000 as reimbursement for costs associated with providing clean water to 50 homes and businesses whose wells were contaminated with 1,4-dioxane as well as additional fines and a court order requiring cleanup (Wall Street Journal, 1988). GSI prevailed against the cost-recovery part of the attorney general's suit and did not have to pay the state's $400,000 bill for hooking up homes to municipal water supplies (Bodwin, 1989).

GSI filed a counter-claim against the state's cleanup suit, claiming that Michigan had never issued criteria for cleaning up 1,4-dioxane. P/GSI again prevailed, but the state won an appeal to a federal court in Detroit, which led to a consent decree. In 2000, Judge Donald Shelton ordered the state to issue cleanup criteria, and P/GSI was ordered to conduct a study of cleanup alternatives and implement a major remediation plan (Kellogg, 2005). Judge Shelton ordered P/GSI to cleanup groundwater to within Michigan's 85 ppb action level within 5 years. That deadline has passed; the goal was not achieved.

GSI's product liability suit against 1,4-dioxane producers and distributors claimed that they failed to alert GSI that 1,4-dioxane is not biodegradable. GSI named eight companies in its suit and claimed that they failed to adequately test 1,4-dioxane to determine its persistence in the environment. The companies named included Dow Chemical Company of Midland, Michigan; PVS-Nolwood Chemicals Inc. of Detroit, Michigan; Ecclestone Industrial Chemical Co. of Warren, Michigan; Chemcentral Detroit Corp. of Romulus, Michigan; Union Carbide Corp. of Danbury, Connecticut; Ashland Chemical Co. of Ashland, Kentucky; McKesson Corp. of San Francisco; and Van Waters & Rogers Inc. of San Mateo, California (Raphael, 1988). The parties settled out of court. Dow Chemical's 1,4-dioxane containers now include a label that advises it is not biodegradable (Kellogg, 2005).

The City of Ann Arbor sued P/GSI over detections of trace levels of 1,4-dioxane in a well that is used by the city as a winter water source (the Montgomery Well). The well was on standby at the

* GSI refers to Gelman Sciences Inc., the name of the firm before it was acquired by Pall Life Sciences in 1997.

time 1,4-dioxane was detected and has not been used since. This well provided less than 5% of the City of Ann Arbor's water. P/GSI noted that the city voluntarily stopped using the well, which had 1,4-dioxane concentrations of 2–4 ppb, whereas the drinking water action level was 85 ppb. Other sources of 1,4-dioxane could have caused the contamination, according to P/GSI, and the well had other water quality problems unrelated to P/GSI's release (Shelton, 2004; Kellogg, 2005). Among the other possible sources, P/GSI noted that three landfills, including the city's closed landfill and the University of Michigan landfill, had detections of 1,4-dioxane in leachate. The city allows untreated groundwater from those landfills to be pumped into the sanitary sewer and discharged into the Huron River because the small contaminant mass is diluted in the voluminous flow of the receiving water (Kellogg, 2005). Leachate from the city's closed landfill contains 1,4-dioxane in excess of 100 ppb (Davis, 2002).

Prior to the litigation over the detection of 1,4-dioxane concentrations of 2–4 ppb in the Montgomery well, the city appealed a decision by MDEQ to allow an increase in the 1,4-dioxane concentration limit to a daily maximum of 60 ppb and a monthly average of 10 ppb for discharge of P/GSI's treated groundwater into Honey Creek, a tributary to the Huron River upstream of the city's water intake. The city also prohibited P/GSI's discharge of treated groundwater into its storm sewer system owing to volume limitations. P/GSI contrasted the city's practice of allowing the discharge of leachate containing 1,4-dioxane at concentrations up to 100 ppb into its sewers, while P/GSI discharges treated groundwater with less than 5 ppb into the same receiving water at a high cost (Davis, 2002). The founder of GSI, Charles Gelman, argued vehemently for consistent application of regulatory standards, and regulation of all sources of 1,4-dioxane, including antifreeze, aircraft deicing fluid, and other sources (Bodwin, 1989).

The delineation of the PZ was also the subject of litigation between the city and P/GSI. P/GSI objected to the city well getting included in the PZ. The court agreed to exclude the well, pending resolution of the separate lawsuit the city filed against P/GSI regarding the contamination of the city's Montgomery well (MDEQ, 2005).

The City of Ann Arbor and P/GSI crafted an agreement that settled three related state and federal lawsuits regarding well contamination and the PZ. Under the terms of the agreement, the city was to receive a $285,000 payment, and P/GSI was required to conduct more groundwater monitoring to safeguard the city water supply against 1,4-dioxane contamination. The agreement provides for P/GSI to pay the city more than $4 million if 1,4-dioxane concentrations exceed the state's 85 ppb drinking water limit (Davis, 2006a).

The longest running lawsuit—lasting 16 years—related to the P/GSI case was between P/GSI and the State of Michigan over cleanup requirements. In addition to the existing 1,300 gpm ozone-peroxide treatment system that P/GSI installed at its main facility, the MDEQ required that P/GSI install a treatment system large enough to accommodate 1,150 gpm in a commercial area. P/GSI argued that such a facility would not be safe because of the large volumes of hazardous chemicals required to treat that volume of contaminated water. The court questioned the feasibility and safety of MDEQ's proposal, because of the challenges of acquiring and rezoning enough land to site both the treatment facility and the required treatment ponds in a congested commercial area (Shelton, 2004). The court also noted that MDEQ's requirements would involve piping millions of gallons of water contaminated with 1,4-dioxane beneath the city to the proposed treatment facility and would require miles of pipelines including at least 1.5 miles of pipelines in residential Ann Arbor neighborhoods. The judge noted that the City of Ann Arbor concurred with MDEQ's requirements, but did not offer a commitment to facilitate the installation of pipelines for contaminated groundwater through Ann Arbor's residential streets (Shelton, 2004).

Judge Shelton determined that MDEQ's requirement that P/GSI comprehensively treat and eliminate 1,4-dioxane was not feasible, was unwarranted, and was not supported by competent, material, and substantial evidence. However, P/GSI designed and positioned a 200-gpm mobile ozone-peroxide system to intercept 1,4-dioxane contamination to the State of Michigan Groundwater-Surface Water Interface criteria of 2,800 μg/L. The court did not offer a different remedial solution to remove contaminants from the plume that spread eastward into the deep Unit E aquifer. Judge Shelton noted, "It will never be possible to extract all of the 1,4-dioxane from this deep aquifer and the geology is such that it will ultimately end up in the Huron River and be diluted to concentrations far below currently acceptable standards. But the goal must be to remove as much of the contaminant as possible, as quickly as possible, so that the ultimate dilution will take place with minimal impact on the water resource" (Shelton, 2004).

The court's unusual solution was to allow the leading edge of the 1,4-dioxane plume to migrate through Ann Arbor unabated. The court ruled that this option is allowed by Michigan law, provided unacceptable exposures to human health and the environment are prevented, including an institutional control to prevent use of the groundwater in the expected path of the Unit E plume (Shelton, 2004). This ruling led to the creation of the PZ, which is the antithesis of the antidegradation policy used for groundwater resources protection elsewhere in the United States. Nevertheless, MDEQ has effectively limited exposure and protected water supply reliability, while P/GSI has continued its progress with remediating this massive 1,4-dioxane plume and monitoring its migration.

In July 2008, Charles and Rita Gelman donated $5 million to the University of Michigan's Risk Science Center, which promotes the application of multidisciplinary skills to the study of the diverse health hazards that people face. The university expects the gift to position its Risk Science Center to become the nation's premier center for assessing, quantifying, and communicating risks to public health. The gift also supports graduate student fellowships and the hiring of two additional faculty members in the field and establishes a named professorship for a senior researcher. The Gelmans announced that their vision is to "help inform industry, government and the public about how to properly assess the benefits and hazards posed by technology (and chemicals in particular) in our society." Mr Gelman believes that settlement of the P/GSI 1,4-dioxane cases would have come far earlier if a neutral entity such as the U-M Risk Science Center had existed (University of Michigan, 2008).

The P/GSI plume also illustrates another feature of investigating and remediating 1,4-dioxane in groundwater: the challenge of assessing and managing geologic complexity. A 2017 presentation by Professor Lawrence D. Lemke of Central Michigan University provides a brilliant progression from the layer-cake geology conceptual model to deterministic and stochastic hydrostratigraphic modeling (Lemke, 2017). Lemke compares monitoring well networks to the game of "Battleship" (by Hasbro—a guessing game to hone in on an opponent's ship coordinates). Another high impact image is a Play-Doh extruder pushing out tubes of Play-Doh, which represent preferential high-conductivity pathways that 1,4-dioxane may follow, but whose continuity and orientation is unknown and unknowable. By coupling particle tracking with multiple realizations of stochastic geologic models, Lemke's team of graduate students have developed a range of plausible outcomes, some of which show that 1,4-dioxane arrival at the Huron River could already have occurred, illustrating that the wide spacing between monitoring wells may be inadequate to detect the advancing front of the 1,4-dioxane plume. As a fully miscible compound, 1,4-dioxane presents an ideal tracer that can inform the hydrostratigraphic interpretation. By employing hybrid models of spatial variability, Lemke's team demonstrates that combined deterministic and stochastic models are better at representing the complexity of glacial outwash plain hydrostratigraphy, and at quantifying the uncertainty associated with a given conceptual model.

9.3 FORMER AMERICAN BERYLLIUM COMPANY, TALLEVAST, FLORIDA

The former American Beryllium Company (ABC) site in Tallevast, Florida, is an example of the impact that the late discovery of 1,4-dioxane in a cleanup can have on-site investigation, remediation, and community relations. The bulk of the information presented herein was obtained from the very comprehensive website for the project (http://www.tallevast.info/) and from the Florida Department of Environmental Protection (FDEP) website (http://www.dep.state.fl.us/).

Located in Florida's Manatee County, Tallevast traces its roots back to the 1890s, when shacks were built there for a community of African-American laborers. The laborers who first lived in

Tallevast worked tapping sap from the local long-leaf "slash" pine forests and boiled it to make turpentine for use in the nation's shipyards and harbors. Residents of this "turp camp" also grew sugarcane, celery, and strawberries in area fields or worked in orange groves or on dairy farms (Lerner, 2008). At present, Tallevast is a blend of single-family residential homes and light commercial and industrial development (Cilek et al., 2004). Tallevast is located between Bradenton and Sarasota and less than 2 miles from Florida's west coast.

9.3.1 Geologic Setting

Tallevast is situated in north-central Manatee County, in the terraced coastal lowlands physiographic province, a featureless plain extending to the Gulf of Mexico. The stratigraphy in the area consists of sandy sediments overlying a sequence of limestones, clays, and sandy limestones. The near-surface geology to a depth of 18 ft in the vicinity of the American Beryllium site consists of an upper unit of undifferentiated sandy and silty sediments of Holocene to Pliocene age and a lower undifferentiated carbonate unit of the Miocene Arcadia Formation extending to a depth of 300 ft below ground surface (bgs). The carbonate unit is predominantly dolostone with thin beds of sand and clay and phosphate deposits (Cilek et al., 2004).

The percentage of silt in the shallow Holocene sediments increases with depth, and the sand grain size decreases from fine to very fine. The transition to the undifferentiated Arcadia Formation occurs at approximately 15–18 ft bgs, where phosphatic sands and pebble-size quartz grains appear. With increasing depth, there is a gradual increase of phosphatic sand and percentage of clay (nearly 20%). At about 30 ft bgs is a thick clay unit, the Venice Clay, described as a mix of gray to olive-green clayey sand, tan lime mud, and limestone or dolostone fragments with quartz and phosphatic sands (Cilek et al., 2004). At approximately 300 ft bgs, the 3,000-ft-thick carbonate Floridan aquifer begins (Campbell and Scott, 1993; Cilek et al., 2004).

The production well nearest the site is an agricultural well located one-quarter mile to the east. The agricultural water supply well is 805 ft deep and cased to 368 ft bgs (TetraTech, 1997). Most of the irrigation and potable wells in the Tallevast community are less than 100 ft deep; they are cased from land surface to between 20 and 30 ft bgs with an open-hole interval to total depth. The majority of irrigation and potable wells at the site produce from the regional intermediate aquifer system.

Karst conditions are suggested by topographic features in the area, including low-lying marshy areas and closed depressions; however, no sinkholes are reportedly present within 2 miles of the site. Groundwater is located approximately 2 ft bgs (TetraTech, 1997). The groundwater hydraulic high within the surficial aquifer is near the south-central part of the ABC site, which also represents the area's topographic high point. Proceeding away from this topographic high, the surficial aquifer's hydraulic head parallels the slope of decreasing ground surface elevation. Consequently, groundwater flows outward from the site, toward the northwest, north, northeast, east, and southeast (Cilek et al., 2004).

In a stunning example of an unfortunate coincidence, the hydraulic mounding was later discovered to be caused by a municipal waterline leak at the facility under a concrete slab adjacent to the former sump location; the sump was the probable source of the contamination described here. The broken pipe was discharging as much as 70,000 gallons/day, possibly for several years, and was probably the principal cause of groundwater and contaminants flowing radially outward and away from the former ABC facility (Arcadis-BBL, 2007a).

9.3.2 Site History

ABC operated on a 5-acre parcel located near the northeast corner of the Sarasota-Bradenton airport. In 1957, a small machine shop called Visioneering opened its factory in Tallevast. In May 1961, Visioneering became the ABC, a subsidiary of Loral Metals Technology, which operated as a machining and metalworking plant until 1996. ABC was one of the largest precision machiners of beryllium, a lightweight, toxic metal that can withstand harsh environments. ABC became a contract manufacturer of components for aerospace guidance systems and nuclear reactors. Because beryllium has a low density and is stronger than steel, it is used by aerospace industry companies. At ABC, beryllium machine parts were milled, lathed, and drilled into various components. Components were finished by electroplating, anodizing, and vapor degreasing and ultrasonic solvent cleaning (Cilek et al., 2004; FDOH, 2007).

Solvent use and wastewater handling are inferred from aerial photographs and from company records. A 1,000-gallon solvent tank was used from 1960 until it was removed in 1986, when it was replaced with drum storage for solvents. Air permit records show that three portable vapor degreasers with 10- to 15-gallon methyl chloroform solvent tanks were used at the site. The first degreaser was acquired in the 1960s; the other two in the 1970s. The methyl chloroform was supplied by Ashland Chemical Company and contained 90–95% methyl chloroform and a proprietary inhibitor package that included 1,4-dioxane (TetraTech, 1997; Ashland Chemical Company, 1986). All three methyl chloroform degreasers were replaced in 1988 with ultrasonic cleaners that used Freon (TetraTech, 1997).

A small wastewater treatment system pumped all the facility's wastewater streams to an evaporation pond before the facility was connected to the public sewer system in 1986. The wastewater treatment system was subsequently removed and the pond backfilled and paved (www.tallevast.info). An unspecified quantity of soil was removed from the evaporation pond at the time of closure (TetraTech, 1997). A 400-ft-deep supply well provided water to the site; it was located near the plating shop where the wastewater sumps were located. This well was reported to have been in use through the 1970s; it was abandoned when the facility was connected to the municipal water supply.

A series of sumps were reportedly located near the wastewater treatment facility (TetraTech, 1997). One sump, dug near the plating shop, was described as an unlined pit filled with gravel and used between 1968 and 1984 for disposal of acid baths. The pit was paved with concrete in 1984 (TetraTech, 1997). The sumps were determined to be the predominant source of groundwater impacts at the site (Cilek et al., 2004).

The first indication of the presence of solvents in groundwater was noted in 1987, when 1,1-dichloroethane was detected at 65 μg/L in a single groundwater sample (TetraTech, 1997). A 1993 investigation of a fuel leak at a nearby fiberglass boat manufacturer revealed TCE in groundwater at concentrations of less than 25 μg/L (Cilek et al., 2004).

In 1996, a major aerospace contractor, referred to hereafter as "the responsible party (RP)," acquired the ABC and closed the Tallevast facility in the same year. In 2000, the RP sold the facility to Wiring Pro International (WPI) but retained the groundwater and soil remediation responsibility (FDOH, 2007). In 2000, the RP identified VOCs during a due diligence investigation to support the land sale to WPI. Manatee County Environmental Management and the FDEP were notified voluntarily of the VOC contamination caused by American Beryllium's past operations in July 2000 by the RP (www.tallevast.info). Between 2000 and 2003, the responsible party performed voluntary assessment and soil removal at the former ABC site, with oversight by the FDEP.

During the course of the voluntary investigations in 2002 and 2003, VOCs, chromium, and beryllium were identified in soil and groundwater. In 2001, 538 tons of impacted soil were removed and landfilled off-site as an interim cleanup measure. In July 2003, FDEP approved the Contamination Assessment Report that delineated impacts from former solvent use, which were largely contained within the site except for a low-concentration plume that extended a short distance off-site to the north and east of the facility. The conclusion was based on data available from monitoring wells installed for the investigation, which were later determined by the FDEP Site Investigation Section to underrepresent the extent of groundwater impacts.

In 2002, FDEP asked the Florida Department of Health (FDOH) to assess potential impacts to off-site private wells. FDOH visited residential areas surrounding the site and tentatively identified several private wells north of the site and outside of the VOC plume as delineated by groundwater samples taken from existing monitoring wells (FDEP, 2004). The RP conducted a survey of water supply wells in the adjacent community and identified several domestic drinking water wells beyond the delineated plume. Because data collected by their consultants suggested that the plume was stable or shrinking, and no domestic wells were identified within or immediately adjacent to the plume, the domestic wells outside of the delineated plume were not sampled (FDEP, 2004).

In the fall of 2003, Tallevast residents living near the site learned of the off-site activities after observing a consultant conducting field work in the area. Although the RP notified the regulatory agencies in 2000, those agencies apparently did not notify the community about the site investigation activities over that 3-year period (there is no regulatory requirement for a RP conducting voluntary investigation and remediation activities to notify the local community). In the fall of 2003, the RP provided information to the community about voluntary site assessment and remediation activities that were being performed. At that time, local residents voiced concerns about the potential for impacts to private wells. In 2004, community leaders provided the RP with additional information regarding the location of domestic wells, and FDEP, FDOH, and Manatee County surveyed the area and located and sampled 17 wells (FDEP, 2004).

FDEP and Manatee convened a community meeting, at which the RP asked residents for help identifying private well locations and agreed to test wells and pay for connection to public water if impacts were detected (FDEP, 2004). Several iterations of sampling detected solvents above MCLs in some domestic wells and irrigation wells, and affected residents were provided with bottled water. Temporary county water hookups were provided to all 17 homes that were not previously connected to the public supply. Some of the affected private wells were located a significant distance from the delineated plume; hence, further investigation of the plume was conducted (FDEP, 2004).

Comparisons of contaminant levels between private and monitoring well data sets are difficult to evaluate owing to the multiple aquifers in this geologic setting and the unknown construction details for most of the private wells that have been tested. Because of these limitations, FDEP generally does not rely on private well data for contamination delineation or remediation monitoring purposes (FDEP, 2006).

Because the impacted private wells were located outside the established VOC plume boundaries, as delineated by the RP's consultant, FDEP conducted an investigation to identify the source(s) of contamination and to resample all monitoring wells (Cilek et al., 2004). FDEP collected 39 groundwater samples from the surficial aquifer; most of the samples were collected near its base. Direct-push groundwater grab samples were collected at locations adjacent to existing monitoring wells, selected irrigation wells, and potential source areas (Cilek et al., 2004). The highest TCE concentration reported in the initial site groundwater survey, 2,050 μg/L, was a direct-push grab sample retrieved near the eastern boundary of the facility (Cilek et al., 2004; FDEP, 2004).

FDEP's results indicated a larger chlorinated solvent groundwater plume with higher concentrations of chlorinated solvents than that delineated by the RP's monitoring wells. For example, a TCE concentration of 550 μg/L was reported in one well, whereas analysis of the nearby grab groundwater sample collected nearby reported a TCE concentration of 29,000 μg/L (35,000 μg/L in a duplicate sample). FDEP interprets this result to be indicative of DNAPL (dense, non-aqueous-phase liquid) because the concentration is greater than 1% of TCE's solubility in water (Cilek et al., 2004). No analyses for 1,4-dioxane were performed in FDEP's or the RP's 2004 investigations.

In May 2004, some Tallevast neighborhood residents reported that they had received soil excavated from the former American Beryllium site to fill in low spots on their properties. Others, concerned that they might be breathing vapors from contaminated groundwater underneath their homes, requested indoor air testing of their homes (FDOH, 2007). During public meetings held in June and July 2004, residents expressed a variety of health concerns they believed American

Beryllium caused, including high rates of cancer, fertility problems, and illnesses that they attributed to exposures to dust and surface soils transferred from American Beryllium to resident yards (FDOH, 2007). In June 2004, the FDEP and the RP's consultants tested neighborhood surface soil for metals, VOCs, semi-VOCs, and hydrocarbons (FDOH, 2007). Although several of the soil samples were found to contain elevated levels of arsenic, lead, and petroleum-related organic compounds* (Cilek et al., 2004), concentrations were generally only slightly above standards, and no pattern of pervasive soil contamination was identified.

In the fall of 2004, FDOH and the Manatee County Health Department tested for 61 VOCs in the indoor air of three homes and in the community center located adjacent to the American Beryllium site. The indoor air-testing sites are located over the highest concentrations of TCE in groundwater. Summa canisters were used to collect three 8-h samples at each location. FDOH mailed letters to each indoor air-testing participant explaining chemical results obtained from their indoor air, including copies of the laboratory results and information on each chemical found. At all of the four locations tested, most of the 61 chemicals analyzed in indoor air were nondetect. FDOH enlisted the assistance of the Agency for Toxic Substances Disease Registry (ATSDR) to prepare a health consultation and exposure investigation report. For those with known comparison guidance concentrations, all levels detected were below ATSDR's guidance concentrations (ATSDR, 2005; FDOH, 2007). 1,4-Dioxane was not detected in any of the four indoor air samples (ATSDR, 2005).† Subsequent vapor-intrusion investigations also tested for 1,4-dioxane, including 23 soil-vapor samples and two ambient-air samples in June 2006; however, analyses did not detect 1,4-dioxane above its 18 μg/m^3 reporting limit (5 ppbv) (Arcadis-BBL, 2007b).

Five years into the site investigation and remediation activities, in February 2005, a Site Assessment Report Addendum presented additional groundwater findings, including the detection of 1,4-dioxane in 24 upper-aquifer monitoring wells and in 16 lower-aquifer monitoring wells. Of these results, 21 upper-aquifer monitoring wells had 1,4-dioxane greater than Florida's Groundwater Cleanup Target Level of 3.2 μg/L, and 13 lower-aquifer monitoring wells exceeded the 3.2 μg/L target level (TetraTech, 2005).

The discovery of 1,4-dioxane substantially increased the plume size because 1,4-dioxane had spread much farther than the methyl chloroform‡ from which it originated and much farther than the TCE and PCE that was also released from the site. Prior to the discovery of 1,4-dioxane, the plume of chlorinated ethenes was mapped at 50 acres; following the discovery of 1,4-dioxane, the plume was mapped at 131 acres and eventually was delineated at about 200 acres.

An independent consultant, Environmental Science & Technology (ES&T), was retained by the local community group (but funded by the RP) to complete an independent sampling study of private wells. ES&T suggested that their findings indicated the plume was moving more rapidly than was first thought. They attributed the plume's movement to changes in hydraulic pressure that occurred when Tallevast residents were connected to county water in the summer of 2004 and stopped pumping their wells. Cessation of pumping near the center of the plume, coupled with continued pumping by surrounding residents and businesses on the periphery of the plume caused an apparent shift in the directions and rates of plume migration (ES&T, 2005).

The RP's February 2005 1,4-dioxane results did not correlate with ES&T's December 2005 reported levels. ES&T's 1,4-dioxane results were obtained by EPA Method 8260 using a reporting limit of 20 μg/L, whereas the RP's consultant used FDEP-approved EPA Method 8270, which was then generally regarded in the industry as a more reliable method. ES&T's EPA 8260 results suffered from quality control problems. For example, one duplicate sample was nondetect (<20 μg/L) in the first analysis and 150 μg/L in the second analysis. Overall, ES&T's EPA

* 1,4-Dioxane was not tested on soil samples, but because of its low sorption, is commonly not found in soil.

† A common laboratory method for analysis of soil vapor and indoor air samples is EPA Method TO-15, which includes 1,4-dioxane; however, 1,4-dioxane was not the focus of the indoor air sampling.

‡ The methyl chloroform is virtually absent from the Tallevast plume; its abiotic and microbial degradation products, 1,1-dichloroethane and 1,1-dichloroethylene, are all that remain.

8260 results showed higher concentrations of 1,4-dioxane. ES&T suggested that the differences between their sample results and those obtained by the RP's consultant may have been related to how deep samples were retrieved. ES&T noticed that samples collected from the deeper, open-cased bore hole had higher concentrations than those from the upper, cased section of the well (ES&T, 2005). The large differences in 1,4-dioxane results from the two sampling surveys, while collected at different times and with different methods, led to public confusion and suspicion of the lower concentration results.

The RP funded a field study in 2006 to collect whole-volume split samples for analysis of 1,4-dioxane by three methods at three laboratories. A new consultant, Arcadis-BBL, collected field samples, including some samples spiked with standards at three different concentrations. Analysis was performed by EPA Methods 8270C, 8260B, and 8270 with isotope dilution ("8270-ID") at three different commercial laboratories. 1,4-Dioxane concentrations in the duplicate samples analyzed by Method 8260B were highest among the three methods; some results were more than an order of magnitude higher than the Method 8270C results. Detailed information regarding this laboratory method comparison study is presented in Chapter 4.

9.3.3 Regulatory Response: Remediation and Water Supply Replacement

The record for the Tallevast site reflects significant effort by the RP to delineate the contamination, report findings to the regulatory agencies, and initiate voluntary remedial actions. Although not required under environmental regulations to communicate directly with the community, the RP made significant attempts to keep them informed, addresses their concerns, and provides financial recompense to the affected community. Substantial additional effort by FDEP, FDOH, and Manatee County Environmental Management was required to oversee the investigation, engage an active and concerned community, address health issues, and respond to a heavy involvement by local, state, and federal elected officials, as well as intensive coverage by the local media. FDEP worked with the RP to aggressively investigate the extent of contamination, which was accomplished by installation and sampling of 250 monitoring wells in the space of only a few years. A soil removal action and construction and operation of an interim remedial action were also used to remove contaminants from the source area. 1,4-Dioxane turned out to be the most widespread contaminant. This finding greatly expanded the scope of the investigation and presented a challenge to both analysis and remediation.

The primary remedial strategy is groundwater extraction and treatment for removal of TCE, perchloroethylene, ds-1,2-DCE, 1,1-DCE, 1,1-dichloroethane, and 1,4-dioxane. The groundwater extraction plans include on-site extraction wells, six extraction trenches in the upper aquifer totaling more than 1 mile of linear trenching to a depth of approximately 25 ft bgs, and 50 off-site extraction wells in several lower-aquifer units (Arcadis-BBL, 2007c). Figure 9.4 shows the configuration of extraction trenches at the former American Beryllium site.

Pilot testing was performed to evaluate treatment technologies—including photo-oxidation, ozone-peroxide oxidation, and biostimulation—to remove 1,4-dioxane and chlorinated solvents. Pilot-test outcomes are described in the following sections. The treatment technology selected at that time for the on-site Interim Remedial Action was the Photo-Cat™ advanced oxidation system coupled with an iron oxidation pretreatment system consisting of a 20,000-gallon influent tank to facilitate contact time with a coagulant and precipitation of iron. A series of bag and cartridge filters provides secondary filtration for removal of iron hydroxide precipitate. Effluent from the Photo-Cat system is further treated by two larger GAC canisters installed in series in a lead/lag fashion that remove any remaining 1,1-DCA and polish the treated groundwater. Following GAC treatment, the groundwater is discharged to the Manatee County wastewater treatment plant (Arcadis-BBL, 2007c).

The Photo-Cat water treatment system destroys VOCs and 1,4-dioxane by mixing titanium dioxide (TiO_2) catalyst with the groundwater and exposing it to UV light in a reactor; the contaminants are converted to carbon dioxide, water, and salts. Pilot testing (described further in the

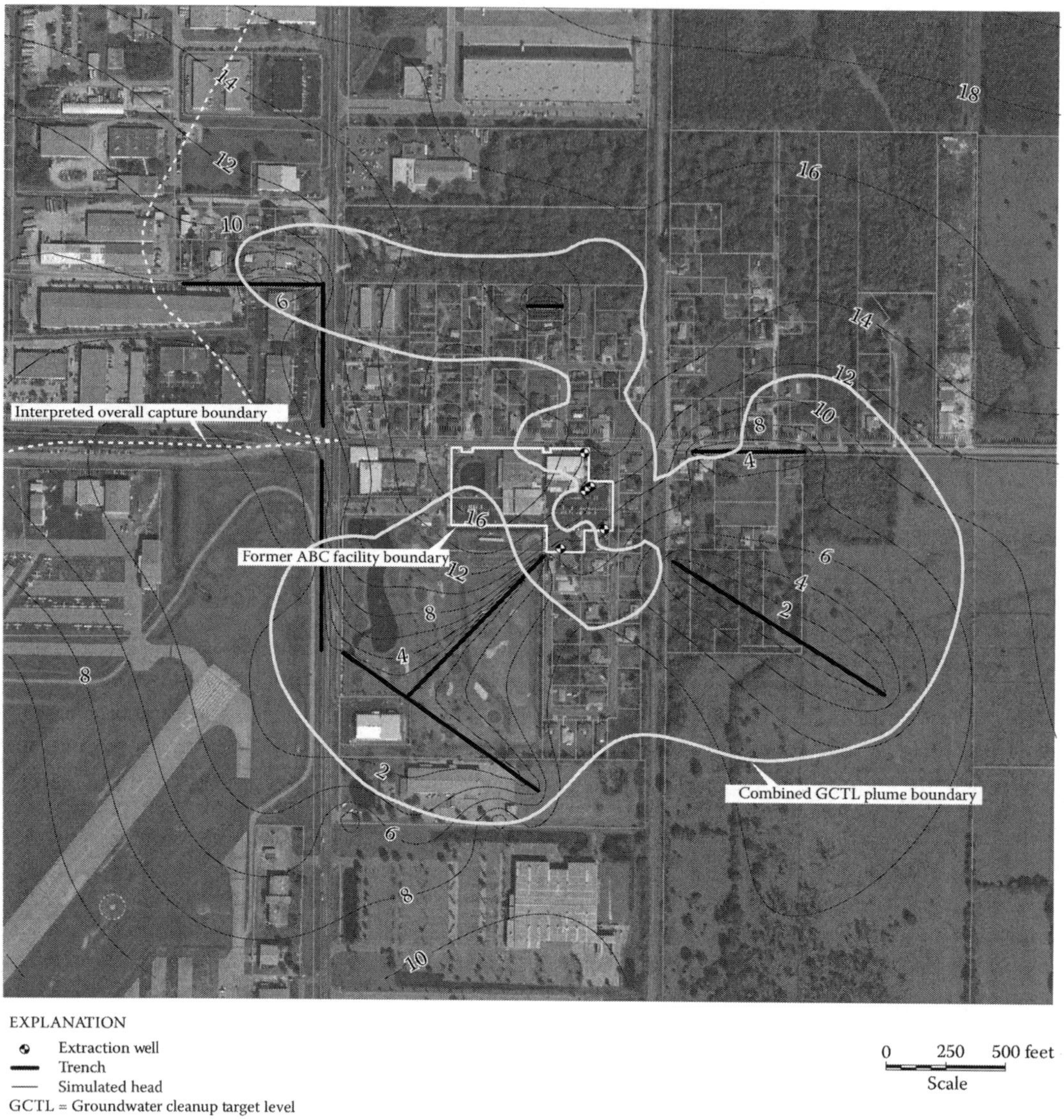

FIGURE 9.4 Configuration of extraction trenches and extraction wells proposed for remediating 1,4-dioxane and VOCs at the former American Beryllium site (Arcadis-BBL, 2007c).

next section) confirmed that the Photo-Cat system reliably treats TCE and 1,4-dioxane to less than 3 μg/L, which is at or below the anticipated discharge limits to be established by Manatee County (Arcadis-BBL, 2007c).

9.3.3.1 UV Oxidation Pilot Treatability Study

Purifics ES, Inc. conducted an on-site pilot test using the Photo-Cat UVOx treatment process, a patented closed-loop TiO_2 slurry-based photocatalytic process to remove 1,4-dioxane and VOCs from groundwater at the former ABC facility.* Bicarbonate ions, abundant in the carbonate aquifers at the Tallevast site, scavenge the hydroxyl radicals generated by the Photo-Cat process and can limit

* In United States, Photo-Cat is sold under licenses from Purifics ES by Basin Water; the technology is NSF-61 certified for drinking water systems.

the effectiveness of the photocatalytically generated hydroxyl radicals. Bicarbonate is controlled by lowering pH with sulfuric acid, nominally to 4.6, where bicarbonate is converted to CO_2; however, to control formation of iron hydroxide precipitates from dissolved iron present in the upper aquifer, pH must be further lowered to 3.0 (Powell, 2005).

To normalize operating parameters between pilot tests, a first-order rate relationship was used to establish a rate constant for the Photo-Cat reaction with the site contaminants. The simple first order equation for the rate constant k is

$$k = \frac{1}{t} \cdot \ln\left(\frac{C_0}{C}\right) \propto \left(\frac{v_f}{P}\right) \cdot \ln\left(\frac{C_0}{C}\right), \tag{9.1}$$

where C_0 is the initial concentration, C is the final concentration, t is the time (in s), k is the first-order rate constant, v_f is the flow rate (in L/min), and P is the power (in kW). Because time is directly proportional to the power used to operate the Photo-Cat system, flow in liters per minute is divided by power in kilowatts, giving units of (L/min)/kW for the first-order rate constant in Equation 9.1.

Influent concentrations in various pilot-test configurations ranged from 250 to 900 μg/L 1,4-dioxane and 2,800–10,000 μg/L TCE. The corresponding rate constant for TCE, averaged from both the upper- and lower-aquifer pilot tests, was 36.8 (L/min)/kW and for 1,4-dioxane, 24.9 (L/min)/kW. The lower aquifer has higher chloride concentration, which also scavenges hydroxides and affects TCE destruction, but has less impact on the destruction of 1,4-dioxane (Powell, 2005).

Nearly 80 million gallons of groundwater per year are now pumped and treated in a centralized filtration and AOP system. Since system startup in November 2013 until August 2019, a total of 458,575,000 gallons have been extracted and treated. Influent 1,4-dioxane concentrations have ranged between 60 and 150 μg/L, with a clear decreasing trend. The RP has already spent $35 million on the cleanup. Pending system enhancements are expected to clean up the majority of the contamination within five years, and all pollution will be removed in approximately 50 years (O'Donnell, 2009).

9.3.3.2 Ozone and Peroxide Oxidation Pilot Study

Samples of contaminated site groundwater were sent to Applied Process Technology, Inc. (APT) in California to evaluate treatability of 1,4-dioxane and chlorinated solvents using APT's HiPOx™ technology. HiPOx, described further in Chapter 8, uses a continuous, in-line, at-pressure AOP to destroy VOCs and 1,4-dioxane. The HiPOx process employs ozone (O_3) and hydrogen peroxide (H_2O_2) chemistry in a uniquely designed oxidation reactor. These reactants are injected under high pressure directly into the treatment water stream at specified ratios and locations, generating hydroxyl radicals that attack the bonds in organic molecules, progressively oxidizing contaminants and their by-products into carbon dioxide, water, and salts. Table 9.1 summarizes influent

TABLE 9.1
Ozone Demand for HiPOx Treatment of 1,4-Dioxane and Trichloroethylene

Influent TCE (μg/L)	Effluent TCE (μg/L)	Influent 1,4-Dioxane (μg/L)	Effluent 1,4-Dioxane (μg/L)	Applied Ozone (mg/L)
2,700	2	500	3.2	15.2
2,700	4.2	300	3.2	13.6
2,700	3	300	2.5	14.3

Source: Herlihy, P. and Bowman, R., 2005, HiPOX™ Technology Lab Testing, Former American Beryllium site, Tallevast, Florida. Pleasant Hill, CA: Applied Process Technologies Inc.

and effluent concentrations for Tallevast groundwater treated with the HiPOx process and the corresponding ozone demand (Herlihy and Bowman, 2005).

The reasons for selecting Photo-Cat instead of HiPOx are not stated in the consultant reports, but the HiPOx system generally has a higher capital cost, but lower long-term operating and maintenance costs, and the higher capital cost may have influenced the decision. Both appear to be effective at removing 1,4-dioxane and chlorinated solvents, and both require the use of hazardous chemicals (sulfuric acid for Photo-Cat and hydrogen peroxide for HiPOx). Systems dependent on UV light may be impeded where nitrate, colloids, or other light-blockers are present, whereas systems using ozone may form undesirable bromates where bromide is present.

9.3.3.3 *In Situ* Biostimulation and Bioaugmentation Treatability Study

Primary and secondary evidence was established for reductive dechlorination of chlorinated solvents, but it was concluded that biodegradation of 1,4-dioxane through biostimulation or bioaugmentation was unlikely to be effective. The primary positive evidence includes the presence of reductive dechlorination by-products, including *cis*-1,2-DCE, 1,1-DCE, 1,1-DCA, and ethene in some groundwater samples; secondary evidence includes reducing geochemical conditions, specifically iron-reducing, sulfate-reducing, and methanogenic conditions (Arcadis-BBL, 2007c). The presence of reducing geochemical conditions in all monitored zones is supported by low dissolved oxygen concentrations (less than 1.0 mg/L), negative ORP (oxidation-reduction potential) values, abundant naturally occurring organic carbon in site groundwater, elevated sulfide concentrations (above 500 mg/L), and elevated methane concentrations (above 50 ppb in some groundwater samples). The lack of further degradation products beyond *cis*-1,2-DCE (i.e., vinyl chloride, ethene) in most monitoring wells suggests that natural biodegradation of chlorinated ethenes may be stalled, possibly owing to the presence of excessive alternative electron acceptors such as sulfate and chloride (Arcadis-BBL, 2007c).

SiREM laboratories used polymerase chain reaction (PCR) tests to determine whether indigenous *Dehalococcoides,* an ethene-degrading microbe, was present. SiREM also performed biostimulation tests on-site groundwater augmented with EOS™ as an electron donor, with and without bioaugmentation using a proprietary bacterial culture, KB-1™.* The fact that the *Dehalococcoides* assay on two tested samples was negative indicated that bioaugmentation would be necessary; however, site data are indicative of active reductive dechlorination. The widespread and persistent occurrence of 1,4-dioxane, in conjunction with evidence of chemically reducing conditions, which are the opposite of the aerobic conditions that 1,4-dioxane-degrading bacteria generally require, suggested that geochemical conditions do not favor natural biodegradation of 1,4-dioxane.

9.3.4 Community Response

Communication with the Tallevast community has been a major challenge for the regulators and the RP due in large part to the dynamic nature of characterizing a plume and the late discovery of 1,4-dioxane. Because some community members felt that they should have been informed of the site investigation earlier, these residents were not inclined to trust the assurances they received that their concerns would be addressed. Local newspaper articles suggest that residents viewed much of the information they received through a lens of distrust. The events that followed the initial discovery of contamination in private wells apparently strengthened the residents' assumptions that they were not being told the full story of the contamination. These events include the chlorinated solvent plume's turning out to be much larger than originally determined, the discovery of 1,4-dioxane in 2006, inconsistent laboratory results, and a health study by FDOH that mistakenly used incorrect data.

* EOS™ is an emulsified edible oil used to stimulate anaerobic biodegradation of chlorinated solvents and other contaminants; KB-1 Dechlorinator (KB-1™) is a natural microbial culture used to introduce Dehalococcoides organisms to sites where they are absent to promote the complete dechlorination of PCE, TCE, DCE, and vinyl chloride in groundwater.

Tallevast residents formed a citizen action group called FOCUS—Family Oriented Community United Strong. FOCUS was successful in obtaining substantial funding from the RP to retain an independent consultant to review and advise community members on the content of technical documents in lay terms. The independent consultant also performed additional soil and groundwater sampling and a supplemental health risk assessment and participated in technical strategy discussions with the RP and regulators. The Vice President of FOCUS, Ms. Wanda Washington, provided the local newspapers with a series of insightful quotations that underscore the challenges that regulators and responsible parties face when plumes affect private wells. These quotes are included here not to malign the FDEP, FDOH, the RP or their consultants; rather, the quotes are provided to emphasize the dilemma faced by regulators, responsible parties, consultants, and the community alike when 1,4-dioxane contamination is discovered late in the cycle of site investigation and remediation.

Regarding the expanding plume, Ms. Washington provided this comment:

> "These experts get up and talk about my safety and welfare and say everybody is OK," said Wanda Washington, vice president of Tallevast's community group FOCUS. "Then they draw this 50-acre plot, and now it is 131 acres? My confidence is just shrinking by the minute. Just when you think it can't get any worse, it gets worse." (*Bradenton-Herald,* Radway and Wright, 2005)

Regarding the discovery of 1,4-dioxane late in the investigative process, an editorial in the *Bradenton-Herald* expressed empathy for Tallevast residents:

> One wonders why it took until now to test for [1,4-] dioxane—or other possible chemicals. Wouldn't it be cheaper to check all potential pollutants at once rather than going back to previous drilling sites? This drip-drip-drip process of discovering and releasing bad news leaves affected residents uneasy. The very existence of a pollution threat was withheld from residents for almost four years. It's no wonder that [Ms.] Washington and members of the community group FOCUS are suspicious of almost everything they're being told. (Fishbein, 2005)

Some members of the Tallevast community are convinced that residents' health was impaired by ABC's historical operations and the resulting groundwater contamination that affected their wells. Some Tallevast residents point out that dozens of their neighbors have taken ill with cancer and other health issues that they attribute to the presence of industrial contaminants in their well water. The FDOH drafted a report on the potential health impacts to the community but found only four cases of cancer, a finding in stark contrast to an informal survey by residents that showed about 90 cases of cancer or other diseases in the Tallevast community (Fishbein, 2008).

When FDOH officials met with FOCUS in March 2008, they acknowledged that their numbers, based on a state database and figures from a local hospital, were "wildly off the mark." FDOH also admitted that they had studied the wrong ZIP code: although Tallevast has a post office, most Tallevast residents live in a Sarasota ZIP code (Fishbein, 2008). By using a statewide database to review health records, FDOH administrators apparently missed important demographic features of the Tallevast community.

Despite the adverse perceptions by the community, newspaper articles and regulatory records clearly show that once the private well contamination was discovered, state and local regulators and the RP provided a comprehensive response to ensure a safe supply of drinking water. An account of the Tallevast case by Steve Lerner posted on the Collaborative on Health and the Environment's website,* describes the response as follows:

> Once news reports of the contamination began to appear, county officials were galvanized into action. They appeared in Tallevast at 8 p.m. one evening handing out five-gallon plastic bottles of water and

* Collaborative on Health and the Environment (www.healthandenvironment.org).

> warning residents not to drink water if it came from a well. Subsequently, all wells in town were capped and above-ground blue plastic pipes were installed as a "temporary hook-up" to county water lines. The plastic pipes remain in place today, four years later. The increase in pressure from the county water hookup caused numerous leaks in faucets and hot water heaters in Tallevast homes and the telephones of Laura Ward and Wanda Washington began ringing with requests for help with plumbing problems. (Lerner, 2008)

The RP initiated a financial incentive program to abandon private water supply wells in the area, and the Manatee County government restricts the construction of new wells in the affected area.* The goals of both programs are to reduce the potential for exposure to affected groundwater and to limit potential cross-connection between vertically distinct groundwater aquifer zones (Arcadis-BBL, 2007c). More than 50 private wells were closed under this program (Arcadis-BBL, 2007a).

The community nevertheless remained dissatisfied, and construction work to install monitoring wells and pilot trenches associated with the groundwater remedial action apparently alarmed residents even more. Residents demanded that Tallevast remediation work stop until the community is moved out of harm's way. Their cause was taken up by the Tallevast representative to the state legislature, Republican Bill Galvano. Representative Galvano proposed that the whole town be moved to a new site and sent the RP a proposal suggesting that they pay for most of the cost of relocating residents to a new site of their selection in exchange for dropping their 2005 lawsuit. Tallevast residents apparently asked for assistance after some of them were declined refinancing of their mortgages because of the plume of contamination from the former American Beryllium plant (Marsteller, 2005).

To address residents' concerns over their property values, the RP established the Tallevast Property Value Protection Program for current homeowners interested in selling or refinancing their residence. The program offers a payment for the reasonable difference (up to 40%), if any, between the actual sale price negotiated and the appraised fair market value if there were no associated groundwater impact. The program also provides assistance with residential seller support services such as appraisals, home inspections, brokerage services, and home marketing assistance (www.tallevast.info).

The Tallevast site provides an excellent example of how to manage dynamic community relations through multiple channels to ensure that relevant information is distributed to community stakeholders in a timely manner. Following the initial missteps described above that eroded the community's trust, the RP hosted frequent and widely announced public meetings to explain the ongoing work and staffed these meetings with the consultant team that was performing the work. In addition, the RP maintained a website with a complete report repository and prepared dozens of informative newsletters that were mailed to the affected residents. Communications have at times been contentious, but the RP ensured that residents had ample opportunity to register their concerns and inquire directly with the parties performing the work.

9.3.5 Discussion

The Tallevast saga continues. Looking back on all that has transpired, it might be tempting to second-guess how the site was addressed. But several features of this case would probably have caused most environmental professionals to follow the same course and arrive at a similar conceptual model, only to discover later that it does not quite fit. By following conventional wisdom and orienting an investigation around biodegradable hydrophobic chlorinated solvents—as is typically done at these types of sites—the persistent and extremely mobile nature of 1,4-dioxane was not effectively assessed. The bias in the sampling and analytical methods for 1,4-dioxane in open-cased

* In this context, "abandon" means to properly seal the well according to a technical standard that will prevent cross-contamination.

irrigation and domestic wells also made a significant difference. The peculiar mounding seen in the site potentiometric surface map was generally consistent with the site location on a relative topographic high; yet, the magnitude of the mounding appeared greater than the local topographic relief. Nevertheless, most hydrogeologists would probably not start out with a conceptual model that included an unknown 50 gallons/min water main leak next to the source area! This feature alone was probably responsible for a much larger mass fraction leaving the source area in multiple directions than could have been anticipated to occur under a natural gradient.

The Tallevast case is a reminder that a conceptual model is always a work in progress, to be revisited and updated frequently as more data become available, and whose assumptions should be periodically questioned. On the basis of the documents reviewed by the author, it appears that the RP consultants, and regulators involved have been quite effective at adapting to changing conditions and have made substantial progress to address a complex and vexing 1,4-dioxane plume.

9.4 1,4-DIOXANE AND ORANGE COUNTY WATER DISTRICT OCWD GROUNDWATER REPLENISHMENT SYSTEM

The unexpected discovery of 1,4-dioxane in advance treated wastewater used in Orange County Water District's (OCWD) Groundwater Replenishment System underscores the challenge of using recycled water for groundwater basin management. Fortunately, most recycled water is not tainted by the large-volume industrial discharge of 1,4-dioxane that caused problems in Orange County; however, as this study shows, 1,4-dioxane is nevertheless ubiquitous at low concentrations in treated wastewater because of the presence of 1,4-dioxane as an impurity of surfactants in household sundries such as shampoos and liquid detergents. This study provides a cautionary tale for planners of recycled water projects and shows why it is important to test for and consider 1,4-dioxane when designing advanced wastewater treatment projects. This case study is also a success story: OCWD is the gold-standard bearer for recycled water projects and is the destination for engineers from across the globe to learn about successfully transforming wastewater into a valuable water resource.

9.4.1 History of the Orange County Water District Groundwater Replenishment System

OCWD was formed in 1933 to manage water supply and groundwater in a 360-square-mile area in the lower Santa Ana River watershed, covering much of Orange County on the southern California coast. OCWD is the water provider to about 2.3 million people in more than 20 cities, including Newport Beach, Huntington Beach, and Seal Beach on the coast, Cypress and Fullerton on the northern end, Anaheim and Yorba Linda to the east, and Santa Ana and Irvine to the south. The groundwater basin is a 3,000-ft-thick wedge of alluvium overlying bedrock formations; water supply wells extend as deep as 1,000 ft. Under normal operating conditions, groundwater is the resource for about 70% of the water provided by OCWD (Deshmukh, 2007). The alluvial basins are vulnerable to seawater intrusion at several "gaps" where bedrock has been eroded by rivers to leave a direct alluvial aquifer interconnection between the main groundwater basin and the Pacific Ocean. The 2.5-mile-wide Talbert Gap between the Newport and Huntington Mesas presents the greatest seawater intrusion challenge to groundwater-basin operations in Orange County (OCWD, 2004).

To sustain the groundwater basin and prevent overdraft while managing seawater intrusion, OCWD actively recharges the groundwater basin, primarily with water from the Santa Ana River and to a lesser extent with imported water. Percolation ponds introduce water into the groundwater basin in the upland cities of Anaheim and Orange. OCWD operates one of the oldest and most sophisticated conjunctive use and groundwater protection programs in the world, sampling more than 700 wells to collect more than 13,000 water samples and conduct more than 300,000 groundwater analyses every year. The OCWD monitoring program analyzes more than 300 constituents, including emerging contaminants beyond regulatory requirements (Yamachika, 2007).

Between 1945 and 1969, the basin went into overdraft because of increased pumping and prolonged drought conditions with only two intervening wet years during these three decades. An annual overdraft of 100,000 acre-feet brought the water table to 15 ft below sea level, leading to seawater intrusion into some aquifers; coastal wells soon began producing brackish water and had to be abandoned. OCWD increased its artificial recharge using Santa Ana River water and, beginning in 1948, water imported from the Colorado River. To pay for the imported water, OCWD assessed a pump tax on all producers of groundwater and began the long-term process of replenishing the overdrafted groundwater basin. By 1956, Orange County's population was booming, and the water level dropped further; the cumulative overdraft grew to 700,000 acre-feet, and salt water invaded aquifers as much as 5 miles inland. By 1964, the replenishment program caught up with extraction and began refilling the basin. During this timeframe, the Santa Ana River became increasingly unreliable as a source of replenishment water. In the face of increasingly expensive imported water and less reliable local surface water, OCWD turned to reclaimed wastewater, that is, recycled water (OCWD, 2008).

In 1965, OCWD began a pilot project to inject treated wastewater to create a hydraulic barrier to seawater intrusion along the Talbert Gap in Huntington Beach. In 1971, officials of the California Department of Health Services approved a full-scale project after reviewing data from the pilot project. OCWD drilled a series of 23 multipoint injection wells distributed across the 2.5-mile-wide Talbert Gap. Construction on an advanced wastewater treatment system called Water Factory 21 began in 1972. Full-scale operation of the project started in 1976 (OCWD, 1996).

Water Factory 21 was dedicated in a 1977 ceremony that included serving punch made with reclaimed water. The plant was very successful and won international acclaim. By 1991, OCWD received a permit to inject 100% recycled wastewater in the Talbert Gap wells without blending, accompanied by intensive monitoring (Sovich, 2001; OCWD, 2008). Over 95% of the injected water flows inland and contributes to basin replenishment. As of 2004, 12 MGD was injected into four aquifers in the Talbert Gap; the injected water consisted of one-third Water Factory 21 recycled water, one-third deep well water, and one-third made up of a blend of imported water and groundwater (OCWD, 2004).

The advanced treatment system consisted of chemical clarification, air stripping (for removal of ammonia), recarbonation, filtration, GAC absorption, and chlorination; the system treated 15 MGD. OCWD treated as much as 5 MGD by blending advanced treated wastewater with water purified and demineralized by reverse osmosis. The injection effort was augmented by drilling and operating seven extraction wells between the injection wells and the shoreline to intercept salt water and return it to the ocean (OCWD, 1996).

9.4.2 1,4-Dioxane Detected at Water Factory 21

In December 2001, 1,4-dioxane was discovered in Water Factory 21 influent and effluent at concentrations ranging from 1 to 75 μg/L, as well as two detections at 150 and 200 μg/L. Because the Talbert Gap injection barrier had been receiving highly treated water from Water Factory 21 for decades, in January 2002 the OCWD staff decided to sample 19 water supply wells that drew water from the affected aquifers. 1,4-Dioxane was detected in nine of these supply wells at concentrations ranging from 4 to 20 μg/L—i.e., above California's then 3 μ/L Notification Level (NL), but below the threshold for shutting down the water supply wells (Mehta, 2002a). OCWD notified the three affected water agencies—the cities of Newport Beach and Fountain Valley, and the Mesa Consolidated Water District—which placed their wells on standby. The nine wells served more than 100,000 people. California's 3 μg/L NL requires notification of consumers when 1,4-dioxane is detected at higher concentrations, but does not require shutting down a water supply unless the concentration exceeds the NL by 100-fold. Local water utility officials nevertheless decided to shut down the nine wells while researching the newly discovered 1,4-dioxane problem, forfeiting approximately 34 MGD of groundwater supply (Woodside and Wehner, 2002). The City of Newport Beach staff decision to shut down all its wells, which supply all of the city's winter drinking water, led it to purchase costly imported water to replace the supply. One-third of the supply to two other cities was similarly shut off after the discovery of 1,4-dioxane (Mehta, 2002a).

OCWD was among the first in California to test for 1,4-dioxane after encountering it while investigating contaminated groundwater in an industrial section of northern Orange County. Although the industrial groundwater contamination by 1,4-dioxane did not threaten drinking water, it led OCWD's analytical chemists (1) to adapt and refine analytical methods to detect 1,4-dioxane and (2) to decide to test for 1,4-dioxane at Water Factory 21 (Mehta, 2002a). OCWD's laboratory analytical experience with 1,4-dioxane is described in a laboratory case study in Section 4.3. Within a month of detecting 1,4-dioxane in 9 out of 19 wells, OCWD proceeded to test 51 more supply wells, revealing one more detection in a Santa Ana supply well that was not in use (Mehta, 2002b).

Water Factory 21 operators and OCWD water quality professionals intensified their monitoring for 1,4-dioxane in the influent and the effluent. The fact that influent concentrations of 1,4-dioxane were highly variable suggested intermittent industrial discharges as the likely source of the observed 1,4-dioxane levels. The Orange County Sanitation District quickly reviewed potential industrial discharges of 1,4-dioxane and identified one significant source, a manufacturer of cellulose acetate membranes, ironically a producer of reverse osmosis filters similar to those used at Water Factory 21. The Orange County Sanitation District worked cooperatively with the discharger, who voluntarily ceased the 1,4-dioxane discharge. Concentrations of 1,4-dioxane subsequently declined to levels associated with domestic wastewater (near 1 μg/L, attributed to shampoos and detergents), as shown in Figure 9.5 (Woodside and Wehner, 2002). According to USEPA's *Toxic Release Inventory,* from 1993 to 1996, the membrane manufacturer discharged 1,4-dioxane to the sewer at rates ranging from 48,000 to 87,000 lbs/year (USEPA, 2008a). This mass of 1,4-dioxane was apparently sufficient to overcome the dilution occurring over more than 5 miles of transit in sewers to Water Factory 21.

9.4.3 Wastewater Treatment and 1,4-Dioxane Removal

The passage of 1,4-dioxane through Water Factory 21 shows that the series of advanced wastewater treatment steps were not configured for 1,4-dioxane removal. The treatment technologies included GAC, air stripping for ammonium removal, and reverse osmosis, all of which are ineffective at removing 1,4-dioxane. In addition, treatment at Water Factory 21 in 2002 included UV light with

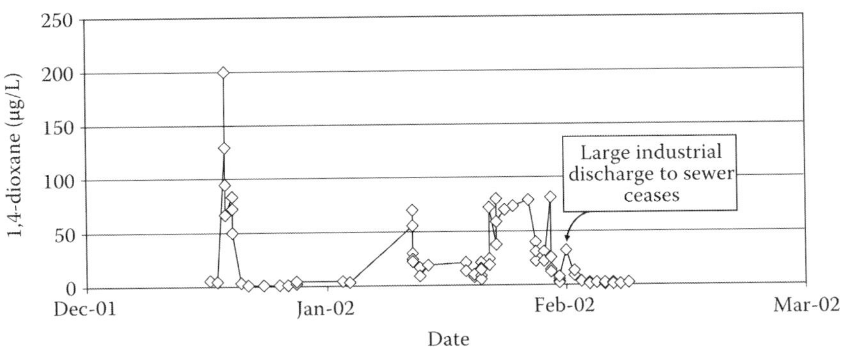

FIGURE 9.5 Influent 1,4-dioxane concentrations at Water Factory 21. (From Woodside, G.D. and Wehner, M.P., 2002, Lessons learned from the occurrence of 1,4-dioxane at Water Factory 21 in Orange County, California. Proceedings of the 2002 Water Reuse Annual Symposium. Alexandria, VA: WateReuse Association. With permission. Courtesy of the WateReuse Association.)

hydrogen peroxide, a technology that is widely used for 1,4-dioxane removal. The UV-peroxide treatment step uses a 70-MGD Trojan UV-Phox low-pressure, high-output lamp system. The Trojan UV-Phox was added in 2001 to remove nitroso-dimethyl amine (NDMA), another potentially carcinogenic contaminant discovered in 2000 (Deshmukh, 2007). Hydrogen peroxide was added at less than 5 mg/L, which was apparently insufficient to remove 1,4-dioxane to below the NL. UV light with hydrogen peroxide concentrations in the range of about 10–15 mg/L would have achieved more significant removal of 1,4-dioxane; however, the Orange County Sanitation District's successful effort to locate and eliminate the discharge made increasing the hydrogen peroxide feed unnecessary (Woodside and Wehner, 2002).

9.4.4 Communicating with Water Utilities and the Public about 1,4-Dioxane in Drinking Water Wells

Removing the source of 1,4-dioxane in Water Factory 21 did not put the local water utilities and the public at ease, as the nine wells remained contaminated with 1,4-dioxane. CDPH officials urged the public not to worry and noted that the local water agencies acted out of an abundance of caution by testing for 1,4-dioxane and closing wells despite relatively low levels of contamination. CDPH staff reminded city officials that the three parts per billion NL is the amount expected to cause one extra case of cancer per million people who drink 2 L of contaminated water every day for 70 years (Mehta, 2002a). CDPH evaluated the relatively low concentrations of 1,4-dioxane detected in the nine supply wells and assured water utility officials that the concentration of 1,4-dioxane did not represent a significant risk to health. On the basis of CPDH's recommendation, the production wells were returned to service (Woodside and Wehner, 2002).

To put the perceived threat of 1,4-dioxane in drinking water in perspective, CDPH staff prepared a "Q&A" sheet to assist water utilities with public outreach. The discovery of 1,4-dioxane in Water Factory 21 effluent was a blow to the campaign to promote the reuse of recycled water in the Talbert Gap intrusion barrier and other recharge projects, coming only 2 years after the discovery of NDMA. To deter the inevitable mentality of paralysis in which recycled water projects are opposed because they *might* introduce some unknown contaminant, CDPH adopted the strategy of acknowledging that recycled water projects indeed *will* introduce contaminants for which tests have not yet been developed. An excerpt from OCWD's Q&A on 1,4-dioxane follows:

> Q—With the discovery of NDMA in 2000, and now 1,4-dioxane this year—both cancer-causing compounds put into the groundwater by reusing sewer water—shouldn't we not reuse sewer water in the future as part of our water supply?
>
> A—No, we will always be finding new contaminants in water, no matter what the source—for example, perchlorate in water from the Colorado River. Finding new possible contaminants, regardless of where the water comes from, is part of keeping the water safe. What is important is the action that is taken when the new compounds are found and determined to be a possible threat. The water industry has a superior record in this area. At OCWD we routinely test our water with state-of-the-art methods to ensure water quality. Remember, we are living longer today [owing] in large part to the new technology that allows us to find these new substances.

The nine contaminated wells remained shut off during the month it took to identify and eliminate the source of the 1,4-dioxane; water utilities' costs for purchasing imported water mounted. The mayor of the City of Newport Beach reported that annual costs to purchase imported water would be close to $4 million, an expense for which the city requested reimbursement from OCWD, because it injected 1,4-dioxane into the groundwater basin. Similarly, the City of Costa Mesa projected an imported water cost of $850,000 annually, for which they too sought reimbursement from OCWD (Mehta, 2002c). In late February, CDPH's Chief of Drinking Water, David Spath, met with water utility officials at OCWD's offices to further explain the role of Notification Levels and why 1,4-dioxane does not pose a risk to water utility operations in the 4–20 μg/L concentration range.

Dr. Spath advised the agencies that, consistent with CDPH guidance, removing the wells from service is not necessary, nor is treatment necessary to remove 1,4-dioxane at the low levels detected. However, continued and frequent monitoring was recommended because the level of 1,4-dioxane exceeded the NL (OCWD, 2002). To reassure the public, Dr. Spath told reporters, "We feel there isn't any significant risk to the public with the use of these wells" (Mehta, 2002b).

CDPH and OCWD officials were confident that well operations could resume in large part because of OCWD's advanced understanding of the groundwater-basin hydrogeology and water balance. OCWD's groundwater flow model showed that the approximate residence time of Water Factory 21 effluent injected in the Talbert Gap injection barrier was 20 years. Because 1,4-dioxane is miscible and resistant to sorption, OCWD's hydrogeologists estimated that the remaining 1,4-dioxane in the aquifer would be pumped out within 20 years. Therefore, the 70-year exposure time upon which the NL for carcinogens is predicated would not be met, and the risk from exposure would diminish correspondingly.* This estimate of 1,4-dioxane's fate may be neglecting its tendency to be stored in fine-grained sediments, as discussed in Section 3.5. Because 1,4-dioxane migrates rapidly and resists sorption and abiotic or biologic degradation, the unintended injection of 1,4-dioxane creates a time marker for tracing the long-term movement of injected water in the groundwater basin.

As of July 2008, two of the four affected Mesa Consolidated Water District wells had decreasing concentrations of 1,4-dioxane, whereas two wells had increasing concentrations of 1,4-dioxane (MCWD, 2008a). The range of detections in Mesa's wells in 2007 was nondetect to 16 μg/L (MCWD, 2008b). The City of Fountain Valley's wells continue to be tested for 1,4-dioxane; in 2007, results ranged from nondetect to 6.6 μg/L (City of Fountain Valley, 2008). Blending with water from uncontaminated wells lowers 1,4-dioxane concentrations such that levels in water delivered to homes may be below the California NL, which in 2010 was lowered to 1 μg/L.

9.4.5 Institutional Measures to Prevent Future Industrial Discharges Adverse to Recycling Wastewater

In order to ensure that Water Factory 21 is not again subjected to unexpected high influent concentrations of 1,4-dioxane, the Orange County Sanitation District entered into an agreement with OCWD to establish a 5 μg/L target concentration for 1,4-dioxane in the wastewater treatment plant effluent (i.e., the influent to the Water Factory 21 advanced treatment system) (Mowbray, 2008). OCWD continues to pioneer new advances in the reuse of treated wastewater, the identification of emerging contaminants, and their treatment. USEPA awarded OCWD a $300,000 grant to research the fate of emerging contaminants including 1,4-dioxane in treatments using a variety of reverse osmosis membranes (Yamachika, 2007).

9.5 AIR FORCE PLANT 44, TUCSON, ARIZONA

The Air Force Plant 44 case study reflects work funded by the United States Air Force (USAF), through the Air Force Civil Engineer Center (AFCEC), formerly the Air Force Center for Engineering and the Environment (AFCEE). However, the following is solely the opinion and assessment of the author and does not reflect the opinion or approval of the USAF.

9.5.1 Background

Air Force Plant (AFP) 44 is located 15 miles southwest of downtown Tucson, Arizona. It is bounded on the east by Tucson International Airport and on the west by Nogales Highway (Route 89), the Union Pacific Tucson-Nogales railroad spur, and the San Xavier Papago Indian Reservation, of the

* Personal communication with Dr. David Spath, Chief of Drinking Water, California Department of Health Services (now CDPH), April 2003. Discussion of CDPH role in Water Factory 21 1,4-dioxane episode.

Tohono O'odham nation. To the south are the Hughes Access Road, vacant land, and light commercial property. AFP 44 covers approximately 1,266 acres and has industrial facilities occupying a total building area in excess of 1.2 million square feet. The government-owned, contractor-operated defense industrial plant was constructed by Hughes Aircraft Company (Hughes) in 1951 for the purpose of manufacturing Falcon air-to-air missiles. Hughes sold the plant to the U.S. government in 1951. Hughes then operated AFP 44 under a series of facility contracts from 1951 to 1997. In December 1997, Hughes merged with Raytheon Company (Raytheon), which has operated the plant since that time. AFP 44 has been used for the production of weapons systems since its inception in 1951 (Earth Tech, 1994).

Since 1981, the USAF, in conjunction with Hughes, has investigated soil and groundwater impacts at AFP 44 under the DOD (U.S. Department of Defense) Installation Restoration Program (IRP). The Tucson International Airport Area (TIAA) was placed on the preliminary USEPA National Priorities List (NPL) in July 1982. It was declared a Superfund Site and placed on the final NPL list in September 1992 (Earth Tech, 1994). AFP 44 is one of six project areas within the TIAA Superfund Site. The USAF is the lead federal agency at AFP 44 and is responsible for environmental response actions that are being conducted in accordance with CERCLA, as amended. In 1983 and 1984, additional USAF/Hughes studies defined an off-site groundwater TCE plume substantially attributable to AFP 44, although contributed to by the Tucson International Airport. The plume encompassed an area containing City of Tucson water supply wells, several of which were closed following verification of solvent impacts through additional sampling. In 1984, the USAF accepted responsibility for remediation activities south of Los Reales Road. An informal agreement was reached among the USAF, Hughes, the USEPA, and the Arizona Department of Health Services (ADHS) to begin groundwater remediation at AFP 44 first, with subsequent historical disposal site/source area investigation and remediation. The USAF was also party to an agreement and provided funding for design, installation, and operation of an extraction well field and treatment plant to replace the lost water supply capacity for the City of Tucson.

9.5.2 Geologic and Hydrogeologic Setting

Groundwater beneath AFP 44 occurs within Tucson Basin alluvial sediments, starting at depths of 90–140 ft bgs. Figure 9.6 presents a conceptual site model intended to schematically portray the various water-bearing units and potential migration pathways for 1,4-dioxane at the site. The unsaturated zone above the regional aquifer is made up of discontinuous sand and gravel lenses interbedded with layers of sandy clay, clayey sand, and clay. The material is unconsolidated except where cemented with caliche. A continuous stratum of reddish brown sandy clay and clay occurs between about 80 and 130 ft deep and is locally referred to as the upper clay unit. In the northwestern portion of AFP 44, perched groundwater is present above and within the upper clay unit in an area referred to as the "Shallow Groundwater Zone (SGZ)."

The regional aquifer beneath the site consists of an upper and a lower-aquifer zone separated by a thick, laterally extensive, clay aquitard. The upper zone extends down to the contact with the clay aquitard, at depths of 180–225 ft. The average thickness throughout the remediation area is 70 ft. The upper zone consists of unconsolidated, highly permeable sand and gravel layers interbedded with layers of clayey, impermeable sediments. The stratigraphy is similar to that of the unsaturated zone. Water within the individual sand and gravel beds occurs under unconfined to semiconfined conditions. In some parts of the area, the individual sand and gravel beds appear to be hydraulically isolated from beds above and below at the same location.

Locally, the upper zone of the regional aquifer is the most productive zone. The upper zone will support sustainable pumping rates of 50–150 gpm in the southeast part of the remediation area and as much as 350 gpm in the northwest (where more coarse-grained, permeable beds predominate). The upper zone also contains the bulk of the solvent contamination. An elongated contaminant plume formed as a result of a continuing migration from source areas (below former disposal sites)

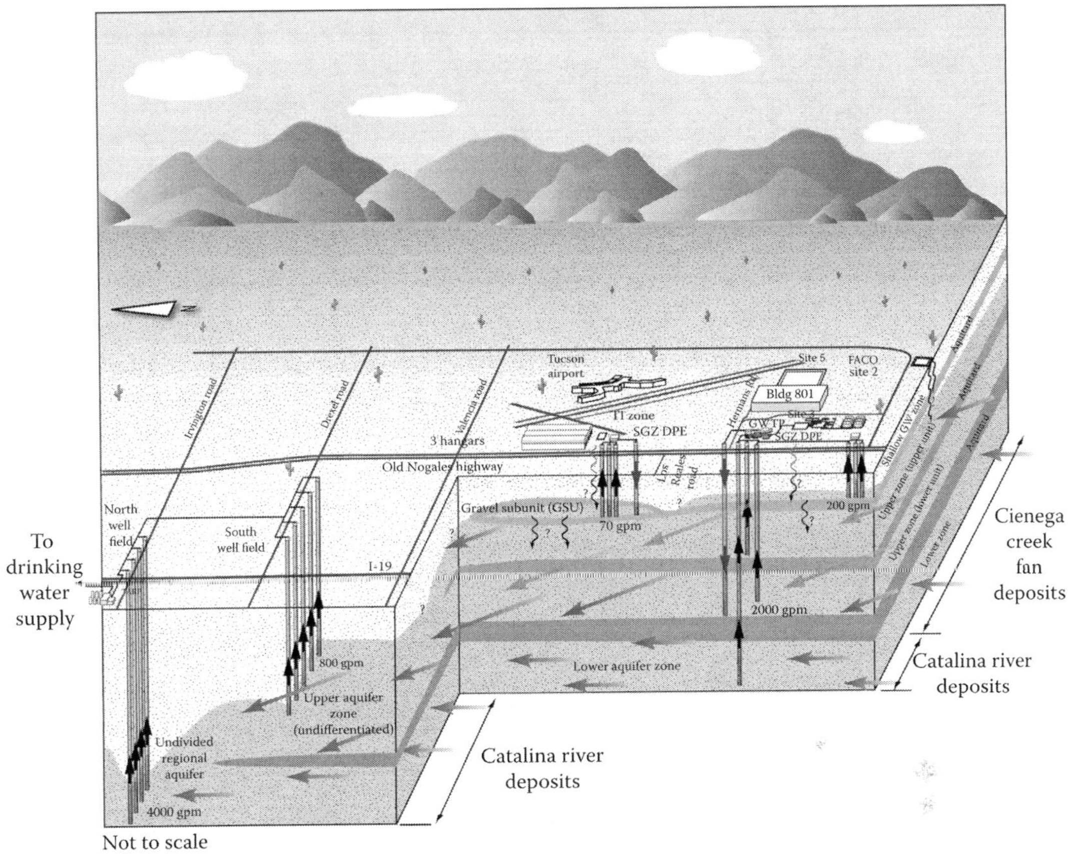

FIGURE 9.6 Schematic conceptual site model for Air Force Plant 44.

combined with relatively high groundwater flow rates in the upper zone. Regional groundwater flow in the upper zone is northward to northwestward across the remediation area.

The lower zone of the regional aquifer begins at about 300–350 ft bgs at the base of the confining clay and continues to an unknown depth. The lower zone contains minor levels of solvent contamination. To the northwest, the upper zone and lower zone combine to become the undivided regional aquifer, which is the principal aquifer for large volumes of drinking water for the City of Tucson.

9.5.3 Chlorinated Solvent Use, Release Mechanisms, and Discovery

Historical industrial processes conducted at AFP 44, including the production, maintenance, and modification of weapons systems, have resulted in wastewater and general industrial waste including solvents, paint sludge, and thinners. Chromium and chlorinated solvents, including TCE and methyl chloroform, were the principal materials used at the facility. Releases occurred primarily from sludge-drying beds, lagoons, degreasers, and landfills. Chlorinated solvents associated with AFP 44 were identified in off-site groundwater to the northwest, as were similar compounds released within other project areas at the TIAA Superfund Site (Figure 9.7). Chlorinated compounds were present in both shallow and deep soil; shallow soils were contaminated primarily by chromium and other inorganic constituents.

TCE was used at AFP 44 from the 1950s until it was replaced by methyl chloroform in 1974. Methyl chloroform was used from 1974 until the mid-1990s although the last decade of usage was at a much lower rate owing to process changes and efforts to minimize waste. The chlorinated solvents

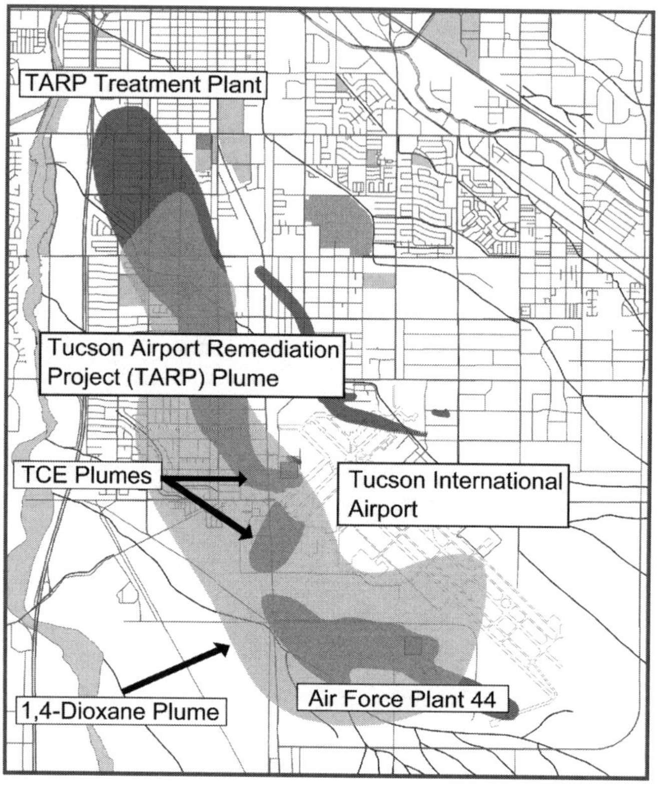

FIGURE 9.7 Groundwater contamination occurrence in the TIAA Superfund Site Project areas. Reprinted with permission from the Arizona Department of Environmental Quality (ADEQ). Copyright 2019 ADEQ.

identified in the groundwater consist of TCE with lesser amounts of DCE present. 1,1-DCE was not used as a solvent on-site, but resulted from the breakdown of methyl chloroform. Other chlorinated daughter products are present on the site at insignificant levels. The primary sources of solvents that contaminated the groundwater, in the order of greatest probable contribution, are Site 2—FACO Landfill, Site 3—Inactive Drainage Channel Disposal Pits, Site 5—Sludge Drying Beds/Building 801, and Site 14—Shallow Groundwater Zone.

The chromium plume was much smaller than the solvent plume. At the start of remediation in 1987, it covered 190 acres and was almost entirely on AFP 44 property. The sources of chromium in the groundwater were plating wastes placed in unlined sludge-drying beds at Site 5, plating wastes placed in disposal pits at Sites 3 and 4, and wastes accumulated in associated drainage channels at Site 6. The chromium plume extent was reduced dramatically once these source areas were excavated in the early to mid-1990s. Chromium is only present above standards in a few wells at the site, which are in areas of known disposal or release. Hexavalent chromium is the more soluble and more toxic oxidation state for chromium; however, at AFP 44, natural aquifer conditions are conducive to the reduction of hexavalent chromium to trivalent chromium, which is the less soluble and less toxic oxidation state.

9.5.4 Community Involvement

In the face of rising community concerns about the TCE discovered in the groundwater on the south side of Tucson, the USAF in the early 1990s developed a facility-wide community involvement program. This included holding meetings of a group that evolved to become the Unified Community

Advisory Board (UCAB), which involves all the parties at the Superfund Site. The USAF participates in the UCAB because it is the lead agent for the cleanup of AFP 44. Negative publicity for the site (measured by media coverage, letters to elected officials, and other activities) decreased by 89% following the onset of this program. In addition to support provided to the UCAB, the USAF has coordinated and documented public meetings, responded to community comments, and provided information about the ongoing cleanup at AFP 44 to community residents. Through this proactive public participation program, the USAF has been able to improve public opinion about the overall cleanup effort.

In addition to its support for the UCAB, the USAF has cohosted an ongoing community outreach program that includes tours of AFP 44 and special events such as ribbon cuttings and environmental fairs for middle schools located near the Superfund Site. The Air Force and the UCAB (with as many as 20 other area organizations) have provided opportunities for more than 1,600 students from the south Tucson area to learn about environmental science issues and technologies. Additionally, the USAF works with three different area middle schools to highlight their environmental science projects at the Tucson Earth Day Festival. The festival has become the region's largest Earth Day event. Most recently, the UCAB sponsored and supported the development of a middle and high school curriculum that introduces local students to the history of the site and examines the social, scientific, and political ramifications of the groundwater contamination.

Despite the progress the USAF has made to build trust among community members, the USAF continues to be challenged by shifting community attitudes about those responsible for the Superfund cleanup. In addition to filing lawsuits over health effects, some community members continue to distrust (or have become distrustful of) the USAF, other government agencies, or others responsible for the cleanup.

9.5.5 Remedial Actions

In 1986, the USAF released a Remedial Action Plan (RAP) that proposed groundwater extraction, treatment, and reinjection to address the contamination at the south of Los Reales Road (USAF, 1986). Construction of the groundwater treatment plant (GWTP) began that same year (Figure 9.8).

FIGURE 9.8 Aerial view of Air Force Plant 44 GWTP.

This "Groundwater Reclamation System" was intended to reduce or eliminate off-site migration of TCE and other chlorinated VOCs from AFP 44. The well field is designed to contain and remediate the southern half of the TIAA Superfund Site contaminant plume, which formed in the upper zone of the regional aquifer. A separate remediation system, referred to as the Tucson Area Remediation Project (TARP), handles the northern half of the plume, which includes solvent contamination emanating from the Tucson Airport property and 1,4-dioxane that may ultimately be determined to be coming from the airport as well as AFP 44. The regional aquifer in the Tucson Basin is a heavily used drinking water source, so restoring the contaminated part of the aquifer is a very high priority for the USAF, local governments, regulatory agencies, and the community.

The AFP 44 well field utilizes extraction and injection wells across a 2-square-mile area to achieve hydraulic containment of the contamination plume by extracting groundwater from the center of the plume and reinjecting it along the outside perimeter of the plume. This reinjection was mandated because the site is within the Tucson Active [groundwater] Management Area, which requires that groundwater extraction does not result in an overall decline in available water resources in the sole drinking water aquifer for the City of Tucson. In April 1987, the GWTP began to extract and treat groundwater at a capacity of approximately 5,000 gpm. This flow rate was reduced over time, based on MODFLOW groundwater modeling results, to optimize capture while minimizing the quantity of groundwater removed. The plant operated between 2,000 and 2,700 gpm throughout most of the 1990s. AFP 44 currently has the capacity to operate approximately 25 active upper-zone and four lower-zone extraction wells, and 10 active upper-zone recharge wells. In addition to the extraction wells, there are 65 upper-zone and 10 lower-zone wells in which water levels and water quality are monitored on a regular basis. The air-stripper system continues to operate today in accordance with the 1986 RAP; however, as will be discussed subsequently, the system is being upgraded with an advanced oxidation treatment (AOT) system specifically to address 1,4-dioxane contamination.

Once the AFP 44 remediation system was installed and demonstrated to be operating as designed, subsequent remedial actions focused on potential groundwater contamination sources, especially those associated with VOCs and chromium waste. SVE systems were installed between 1994 and 1996 at solvent contamination Sites 1, 2, and 3. Soil excavation and off-site disposal was conducted at shallow metals-contaminated Sites 4, 5, and 6. Additionally, dual-phase extraction (DPE) was conducted at Site 5 to address deep vadose zone VOC contamination. Most of these remediation systems successfully reduced contaminant levels to below agreed-upon standards, and the sites were documented as having remedial action complete.

An additional set of remedial actions involved *in situ* treatment methods including chemical oxidation (ISCO) and bioremediation/chemical reduction. The ISCO was focused on TCE and 1,1-DCE and utilized potassium permanganate ($KMnO_4$). This oxidizer has been effectively demonstrated at numerous solvent sites and was pilot-tested in a series of large-scale injection programs in the regional groundwater beneath soil Sites 2 and 3. Over 1 million gallons of water containing more than 40,000 pounds of $KMnO_4$ at concentrations ranging from 0.5% to 2% were injected at these two sites over the course of the multiyear program. VOC destruction has been encouraging: less than 10% of the maximum-measured VOC concentrations in groundwater remained after this massive $KMnO_4$ injection. Monitoring and evaluation of the program is ongoing. An *in situ* bioremediation pilot study was performed in an area with chlorinated solvent and hexavalent chromium contamination in the deep vadose zone and groundwater. Because hexavalent chromium was present in this area, oxidation technologies could not be used. An emulsified mixture of vegetable oil and sodium lactate was injected to create chemically reducing conditions, which would lead to chemical reduction of hexavalent chromium to the less mobile and less toxic form, trivalent chromium. Additionally, the injection would create favorable conditions for the growth of indigenous bacteria capable of consuming the chlorinated contaminants, either directly or cometabolically. This 2006–2008 study yielded mixed results.

The TARP plant was constructed near Irvington Road and Interstate 19, northwest of AFP 44, and placed in operation in September 1994 to contain the leading edge of the TCE plume. The system has been operating continuously since that time, with the exception of brief maintenance-related shutdowns. The system includes two separate well fields, the south well field, extracting approximately 1,000 gpm from five wells in the upper zone of the regional aquifer, and the north well field, extracting 4,000 gpm from four wells in the undivided regional aquifer. The combined extraction of approximately 4,900 gpm was treated by using air strippers, with GAC for off-gas treatment, since 1994. The system was upgraded to treat 1,4-dioxane in 2009. Treated water from the system is blended with water from other sources and distributed to the community. An estimated 3,500 pounds of chlorinated VOCs was removed from the drinking water aquifer between 1994 and 2008 (ADEQ, 2008). Because this water is provided to the community, a significant amount of public scrutiny has been directed at the pumping, blending, and distribution, especially with respect to concentrations of 1,4-dioxane in the groundwater.

9.5.6 1,4-Dioxane Discovery and Regulation

In 2002, 1,4-dioxane was discovered in influent and effluent groundwater at the AFP 44 GWTP at concentrations of approximately 10 μg/L. Ultimately, the 1,4-dioxane plume was determined to extend over 5 miles and to have a slightly wider footprint than the chlorinated VOC plume. The identification of 1,4-dioxane at the site led to the supposition, and later empirical demonstration, that the existing treatment system was not adequate to address this new-found recalcitrant contaminant. In response to this finding, the extraction well-field operation was reconfigured to containment-only at minimal extraction rates, resulting in a reduction in 1,4-dioxane levels in the effluent/recharge water from approximately 10 μg/L to below 4 μg/L. Additionally, the USAF conducted a Technology Evaluation (Earth Tech, 2004) that determined that an AOT system could treat both the VOCs and the 1,4-dioxane for about the same annual cost as the present system (when taking into account costs for chemicals, GAC, natural gas, evaporative water loss, electricity, and ongoing maintenance of the 20-year-old stripping towers).

After the USAF decided to install an AOT system, the EPA issued an Administrative Order to the USAF in July 2007 under Section 1431 of the Safe Drinking Water Act (SDWA), stating that the 1,4-dioxane in the regional aquifer posed an "imminent and substantial endangerment" to the public (USEPA, 2007d). The Administrative Order required installation of an AOT system at the AFP 44 GWTP to address 1,4-dioxane identified in the groundwater above a Region 9 Preliminary Remediation Goal (PRG) of 6.1 μg/L and to ensure that recharge water contained less than 3 μg/L. A proposed Settlement Agreement separate from the order requested that the USAF conduct a focused remedial investigation (FRI) of 1,4-dioxane in the entire plume area. The USAF believed that the SDWA Administrative Order was invalid, because the remediation was occurring under CERCLA, and therefore other regulatory enforcement mechanisms are not applicable, among other reasons. In spite of this disagreement, the USAF expedited the design and installation of the AOT system, which was programmed for installation in the fiscal year 2009. The startup of the system occurred in September 2009 as detailed below. The USAF also initiated a Phase I FRI, which addressed the 1,4-dioxane concerns expressed by the USEPA in the proposed Settlement Agreement submitted in April 2007. The scope of the Phase I FRI was (1) to evaluate the geology, hydrogeology, and water quality with data gathered from all the TIAA Superfund Site parties, (2) to assess the groundwater model that had been used by the Environmental Protection Agency, and (3) to provide recommendations to fill any data gaps identified. The Phase I FRI was completed in November 2008. In July 2008, while the Phase I FRI was being conducted, the EPA issued a Special Notice Letter to the TARP parties, which required a full remedial investigation to be performed (to include drilling, well installation, well abandonment, and other tasks) (USEPA, 2008b). This scope included tasks from Phase I and Phase II, which was funded by the USAF in March 2009 and was completed in the spring of 2016. The requirements of the Special Notice Letter did not materially change the USAF activities that were funded and ongoing.

In August 2008, the USEPA issued a request to the Tucson Airport Authority that a specific area near their Three-Hangars building be investigated as a potential source of 1,4-dioxane. This request was triggered by sampling conducted by the USGS on the Tucson Airport property that identified elevated 1,4-dioxane in a location and geologic unit that could not have been related to AFP 44 contaminant migration. Ongoing investigations by several parties focus on whether additional potential sources, besides AFP 44, exist for the 1,4-dioxane in the regional aquifer.

9.5.7 Advanced Oxidation Treatment System Upgrade

The USAF accelerated the design and installation of an AOT system to address both chlorinated VOCs and 1,4-dioxane. The system selected (following an extensive review of available technologies) was APT's HiPOx™ system. The AOT system utilizes ozone and hydrogen peroxide to form hydroxyl radicals. This application of peroxone chemistry is well documented and is further detailed in Chapter 8. Hydroxyl radicals are second only to fluorine in their oxidizing potential. Hydroxyl radicals react very rapidly to oxidize organic compounds into nonhazardous compounds including carbon dioxide and water. This technology has been demonstrated to be effective on recalcitrant compounds, including 1,4-dioxane. The oxidation process occurs in the aqueous phase and does not increase the temperature or pressure of the water because it usually occurs at very low concentrations (<1%). The fundamental difference between other peroxone technologies and HiPOx is in the associated mass transfer. In "traditional" commercial processes, air containing 2–3% ozone by weight is injected through a diffuser and bubbles upward through a contactor at or near atmospheric pressure. Mass transfer of ozone from the gas phase to the liquid (and diffusion/mixing within the liquid) is a very slow and inefficient process. As a result, destruction requires contact or residence times as long as 20 min. HiPOx incorporates high-pressure, high-precision distribution of ozone and reagents and high-efficiency mixing in order to maximize mass transfer, manage localized reagent and ozone concentrations, and reduce the mass-transfer limitation on the overall oxidation rate by orders of magnitude. Figure 8.11 illustrates how higher ozone concentrations (8–10% by weight in air), higher operating pressure (25–30 pounds per square inch gauge [psig] versus ambient pressure), and efficient, in-line mixing are combined to maximize mass transfer and reaction efficiency. The dosed fluid flows immediately through a static mixer followed by a reaction zone specifically designed for the required residence time. A critical design feature of the distributed-injection approach is the low, local concentrations of ozone and hydrogen peroxide.

While the AOT system was being designed and fabricated, the existing treatment system piping was modified to accommodate the system upgrade. The piping is configured with valves to allow bypass of the existing air-stripper systems or bypass of the HiPOx system, or operation of both systems in series, as a safety precaution. Some of the existing piping and pipe racks were utilized in the final installation configuration.

The AOT system was delivered to AFP 44 in June 2008 on a flatbed truck trailer. The AOT unit, which is installed in a steel CONEX box, was placed on an existing, unused concrete slab with a crane (Figure 9.9). The unit is connected to the power supplies utilizing existing transformers. Fiber-optic cables are connected to allow automated operation and control from the integrated control systems in place for the GWTP operations. These systems allow remote monitoring and operation of the system. Influent and effluent water samples were collected and analyzed to provide assessment of the system's operational effectiveness during the 30-day start-up phase in the summer of 2009. Optimization of the system occurred during the initial week of operations, while both treatment systems were operated in series. Once the HiPOx system performance was demonstrated, the air strippers were bypassed and mothballed, and the AOT system formally put into full-time service in September 2009.

Approximately 31 billion gallons of groundwater have been extracted and recharged to the aquifer as of December 2017 (USEPA, 2018). Over 25,000 lbs of VOCs have been removed since 1987, when the groundwater extraction and treatment system was started up. Since 2009, when the AOP

FIGURE 9.9 Advanced oxidation unit delivery and placement.

system was installed, approximately 120 lbs of 1,4-dioxane have been removed. The average pumping and injection rate is presently about 1,600 gpm. Monitoring data show that contaminant levels of TCE and 1,4-dioxane are either slowly decreasing or stable. The groundwater plume is being hydrologically controlled and the migration of both TCE and 1,4-dioxane contained. Concentrations of 1,4-dioxane from within the plume show consistent decreases over time.

9.5.8 Cost Analysis

Operating the new AOT system instead of using the existing air strippers results in an annual electrical cost reduction of $25,000 over the previous air stripper system. Additionally, the elimination of the air-stripper systems saves $25,000 in sulfuric acid neutralization costs, $10,000 in GAC cost, and $30,000 in evaporative water loss per year. However, operating the HiPOx system adds $60,000 in peroxide cost and $5,000 in the generation of oxygen for ozone production. The AOT upgrade represents a net overall savings of approximately $25,000/year in materials and electrical costs for water treatment. In addition, the new treatment plant replaces the aging air strippers with new equipment and has fewer moving parts; hence, future maintenance costs will be reduced. The existing air strippers are over 20 years old, and maintenance costs are increasing over time as the system has exceeded its 15-year designed life expectancy. The new system is also modular and therefore capable of being relocated and reused at another location in the future. Lastly, advanced oxidation is a more aggressive treatment technology, which could be capable of treating other organic contaminants that may be discovered later.

9.6 CAPE FEAR RIVER BASIN, NORTH CAROLINA

The Cape Fear River Basin in North Carolina covers 26 counties, 9,164 square miles, more than 21,000 miles of rivers and creeks, and supplies about 1.5 million people with drinking water in 120 public water systems. Detections of 1,4-dioxane, GenX, PFAS, arsenic, and selenium in raw water that supplies drinking water has alarmed regulators and the public, prompting a number of responses described in this case study. About 1 million people have been affected by wastewater discharges containing high levels of industrial contaminants (Knappe, 2017). The Cape Fear River Basin has the unfortunate distinction of having six of the 12 highest 1,4-dioxane concentrations in the Unregulated Contaminant Monitoring Rule (UCMR) database, as described below.

The seasonal variations in river flows in the watershed, coupled with fairly constant rates of wastewater discharge, have produced relatively high wastewater proportions in river flows, which partly explains the high 1,4-dioxane concentrations in the Cape Fear River watershed. For example, the wastewater percentage in the Haw River at Bynum, where the town of Pittsboro (population 4,200) draws its drinking water source, reaches more than 40% and can approach 60% when streamflows drop to 100 cubic feet per second between June and September (Knappe, 2017). The wide swings in flows and wastewater percentages are further illustrated in Fayetteville (population 204,000), where the average wastewater percentage was estimated as 2.2%, but low flow wastewater percentage was estimated as 92% (Swayne et al. 1980).

Contamination in the Cape Fear River Basin highlights the necessity to properly treat industrial and manufacturing waste streams containing 1,4-dioxane to prevent discharge into drinking water supplies. Municipal wastewater treatment facilities were found directly upstream from three of the four reaches in the basin identified with concentrations of 1,4-dioxane exceeding 0.35 μg/L, the North Carolina standard for surface water supply (NCDWR, 2016). The 1,4-dioxane surface water supply standard was exceeded along more than 200 hundred miles of river in the Haw, Deep, and Cape Fear Rivers. The fourth area exceeding 1,4-dioxane standards is slightly further downstream from a wastewater treatment facility and an inactive textile manufacturing site. Since 1,4-dioxane is not used, generated, or effectively treated by conventional wastewater treatment practices, it is likely that upstream industrial and manufacturing wastewater discharges to the Cape Fear River or its tributaries result in 1,4-dioxane in raw water at drinking water treatment plants that are not equipped for 1,4-dioxane removal. Due to the prohibitive cost of removing 1,4-dioxane from wastewater, it is unlikely that these municipal wastewater treatment facilities, and others around the country, will be able to properly treat 1,4-dioxane.

9.6.1 Detection in the Cape Fear River Basin

As part of the UCMR3, all water utility systems serving more than 10,000 people were required to test finished drinking water for over 20 chemicals, including 1,4-dioxane, between January 2013 and December 2015). UCMR3 data indicated the finished drinking water supplied by the Cape Fear River Basin contained some of the highest 1,4-dioxane concentrations in the United States (NCDWR, 2016). In response to these results, the North Carolina Department of Water Resources launched a monitoring study to delineate the high 1,4-dioxane concentrations (NCDWR, 2016; Knappe et al., 2016). Surface waters in the basin were sampled for 1,4-dioxane during all seasons to determine areas with consistently elevated 1,4-dioxane concentrations. The study included sampling upstream and downstream from municipal wastewater treatment plants that received industrial effluents, which revealed up to two orders of magnitude increase in 1,4-dioxane concentrations (Sun et al., 2016). Results from the study helped identify sources, the identification of which was used to inform appropriate parties. Finally changes in concentrations were documented in response to actions to reduce 1,4-dioxane discharges. While these objectives were accomplished, the results prompted an extension of the study to further evaluate 1,4-dioxane concentrations and distributions in the Cape Fear, Yadkin, and Neuse River Basins.

9.6.2 Phase I Monitoring Results and the Plan for Further Monitoring

To support 1,4-dioxane monitoring, North Carolina State University developed a new analytical method using heated purge-and-trap preconcentration and gas chromatography–mass spectrometry. The method achieved the lowest method reporting limit, following USEPA protocols, yet reported for purge and trap techniques, at 0.15 μg/L (Sun et al., 2016).

During the first year of monitoring, four areas within the watershed greatly exceeded the USEPA's health advisory of 0.35 μg/L. The maximum concentrations from these four areas ranged between 171 and 1,030 μg/L, with mean concentrations (43 and 351 μg/L) two to three orders of magnitude higher than the health advisory. Findings were communicated to the municipalities, project partners, and interested public (NCDWR, 2016).

During the second year of the study (October 2015 to October 2016), 12 monitoring stations were sampled quarterly to document declines in concentrations. While all concentrations declined, they remained well above the health advisory and North Carolina Groundwater Standard. Maximum 1,4-dioxane concentrations ranged from 20 to 614 μg/L, and mean concentrations at the 12 monitoring stations ranged from 11 to 260 μg/L (NCDWR, 2017a).

Following the completion of this study, a second study was proposed for 2017–2018 to evaluate 1,4-dioxane trends in the Cape Fear River Basin and extend the study to evaluate 1,4-dioxane transport in surface water for two other basins supplying drinking water, the Yadkin and Neuse River Basins. This study will be focused on areas that returned UCMR3 results for 1,4-dioxane above the calculated human health criteria for water supply bodies (NCDWR, 2017b).

9.6.3 Discussion

UCMR3 data can identify other areas of the country where finished drinking poses a human health risk due to high concentrations of 1,4-dioxane. Where applicable, industrial and manufacturing waste streams entering wastewater treatment facilities should be evaluated. Municipal water and wastewater treatment facilities are not currently configured to remove 1,4-dioxane, and many facilities have reported effluent 1,4-dioxane concentrations equal to influent concentrations. At two wastewater treatment plants in the Cape Fear River Basin, average influent 1,4-dioxane concentrations of 8.8 μg/L and 2.8 μg/L were almost unchanged in effluent concentrations, whose averages were 8.7 μg/L and 2.6 μg/L, respectively (Hopkins, 2018). Removal of 1,4-dioxane from drinking water can be done with advanced oxidation processes applying a combination of hydrogen peroxide and ultraviolet light or ozone, or UV with hypochlorite (Stepien et al., 2014).

One of the few drinking water treatment plants in the Cape Fear River Basin equipped to achieve partial removal of 1,4-dioxane, the Sweeney plant in Wilmington, achieves an average 1,4-dioxane concentration reduction of 67% using ozone and granular activated carbon operated in biological mode. Figure 9.10 profiles influent and effluent 1,4-dioxane concentrations at the Sweeney drinking water treatment plant. However, ozone by itself is far less effective than AOP technologies that generate hydroxyl radicals, as described in Chapter 8. Direct oxidation by ozone has a degradation rate constant of 0.32 $M^{-1}s^{-1}$, whereas hydroxyl radical oxidation has a degradation rate constant in the range of $2.5–3.1 \times 10^9$ $M^{-1}s^{-1}$. Ozonation to remove 1,4-dioxane from Cape Fear River drinking water is slightly sensitive to pH within the first 10 minutes of contact time (less oxidation at lower pH), and more sensitive to alkalinity, with a 10% reduction of 1,4-dioxane oxidation occurring when alkalinity is 150 mg/L versus 21 mg/L. 1,4-Dioxane oxidation is most sensitive to the ratio of ozone to total organic carbon, and wastewater effluent dominated dissolved organic matter suppressed ozone oxidation rates to only 10% (Hopkins, 2018). Other constituents in Cape Fear River water compete with 1,4-dioxane during ozonation. For example, bromide inhibits 1,4-dioxane oxidation by ozone in Cape Fear River water, with a concomitant production of bromate, another undesired drinking water contaminant (Hopkins, 2018).

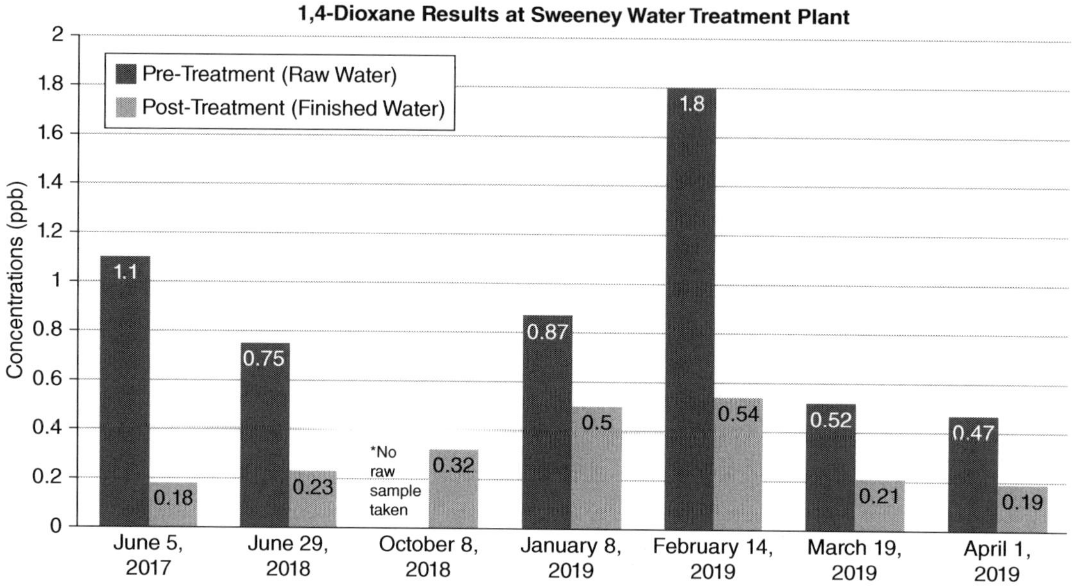

FIGURE 9.10 1,4-Dioxane Removal at the Sweeney Water Treatment Plant. Reprinted with permission from the Cape Fear Public Utility Authority. Copyright 2019 Cape Fear Public Utility Authority. https://nc-cfwu3.civicplus.com/ImageRepository/Document?documentId=12269

Other utilities have evaluated low pressure reverse osmosis (LPRO) to remove 1,4-dioxane from Cape Fear River water. Brunswick County commissioned a feasibility analysis for treatment technologies capable of removing both GenX and the suite of PFAS compounds particular to the Cape Fear River Basin, as well as 1,4-dioxane. The recommended technology, subject to pilot testing verification, was LPRO, for which 1,4-dioxane removal rate was forecasted as 90% or greater, based on earlier studies of RO Membrane performance (Linares, 2011; CDM Smith, 2018).

For most water utilities, however, installation and operation of advanced oxidation and other 1,4-dioxane removal technologies is prohibitively expensive for local governments. Therefore, the most effective strategy is to require elimination of 1,4-dioxane at the industrial or manufacturing source. To facilitate this strategy, in May 2019 the North Carolina Department of Environmental Quality established requirements for 25 municipalities in the Cape Fear River Basin with industrial pretreatment monitoring programs to monitor for a set of emerging compounds starting in the summer of 2019 (NCDEQ, 2019).

9.7 1,4-DIOXANE SUPPLY WELL CONTAMINATION ON LONG ISLAND

The 2.86 million residents of Suffolk and Nassau counties, which account for 95% of Long Island's area, rely solely on groundwater for their water supply. Urbanization on Long Island has resulted in increased groundwater withdrawals, about half of which is discharged to the ocean as treated wastewater, causing groundwater levels to decline, and risking upconing or landward migration of marine saltwater. Increased build-out has led to a higher proportion of impervious surfaces, increasing runoff, and decreasing recharge. The challenge of managing the delicate balance of water supply for a growing population is made more difficult by a wide range of groundwater quality problems found in Long Island's aquifers.

Following Charles Lindbergh's historic transatlantic flight, Long Island became the "cradle of aviation", with numerous civil and defense aircraft companies establishing production plants there. Due to highly transmissive soils and aquifers and decades of widespread use of agricultural and industrial chemicals, Long Island supply wells have experienced significant contamination from chlorinated solvents, aldicarb (a pesticide used in production of potatoes and other crops), nitrate, perchlorate, and as identified more recently, per- and poly-fluoroalkyl substances (referred to as "PFAS") and 1,4-dioxane.

Long Island's groundwater supply is recharged entirely by precipitation, which is plentiful, with an island-wide average of 44 inches/year, about half of which recharges groundwater (Cartwright, 1997). Groundwater occupies multiple aquifers whose lithology and stratigraphy vary by location. Long Island's aquifers are comprised of unconsolidated deposits overlying crystalline bedrock, which dips slightly to the southeast. Bedrock outcrops at the northwestern shore, but is buried by about 2,700 ft of aquifer sediments beneath the south-shore barrier islands. The unconsolidated units, in descending order, are the Pleistocene-aged glacial deposits, which form the upper glacial aquifer and the Gardiners Clay, the Cretaceous-aged fluvial and deltaic Magothy aquifer, the Raritan confining unit (the Raritan clay), and the Lloyd sand aquifer. These aquifers are hydraulically connected throughout, and the glacial and Magothy aquifers act as recharge for underlying units (Franke and Cohen, 1972).

Geologically, Long Island was formed very recently—11,000 years ago—when the last advance of the Laurentide Ice Sheet in the Wisconsin Stage retreated. The retreating, ~1,000 foot-tall glacier deposited the Ronkonkoma moraine (a hilly ridge of rock mixed with sand and gravel pushed by glacier and deposited as the glacier retreated). The moraine comprises the rocky, higher elevation northern half of Long Island, and the glacial outwash plain, forms the lower, flatter southern half of Long Island. The landscape and buried aquifers are shaped in part by kettle hole depressions that formed lakes and sand and silt deposits.

Precipitation percolating into the ground either stays in the shallow, unconfined water-table aquifer, or percolates deeper, into confined aquifers. Water in the water-table aquifer flows horizontally and discharges to streams or the shore. Water in the glacial aquifer recharges the deeper Magothy aquifer over large areas, particularly where Pleistocene deposits or the Gardiner Clay are absent. North of the groundwater divide, water that reaches the water table flows northward toward Long Island Sound, and water that reaches the water table south of the divide flows southward toward the Atlantic Ocean. There are a number of factors that may alter this flow pattern, however, particularly in the deeper Magothy Aquifer. These include the effects of large artificial recharge basins, municipal well pumping, and, in some areas, the injection of water through diffusion wells. There is a deep recharge zone that runs through the ridge of highest water table elevation along Long Island's "groundwater divide," an approximately five-mile wide spine that bisects the center of the island. Precipitation that falls in that zone percolates deeper, into the confined Magothy, Raritan, and Lloyd aquifer units.

9.7.1 Contamination

The Old Bethpage, Port Washington, and Syosset Landfills are known sources of groundwater contamination listed as federal Superfund sites, and many of the other, smaller "town dumps" among Long Island's 35 operating and inactive landfills may also be sources of contamination. Many of these landfills legally received large quantities of chemical waste until late 1980, when federal law required disposal of solvent and other chemical wastes in registered hazardous waste facilities. Municipal landfills receiving only municipal refuse, construction debris, and septic waste were nevertheless sources of chlorinated solvent contamination, e.g., North Sea's landfill in Southampton Town. The Town of Babylon's landfill generated a very large municipal leachate plume, two miles long, and hundreds of feet wide. Local legislation (the Long Island Landfill Law) required closure of landfills by 1993. Even so, landfills nevertheless represent a significant potential contribution to Long Island well contamination with 1,4-dioxane. This is in part because some landfills received septic waste, in which septic cleaning solvents containing 1,4-dioxane stabilized methyl chloroform were used before such cleaners were banned in 1981. In 1977, a survey of septic cleaner products usage showed annual usage was 67,500 gallons of solvent-based septic tank and drain opening cleaners in Nassau County alone, of which 18,600 gallons was methyl chloroform (Nassau County Department of Health, 1977). At that time, 7% of 228 community wells tested had detections of methyl chloroform, as did 6% of non-community water systems (maximum detection 85 μg/L), with a maximum concentration of 310 μg/L (NYDOH, 1977). Cess-pool (septic tank) cleaners and drain openers that contained methyl chloroform included "DRAINZ" (Jan Cyn Manufacturing, Central Islip, NY—31% methyl chloroform) and "Kitchen Drano" (The Dracket Company, Clifton, NJ—75% methyl chloroform).

A survey of 93 streams in Suffolk County sampled for 61 volatile organic compounds, and found methyl chloroform most frequently among all VOCs, at 32% of the streams. The streams in question drain the glacial deposits of sand, gravel, and clay, which also comprise the water-table aquifer in Suffolk County (USGS, 1997). The detection rate varied seasonally, with methyl chloroform detected in 26% of streams in samples collected between April and September, and 40% of streams in samples collected between October and March. The maximum concentration was 4 μg/L. The biodegradation product of methyl chloroform, 1,1-dichloroethane, was detected in 15% of stream samples, with a maximum concentration of 3 μg/L. The majority of detections were located in the more populated western end of Suffolk County, with more detections on the Atlantic Ocean (south) side than in streams that drain northerly to Long Island Sound (USGS, 1997).

9.7.2 Supply Well Contamination

Between 1989 and 1998, major industries in Nassau County alone used more than 1,000,000 pounds of methyl chloroform per year, as reported to EPA's Toxic Release Inventory database. Approximately 30% of supply wells in Nassau County are treated for VOCs, with 70 air strippers treating 100 wells. The specific sources of methyl chloroform groundwater contamination are unknown for 85% of treatment facilities.

Nassau and Suffolk counties have of necessity led the nation in detailed aquifer characterization and groundwater quality assessment, installing extensive networks of multilevel monitoring wells to gage conditions in each of the aquifers serving the island. Two multi-year surveys for total volatile organic compounds (TVOCs) were conducted in Nassau County between 1985–1987 and 2000–2003; results are summarized in Table 9.2. The survey for TVOCs does not differentiate between methyl chloroform, to which 1,4-dioxane was added, and other solvents. Nevertheless, the extensive detections of TVOCs in Nassau and Suffolk County monitoring wells sets the stage for widespread 1,4-dioxane contamination.

TABLE 9.2
Inter-Decadal Monitoring for Total Volatile Organic Compounds in Regional Monitoring Wells, by Aquifer—Nassau County

Aquifer	Period	>100 μg/L TVOC	5–100 μg/L	ND to 5 μg/L	Wells Sampled
Magothy and Jameco Aquifers	1985–1987	4 samples (7%)	24 samples (43%)	28 samples (50%)	56 samples = 53 wells
	2000–2003	10 samples (8%)	8 samples (6%)	107 samples (86%)	125 samples = 94 wells
Upper Glacial Aquifer	1985–1987	49 samples (17%)	95 samples (33%)	148 samples (51%)	292 samples = 222 wells
	2000–2003	11 samples (4%)	34 samples (11%)	287 samples (87%)	302 samples = 227 wells
Lloyd And North Shore Aquifers	2000–2003	0 samples	4 samples (4%)	91 samples (96%)	95 samples = 64 wells

Source: Nassau County Department of Public Works

TABLE 9.3
Inter-Decadal Frequency of Detection: Methyl Chloroform and Breakdown Products

1990–1999		2000–2003	
Compound	**Frequency**	**Compound**	**Frequency**
1,1,1-Trichloroethane	41%	1,1-Dichloroethane	30%
1,1-Dichloroethane	25%	1,1,1-Trichloroethane	26%
1,1-Dichloroethene	15%	1,1-Dichloroethene	19%
2015 and 2016 Detections of Methyl Chloroform and Its Breakdown Products in Nassau County Supply Wells (Source: WaterTraq)			
Methyl Chloroform	49 wells	15 Water Systems	
1,1-dichloroethane	21 wells	9 Water Systems	
1,1-dichloroethylene	12 wells	6 Water Systems	

Table 9.3 provides historical and recent rates of detection for methyl chloroform and its breakdown products, 1,1-dichloroethane and 1,1-DCE, in Nassau County production wells. The high rates of methyl chloroform detection indicate a strong likelihood for 1,4-dioxane detection, if one assumes that the predominant grades of methyl chloroform used on Long Island were stabilized with 1,4-dioxane. Some grades of methyl chloroform used other stabilizers to prevent reaction with aluminum chloride in vapor degreasing, e.g., 1,3-dioxolane and dimethoxymethane. In 1972, the majority of the methyl chloroform produced in the United States (64%) was stabilized with 1,4-dioxane, based on market share, stabilizer patents, and licensing arrangements. By 1978, 75% of the methyl chloroform was produced by the patent holder or licensee for 1,4-dioxane as a stabilizer for methyl chloroform (USEPA, 1982).

A surprising data point in Table 9.3 is the continuing detection of methyl chloroform in supply wells, 20 years after its use was phased out by the Montreal Protocol. As discussed in Chapter 3, biodegradation of methyl chloroform under anaerobic conditions will produce 1,1-dichloroethane. Degradation of methyl chloroform also occurs under iron and sulfate reducing conditions. Biotic reduction of iron and sulfate forms FeS (iron sulfide), which can then abiotically degrade methyl chloroform. The abiotic degradation of methyl chloroform, mediated by biotic FeS formation, has been confirmed by measuring ratios of carbon isotopes (Broholm et al, 2014).

Hydrolysis of methyl chloroform yields acetic acid and 1,1,-DCE, with a half-life of approximately 2.9 years, as discussed in Chapter 10. In the 16–20 years that have passed since methyl chloroform was last actively used for vapor degreasing, 98.4–99.2% of methyl chloroform is expected to have hydrolyzed to acetic acid and 1,1-DCE. Hydrolysis rate measurements are typically conducted at 25°C. The 2.9-year hydrolysis half-life for methyl chloroform was deduced from analysis of field data in Santa Clara, California, where ambient groundwater temperature is 15°C. Groundwater on Long Island typically ranges between 10° and 13°C (LIAC, 2018). While there is a significant dependency of hydrolysis rate on system temperature, a more likely explanation for the persistence of methyl chloroform is the slow dissolution of pure phase methyl chloroform. Hydrolysis is limited to the surfaces of the DNAPL; methyl chloroform in pooled DNAPL or ganglia are isolated from water and therefore immune to hydrolysis (Morrison and Murphy, 2013).

UCMR 3 testing results found 1,4-dioxane in nearly 87 of 145 supply wells sampled in Nassau County (60%), 54 of which exceed the 2012 US EPA Health Advisory Level of 0.35 ug/L (37%). The population served by wells with 1,4-dioxane detections is approximately 400,000, or about 30% of Nassau County's population. Water utilities employ blending strategies to prevent or limit exposure to 1,4-dioxane, however, in 9 systems, most or all wells (≥75%) have detections. UCMR 3 sampled only large water systems (10,000 or more connections), so these statistics exclude smaller water

TABLE 9.4
UCMR 3 Results for 1,4-Dioxane Sampling in Nassau County

Nassau County Water Utility Name	Number of Wells Tested	Percent of Wells w/Dtxns	Percent of Wells w/Dtxns > 0.35 ppb	Max Detection, ug/L (bold where >0.35 ppb)	Population Served
Plainview Water District	7	100	86	***4.40***	32,000
Village of Garden City	6	100	83	***3.90***	22,371
Garden City Park Water District	5	100	80	***3.90***	20,000
Hicksville Water District	9	89	78	***33.00***	47,800
Port Washington Water District	8	75	63	***2.00***	30,000
Westbury Water District	10	70	60	***1.10***	6,000
Village of Hempstead	2	50	50	***4.50***	56,000
Manhasset Lakeville Water District	9	56	44	***0.68***	45,000
Village of Mineola	5	100	40	***0.65***	19,245
Locust Valley WD	5	60	40	***0.90***	3,400
Jericho Water District	11	91	36	***3.50***	13,600
South Farmingdale Water District	7	57	29	***1.30***	14,500
Water Authority of Great Neck North	8	88	25	***1.24***	10,200
Aqua NY, Inc.	5	40	20	***1.35***	177,000
Roslyn Water District	7	29	14	***0.86***	3,000
Long Island American Water Corporation	21	14	10	***0.92***	230,300
Massapequa Water District	9	0	0	***<0.07***	43,000
Village of Rockville Centre	4	25	0	0.21	24,000
City of Glen Cove	3	100	0	0.11	27,000
West Hempstead Water District	2	50	0	0.31	32,000
Aqua NY - Sea Cliff	2	50	0	0.10	15,000

systems (17 water systems on Long Island serve fewer than 10,000 customers, with a still smaller number of connections, altogether serving 71,000 people). Table 9.4 summarizes the 2013–2015 UCMR 3 results for Nassau County. Some of the systems in Table 9.4 serve fewer than 10,000 customers, and may have been among the 800 smaller water utilities pre-selected for inclusion in UCMR3.

In Suffolk County (2,373 mi²), 1,4-dioxane has been detected in at least 252 of the Suffolk County Water Authority's (SCWA) 586 active wells (43%), at locations throughout its service area. At least 108 of SCWA's wells (18%) have detections exceeding the U.S. EPA Health Advisory Level of 0.35 parts per billion (ppb). In 2016, SCWA detected methyl chloroform in 106 wells, of which 5 exceed New York's principal organic contaminant (POC) groundwater standard of 5 ug/L (6 NYCRR 703.5). The breakdown products of methyl chloroform, 1,1-dichloroethane and 1,1-DCE, were detected in 120 wells (5 exceeding the 5 ppb POC) and 88 wells (7 exceeding the 5 ppb POC), respectively. Significantly, every well in which there is a detection of methyl chloroform or one of its breakdown products also had a detection of 1,4-dioxane in 2016, as summarized in Table 9.5 and Figure 9.11.

As indicated in the table, TCA and/or its degradation products (DCE and DCA) were detected in slightly more than half of SCWA's 252 wells with 1,4-dioxane detections. Many wells in which 1,4-dioxane occurs alone are also likely the result of methyl chloroform releases, because 1,4-dioxane in the mobile phase of groundwater (in highly transmissive sands and gravels) is more mobile, less susceptible to adsorption, and more persistent than methyl chloroform and its degradation products.

TABLE 9.5
Frequency of Co-Detection of 1,4-Dioxane with Methyl Chloroform and Degradation Products in SCWA Production Wells

Category	Number of Wells	Percent of Wells with Dioxane Detections (%)
Wells with Dioxane Only	123	49
Wells with Dioxane and TCA, DCE and/or DCA*	129	51

* Percentages based on 2016 data posted to SCWA's WaterTraq online data mapping tool.

Some of the detections of 1,4-dioxane in Long Island supply wells occur at the eastern end of the island, e.g., Montauk, where there has been no industrial operations. In addition to the septic tank cleaners previously described, one potential source of 1,4-dioxane in such locations is detergents, which may carry 1,4-dioxane into septic leach fields or leaking sewer lines. Other incidental uses of products containing methyl chloroform, as well as historical disposal of household hazardous waste to unlined landfills and town dumps may also comprise sources of 1,4-dioxane. Forensic tools such as identification of marker chemicals associated with wastewater may be needed to identify 1,4-dioxane source(s).

Long Island water suppliers have been challenged by VOCs such as methyl chloroform and its breakdown products in wells since the 1970s. As described in Chapter 8, the conventional treatment technologies effective for removing VOCs from municipal supply wells are not effective for removing 1,4-dioxane, which requires treatment with advanced oxidation processes (AOP). Wellhead treatment employing AOP for dioxane generically runs about $2 per thousand gallons—excluding the significant costs of property acquisition for treatment facilities—or, in very round terms, well over $1 million per well over 5 years of treatment operations. GAC costs to treat methyl chloroform,

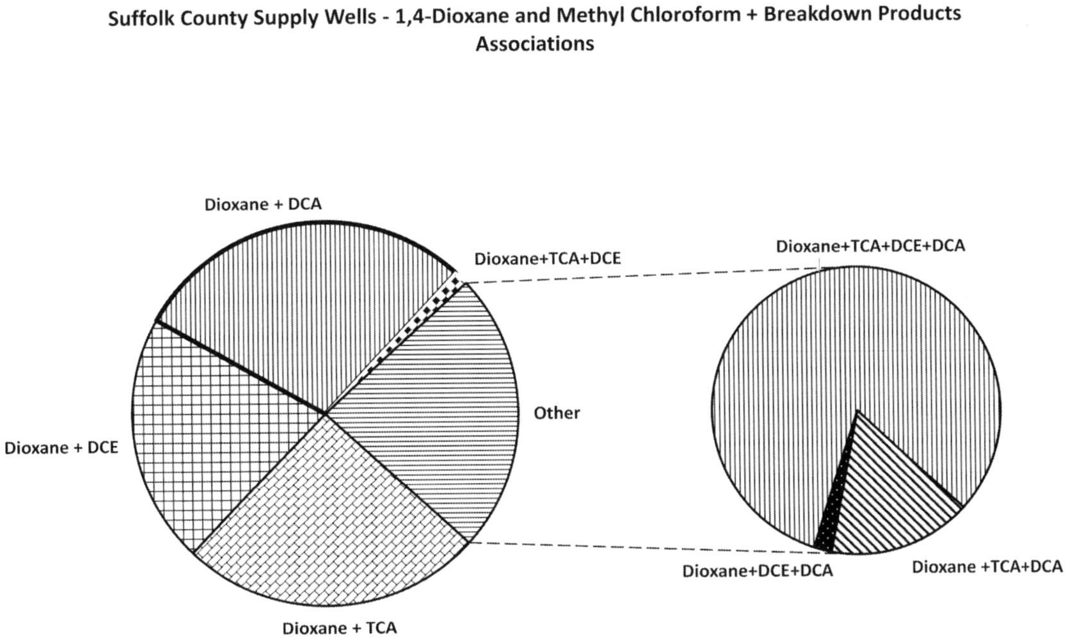

FIGURE 9.11 Association of 1,4-dioxane with Methyl Chloroform and Breakdown Products in Suffolk County Supply Wells (Source: WaterTraq).

TABLE 9.6
Prospective Costs for Treating New York Supply Wells (Statewide) to Remove 1,4-Dioxane to Draft MCL Concentrations

If MCL Is Set at:	Number of Wells Requiring Treatment	Total Capital Cost ($)
0.07	2,663	3,995,000,000
0.35	1,685	2,527,000,000
1.0	1,087	1,631,000,000
3.5	598	897,000,000

Source: Dooley, 2018. *Assumed treatment cost per well is $1.5 million; excluding annual operating costs, which run from 7% to 10% of capital costs.*

1,1-dichloroethane, and/or 1,1-DCE can vary widely. Site-specific factors such as water chemistry, the presence of multiple compounds requiring treatment (e.g., VOCs, PFAS, nitrate), property acquisition, centralized versus individual treatment, and many other factors will affect both capital costs and operating and maintenance (O&M) annual and periodic expenses.

The multi-million dollar question becomes the treatment target, i.e., the level at which New York sets its 1,4-dioxane MCL. While all water utilities would strongly prefer to serve their customers water free of any contaminants at detectable levels, the cost of doing so is prohibitive and impractical, and beyond the economic means and acceptability threshold for most ratepayers. Water utility operators must therefore operate their systems by putting contaminated wells on standby, blending water from different wells to achieve compliance with legal and advisory drinking water standards, or arranging for clean supply from neighboring systems, which also rely on groundwater. In some cases, municipal utilities in Nassau County with most or all wells impacted by 1,4-dioxane cannot import water from neighboring systems, because they are in the same position. Incorporating AOT for 1,4-dioxane can be prohibitively expensive, with costs ranging from $1.5 to $5 million per well. Municipal utilities are pursuing alternative funding models, relying on rate increases and bond measures for near-term financing, and litigation for long-term cost recovery.

The New York State Drinking Water Quality Council has evaluated 1,4-dioxane concentration thresholds for setting a New York Maximum Contaminant Level for 1,4-dioxane in drinking water. The Council's recommended MCL is 1 μg/L. In testimony regarding the draft MCL, the New York Office of Public Health provided the prospective costs for treating to different MCLs, as summarized in Table 9.6. Based on the assumption that the MCL will remain at 1 μg/L, Long Island water providers estimate that $840 million will be needed to add treatment systems to 185 drinking water wells contaminated by 1,4-dioxane, considerably more than the $317.7 million estimate provided by the State Department of Health. Annual operating costs are estimated at $30 million (Schwartz, 2019).

9.8 CONFIDENTIAL LANDFILL CASE STUDY

A case study using multiple lines of evidence (MLOE) to demonstrate the effectiveness of intrinsic bioremediation/monitored natural attenuation of 1,4-dioxane was presented in the *Remediation Journal* in Winter 2016 (Gedalanga et al., 2016). The project team used MLOE to demonstrate the 1,4-dioxane biodegradation potential in groundwater at a landfill site in the Midwest and to develop a strategy or framework for demonstrating the intrinsic biodegradation potential. The evaluation used the traditional approach of documenting 1,4-dioxane concentration and mass trends over time, spatial distribution of 1,4-dioxane and co-contaminants to document declining mass and concentrations; and geochemistry to demonstrate conditions favorable to biological activity, coupled with leading edge microbiological tools, including compound-specific isotope analysis (CSIA) to confirm degradation and biomarker analysis to confirm the presence and activity of degrading organisms.

The presence of tetrahydrofuran in groundwater was considered a significant factor since it is a known cometabolite of 1,4-dioxane. The result was a comprehensive approach that presented substantial evidence that 1,4-dioxane was biodegrading within the plume.

Results indicated that the lateral extent of 1,4-dioxane had decreased over the study period and declines in concentration were significant on the margins of the plume. 1,4-Dioxane concentrations were also declining immediately down-gradient of the landfill, likely in response to capping activities there, which would have reduced infiltration and associated leaching of contaminants out of the landfill. Fate and transport modeling indicated that the plume should be far larger if only non-destructive processes were at work. Between 2004 and 2011, the total 1,4-dioxane mass in the plume declined by 82 percent. The decline in 1,4-dioxane mass seemed to correspond with a substantial decrease in THF mass; over the same period THF mass declined by 99 percent. Geochemical analysis indicated that the core of the plume was still largely anaerobic, which is not conducive to 1,4-dioxane biodegradation, but the fringes of the plume were aerobic and therefore highly favorable for 1,4-dioxane biodegradation.

The applied microbiological tools supported the conclusions gleaned using the more traditional data and provided confirmation of the mechanism for attenuation of the 1,4-dioxane plume over time. CSIA data exhibited a trend of increasing ^{13}C with declining concentration, suggesting biological degradation. Biomarker analysis detected a number of markers associated with biological degradation of 1,4-dioxane, including: dioxane monooxygenase enzyme (DXMO, associated with the *dxmB* gene), aldehyde dehalogenase enzyme (ALDH, associated with the *aldH* gene), and soluble methane monooxygenase enzyme (sMMO). The locations exhibiting DXMO presence (Figure 9.12), which would be specific to 1,4-dioxane, tended to correlate to 1,4-dioxane concentrations, while ALDH, which can degrade more compounds than DXMO was more broadly distributed and sMMO, which is non-specific to contaminants, was ubiquitous.

Field-scale pilot testing demonstrated addition of propane, oxygen and nutrient (i.e., diammonium phosphate) to site groundwater enhanced intrinsic co-metabolic degradation of 1,4-dioxane by indigenous microorganisms. Following two months of one day per week groundwater augmentation with propane, oxygen and nutrients, 1,4-dioxane levels decreased in the pilot treatment area by up to 90% and achieved compliance with regulatory standards. The observed 1,4-dioxane degradation

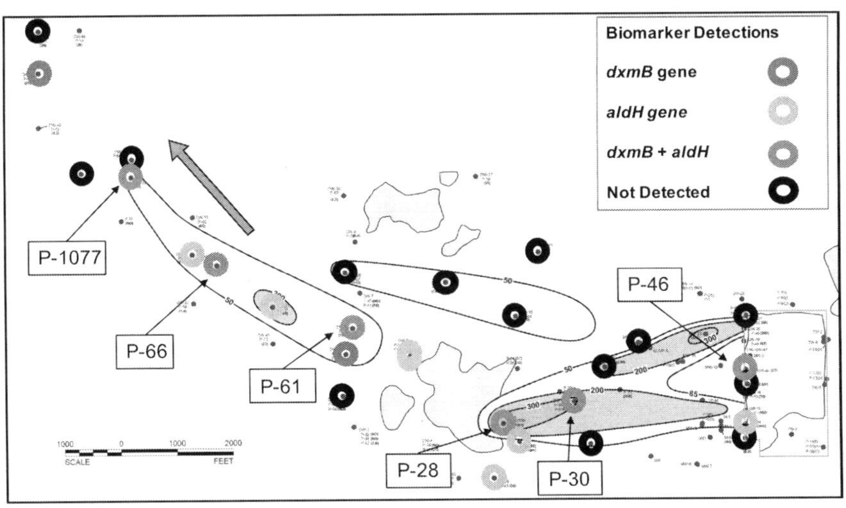

FIGURE 9.12 DXMO and ALDH biomarker detections in groundwater collected September 2015 track well within the 1,4-dioxane plume. (From Gedalanga, P., Madison, A., Miao, Y., Richards, T., Hatton, J., DiGuiseppi, W.H., Wilson, J., and Mahendra, S., 2016, A multiple lines of evidence framework to evaluate intrinsic biodegradation of 1,4-Dioxane. *Remediation Journal* 27(1):93–114. With permission.)

by indigenous microorganisms during pilot testing aligns with positive detections of 1,4-dioxane biodegradation biomarkers within the plume and indicates 1,4-dioxane mass loss can be attributed to biodegradation by multiple mechanisms.

The MLOE approach proved to be a useful tool for demonstrating natural attenuation of this plume:

- Contaminant analysis and spatial distribution demonstrated declining mass and extent of 1,4-dioxane and THF.
- Geochemical analysis delineated oxidizing conditions in the plume, helping to understand the environments conducive to 1,4-dioxane degradation.
- The groundwater model showed that a plume with no degradation would be far larger than the observed plume.
- CSIA provided strong evidence that the losses were due to biodegradation, and the biomarker analysis found a distribution of biomarker detections that made sense in the plume context.
- Biomarker analysis also found evidence that both THF-driven cometabolic biodegradation and catabolic 1,4-dioxane biodegradation were active at the site, providing multiple mechanisms for degradation.

A detailed assessment of the available data suggests that the best explanation for the loss of 1,4-dioxane mass was biodegradation by multiple mechanisms.

REFERENCES

ADEQ, 2008, Tucson Airport Remediation Project (TARP). http://www.azdEquationEquationgov/environ/waste/sps/download/tucson/tucsona.pdf (accessed September 24, 2008).

Arcadis-BBL, 2007a, *Vapor Intrusion Assessment Report: Former American Beryllium Company Site, Tallevast, Florida*. Cranbury, NJ: Arcadis-Blasland Bouck and Lee, Inc.

Arcadis-BBL, 2007b, *Interim Data Report*. Cranbury, NJ: Arcadis-Blasland Bouck and Lee, Inc.

Arcadis-BBL, 2007c, *Remedial Action Plan for the Former American Beryllium Company (ABC) Facility and Site, Tallevast, Florida*. Tampa, FL: Arcadis-Blasland Bouck and Lee, Inc.

Ashland Chemical Company, 1986, MSDS for 1,1,1-trichloroethane degrease cold vapors, Columbus, Ohio. MSDS No. 6810–00–476–5613.

ATSDR, 2005, *Health Consultation: Exposure Investigation Report—Indoor Air Testing*. Former American Beryllium site, Tallevast, Manatee County, FL: Agency for Toxic Substances and Disease Registry.

Bodwin, A., 1989, Lonesome cowboy: Gelman vs. DNR. Crain's Detroit Business, November 20, p. 8.

Brode, J., Fotouhi, F., and Kolon, S., 2005, Ultraviolet and hydrogen peroxide treatment removes 1,4-dioxane from multiple aquifers. *US EPA's Technology News and Trends*, (16): 3–4. January.

Broholm, M.M., Hunkeler, D., Tuxen, N., Jeannottat, S., and Scheutz, C., 2014, Stable carbon isotope analysis to distinguish biotic and abiotic degradation of 1,1,1-trichloroethane in groundwater sediments. *Chemosphere* (108): 265–273.

Campbell, K.M. and Scott, T.M., 1993, *Geologic Map of Manatee County, Florida*. Tallahassee, FL: Florida Geological Survey.

Cape Fear Public Utility Authority, 2019, https://nc-cfwu3.civicplus.com/ImageRepository/Document?documentId=12269. (accessed September 20, 2019).

Cartwright, R.A., 1997. Hydrogeologic-Setting Classification for Suffolk County, Long Island, New York. U.S. Geological Survey Open-File Report 96-457. Prepared in cooperation with the Suffolk County Water Authority and the Suffolk County Department of Health Services, Coram, New York. 22 pages.

CDM Smith, 2018, Northwest Water Treatment Plant Treatment Evaluation Draft Report. March 19, 2018.

Cilek, R., Jensen, D.A., Meyers, D., Martin, W.A., McClain, K., Newton J.L., and Winters, J. 2004, Tallevast Community; Tallevast, Manatee County Florida: Preliminary contamination assessment report. Florida Department of Environmental Protection, Site Investigation Section.

City of Ann Arbor, 2006, A brief history of the Ann Arbor groundwater contamination problem. http://www.a2gov.org/government/publicservices/systems_planning/Environment/pls/Pagestimeline.aspx (accessed June 9, 2006).

City of Fountain Valley, 2008, *2008 Water Quality Report*. City of Fountain Valley, CA: City of Fountain Valley Water Department.

Davis, T., 2002, Delay, denial part of battle over dioxane; First well contamination found in 1986. *Ann Arbor News,* Ann Arbor, MI, April 7, 2002.

Davis, T., 2006a, City, Pall reach deal: water quality still being monitored. *Ann Arbor News,* Ann Arbor, MI, November 21, 2006.

Davis, T., 2006b, Contaminant traced to [Ann Arbor] Inn well. *Ann Arbor News,* Ann Arbor, MI, June 29, 2006.

Davis, T., 2007, New wells track dioxane. *Ann Arbor News,* Ann Arbor, MI, May 1, 2007.

Deshmukh, S., 2007, Emerging contaminants and the groundwater replenishment system. Presented at April Water Issues Meeting, Southern California Alliance of Publicly Owned Treatment Works.

Dooley, E., 2019, Panel: Billions for dioxane's removal. Newsday, February 27, 2018.

Earth Tech, 1994, Remedial investigation report, U.S. Air Force Plant 44, Tucson, Arizona, January 1994.

Earth Tech, 2004, Final technology evaluation for treatment of 1,4-dioxane in groundwater. December 2004.

ES&T, 2005, Sampling and analysis report, Tallevast Community. Prepared for FOCUS Inc. Lakeland, FL: Environmental Sciences & Technologies, Inc.

FDEP, 2004, Letter from Florida Department of Environmental Protection Waste Management Division Director Michael W. Sole to U.S. EPA Waste Management Division Director Winston Smith. Florida Department of Environmental Protection.

FDEP, 2006, SARA #3, response to Florida Department of Environmental Protection comments, former American Beryllium site, Tallevast, Manatee County, Florida, Florida Department of Environmental Protection.

FDOH, 2007, Department of Health timeline of activities around former American Beryllium site, Tallevast Community. http://www.doh.state.fl.us/Environment/community/SUPERFUND/hazwastework/currenttimeline.htm (accessed January 21, 2008).

Feldman, S.M., 2000, *Principal Scientist/Hydrogeologist*. Melville, NY: ARCADIS Geraghty & Miller, Inc.

Fishbein, N., 2005, Level with Tallevast pollution victims. *Bradenton Herald*. Bradenton, FL, April 20.

Fishbein, N., 2008, State admits Tallevast pollution study way off mark. *Sarasota Herald-Tribune*. Sarasota, FL, March 31.

Fotouhi, F., Tousi, S., and Brode, J.W., 2006, Managing a significant release of 1,4-dioxane into a complex glacial depositional environment: The integration of hydrogeology, remedial engineering and politics. Presented at Emerging Contaminants in Groundwater: A Continually Moving Target, Groundwater Resources Association of California, Concord, CA, June 7–8, 2006.

Franke, O.L. and Cohen, P., 1972, Regional rates of ground-water movement on Long Island, New York, in Geological Survey Research 1972: U.S. Geological Survey Professional Paper 800-C, p. C271–277.

Gedalanga, P., Madison, A., Miao, Y., Richards, T., Hatton, J., DiGuiseppi, W.H., Wilson, J., and Mahendra, S., 2016, A multiple lines of evidence framework to evaluate intrinsic biodegradation of 1,4-Dioxane. *Remediation Journal* 27(1):93–114.

Herlihy, P. and Bowman, R., 2005, *HiPOX-TM Technology Lab Testing, Former American Beryllium site, Tallevast, Florida*. Pleasant Hill, CA: Applied Process Technologies Inc.

Hopkins, Z., 2018, Treatment Approaches for 1,4-Dioxane: Consideration of Background Matrix Constituents and Co-contaminants. American Water Works Association Webinar, Emerging Contaminants in Drinking Water Update. September 19, 2018.

Kellogg, K., 2005, On-going cleanup by Pall Life Sciences: A continuing controversy. *Ann Arbor Business News,* July, 2005.

Knappe, D., 2017, Emerging Contaminants: A Case Study for the Cape Fear River Basin. Presentation to the House Select Committee on NC River Quality, Raleigh, NC, October 26, 2017.

Knappe, D., Lopez-Velandia, C., Hopkins, Z., and Sun, M., 2016, Occurrence of 1,4-Dioxane in the Cape Fear River Watershed and Effectiveness of Water Treatment Options for 1,4-Dioxane Control. Water Resources Research Institute of the University of North Carolina, Report No. 478.

Lemke, L.D., 2017, 1,4-Dioxane Beneath Ann Arbor, Michigan: Investigating Groundwater Contamination in a Complex Glacial Aquifer System. Presentation to Michigan Occupational and Environmental Medicine Association, October 7, 2017.

Lerner, S., 2008, Tallevast, Florida: Rural residents live atop groundwater contaminated by high-tech weapons company. http://www.healthandenvironment.org/articles/homepage/3860 (accessed August 16, 2008).

LIAC, 2018, Groundwater Resources Management Plan. Long Island Aquifer Commission, December 13, 2018. www.liaquifercommission.com (accessed September 20, 2019).

Linares, R.V., 2011, Forward Osmosis/Low Pressure Reverse Osmosis for Water Reuse: Removal of Organic Micropollutants, Fouling and Cleaning. Thesis, King Abdullah University of Science and Technology, Thuwal, Kingdom of Saudi Arabia.

Marsteller, D., 2005, Tallevast home values still in limbo. *Bradenton Herald*. Bradenton, FL, July 24.

MCWD, 2008a, *Water Quality Program Update*. Costa Mesa, CA: Mesa Consolidated Water District.

MCWD, 2008b, *The 2008 Water Quality Report*. Costa Mesa, CA: Mesa Consolidated Water District.

MDAG, 2019, State of Michigan Department of Attorney General, March 1st, 2019. Letter from Attorney General's Office to ZAC PC, Counsel for Gelman/Pall/Danaher, re: 85 µg/L threshold for contingency plan.

MDEQ, 2000, *Information Bulletin: Gelman Sciences, Inc. site, Scio Township, Washtenaw County, Michigan, July 31*. Lansing, MA: Michigan Department of Environmental Quality.

MDEQ, 2004, *Operational Memorandum no. 5—Groundwater-Surface Water Interface Criteria*. Lansing, MA: Michigan Department of Environmental Quality, Remediation and Redevelopment Division.

MDEQ, 2005, *Fact sheet—Prohibition Zone to Restrict Use of Groundwater, Gelman Sciences, Inc., Unit E Aquifer, Groundwater Contamination, Washtenaw County*. Lansing, MA: Michigan Department of Environmental Quality, Remediation and Redevelopment Division.

MDEQ, 2007, *Fact Sheet—Prohibition Zone, Groundwater Use Restrictions, Gelman Sciences, Inc., Unit E Aquifer, 1,4-Dioxane Groundwater Contamination, Washtenaw County, March*. Lansing, MA: Michigan Department of Environmental Quality, Remediation and Redevelopment Division.

MDEQ, 2016, Department of Environmental Quality Remediation and Redevelopment Division. Cleanup Criteria Requirements for Response Activity, Rule 299, May 24, 2016.

MDEQ, 2018, 1,4 Dioxane in Ann Arbor: October 27, 2016 Town Hall Meeting Questions & Answers. Michigan Department of Environmental Quality and Michigan Department of Human Health Services, Division of Environmental Health.

Mehta, S., 2002a, 9 O.C. drinking wells closed. *Los Angeles Times,* January 30, 2002.

Mehta, S., 2002b, No health threat from dioxane-tainted wells, officials say. *Los Angeles Times,* March 7, 2002.

Mehta, S., 2002c, Cities seek restitution for tainted well water. *Los Angeles Times,* February 22, 2002.

Morrison, R.D. and Murphy, B.L., 2013, *Chlorinated Solvents—A Forensic Evaluation*. Cambridge, United Kingdom: RSC Publishing.

Mowbray, S.L., 2008, Removal of NDMA and 1,4-dioxane during secondary wastewater treatment at OCSD. Orange County—American Chemical Society Environmental Program, Chapter Meeting, Kennedy/Jenks Consultants, Irvine, CA: American Chemical Society.

Nassau County Department of Health, 1977, Assessment of the Impact of Organic Solvent Cesspool Cleaners and Drain Openers on Nassau County Drinking Water. December 8, 1977, Mineola, NY.

North Carolina Department of Environmental Quality (NCDEQ), 2019, State Takes Action to Manage Emerging Compounds in Wastewater. Press Release, May 6, 2019.

North Carolina Division of Water Resources (NCDWR), 2016, 1,4-Dioxane in the Cape Fear River Basin of North Carolina: An Initial Screening and Source Identification Study. Raleigh, North Carolina: North Carolina Department of Environmental Quality.

North Carolina Division of Water Resources (NCDWR), 2017a, 1,4-Dioxane Monitoring in the Cape Fear River basin of North Carolina: An Ongoing Screening, Source Identification, and Abatement Verification Study. Raleigh, North Carolina: North Carolina Department of Environmental Quality. February, 2017.

North Carolina Division of Water Resources (NCDWR), 2017b, 1,4-Dioxane and Bromide Monitoring Plan in the Cape Fear River Basin of North Carolina. Raleigh, North Carolina: North Carolina Department of Environmental Quality. March, 2017.

NYDOH, 1977, Letter to Francis Padar, Nassau County Department of Health, from David Axelrod, M.D., New York Department of Health, discussing drinking water analysis for volatile organic compounds. April 5, 1977.

Nyer, E.K., Kramer, V., and Valkenburg, N., 1991, Biochemical effects on contaminant fate and transport. *Ground Water Monitoring Review* 11(Spring): 80–82.

OCWD, 1996, *Water Factory 21*. Fountain Valley, CA: Orange County Water District.

OCWD, 2002, *Press Release: OCWD Receives Advice from Department of Health Services on Wells with Low Levels of 1,4-Dioxane*. Fountain Valley, CA: Orange County Water District.

OCWD, 2004, *Orange County Water District Groundwater Management Plan*. Fountain Valley, CA: Orange County Water District.

OCWD, 2008, Orange County Water District web site. http://www.ocwd.com (accessed March 22, 2008).

O'Donnell, C., 2009, Tallevast cleanup may take 50 years. *Sarasota Herald Tribune*.

Powell, T., 2005, *Test Report: Photo-Cat Treatment of Groundwater at the Former ABC Site, Sarasota*. London, Ontario: Purifics ES Inc.

Radway, S. and Wright, D., 2005, Tallevast plume reaches 131 acres. *Bradenton Herald*. Bradenton, FL, April 16.

Raphael, S., 1988, Gelman sues chemical giants. *Crain's Detroit Business* 4(8): 2.

Schwartz, D., 2019, Water providers put cost for 1,4-dioxane treatment systems at $840M. Newsday, February 14, 2019.

Shelton, D.E., 2004, Opinion and order regarding remediation of the contamination of the Unit E aquifer. Circuit Court for the County of Washtenaw, MI, Honorable Donald E. Shelton, State of Michigan Case No. 88–34734-CE.

Sovich, T., 2001, Operation and maintenance of the OCWD Talbert barrier injection well system. Presented at Seawater Intrusion Barrier Workshop—Operation, Maintenance & Repairs: Lessons Learned in Southern California, American Water Resources Association, Loyola Marymount University, Los Angeles, October 13, 2001.

SRSW, 2006, Scio residents for safe water web site: History of the Pall/Gelman site. www.http://srsw.org/ (accessed June 9, 2006).

Stanton, R., 2019, Ann Arbor considering worst-case scenario for Gelman dioxane plume. May 31, 2019. https://www.mlive.com/news/ann-arbor/2019/05/ann-arbor-considering-worst-case-scenario-for-gelman-dioxane-plume.html (accessed September 20, 2019).

Stepien, D.K., Diehl, P., Helm, J., Thomas, A., and Püttmann, W., 2014, Fate of 1,4-dioxane in the aquatic environment: From sewage to drinking water. *Water Research* 48: 406–419.

Sun, M., Lopez-Velandia, C., and Knappe, D.R.U., 2016, Determination of 1,4-dioxane in the cape dear river watershed by heated purge-and-trap preconcentration and gas chromatography–mass spectrometry. *Environmental Science and Technology* 50: 2246–2254.

Swayne, M.D., Boone, G.H., Bauer, D., and Lee, J.S., 1980, Wastewater in Receiving Waters at Water Supply Abstraction Points. Municipal Environmental Research Laboratory Office of Research and Development, U.S. Environmental Protection Agency, Cincinnati, Ohio EPA-600/2-80-044.

TetraTech, 1997, *Phase I Environmental Assessment—Former American Beryllium Company, 1600 Tallevast Road, Tallevast, Florida*. Pasadena, CA: Tetra Tech Inc.

TetraTech, 2005, *February 2005 Site Assessment Report Addendum, Former American Beryllium Company, 1600 Tallevast Road, Tallevast, Florida*. San Diego, CA: Tetra Tech Inc.

U.S. Congress, 1988, *Are We Cleaning Up? 10 Superfund Case Studies—Special Report*. Washington, DC: Office of Technology Assessment.

University of Michigan, 2008, Gift positions U-M as national leader in studying risks to public health. University of Michigan News Service Press Release, July 1, 2008. http://www.ns.umich.edu (accessed December 6, 2008).

USAF, 1986, Remedial Action Plan, Air Force Plant 44, Tucson, AR.

USEPA, 1982, Health Assessment for 1,1,1-Trichloroethane (Methyl Chloroform). Revised Draft. Office of Health and Environmental Assessment. EPA-600/8-82-003

USEPA, 1983, NPL site narrative for Seymour Recycling Corp. Seymour, Indiana. *Federal Register* Notice, September 8, 1983.

USEPA, 1986, EPA Superfund record of decision: Seymour Recycling Corp. EPA ID: IND040313017 Operable Unit 1 Seymour, Indiana, 9/30/1986. U.S. Environmental Protection Agency ROD/R05–86/046.

USEPA, 1987, EPA Superfund record of decision: Seymour Recycling Corp. EPA ID: IND040313017 Operable Unit 02 Seymour, Indiana, 9/30/1987. U.S. Environmental Protection Agency ROD/R05–87/050 1987: 31.

USEPA, 1997, *Five-year Review Report Seymour Superfund Site, Seymour, Indiana*. Region 5, Chicago, IL: U.S. Environmental Protection Agency.

USEPA, 2002, *Five Year Review, Seymour Recycling Corp Superfund Site, Seymour, Indiana*. Region 5, Chicago, IL: U.S. Environmental Protection Agency.

USEPA, 2007a, Superfund information systems. http://www.epa.govR5Super/npl/indiana/IND040313017.htm (accessed February 12, 2007).

USEPA, 2007b, *Third Five Year Review Report—Seymour Recycling Superfund Site, Seymour, Jackson County, Indiana*. Region 5, Chicago, IL: U.S. Environmental Protection Agency.

USEPA, 2007c, September 19th, 2007, Superfund 20th anniversary report—Making the program faster, fairer, and more efficient. http://www.epa.gov/superfund/20years/ch4pg9.htm (accessed November 11, 2007).

USEPA, 2007d, Administrative settlement agreement and order on consent for focused remedial investigation to address 1,4-dioxane. Region 9, San Francisco, CA: U.S. Environmental Protection Agency.

USEPA, 2008a, Toxic release inventory database. http://www.epa.gov/triexplorer/ (accessed January 17, 2008).

USEPA, 2008b, Special notice letter for the upcoming RI/FS activities, Tucson International Airport Area Superfund Site, Pima County, AR, July, 2008.

USEPA, 2018, Second Five-Year Review Report for Tucson International Airport Area Superfund Site Pima County, Arizona, September, 2018. https://semspub.epa.gov/work/09/100010552.pdf (Accessed March 18, 2019).

USGS, 1997, Occurrence of Volatile Organic Compounds in Streams on Long Island, New York, and New Jersey: Overview of available data and reconnaissance sampling.

Wall Street Journal, 1988, Gelman Sciences Inc. accused by Michigan of water pollution. *Wall Street Journal,* March 1, 1988.

Washtenaw County Health Department, 2019, Gelman Sciences Inc. 1,4-Dioxane Groundwater Plume.

Woodside, G.D. and Wehner, M.P., 2002, Lessons learned from the occurrence of 1,4-dioxane at water factory 21 in orange county, California. *Proceedings of the 2002 Water Reuse Annual Symposium.* Alexandria, VA: WateReuse Association.

Yamachika, N., 2007, Orange County Water District. Presented at Workshop on Trace Organics: Mapping a Collaborative Research Roadmap, San Francisco, CA, May 17–18, 2007.

10 Forensic Applications for 1,4-Dioxane and Solvent Stabilizers

Thomas K.G. Mohr

CONTENTS

In the town of Solventville, a printing shop, a former dry cleaner, an electrical utility substation with several large transformers, and a printed circuit board manufacturer are all located within a short distance of each other. The neighboring gas station investigates a fuel tank leak and discovers perchloroethylene (PCE) in groundwater samples. The regulator suspects that the PCE originated from the printed circuit board manufacturer, which is located up-gradient from the gas station. The electronics firm retains a consultant, namely you, to advise it of its potential liability in the PCE spill. Your initial review shows that PCE was used in varying amounts at different times at all the four facilities, and even though two facilities are cross-gradient of the current known occurrence at the gas station's monitoring wells, sewer lines, and storm drains may have provided a horizontal conduit through which PCE could have migrated. How will you proceed? It is time to open your environmental forensics toolbox and see what you have got to work with. Solvent stabilizers are a potentially useful forensic tool; however, very little work has been published documenting their application in completed studies.

An all-encompassing definition for the widely used term "environmental forensics" is found in the definitive primer from two leading authors on the subject:

"Environmental forensics is the systematic and scientific evaluation of physical, chemical, and historical information for the purpose of developing defensible scientific and legal conclusions regarding the source or age of a contaminant release into the environment."*

Forensic investigations into environmental contamination are usually oriented toward attributing responsibility to parties who used and released a contaminant and must therefore share in the burden

* From the foreword to *Environmental Forensics—Contaminant Specific Guide* by Robert D. Morrison and Brian L. Murphy (Morrison and Murphy, 2006).

of investigation and remediation (Murphy and Morrison, 2007). Environmental forensic investigations often focus on cost allocation; however, contaminant hydrogeologists can derive many other dividends by applying environmental forensic techniques to verify and enhance their site conceptual model.

Chlorinated solvents have been the subject of many forensic investigations. Methods to distinguish sources of solvents include the following:

- Mapping the molar ratios of the solvent to its breakdown products
- Correlating historical operations to the general pattern of solvent usage constrained by regulations, equipment developments, and markets
- Back-calculating the timing of a release from the length of a plume, rates of groundwater flow, and contaminant retardation
- Compound-specific isotope analysis (CSIA) to differentiate solvents by use or supplier
- Use of impurities of manufacturing processes used in specific timeframes, and solvent stabilizers as marker chemicals to distinguish two sources of the same solvent (Morrison et al., 2006)

This chapter focuses on the potential for solvent stabilizers to persist and be detected in solvent plumes in a manner useful for various methods of analysis that can be applied toward deconvoluting commingled plumes of a common solvent. Additional tools can be leveraged to confirm the biotransformation of 1,4-dioxane, such as recent improvements in CSIA of 1,4-dioxane.

10.1 LEVERAGING 1,4-DIOXANE DATA FOR PLUME ANALYSIS AND SOURCE APPORTIONMENT

Forensic analysis of environmental data is hampered by a high degree of uncertainty regarding many of the parameters that can be leveraged to interpret the source and timing of a release. Uncertainty can be reduced by collecting more data from more locations, using more accurate measurement techniques, and adding additional types of data to the analysis. When multiple lines of evidence supported by a variety of data types point predominantly in the same direction, the forensic scientist begins to get a warm and fuzzy feeling, and with a little bit of luck, she or he may actually attain the much sought after "Aha!" moment and reach a conclusion.

Perhaps the most important and least available features of solvent contamination sites are the mass, timing, and duration of the release. It is the rare exception in which these are known with certainty, for example, when an accident such as a truck overturn, train derailment, or above-ground tank failure is documented. The potential utility of solvent stabilizers to assist in establishing the mass released and timing (or rate and duration of the release) depends on whether the mass fraction of the stabilizer is sufficient to be detected in environmental samples and whether the stabilizer compounds persist in the subsurface. The ability to reliably identify and quantify solvent stabilizers through laboratory analysis is of course essential to their use in forensic investigations. Detailed site operation records and firsthand accounts of operating practices are also very useful in reconstructing the manner in which solvents were used at the site. Finally, the study of solvent stabilizers at a release site cannot by itself answer source apportionment questions. The incorporation of solvent stabilizers should be viewed as an additional tool in the forensic toolbox whose usefulness will depend on the unique circumstances of the site's history and hydrogeology. This section examines some of the factors that may contribute to or detract from the utility of solvent stabilizers in forensic investigations.

10.1.1 Estimating the Mass of the Release

Methods for estimating the release mass generally utilize a combination of the site's operating history—for example, the timeframe during which a leaking underground solvent-waste tank was used and the portion of that timeframe that the tank is assumed to have leaked—and the interpolation of subsurface data points. The site history analysis is based on a study of tank construction, photographs, and field notes that can be helpful in determining the mode of failure. Common modes of failure include

- Corrosion from an external galvanic cell in the soil
- Acidic corrosion from within the tank due to mixtures of acid wastes or solvent breakdown to hydrochloric acid
- Physical weakening of the tank due to the practice of dropping a tank dipstick used to measure product volume

Subsurface data points encompass soil core and groundwater data to delineate the mass concentration at different points within the source area and various interpolation routines to allow the investigator to infer the amount of pore space in the heterogeneous media occupied by the solvent. The interpolation analysis is made difficult by the heterogeneity of the subsurface. The percentages and interconnected porosity of gravels, sands, silts, and clays can be estimated from soil borings; however, the solution to this inverse problem lies more in the domain of probabilistic conditional simulation than in the more available "back of the envelope" approach using spreadsheet software.

Once an estimate of percentages of different geologic materials and their respective porosities and permeabilities is made, the investigator must also infer the propensity for the solvent to migrate into these media. The ability of dissolved solvent to migrate into the pore space of the analyzed soil core depends not only on advective flow and diffusion, but also on sorption of the solvent to soil organic matter and mineral surfaces, and its abiotic and biologic breakdown.

Where 1,4-dioxane is present, the uncertainty due to sorption and abiotic and biologic transformation is reduced, because 1,4-dioxane is presumed to be resistant to these fates. While intrinsic biodegradation of 1,4-dioxane occurs, the rate is slow, and often the concentrations of 1,4-dioxane in groundwater are too low to sustain significant mass reduction from biodegradation. If the starting concentration of 1,4-dioxane in the release is known or estimated, the 1,4-dioxane concentration measured in down-gradient wells can be useful for inferring the mass released and the mass remaining in the source area. For example, 1,4-dioxane could be used to establish an order-of-magnitude estimate to the upper limit of the volume of the subsurface contaminated by the release, because, in the mobile fraction of groundwater (sands and gravels) 1,4-dioxane is expected to migrate farthest among the contaminants released at facilities using methyl chloroform for vapor degreasing. However, 1,4-dioxane also has a propensity to diffuse into fine-grained pore spaces in silts and clays, which can retain mass near the point of release and decrease the overall mass migrating downgradient. Therefore, an accurate site conceptual model delineating the proportion of coarse versus fine-grained aquifer materials is essential. To this end, high-resolution site characterization tools such as closely spaced borings with continuous cores from sonic drilling, or closely spaced CPT logs coupled with continuous cores are very useful. As described in Chapter 3, the role of matrix diffusion is critical for developing viable 1,4-dioxane conceptual site models.

The range of 1,4-dioxane concentrations present in the solvent wastes released at a vapor degreasing site can be estimated from the vapor degreaser's operating history or from measurements made at other facilities that used methyl chloroform for vapor degreasing in a similar fashion. As discussed in Section 1.2.7.1, the boiling-point difference between 1,4-dioxane and methyl chloroform (101°C versus 78°C) causes 1,4-dioxane to become concentrated in vapor degreaser still bottoms. Experiments have measured the 1,4-dioxane liquid-vapor partitioning factor at methyl chloroform's boiling point. In the operating vapor degreaser, 27% of the 1,4-dioxane will partition to the vapor phase, while 73% will remain in the liquid phase (Spencer and Archer, 1981). Through continued use, a vapor degreaser iteratively partitions 1,4-dioxane such that the proportion of 1,4-dioxane in the sump will increase over time. In several weeks of daily use, the composition of the liquid solvent can evolve to contain as much as 10–20% of 1,4-dioxane. For example, the measured 1,4-dioxane concentration accumulated in the solvent sump in a vapor degreaser used for 24 days was 7.5%, whereas the originally supplied methyl chloroform had been stabilized with only 2.8% 1,4-dioxane (Spencer and Archer, 1981). Laboratory analysis of new and spent methyl chloroform sampled from a vapor degreaser at the Hayes International Corporation showed a 68% increase in 1,4-dioxane concentration, from 1.7% to 2.9% (by mass) (Tarrer et al., 1989).

It is possible to estimate the probable concentration of 1,4-dioxane in vapor degreaser still bottoms if the operating practices are known from records or personnel interviews. For example, a Model MLW-120 stainless steel Baron-Blakeslee vapor-immersion degreaser unit has a 45.4 L solvent-sump capacity. In this estimation, 45.4 L of methyl chloroform solvent with 3% 1,4-dioxane fills the sump, and 73% of the 1,4-dioxane remains in the liquid phase. Moreover, once in each 5-day work week, 6 L of makeup solvent was added to the boiling sump. Solvent vapor carryout, 1,4-dioxane removal in the water trap (see Section 1.1.1.3), and accumulation of dirt and oils in the sump are ignored. The calculations using these assumptions indicate that the 1,4-dioxane content of the solvent waste could build up to 10% by mass within 20 weeks.

Calculating the 1,4-dioxane content of still bottoms in this manner produces lower estimates than the measured data, most likely because of the significant role that vapor drag-out and liquid solvent carryout plays in removing about three times more solvent from the degreaser than 1,4-dioxane. As discussed in Section 1.1.1.3, solvent drag-out rates have been estimated at 1–2 gallons of solvent per ton of small parts cleaned when a covered, in-line conveyor system is used, whereas small parts cleaned at a similar rate in manually operated open-top degreasers lost 20% more of the solvent through vapor drag-out (ASTM, 1962). The average annual solvent losses of three methyl chloroform vapor degreasing operations at southeastern U.S. military installations (summarized in Table 1.8) was 65%.

A broad estimate of the rate and concentration of 1,4-dioxane accumulation in solvent waste can be inferred from solvent-waste recovery studies. The vapor degreasing operations using methyl chloroform at the southeastern air force bases listed in Table 1.8 are profiled more extensively in Table 10.1 for their potential to accumulate 1,4-dioxane in liquid solvent waste.

Estimates of the total mass of 1,4-dioxane at release sites are not widely available. Approximately 16,000 gallons of methyl chloroform from the Rocky Flats Closure Project were buried at the Idaho National Engineering Laboratory (INEL). Engineers at INEL estimated the 1,4-dioxane content of the methyl chloroform from the industrial literature because the methyl chloroform in question was not used for vapor degreasing where 1,4-dioxane would become progressively concentrated; rather, the methyl chloroform was used in uranium processing (Shuckrow et al., 1982). On the basis of an average 1,4-dioxane content in methyl chloroform of 2.75 vol% (2.1 wt%) and a maximum 1,4-dioxane content of 3.5 vol% (2.67 wt%), the corresponding average and maximum 1,4-dioxane masses in the buried methyl chloroform were 1,720 kg and 5,900 kg, respectively. The only documented value for 1,4-dioxane in methyl chloroform at the site of origin was 1.9% (INEL, 2006).

A major Midwest manufacturer studied the stabilizer content of solvents to improve the quality of solvents recovered by distillation for reuse. The firm's industrial engineers sought to identify the solvent stabilizers in methyl chloroform, TCE, and PCE to restore the original composition following on-site distillation. The stabilizers in TCE were identified as 1,2-butylene oxide, cyclohexene oxide, *p-tert*-butyl phenol, and 1-propanol. Stabilizers in new methyl chloroform supplied by two solvent producers were identified. The first contained the familiar formula of 1,4-dioxane (3.17 wt%), nitromethane (0.35%), and 1,2-butylene oxide (0.45%), whereas the second contained 1,3-dioxolane (2.45%), nitromethane (0.3%), and 1,2-butylene oxide (1.45%) (Bohnert and Carey, 1991).* The concentration of stabilizers in the solvent that was removed from vapor degreasers and stored in drums was not measured, as the study objective concerned refining distillation techniques.

* The report did not identify the solvent producers. Dow held the first patent for use of 1,4-dioxane in methyl chloroform, and Vulcan held an early patent for use of 1,3-dioxolane to stabilize methyl chloroform (Bursack et al., 1963). Both Dow and Vulcan produced methyl chloroform stabilized by 1,4-dioxane or 1,3-dioxolane. The third major producer of methyl chloroform, PPG, often used dimethoxymethane and *tert*-butyl alcohol instead of either 1,4-dioxane or 1,3-dioxolane (PPG, 1998).

TABLE 10.1
Annual Operating Parameters for Methyl Chloroform Vapor Degreasers at Southeastern U.S. Air Force Bases and Potential Accumulation of 1,4-Dioxane in Solvent Waste

Vapor Degreaser Number	Vapor Degreaser Dimensions (ft)	Annual Number of Batch Changes	Reservoir Volume (gallons)	Annual Solvent Usage (gallons)	Annual Solvent Waste Generation (gallons)	Annual Emissions Loss (%)	Annual 1,4-Dioxane Consumption (lb/kg)	Annual 1,4-Dioxane in Solvent Waste (lb/kg)	Annual 1,4-Dioxane in Solvent Waste (wt%)	Annual 1,4-Dioxane Lost to Vapor (lb/kg)
1	4.5 × 3 × 5.6	10	50	654	275	58	216/98	158/72	6.9	34/15
2	8 × 4 × 8	6	110	3,850	990	74.3	1,271/578	927/422	11.2	255/116
3	10 × 3 × 10	8	110	4,180	1,320	68.4	1,379/627	1,007/458	9.1	255/116

Source: Tarrer, A.R, et al., 1989, *Reclamation and Reprocessing of Spent Solvent*. Park Ridge, NJ: Noyes Data Corporation.

Notes: 1,4-Dioxane consumption was based on an assumed initial content of 3 wt%. 1,4-Dioxane content in liquid waste applies the 73% 1,4-dioxane partitioning factor for 1,4-dioxane at methyl chloroform's 78°C boiling point (Spencer and Archer, 1981). Weight percent 1,4-dioxane in liquid waste was obtained as the ratio of calculated kilograms of 1,4-dioxane in the waste to the volume of the waste in liters. This screening-level calculation neglects 1,4-dioxane losses due to removal in the water trap and does not account for any addition of make-up solvent to the vapor degreaser between solvent changes. See Section 1.1.1.3 for more on vapor degreasing.

TABLE 10.2
Stabilizer Content in Distilled Methyl Chloroform Solvent Waste at a Major Midwest Manufacturer

Stabilizer	Amount in New Methyl Chloroform (wt%)	Amount in Distilled Methyl Chloroform Solvent Waste (wt%)	Difference
1,4-Dioxane	3.17	0.92–2.79	−12% to −71%
1,2-Butylene oxide	0.45	0.26–1.59	−42% to +253%
Nitromethane	0.35	0.26–0.54	−26% to +54%

Source: Bohnert, G.W. and Carey, D.A., 1991, Scale-up of recovery process for waste solvents. Technical Communications. Kansas City, Missouri: Allied Signal Aerospace Company.

Following a filtering, desiccation, and distillation process, the distilled methyl chloroform solvent waste had a lower weight percent of 1,4-dioxane, as shown in Table 10.2.

This finding of less 1,4-dioxane in the distilled solvent waste is consistent with the expectation of increased 1,4-dioxane concentrations in vapor degreasing waste due to boiling-point differences. Distillation retains only the vapor fraction of 1,4-dioxane (~27%). The Tarrer et al. study of southeastern air force base vapor degreasers also observed a decrease in 1,4-dioxane concentration following distillation of methyl chloroform solvent waste. The waste had 2.9% 1,4-dioxane initially and 1.96% after distillation, a decrease of 32% (Tarrer et al., 1989). A second study by the same Midwest manufacturer analyzed the amount of 1,4-dioxane in distilled methyl chloroform and found a 65% decrease in 1,4-dioxane concentration (Holt, 1990) (see Table 10.7).

The solvent distillation process profiled in the Bohnert and Carey report used a preparatory drying step to return the solvent to required moisture conditions (less than 100 ppm water in methyl chloroform and TCE). Drying was achieved with a recirculating pump and a 3 Å (angstrom) (0.3 nm) molecular sieve desiccant bed. Solvent waste was recirculated through the sieve for 24–48 hours or as needed to decrease the moisture level to the target of 100 ppm water (Bohnert and Carey, 1991).

While removing the water that accumulated in the solvent, water-soluble stabilizers are also likely to be removed, particularly 1,4-dioxane. The fate of the removed water then becomes important to the forensic investigation. If the desiccant was regenerated by using a centrifuge or gravimetric technique, then the handling of the wastewater should be investigated to confirm whether it was discharged on site or hauled away as hazardous waste. If the desiccant was regenerated through an evaporation step, then it is likely that the 1,4-dioxane was emitted to the atmosphere, where it is expected to photo-oxidize, disperse, or return to the surface through precipitation or condensation at inconsequential concentrations (see Section 3.1.4).

10.1.2 Impurities*

The impurities in chlorinated solvents mentioned in Section 1.2.7 can also be important to forensic investigations and can be used together with stabilizers for determining whether iterative concentration in vapor degreasing cycles due to boiling-point differences could produce enough impurity mass to be used as a marker chemical to trace the plume. Studies of the occurrence and fate of solvent impurities provide a model for the analysis of "high-boiler" stabilizers that boil at temperatures higher than the solvent. The presence of impurities in solvents is difficult to document as most producers list their products in terms of their purity, attributing the fraction that is not the solvent to unspecified proprietary stabilizers. Nevertheless, some documentation of impurities in solvents is available, as tabulated in Table 10.3.

* From Gauthier, T.D. and Murphy, B.L., 2003, Age dating groundwater plumes based on the ratio of 1,1-dichloroethylene to 1,1,1-trichloroethane: An uncertainty analysis. *Environmental Forensics* 2003(4): 205–213. With permission.

TABLE 10.3
Documented Impurities in Chlorinated Solvents and Their Potential for Enrichment During Vapor Degreasing

Impurities in Specific Chlorinated Solvents	Impurity Amounts (Total)	Potential Concentration after Vapor Degreasing Cycle	References
	Perchloroethylene (b.p. = 121°C)		
Carbon tetrachloride, methyl chloroform, dichloromethane, trichloroethylene, and water	<1%	All would evaporate	European Chemicals Bureau (2004)
Trichloroethylene	1.6%	Evaporates	Parsons et al. (1984)
"Dichloromethane or 1,1-dichloroethylene"	0.13%	Evaporates	Parsons et al. (1984)
Vinyl chloride	0.07%	Evaporates	Parsons et al. (1984)
1,2-Dichloroethane	<0.3%	Evaporates	Borror and Rowe (1981)
Methyl chloroform		Evaporates	
1,1,2-Trichloroethane		Evaporates	
Symmetrical Tetrachloroethane (1,1,2,2)		*Concentrates*	
Pentachloroethane		*Concentrates*	
Hexachloroethane		*Concentrates*	
Hexachlorobenzene may be present as a by-product	NA	Not calculated	UNEP (2005)
Methyl chloroform, dichloromethane	See note[a]		ASTM (2006)
	Trichloroethylene (b.p. = 87°C)		
Methyl chloroform, dichloromethane	See note[a]		ASTM (2006)
Hexachlorobenzene may be present as a by-product	NA	Not calculated	UNEP (2005)
Perchloroethylene, 1,1,2-trichloroethane, methyl dichloroacetate, 1,1-dichloroethylene, and chloroform[b]	<2%	Perchloroethylene and 1,1,2-trichloroethane will concentrate—see text	Totonidis (2005)
Carbon tetrachloride[c]	0.05%	All would evaporate	Henschler et al. (1977)
Methyl chloroform[c]	0.035%		
Chloroform[c]	0.01%		
Tetrachloroethylene[d]	<0.03%	*Concentrates*	European Chemicals Bureau (2004)
1,1-Dichloroethylene[d]	<0.01%	Evaporates	
Methyl chloroform[d]	<0.01%	Evaporates	
Chloroform[d]	<0.01%	Evaporates	
Carbon tetrachloride[d]	<0.005%	Evaporates	
Dichloromethane[d]	<0.001%	Evaporates	
Bromodichloromethane[d] (b.p. = 90°C)	<0.1%	*Concentrates*	
	Methyl chloroform (b.p. = 73.8°C)		
1,1-Dichloroethylene (in 18 of 22 samples)[e]	30–900 mg/L; avg 300 mg/L	Evaporates	Henschler et al. (1980)
1,1-Dichloroethane[e] (in 11 of 22 samples)	680 mg/L	Evaporates	Henschler et al. (1980)

(*Continued*)

TABLE 10.3 *(Continued)*
Documented Impurities in Chlorinated Solvents and Their Potential for Enrichment During Vapor Degreasing

Impurities in Specific Chlorinated Solvents	Impurity Amounts (Total)	Potential Concentration after Vapor Degreasing Cycle	References
Trichloroethylene[e] (in 12 of 22 samples)	1,885 mg/L	*Concentrates*	
1,1,2-Trichloroethane[e] (in 9 of 22 samples)	1,340 mg/L	*Concentrates*	
1,2-Dichloroethane, 1,1-dichloroethane, chloroform, carbon tetrachloride, trichloroethylene, 1,1,2-trichlorocthanc, and 1,1-dichloroethylene			Stewart et al. (1969)
Chloroform[f]	100 ppm	Evaporates	Maltoni et al. (1986)
Carbon tetrachloride[f] (b.p. = 76.8°C)	250 ppm	*Concentrates*	
1,1-Dichloroethane[f]	426 ppm	Evaporates	
1,2-Dichloroethane[f] (b.p. = 83.5°C)	2,300 ppm	*Concentrates*	
1,1-Dichloroethylene[f]	398 ppm	Evaporates	
Trans-1,2-dichloroethylene[f]	50 ppm	Evaporates	
Trichloroethylene[f] (b.p. = 87°C)	200 ppm	*Concentrates*	
Tetrachloroethylene[f] (b.p. = 121°C)	475 ppm	*Concentrates*	
Dichloromethane technical grade (b.p. = 39.7°C)			
May contain methyl chloride		Evaporates	IARC (1986)
Chloroform (b.p. = 61.2°C)		*Concentrates*	
1,1-Dichloroethane (b.p. = 57.2°C)		*Concentrates*	
1,2-Dichloroethane (b.p. = 83.5°C)		*Concentrates*	

[a] Both would evaporate in PCE & TCE; ASTM test method to determine MC and DCM in TCE & PCE suggests these impurities were present.
[b] Industrial grade TCE.
[c] Industrial grade TCE (1977).
[d] Trichloroethylene (Europe, pre-1995).
[e] German technical grade.
[f] Technical grade.
NA = Not available.

Two published studies describe the potential for PCE to be present at TCE release sites where PCE was never used, through the presence of PCE as an impurity of TCE production and the concentration of PCE during vapor degreasing because of boiling-point differences. The key distinction between the two profiles is the assumption of the starting PCE concentration that may be present as an impurity of TCE.

The first profile of PCE as an impurity of TCE is provided in *Environmental Forensics—Contaminant Specific Guide* (Morrison and Murphy, 2006). Morrison, Murphy, and Doherty have noted that these solvents are produced by the same process and then separated by fractional distillation (Morrison et al., 2006), which is facilitated by the large boiling-point difference between TCE and PCE, that is, 87.2°C and 121.2°C, respectively (Shepherd, 1962). Morrison et al. (2006) cited a chemical supplier's specifications website for impurity content in degreasing grade TCE of up to 3–4% chlorinated compounds.* As described subsequently in this section, this estimate is probably too high for most grades of TCE.

The second profile of PCE as an impurity of TCE is provided by Lane and Smith (2006),† who have cited Dow Chemical's records showing that Dow's TCE had no detectable "perchloroethylene as a manufacturing impurity" (PCEMI) of TCE since about 1990. The reporting limit for analyses performed by Dow Chemical was 1 ppm PCE in TCE. Dow has been one of the leading providers of TCE in the United States since the 1960s. PCEMI may have been present in TCE produced by Dow before the 1990s, or in TCE from other solvent producers, as suggested by older MSDS and other citations. It is also important to bear in mind that many industrial operations for non-precision applications of vapor degreasing used recycled solvent, in which impurities may be enhanced by distillation processes, carryover for distillation equipment used for different solvents, and other causes.

Understanding the history of methods of solvent production can be leveraged to utilize impurities of production as time-markers. For example, PCE was produced using the acetylene method before 1978 (Morrison and Murphy, 2013). An impurity of the acetylene method for producing PCE is 1,1,2,2-tetrachloroethane. Identifying 1,1,2,2-tetrachloroethane can distinguish PCE produced before 1978, while 1,2-DCA, where identified in the absence of other sources (such as older gasoline formulations in which it was used as an anti-knock compound), can be leveraged as a marker of the ethylene method for production of PCE, which was used exclusively after 1978. Analyzing for time-marker chemicals at lowered detection limits (e.g., running GC/MS in selected ion monitoring mode), can be useful for bracketing the timeframe in which releases occurred. The timing of a release is valuable information that can assist with cost allocation and prove that a release occurred while an insurance policy included pollution liability (generally pre-1986 for comprehensive general liability policies). Identifying the time-frame of release can also assist with estimating the probable extent of contaminant migration for planning subsurface investigation to delineate a plume.

One of the attributes that producers of solvents marketed for a competitive edge was the quality and reliability of their products. There was an inherent business interest in producing high-quality, high-purity TCE; hence, it is likely that the purity improved over time and as end-use demands for TCE became more specialized and sensitive to impurities, for example, in the electronics industry.

PCE may occur as a coproduct when TCE is produced by the chlorination of 1,2-dichloroethane or the oxychlorination of ethylene (Doherty, 2000a). As noted in Section 1.2.3, typical technical or industrial grade TCE has been known to include about 0.15% carbon tetrachloride as the major chlorinated hydrocarbon impurity since the 1950s; however, PCEMI was not mentioned (Willis and Christian, 1957). PCEMI was not reported in analyses of pre-1977 American technical grade TCE (Henschler et al., 1977) or in pre-1990 American technical grade TCE (Holt, 1990). A Japanese analysis of pre-1983 industrial grade TCE detected several impurities, but PCE was not among them (Tsuruta et al., 1983). TCE produced in Europe before 1995 had up to 0.03% PCE as an impurity (European Chemicals Bureau, 2004). Indian TCE and British reagent grade TCE both contained detectable PCEMI in more recent analyses (Totonidis, 2005). In America, a widely used grade of vapor degreasing TCE, Dow Chemical's Neu-Tri™, had a purity of 99.4% in 1981, as indicated in Table 1.14, with stabilizers composing the balance. However, a technical grade of TCE, meeting military specification O-T-634 and distributed by Ashland Chemical in 1986, indicated a TCE content of 95% or more. Dow's technical grade TCE for military specification O-T-634 in 1981 was the same as its Neu-Tri™, reported as 99.4% TCE and 0.6% 1,2-butylene oxide with identical

* At time of publication, the cited website is no longer operational.

† See Lane and Smith (2006).

TABLE 10.4
MSDS Descriptions of Trichloroethylene Purity in the 1980s and 1990s

Company	Date	Id Number	Name/Info	Trichloroethylene (%)	1,2-Butylene Oxide (%)
Dow Chemical USA	12/14/81	FSC: 6810 NIIN: 00–924–7107	Trichloroethylene, Technical O-T-634	99.4	0.6
Dow Chemical USA	12/14/81	FSC: 6810 NIIN: 00–924–7107	Neu-Tri Trichloroethylene #56530	99.4	0.6
Dow Chemical USA	1/1/85	FSC: 6810 NIIN: 00–754–2813	Trichloroethylene-E	>95	0.6
Allied Chemical Corp.	1/1/85	FSC: 6810 NIIN: 01–031–5512	Trichloroethylene Low Mobile Ion	>97	—
Ashland Chemical Company	3/1/86	FSC: 6810 NIIN: 00–678–4418	Trichloroethylene Deg Cold/ Vap; Trichloroethylene, Technical O-T-634; ASTM D 4080	95	—
Baxter Healthcare Corp/Burdick and Jackson Div	12/1/89	FSC: 6505 NIIN: LIIN: 00f038115	Trichloroethylene	99	1
Chem Central Kansas City	12/1/94	FSC: 6810 NIIN: 00–184–4800	Trichloroethylene, Technical	99.4	0.5

composition listed in 1991, 1994, and 1999 (Lane and Smith, 2006). Table 10.4 lists some of the MSDS descriptions of TCE in the 1980s and 1990s.

The purity listing on an MSDS does not necessarily imply that the balance of the solvent may include impurities; solvent stabilizers often composed the balance. It is only through the laboratory analysis of solvents that PCEMI and other impurities are definitively revealed.

Whether or not PCEMI occurs in the part per million range or in the percent range, if initially present, it is likely to become concentrated in TCE through use. In vapor degreasing operations, TCE volatilizes more rapidly than PCE, and PCE is concentrated in the remaining liquid TCE. Morrison et al. (2006) explained that even at sites where PCE was allegedly never used, it may be found where spent TCE vapor degreasing waste was discharged because of the concentration of PCE as a high-boiling impurity in TCE.

Morrison et al. (2006) estimated the accumulation of PCEMI in TCE by assuming a starting ratio of 1% PCEMI, calculating the total volume of PCEMI in the volume of TCE present in the vapor degreaser solvent reservoir, and then estimating how much PCEMI volatilizes during operations. They also accounted for the addition of PCEMI in new TCE added to the degreaser between batch changes. In their example evaluation of the potential for PCEMI to become concentrated, they assumed a degreaser with a 350-gallon solvent capacity and the addition of 55 gallons of new TCE per week to make up the volume lost to volatilization, drag-out, and carryout. Morrison et al. (2006) further assumed that the degreaser is cleaned out four times per year and that a total of 100 gallons of TCE and accumulated oils and other waste are removed per year.

To estimate the rate of volatilization of PCEMI in TCE, the vapor pressure of PCE at TCE's boiling point is determined by using the Clausius-Clapeyron equation, shown here in two forms:

$$\ln P_{\text{vap}} = k\left(\frac{1}{T_{\text{B}}} - \frac{1}{T}\right), \tag{10.1a}$$

$$\ln\left(\frac{P_1}{P_2}\right) = \frac{\Delta H_{\text{vap}}}{R}\left(\frac{1}{T_2} - \frac{1}{T_1}\right), \tag{10.1b}$$

where P_{vap} is the vapor pressure of PCE at the TCE boiling point in atmospheres, T_{B} is the boiling point of PCE (394.35 K), T is the TCE boiling point (360.35 K), and k is a constant relating the ideal gas constant (R) and the enthalpy of vaporization of PCE (ΔH_{vap}). The relative rate of solvent loss is proportional to the vapor pressures, under the assumption that PCE is infinitely soluble in TCE (Morrison et al., 2006).

Lane and Smith began their analysis of PCEMI in TCE with a starting concentration between 1 and 10 ppm; they cited communication with Dow officials to support that assumption. PCEMI will accumulate owing to boiling-point differences in the still bottoms. Lane and Smith cited a typical still bottom composition of 20% TCE and 80% oils and other waste. As discussed in Section 1.1.1.3 and Table 1.10, operators were motivated to remove solvent from service when oil and grease reached 25% to avoid solvent breakdown due to overheating; hence, the inverse of this ratio is more likely to have been the composition of solvent wastes removed from vapor degreasers. The American Society for Testing Materials (ASTM) manual for vapor degreasing recommends that solvent be removed from service at 25% oil (ASTM, 1962).

Starting with an assumed PCEMI content of 5 ppm, Lane and Smith assumed a 500-fold increase in PCEMI content through iterative concentration in vapor degreasing cycles and so ended up with 2,500 ppm (or 0.25%). Their result is a substantially higher degree of concentration than the sevenfold increase obtained in the analysis by Morrison et al. (2006) to obtain 7% PCEMI from their starting value of 1% PCEMI. Even with the conservative assumption of tremendous concentration, the ending ratio in the Lane and Smith evaluation gives 400 times more TCE than PCEMI. Given that the solubility of TCE in water is substantially higher than that of PCE (1,100 versus 150 mg/L, i.e., a solubility ratio of 7.33),* TCE will preferentially dissolve out of the solvent waste once it has migrated into the saturated zone. With a 400-fold greater concentration and a 5.5 times greater solubility, the concentration of TCE in groundwater from the solvent waste under discussion is likely to be 2,200 times greater than the concentration of PCEMI (not accounting for other fate and transport properties such as sorption).

Lane and Smith pointed out that the effective solubility must be considered when estimating the likely concentrations in groundwater for TCE and PCEMI. By way of review, $S_{\text{eff}} = X_i S_{\text{water}}$, where S_{eff} is the effective solubility of a compound in a mixture (lower than its solubility when measured alone in water), X_i is the mole fraction of a compound in the mixture, and S_{water} is the compound's single-component aqueous solubility. Table 10.5 provides the terms for calculating and comparing the effective solubility of TCE and PCEMI.

A quantity of solvent in still bottoms released to groundwater may completely dissolve over time. As the Lane and Smith analysis shows, TCE will initially be a great deal more soluble in water than PCEMI owing to its much higher effective solubility. As the residual mass of solvent waste shrinks, the fraction that is PCEMI will increase, and it will eventually dissolve into water as well. The mass of PCEMI is probably small to begin with; therefore, detection of PCE at sites where TCE was used

* These solubility values differ from those used by Lane and Smith and from those cited in Table 3.11. Solubility of the chlorinated solvents has a range of values in the literature. Solubility values used here were obtained from Dow Chemical Company (2006).

TABLE 10.5
Comparison of Effective Solubility and Corresponding Groundwater Concentrations of Trichloroethylene and Perchloroethylene as a Manufacturing Impurity in Trichloroethylene Vapor Degreasing Waste

Parameter	Trichloroethylene	Perchloroethylene	Ratio
Molecular weight (Da)	131.4	165.8	—
Aqueous solubility (mg/L)	1,100	150	7.33
Mass fraction in still bottoms[a]	99.75%	0.25%	400
Mole fraction in still bottoms[a]	0.998	0.001982	503.5
Effective solubility (mg/L)	1,097.8	0.2974	3,692
Probable groundwater concentration (μg/L)	15,000	4	—

Source: Lane, V. and Smith, J., 2006, Fact or fiction? The source of perchloroethylene contamination in groundwater is a manufacturing impurity in chlorinated solvents. Presented to the National Groundwater Association Ground Water and Environmental Law Conference, July 7, 2006. Chicago, Illinois: National Groundwater Association. http://www.ngwa.org (accessed August 2, 2006).

Notes: This analysis ignores volatilization that may occur at the surface of the spill site or in the unsaturated zone, differential sorption in transit to the saturated zone, diffusion, biodegradation, and other fate processes that would alter the probable groundwater concentration. The 15 ppm concentration of trichloroethylene in groundwater is an assumed value found at more highly contaminated trichloroethylene release sites; the 4 ppb concentration of perchloroethylene is calculated from the trichloroethylene value.

[a] After assumed 500-fold concentration.

exclusively is more likely to reflect migration from other sources of PCE such as dry cleaners, brake shops, and printing operations, or such detection may reflect incomplete site operations records, obscuring some period during which PCE was used at the facility.

The analysis of PCEMI suggests that other solvent-stabilizer compounds present in solvents in small quantities would have low potential to concentrate to detectable levels even when boiling-point differences favor their concentration. Table 10.6 summarizes the potential for the high-boiling stabilizers listed in Tables 1.27 through 1.30 to become concentrated, as calculated by using the Clausius-Clapeyron equation to determine vapor pressures and the resulting mass fraction and Raoult's law to determine effective solubilities.

Calculating the iterative enrichment due to boiling-point and vapor-pressure differences and the resulting effective solubilities focuses attention on the probable composition of the still bottoms and other solvent wastes that are the source of the many chlorinated solvent plumes and the subject of ongoing cleanup. The resulting high effective solubility of 1,4-dioxane explains the elevated concentrations at which it is sometimes encountered at methyl chloroform release sites. Table 10.6 also profiles the potential for other solvent stabilizers to become concentrated in waste and preferentially dissolve into groundwater. For example, 1,3-dioxolane is rarely analyzed in groundwater samples; yet it may be present at some dichloromethane release sites. However, where 1,3-dioxolane was a stabilizer of methyl chloroform, the small boiling-point difference (only 4°C) does not favor enrichment: a starting formulation of 2 wt% 1,3-dioxolane in methyl chloroform will have an ending 1,3-dioxolane concentration of 4.5 wt% after 7 weeks and an effective solubility only seven times higher than methyl chloroform. This small increase could still be sufficient to produce detections useful as a marker chemical, but it is unlikely that a 1,3-dioxolane plume would develop at a methyl chloroform release site.

The numerous assumptions necessary to estimate partitioning and calculate stabilizer concentrations will not provide a technically rigorous or legally defensible analysis of the likely origin of solvents, as indicated by detections of stabilizers in groundwater. The most reliable evidence for

TABLE 10.6
Summary of the Potential for High-Boiling Stabilizers to Become Concentrated

Operating Parameters for Vapor Degreasers Used in Stabilizer-Enrichment Calculations[a]

Vapor Degreaser Type	Dimensions (ft)	Solvent	Batch and Reservoir Volume (gallons)	Solvent Added per Week (gallons)	Batch Changes per Year	Annual Solvent Usage (gallons)	Annual Solvent Waste (gallons)	Emissions Losses (%)
Open top	5 × 2.5 × 6.5	Perchloroethylene	70/50	27	17	1,824	420	77
Open top	8 × 4 × 8	Methyl chloroform	165/110	55	6	3,850	990	74.3
Open top	12 × 4.5 × 4.5	Trichloroethylene	280/180	35	12	7,200	1,800	75
Open top	5 × 3 × 3	Dichloromethane	75/50	30	12	2,200	280	85

Vapor Pressures of "High-Boiling" Stabilizers at Solvent Boiling Point

	Boiling Point (K)	Concentration (wt%)	Enthalpy of Vaporization[b] (ΔH_{vap}) (J/mol)	Calculated Stabilizer Vapor Pressure at Solvent Boiling Point (atm)
***Stabilizers of Methyl Chloroform* (b.p. = 346.8 K)**				
1,4-Dioxane	374.1	3	36,500	0.40
Nitromethane	374.2	0.5	34,000	0.42
2-Methyl-3-butyn-2-ol	376.5	2.5	42,000	0.32
Nitroethane	387	0.5	38,270	0.25
***Stabilizers of Trichloroethylene* (b.p. = 360.2 K)**				
Pyridine	388.2	0.5	37,300	0.41
Epichlorohydrin	390.9	0.15	42,900	0.32
p-tert-Amyl phenol	535.5	0.01	58,200	0.0017
***Stabilizers of Perchloroethylene* (b.p. = 394.2 K)**				
Butoxymethyl oxirane	437	0.0043	NA	*Not calculated*
Cyclohexene oxide	402.5	0.0011	NA	*Not calculated*
***Stabilizers of Dichloromethane* (b.p. = 346.8 K)**				
Nitromethane	374.2	0.43	34,000	0.12
1,4-Dioxane	374.1	0.8	36,500	0.10
Cyclohexane	353.7	0.01	32,300	0.24
1,3-Dioxolane	351	2	34,600	0.23

(Continued)

TABLE 10.6 *(Continued)*

Summary of the Potential for High-Boiling Stabilizers to Become Concentrated

	Iterative Enrichment of High-Boiling Stabilizers through Emission Losses and Solvent Addition								
	Initial Mass Fraction (wt%)	Vapor Pressure Ratio	Mass Fraction of Stabilizer in Liquid Solvent (wt%) Weeks of Operation with Additions before Clean out						
			1	2	3	4	5	6	7…
***Stabilizers of Methyl Chloroform* (b.p. = 346.8 K; total solvent volume = 120 gallons; 45 gallons added per week; 55 gallons of emission losses per week; 9 weeks per batch)**									
1,4-Dioxane	3.00	0.40	3.6	4.3	5.4	6.8	8.8	11.9	16.6
Nitromethane	0.50	0.42	0.6	0.7	0.9	1.1	1.4	1.9	2.7
2-Methyl-3-butyn-2-ol	2.50	0.32	3.1	3.8	4.8	6.1	8.0	10.9	15.3
Nitroethane	0.50	0.25	0.6	0.8	1.0	1.3	1.7	2.3	3.3
***Stabilizers of Trichloroethylene* (b.p. = 360.2 K; total solvent volume = 275 gallons; 90 gallons added per week; 104 gallons of emission losses per week; 4 weeks per batch)**									
Pyridine	0.50	0.41	0.59	0.69	0.80	0.94	—	—	—
Epichlorohydrin	0.15	0.32	0.18	0.22	0.26	0.30	—	—	—
p-tert-Amyl phenol	0.01	0.00	0.013	0.017	0.021	0.026	—	—	—
***Stabilizers of Perchloroethylene* (b.p. = 394.2 K; total solvent volume = 460 gallons; 25 gallons added per week; 28 gallons emission losses per week; 3 weeks per batch)**									
Butoxymethyl oxirane (CF: 1.75)[c]	0.0043	0.0075	0.013	0.023	—	—	—	—	—
Cyclohexene oxide (CF: 0.94)[c]	0.0011	0.0010	0.00097	0.0009	—	—	—	—	—
***Stabilizers of Dichloromethane* (b.p. = 346.8 K; total solvent volume = 125 gallons; 30 gallons added per week; 35 gallons emission losses per week; 4 weeks per batch)**									
Nitromethane	0.43	0.12	0.52	0.63	0.79	0.99	—	—	—
1,4-Dioxane	0.80	0.10	0.97	1.19	1.47	1.87	—	—	—
Cyclohexane	0.10	0.24	0.12	0.14	0.17	0.21	—	—	—
1,3-Dioxolane	2.00	0.23	2.35	2.81	3.43	4.30	—	—	—

Effective Solubilities of Stabilizers Concentrated in Still Bottoms

	Mass Fraction in Still Bottoms (wt%)	Mole Fraction	Aqueous Solubility (mg/L)	Effective Solubility (mg/L)	Solubility Ratio (Solvent: Stabilizer)	Stabilizer Concentration[d] (µg/L)
Stabilizers of Methyl Chloroform						
1,4-Dioxane	16.6	0.20	10,000,000	2,045,261	0.00017	>100,000
Nitromethane	2.7	0.048	11,100	530	0.67	>10,000
2-Methyl-3-butyn-2-ol	15.3	0.20	10,000,000	1,963,840	0.00018	>100,000
Nitroethane	3.3	0.047	45,000	2,127	0.17	>10,000
Methyl chloroform	62.1	0.50	700	353	—	—
Stabilizers of Trichloroethylene						
Pyridine	0.94	0.015	10,000,000	154,850	0.007	>100,000
Epichlorohydrin	0.30	0.0043	65,900	281	3.83	>1,000
p-tert-Amyl phenol	0.026	0.00022	168	0.037	28,911	<1
Trichloroethylene	98.6	0.98	1,100	1,076	—	—
Stabilizers of Perchloroethylene						
Butoxymethyl oxirane	0.023	0.00029	20,000	5.87	25.549	<100
Cyclohexene oxide	0.0009	0.000015	—	—	—	<1
Perchloroethylene	99.98	0.9997	150	150	—	—
Stabilizers of Dichloromethane						
Nitromethane	1.0	0.0137	11,100	152	103	<100
1,4-Dioxane	1.9	0.0178	10,000,000	178,312	0.087	>10,000
Cyclohexane	0.21	0.0021	55	0.12	132,346	<1
1,3-Dioxolane	4.3	0.0488	10,000,000	487,945	0.032	>100,000
Dichloromethane	92.6	0.9175	17,000	15,598	—	—

Notes: The values given here are rough approximations; the calculations that yielded these estimates do not account for the ~25% of oily wastes that accumulate in vapor degreaser still bottoms.

[a] Operating parameters for perchloroethylene and methyl chloroform vapor degreasers are from Tarrer et al. (1993); parameters for trichloroethylene and dichloromethane vapor degreasers are inferred from data in USEPA (1977).

[b] Enthalpy data are from Chickos and Acree (2003). Enthalpies of vaporization were not located for butoxymethyl oxirane or cyclohexene oxide. Where enthalpy data are not available, enthalpy of vaporization can be calculated by using Antoine coefficients with the method presented in the Handbook of Chemical Property Estimation Methods (Lyman et al., 1990); however, Antoine coefficients were also not located for these two compounds.

[c] CF = Concentration factor, taken as the ratio of stabilizer concentration in spent versus new solvent.

[d] Where solvent concentration is 10 mg/L. Final stabilizer concentration in groundwater resulting from a spill of still bottoms is estimated to an order of magnitude without considering fine-grained storage or other fates of stabilizers in the subsurface.

TABLE 10.7
Gas Chromatographic Measurements of Solvents, Stabilizers, and Impurities

Solvents with Stabilizers (S) and Impurities (I)	Fraction in Solution (wt%)		
	Before Use	In Sump	Distilled
Trichloroethylene	96.24	98.89	95.41
1,2-Butylene oxide (S)	0.5	0.22	0.78
Methyl chloroform (I)	0.022	0.015	0.089
1,1,2-Trichloroethane (I)	0.099	0.15	0.046
Unknowns	3.14	0.72	3.67
Methyl chloroform	96.25	96.34	96.21
sec-Butanol (S)	0.97	0.80	1.12
1,4-Dioxane (S)	0.91	1.50	0.80
1,3-Dioxolane (S)	0.84	0.75	0.95
Nitromethane (S)	0.57	0.34	0.77
1,2-Butylene oxide (S)	0.32	0.25	0.38
1,2-Dichloroethane (I)	0.12	0.13	0.11
Trichloroethylene (I)	0.0087	0.013	0.0074
Cyclohexane (S)	0.0049	0.0046	0.0088
Nitroethane (S)	0.0047	0.0063	0.0027

Source: Holt, R.D., 1990, Physical properties of contaminated TCE and 1,1,1-trichloroethane. *Technical Communications*. Kansas City, Missouri: Allied Signal Aerospace Company.

stabilizer enrichment in vapor degreasing is laboratory analyses done in the context of industrial engineering studies. The data from Bohnert and Carey (1991) and Tarrer et al. (1989, 1993) are based on such studies, as are the data from Holt (1990), presented in Table 10.7. The Holt study measured a 65% increase in 1,4-dioxane concentration in a vapor degreasing sump, which corroborates the analyses by Tarrer et al. in which a 68% increase in 1,4-dioxane in the sump was measured. The decrease in 1,4-dioxane concentration in solvent recovered by distillation (the last column in Table 10.7) is due once again to the boiling-point difference; 1,4-dioxane remains behind in the distillation process, as discussed in Section 10.1.1.

10.2 RATIO ANALYSIS

At many large-scale solvent release sites, passive remediation approaches such as monitored natural attenuation (MNA) are favored where risk can be managed and where processes that degrade or destroy contaminants can be demonstrated. The MNA approach is often a cost-effective and preferred remedial strategy where the contaminated property has a single owner or where long-term passive remediation will not interfere with beneficial land and groundwater uses.

To demonstrate MNA, remediation consultants must prove that natural processes, whether abiotic or microbially mediated, are transforming the solvents and associated chemicals into breakdown products that are less harmful, thereby decreasing risk and reducing contaminant mass. This demonstration is often attempted by plotting changes in ratios of the molar concentrations of the contaminant and its breakdown products with distance from the source. The ratio can serve as an indicator of the degree to which the breakdown reaction or microbial process has progressed. However, the original contaminant and its breakdown product may behave differently in the subsurface, leading to different rates of migration. For example, the molar ratio in a down-gradient monitoring well may reflect differences in the rate of adsorption due to the organic carbon partition coefficients for these compounds (TCE has an average K_{oc} value of 101, while *cis*-1,2-dichloroethylene has an estimated

K_{oc} value of 240). Consequently, the molar ratios of these two compounds may reflect both the rate of biotransformation and differences in rates of migration. Because the more mobile compound is likely to migrate farther down-gradient, the ratio of the more mobile compound to the less mobile compound will decrease in monitoring wells located near the source zone and increase farther from the source zone. Similarly, the ratio of the compound more resistant to degradation to the compound more prone to degradation will increase with time, especially when the compound more prone to degradation produces the more resistant compound as a daughter product (McNab and Narasimhan, 1994). Determining rates of transformation from groundwater samples is especially challenging—because the groundwater is moving, the solutes are subject to dispersion, diffusion, and potential multiple pathways of biodegradation or abiotic degradation. Contaminant migration is often retarded owing to sorption and matrix diffusion.

To discern the rates of biodegradation that are independent of variable migration rates among parent and breakdown products, a conservative tracer can be useful to confirm the rate of groundwater flow. Where 1,4-dioxane was released together with methyl chloroform, the 1,4-dioxane can serve as a tracer, because monitoring wells are generally completed in coarse-grained strata. Molar concentration ratios of 1,4-dioxane to methyl chloroform and its elimination reaction product, 1,1-dichloroethylene, can reveal patterns in the abiotic breakdown of methyl chloroform. Similarly, ratios of 1,4-dioxane to methyl chloroform and its reductive dechlorination products, 1,1-dichloroethane and chloroethane, can profile the rate of biologic transformation of methyl chloroform. As a conservative tracer, 1,4-dioxane can demonstrate the degree to which physical fates such as transverse dispersion or matrix diffusion may complicate the interpretation of molar ratios to discern degradation rates. With sufficient monitoring points, sampling for 1,4-dioxane can reveal the degree to which subsurface heterogeneity at a site introduces uncertainty into the degradation rates inferred from constituent ratios. This section examines the utility of ratio comparisons to demonstrate MNA with 1,4-dioxane as a conservative tracer.

Until recently, 1,4-dioxane had not been directly observed to undergo breakdown at field sites; therefore, its presence at a site was conflicted with the requirements for demonstrating MNA to the satisfaction of the U.S. Environmental Protection Agency's (USEPA's) Directive on MNA (USEPA, 1999). The USEPA Directive notes that the fuel-oxygenate methyl *tert*-butyl ether (MTBE) migrates farther and threatens down-gradient water supplies at the same sites where the migration of fuel constituents has either reached steady state (i.e., plume stabilization) or diminished by natural attenuation. 1,4-Dioxane released at methyl chloroform sites presented a similar dilemma. USEPA's Directive cautions that compounds that are not *readily* degradable in the subsurface, such as MTBE and 1,4-dioxane, may represent an actual or potential threat to the beneficial uses of groundwater and should be assessed when evaluating the appropriateness of MNA remedies (USEPA, 1999). However, there have now been several publications that have confirmed intrinsic biodegradation of 1,4-dioxane in the subsurface using stable isotope probing and molecular biological tools, as described in Chapter 9. The extent to which confirmed intrinsic biodegradation of 1,4-dioxane will satisfy MNA requirements will vary by the strength of the evidence and the state or federal regulator's requirements, as well as the proximity to drinking water supply wells. To the authors' knowledge, MNA has not been formally accepted by the regulatory community as a permanent solution for any 1,4-dioxane sites.

10.2.1 Ratio Analysis of Methyl Chloroform Breakdown Using 1,4-Dioxane

When in use, methyl chloroform is usually stabilized against oxidative attack by and reaction with metals and acids; once released to the subsurface, it breaks down fairly quickly due to hydrolysis and biodegradation. Elimination reactions remove HCl from methyl chloroform to produce 1,1-dichloroethylene, and hydrolysis of methyl chloroform produces acetic acid. Reductive dechlorination transforms methyl chloroform to 1,1-dichloroethane (Vogel and McCarty, 1987a, 1987b). The hydrolysis of methyl chloroform to acetic acid occurs about five times more rapidly than the elimination reaction that produces 1,1-dichloroethylene (Vogel and McCarty, 1987b). Acetic acid (vinegar) is not a contaminant of interest and is not usually monitored because it is quickly mineralized by soil microbes. One can expect the detection of 1,1-dichloroethylene at concentrations of at least 5 μg/L where methyl chloroform has been in groundwater at concentrations greater than 120 μg/L for more than 1 year (Vogel and McCarty, 1987a).

The reported half-life of methyl chloroform in the elimination reaction has been variously cited as ranging from 1.7 to 3.8 years at 20°C, whereas the yields of the abiotic elimination and hydrolysis reactions of methyl chloroform range from 20% to 27% 1,1-dichloroethylene and 73% to 80% acetic acid, respectively (literature survey summarized in Wing, 1997). For each molecule of 1,1-dichloroethylene generated by the elimination of HCl from methyl chloroform, approximately three molecules of acetic acid are concurrently generated by the hydrolysis of methyl chloroform (Wing, 1997). The fact that 1,1-dichloroethylene is itself resistant to both hydrolysis and biodegradation permits its use in ratio analysis to assess the rates of methyl chloroform breakdown. In a study that analyzed groundwater data collected over a 10-year period from a methyl chloroform release site, methyl chloroform exhibited first-order kinetics and a half-life of 2.9 years at 15°C for the abiotic transformation to 1,1-dichloroethylene and acetic acid (Wing, 1997). This reaction is described further in Section 10.2.2.

Methyl chloroform is less soluble and has a higher affinity for soil organic matter than 1,1-dichloroethylene; therefore, its migration is relatively retarded. Because 1,1-dichloroethylene is less prone to degradation, ratios of 1,1-dichloroethylene to methyl chloroform should increase with distance from the source zone. The superimposed effect of retardation may complicate the interpretation of methyl chloroform degradation. To determine the degree to which retardation hampers interpretation of degradation rates, McNab and Narasimhan (1994) modeled data from the Lawrence Livermore National Laboratory (LLNL) site by using a transport code that accounts for sequential decay chains. The modeling effort led to the conclusion that the effect of degradation of methyl chloroform to 1,1-dichloroethylene overwhelmed the effect of retardation at the LLNL site (McNab and Narasimhan, 1994). This conclusion suggests that sorption, dispersion, and matrix diffusion play only a minor role in the ratios of methyl chloroform to 1,1-dichloroethylene and so permit their use to interpret degradation rates. Their modeled methyl chloroform abiotic degradation rate (half-life of 2 years) agreed well with literature values. It may therefore be unnecessary to use 1,4-dioxane as a control on migration rates when using ratios to interpret abiotic degradation of methyl chloroform; however, 1,4-dioxane may still be a useful tracer for interpreting rates of reductive dechlorination of methyl chloroform to 1,1-dichloroethane.

10.2.2 Using Ratios to Approximate the Date of a Release

Ratios of 1,1-dichloroethylene to methyl chloroform can be used to infer the timing of a release when the release was episodic in nature. "Dating" a plume applies knowledge of the degradation half-life for the chemical and the source-zone concentration to estimate when the chemical was released into the subsurface (Morrison, 2006). The ratio approach to determining the timeframe of a methyl chloroform release requires knowledge of (1) the rate of chemical hydrolysis of methyl chloroform to acetic acid and (2) the rate of the elimination reaction to form 1,1-dichloroethylene based on laboratory data or field measurements. Several assumptions are also required (Gauthier and Murphy, 2003):*

- The rate of hydrolysis is constant over time.
- Because the hydrolysis of methyl chloroform exhibits first-order kinetics and is relatively independent of pH between 4 and 9, its rate depends only on temperature. Assuming a constant rate of hydrolysis implies an assumption of constant groundwater temperature.

* The ratio approach to establishing the timeframe of a release described here summarizes the 2003 Gauthier and Murphy original article in the *Journal of Environmental Forensics*, to which the reader is referred for a more complete treatment of the topic: "Age Dating Groundwater Plumes Based on the Ratio of 1,1-Dichloroethylene to 1,1,1-Trichloroethane: An Uncertainty Analysis," Thomas D. Gauthier and Brian L. Murphy, Exponent, Tampa, FL, USA. Used with permission.

- Biodegradation of methyl chloroform and 1,1-dichloroethylene is either unimportant or can be adequately modeled.
- Chromatographic separation effects such as differing volatilization rates or sorption rates do not skew the ratio of 1,1-dichloroethylene to methyl chloroform, that is, they are either negligible or can be modeled.

The rate of the elimination reaction is first order; therefore, the Arrhenius equation can be used to describe the temperature dependence of the rate of methyl chloroform degradation (Gauthier and Murphy, 2003). The general form of the Arrhenius equation is shown as follows:

$$k = Ae^{E_a RT} \tag{10.2}$$

where k is the first-order abiotic rate constant for methyl chloroform loss, A is the Arrhenius constant, E_a is the activation energy (in kJ/mol), R is the gas constant [8.3145×10^{-3} kJ/(mol K)], and T is the absolute temperature (in K). For the abiotic degradation of methyl chloroform, the rate constant k is the lumped sum of the reaction rates for the elimination reaction, k_e, and the hydrolysis or substitution reaction, k_s: $k = k_e + k_s$. The elapsed time of reaction corresponding to the ratio of the concentration of the elimination product, 1,1-dichloroethylene, to that of its parent, methyl chloroform, can be determined by applying the literature values of the reaction rates, as shown in Equation 10.3 (Gauthier and Murphy, 2003).

$$t = \frac{1}{k}\ln\left(1+\frac{k\left[\text{1,1-dichloroethylene}\right]}{k_e\left[\text{methyl chloroform}\right]}\right). \tag{10.3}$$

A "field-estimate" of the abiotic rate constant k can be derived by plotting time-series data as the natural logarithm of the molar concentration ratios from monitoring points along the longitudinal axis of the plume. The natural logarithm of the molar concentration ratio in Equation 10.3 is plotted on the y-axis and time on the x-axis as dates or Julian days. A best-fit straight line then has a slope equal to the estimated k, and the x-intercept provides an estimate of the date at which hydrolysis began, that is, the approximate date when dissolved methyl chloroform was first subjected to abiotic degradation. This date does not account for the time since the actual spill, disposal, or tank leak, but reflects the date when methyl chloroform entered groundwater as a dissolved phase. Pure (nonaqueous phase) methyl chloroform does not participate in abiotic degradation reactions; however, condensed water vapor in vapor degreasers is likely to facilitate abiotic reactions that form 1,1-dichloroethylene. The molar ratios measured in down-gradient monitoring points within the plume can only represent a travel time from the source zone and not the date when the spill occurred (Gauthier and Murphy, 2003). In order for k to be determined by time-series data from a single well, that well must be at the plume front. If the methyl chloroform was released in the dissolved phase, for example, from wastewater discharge out of a vapor degreaser water trap, measurement of the ratio of 1,1-dichloroethylene to methyl chloroform at any location or time can be used to determine the date when hydrolysis began (Gauthier and Murphy, 2003).

In performing ratio analysis to discern the timeframe of a release, uncertainty due to variable rates of migration among the constituents should be considered. The primary transport parameter that may affect the molar ratio of 1,1-dichloroethylene to methyl chloroform is sorption, which is governed by the fraction of organic carbon in soil (f_{oc}) and the compound's organic carbon partition coefficient, K_{oc}. The K_{oc} for methyl chloroform is more than twice that of 1,1-dichloroethylene (152 and 65 mL/g, respectively; Pankow and Cherry, 1996). The degree to which retardation of methyl chloroform relative to 1,1-dichloroethylene affects the ratio depends on the f_{oc} and can be

TABLE 10.8
Retardation Coefficients and Relative Velocities as a Function of Fraction of Organic Carbon

Fraction of Organic Carbon, Oc (%)	Retardation Coefficient/[a] *R* Methyl Chloroform	Retardation Coefficient/[a] *R* 1,1-Dichloroethylene	Ratio of 1,1-Dichloroethylene Velocity to Methyl Chloroform Velocity
0.01	1.07	1.03	1.04
0.02	1.14	1.06	1.08
0.05	1.35	1.15	1.17
0.1	1.69	1.30	1.31
0.2	2.39	1.59	1.50
0.5	4.47	2.49	1.80
1.0	7.95	3.97	2.00

Source: Gauthier, T.D. and Murphy, B.L., 2003, Age dating groundwater plumes based on the ratio of 1,1 dichloroethylene to 1,1,1-trichloroethane: An uncertainty analysis. *Environmental Forensics* 4: 205–213.

[a] These calculations of retardation coefficient $R = 1 + (\rho_b K_d/\Theta_e)$ assume bulk soil density $\rho_b = 1.6$ g/cm^3, water-filled porosity $\Theta = 0.35\%$, and partition coefficient $K_d = K_{oc} \times f_{oc}$. Because f_{oc} can vary within an aquifer and may be higher in fine-grained deposits where hydraulic conductivity is lower, its heterogeneous distribution may impart uncertainty to the analysis of ratios to estimate the timeframe of a release. Overall retardation can be estimated by comparing the distance that methyl chloroform and its breakdown products have migrated compared to that of 1,4-dioxane.

significant where f_{oc} is greater than 0.1%, as summarized in Table 10.8 (Gauthier and Murphy, 2003). At higher f_{oc} values, the migration rate of 1,1-dichloroethylene may substantially exceed that of methyl chloroform and so affect the molar concentration ratios. This possibility is seen by the measured and predicted range of volume concentration ratios observed with increasing distance from the methyl chloroform source, as shown in Figure 10.1.

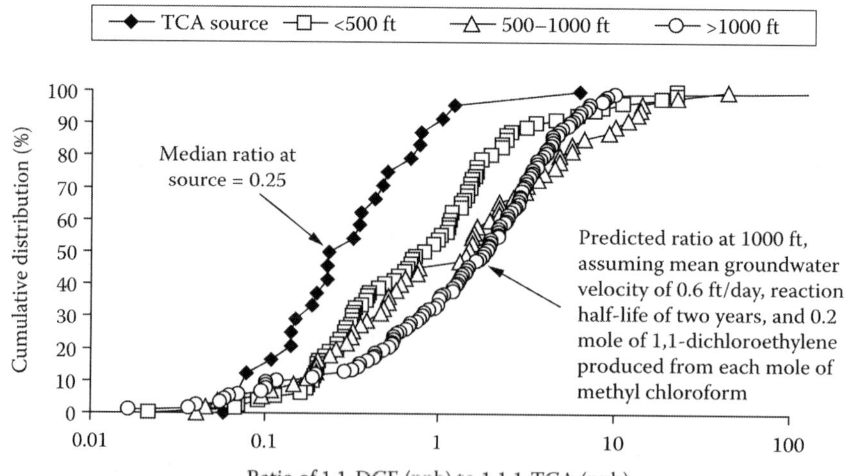

FIGURE 10.1 Field and modeled volume concentration ratios of methyl chloroform (1,1,1-TCA) and 1,1-dichloroethylene (1,1-DCE). Predicted volume concentration ratio is given at 1,000-ft distance from plume source. The modeling assumes a mean velocity of 0.6 ft/day, reactant half-life of 2 years, and 0.2 mol of 1,1-dichloroethylene produced from each mole of methyl chloroform. (From McNab, W.W. et al., 1999, Historical case analysis of chlorinated volatile organic compound plumes. Lawrence Livermore National Laboratory report UCRL-AR-133361. With permission.)

Where biologic transformation of methyl chloroform is also evident from detection of 1,1-dichloroethane and chloroethane, Equation 10.4 can be used:

$$t = \frac{1}{k_T} \ln\left(1 + \frac{k}{k_e}\left(\frac{[\text{1,1-dichloroethylene}]}{[\text{methyl chloroform}]}\right) + \frac{[\text{1,1-dichloroethane}]}{[\text{methyl chloroform}]}\right), \tag{10.4}$$

where the total rate constant $k_T = k + k_b$, and k_b is the biodegradation rate constant for the reductive dechlorination of methyl chloroform to 1,1-dichloroethane (Gauthier and Murphy, 2003).

One case study provides a remarkable example of the application of ratio analysis to interpret the timeframe of a release. The estimate predicts the date of release to within 1 week of the actual date (Wing, 1997). On the morning of August 14, 1984, a pipe supplying a vapor degreaser at a precision medical instrument plant in Santa Clara, California, failed and released as much as 20 gallons of methyl chloroform. Three hundred cubic yards of contaminated soil were removed within a year of the spill. Methyl chloroform concentrations in groundwater were up to 11,000 μg/L; a pump-and-treat system removed 41 million gallons of contaminated water until 1994, when the system was shut down because cleanup goals had been attained at most monitoring wells. By 2003, methyl chloroform concentrations had decreased to 2.3 μg/L, with 5.1 μg/L 1,1-dichloroethylene and 4.7 μg/L 1,1-dichloroethane. The results of analyses for 1,4-dioxane at the site were nondetect at less than 2 μg/L in all monitoring wells, based on one-time sampling in 2002, 18 years after the initial release (Cal EPA, 2003).

Using data from monitoring wells placed within the 500-feet-long methyl chloroform plume at the Santa Clara site, Wing (1997) plotted time-series concentration data from two groundwater extraction trenches as the log of the molar ratios of the sum of 1,1-dichloroethylene and 1,1-dichloroethane to methyl chloroform against sample dates. Because no data for acetic acid was available, Wing applied the approximate kinetic ratio of 3:1 (molecules acetic acid to 1,1-dichloroethylene produced from methyl chloroform) and multiplied the molar concentrations of 1,1-dichloroethylene by four. The specific ratio plotted to interpolate the timeframe of release was (Wing, 1997)

$$\ln\left(\frac{4[\text{1,1-dichloroethylene}] + [\text{1,1-dichloroethane}]}{[\text{methyl chloroform}] + 1}\right). \tag{10.5}$$

Wing's analysis corroborated the finding by McNab and Narasimhan which showed that degradation effects overwhelm retardation effects. The slope of the linear regression line (correlation coefficient $r^2 = 0.81$) was 0.298 year^{-1}, from which an overall half-life of methyl chloroform of 2.3 years was obtained. To account for only the elimination reaction that transforms methyl chloroform to its field-measured degradation product, 1,1-dichloroethylene, Wing removed 1,1-dichloroethane and the factor of four that accounted for acetic acid to obtain a slope of 0.243 year^{-1}. The resulting half-life of methyl chloroform for the elimination reaction was determined to be 2.9 years (Wing, 1997).

The study of field data by Wing produced unexpectedly accurate results, providing encouragement for applying this technique where an episodic release is suspected (e.g., a plume of unknown origin that may have originated as a "midnight dumping" episode). However, caution is in order despite the high degree of accuracy obtained by Wing. Sources of error and uncertainty include hydrodynamic dispersion and molecular diffusion into fine-grained sediments, which can be expected to decline with distance from a continuous source (McNab and Dooher, 1998). The number, placement, and construction of monitoring wells may invalidate assumptions of a continuum between data points where anisotropy creates preferential pathways (e.g., paleochannels) affecting some monitoring wells and not others. Heterogeneity of hydraulic conductivity, redox conditions, and temperature may all affect the ratios used to interpret the timeframe of release. Seasonal variations in hydraulic conditions such as water level fluctuations or seasonal recharge causing variation in groundwater temperature and groundwater flow direction may also confound the application of ratios to estimating the timing of a release. To determine the extent to which these factors may be influencing the ratio data at a site, contrasting the concentrations of methyl chloroform and its degradation products to the concentrations of 1,4-dioxane where it has been established as a co-contaminant is recommended.

The application of ratio analysis to determine the timing of an episodic release may be strengthened by including one or more additional lines of evidence, such as analysis for time marker chemicals (e.g., MTBE, symmetrical tetrachloroethane), analysis for atmospheric tracers (e.g., chlorofluorocarbons 11, 12, and 113, and sulfur hexafluoride), and decay of tritium, the radioisotope of hydrogen from open-air testing of hydrogen bombs, to helium.

10.2.3 Modeling 1,4-Dioxane Migration to Approximate the Date of a Release

Two modeling studies used the 1,4-dioxane data set from the Gloucester Landfill (described in Section 3.4.2) to refine the methods of solving the inverse problem to estimate the probable timing of a 1,4-dioxane release to groundwater in a special waste unit. Laboratory wastes, presumably including scintillation wastes, were discharged to a trench and periodically burned. Wastes containing 1,4-dioxane leached from the trench to groundwater and migrated about 1,600 ft down-gradient. The modeling studies sought to solve the linear inverse problem by using assumed values for contaminant transport parameters such as lateral and transverse dispersivity.

The first study by Woodbury et al. (1998) used the minimum relative entropy (MRE) method to recover the source-release history of the plume in three dimensions. The MRE method applies a probabilistic approach that treats parameters as random variables rather than as fixed deterministic values. Assessing the statistical properties of unknown model parameters is a significant challenge when using probabilistic models. Although most probabilistic studies usually assume Gaussian distributions, Woodbury et al. used the MRE approach to assign truncated exponential distributions where only the lower and upper bounds and some knowledge of the mean are given. The MRE approach supports a more robust reconstruction of the probability density function associated with the model parameters. Woodbury et al. applied hydrogeologic parameters discerned from detailed field studies conducted by Jackson et al. (1991) and inferred transport parameters from the distribution of 1,4-dioxane measured in a dense network of multilevel monitoring wells. The retardation factor assumed for 1,4-dioxane was 1.0, that is, 1,4-dioxane was assumed to move at the rate of the groundwater flow without retardation. The sample data used were from the early and late 1980s (Woodbury et al., 1998).

The recovered or modeled release history suggests a fairly narrow window of time during which 1,4-dioxane releases were likely to have occurred; two peaks were evident, coinciding with times of 8.4 and 9.7 years. The latter time corresponds to the year 1978, when approximately 1 ton of organic solvents was reported to have been disposed of within the special waste compound, whereas the former time estimate does not correspond to a known event (Woodbury et al., 1998).

The second modeling study of the Gloucester Landfill 1,4-dioxane data set used to refine techniques for back-dating the release was published by Michalak and Kitanidis in 2002. Their study used a geostatistical approach to inverse modeling, applying a Bayesian stochastic inference method to estimate the release history for 1,4-dioxane, and a novel approach to enforce non-negative concentration values. The advantage of using a stochastic approach such as Bayesian inference is that it provides a best estimate together with the associated confidence intervals that represent the information contained in the available data (Michalak and Kitanidis, 2002). The conditional simulations performed by Michalak and Kitanidis generated 50,000 realizations, which were summarized to obtain a range of probable outcomes. The estimated timeframe of release successfully reproduced the observations to within the estimated measurement error variance, which accounted for both analytical measurement error and the limitations of the physical model (Michalak and Kitanidis, 2002). The modeling results indicated that the 1,4-dioxane release from the special waste compound continued after the end of landfill operations and identified the probable occurrence of multiple release periods spanning a 20-year timeframe.

10.2.4 Compound-Specific Isotope Analysis

CSIA assesses the degree of isotopic enrichment or depletion that may occur during biotransformation, phase transitions, or chemical reactions involving a contaminant of interest. CSIA leverages variations in the naturally occurring stable isotopes of elements incorporated in the molecular structure of the contaminant under study, for example, carbon, hydrogen, and chlorine in TCE (C_2HCl_3). Carbon has six protons and six neutrons in 98.89% of all carbon atoms in nature, whereas its stable isotope (^{13}C) with six protons and seven neutrons, forms 1.11% of all carbon atoms. Hydrogen most often has one proton, but 0.015% of natural hydrogen atoms have one proton and one neutron (that isotope is called deuterium, ^2H or "D"). Chlorine in its most abundant state has 17 protons and 18 neutrons, but 31.978% of chlorine atoms have 17 protons and 20 neutrons (^{37}Cl). Oxygen with eight protons and eight neutrons is most abundant; 0.204% of oxygen atoms have eight protons and 10 neutrons (^{18}O), and 0.037% of oxygen atoms have eight protons and nine neutrons (^{17}O).

Very sensitive isotope ratio mass spectrometers can measure the small differences in molecular mass to determine the ratios of the isotopes present, for example, the inclusion of heavier chlorine isotopes in a TCE molecule. Physical, chemical, or biological processes may favor heavier or lighter molecules and enrich or deplete the ratio of the rare to the most abundant isotope. The natural variations in the mass of the oxygen, hydrogen, and carbon atoms on the 1,4-dioxane molecule can potentially be used to interpret its fate in the subsurface, as discussed further in Section 10.2.5.

Ratios are referenced to an international standard material, such as Vienna Standard Mean Ocean Water for oxygen and hydrogen; the Peedee Belemnite (PDB), a South Carolina limestone formation, for carbon; and seawater for chlorine (Fritz and Clark, 1997). By convention, the ratio of two isotopes of the same element is described by δ notation and expressed in units of per mil (‰), as shown for carbon:

$$\delta^{13}C = \left(\frac{R_{sample}}{R_{std}} - 1\right) \times 1{,}000, \tag{10.6}$$

where R_{sample} is the $^{13}C/^{12}C$ ratio for a compound in a sample and R_{std} is the $^{13}C/^{12}C$ ratio for the international CO_2 standard (PDB) (Fritz and Clark, 1997). A detailed treatment of isotope hydrology and CSIA is beyond the scope of this book; the reader is directed to the leading texts and online resources on this subject.*

CSIA is accomplished with in-line gas chromatography-isotope ratio mass spectrometry (GC-IRMS). CSIA can be effective for differentiating solvents that have been through processes causing isotopic fractionation (i.e., enrichment or depletion, shifting the initial isotope ratio). Applications include characterizing, monitoring, and evaluating natural and engineered bioremediation (Sturchio et al., 1998). For example, the analysis of deuterium (^2H) in TCE has been used to distinguish two sources of TCE: the first due to biodegradation of PCE and the second due to a direct release of TCE. This differentiation is possible because the deuterium isotope ratio (δ^2H) in manufactured TCE is typically in the range of +466.9‰ to +681.9‰, whereas TCE that has

* Texts: (1) Fritz and Clark (1997). (2) Immanuel Mazor, 1996, *Chemical and Isotopic Groundwater Hydrology*. New York: Marcel Dekker. (3) Pradeep K. Aggarwal, Joel R. Gat, and Klaus F.O. Froehlich, 2005, *Isotopes in the Water Cycle: Past, Present and Future of a Developing Science*. New York: Springer-Verlag. Websites: (1) Compound Specific Isotope Analysis (CSIA): https://www.microbe.com/compound-specific-isotope-analysis-csia; (2) CSIA Fundamentals and Application Guide https://www.pacelabs.com/environmental-services/energy-services-forensics/csia/fundamentals-of-csia.pdf; (3) A Guide for Assessing Biodegradation and Source Identification of Organic Ground Water Contaminants using Compound Specific Isotope Analysis (CSIA) [US EPA 600/R-08/148 | December 2008] https://clu-in.org/download/contaminantfocus/dnapl/Treatment_Technologies/600r08148.pdf

been dechlorinated from PCE typically has a deuterium isotope ratio in the range of –351.9‰ to –320.0‰ (Shouakar-Stash et al., 2003; Morrison et al., 2006).

Depletion of heavy isotopes in the solvent can also be expected from vaporizing solvents such as will occur in a vapor degreaser or dry-cleaning operation. The enrichment of heavy isotopes in the vapor phase is due to the higher vapor pressure for heavier compounds. This "inverse isotope effect" is also attributed to a greater surface tension (i.e., intermolecular attraction) for lighter compounds. The smaller volume and heavier mass in the liquid phase confer greater motion energy for ^{13}C molecules in the liquid phase, which together with lower intermolecular cohesive forces, produces a higher volatility for the molecules with the heavy isotopes (Bouchard, 2007). Studies have also demonstrated a large enrichment of ^{13}C in the headspace above the liquid phase contained in closed vessels for chlorinated solvents (e.g., Slater et al., 1999; Hunkeler and Aravena, 2000).

CSIA has also been applied to differentiate the manufacturer of origin for chlorinated solvents. Isotopic differentiation between two grades of the same solvent produced at different manufacturers may occur because of variation in the method of solvent production, such as dehydrochlorination and dehydrogenation reactions, temperatures, and durations of the different stages of production, catalysts used, and isotopic variation in the original feedstock used to produce the solvent (Morrison and Murphy, 2006). The utility of this isotopic differentiation may be limited because carbon isotope ratios for the same solvent produced by the same manufacturer in different years have shown substantial variation, and because feedstocks and methods of production may vary over time. A further limitation to using CSIA for solvent source identification is that once delivered to the site of use, the solvent may undergo additional isotopic depletion during vapor degreasing, dry cleaning, on-site recycling including distillation and carbon filtering, and evaporation during cold-cleaning operations. Carbon isotope ratios for solvents from different manufacturers in different years were analyzed by Shouakar-Stash et al. (2003), Van Wamerdam et al. (1995), and Beneteau et al. (1999) and compared in Morrison et al. (2006). Source identification applications of CSIA are best suited to the rare circumstance in which a sample of the original solvent waste is available. A source-zone sample may be a reasonable surrogate, particularly for studying biodegradation of solvents with CSIA.

One challenge to using CSIA has been the mass of the contaminant required on the gas chromatography column in GC-IRMS, which previously limited or even prevented CSIA from contaminant fate studies and the differentiation between sources of contaminants based on their isotopic signature. Because of the large mass required for CSIA, studies of *in situ* biotransformation of contaminants were previously restricted to source-zone samples at sites with high aqueous concentrations of contaminants, to avoid collecting and processing very large volumes of samples in which the subject contaminant is present at low concentrations (Blessing et al., 2008). New methods such as SPME have been developed to improve the isotopic resolution of GC-IRMS on low-concentration samples (Zwank et al., 2003; Jochmann et al., 2006; Blessing et al., 2008). Currently, most CSIA laboratories use purge and trap as the GC-IRMS concentration method to achieve detection limits of $\delta^{13}C$ of about 1 ppb for chlorinated solvents and MTBE.

The limitations and uncertainty of CSIA for fingerprinting sources of chlorinated hydrocarbons may discourage its use in forensic investigations; however, it may nevertheless prove useful if approached in a stepwise fashion, beginning with a screening-level evaluation. To ascertain whether CSIA has the potential to be a useful tool for a particular investigation, the expected isotopic "end-members" of the system should be identified and sampled as a screening-level evaluation of the potential viability of CSIA. For example, where groundwater analyses show a sequence of geochemical conditions suggestive of a reducing environment and reductive dechlorination of chlorinated solvents, samples of the original solvent and its primary breakdown product(s) can be analyzed by using CSIA to confirm whether there is a sufficient shift in carbon, hydrogen, or chlorine isotope ratios to justify additional investigation. Similarly, if CSIA is to be useful for solvent-source identification for allocating cost in a commingled plume of the same solvent from two or more sources, sampling the different source areas and a few monitoring wells in the commingled plume should be the first step. If the source-area isotope ratios show a wide enough differentiation from the isotope ratios in the commingled plume, it may be possible to apply this method to establish mixing patterns. The range in values of carbon isotope ratios ($\delta^{13}C$) should be larger than the margin of analytical error, that is, typically 1.0–0.5‰ (Morrison et al., 2006).

One effort to confirm the biodegradation of TCE in a plume from a landfill in Switzerland produced the surprising result that the presence of *cis*-1,2-DCE was not the result of reductive dechlorination of TCE. Isotopic signatures showed the absence of *in situ* degradation of TCE, despite the presence of *cis*-1,2-DCE, a known metabolite of TCE (Zwank et al., 2003).

To constrain the uncertainty or ambiguity that a CSIA investigation may produce, investigators may include additional lines of evidence, such as the presence of proprietary solvent-stabilizer formulations, which together with the basic geochemical and hydrogeological framework may result in an improved conceptual model.

10.2.5 Compound-Specific Isotope Analysis of 1,4-Dioxane Biodegradation

CSIA has recently been applied to assess the isotopic fractionation that occurs during the biodegradation of 1,4-dioxane. CSIA provides a direct means of demonstrating the occurrence of biodegradation in the field. Biodegradation reactions preferentially open the dioxane ring at bonds between the highly abundant lighter isotopes (e.g., $^{12}C–^{1}H$) over the bonds that include the rarer heavier isotopes ($^{13}C–^{1}H$ or $^{12}C–^{2}H$). Consequently, the remaining, unbiodegraded mass of 1,4-dioxane will become enriched with molecules that incorporate the heavier isotopes of carbon and hydrogen. This enrichment process is referred to as the kinetic isotope effect (KIE) of degradation reactions. As described below, CSIA involves measurement of the stable isotope ratios in isolated 1,4-dioxane (e.g., $^{13}C/^{12}C$ and $^{2}H/^{1}H$).

Reaction rates for different isotopes vary because of mass-dependent differences in activation energies for the respective reactions. This variation is called the KIE, which is used to predict the degree of both carbon and hydrogen isotopic fractionation during biodegradation. KIE describes the ratio of the reaction rates for the heavy, ^{H}k, and light, ^{L}k, isotopes

$$\mathrm{KIE} = \frac{^{L}k}{^{H}k}. \tag{10.7}$$

Published KIE values from known reactions can be used to estimate the degree of fractionation for biodegradation of a compound whose KIE has not yet been measured (Elsner et al., 2005).

Isotopic fractionation resulting from biodegradation follows the Rayleigh model, in which the extent of measured fractionation is proportional to the extent of biodegradation by the following relationship:

$$1{,}000 \times \ln\left(\frac{\delta^{13}C_x + 1{,}000}{\delta^{13}C_0 + 1{,}000}\right) = 1{,}000 \times (\alpha - 1)\ln F = \varepsilon \ln F, \tag{10.8}$$

where $\delta^{13}C_x$ and $\delta^{13}C_0$ are the $^{13}C/^{12}C$ isotope ratios determined from microcosm experiments at time x and time 0, ε is the carbon isotope enrichment factor, α is the carbon fractionation factor, and F is the ratio of contaminant concentrations at time x and 0 (e.g., the fraction remaining). Shifts in contaminant isotopic ratios are useful as indicators of biodegradation (Slater et al., 2000).* Nondegrading physical- or phase-transfer processes such as dissolution, vaporization, or sorption for *dissolved* TCE and PCE Carbon isotopic fractionation is not significant at equilibrium (i.e., <0.5‰).

* The negligible isotopic shift for dissolved chlorinated ethenes is in contrast to the expected enrichment in the vapor phase for these solvents in the vapor degreaser or dry-cleaner setting, where solvents are not in aqueous solution and a significant isotopic shift is expected.

The extent of 1,4-dioxane biodegradation can be calculated using the Rayleigh isotopic enrichment model (described below) if both the isotopic ratio of the undegraded contaminant and the relevant carbon and hydrogen enrichment factors (ε_C and ε_H, respectively) are known. Selection of appropriate values for ε_C and ε_H is critical, and for biodegradation reactions, ε_C and ε_H are typically established with pure cultures of bacteria with known kinetic and enzymatic features (Bennett, et al., 2018). Recent commercial laboratory improvements now allow CSIA to be reliably conducted for 1,4-dioxane at concentrations in the range of 1–10 µg/L (Wang, 2016). Research funded by SERDP led to the development of a different method from Wang (2016), has been applied to 1,4-dioxane concentrations as low as 1 µg/L and applied at multiple sites.

Sample collection for CSIA analysis can be accomplished by collecting a larger volume sample in the field, then concentrating 1,4-dioxane on Ambersorb 560 in the lab. For example, samples collected from McClellan Air Force Base for the 2017 study by Bennett and Aravena retrieved 1 L of groundwater in amber glass bottles. Laboratories that have completed 1,4-dioxane CSIA analysis and may offer commercial analytical services or collaboration opportunities include Pace Analytical, Microbial Insights, University of Waterloo Environmental Isotope Laboratory, and UCLA Civil Engineering, and others.

The accuracy of CSIA of 1,4-dioxane is reportedly within 0.6% relative percent difference; however, high concentrations of co-contaminants and other matrix interference issues can constrain CSIA accuracy. A single source of neat 1,4-dioxane was used to prepare a dilute aqueous solution, which was analyzed at two different laboratories, 4 months apart, using two different analytical approaches. The University of Waterloo determined a $\delta^{13}C$ value of 33.3‰ using elemental analysis coupled with isotope ratio mass spectrometry, while Pace Analytical determined a $\delta^{13}C$ value of 33.1‰ using GC-IRMS. The results show the accuracy achieved by independent analyses, and the stability of isotope ratios over time in neat 1,4-dioxane, when samples are preserved against ongoing biodegradation (Bennett and Aravena, 2017).

Lowering the threshold of detection for CSIA is a significant advancement, because many contaminated sites have 1,4-dioxane concentrations in the range of 1–10 µg/L. Achieving a 1 µg/L threshold for 1,4-dioxane CSIA can be useful for water utilities addressing 1,4-dioxane in the low or sub parts per billion range. However, without published ε_C and ε_H values for 1,4-dioxane biodegradation, CSIA results can be inconclusive to confirm biodegradation is occurring (Bennett, et al., 2018). Until Bennett and Aravena's 2017 publication, only one study, conducted by UCLA scientists Peerapong Pornwongthong and Shaily Mahendra with the 1,4-dioxane-metabolizing strain *Pseudonocardia dioxanivorans* CB1190, had reported an ε_C value for 1,4-dioxane biodegradation (−1.73 ± 0.14‰) (Pornwongthong, 2014). The Bennett and Aravena study provides three additional sets of both ε_C and ε_H, shown in Table 10.9.

TABLE 10.9
Isotopic Enrichment Factors and Apparent Kinetic Isotope Effects (AKIE) During 1,4-Dioxane Biodegradation in Microcosms (Bennett et al., 2018, with permission)

Strain	Growth Substrate	ε_C (‰)	$AKIE_C$	ε_H (‰)	$AKIE_H$	Δδ2H/Δδ13C
R. rhodochrous ATCC 21198	propanc	−2.7 ± 0.3	1.01	−21 ± 2	1.2	7.5 ± 1.1
R. rhodochrous ATCC 21198	isobutane	−2.5 ± 0.3	1.01	−28 ± 6	1.3	10.9 ± 2.2
P. tetrahydrofuranoxidans K1	tetrahydrofuran	−4.7 ± 0.9	1.02	−147 ± 22	>10	37.2 ± 2.6

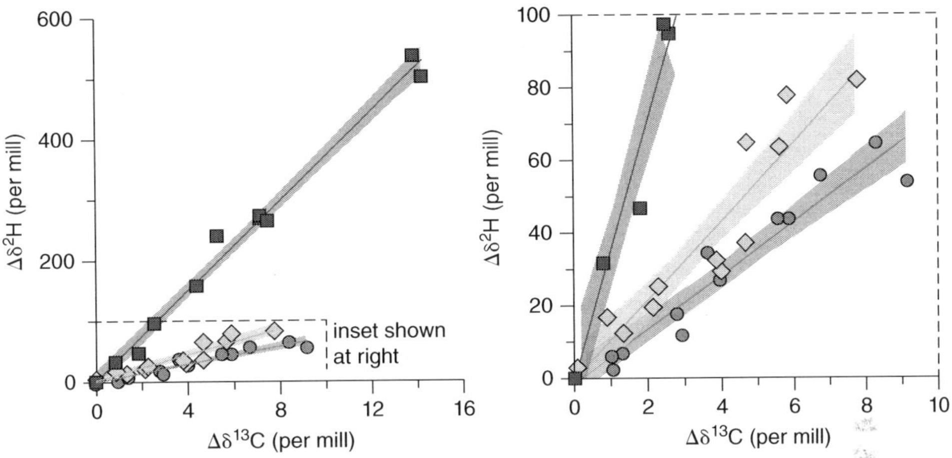

FIGURE 10.2 Carbon and hydrogen enrichment during 1,4-dioxane biodegradation. Two-dimensional plot highlighting the enrichment with ^2H relative to ^{13}C during 1,4-dioxane degradation by strain 21198 grown on propane (circles) and isobutane (diamonds) and for strain K1 grown on THF (squares). Solid lines represent the linear best fits of the slope and R^2: for propane, slope = 7.5 and $R^2 = 0.94$; for isobutane, slope = 10.9 and $R^2 = 0.91$; for THF, slope = 37.2 and $R^2 = 0.99$. Shaded areas represent 95% confidence intervals of the slope of the regression line. *Source*: Bennett et al., 2018, with permission.

The study concludes that two-dimensional CSIAs can be used not only to determine the extent of 1,4-dioxane biodegradation but also to discriminate between different types of 1,4-dioxane oxidizing monooxygenases (Bennett and Aravena, 2017; Bennett et al., 2018). The two-dimensional CSIA results are shown in Figure 10.2.

The isotopic fractionation factors (α) for biodegradation processes can be calculated by posting the data for laboratory degradation experiments on a plot of ln F versus $1{,}000 \times \ln(\delta^{13}C_x/\delta^{13}C_0)$. Through the use of least-squares regression, the fractionation factor, α, is determined as the slope of the plotted data. Researchers applying CSIA in biodegradation studies use fractionation enrichment factors from laboratory microcosm experiments and the KIE. Measurements of isotope ratios for contaminants in field samples, when combined with the laboratory-determined enrichment factors, permit the calculation of F, the ratio of contaminant remaining relative to the initial concentration, which is, in theory, a direct measure of the extent of biodegradation of the contaminant. However, such quantitative measurements may not be applicable to field sites where the isotopic composition of the 1,4-dioxane source may vary, and because of the masking of isotopic enrichment due to blending in monitoring wells (EPA, 2008).

KIE can be classified as primary (a large ratio, resulting from the atom being directly involved in the reaction) or secondary (a small ratio, resulting from the atom being bonded to the reactive center, but not immediately involved in the reaction). On the basis of KIE values for the hydroxylation reaction in biodegradation of other contaminants, only a moderate secondary carbon isotope effect (KIE = 1.015) is expected for 1,4-dioxane, as the carbon atom is not immediately involved in the reaction. A substantial primary hydrogen isotope effect (KIE = 3–8) is expected as the hydrogen atom is thought to be directly involved in the reaction. Hydroxylation is proposed as the initial degradation step in one possible pathway for 1,4-dioxane degradation, in which hydrolysis to diethylene glycol occurs (see Figure 3.7) (Steffan et al., 2007). Other degradation pathways may be present and produce greater or lesser degrees of fractionation.

Given the established relationships between reaction pathway and KIE, it is possible to hypothesize reaction mechanisms from the degree of isotopic fractionation observed for a given atom.

1,4-Dioxane biodegradation may undergo an initial step of proton abstraction. If proton abstraction is involved, a substantial primary hydrogen isotope effect is expected with a theoretical KIE ranging from 3 to 8 (Elsner et al., 2005). The large difference in KIE between carbon and hydrogen results from the hydrogen atom undergoing a primary isotope effect whereas the carbon atom fractionation is a secondary isotope effect.

At sites contaminated with 1,4-dioxane from methyl chloroform releases, it is likely that 1,4-dioxane is present in the highly anoxic or anaerobic portions of the release near the source zone, at the oxygenated edges of the solvent plume, and at more distant locations where 1,4-dioxane has migrated beyond the solvent plume. It is therefore relevant to use CSIA to determine the biodegradability of 1,4-dioxane under both aerobic and anaerobic conditions. Both pure and mixed cultures that possess the ability to degrade 1,4-dioxane aerobically and one study of anaerobic biodegradation are described in Chapter 7.

The rate of anaerobic biodegradation of 1,4-dioxane is expected to be slow because the anaerobic biodegradation of structurally similar compounds, such as MTBE, is slow (Kuder et al., 2005). Under anaerobic systems, the addition of proportionately larger concentrations of another biodegradable ether may be a viable strategy for promoting anaerobic cometabolism of 1,4-dioxane. For example, water-soluble polyethylene glycol (PEG) is degraded by various bacteria in the absence of molecular oxygen (Yeh and Pavlostathis, 2005). The addition of PEG may stimulate "etherase" activity in the subsurface capable of cleaving the 1,4-dioxane ring. Nakamiya et al. (2005) (see Chapter 3) reported that a filamentous fungus degrades 1,4-dioxane via etherase-type reactions to sequentially produce ethylene glycol, glycol-aldehyde, glycolic acid, and oxalic acid.

By combining isotope ratio measurements and biodegradation studies under aerobic and anaerobic conditions, CSIA methods can be used to assess *in situ* 1,4-dioxane biodegradation at enhanced bioremediation sites and MNA sites. This information can be highly useful for risk-based remediation decision-making by facility owners, regulators, and the environmental engineering community. Numerous published studies have shown that CSIA is an excellent tool for providing the evidence of *in situ* biodegradation and to quantify the extent of *in situ* biodegradation of chlorinated solvents at field sites. Recent improvements to lower the threshold of detection in GC-IRMS show that CSIA is now a viable tool for 1,4-dioxane as well.

10.3 SOLVENT STABILIZERS AS AGE MARKERS

The date when a chemical first became commercially available is important for establishing the earliest possible date that a release could have occurred. Knowledge of the dates that chemicals were produced must be coupled with evidence that the facility used the chemical in question. An excellent compilation of dates of production and use patterns for chlorinated solvents is found in Richard Doherty's solvent history articles in the *Journal of Environmental Forensics* (Doherty, 2000a, 2000b). A brief summary of the history of solvent use is profiled in Section 1.1. A tabulation of the dates of introduction of the most widely used solvent-stabilizer packages for the major chlorinated solvents could be similarly useful. Section 1.2.6 reviews the available sources of information on solvent-stabilizer packages, and Tables 1.21 through 1.24 list the formulations. The dates listed in these tables unfortunately do not correspond to when the stabilizer was first used; rather, the dates refer to the date of publication of the citation for the inclusion of the listed stabilizer in the solvent.

In the case where there is documentation available for stabilizer use in a particular solvent, then the earliest patent claiming its use can be helpful for bracketing its introduction for commercial application. However, this information must also be integrated with the general pattern of commercial use for the major chlorinated solvents. For example, 1,4-dioxane was first patented as a stabilizer for methyl chloroform in 1957, and unpublished sources indicate that 1,4-dioxane was in commercial use as a stabilizer of methyl chloroform as early as 1953. But it was not until the early 1960s that methyl chloroform found somewhat wider use in cold-cleaning and vapor degreasing applications,

TABLE 10.10
Stabilizers and Impurities in 1965 Technical Grade Methyl Chloroform from Four Suppliers

	Octagon Process Company (vol%)	Fisher Chemical Company (vol%)	Phillips-Jacobs Chemical Company (vol%)	Eastman Kodak Company (vol%)
Methyl chloroform (%)	65	94	95.5	95
Stabilizers				
1,4-Dioxane	—	3.6	3.4	4.5
1,2-Dimethoxymethane	0.24	—	—	—
tert-Butyl alcohol	3.64	—	—	—
Nitromethane	1.3	—	—	—
Diethyl ether	Trace	—	—	—
sec-Butyl alcohol	—	0.2	0.2	0.3
Impurities				
Trichloroethylene	7.6	—	—	—
1,2-Dichloroethane	22	2.1	0.8	—
1,1-Dichloroethylene	0.06	Trace	Trace	Trace
1,1,2-Trichloroethane	Trace	—	—	—
1,2-Dichloroethylene	Trace	—	Trace	Trace
2-Chlorobutane	—	Trace	—	—

Source: Saunders, R.A., 1965, *The Composition of Technical Grade Methyl Chloroform*. Washington, DC: Physical Chemistry Branch, U.S. Navy Research Laboratory.

and not until the mid-1970s that it displaced TCE as the most widely used vapor degreasing solvent. In 1970, TCE accounted for 82% of all chlorinated solvents used in vapor degreasing (Morrison, 2006). By 1974, TCE accounted for only 46% of vapor degreasing solvent consumption (USEPA, 1977), and by 1976, TCE's share had declined to 42% (Morrison, 2006).

By the mid-1960s, the Navy and NASA were already using 1,4-dioxane-stabilized methyl chloroform. The Navy analyzed four commercially available methyl chloroform products and identified 16 compounds, including stabilizers and impurities. The report notes that most of the compounds (listed in Table 10.10) were previously found in submarine, spacecraft, and environmental test chamber atmospheres (Saunders, 1965). A chromatogram from the report (Figure 10.3) shows a distinct peak for 1,4-dioxane.

Similarly detailed composition profiles of technical grade methyl chloroform can be found in the toxicological literature in which scientists analyzed the composition of technical grade solvents to determine all of the toxic agents to which laboratory animals are exposed. For example, Table 10.11 lists the analysis of pre-1986 technical grade methyl chloroform from a study on the toxicity of methyl chloroform to male and female Sprague Dawley rat pups (Maltoni et al., 1986).

When the date a solvent stabilizer was discontinued is known, the stabilizer may serve as a time-marker. For example, federal specification O-T-634C (April 25, 1984) required that solvent stabilizer compounds may be no more toxic than the solvent to which they are added. Major solvent producers supplying federal contractors in the defense and other industries complied by removing epichlorohydrin as an acid acceptor solvent stabilizer for PCE and TCE. Epichlorohydrin is therefore a potentially useful time-marker chemical for forensic investigations. However, epichlorohydrin is relatively insoluble in water (66 mg/L), and also prone to biodegradation and rapid hydrolysis, so it is not expected to be sufficiently recalcitrant and persistent to allow its detection decades after it was released to the subsurface. Because epichlorohydrin is infinitely soluble in solvents, it may nevertheless be a viable time-marker compound in source zones harboring residual pure phase solvent (non-aqueous phase liquid [NAPL]). Owing to its low aqueous solubility, it will remain

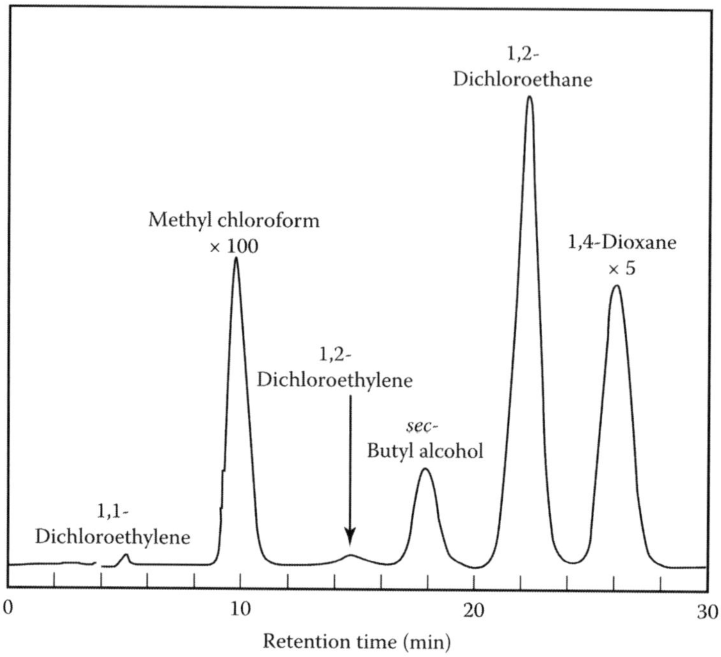

FIGURE 10.3 Chromatogram of 1965 methyl chloroform sold by Phillips-Jacobs Chemical. (From Saunders, R.A., 1965, *The Composition of Technical Grade Methyl Chloroform*. Washington, DC: Physical Chemistry Branch, U.S. Navy Research Laboratory.)

TABLE 10.11
Stabilizers and Impurities in Pre-1986 Technical Grade Methyl Chloroform

	Concentration
Stabilizers	
1,4-Dioxane	3.8 wt%
1,2-Epoxybutane	0.47 wt%
Nitromethane	0.27 wt%
n-Methyl pyrrole	<0.0001 wt%
Impurities	
Chloroform	100 ppm
Carbon tetrachloride	250 ppm
Trichloroethylene	200 ppm
1,2-Dichloroethane	2,300 ppm
1,1-Dichloroethylene	398 ppm
1,2,3-Trichloroethane	41.8 ppm
trans-1,2-Dichloroethylene	50 ppm
Tetrachloroethylene	475 ppm

Source: Maltoni, C., et al., 1986, Results of long-term carcinogenicity bioassays on Sprague-Dawley rats of methyl chloroform administered by ingestion. *Acta Oncology* 7: 101–117.

dissolved in pure phase TCE or PCE and dissolve into groundwater only slowly and after its relative mass proportion has increased due to TCE or PCE dissolution over many years.

The best opportunity to leverage epichlorohydrin as a time marker is likely obtained by centrifuging fine grained soil cores retrieved from the source zone, in which PCE or TCE concentrations are well in excess of their aqueous solubilities. Similarly, a synthetic precipitation leaching procedure test can be used to dissolve into water any hydrophobic residual stabilizers or impurities from pure-phase solvent bound to a fine-grained soil matrix. Epichlorohydrin has many other uses; however, other industrial and commercial uses of epichlorohydrin react it to form a resin or coating, such that it is consumed in reaction and remains essentially insoluble thereafter. Leveraging epichlorohydrin as a time-marker chemical requires a thorough investigation to ensure no other explanation for its presence is plausible.

While epichlorohydrin may be consumed by hydrolysis or biodegradation, it is likely to leave behind its own recalcitrant marker, 1,2,3-trichloropropane (123TCP). 123TCP occurs as an impurity of epichlorohydrin in the range of 0.1–0.4% weight, but has a substantially higher boiling point than either TCE of PCE (156.8°C, compared to 121°C for PCE and 86.7°C for TCE). Where TCE or PCE was used in vapor degreasing, 123TCP may become iteratively concentrated due to its higher boiling point. Because 123TCP is far more persistent than epichlorohydrin, it can serve as a surrogate marker of epichlorohydrin. In 2018, California set a drinking water maximum contaminant level (MCL) for 123TCP at 0.005 μg/L. Commercial laboratories routinely report 123TCP in the low parts-per trillion range, which can assist with detecting this potential time-marker at solvent release sites.

123TCP is also an impurity of the soil fumigant nematocide compound 1,3-dichloropene. It is therefore critical that a forensic investigation relying on 123TCP as a surrogate for epichlorohydrin confirm the land use history, and check for 1,3-dichoropropene and its breakdown products, as well as for the soil fumigant 1,2-dibromo-3-chloropropane (DBCP), which was also stabilized using epichlorohydrin (EPA, 1995).

10.4 SOLVENT STABILIZERS AS MARKERS OF SOLVENT APPLICATIONS

Solvent stabilizers were tailored to prevent solvent breakdown in the demanding environments in which solvents were used. The use of certain stabilizers to prevent solvent-breakdown reactions that occur in unique applications—such as PCE used in phosphatizing baths for automotive coatings or high temperatures encountered by PCE used in transformer fluids—can provide unique associations that may allow the interpretation of solvent origins. The following circumstances occurring together would favor the use of solvent stabilizers as marker chemicals to distinguish the origins of the same solvent used in different applications:

- Stabilizer is unique to a specific application employed by only one of the potentially responsible parties assumed to have contributed to the solvent spill(s)
- Stabilizer is added in sufficient quantity to favor detection, or stabilizer is concentrated in the application so that the mass released is sufficient to favor detection
- Stabilizer that would serve as a unique marker is sufficiently mobile and persistent to favor its migration to the region in which plumes of the common solvent have commingled
- Analytical methods are available to detect the marker stabilizer
- Stabilizer compound intended to serve as a unique marker of the solvent is not commonly used in other applications that might be released in neighboring industrial operations

The combined occurrence of these circumstances is unlikely because stabilizers generally compose only a small percentage of the solvent mixture. The following examples profile the potential for solvent stabilizers to serve as marker chemicals.

10.4.1 PCE as a Dielectric Fluid

As described in Chapter 1, PCE is stable up to 150°C in the presence of air and moisture, and in the sealed environment of an electrical transformer, but in the absence of catalysts, PCE may be stable up to 500°C. The desired characteristics of a dielectric fluid include long-term resistance to decomposition at high temperatures in the presence of oxygen, to prevent formation of electrically conductive or corrosive materials. PCE was used in transformers as a replacement for the more toxic class of dielectric fluids containing polychlorinated biphenyl (PCB) compounds. Commercial grades of PCE may contain manufacturing impurities including any of the chlorinated ethanes (e.g., 1,1,2,2-tetrachloroethane in grades produced before 1978) at levels up to 0.3 wt% (Borror and Rowe, 1981). When PCE is used as a dielectric fluid in an electrical transformer, any chlorinated ethanes present can be expected to undergo dehydrochlorination, producing hydrochloric acid and causing corrosion. To function well as a dielectric fluid, PCE must have less than 50 ppm total chloroethanes (Borror and Rowe, 1981). In particular, dichloroethane, methyl chloroform, symmetrical tetrachloroethane, pentachloroethane, and hexachloroethane should not be present in PCE used as a dielectric fluid at masses totaling more than 10 ppm by weight, and unsymmetrical tetrachloroethane (1,1,1,2-tetrachloroethane) should not exceed 30 ppm by weight. Even small amounts of chloroethanes present as impurities will react with aluminum chloride or other catalysts and form acid, which will in turn begin to attack PCE. PCE destined to be used as a dielectric fluid was subjected to extra refining, including scrubbing and distillation, and was stabilized with *n*-methyl pyrrole and pentaphen (*p*-*tert*-amyl phenol).

Some transformers require a dielectric fluid that will vaporize at higher temperatures. PCE will react with oxygen at high temperatures, particularly in the vapor phase; therefore, the stabilizers added to PCE must be equally effective in the liquid and vapor phases. *N*-methyl pyrrole boils at 112°C (compared with 121°C for PCE) and will be carried into the vapor phase, whereas pentaphen boils at 266°C but has sufficiently high vapor pressure in the system to be partially carried into the vapor phase at the boiling point of PCE (Borror and Rowe, 1981).

N-methyl pyrrole is not unique to the stabilization of PCE in transformers; it is also used to stabilize methyl chloroform and TCE, as described in Chapter 1. Similarly, pentaphen is also used to stabilize TCE against oxidation. Neither stabilizer provides a unique marker if TCE was among the solvents released at a site that also had a PCE release from a transformer. However, if the task is to differentiate PCE from a transformer versus PCE from a dry cleaner, then there may be some potential to use these stabilizers as marker chemicals. Dry-cleaning grades of PCE apparently did not use *n*-methyl pyrrole or pentaphen. The patent literature includes citations of commercial PCE containing 0.022–0.028% *n*-methyl pyrrole as an antioxidant (Copelin, 1959); however, citations referring to pyrrole as a stabilizer of PCE are presented in the vapor degreasing/metal-cleaning context and not in the dry-cleaning context.

Dow Chemical's dry-cleaning grade of PCE, DowPer, used 4-methyl morpholine, Vulcan's PerSec used diallylamine and tripropylene compounds, and PPG's dry-cleaning grade used cyclohexene oxide and β-ethoxypropionitrile. A stabilizer concentrate for PPG's dry-cleaning grade of PCE adds cyclohexene oxide, β-ethoxypropionitrile, *n*-methyl morpholine, and 4-methoxyphenol (see Table 1.21).

The comparative fate of these stabilizers is fundamental to establishing their utility as marker chemicals. Table 10.12 contrasts the fate and transport properties of stabilizers of dry-cleaning versus transformer grades of PCE. The fate and transport properties rule out some stabilizers as marker chemicals; for example, cyclohexene oxide has low solubility and hydrolyzes easily, and both diallylamine and 4-methyl morpholine are susceptible to sorption in the ionized form they assume once dissociated in water. In the BIOWIN GCM, all of the PCE stabilizers are estimated to be easily biodegradable in a matter of weeks.

As discussed in Chapter 3, BIOWIN estimates are less reliable than some of the other estimation tools in the EPIWIN suite, and field evidence suggests that at least one of the compounds

TABLE 10.12
Comparison of Stabilizer Fate and Transport Properties for Electrical Transformer and Dry-Cleaning Grades of Perchloroethylene

Stabilizer	Solubility (mg/L)	log K_{oc}	Hydrolysis Potential	Acid Dissociation Constant[a] (pK_a) at pH 7	BIOWIN-Estimated Biodegradation Potential	Subsurface Mobility
Stabilizers of dielectric fluid grade perchloroethylene for transformers						
p-tert-Amyl phenol (pentaphen)	168	2.41	Low	0.00167%	Weeks to months	Minimal
n-Methyl pyrrole	71,900	*2.16*[b]	—	−2.9	Weeks to months	High
Stabilizers of dry-cleaning grade perchloroethylene						
4-Methyl morpholine	*Miscible*	*0.719*	—	8.49, 7.38	Weeks to months	High
Diallylamine	86,000	1.11	—	9.29	Weeks	High
Tripropylene compounds[c]	*196,000*	1	—	NA	Weeks to months	High
Cyclohexene oxide	*4,530*	*1.2*	High	NA	Weeks	Minimal
3-Methoxy propionitrile	*732,000*	*0.186*	—	NA	Weeks	High
n-Butyl glycidyl ether	20,000	*0.451*	Very low	4.89	Weeks	High
Perchloroethylene[d]	150	2.32	—	—	Months	Moderate

Notes: See Chapter 3 for description of these parameters and BIOWIN estimates.

[a] A high pK_a indicates that the compound may exist in a protonated form in the environment; cations generally adsorb to organic carbon and clay more strongly, thereby reducing mobility compared to that of neutral compounds.

[b] Italicized values were estimated by using EPIWIN.

[c] "Tripropylene compounds" are assumed to be tripropylene glycol monomethyl ether, also known as DOWANOL® TPM glycol ether and ARCOSOLV® TPM; tripropylene glycol-mono-*n*-propyl ether, also known as DOWANOL® TPnP; or tripropylene glycol mono-*n*-butyl ether, also known as ARCOSOLV® TPNB or DOWANOL® TPNB; listed values are for tripropylene glycol monomethyl ether.

[d] Mass fraction = 99%.

in Table 10.12 is resistant to biodegradation because it has been detected years after the probable date of release, or that its dissolution is delayed. For stabilizer compounds that are readily biodegraded or hydrolyzed after dissolution into groundwater, it is important to compare their aqueous solubility to their solubility in the host solvent. When the solubility of the stabilizer in the solvent is orders of magnitude higher than its aqueous solubility, the stabilizer may remain in the solvent as the solvent slowly dissolves. As the solvent mass diminishes, the proportion of the stabilizer will increase, and the stabilizer will eventually dissolve into groundwater as well (to the extent it is water-soluble), whereafter it may undergo rapid degradation. Therefore, stabilizers that remain dissolved in solvent will not immediately enter groundwater; rather, their release is delayed, thereby extending their potential utility as marker chemicals in forensic investigations by many years after their release. Online chemical property databases show that all of the compounds listed in Table 10.12 are miscible in organic solvents, whereas several are relatively insoluble in water.

Butoxymethyl oxirane, also called *n*-butyl glycidyl ether or *n*-BGE, was found in soils at a PCE release in Mountain View, California. The *n*-BGE was detected in soils adjacent to a chemical packaging facility that packaged paint thinners and related products, but never handled PCE. A PCE plume of unknown origin was identified adjacent to the property, and in soil samples retrieved beneath a stormwater drainage ditch, where it occurs together with *n*-BGE. The adjacent property housed a printing operation, which may have used PCE to clean printing presses and could be a source of PCE; the stormwater drainage ditch drains from the printing facility toward the chemical packaging facility. Sewer lines also traverse the area and collect wastewater from an area that included several former dry-cleaning operations. A complicating factor for interpreting the presence of *n*-BGE co-located with PCE is its possible presence in printing inks. Butoxymethyl oxirane is used as an epoxy resin diluent in some ink formulations (NIEHS, 2004). The detection of butoxymethyl oxirane did not factor into the determination of responsibility for the PCE plume.

For the hypothetical situation in which PCE was released from both a dry cleaner and an electrical transformer, the fate and transport properties suggest that it may be possible to track the movement of the transformer solvent with *n*-methyl pyrrole and the dry-cleaner solvent with *n*-butyl glycidyl ether. The scenario used for this example would be uncommon if it has occurred at all, so of course no field data are available to test the viability of leveraging these stabilizers to differentiate commingled PCE plumes. Nevertheless, there is a reasonably good potential that some combinations of stabilizers and solvents in addition to 1,4-dioxane and methyl chloroform will prove useful for distinguishing the origins of commingled plumes.

REFERENCES

ASTM, 1962, *Handbook of Vapor Degreasing*. American Society for Testing and Materials Special Technical Publication No. 310.

ASTM, 2006, Standard test methods for determination of 1,1,1-trichloroethane and methylene chloride content in stabilized TCE and tetrachloroethylene. D5320–96 (2000) (Withdrawn 2004; replaced by D6806), *American Society for Testing and Materials*.

Beneteau, K.M., Aravena, R., and Frape, S.K., 1999, Isotopic characterization of chlorinated solvents—laboratory and field results. *Organic Geochemistry* 30(8 Part 1): 739–753.

Bennett, P. and Aravena, R., 2017, Extending the applicability of compound-specific isotope analysis to low concentrations of 1,4-dioxane. *SERDP Project* ER-2535 February 2017.

Bennett, P., Hyman, M., Aravena, R., et al., 2018, Enrichment with carbon-13 and deuterium during monooxygenase-mediated biodegradation of 1,4-dioxane. *ES&T Letters* 5(3): 148–153.

Blessing, M., Jochmann, M.A., Haderlein, S.B., and Schmidt, T.C., 2008, Compound-specific isotope analysis at trace concentrations: Evaluation of a large volume injection (LVI) based method. *Geophysical Research Abstracts* 10(EGU2008-A-08981).

Bohnert, G.W. and Carey, D.A., 1991, Scale-up of recovery process for waste solvents. *Technical Communications*. Kansas City, Missouri: Allied Signal Aerospace Company.

Borror, J.A. and Rowe, E.A.J., 1981, United States Patent 4,293,433: PCE dielectric fluid containing pyrrole and phenol. Diamond Shamrock Corporation, Dallas, TX.

Bouchard, D., 2007, Use of stable isotope analysis to assess biodegradation of volatile organic compounds in the unsaturated subsurface. PhD dissertation. Centre D'Hydrogéologie. Neuchâtel, University of Neuchâtel, Switzerland.

Bursack, K.F., Rains, J.H., and Bussard, P.A., 1966, British Patent 1,042,076: Stabilization of methyl chloroform. Assignee: Vulcan Materials Company, Birmingham, AL.

California Environmental Protection Agency (Cal EPA), 2003, Letter from San Francisco Bay Regional Water Quality Control Board to Varian Medical Systems: Request for Site Closure at 3100 Jay Street, Santa Clara, CA, San Francisco Bay Regional Water Quality Control Board, May 5th.

Chickos, J.S. and Acree, W.E., Jr., 2003, Enthalpies of vaporization of organic and organometallic compounds, 1880–2002. *Journal of Physical and Chemical Reference Data* 32(2): 519–878.

Copelin, H.B., 1959, United States Patent 2,904,600: Stabilization of chlorinated hydrocarbons. Assignee: E.I. du Pont de Nemours and Company, Wilmington, DE.

Doherty, R.E., 2000a, A history of the production and use of carbon tetrachloride, tetrachloroethylene, TCE and 1,1,1-trichloroethane in the United States: Part 1-Historical background; carbon tetrachloride and tetrachloroethylene. *Journal of Environmental Forensics* 1(2): 69–81.

Doherty, R.E., 2000b, A history of the production and use of carbon tetrachloride, tetrachloroethylene, TCE and 1,1,1-trichloroethane in the United States: Part 2-TCE and 1,1,1-trichloroethane. *Journal of Environmental Forensics* 1(2): 83–93.

Dow Chemical Company, 2006, Chlorinated solvents-physical properties. Form No. 100–06358. http://www.chlorinatedsolvents.co (accessed May 29, 2006).

Elsner, M., Zwank, L., Hunkeler, D., and Schwarzenbach, R.P., 2005, A new concept linking observable stable isotope fractionation to transformation pathways of organic pollutants. *Environmental Science and Technology* 39(18): 6896–6916.

EPA, 1995, Decisions of the United States Environmental Protection Agency, March 1972 to March 1985. See page 683. ISBN 0-16-048230-5.

European Chemicals Bureau, 2004, European Union risk assessment report: Trichloroethylene. 348. Environment Agency Chemicals Assessment Section, Ecotoxicology and Hazardous Substances National Centre, Isis House, Howbery Park Wallingford, Oxfordshire. European Commission Joint Research Centre, Institute for Health and Consumer Protection, European Chemicals Bureau.

Fritz, I.D. and Clark, P., 1997, *Environmental Isotopes in Hydrogeology*. Boca Raton, Florida: CRC Press.

Gauthier, T.D. and Murphy, B.L., 2003, Age dating groundwater plumes based on the ratio of 1,1-dichloroethylene to 1,1,1-trichloroethane: An uncertainty analysis. *Environmental Forensics* 4: 205–213.

Henschler, D., Eder, E., Neudecker, T., and Metzler, M., 1977, Carcinogenicity of TCE: Fact or artifact? *Archives of Toxicology* 37(3): 233–236.

Henschler, D., Reichert, D., and Metzler, M., 1980, Identification of potential carcinogens in technical grade 1,1,1-trichloroethane. *International Archives of Occupational and Environmental Health* 47(3): 263–268.

Holt, R.D., 1990, Physical properties of contaminated TCE and 1,1,1-trichloroethane. *Technical Communications*. Kansas City, Missouri: Allied Signal Aerospace Company.

Hunkeler, D. and Aravena, R., 2000, Determination of stable carbon isotope ratios of chlorinated methanes, ethanes and ethenes in aqueous samples. *Environmental Science and Technology* 34(13): 2839–2844.

INEL, 2006, Mass estimate of organic compounds in 743-series waste buried in the subsurface disposal area for operable units 7–08 and 7–13/14. U.S. Department of Energy, Idaho National Engineering and Environmental Laboratory. Idaho cleanup project report 23256.

International Agency for Research on Cancer (IARC), 1986, Monographs on the evaluation of the carcinogenic risk of chemicals to humans. *Some Halogenated Hydrocarbons and Pesticide Exposures*, Vol. 41. Lyon, France: International Agency for Research on Cancer.

Jackson, R.E., Lesage, S., Priddle, M.W., Crowe, A.S., and Shikaze, S., 1991, *Contaminant Hydrogeology of Toxic Organic Chemicals at a Disposal Site, Gloucester, Ontario: 2. Remedial Investigation*. Burlington, Ontario, Canada: National Water Research Institute.

Jochmann, M.A., Blessing, M., Haderlein, S.B., and Schmidt, T.C., 2006, A new approach to determine method detection limits for compound-specific isotope analysis of volatile organic compounds. *Rapid Communications in Mass Spectrometry* 20(24): 3639–3648.

Kuder, T., Wilson, J.T., Kaiser, P., Kolhatkar, R., Philp, P., and Allen, J., 2005, Enrichment of stable carbon and hydrogen isotopes during anaerobic biodegradation of MTBE: Microcosm and field evidence. *Environmental Science and Technology* 39(1): 213–220.

Lane, V. and Smith, J., 2006, Fact or fiction? The source of perchloroethylene contamination in groundwater is a manufacturing impurity in chlorinated solvents. Presented to the National Groundwater Association Ground Water and Environmental Law Conference, July 7, 2006. Chicago, Illinois: National Groundwater Association. http://www.ngwa.org (accessed August 2, 2006).

Lyman, W.J., Reehl, W.F., and Rosenblatt, D.H. (Eds), 1990, *Handbook of Chemical Property Estimation Methods: Environmental Behavior of Organic Compounds*. Washington, DC: American Chemical Society.

Maltoni, C., Cotti, G., and Patella, V., 1986, Results of long-term carcinogenicity bioassays on Sprague-Dawley rats of methyl chloroform administered by ingestion. *Acta Oncology* 7: 101–117.

McNab, W.W. and Dooher, B.P., 1998. A critique of a steady-state analytical method for estimating contaminant degradation rates. *Ground Water* 36(6): 983–987.

McNab, W.W. and Narasimhan, T.S., 1994, Degradation of chlorinated hydrocarbons and groundwater geochemistry: A field study. *Environmental Science and Technology* 28(5): 769–778.

McNab, W.W., Rice, D.W., Bear, J., Ragaini, R., Tuckfield, C., and Oldenburg, C., 1999, Historical case analysis of chlorinated volatile organic compound plumes. Lawrence Livermore National Laboratory report UCRL-AR-133361.

Michalak, A.M. and Kitanidis, P.K., 2002, Application of Bayesian inference methods to inverse modeling for contaminant source identification at Gloucester Landfill, Canada. XIVth International Conference on Computational Methods in Water Resources, June 23–28, 2002, Delft, Netherlands: Elsevier.

Morrison, R.D., 2006, Forensic techniques for establishing the origin and timing of a contaminant release, Environmental Data Pages. http://www.techstuff.com/bmorrep.htm (accessed July 29, 2006).

Morrison, R.D. and Murphy, B.L., 2006, *Environmental Forensics—Contaminant Specific Guide*. San Diego: Elsevier.

Morrison, R.D. and Murphy, B.L., 2013, Contaminant Transport Models. In: R.D. Morrison and B.L. Murphy (Eds), *A Forensic Analysis*. Royal Society of Chemistry, Cambridge, UK, pp. 426–468.

Morrison, R.D., Murphy, B.L., and Doherty, R.E., 2006, Chlorinated solvents. In: R.D. Morrison and B.L. Murphy (Eds), *Environmental Forensics—Contaminant Specific Guide*. San Diego: Elsevier.

Murphy, B.L. and Morrison, R.D., 2007, *Introduction to Environmental Forensics*. San Diego: Elsevier Academic Press.

Nakamiya, K., Hashimoto, S., Ito, H., Edmonds, J.S., and Morita, M., 2005, Degradation of 1,4-dioxane and cyclic ethers by an isolated fungus. *Applied and Environmental Microbiology* 71(3): 1254–1258.

National Institute of Environmental Health Sciences (NIEHS), 2004, n-Butyl glycidyl ether (BGE)—[CAS No. 2426–08–6]: Review of toxicological literature. National Toxicology Program (NTP), National Institute of Environmental Health Sciences (NIEHS), Research Triangle Park, North Carolina.

Pankow, J.F. and Cherry, J.A., 1996, *Dense Chlorinated Solvents and Other DNAPLs in Groundwater: History, Behavior, and Remediation*. Portland, Oregon: Waterloo Press.

Parsons, F., Wood, P.R., and DeMarco, J., 1984, Transformations of tetrachloroethene and trichloroethene in microcosms and groundwater. *Journal of the American Water Works Association* 76(2): 56–59.

PPG (Pittsburgh Plate and Glass), 1998, Material Safety Data Sheet for tri-ethane 348, issued March 2, 1998, Pittsburgh, Pennsylvania.

Pornwongthong, P., 2014, Stable Isotopic and Molecular Biological Tools to Validate Biodegradation of 1,4-Dioxane. Dissertation, UCLA Civil Engineering.

Saunders, R.A. 1965, *The Composition of Technical Grade Methyl Chloroform*. Washington, DC: Physical Chemistry Branch, U.S. Navy Research Laboratory.

Shepherd, C.B., 1962, Perchloroethylene and trichloroethylene. In: S.J. Sconce (Ed), *Chlorine: Its Manufacture, Properties and Uses, American Chemical Society Monograph Series, No. 154*, Chapter 13. Washington, DC: American Chemical Society.

Shouakar-Stash, O., Frape, S., and Drimmie, R., 2003, Stable hydrogen, carbon, and chlorine isotope measurements of selected organic solvents. *Journal of Contaminant Hydrology* 30(3–4): 211–228.

Shuckrow, A.J., Pajak, A.P., and Touhill, C.J., 1982, *Hazardous Waste Leachate Management Manual*. Park Ridge, NJ: Noyes Data Corporation.

Slater, G.F., Ahad, J., Sherwood-Lollar, B., Allen-King, R., and Sleep, B., 2000, Carbon isotope effects resulting from equilibrium sorption of dissolved VOCs. *Analytical Chemistry* 72(22): 5669–5672.

Slater, G.F., Dempster, H.S., Sherwood-Lollar, B., and Ahad, J., 1999, Headspace analysis: A new application for isotopic characterization of dissolved organic contaminants. *Environmental Science and Technology* 33(1): 190–194.

Spencer, D.R. andArcher, W.L., 1981, British Patent 1,582,803: Stabilization of 1,1,1-trichloroethane. Assignee: Dow Chemical Company, Midland, MI.

Steffan, R.J., McClay, K.R., Masuda, H., and Zylstra, G.J., 2007, Biodegradation of 1,4-Dioxane. Arlington, Virginia: SERDP—Strategic Environmental Research and Development Program.

Stewart, R.D., Gay, H.H., Schaffer, A.W., Erley, D.S., and Rowe, V.K., 1969, Experimental human exposure to methyl chloroform vapor. *Archives of Environmental Health* 19(4): 467–472.

Sturchio, N.C., Clausen, J.C., Heraty, L.J., Huang, L., Holt, B.D., and Abrajano, T., 1998, Stable chlorine isotope investigation of natural attenuation of trichloroethene in an aerobic aquifer. *Environmental Science and Technology* 32(20): 3037–3042.

Tarrer, A.R., Donahue, B.A., Dhamavaram, S., and Joshi, S.B., 1989, *Reclamation and Reprocessing of Spent Solvent*. Park Ridge, NJ: Noyes Data Corporation.

Tarrer, A.R., Sanjay, H.J., and Howell, G.S., 1993, Minimizing pollution in cleaning and degreasing operations—Armstrong Laboratory, Tyndall Air Force Base, Florida. U.S. Environmental Protection Agency, Cincinnati, Ohio. EPA/600/R-93/191.

Totonidis, S., 2005, A role for TCE in developing nation anaesthesia. *Kathmandu University Medical Journal* 3(2): 181–190.

Tsuruta, H., Iwasaki, K., and Fukuda, K., 1983, Analysis of trace impurities in reagent and technical grade TCE. *Industrial Health* 21(4): 293–295.

U.S. Environmental Protection Agency (USEPA), 1977, Control of volatile organic emissions from solvent metal cleaning. Research Triangle Park, North Carolina, U.S. Environmental Protection Agency, Office of Air Quality Protection Standards.

U.S. Environmental Protection Agency (USEPA), 1999, Use of monitored natural attenuation at Superfund, RCRA corrective action, and underground storage tank sites. U.S. Environmental Protection Agency, Office of Solid Waste and Emergency Response. Directive 9200.4–17P.

U.S. Environmental Protection Agency (USEPA), 2008. A Guide for Assessing Biodegradation and Source Identification of Organic Ground Water Contaminants using Compound Specific Isotope Analysis (CSIA). Office of Research and Development, National Risk Management Research Laboratory, Ada, Oklahoma. EPA 600/R-08/148 December 2008.

United Nations Environment Programme (UNEP), 2005, *Draft technical guidelines for the environmentally sound management of wastes consisting of, containing, or contaminated with hexachlorobenzene (HCB)*. Open-ended Working Group of the Basel Convention on the Control of Transboundary Movements of Hazardous Wastes and Their Disposal, Nairobi, Kenya. United Nations Environment Programme.

van Wamerdam, E.M., Frape, S.K., Aravena, R., Drimmie, R.J., Flatt, H., and Cherry, J.A., 1995, Stable chlorine and carbon isotope measurements of selected chlorinated organic solvents. *Applied Geochemistry* 10(5): 547–552.

Vogel, T.M. and McCarty, P.L., 1987a, Abiotic and biotic transformations of 1,1,1-trichloroethane under methanogenic conditions. *Environmental Science and Technology* 21(12): 1208–1213.

Vogel, T.M. and McCarty, P.L., 1987b, Rate of abiotic formation of 1,1-dichloroethylene from 1,1,1-TCE in groundwater. *Journal of Contaminant Hydrology* 1(3): 299–308.

Wang, Y., 2016, Breakthrough in 2D-CSIA technology for 1,4-dioxane. *Remediation* 27(1): 61–70.

Willis, G.C. and Christian, C.A., 1957, United States Patent 2,803,676: TCE stabilized with propargyl alcohol and pyrrole. Assignee: Dow Chemical Company, Midland, MI.

Wing, M.R., 1997, Apparent first-order kinetics in the transformation of 1,1,1-trichloroethane in groundwater following a transient release. *Chemosphere* 34(3): 771–781.

Woodbury, A., Sudicky, E., Ulrych, T.J., and Ludwig, R., 1998, Three-dimensional plume source reconstruction using minimum relative entropy inversion. *Journal of Contaminant Hydrology* 32: 131–158.

Yeh, D. and Pavlostathis, S., 2005, Anaerobic biodegradability of tween surfactants used as a carbon source for the microbial reductive dechlorination of hexachlorobenzene. *Water Science and Technology* 52: 343–349.

Zwank, L., Berg, M., Schmidt, T.C., and Haderlein, S.B., 2003, Compound-specific carbon isotope analysis of volatile organic compounds in the low-microgram per liter range. *Analytical Chemistry* 75(20): 5575–5583.

Appendix 1: Synopses of Selected Patents from the Art of Solvent Stabilization

1,925,602, 1933; PITMAN, FOR E.I. DU PONT DE NEMOURS & CO., INC., USA

Notes that TCE, PCE, and DCM form acids upon decomposition that corrode metal storage tanks. Gasoline has been used to stabilize TCE—but large volumes are required. Notes other stabilizers have boiling point differences causing fractionation upon distillation. Seeks a stabilizer that boils with TCE and is not malodorous. Claims that triethylamine meets these criteria, at 0.0001–0.1% by volume. If TCE is not exposed to light, only 0.0005% triethylamine is needed, more if storage includes exposure to light. Presents data for storage stability of PCE, chloroform, TCE. No discussion of solvent breakdown in vapor degreasing.

1,971,318, 1934; SMITH AND DE PREE, FOR DOW CHEMICAL CO., USA

Stabilizes CCl_4 and PCE with small amounts of natural resins, such as gum mastic, sandarac,* and rosin. Adds 0.006–0.13% to protect solvents during storage when in contact with iron or copper.

2,108,390, 1938; PRICE, FOR WESTVACO CLORINE PRODUCTS CORP., USA

Stabilizes TCE with thiocyanate esters at 0.001–1.0% (preferably methyl thiocyanate). Notes that the stabilizer itself must be stable, and stand up to repeated phase transitions from liquid to vapor and vapor to liquid.

2,111,253, 1938; STOESSER AND ALQUIST, FOR DOW CHEMICAL CO., USA

Observes that all chlorinated aliphatic hydrocarbons decompose in the presence of moisture, leading to corrosion of metal surfaces. Proposes the addition of a piperazine compound such as tetramethyl piperazine or piperazine hydrate to stabilize chlorinated aliphatic hydrocarbons (PCE in particular) against breakdown from exposure to light and against formation of acid.

2,155,723, 1939; LEVINE AND CASS, FOR E.I. DU PONT DE NEMOURS & CO., INC., USA

Notes that TCE decomposes during recovery (distillation, condensation) if not restabilized at this point in the solvent life cycle. "In order to combat this tendency to decompose during storage, use, and recovery, it has been usual to dissolve in the TCE an agent [*p-tert*-butyl phenol] which will stabilize the solvent under the conditions at which these operations are normally carried out."

2,371,644, 1945; PETERING AND AITCHISON, FOR WESTVACO CHLORINE PRODUCTS CORP., USA

Emphasizes the addition of oxygen-containing organic compounds such as organic oxides, oximes, ethers, and alcohols. The more alcohol or other oxygen compound present, the more effective the

* Sandarac is a resin obtained from the small cypress-like tree *Tetraclinis articulata* of northern Africa.

composition in restraining the "metal-induced" decomposition of solvents. Recommends using 1 mol oxygen compound per 99 mol chlorohydrocarbon. Emphasizes that alcohols boil with the solvent. A minor amount of oxygen-containing compound results in a marked decrease in metal-induced decomposition. Claims *n*-butyl alcohol and isoamyl alcohol to stabilize TCE.

2,435,312, 1948; KLABUNDE, FOR E.I. DU PONT DE NEMOURS & CO., INC., USA

Stabilizes TCE with diisobutylene to inhibit the condensation decomposition reaction catalyzed by aluminum or iron. Notes that when exposed to air, light, or heat, TCE degrades by oxidation to form dichloroacetyl chloride, HCl, and phosgene. At higher temperatures, TCE undergoes a condensation type decomposition to form HCl together with a resinous material. Describes diisobutylene as an unsaturated organic compound boiling at about 102°C. Uses diisobutylene to stabilize TCE at 0.1–1.0%. Recommends addition of an antioxidant.

2,517,894, 1950; LARCHAR, FOR E.I. DU PONT DE NEMOURS & CO., INC., USA

Proposes adding cyclohexene at 0.02–0.25% to TCE to stabilize against reactions with metals. Notes that cyclohexene boils with TCE, allowing recovery on distillation. Finds that cyclohexene protects against solvent breakdown in the presence of alkali metals 17 times more effectively than cyclohexane. Also finds that cyclohexene in TCE is three times more effective than aniline.

2,721,883, 1955; STEVENS, FOR COLUMBIA-SOUTHERN CHEMICAL CORP., USA

The decomposition of perchloroethylene is more pronounced when in contact with iron or copper. Products of PCE decomposition materially injure its beneficial effect, causing metal corrosion. Notes that perchloroethylene shows poor light stability; fluorescent and ultraviolet light cause PCE to decompose, producing a measureable decrease in pH. Compounds previously used to stabilize PCE, such as benzaldehyde, cyclohexane, toluene, benzyl chloride, amylene hexylresorcinol, butyl mercaptan, ethyl acetate, guanidine, and pyridine, have been found to be unsatisfactory because they are ineffective or their odors or color are objectionable. Proposes *N*-methylmorpholine as a stabilizer to protect PCE against UV-light attack.

2,751,421, 1956; STAUFFER, FOR E.I. DU PONT DE NEMOURS & CO., INC., USA

Claims that *tert*-butyl-4-hydroxyanisole (BHA, the food preservative) stabilizes TCE in storage, which otherwise oxidizes and decomposes resulting in the formation of acids. This decomposition is accelerated by heat, light, moisture, and contact with metals. The use of amines is disadvantageous because they are dissipated by reaction with the acids that the solvent may encounter during normal use. Amines have a further disadvantage: the production of amine salts accelerates the corrosion of metals.

2,775,624, 1956; SKEETERS AND BALDRIDGE, FOR DIAMOND ALKALI CO., USA

Claims that acetylenic alcohols, particularly the methyl butynol and methyl pentynols (e.g., 2-methyl-3-butyne-2-ol, and 3-methyl-1-pentyne-3-ol, and propargyl alcohol), stabilize PCE against oxidation. Provides boiling ranges for several acetylenic alcohols that fall within or slightly below PCE's boiling range of 119–122°C. Discusses methods for purifying PCE to remove trace lower aliphatics that lead to decomposition. Skeeters and Baldridge assert that acetylenic alcohols perform better than all other stabilizers for metal degreasing and wood dehydration using PCE.

2,795,623 1957; STARKS, FOR F.W. E.I. DU PONT DE NEMOURS & CO., INC., USA

Starks presents a stabilizer for a stabilizer: *N*-methyl pyrrole. Pyrroles added to TCE and PCE at 0.001–1% by weight are somewhat unstable and may decompose when exposed to light, heat, or oxygen. TCE containing *N*-methyl pyrrole will show discoloration on exposure to bright sunlight in as little as 2–4 hours. Solvent in which *N*-methyl pyrrole has become unstable will itself destabilize and deposit black tar. This invention uses organometallic chelate compounds to stabilize *N*-methyl pyrrole added at 0.001% by weight. Claims copper butylacetoacetate, copper salicylate, aluminum methylsalicylate, and aluminum ethylacetoacetate.

2,797,250, 1957; COPELIN, FOR E.I. DU PONT DE NEMOURS & CO., INC., USA

Proposes use of an organic oxide, preferably epichlorohydrin, in combination with a basic amine, such as triethylamine, to act as antacids for TCE and PCE. Basic amines are commonly used to impart alkaline conditions to chlorinated hydrocarbons. HCl formation is a problem because it catalyzes the rusting of iron or steel workpieces cleaned in the machine or in the degreasing apparatus itself. Even small amounts of HCl will catalyze the rusting of iron or steel workpieces put through the degreasing machine. If HCl formation occurs, or if metallic salts—particularly aluminum chloride—are formed, the operator must shut down the degreasing production line to neutralize the solvent.

2,811,252, 1957; BACHTEL, FOR DOW CHEMICAL COMPANY, USA

Explains that methyl chloroform is inherently unstable and reactive with metals, particularly aluminum, and that none of the metal-stabilizing inhibitors added to TCE and other solvents are effective at preventing this reaction in methyl chloroform. This patent is the first to claim the discovery of 1,4-dioxane's effectiveness at inhibiting the reaction between methyl chloroform and aluminum, even at boiling temperature, when added at 4% by weight. Recommends including *sec*-butyl alcohol or *tert*-amyl alcohol to keep the solvent water white. Experiments show that the inhibiting effect of 1,4-dioxane on methyl chloroform is not lost as the methyl chloroform evaporates, the unevaporated residue being still stabilized.

2,818,446, 1957; STARKS, FOR E.I. DU PONT DE NEMOURS & CO., INC., USA

Notes that many other stabilizers increase the solubility of HCl. Invention claims synergistic combination of an epoxide and an ester to stabilize TCE. Preferred combination is 0.2% by weight of ethyl acetate and 0.3% by weight of epichlorohydrin. Claims this combination provides a stabilizing effect from 7- to 20-fold greater than that of either component used singly. Notes that it may still be necessary to add a stabilizing amount of a phenolic antioxidant.

2,838,458, 1958; BACHTEL, FOR DOW CHEMICAL COMPANY, USA

Amends findings in earlier patent, noting that 1,4-dioxane mixed with alcohol does not entirely prevent corrosion of iron or aluminum when exposed to condensing vapors of methyl chloroform at boiling temperatures. Proposes addition of aliphatic monohydric acetylenic alcohols such as 2-methyl-3-butyn-l-ol at 0.1% with 1,4-dioxane at 3.5% to prevent this reaction.

2,947,792, 1960; SKEETERS, FOR DIAMOND ALKALI CO., USA

Lists numerous stabilizers of PCE and suggests they have light stabilizing properties. This invention claims isoeugenol (4-hdyroxy-3-methoxy-1-propenylbenzene) as a light stabilizer of PCE, and recommends it should be used in combination with methyl pentynol.

3,049,571, 1962; BROWN, FOR DOW CHEMICAL CO., USA

Notes that methyl chloroform (TCA) will react with zinc in the boiling vapor when stabilized with 1,4-dioxane, nitromethane, and *sec*-butyl alcohol. It is the stabilizers themselves that catalyze the reaction with zinc. This is a problem for reclaiming spent solvent from cold degreasing using TCA, because much solvent recycling equipment has galvanized iron containing zinc. Zinc-lined stills commonly used in solvent recycling would be damaged. Zinc causes degradation of the solvent vapor and evolves acid, which attacks and liberates more zinc. This patent claims the addition of vicinal monoepoxides to methyl chloroform (aliphatic epoxides containing an ethylene oxide group). 1,2-Butylene oxide is the example tested.

3,133,885, 1964; PETERING AND CALLAHAN, FOR DETREX CHEMICAL INDUSTRIES, INC., USA

Claims epoxides, for amine-stabilized PCE and TCE, noting these formulations will react with iron and copper in the absence of water. Notes most stabilizers have high boiling points, but the wide difference in boiling points between the solvent and the stabilizers is undesirable because of the difficulty of stabilizing the vapor phase. Distillation of solvent may result in complete loss of stabilizer if boiling point differences are large.

990,302 (GREAT BRITAIN), 1965; TECT, INC., USA

Notes that utility of methyl chloroform (TCA) as a cold cleaning agent was recognized as a preferable replacement to the more toxic carbon tetrachloride (CT) since 1950, but it was not until 1954 that addition of stabilizers enabled the use of TCA for cold cleaning without formation of a tarry mass or releasing phosgene gas and HCl. This 1965 patent notes that until recently TCA was not considered stable for hot solvent vapor degreasing. Invention was successful to stabilize TCA for degreasing for up to 7,200 hours.

3,326,988, 1967; STACK, FOR PITTSBURG PLATE GLASS COMPANY, USA

Alcohol-containing compositions of methyl chloroform (TCA) can be better stabilized by adding perchloroethylene to form an azeotrope. Addition of PCE at 0.5–5% by weight increases the rate of TCA evaporation. Alcohol-stabilized TCA will leave spots on the work if PCE is not added. Most common TCA formulations contain 1–2% alcohol (saturated- and unsaturated-monohydric aliphatic alcohols, e.g., *sec*-butyl alcohol).

3,878,256, 1975; RICHTZENHAIN AND RUDOLF, FOR DYNAMIT NOBEL AG, GERMANY

Tested different proportions of a four-component system using MIL-T-7003 metal strip method with aluminum, zinc, and iron. Found that the best mix stabilizing mixture was methoxyacetonitrile, 0.5%; 1,2-epoxybutane, 0.8%; 1,4-dioxane, 2%; and nitromethane, 1%. This mixture caused no corrosion of metal strips in a 168-hour test.

4,032,584, 1977; IRANI, FOR STAUFFER CHEMICAL CO., USA

Notes dichloromethane is the least toxic of the chloromethanes, and it is more stable than PCE, TCE, and MC, and has a lower boiling point. Dichloromethane is also resistant to photochemical activity and therefore does not produce smog. Dichloromethane can react with aromatic and aliphatic compounds in the presence of metals, metal halides, and combinations thereof, including aluminum, zinc, and iron, and their halides. The reaction product is generally an objectionable, high-boiling tarry substance that renders the dichloromethane unsuitable for further use. Compounds that react with methylene chloride are generally introduced from various cutting oils and lubricants used in metal fabricating operations. Stabilizers are added to dichloromethane to prevent degradation and other types of deterioration such as oxidation, hydrolysis, and pyrolysis. Stabilizing amounts of mixed amylenes, propylene oxide, butylene oxide, and tertiary butylamine are added.

Appendix 2: Fate and Transport Properties of Solvent-Stabilizer Compounds

Stabilizer Compound	Chemical Abstract Service Registry Number	Boiling Point (°C @ 760 mm Hg)	Water Solubility (mg/L @ 25°C)	Solubility Measurement Temperature (°C)	log K_{ow}	log K_{oc}	Henry's Law Constant [(atm·m³)/mol]	Vapor Pressure (mm Hg)	Vapor Temperature (°C)	Vapor Density (@ 25°C) Relative to Air (unitless)
1,2-Butylene Oxide	106-88-7	63.3 *[1]*	95,000	25 *[12]*	0.9 *[22]*	0.9 *[12]*	0.00018 *[32]*	180	25 *[32]*	2.2 *[32]*
1,3,5-Trioxane	110-88-3	114.5 *[2]*	212,000	25 *[3]*	−0.47 *[23]*	−0.43 *[31]*	– *[23]*	10	20 *[23]*	– –
1,3-Dioxolane	646-06-0	78 *[1]*	miscible	25 *[1]*	−0.37 *[24]*	1.17 *[32]*	0.000024 *[23]*	70	20 *[6]*	2.6 *[51]*
1,4-Dioxane	123-91-1	101.1 *[3]*	miscible	25 *[13]*	−0.27 *[24]*	1.23 *[22]*	4.29×10^{-6} *[57]*	38.09	25 *[38]*	3.03 *[6]*
2,6-di-*t*-Butyl-*p*-Cresol (BHT)	128-37-0	265 *[4]*	0.4	20 *[14]*	5.1 *[25]*	– –	0.00059 *[25]*	–	– –	7.6 *[6]*
2-Butanone	78-93-3	79.6 *[4]*	353,000	10 *[14]*	0.29 *[24]*	1.5 *[23]*	0.000047 *[35]*	91	25 *[45]*	2.41 *[14]*
2-Methyl-3-Butyn-2-Ol	115-19-5	103.5 *[5]*	miscible	20 *[5]*	0.318 *[5]*	1.55 *[5]*	9.9×10^{-7} *[5]*	15	20 *[5]*	2.49 *[52]*
Acetonitrile	75-05-8	81.6 *[4]*	miscible	25 *[15]*	−0.34 *[24]*	1.2 *[32]*	0.000035 *[36]*	88.8	25 *[46]*	1.42 *[46]*
Butoxymethyl Oxirane	2426-08-6	164 *[6]*	20,000	20 *[16]*	0.63 *[24]*	0.24 *[33]*	0.000027 *[33]*	3.2	25 *[33]*	3.78 *[16]*
Cyclohexane	110-82-7	80.7 *[4]*	0	25 *[4]*	3.44 *[24]*	2.2 *[32]*	0.15 *[37]*	97	25 *[47]*	2.98 *[53]*
Diisobutylene	107-40-4	104.9 *[1]*	0	25 *[1]*	4 *[25]*	2.44 *[32]*	0.88 *[32]*	35.9	25 *[32]*	3.8 *[32]*
Diisopropylamine	108-18-9	84 *[7]*	11,000	25 *[10]*	1.4 *[24]*	2.15 *[32]*	0.000096 *[38]*	79.4	25 *[38]*	3.5 *[51]*
Epichlorohydrin	106-89-8	117.9 *[4]*	65,900	25 *[17]*	0.45 *[27]*	1.6 *[32]*	0.00003 *[32]*	16.4	25 *[32]*	3.29 *[6]*
Ethanol	108-01-0	135 *[4]*	miscible	25 *[4]*	−0.31 *[24]*	– –	– –	–	– –	3.03 *[6]*
Ethyl Acetate	141-78-6	77 *[4]*	64,000	25 *[18]*	0.73 *[29]*	1.77 *[29]*	0.00013 *[29]*	79.4	25 *[38]*	3.04 *[4]*
Nitroethane	79-24-3	114 *[4]*	45,000	20 *[17]*	0.18 *[24]*	1.48 *[32]*	0.000048 *[32]*	20.80	25 *[32]*	2.58 *[6]*
Nitromethane	75-52-5	101.2 *[4]*	11,100	25 *[15]*	−0.35 *[24]*	0.28 *[32]*	0.000026 *[32]*	35.8	25 *[38]*	2.11 *[6]*
N-Methylmorpholine	109-02-4	113 *[8]*	miscible	25 *[8]*	−0.35 *[26]*	0.791 *[26]*	2.5×10^{-7} *[26]*	18.7	25 *[26]*	3.5 *[8]*
N-Methyl Pyrrole	96-54-8	112 *[9]*	0	25 *[9]*	0.75 *[24]*	2.162 *[26]*	0.00019 *[26]*	19.5	25 *[26]*	2.8 *[9]*
Phenol	108-95-2	181.75 *[10]*	82,800	25 *[19]*	1.46 *[24]*	1.2 *[32]*	3.3×10^{-7} *[36]*	0.35	25 *[32]*	3.24 *[54]*
1,2-Propylene Oxide	75-56-9	34.23 *[4]*	590,000	25 *[12]*	0.03 *[24]*	1.4 *[32]*	0.000070 *[32]*	538	25 *[48]*	2 *[16]*
p-*t*-Amyl Phenol	80-46-6	262.5 *[4]*	168	25 *[17]*	4.03 *[28]*	2.41 *[22]*	2.0×10^{-6} *[39]*	0.008	25 *[38]*	– –
Pyridine	110-86-1	115.2 *[4]*	miscible	20 *[10]*	0.65 *[24]*	1.7 *[22]*	0.00001 *[40]*	20.8	25 *[32]*	0.982 *[55]*
Resorcinol	108-46-3	280 *[4]*	71,700	25 *[17]*	0.8 *[24]*	1.81 *[32]*	1.1×10^{-11} *[32]*	0.00049	25 *[49]*	3.79 *[16]*
sec-Butyl Alcohol	78-92-2	99.5 *[11]*	181,000	25 *[10]*	0.61 *[24]*	1.7 *[32]*	9.1×10^{-6} *[41]*	18.3	25 *[38]*	2.6 *[51]*
t-Amyl Alcohol	75-85-4	102.4 *[1]*	110,000	25 *[20]*	0.89 *[24]*	0.46 *[32]*	0.00073 *[42]*	16.8	25 *[38]*	3.0 *[56]*
t-Butyl Alcohol	75-65-0	82.41 *[4]*	miscible	25 *[15]*	0.35 *[24]*	1.57 *[32]*	9.1×10^{-6} *[43]*	40.7	25 *[38]*	2.55 *[32]*
t-Butyl Phenol	128-37-0	265 *[4]*	0.6	25 *[21]*	6.2 *[30]*	2.77 *[30]*	0.0030 *[21]*	0.18825	25 *[30]*	7.6 *[6]*
Thymol	89-83-8	233 *[6]*	1,000	25 *[4]*	3.3 *[24]*	– –	– –	0.095	40 *[32]*	– –
Triethylamine	121-44-8	89.3 *[1]*	15,000	20 *[14]*	1.45 *[24]*	2.16 *[29]*	0.00015 *[44]*	57.1	24 *[50]*	3.49 *[16]*

Notes: References are listed by number in brackets and italics to the right of each value; see the list of references on the pages.

REFERENCES

1. Lide, D.R. (Ed.), 2000, *CRC Handbook of Chemistry and Physics*. Boca Raton, Florida: CRC Press LLC.
2. Windolz, M. (Ed.), 1983, *The Merck Index*. Rahway, New Jersey: Merck & Co., Inc.
3. Budavari, S. (Ed.), 1989, *The Merck Index—An Encyclopedia of Chemicals, Drugs and Biologicals*. Rahway, New Jersey: Merck and Co., Inc.
4. Budavari, S. (Ed.), 1996, *The Merck Index—An Encyclopedia of Chemicals, Drugs, and Biologicals*. Rahway, New Jersey: Merck and Co., Inc.
5. UNEP (United Nations Environment Programme), 1998, Screening information data set (SIDS) for 2-methyl-3-yn-2-ol. Nairobi, Kenya: United Nations Environment Programme Report 115-19-5. http://www.chem.unep.ch/irptc/sids/OECDSIDS/115195.pdf (accessed June 3, 2007).
6. Sax, N.I. and Lewis, R.J.S., 1987, *Hawley's Condensed Chemical Dictionary*. New York: Van Nostrand Reinhold Company.
7. Lide, D.R. and Kehiaian, H.V., 1994, *CRC Handbook of Thermophysical and Thermochemical Data*. Boca Raton, Florida: CRC Press.
8. Huntsman, 2005, *Technical Bulletin: N-Methylmorpholine (NMM) CAS 109-02-4*. The Woodlands, Texas: Huntsman Corporation.
9. Fisher Scientific, 1998, Material Safety Data Sheet for *n*-methyl pyrrole for 96-54-8, Fair Lawn, New Jersey. https://fscimage.fishersci.com/msds/43291.htm (accessed on February 4, 2007).
10. Kirk, R. and Othmer, D., 1982, *Kirk-Othmer Encyclopedia of Chemical Technology*. New York: John Wiley and Sons, Inc.
11. Snyder, R. (Ed.), 1992, *Ethel Browning's Toxicity and Metabolism of Industrial Solvents*, 2nd ed., Vol. 3: Alcohols and Esters. New York: Elsevier.
12. Bogyo, D.A., Lande, S.S., Meylan, W.M., Howard, P.H., and Santodonato, J., 1980, *Investigation of selected potential environmental contaminants: Epoxides—Final technical report*. Prepared by Syracuse Research Corporation for the U.S. Environmental Protection Agency Office of Toxic Substances.
13. Hawley, G.G., 1977, *The Condensed Chemical Dictionary*. New York: Van Nostrand Reinhold Company.
14. Verschueren, K., 1996, *Handbook of Environmental Data on Organic Chemicals*, 3rd ed. New York: Van Nostrand Reinhold Company.
15. Riddick, J.A., Bunger, W.B., and Sakano, T.K., 1985, *Techniques of Chemistry*, 4th ed., Vol. II: Organic Solvents. New York: John Wiley and Sons, Inc.
16. Patty, F., 1963, *Industrial Hygiene and Toxicology*, Vol. II: Toxicology. New York: Wiley Interscience Publishers.
17. Yalkowsky, S.H. and Dannenfelser, R.M., 1992, *The AQUASOL Database of Aqueous Solubility*. Tucson: University of Arizona, College of Pharmacy.
18. Wasik, S.P., Tewari, Y.B., Miller, M.M., and Martire, D.E., 1981, *Octanol/Water Partition Coefficients and Aqueous Solubilities of Organic Compounds*. Washington, D.C.: U.S. Department of Commerce, National Bureau of Standards. NBSIR 81-2406
19. Southworth, G.R. and Keller, J.L., 1986, Hydrophobic sorption of polar organics by low organic carbon soils. *Journal of Water, Air, & Soil Pollution* 28(3–4): 239–247.
20. Barton, A.F.M., 1984, *Alcohols with water*. In: *Solubility Data Series*, Vol. 15. Oxford: Pergamon.
21. Inui, H., Akutsu, S., Itoh, K., Matsuo, M., and Miyamoto, J., 1979, Studies on biodegradation of 2,6-di-*tert*-butyl-4-methylphenol (BHT) in the environment. Part IV: The fate of ^{14}C-phenyl-BHT in aquatic model ecosystems. *Chemosphere* 8(6): 393–404.
22. Lyman, W.J., Reehl, W.F., and Rosenblatt, D.H., 1990, *Handbook of Chemical Property Estimation Methods: Environmental Behavior of Organic Compounds*. New York: McGraw-Hill.
23. U.S. Environmental Protection Agency (USEPA), 2000, High production volume challenge program submission for 1,3,5-trioxane by Trioxane Manufacturer's Consortium. Prepared by Toxicology and Regulatory Affairs, Flemington, New Jersey. http://www.epa.gov/hpv/pubs/summaries/triox/c12863.pdf (accessed January 4, 2009).
24. Hansch, C. and Leo, A.J., 1985, Medchem Project issue no 26. Claremont, California, Pomona College, Medchem Project.
25. UNEP (United Nations Environment Programme), 2002, Screening information data set (SIDS) for 2,6-di-*tert*-butyl-*p*-cresol [betahydroxy toluene or BHT]: Initial assessment report for SIDS initial assessment meeting (SIAM), 14th meeting, Paris, France, 26–28 March 2002.
26. U.S. Environmental Protection Agency (USEPA), 2007, Estimations programs interface for Windows (EPI Suite), v. 3.20. http://www.epa.gov/oppt/exposure/pubs/episuite.htm (accessed November 3, 2007).

27. Deneer, J.W., Sinnige, T.L., Seinen, W., and Hermens, J.L.M., 1988, A quantitative structure-activity relationship for the acute toxicity of some epoxy compounds to the guppy. *Aquatic Toxicology* 13(3): 195–204.
28. Schultz, T.W., 1986, The use of the ionization constant (pK_a) in selecting models of toxicity in phenols. *Ecotoxicology and Environmental Safety* 14(2): 178–183.
29. Swann, R.L., Laskowski, D.A., McCall, P.J., and Vander Kuy, K., 1983, A rapid method for the estimation of the environmental parameters octanol/water partition coefficient, soil sorption constant, water to air ratio and water solubility. *Residue Reviews* 85: 17–28.
30. NICNAS (National Industrial Chemicals Notification and Assessment Scheme), 2003, Full report: 2,6-di-*t*-butyl-4-methyl phenol. Canberra, Commonwealth of Australia: N. I. C. N. A. S. (NICNAS), National Occupational Health and Safety Commission.
31. U.S. Environmental Protection Agency (USEPA), 2007, High Production Volume Information System (HPVIS) online database. http://www.epa.gov/chemrtk/hpvis/index.html (accessed November 3, 2007).
32. NLM (National Library of Medicine), 2006, Hazardous substance data bank: MEDLARS Online Information Retrieval System. National Library of Medicine. http://toxnet.nlm.nih.gov/ (accessed April 12, 2006).
33. ERSTG (Epoxy Resin Systems Task Group), 2002, *High production volume (HPV) challenge program test plan and robust summaries for* n-*butyl glycidyl ether.* U.S. Environmental Protection Agency, AR201-1400A.
34. Park, J.H., Hussam, A., Couasnon, P., Fritz, D., and Carr, P.W., 1987, Experimental reexamination of selected partition coefficients from Rohrschneider's data set. *Analytical Chemistry* 59(15): 1970–1976.
35. Bhattacharya, S.K., Qu, M., and Madura, R.L., 1996, Effects of nitrobenzene and zinc on acetate utilizing methanogens. *Water Research* 30(12): 3099–3105.
36. Gaffney, J.S., Streit, G.E., Spall, W.D., and Hall, J.H., 1987, Beyond acid rain. Do soluble oxidants and organic toxins interact with SO_2 and NO_x to increase ecosystem effects? *Environmental Science and Technology* 21(6): 519–524.
37. Bocek, K., 1976, Relationships among activity coefficients, partition coefficients and solubilities. *Experientia Supplementum* 23: 231–239.
38. Daubert, T.E. and Danner, R.P., 1985, *Data Compilation Tables of Properties of Pure Compounds*. New York: American Institute of Chemical Engineers.
39. Meylan, W.M. and Howard, P.H., 1991, Bond contribution method for estimating Henry's law constants. *Environmental Toxicology and Chemistry* 10(10): 1283–1293.
40. Hawthorne, S.B., Slevers, R.E., and Barkley, R.M., 1985, Organic emissions from shale oil wastewaters and their implications for air quality. *Environmental Science and Technology* 19(10): 992–997.
41. Snider, J.R. and Dawson, G.A., 1985, Tropospheric light alcohols, carbonyls, and acetonitrile: Concentrations in the southwestern United States and Henry's law data. *Journal of Geophysical Research* 90(D2): 3797–3805.
42. Butler, J.A.V., Ramchandani, C.N., and Thompson, D.W., 1935, The solubility of non-electrolytes: Part I. The free energy of hydration of some aliphatic alcohols. *Journal of the Chemical Society, Faraday Transactions* 58: 280–285.
43. Altschuh, J., Berggemann, R., Santl, H., Eichinger, G., and Piringer, O.G., 1999, Henry's law constants for a diverse set of organic chemicals: Experimental determination and comparison of estimation methods. *Chemosphere* 39(11): 1871–1886.
44. Christie, A.O. and Crisp, D.J., 1967, Activity coefficients of the *n*-primary, secondary and tertiary aliphatic amines in aqueous solution. *Journal of Applied Chemistry* 17: 11–14.
45. Alarie, Y., Nielsen, G.D., Andonianhaftvan, J., and Abraham, M.H., 1995, Physicochemical properties of nonreactive volatile organic chemicals to estimate RD_{50}: Alternatives to animal studies. *Toxicology and Applied Pharmacology* 134(1): 92–99.
46. Clayton, G.D. and Clayton, F.E., 1994, *Patty's Industrial Hygiene and Toxicology*. New York: John Wiley and Sons, Inc.
47. Chao, J., Lin, C.T., and Chung, T.H., 1983, Vapor pressure of coal chemicals. *Journal of Physical and Chemical Reference Data* 12(4): 1033–1063.
48. Boublik, T., Fried, V., and Hala, E., 1984, *The Vapor Pressures of Pure Substances: Selected Values of the Temperature Dependence of the Vapor Pressures of Some Pure Substances in the Normal and Low Pressure Region*. New York: Elsevier Science Publishing Company.
49. Yaws, C.L., 1994, *Handbook of Vapor Pressure*, Vol. 2: C5 to C7 Compounds. Houston, Texas: Gulf Publishing Company.
50. U.S. Coast Guard, 1984, *CHRIS–Hazardous Chemical Data*. Vol. II. Washington, D.C.: Department of Transportation, U.S. Government Printing Office.

51. NFPA (National Fire Protection Association), 1991, *National Fire Protection Guide: Fire Protection Guide to Hazardous Materials*. Quincy, Massachusetts: National Fire Protection Association.
52. BASF, 2006, *Safety Data Sheet for 2-methylbut-3-yn-2-ol*, Florham Park, New Jersey.
53. Mackison, F.W., Stricoff, R.S., and Partridge, L.J., Jr. (Eds.), 1981, *NIOSH/OSHA—Occupational Health Guidelines for Chemical Hazards*. Washington, D.C.: National Institute of Occupational Safety and Health (NIOSH).
54. Bingham, E., Cohrssen, B., and Powell, C.H., 2001, *Patty's Toxicology*. New York: John Wiley and Sons, Inc.
55. Browning, E., 1965, *Toxicity and Metabolism of Industrial Solvents*. New York: American Elsevier.
56. Mallinckrodt Baker, 2008, *Material Safety Data Sheet for tert-amyl Alcohol—75-85-4*. Phillipsburg, New Jersey.
57. Hiatt, M.H., 2013, Determination of Henry's law constants using internal standards with benchmark values. *Journal of Chemical & Engineering Data* 58: 902–908.

Appendix 3: Compilation of Instrumental Parameters in Literature Studies of 1,4-Dioxane Analyses of Various Media

Matrix	MDL	Preparation	Standard and Carrier Gas	Solvent	Drying	GC
Water	1 ppb	SPE coconut shell charcoal: SKC 226–26 Tekmar LSC-2 P&T	1,4-Difluorobenzene	Methanol + CS_2 5% v/v	3.5 M Na_2SO_4	CG-FID: HP5580
Water	GC-FID: 2.5 ppb GC–MS: 0.25 ppb	SPME on CAR-PDMS with headspace extraction	1,4-Dioxane-d_8, isopropanol, methyl chloroform Helium	None	Desorption at 310°C	GC-FID: Varian 3400 MS: Varian Saturn 2200 ion trap
Water	LLE: 0.2 ppb SPE: 6 ppb P&T: 25 ppb	LLE separatory funnel compared with J&W SPE using AC-2 on SDB and with P&T using Tekmar 3000 P&T	1,4-Dioxane-d_8, 2-bromo-1-chloropropane Helium	DCM, MTBE, NaCl, acetone for AC-2 cartridge	Centrifuge, air	GC: Agilent 6890 MS: 5973N
Water	SPE with LLE: 0.13 ppb LOQ = 0.31 ppb	Sorption on carbon disks	1,4-Dioxane-d_8 THF-d_8	Acetone	Laboratory airflow for 1 h	GC: HP 5890 Series II MS = Finnigan Mat TSQ 700
Water	SPE: 50 ppb	LLE in hexane-DCM solvent followed by SPE using C_{18} cartridges	1,4-Dioxane-d_8 Helium	Hexane: DCM 80:20 v/v	Eluted from cartridge with acetonitrile	GC: HP 5890 Series II MS: HP 5989
Water	P&T: 1 ppb	20-min purge and 20-min desorb Tekmar 3100 with Vocarb 4000 Trap	1,4-Dioxane-d_8 Helium	None	Air dried for 2–3 min under 2.7 kPa; centrifuged at 3000 rpm (1700 g) for 10 min	GC: Varian 3800 MS: Varian Saturn 2000 GC–MS/MS system
Water	SPE: 0.02 ppb	SPE on carbon felt, 1500 m^2/g	1,4-Dioxane-d_8 Helium	Acetone, DCM N_2 gas blowdown	Air dried for 2 min, then centrifuged at 3000 rpm for 10 min	Shimadzu GC–MS QP5050A

Injection Temperature (°C)	Column	Column Material	Final Column Temperature (°C)	Gas Flow, Pressure at Head	Monitored Ions *m/z*	References
50	J&W DB-5: 6 ft × 1/4 in.	Supelco: Glass 1% SP-1000 on Carbopack B	50 → 100	30 mL/min	88, 58 for 1,4-dioxane 114 for 1,4-difluorobenzene	Epstein et al. (1987)
45	Supelco PDMS: 30 m × 0.32 mm × 4 μm	Bonded polydimethyl siloxane (PDMS)	45 → 230	GC: He at 13 psi and 40 cm/s	88–98 57–67	Shirey and Linton (2006)
200	Ultra-2: 50 m × 0.2 mm id; 0.33 μm film thickness	Ultra-2: 5% diphenyl, 95% methylpolysiloxane stationary phase	40 → 200	GC: He at 0.5 mL/min	88 58, 57, 87 for 1,4-dioxane 96, 64 for 1,4-dioxane-d_8 77, 79 for 2-bromo-1-chloropropane	Park et al. (2005)
150	Supelco SPB-1: 30 m × 0.32 mm × 4 μm	Supelco SPB-1: Nonpolar methylsilicone (bonded polydimethylsiloxane)	40 → 120	77 kPa	45 [C_2H_4O + H] + for dioxane 49 [C_2D_4O + H] + for dioxane-d_8 62 [C_4D_7] + for THF-d_8	Isaacson et al. (2006)
130	HP-5: 25 m × 0.2 mm × 0.33 μm	Agilent HP-5: Nonpolar (5% phenyl) methylpolysiloxane	40 → 100	GC: He at 1 mL/min	88 for 1,4-dioxane 96 for 1,4-dioxane-d_8	Song and Zhang (1997)
125	Varian VRX: 60 m × 0.32 mm × 1.8 μm	J&W DB-WAX: Fused silica with polyethylene glycol (PEG)	35 → 220	GC: He at 1 mL/min	40–100	Kawata et al. (2001)
170	J&W DB-WAX: 30 m × 0.25 mm × 0.5 μm	J&W DB-WAX: Fused silica with polyethylene glycol (PEG)	35 → 190	GC: He at 1 mL/min	1,4-Dioxane: 88, 58	Song and Zhang (1997)

(Continued)

Matrix	MDL	Preparation	Standard and Carrier Gas	Solvent	Drying	GC
Water	SPDE: 0.03 ppb	SPDE: CTC-CombiPAL autosampler by Chromtech (Idstein, Germany)	1,4-Dioxane-d_8 (99 + %)	Acetone, dichloromethane	Cartridges air dried with aspirator and centrifuged at 3000 rpm (1700 g) for 10 min	GC: TraceGC 2000 MS: TraceDSQ ThermoFinnigan
Water	P&T: 20 ppb	Heated purge and trap with sodium sulfate salt; 2 µL of 250 µg/mL 1,4-dioxane-d_8 solution added to the syringe	1,4-Dioxane-d_8 Helium	Purge at 85°C	Na_2SO_4	GC–MS Hewlett-Packard 5970 and Tekmar LSC-1000 in scan mode
Water	SPE: 5 ppb	SPE: SDB + AC	1,4-Dioxane-d_8 Helium	Acetone N_2 gas blowdown	Cartridge air dried	GC–MS: QP2010 Shimadzu MS in SIM
Water	1.6 ppb	Frozen Micro-Extraction	1,4-Dioxane-d_8	DCM	Freezing	GC-MS: Agilent 5973, 6890
Water	GC-FID: 2.5 ppb GCMS: 0.25 ppb	SPME	1,4-Dioxane-d_8 Isopropanol	SPME: CAR-PDMS	None	GC: 3400 GC (Varian) GCMS: 3800 GC (Varian); 2200 Saturn ion trap (Varian)
Synthetic leachates		Solvent extraction	1,4-Dioxane-d_8 Helium	DCM	Dry Na_2SO_4 centrifuge	JEOL Automass #50
Landfill leachates		SPE on Sep-Pak PS-2 and AC-2	1,4-Dioxane-d_8 Helium	Methanol and methyl acetate	Airflow, dry Na_2SO_4	GC: Agilent 6890A MS: Agilent 5973

Injection Temperature (°C)	Column	Column Material	Final Column Temperature (°C)	Gas Flow, Pressure at Head	Monitored Ions *m/z*	References
200	Restek Corp.: 60 m × 0.32 mm × 0.5 µm	Restek Stabilwax: Fused-silica capillary column	40 → 180	GC: He at 1.5 mL/min	71, 55 for acrylamide 87, 44 for *N,N*-dimethylacetamide 73, 44 for *N,N*-dimethylformamide 88, 58 for 1,4-dioxane 96, 95 for furfural 98, 81 for furfuryl alcohol 102, 56 for *N*-nitrosodiethylamine 74, 42 for *N*-nitrosodimethylamine	Kawata et al. (2001)
160	ID DB-624: 75 m × 0.53 in. with a 3 µm film thickness	J&W DB-624: Medium-polarity column; equivalent to VOCOL	40 → 160at 20°C/min	GC: He at 120 mL/min	88, 58 for 1,4-dioxane 96, 64 for 1,4-dioxane-d_8	Rudinsky et al. (1997)
200	Rtx-1701: 30 m × 0.25 mm × 1 µm	14% cyanopropylphenyl and 86% dimethylpolysiloxane	40 → 250	GC: He at 45 cm/s	88, 58 for 1,4-dioxane 96, 64 for 1,4-dioxane-d_8	Jochmann et al. (2006)
200	HP-5: 30 m × 0.25 nm × 0.25 µm	Agilent HP-5: Nonpolar (5% phenyl) methylpolysiloxane	35 → 275	He at 1.3 mL/min	88, 58 for 1,4-dioxane 96, 64 for 1,4-dioxane-d_8	Li et al. (2011)
310	GC-FID: PDMS (Supelco): 30 m × 0.32 mm × 4.0 µm GCMS: PDMS (Supelco): 60 m x 0.25 mm x 3.0 µm	Polydimethylsiloxane	GC-FID: 80 → 230 GCMS: 50 → 230	He at 13 psi	86 to 98 and 57 to 67 for 1,4-dioxane	Shirey and Linton (2006)
200	DB-1: 30 m × 0.53 mm id	DB-1: 100% nonpolar dimethyl polysiloxane	30 → 100 → 200	7 psi	88, 58 for 1,4-dioxane 96, 64 for 1,4-dioxane-d_8	Yasuhara et al. (1999)
250	FFAP: 30 m × 0.25 mm id	FFAP: Nitroterephthalic acid modified polyethylene glycol (PEG)	35 → 250	1.8 mL/min 97.4 kPa	88, 58 for 1,4-dioxane 96, 64 for 1,4-dioxane-d_8	Yasuhara et al. (2003)

(Continued)

Matrix	MDL	Preparation	Standard and Carrier Gas	Solvent	Drying	GC
Aspirin, acetaminophen	1.5 ppm	Static headspace	Benzene, trichloroethylene, chloroform, dichloromethane, 1,4-dioxane Hydrogen	Water	None—aqueous headspace	GC-FID: 9001, Finnigan
Tween 20 Tween 80	50 ppb 100 ppb	Thermal desorption	Nitrogen	None—thermal desorption	None—thermal desorption	GC-FID SP-6800
Cosmetics and nonionic surfactants	300 ppb 60 ppb by GC–MS	SPME: 75 μm Carboxenpoly (dimethylsiloxane) (PDMS–CAR), 85 μm poly(acrylate) (PA), and 100 μm poly (dimethylsiloxane) (PDMS) fiber assembly (Supelco)	1,4-Dioxane, THF Nitrogen	Methanol, acetonitrile, or acetone	Salting out with 5 mL of 20% NaCl	GC-FID: HP 6890 and Varian Saturn 3800 GC coupled with a Varian 2000 ion-trap MS

Notes: Temp. = temperature; AC2 = activated carbon; C_{18} = 18% activated carbon loading; CAR-PDMS = 100 μm carboxen extraction; LOQ = Limit of quantitation; MDL = method detection limit; MTBE = methyl *tert*-butyl ether; NaCl = sodium SPE = solid-phase extraction; SPME = solid-phase microextraction; THF = tetrahydrofuran.

Injection Temperature (°C)	Column	Column Material	Final Column Temperature (°C)	Gas Flow, Pressure at Head	Monitored Ions *m/z*	References
250	Restek Rtx-1701: 30 m × 0.53 mm × 1.0 μm Restek Rtx-Volatiles: 30 m × 0.53 mm × 2.0 μm	G43: 6% cyanopropylphenyl and 94% dimethyl polysiloxane	35 → 170	GC: He at 80 cm/s	GC-FID configuration: Precolumn and two megabore columns to analyze on three detectors	Rankin (1996)
155	60 cm SS column 2 mm id packed with 3% DEGS (polydiethylene glycol succinate)	3% DEGS and 150–200 mesh SS-101 white supporter	105	0.32 MPa 42 mL/min	Not provided	Guo and Brodowsky (2000)
To GC: 40 To MS: 240	HP-1 (Agilent Technologies) column: 30 m × 0.25 mm id	HP-1 column; nonpolar, 100% dimethyl polysiloxane	160	GC: N_2 at 1 L/min FID: N_2 at 45 L/min; H_2 at 45 mL/min Air at 450 mL/min	88 for 1,4-dioxane 71 for THF	Fuh et al. (2005)

polydimethylsiloxane-coated wire adsorbent; DCM = dichloromethane; dry Na_2SO_4: anhydrous sodium sulfate; LLE = liquid-liquid chloride solution; P&T = purge and trap; PEG = polyethylene glycol; SDB = styrene divinyl benzene; SIM = selective ion monitoring;

Index

B

D

E

N

O

P

Q

R

S